HANS CHRISTIAN ØRSTED

Hans Christian Ørsted

Reading Nature's Mind

DAN CHARLY CHRISTENSEN

OXFORD
UNIVERSITY PRESS

Great Clarendon Street, Oxford, OX2 6DP,
United Kingdom

Oxford University Press is a department of the University of Oxford.
It furthers the University's objective of excellence in research, scholarship,
and education by publishing worldwide. Oxford is a registered trade mark of
Oxford University Press in the UK and in certain other countries

© Dan Charly Christensen 2013

The moral rights of the author have been asserted

First Edition published in 2013

Impression: 1

All rights reserved. No part of this publication may be reproduced, stored in
a retrieval system, or transmitted, in any form or by any means, without the
prior permission in writing of Oxford University Press, or as expressly permitted
by law, by licence or under terms agreed with the appropriate reprographics
rights organization. Enquiries concerning reproduction outside the scope of the
above should be sent to the Rights Department, Oxford University Press, at the
address above

You must not circulate this work in any other form
and you must impose this same condition on any acquirer

British Library Cataloguing in Publication Data
Data available

ISBN 978–0–19–966926–4

Printed in China by
C&C Offset Printing Co., Ltd

Publication supported by the Carlsberg Foundation

TABLE OF CONTENTS

List of Abbreviations	*xi*
List of Illustrations	*xv*
Dedication	*xx*

PART I: THE STUDENT

1 1777–1851
    Introduction — 3

2 1777–94
    A Childhood without Playing — 15

3 1794–6
    A University without Science — 25

4 1796–8
    Two Philosophical Minds — 40

5 1796–7
    Hans Christian's Gold Medals — 47

6 1798
    Anders's Gold Medal — 52

7 1798–1800
    Editors for Kant — 59

8 1798–9
    Doctoral Thesis on the Dynamical System — 66

9 1800–01
    Pharmacy Manager and Fiancé — 72

10 1799–1801
    Galvanism — 80

PART II: THE COSMOSPOLITAN

11 1801–2
    First Grand Tour
    Tourist far away from Sophie — 97

12 1801–2
    Encountering Ritter and Winterl — 108

13 1801–2
    Jena: Romanticism, Salons, and Societies — 122

| TABLE OF CONTENTS

14 1802–3
    Post-revolutionary Paris   135

15 1802–3
    Ritter and the Napoleon Prize   147

16 1802–3
    The Double Game   156

PART III: THE RESEARCHER AND TEACHER

17 1804
    Alone and Abandoned in Copenhagen with a Collection of Instruments   165

18 1805
    Rivalry and Love   177

19 1804–9
    Textbook Writer and Professor   187

20 1807
    Fichte's Idealism and Napoleon's Wars   196

21 1808
    Sonorous Figures   207

22 1808
    The Art of Music   214

23 1808
    The Royal Danish Society of Sciences and Letters   221

24 1809
    Family and Friends   227

25 1810
    Dialogue on Mysticism Ritter's Death   237

PART IV: THE SPOUSE

26 1811
    Career and Brothers Working Together   247

27 1812–13
    Second Journey Abroad
    Berlin and Paris   254

28 1812–13
    The Major Work   264

29 1813
    Controversy on Pantheism   274

30 1814
    Love and Marriage   281

31	1812–15 The Prime Mover of Danish Science, and Gitte's First-Born Child	288
32	1815–17 Dynamical Research A.S. Ørsted's Dissent	298
33	1818–19 Shadows of Death Expedition to Bornholm	312

PART V: THE TRIUMPHATOR

34	1820 The Happiest Year	327
35	1820 A Discovery by Chance?	336
36	1820–21 Domestic and Foreign Reactions	350
37	1822–3 Ørsted's Triumphal Progress Germany	360
38	1822–3 The Triumphal Progress Paris	376
39	1823 The Triumphal Progress Britain	390

PART VI: THE ORGANISER

40	1823–4 The Society for the Dissemination of Science in Denmark	407
41	1824 The Ørsted Brothers in the Howitz Controversy	415
42	1825 Aluminium Priority and Nationalism	424
43	1826–32 The Downfall of A.S. Ørsted The Millennium of Christianity The Tercentenary of the Reformation	431
44	1827–8 Family Life and Conferences Abroad	440

| TABLE OF CONTENTS

45 1828–9
    The Polytechnic Institute   453

46 1829–1833
    The Literary Critic
    The Airship   465

47 1831–9
    The Awakening of Political Life   481

PART VII: FAME AND TRIBULATIONS

48 1831–9
    Technology and Industry   497

49 1833–9
    The Natural Laws of General Education   506

50 1839–47
    Scandinavian Science Conferences   517

51 1839–46
    Politics and Nationalism   534

52 1842–8
    The Centenary of the Royal Danish Society
    Magnetischer Verein
    Henrik Steffens
    L.A. Colding   544

53 1843
    Homage in Berlin   560

54 1843–6
    Aesthetics of Nature   569

55 1846
    Homage in Britain   585

56 1840–50
    Polytechnic Criticism   600

57 1848–9
    Civil War and Free Constitution   609

58 1849–50
    The Soul in Nature   621

59 1850–1
    Big and Little Hans Christian's Modern Turning Point   634

60 1849–51
    Jubilee and Death   643

61 Hans Christian Ørsted and the Golden Age in a Wider Perspective	653
Notes	665
Archival Material & Bibliography	710
Index of Names	733

LIST OF ABBREVIATIONS

ADB	Allgemeine Deutsche Biographie
AOe	Adam Oehlenschläger
ASØ	Anders Sandøe Ørsted
BAAS	British Association for the Advancement of Science
BAS	Bibliotheca Academia Sorana, Sorø [Library of Sorø Academy]
BPK	Bildarchiv Preussischer Kulturbesitz, Berlin [Prussian Picture Archive]
BT	Berlingske Tidende
C I-II	*Correspondance de H.C. Örsted avec divers savants* [HCØ's Correspondence with sundry scholars], vols.i–ii, ed. by M.C. Harding, Copenhagen 1920
CCA	Copenhagen City Archive
CCM	Copenhagen City Museum
CF	Prince Christian Frederik, (King CVIII, 1839–48)
CH	Christopher Hansteen
CM	Christian Molbech
DBL	Dansk Biografisk Leksikon
DCC	Dan Ch. Christensen
DFO	Den danske Frimurerorden [Danish Masonic Order]
DHS	Den Hirschsprungske Samling, Copenhagen
DNB	Dictionary of National Biography
DPB	Danske Politiske Breve [Danish Political Letters]
DSB	Dictionary of Scientific Biography
DTM	Danmarks Tekniske Museum, Ellsinore
DTU	Danmarks Tekniske Universitet, Lundtofte
EA	Erhvervsarkivet, Aarhus
EAS	E.A. Scharling
FVI	King Frederik VI (1808–1839)
FMA	Den store danske Frimurerordens arkiv [Masonic Archive]
FRM	Frederiksborgmuseet, Hillerød
FCS	F.C. Sibbern
FPL	Frederiksberg Public Library
FRS	Fellow of Royal Society
FWJS	F.W.J. Schelling
GDNÄ	Gesellschaft Deutscher Naturforscher und Ärtzte
HCA	Hans Christian Andersen
HCA SV	Hans Christian Andersen, *Samlede Værker 1–17*, Cph. 2003–2007
HCØ	Hans Christian Ørsted

LIST OF ABBREVIATIONS

IBØ	Inger Birgitte Ørsted, née Ballum
JC	Jonas Collin
JFS	Joakim Frederik Schouw
JGF	Johan Georg Forchhammer
JJB	Jöns Jacob Berzelius
JLH	Johan Ludvig Heiberg
JWR	Johann Wilhelm Ritter
KDVS	Det Kongelige Danske Videnskabernes Selskab [The Royal Danish Society of Sciences and Letters, Copenhagen]
KLE	Kjøbenhavns Lærde Efterretninger [Copenhagen Learned News]
KU	Københavns Universitet
KUJ	Kjøbenhavns Universitetsjournal
LAB	Landesarchiv, Berlin
LAS	Landsarkivet for Østifterne, Copenhagen
LHS	Landhuusholdningsselskabet [Royal Danish Agricultural Society]
LM	Ludvig Manthey
MØI-II	Breve til og fra HCØ, [Letters to and from HCØ], ed. by Mathilde Ørsted, vols. i–ii, Copenhagen 1870
NBD	Nyere Brevsamling Dansk [New Collection of Letters, RL]
NFSG	N.F.S. Grundtvig
NKS	Nye Kongelige Samling [New Royal Collection, RL]
NPG	National Portrait Gallery, London
NS I-III	HCØs Naturvidenskabelige Skrifter, [HCØ's Scientific Writings], vols. i–iii, ed. by K. Meyer, Copenhagen, 1920
OBM	Odense Bys Museer [Museums of Odense]
PLA	Polytechnisk Læreanstalt [Polytechnic Institute, Copenhagen]
PRO	Rigsarkivet [Public Record Office, Copenhagen]
RL	Royal Library, Copenhagen
RMN	Agence photographique de la réunion des musées nationaux
RS	Royal Society, London
RUB	Roskilde Universitetsbibliotek
SAK	Søren Aabye Kierkegaard
SCØ	Søren Christian Ørsted
SES	Samlede og Efterladte Skrifter af H.C. Ørsted [Collected and Posthumous Works by HCØ], vols. i–ix, ed. by M. Ørsted, Copenhagen 1852–53
SLS	Skandinavisk Litteraturselskab [Scandinavian Literary Society]
SMK	Statens Museum for Kunst, Copenhagen
SNU	Selskabet for Naturlærens Udbredelse i Danmark [The Society for the Dissemination of Science in Denmark]
SP	Sophie Probsthein

SSW	*Selected Scientific Works of Hans Christian Ørsted*, transl. and ed. by K. Jelved, A.D. Jackson, and O. Knudsen with an Introduction by A.D. Wilson, Princeton 1997
SØ	Sophie Ørsted
TL	*The Travel Letters of H.C. Ørsted*, ed. and transl. by K. Jelved and A.D. Jackson, Copenhagen 2011
WCZ	W.C. Zeise
ØC	The Ørsted Collection, RL

LIST OF ILLUSTRATIONS

Fig. 1	Ørsted statue by A. Munro at the Oxford University Museum, 1885.	12
Fig. 2	Ørsted statue by J.A. Jerichau in Copenhagen, 1861.	13
Fig. 3	Silhouette of S.G. Ørsted (father) and A.D. Borring (stepmother).	18
Fig. 4	Ørsted's pedigree (family tree).	19
Fig. 5	Drawing of the pharmacy of Rudkøbing by J.K. Jauch.	21
Fig. 6	Colour drawing of Vester Port, Copenhagen, by C.W. Eckersberg.	26
Fig. 7	The great fire of Copenhagen, drawing by C.F.F. Stanley.	27
Fig. 8	Drawing of the Ørsted brothers by E.L. Hemmingsen.	28
Fig. 9	Map of Copenhagen by Sterm, 1839.	29
Fig. 10	Map of the University of Copenhagen by C. Gedde.	33
Fig. 11	*The Clover Leaf or the Masked Jesuits* by O.P. Gram.	38
Fig. 12	Portrait of I. Kant by an unknown artist, 1790.	43
Fig. 13	The first known writing by eighteen-year-old student Ørsted.	55
Fig. 14	Brunonianism by C.H. Pfaff, 1796.	57
Fig. 15	Portrait of A.W. Hauch.	61
Fig. 16	A.W. Hauch's apparatus to prove that water is no element.	62
Fig. 17	A. Oehlenschläger reciting in Mrs Møller's dye-works in Vestergade c. 1799, by C.C.F. Thomsen.	70
Fig. 18	Portrait of J.G.L. Manthey by an unknown artist.	73
Fig. 19	The entrance of the Lion Pharmacy at the corner of Østergade and Hyskenstræde.	74
Fig. 20	Painting of N. Bonaparte watching A. Volta's experiment by an unknown artist.	82
Fig. 21	Diagram of contact electricity, 1816.	83
Fig. 22	Voltaic pile, 1802.	84
Fig. 23	Drawing of Ritter's experiment on the separation of water.	85
Fig. 24	Ritter's experiment on the separation of water.	86
Fig. 25	Ritter's subsequent experiment on the separation of water.	87
Fig. 26	Hauch's control experiment.	89
Fig. 27	Hauch's control experiment.	90
Fig. 28	Hauch's control experiment.	90
Fig. 29	Ørsted's portable galvanic battery.	92
Fig. 29a	HCØ as a twenty-five-year old. Drawing after G.L. Chrétien's physionotrace.	92
Fig. 30	Map of Europe showing HCØ's first three itineraries.	99
Fig. 31	'Der Stein zu Wörlitz' (the Rock of Wörlitz) by K. Kunz.	105
Fig. 32	*Utsikt mot Pillnitz gjennom ett vindu* (View towards Pillnitz through a window) by J.C. Dahl.	106

LIST OF ILLUSTRATIONS

Fig. 33	Woodcut of J.W. Ritter.	111
Fig. 34	Map of Berlin, 1820.	113
Fig. 35	The Chemical Institute of the Prussian Industrial College, photo by A. Matschenz.	115
Fig. 36	Drawing of J.J. Winterl.	120
Fig. 37	Map of Jena with city wall.	123
Fig. 38	*Versuch auf den Parnass zu gelangen* (Attempt to climb the Parnassus).	125
Fig. 39	Portrait of H.J. Herz by A.D. Therbusch.	127
Fig. 40	The Royal Theatre on Gendarmenmarkt by L.E. Lütke.	129
Fig. 41	Map of Paris, 1840, belonging to Ørsted.	136
Fig. 42	Drawing of L.N. Vauquelin.	141
Fig. 43	Drawing of J.A.C. Charles.	142
Fig. 43a	Oil painting of P.-S. de Laplace (1745–1827).	146
Fig. 44	J.W. Ritter's letter of 2 May 1803 to Ørsted.	149
Fig. 45	Physionotrace of Ørsted by G.L. Chrétien.	160
Fig. 46	Drawing of a recital of A. Oehlenschläger's drama Hakon Jarl by C.C.F. Thomsen.	170
Fig. 47	Pastel portrait of S.C. Ørsted by P. Copmann.	171
Fig. 48	Oil painting of Collin's Court, 156, Norgesgade (today's Bredgade) by P.F.N. Grove.	175
Fig. 49	Caricature of a chemical experiment (where the chemist looks unlike HCØ) by C.W. Eckersberg.	180
Fig. 50	Ørsted's electrical figures.	183
Fig. 51	Oil painting of the yard behind Thott's Palace on Kongens Nytorv by F. Vernehren.	194
Fig. 52	Oil painting of S. Ørsted by J.L.G. Lund.	199
Fig. 53	Drawing of J.G. Fichte lecturing by an anonymous artist.	200
Fig. 54	Copperplate engraving of the sally of a corps of volunteers from Classen's Have.	204
Fig. 55	Illustration of the British bombardment of Copenhagen.	205
Fig. 56	Acoustic figures drawn by E.F.F. Chladni.	209
Fig. 57	Ørsted's 'Experiments on Acoustic Figures'.	210
Fig. 58	Oil painting of a soirée of chamber music at the home of Chr. Waagepetersen by N.W. Marstrand.	216
Fig. 59	Portrait of A.S. Ørsted by C.W. Eckersberg.	250
Fig. 60	Coloured print of a drawing of F.C. Sibbern by C. Købke.	257
Fig. 61	Lithography of a drawing of B.G. Niebuhr by S.v. Karolsfeld.	258
Fig. 62	The galvanic chain according to Ørsted's *Ansicht*.	269
Fig. 63	Caricature of N.F.S. Grundtvig by C. Hansen.	280
Fig. 64	Copperplate print of Frederik VI's ceremony of anointing by W. Heuer.	293
Fig. 65	Copperplate engraving of Trinitatis Church by J.C.E. Walter, 1826.	301
Fig. 66	The electrometer, or torsion balance, 1785.	302

Fig. 67	Ørsted's galvanic trough apparatus.	305
Fig. 68	Ørsted's piezometer.	307
Fig. 69	Oil painting of the Rasphouse Prison on fire by M. Bang.	309
Fig. 70	Portrait of W.C. Zeise by F.F. Helsted.	314
Fig. 71	Portrait of G. Forchhammer by an unknown artist.	315
Fig. 72	Technical drawing of a coalmine at Sorthat on Bornholm.	317
Fig. 73	Lithograph of the SS Caledonia by an unknown artist.	319
Fig. 74	Pingel's house, 35 Nørregade.	320
Fig. 75	Groundplan of Ørsted's premises at 35 Nørregade.	321
Fig. 76	Drawing of Regensen College around 1830.	332
Fig. 77	Dyrehavsbakken, the amusement park of the Royal Deer Park at Klampenborg.	337
Fig. 78	Ørsted's electromagnetical experiment, 1820.	338
Fig. 79	Diagram showing the lines of magnetic force and the direction of the magnetic field of a current-carrying wire.	344
Fig. 80	Ørsted's notes of 15.07.1820.	346
Fig. 81	Further notes.	347
Fig. 82	Ørsted's *Experimenta circa effectum conflictus electrici in acum magneticam*.	348
Fig. 83	A.-M. Ampère's contradictory experiment.	353
Fig. 84	D.F.J. Arago's corroboratory experiment.	354
Fig. 85	The Copley medal.	356
Fig. 86	Portrait of HCØ by C.W. Eckersberg.	363
Fig. 87	Portrait of Prince Carl of Hessen by P. Copmann.	367
Fig. 88	Lithograph of N.J. Conté, head of the Egyptian Institute in Cairo/Paris by Louis-Pierre Baltard de la Fresce.	368
Fig. 89	Thomas Johann Seebeck's experiments with thermoelectricity.	370
Fig. 90	Portrait of J.W. Goethe by J. Darbes.	372
Fig. 91	Copperplate of the Cathedral of Cologne.	373
Fig. 92	A.-M. Ampère and D.F.J. Arago discussing an experiment.	379
Fig. 93	Portrait of Pierre-Simon, Marquis de Laplace by an unknown artist.	383
Fig. 94	Caricature of the mathematician P.S. de Laplace and the chemist C.L. Berthollet by L.-L. Boilly.	384
Fig. 95	The order of the French Legion of Honour.	385
Fig. 96	Copperplate of the Hall of Caryatids in the Louvre by Berthault.	388
Fig. 97	Oil painting of Sir H. Davy by Sir T. Lawrence.	391
Fig. 98	The Royal Society's assembly hall in Somerset House, The Strand, London.	392
Fig. 99	Lithograph of W. Whewell by E. Upton.	396
Fig. 100	M. Faraday's instrument to prove the rotation of the electromagnetic force.	399
Fig. 101	Oil painting of Sir W. Scott by E.H. Landseer.	401
Fig. 102	Reconstruction of Ørsted's study in Studiestræde.	409
Fig. 103	a. C.W. Gluck, b. his skull, c. his brain.	421
Fig. 104	Skull with marked phrenological bumps.	422

LIST OF ILLUSTRATIONS

Fig. 105	Lithograph of J.J. Berzelius by A. Tardieu.	427
Fig. 106	Lithograph of F. Wöhler by an unknown artist.	428
Fig. 107	Ørsted's bust designed by M.S. Elo (1887–1948) and cast by N. Aluminiumindustri, 1937.	429
Fig. 108	Lithograph of A.S. Ørsted by E. Bærentzen.	434
Fig. 109	Etching of HCØ by E. Eckersberg.	435
Fig. 110	Etching of the University of Copenhagen by L.A. Winstrup.	436
Fig. 111	HCA, paper collage of ballerinas dancing in a clearing presented to M. Ørsted.	444
Fig. 112	Map of Europe showing HCØ's last four itineraries.	445
Fig. 113	Ørsted's crucial experiment.	449
Fig. 114	Ground plan of the Polytechnic Institute.	461
Fig. 115	HCA at the examination table.	468
Fig. 116	Painting of J.L. Heiberg, actress, with her husband J.L. Heiberg, the playwright and literary critic, by N.W. Marstrand.	470
Fig. 117	Caricature of F. Paludan-Müller by C. Hansen.	472
Fig. 118	Painting of the ascent of the hot air balloon of the Montgolfier brothers from the Tuileries, Paris, by M. Carnavalet.	477
Fig. 119	HCA's paper collage of an airship.	479
Fig. 120	Caricature of 'the triumphal procession of censorship' by N.W. Marstrand.	488
Fig. 121	The Assembly of the Estates in the Yellow Palace.	492
Fig. 122	Work table for the artesian well boring in Nyholm 1831–3.	503
Fig. 123	Portrait of HCØ by C.A. Jensen.	507
Fig. 124	Portrait of N.F.S. Gruntvig by C.A. Jensen.	508
Fig. 125	Painting of the Sorø Academy by H.G. Harder.	514
Fig. 126	Etching of King Christian VIII's anointing in Frederiksborg Church by J.W. Gertner.	522
Fig. 127	Ørsted welcomes Berzelius to the Scandinavian Science Conference in Copenhagen.	524
Fig. 128	Lithograph of Christopher Hansteen by E. Bærentzen.	527
Fig. 129	Lithograph of the Tivoli Gardens by E. Bærentzen.	530
Fig. 130	The Casino Theatre.	531
Fig. 131	The party for the Scandinavian scientists, 12 July 1847, in the Yellow Palace.	532
Fig. 132	Lithograph of J.C. Hauch, zoologist, poet, and novelist by E. Bærentzen.	540
Fig. 133	Pencil drawing of the Christian-Albrecht University of Kiel by A. Burmester.	542
Fig. 134	Lithograph of H.C. Schumacher by Ausborn.	547
Fig. 135	Drawing of K.F. Gauss, German mathematician and astronomer, by an unknown artist.	548
Fig. 136	Lithograph of H. Steffens Norwegian-Danish-German *Naturphilosoph* by an unknown artist.	550
Fig. 137	Portrait of L.A. Colding independent discoverer of the principle of the conservation of energy.	553

Fig. 138	L.A. Colding's apparatus for the discovery of the law of the conservation of energy.	555
Fig. 139	Watercolour of A.v. Humboldt's study by E. Leist.	562
Fig. 140	Neues Palais, Potsdam, 1820.	563
Fig. 141	Ørsted's symmetrical figures formed as the names of his great examples written in ink and immediately folded to make a symmetrical mirror image.	571
Fig. 142	HCA's symmetrical ink blot.	571
Fig. 143	HCØ's colour circle showing the polarisation of sunlight through a prism.	573
Fig. 144	HCA's clip of a swan with an added rhyme.	574
Fig. 145	Oil painting of the Sarps Force waterfall by E. Pauelsen.	576
Fig. 146	C.W. Eckersberg's drawing of himself, Ørsted, J.P. Møller, landscape painter, and A. Wallick, scene painter.	579
Fig. 147	'A Party of Danish Artists in Rome' 1837, by C. Hansen.	580
Fig. 148	Caricature of the model school of the Academy of Arts, Charlottenberg, by C.W. Eckersberg.	581
Fig. 149	'Frederiksborg Castle in Evening Glow' by C. Købke.	583
Fig. 150	Daguerreotype of Mathilde Elisabeth Ørsted, Ørsted's daughter, 1870.	586
Fig. 151	Drawing of the Adelaide Gallery, London, by T. Kiernan.	589
Fig. 152	Ørsted at sixty-nine.	592
Fig. 153	Lithograph of M. Faraday lecturing at the Royal Institution, 1856.	593
Fig. 154	M. Faraday's great electromagnet.	594
Fig. 155	Faraday's box.	595
Fig. 156	Leonard Horner, chemist and mineralogist, and his wife.	598
Fig. 157	H. Olde's draft for an oil painting of the proclamation of the Provisional Government of Schleswig-Holstein.	613
Fig. 158	The Brave Soldier, a coloured lithograph.	618
Fig. 159	Daguerreotype of the Ørsted family, 1849.	619
Fig. 160	Daguerreotype of Sir J.F.W. Herschel, astronomer and mathematician.	623
Fig. 161	J.P. Mynster, Chaplain-in-Ordinary, Bishop of Zealand.	627
Fig. 162	Daguerreotype of HCA.	630
Fig. 163	Ørsted's doctoral ornament framed by diamonds.	645
Fig. 164	Glass containing Spanish flies.	647
Fig. 165	Ørsted's death mask in gypsum made by the sculptor J.A. Jerichau.	648
Fig. 166	Ørsted's tomb at Assistens Cemetery.	650
Fig. 167	Ørsted's Masonic coat of arms with his motto 'Truth in Love'.	651
Fig. 168	Bust of Ørsted by H.W. Bissen.	661

To Karin Bastian

PART I
THE STUDENT

> He was a man carved out of one piece,
> he did not subscribe to one belief in his writings
> and another one at home for everyday use.[1]

1 | 1777–1851
Introduction

Ørsted is probably one of only three internationally well-known Danish figures in the history of science, the others being Tycho Brahe and Niels Bohr. So though his career invites a painstaking and comprehensive biography, strange as it may seem there has been no profound study of his life and works, no biography that takes full advantage of the enormous, though scattered, body of source materials he left behind. Of course there are separate studies of his research, his discovery of electromagnetism in particular, of his achievements in the field of scientific education, and of his influential position in Danish cultural life. Still, there is no biography that tries to understand the complete human being and to assess his importance as a scholar and as an organiser, philosopher, poet, and aesthete. As we can tell from what we find out about him, his life and works formed a coherent whole and took place in a close and continuous collaboration with his brother Anders Sandøe. It is hardly an overstatement to suggest that the two Ørsted brothers headed the nation's intellectual elite during the dramatic period from the turn of the eighteenth century to the middle of the nineteenth, the period posterity has dubbed the Danish Golden Age. His fame was the fruit of an unfailing and life-long store of energy which was channeled to benefit Danish science and culture in the broadest sense, and which did not diminish with the many honours heaped upon him. His biographer does not need to build him up with flattering words. He was a man of sterling character marked by a rare harmony of conviction and action. In all its aspects Ørsted's life story is an edifying tale.

It is hard not to admire his accomplishments in scientific research and education. When he matriculated in 1794, there was no degree in physics, chemistry, or any other individual scientific discipline on offer at the University of Copenhagen. By the time of his death the Polytechnic Institute as well as a Faculty of Science at the University had been established thanks to his efforts. Although his career was launched in direct opposition to the sparse research activities of individual fellows of the Royal Danish Society of Sciences and Letters, who kept all doors closed to the young and critical undergraduate, by the end of his life he succeeded in occupying one of the highest positions in Denmark.

In his dissertation Ørsted took as his point of departure Kant's metaphysical philosophy of nature, which offered a critical analysis of Newton's laws of motion. The small scientific community in Copenhagen was oriented towards Paris, which was generally recognised as the universal headquarters of mechanical and mathematical physics. No other Danish philosopher of nature had immersed himself in Kant's critical epistemology (nor had any been attracted to it), apart from Hans Christian Ørsted, whose brother had equally thrown himself into Kant's moral philosophy. Inspired by Kantian metaphysics he worked out a research programme for a dynamical physics that focused on the study of immaterial phenomena such as light, heat, electricity, and magnetism. Against the technological utility conventionally demanded by Denmark's absolutist regime, Ørsted maintained that the study of natural philosophy should be justified for its cognitive and aesthetic benefits. Technological gains were allocated only third place. As a teacher of science he was convinced that natural philosophy equips its students with the best method to distinguish between true and false. To him the laws of nature enshrined a particular beauty and harmony—an aesthetic dimension that would open the eyes of artists and art-lovers to the beautiful in nature and in art. Ørsted visualised an intimate relationship between science and the arts. A natural philosopher would take his point of departure a priori in the metaphysics of nature and would demonstrate the empirical reality of the laws of nature, while an artist would embrace phenomena with his sensory organs in order to reveal subtle ideas to the art-lover. He believed scientific research per se to be closely akin to religious worship. For some people miracles serve as proofs of the existence of God. For Ørsted, however, the absence of miracles reflected divine intelligence. The fact that nature is intelligible and that she abides by laws was awe-inspiring to him. To Ørsted the rationality of the laws of nature and the harmony of all existence revealed God. Moreover, the rationality of laws of nature, which every research effort takes for granted, must be equivalent to human rationality as but a small part of divine rationality. Ørsted's character embodied this awe and a self-confidence which made him consider the laws of nature, including electromagnetism of course, to be 'nature's thoughts' or 'nature's mind'. It is this which is reflected in the subtitle of this biography: *Reading Nature's Mind*.[2]

*

Hans Christian Ørsted revolutionised natural science when in 1820 he observed the deflection of a magnetic needle caused by the electrical force in a wire. His discovery flabbergasted the international community of physicists, particularly the French Académie des Sciences, who had long been convinced that the two phenomena, magnetism and electricity, are embodied in two different, imponderable fluids and are therefore independent. Consequently, all attempts at discovering interaction between them must be doomed to failure. Moreover, everybody, including Ørsted himself, was utterly surprised to learn that the interaction took place in a snail-like orbit which deviated from the rectilinear attraction/repulsion observed in Coulomb's electrometer.

Ørsted's discovery opened the gates to an entirely new field of scientific research. One decade later Michael Faraday discovered that a magnet wound with an electrical wire would induce the inverse effect. Five decades later James Maxwell gave a mathematical treatment to Ørsted's and Faraday's theories of electrical and magnetic forces. The interaction of these forces was subsequently exploited in a range of technological innovations that fundamentally changed everyday life for ordinary people all over the world.

Initially, let me briefly sketch some main points in the historiography of our protagonist. An appropriate first question is this: Can we be sure that the discovery was actually made by Ørsted? His priority has been contested by a few historians of science, notably J.J. Hamel writing in 1859 about early telegraphy. They have argued that two Italians, Giuseppe Mojon and Gian Domenico Romagnosi deserve the honour bestowed upon Ørsted or at least came very close to discovering electromagnetism. These claims have subsequently been repudiated.[3] Although Ørsted owned a book published in Paris 1804 by J. Aldini briefly reporting that 'Romagnosi has made experiments with the magnetic needle', in all probability Ørsted remained unaware of them, hidden away as they are on page 340.[4] Romagnosi never came anywhere near discovering electromagnetism and stated in a letter of 1827 'I was only an amateur physicist and I do not want the honour due to Ørsted'.[5]

Having ascertained Ørsted's priority I shall turn to various views that remain etched in current historiography. Some argue that Ørsted stumbled over the discovery by chance. Wilhelm Ostwald for one labelled Ørsted a *Naturphilosoph* and remarked condescendingly, 'Sometimes nature whispers its secrets into the ears of researchers in the most absurd way'. In Ostwald's eyes 'Naturphilosophie ravaged Germany like a plague in the first years of the 19th century', and he considered Ørsted to be contaminated by this disease, thus implying that he was merely a lucky man, not a sound scientist.[6]

Conversely, Faraday, who redirected his field of research for the rest of his life in consequence of the discovery, found in 1821 already that Ørsted, due to 'constancy in his pursuit of his subject—both reasoning and experiment—was well rewarded...by the discovery of a fact of which not a single person beside himself had the slightest suspicion.'[7] In other words Faraday, as opposed to Ostwald, stressed the singlemindedness and originality of Ørsted's epochmaking discovery. A similar view was entertained by John Herschel who compared Ørsted's perseverance to that of Columbus. Both discoverers obstinately anticipated the necessary existence of hitherto unknown phenomena, and when they found them they realised at the same time that what they eventually found differed not only from their initial expectations, but also from contemporary doctrines.[8]

Oswald's degrading view that the discovery was a lucky accident is a world apart from Faraday's and Herschel's unreserved acknowledgement of originality. Still they have one thing in common. Ørsted acted single-handedly. Their explanations require there to be no impact from the scientific community. But for almost a century other historians of science have been on the look-out for sources of influence, taking it for granted that scientists do not act in a vacuum, but are in fact intellectually influenced by encounters with other scientists or their works. This may be true, but leaves open the question of degree of influence, varying from imitation or even plagiarism to relative independence or even originality.

For the past five decades controversy among historians of science has revolved around the impact from Kantian metaphysics and/or *Naturphilosophie*, although in ambiguous ways, because attitudes towards the two sources of inspiration—sometimes perceived as lying outside the field of proper scientific investigation, sometimes not—have oscillated between two extremes. K. Meyer claimed (1920) that Ørsted was deeply influenced by Kant or by Schelling or by both at least in his younger days, but also that he came to realise the futility of their speculative approach to science as he matured and headed for empirical research.[9]

B. Gower in his penetrating article 'Speculation in Physics' (1973) aimed at demonstrating that *Naturphilosophie* represents a cul-de-sac in the history of science. Schelling's 'baffling abstruseness' cannot possibly have been conducive to enduring scientific knowledge, and in any case Ørsted's enthusiasm for *Naturphilosophie* waned before his discovery. T. Shanahan in his reassessment (1989) of Schelling's intellectual influence raised equally strong objections and to underpin his scepticism quoted Ørsted as follows: 'He [Dr J.J. Wagner, a German philosopher] wants to give us a complete philosophical system of physics, but without any knowledge of nature except from text-books and without possessing the same rigorousness of philosophical construction as Kant, exactly like his master [Schelling]…These people all bring to market lame philosophical proofs and lopsided physical theories, and then they grumble when others will not accept them.'[10] Shanahan argued that Ørsted drew his inspiration from Kantian epistemology and metaphysics, which was fully sufficient to have inspired his great discovery. G. Buchdahl(1986), finally, stated that Ørsted's discovery was probably the only conspicuous example to show that Kantian metaphysics was to some extent instrumental, thus proving its scientific fruitfulness.[11]

However, the pendulum had already started to swing to the opposite extreme from the 1950s. The virtues of *Naturphilosophie* were vindicated by a number of American historians of science. R. Stauffer (1953 and 1957) was perhaps the first to claim that Ørsted had been influenced (not only by Kant, but also) by 'Schelling's beautiful and great ideas…(that) should be recognized as factors involved in a major discovery in physics'.[12] P.L. Williams followed suit in his authoritative article in the *Dictionary of Scientific Biography* (1974) stating, 'Since both Oersted and the *Naturphilosophen* drew their inspiration and basic ideas from Kant, it is no coincidence that Oersted's later [sic!] philosophy closely resembled *Naturphilosophie*.'[13]

The next generation of American historians pursued this line, but stopped to dig a little deeper. Reasonably, K.L. Caneva(1997, 2007), A.D. Wilson 1998, 2007), and M. Friedman (2007) inquired:[14] 'if it is true that Schelling's ideas had an impact on Ørsted's discovery it must be possible to clarify more precisely what idea helped him'. In looking for an answer to this question they point out Schelling's tripartite scheme, and Friedman calls special attention to the publication of this scheme in the 1803 edition of Schelling's *Ideas*. So, we have now moved from Stauffer's suggestion that Kant's and Schelling's ideas in general influenced Ørsted to the assertion that Ørsted's discovery was even conditioned by Schelling's specific scheme, without which there would be no path leading from Kant's metaphysics to the discovery of electromagnetism. Friedman concluded: 'The crux of the matter, of course, is that Schelling's *Naturphilosophie* was in fact an intelligent, perceptive, and appropriate response to both the tensions in Kant's system and the new empirical results. Viewed in this particular context, I believe, there is indeed much of "philosophical value" to be gained by studying it.'[15]

Since the discovery of electromagnetism was so completely at odds with contemporary doctrines, one wonders why other physicists or chemists under the spell of *Naturphilosophie* did not make scientific discoveries of any significance. In 1829 Steffens, perhaps Schelling's most ardent disciple and once Ørsted's rival, congratulated Ørsted on his scientific success while regretting that *Naturphilosophie* had achieved so very little (ch. 52).

This introduction is not the right place to comment on this multipolar, complicated, and ongoing controversy. Throughout this biography I shall aim at elucidating my claim that Ørsted was open-minded, cosmopolitan, and under a wide range of influences, but never became a partisan of any established intellectual movement let alone of *Naturphilosophie*. I believe that this debate has strongly exaggerated the issue of 'influence'. Furthermore, the stronger the dependency on a particular authority the more the autonomy and originality (qualities Ørsted valued highly) of our protagonist are dwarfed. This study will explore practically all available source material illuminating Ørsted's persona and works by focusing on the epistemology and experimentation of his scientific work, by scrutinizing his many personal relationships with the scientific community at home and abroad, by bringing to light his many activities as an organizer of academic and political institutions, and finally by unravelling his particular interests relating to religion, ethics, and aesthetics.

*

Since Ørsted was the only chemist/physicist in Denmark who considered himself a Kantian, it is hardly surprising that he received very limited response from domestic academia. Court Steward Adam Hauch, who had studied under Lavoisier, was an independent amateur scientist, open-minded and of high renown, and he was one of the very few with whom young Ørsted entered into a fruitful dialogue. Professor Manthey, the pharmacist, protected him and provided the travel grant that enabled his protegé to go abroad. But Thomas Bugge, professor of mathematics and astronomy found Kantian metaphysics irrelevant, if not harmful for science, so Ørsted's early career at the University was most uncertain. What mattered most of all to him was the opportunity to meet natural philosophers in the German states and in Paris, to escape provincialism, and to encounter philosophers and experimenters who might become collaborators on his dynamical project. In the five decades from 1800 to 1850 Ørsted spent more than five years abroad, that is more than ten per cent of his adult life. And he did make friends as well as enemies. The first journeys were primitive *Wanderjahre*, years of apprenticeship, whereas his travels after his discovery were veritable triumphal entries into Berlin, Paris, and London.

Ørsted was well equipped to engage with foreign communities of scholars. His German which he had picked up in childhood was fluent, and when he arrived in Paris he paid a tutor to give him instruction in French every morning. Soon he would write articles for scientific journals in French and there seems to have been no problem for him in conversing with the members of the French Académie. His first visit to Britain took place as late as 1823, and his problems with English were considerable. At that time English ranked third as a foreign language in Denmark, and in the beginning he was utterly confused by the relationship between English spelling and pronunciation, and struggled to spell what he heard and to pronounce what he read. His speeches were flat and formal for want of vocabulary and the letters he sent John Herschel in the late '40s testify that he never quite overcame these initial difficulties.

His journeys abroad meant everything to him, because now he encountered his peers and found role-models. He gave special attention to the styles in which lectures were delivered and experiments made. He saw laboratories and collections of instruments he had never seen before, and he picked up useful social competences in the salons and societies he joined.

He experienced the enormous social differences between natural philosophers in Germany and France. The worlds of Ritter and Laplace could hardly be further apart. But of course he also noticed the different modes of thinking about and practising science in Germany where *Naturphilosophie* had a strong foothold and France where Laplacean pre-positivism ruled supreme. Ørsted could not help observing the opportunism that impregnated French science in the post-revolutionary period and as a Nordic puritan he was almost bound to criticise the well rewarded members of the French Académie who unlike his friend Ritter—did not have to work hard to make a living. Moreover, their philosophies were so different. In France it was as if Kant had lived on another planet, whereas the Germans were better informed about their neighbours thanks to their ability to read French.

Crossing the borders between these diverse scientific cultures alerted Ørsted to the particularities of paradigmatic differences across Europe. Representatives of these cultures were part of his vast network of colleagues, and they did not speak the same scientific language. Humboldt and Ørsted were probably the two natural philosophers who communicated with the greatest number of scientists across Europe. I have found it of great interest to study how Ørsted managed to cope with this state of affairs. He found it necessary to adapt his major work to the two contending cultures, the German and the French, by publishing it in two different versions. Having made this discovery he immediately saw its potential in supporting the dawning opposition to Laplacean supremacy in French science. In the end Ørsted's electromagnetism was certainly a contributory factor in bringing about the fall of Laplace. In Britain his discovery was immediately acclaimed as a revolutionary conquest, but his interpretation of it was met with some reservation, as being less interesting. Towards the end of his life he urged Herschel to help him find a publisher for his aesthetic and religious philosophy (*The Soul in Nature*), but to no avail, and when the book finally landed on the British market it fell on barren ground. Darwin found it 'dreadful'.

*

Ørsted's achievement, however, went further than taking advantage of his discovery of electromagnetism to secure a place for the sciences at the University. Already as undergraduates Hans Christian and his brother Anders had immersed themselves in Kantian philosophy, the elder in the metaphysics of science, the younger in the philosophy of morality. Belonging to two different camps, science and moral philosophy, this interdisciplinary collaboration was from first to last obviously attracted by one of Kant's most essential problems: the apparent antinomy between the determinism of the laws of nature ('the starry heaven above me') and the freedom of the will ('the moral law within me') so crucial to upholding the dignity of man. Like their master who sought after the desirable possibility of joining his three critiques into one philosophy, both brothers were engaged in trying to find out whether these realms, so opposite at the outset, could be unified.

One of the reasons that some historians of science have included Ørsted in the movement of *Naturphilosophie*, I think, stems from the observation that this movement embarked on a similar endeavour. So what is the difference between Schelling's solution and Ørsted's? The decisive difference, as I shall try to show, is that Schelling already from the beginning founded a philosophy of identity according to which he justified the imposition of speculative ideas upon nature in order to invest it with consciousness. If Ørsted was sometimes tempted to embrace Schelling's philosophy of identity (objective idealism) or parts of it, he more often rebuked it for ignoring the importance of empirical investigation. He never adopted Schelling's philosophy of identity, but remained faithful to Kant's dualism, and in particular to his distinction between nature and mind, and between the phenomenal and the noumenal realms (chs 25 and 52).

Nevertheless, Kant did bridge the gulf between the determinism of natural laws and the freedom of morality by the following chain of arguments: the laws of nature, apparently, are endowed with reason in so far as—according to human reflection—there seems to be a plan in nature and consequently design and plot. Also the categorical imperative—based on human reason—presupposes that it is at least possible to realise the moral objectives in freedom in this phenomenal world as otherwise it would lose its meaning. Hence, the freedom (to intend at least) to live a morally good life in an ambience determined by laws of nature is open to us as a possibility. This view, of course, is a product of practical philosophy. It is not knowledge, but an idea fostered by human reason. And it is different from the bridge established by Schelling's *Identitätsphilosophie* or objective idealism, which eliminated the possibility of moral freedom by ascribing a theoretical identity between nature and mind to an Absolute Godhead, to which human morality (and empirical science) is irrelevant.[16]

The practical lives of the brothers seem to have been governed by Kantian and Fichtean ideas. Apart from his many scientific activities Hans Christian dedicated most of his time after 1820 to cultural activities in the broad sense: 1, dissemination of scientific knowledge to the general public; 2, aesthetics of nature; 3, religion; and 4, politics and the rule of law.

1. To start with, Ørsted gave public lectures because he had to make a living. But they continued beyond that need for the rest of his life. The scientific culture for Ørsted was the challenge of disseminating scientific knowledge to the entire population. He was convinced that knowledge of the laws of nature would help people understand the importance of distinguishing between true and false. He started in 1815 to lecture to the scientific laity on Sundays. Inspired by voluntary associations he visited in London, he founded in 1824 the Society for the Dissemination of Science in Denmark; this provided the leverage for the establishment of the Polytechnic Institute five years later, which in turn facilitated the establishment of a Faculty of

Science at the University in 1850. Ørsted's lectures enjoyed large audiences. Series of lectures he delivered to fora of young artists in collaboration with the Academy of Arts took as their theme the significance of knowledge of laws of nature for the creative arts.

2. Although Ørsted was burdened with technological assignments laid upon his shoulders by the government, he opposed the public demand for the sciences to take on the role of a servant to technology. Aside from the intrinsic value of the sciences, he considered them basic to the arts. For to him art imitates nature, so how can the artist paint the idea of a natural landscape without understanding at least the fundamentals of optics and the theory of light and colours? He also formed an editorial board from members of the learned republic to publish a critical journal of the arts, theology, new and disputed sciences such as phrenology, and literature. The intention was to set a course to influence public taste and inspire cultural life.

3. As boys the two brothers had already shown an active interest in theological issues. Anders wrote sermons and planned to study divinity. As students they frequented the services of Dr Marezoll, the German and Kantian vicar at St Petri. Grundtvig, the theologian, mythologist, and hymn writer so provoked the Ørsted brothers by launching an impetuous attack on the new German school of philosophy that Hans Christian engaged him in a long debate on pantheism. Both took part in theological clashes between the state church and dissenting movements, asserting their opposition to smug dogmatism and advocating instead a minimalist creed and an unorthodox concept of God. Here, too, Hans Christian thought like a Kantian. God cannot be an object of theoretical knowledge, but it may accord with human reason to imagine God as the creator of nature. As such we may perceive him as revealed in nature, but nothing can be deduced from theology to science. In the end his unorthodox concept of God brought him on a collision course with the supreme authority of the state church.

4. When eventually the absolutist regime opened its doors to popular participation in politics he became a 'fiery soul' (a term he invented) in The Society for Liberty of the Press, seeking to influence the government in order to advance the protection and extension of the freedom of the press and the rule of law. He worked closely with his brother Anders who in 1826 had been forbidden to take part in public debate, so they agreed on a secret plan to work closely together. Anders would be the mole in the government and feed Hans Christian with the insider's knowledge necessary for him to act as a mouthpiece for both. Much to Hans Christian's annoyance, however, the National Liberal movement gradually succeeded in dominating the society, and he resented their nationalistic programme and power-seeking aspirations. He ranked the rule of law above democratic rule and ended up disillusioned with the political events taking place in Denmark in 1848, leading to a premature constitutional change and a devastating civil war.

*

Given this background, I endorse the image of Ørsted as a man of two cultures. This is probably not his image outside his native country. I think one of the reasons that Ørsted's persona has been split into two is that his works have been published in that way. His daughter Mathilde Ørsted edited a posthumous collection of her father's cultural works in nine volumes [*Samlede og Efterladte Skrifter*]. Most of the first two volumes were published in English under the title *The Soul in Nature*. They contain his philosophical, religious, aesthetical, pedagogical, and political thoughts. At the end of the ninth volume comes Hauch's unsurpassed biography, carefully depicting his protagonist as one who bridged the gap between two cultures and distinguished himself in both.

By 1920, the centenary of his discovery, the dualism of the Golden Age had long lost general support in favour of positivism. The celebration of Ørsted, the national icon of Danish science, that year (which coincided with the reunification of part of Schleswig with the Danish monarchy), was used as an opportunity to harness him to a political project to promote engineering and scientific education. The Royal Danish Society of Sciences and Letters asked Dr Kirstine Meyer to edit and introduce Ørsted's scientific writings, while completely ignoring the cultural parts of his oeuvre. Meyer's publication of Ørsted's collected scientific writings [*Naturvidenskabelige Skrifter*, translated as *Selected Scientific Works of H.C.Ørsted* (1998)] widened the gap in his life, a gap that was more than simply formal. According to positivism, the prevailing paradigm of the 1920s, science must be cleansed of values. Hence, complete separation of science from metaphysics and cultural values was considered not only reasonable but virtuous.

The gap between the sciences and the arts reflects C.P. Snow's well-known division between two cultures as having two languages that are mutually incomprehensible. Piet Hein, the Danish architect and poetic philosopher, has amused his readers by calling the scientific and technological camp 'technocists' and their adversaries 'cultists'. The two terms deliberately echo 'idiots' and 'occultists', hinting at the reciprocal prejudices of the two camps. According to this animosity 'technocists' only take an interest in exact and material problems and their concrete solutions whereas they lack aesthetic education and taste. 'Cultists' on the other hand wander aimlessly about engaged in metaphysical speculation and even approach the occult, while boasting of their incapacity to appreciate technology at the same time as making use of high-tech gadgets. Sadly, Snow claimed, this cultural gap is widening. Ørsted, however, exerted a strong influence in both camps insisting on uniting the scientific and cultural elite. Members of the learned republic of the Golden Age were still on speaking terms across cultural borders. For Ørsted human reason bridged the two camps.

Ørsted was a formidable, original, and ambitious character in Danish science and culture. Nevertheless he maintained a child-like, almost naive and trustful attitude that endeared him even to people who did not share his convictions. Am I depicting some kind of saint and are you about to read a hagiography? In fact it is hard to find examples of unpleasantness or accusations of selfishness being levelled at Ørsted, and I have searched meticulously for them in the archives. What personal benefit could he possibly hope to gain from fostering a fourteen-year-old chemist's son and a hypochondriac in his own house, or from supporting Hans Christian Andersen, an anxious cobbler's son and budding poet of sixteen? What ulterior motive could he possibly have in discreetly but open-handedly sharing his money with Ritter, a German pauper and amateur scientist, or with Anders Sandøe, his sick brother, or in exercising his influence in favour of young talents such as Oehlenschläger, Heiberg, and Carsten Hauch to promote their careers? His readiness to help reflected a well-developed ability to spot talent and a conviction that just as he had previously sought and found shelter under the protective wings of others, so it was now his obligation to pay off that debt. According to Ørsted it was a

Fig. 1. Oxford University Museum had planned the Ørsted statue to be carved in marble by A. Monro and to be paid for by Queen Victoria, like the other statues. Unfortunately, the sculptor had to give up for want of a model and he was unable to get hold of Ørsted's death mask. This was no wonder, because the requisition had been sent to Sweden (Phillip's letter of 10.07.1861, The Oxford University Museum Archive). Meanwhile Monro had already been paid a sum in advance according to the contract of 1859, so a protracted dispute ensued and the statue was only erected in 1885 after a lot of fuss. According to archival material kept by the museum the statue is modelled by a certain K. Jobhen (B. Haward 1991, F. O'Dwyers.a. 257, and personal information from curator Stella Brecknell). 'K. Jobhen' is a sloppy transcription of a Danish hand. Actually, the artist was J.A. Jerichau (1816–83), who had been commissioned to model and cast a bronze statue for erection in the Ørsted Park, where it was unveiled in 1876 (fig. 2). When J.C. Jacobsen, founder of the Carlsberg Breweries, became aware of the problem at the Museum he took the trouble to buy an unused design in clay from Jerichau's estate and had it shipped to Oxford (Nos. 103a and b in *Fortegnelse over de af afdøde Professor Jerichau efterladte Kunstværker* [Catalogue of posthumous artworks by the deceased Professor Jerichau], 1885, Studiesamlingen, SMK). Obviously, both statues are made by the same artist (similar facial features, but different positions of hands). Jacobsen had been attending Ørsted's lectures in the Society of the Dissemination of Science for many years. Photo by Oxford University Museum. I am grateful to Stella Brecknell for her help.

Fig. 2. In 1861 a committee consisting of Hans Christian Andersen, the storyteller, G. Forchhammer, geologist, O.B. Suhr, merchant, and F. Tillisch, Privy Councillor organized a collection of voluntary contributions to erect a monument in honour of Ørsted. The committee soon approached Professor Jerichau, whose first design showed four bas-reliefs representing subterranean forces of nature and allegorical women from antiquity, with Ørsted dressed in a Greek chiton on top of the pedestal. The committee rejected this classical design, demanding Nordic norns and a modern scientist dressed in contemporary clothes. Jerichau's new design shows Ørsted in a didactic position and with a magnetic needle on a pillar and at his feet a galvanic apparatus. The statue rests on a base of granite on a hill in the new park established on the site of the former moat surrounding Copenhagen. (Anne Christiansen, *Skønhed og skrøbelighed. Værker af billedhuggeren J.A. Jerichau fra museets samling* [Beauty and Frailty. Works by the Sculptor J.A. Jerichau from the Museum's Collections], OBM 2003.) Photo by the author.

personal duty to develop a talent bestowed by nature for the common good. His trust in individual perfectibility was strong and emanated from reason, for it would be contrary to reason to let gifts of nature vanish. Genius (like noblesse) obliges.

In his biography of Niels Bohr, Abraham Pais notes that a reproduction of a painting of Ørsted speaking at the meeting of Scandinavian Scientists at Roskilde 1847 (fig. 131) decorated Bohr's office at The Institute of Theoretical Physics in Copenhagen. Pais regrets that he never asked Bohr what precisely this picture meant to him, but assumes that Bohr felt a certain affinity with Ørsted. Both had done epoch-making scientific research, both were internationally recognised and widely travelled; both had leading roles in The Royal Danish Society of Sciences and Letters and in The Society for the Diffusion of Science, both were home-loving persons, and both had lifelong ties to brothers who also had important roles in academia. Finally, both of them had something boyish about them, some trait of straightforward innocence and plain modesty.

Ørsted's international image resembles a sketch or a draft, not unlike the statue of him in the Oxford University Museum of Natural History, which is also symptomatic of his position in the history of science. In this great hall thirty-four statues of famous scientists from history have been erected. There are three Greeks from Antiquity (Aristotle, Hippocrates, and Euclid), twenty-six Britons, and four Europeans (Galileo, Leibniz, Linnaeus, and Ørsted) as well as Albert, the Prince Consort, who was neither quite British nor quite a scientist. It is a mark of merit, of course, that Ørsted is categorised on an equal footing with Newton and Darwin; but it is hard to escape the notion that the British have been over-generous towards their fellow countrymen as, for instance, no French scientist has been given the honour. Most conspicuous, however, is the fact, that whereas all other statues are carved in stone, Ørsted's is modelled in plaster. He stands out as an emergency solution. Thus Ørsted is incorporated into the international company of immortal scientists in this hall of fame, but nevertheless he is represented as an outsider.

Most of the questions raised in this introductory chapter are deliberately left open. They will find answers in the following chapters, which are basically structured chronologically, but also, where appropriate, with due regard to thematic coherence. The epilogue is particularly intended to discuss Ørsted's position in international historiography. This life of Ørsted endeavours to put him in his proper niche in the history of science by elucidating all aspects of his many faceted life: scientific, religious, aesthetic, educational, and political, private and public, as well as national and international.

What is the most distinguished thing one can be in the world?[1]

2 | 1777–94
A Childhood without Playing

BEFORE THE boy was seven years old, his private teacher, a German wig-maker living in Rudkøbing, asked him the standard question:

'What do you want to be when you grow up?'

Immediately, the reply came in the form of another question: 'What is the most distinguished thing one can be in the world?'

'The Emperor of Rome!' the teacher replied.

'All right, then I'll be the Emperor of Rome.'

'Impossible!' the wig-maker objected: 'It's hereditary.'

The boy found the objection unreasonable and asked what other highly distinguished things one could be.

'General-superintendent in Lybeck!' said the teacher who had been born in Lybeck himself. 'He is an exceedingly important man.'

'Well. Then I want to be the Bishop of Lybeck!' the boy burst out enthusiastically, knowing that a general-superintendent in Germany is equal to a Danish bishop.

'Then you have to learn a great deal first', the teacher exhorted.

'Of course' the boy said, determined to become a student of theology.[2]

Hans Christian was the boy's name. His parents were terribly busy taking care of the pharmacy at Rudkøbing, far too busy to educate their offspring, who were increasing by nearly one a year. So they sent their first-born and his siblings to private education at Oldenburg's, the wig-maker, and his wife living just across from the pharmacy. But let us begin with the boy's earliest memories.

Hans Christian recalled a summer night when he and his exhausted family landed at Stigsnæs, the nearest point on the coast of Zealand, having sailed from Hov at the northern end of Langeland. The head of the family was the chemist of Rudkøbing, Søren Christian Ørsted. The other members of the family were Karen, his wife, and their three sons, Hans Christian (aged 4), Anders (3) and Jacob (1). The sea had been rough, waves had dashed their boat hither and thither, the children were drenched to the skin and seasick, and everybody yearned for a bed. Lightning flashed across the sky, and thunder and pouring rain made them feel even worse.

From how far back in our childhood can we remember things? *Really* remember things, that is, not 'recall' episodes that have been recounted to us so many times by our parents that it seems as if we remember them ourselves. The eldest of the young boys claimed to remember this landing, and that his father knocked with his stick on a door to wake up people at the guesthouse near the jetty. Inside, the soaked family was offered beer, but after the first pull Hans Christian pushed the mug away insisting that the beer had gone flat and was distasteful.[3]

The Ørsteds had come to visit branches of their family on Zealand that summer. First they saw Karen's relatives in Holbæk and then went on to see Søren Christian's in Slagelse and Copenhagen. Hans Christian retained a few scattered memories from these visits.. His mother was the daughter of a merchant, the well-to-do Herman Hansen, and her wedding to Søren Christian in Saint Nicolai's Church, Holbæk, had been followed by a celebration in Herman Hansen's general store on 24th July 1776, little more than a year before Hans Christian was born. The couple had fallen in love in Holbæk, where the groom had been in charge of the Elephant Pharmacy at Ahlgade, assisting the widow of its former owner until the pharmacy was sold to another chemist.[4] Unfortunately Søren Christian had not been able to afford to buy it.

Karen Hermansdatter came from a large family, since her father had been married three times and had fifteen children, five with each wife. At Holbæk the Ørsteds saw Aunt Bodil who was married to another merchant, Anders Sandøe. Bodil and Karen were the first and last children respectively of their father's second marriage.

Hardly anything about Karen has survived among the memories of the two elder brothers. Yet young Hans Christian wrote her a poem for her birthday:

> The day you saw light for the first time
> We shall honour and celebrate.
> It is a festive day for your young ones
> And will ever remain so.
>
> To make us children happy is your pleasure
> You wish to teach us all that is best.
> You'll guide us towards duty and virtue,
> And we shall always remain obedient to you.

> We wish you joy, we wish you luck!
> Let far more than we can ask for
> Descend upon you from above.
> May every day be a pleasure to you.'[5]

This poem hardly goes beyond what many other well-brought-up children could have written for their mother. Søren Christian was only allowed to keep her for fifteen years. Karen died in 1791, at 46 years of age, when Hans Christian was fourteen. By that time she had already presented the chemist with eight children and Hans Christian with seven younger siblings.[6]

The Ørsted family went on to Slagelse where they paid a visit to Søren Christian's mother, Barbara Albertine (née Witth) who had been a widow for twenty years after the death of Christian Sørensen Ørsted, pastor at Saint Peter's Church and Antvorskov Castle.

Christian and Barbara had thirteen children in fifteen years, and at the time she was widowed seven of them between the age of three and seventeen had survived, among them Søren Christian. When he visited his mother and proudly introduced his three sons, she lived in her own house in Slagelse, which her considerate husband, who was older than her, had bought and refurbished after a fire. Hans Christian recalled that he was offered sandwiches by his grandmother in the basement of her house, but refused them and demanded to be served upstairs, not in the company of servants. Even as a child he held a high opinion of himself, he later admitted.

In this house Søren Christian grew up with his siblings—aunts and uncles to Hans Christian. Those that the children knew best were the aunts Engelke and Benedikte and the uncles Lauritz Gerhard and Jacob Albert. In 1778 Engelke moved to Copenhagen and married a dyer who died the same year, after which she married another dyer, and when after a few years he died as well she probably concluded that enough was enough and took charge of the dye-works at Vestergade herself. She had no children, and nor did her sister, Benedikte, who had stayed at home to help her mother. After their mother died Benedikte moved to her sister Engelke's to take care of the big household of the dye-works. By common consent she was not a complete success because she suffered from somnolence, which was somewhat of a handicap (particularly when falling asleep in the kitchen). Engelke's dye-works eventually became the rendezvous where the two brothers met with other students.

Søren Christian and his brothers began their education at home and continued at Slagelse Grammar School. Jacob Albert became a student of divinity while Lauritz Gerhard and the youngest brother did not make it into university. Søren Christian was apprenticed to a chemist at Aabenraa and later worked as a pharmacist at Faaborg and Holbæk until eventually he passed the pharmaceutical exam at the University of Copenhagen. Their mother supported the boys to the best of her ability selling off parcels of land and buildings of her estate. In 1776 Søren Christian borrowed 250 rixdollars from her to buy the pharmacy at Rudkøbing, and Lauritz Gerhard spent his advance on setting up a silk and cloth store on Købmagergade in Copenhagen.[7] The Ørsted family maintained an attitude of mutual help.

Fig. 3. HCØ's father, Søren Christian Ørsted, the pharmacist (1750–1822) and stepmother Anne Dorothea Borring (1764-c.1817). There is no picture of Karen, his mother, who died in 1791, three years before the two brothers moved to Copenhagen. Silhouette by unknown artist. DTM.

From Slagelse the family proceeded to Copenhagen to Aunt Engelke Møller, by then a widow and running the dye-works close to Vester Port, one of the gates through the rampart. Hans Christian remembers sailing about in a wooden tub in one of the large vats used for dying the cloth woven by peasants at home and brought in to the city. Another incident testifies to his childish conceit. He was put to bed in a clothes basket, but he refused to sleep so primitively and demanded a proper alcove with a curtain. Only when he was persuaded that a clothes basket was most suitable for a nobleman did he acquiesce.

From Copenhagen the family's itinerary continued to Northern Zealand to see Uncle Jacob Albert, minister to the congregations of Kirke Helsinge and Drøsselberg. 'Who are you?' Jacob Albert asked Hans Christian. As a young student of divinity he had carried the little boy to the baptismal font at Rudkøbing six days after his birth on 14th August 1777.[8] 'I am Hans Christian

A CHILDHOOD WITHOUT PLAYING

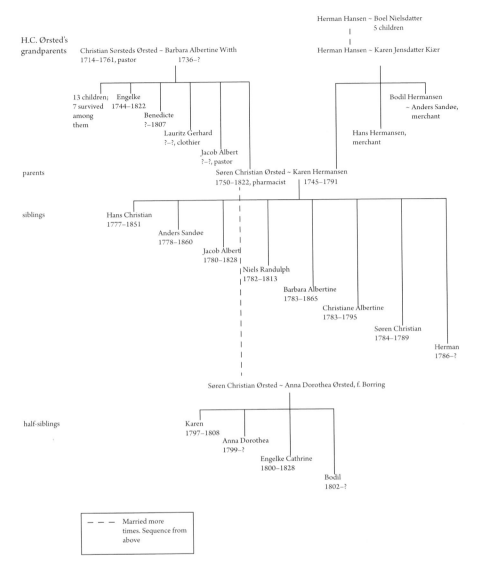

Fig. 4. HCØ's pedigree.

Bravkarl [Brave Fellow]!' the boy answered smartly.[9] This honorific nickname became a firm attachment to his name. Thirteen years later when as a student he would sit down at Aunt Engelke's dinner table, his friends, Adam Oehlenschläger for one, would often address him jokingly as 'Hans Christian Bravkarl'.

The family visits that summer had a purpose. Karen and Søren were proud to show off their boys and give them an impression of belonging to the wider family into which they were born.

19

This family extended from their mother's parents and siblings in Holbæk to their father's mother in Slagelse, Aunts Engelke and Benedikte in Copenhagen, and Uncle Jacob Albert at Kirke Helsinge. Afterwards the Ørsted family returned to their everyday life at Rudkøbing, Søren Christian to his pharmacy, Karen to more pregnancies and births, and the boys to the wig-maker. Between 1782 and 1786 Hans Christian acquired five more siblings, two sisters and three brothers. Mother Karen was not strong enough to survive this and died in 1791. By then one of the children had died aged 5. Their father could not possibly look after all these children as well as his pharmacy and guesthouse on his own, and three years later he married Anna Dorothea Borring, his junior by fourteen years.

These family relations may be difficult to take in, but it becomes even harder to grasp the individual names because Søren Christian and his two wives chose to recycle the names of their forebears when their own children were baptised. For instance the double name of our protagonist probably derives from his uncle, Hans Hermansen, and his grandfather, Christian Ørsted.

The Ørsteds were not alone in following this custom and the next generation continued this recycling of names. It served the purpose of strengthening family ties at a time when the different branches of the family were difficult to keep together for geographical reasons, because well-educated fathers had to take up jobs as medical doctors, pastors, and chemists wherever vacancies were available. Secondly, parents were often replaced by step-parents owing to the early death of the father or (more often) the mother. Family ties were useful, even indispensable, in a society without public social security, when mothers passed away in childbirth or breadwinners died young, making it impossible for a single parent to manage both to work and to bring up a crowd of children. Indeed, the summer's tour of Zealand was precisely an attempt to maintain and strengthen the ties of kinship between the scattered sections of the family.

The family name refers to the village of Ørsted outside the town of Randers, where Søren Olufsen had been a pastor at the time of King Christian IV (1588–1648), a post held both before and after by many members of the family. So the Ørsteds could look back on a pedigree of Lutheran priests. Among their forebears were bishops in both Norway and Denmark: a Randulph, a Wandal and a Winstrup. Thanks to their stepmother's descent from bishop Wandal, Hans Christian and Anders Sandøe enjoyed the obvious privilege of a scholarship at Elers' College in Copenhagen founded by the bishop.[10]

In 1776 Søren Christian had bought the small pharmacy in Rudkøbing, when the court ordered it to be sold for the fourth time. The price was a trifling 600 rixdollars, less than the insured value. Rudkøbing was a desolate place, but its pharmacy—one of the cheapest in the kingdom and surviving as a distillery rather than a pharmacy—was as much as the newly-wed chemist could afford. The young couple had their hands full just keeping it afloat. At first the employment of servants was out of the question, so Karen had to take care of the guesthouse that soon turned out to be the most popular accommodation in the town. The pharmacy and guesthouse were close to the harbour just a few steps up the hill. Søren was an industrious chemist, he set up a distillery and a packing room and gradually hired women to collect and dry medicinal plants and aromatic herbs; he also built a small lab to carry out simple chemical

Fig. 5. The pharmacy of Rudkøbing which Søren Christian Ørsted bought in 1776 in a desolate state, but soon modernised by establishing a lab in the wing to the right of the courtyard. Drawing by Jens Kortermann Jauch, Museum of Langeland.

analyses and where he would introduce his sons to the enchanting world of chemistry. The Royal Agricultural Society awarded him their gold medal in 1785 for his initiative in collecting plants and herbs and thus creating new jobs on the island.[11]

Even though Rudkøbing had been granted a municipal charter in the Middle Ages, its population of less than a thousand inhabitants was insufficient for it to have its own grammar school. A former grammar school had been closed in 1739 and its premises made available to the parish clerk for the instruction of confirmands. Subsequently, the parish clerk had been persuaded to establish a primary school where no Latin was taught. However, though the town had appointed a schoolmaster, what kind of education could be expected from a school that was only able to charge two to four skillings a week per pupil? 'What dire wages for such dismal work', the district judge noted resignedly, suggesting that the posts of schoolmaster and undertaker were merged as soon as they became vacant. For want of a reasonable livelihood it would be impossible to obtain a schoolmaster who would teach 'without sighing [for poverty]'. The district judge demanded regular examinations in public, 'because true enlightenment can only be expected where honour is spurred on by diligence'.[12] The chemist craved a good school for his sons, but no such thing was available in Rudkøbing.

Hence, on the family's return from Zealand, Hans Christian and Anders Sandøe were sent to the infant school run by Christian Oldenburg and his wife just a few steps away.[13] They were not the only untrained teachers offering private education in the market place. Oldenburg was an immigrant from Lybeck and a wig-maker whose clientele in a town with two competing wig-makers was dwindling. As is often the case with immigrants who are no longer in a position to take advantage of their native language and professional skills, they cling to the values they brought with them to safeguard their threatened identity. Oldenburg, as we can imagine, tried to get to grips with his disappointment at the lack of demand for his noble craft by sticking to the indestructible values of his past: the grandeur of Lybeck compared to the mediocrity of Rudkøbing, the genuine spirituality of German Pietism compared to the stolid rationalism on Langeland, and above all the superiority of the German language and culture over plain and boorish Danish and the local cultural wasteland. His know-all attitude was a source of irritation to his Danish wife whose language he refused to learn.

What then could be more conducive to his comfort than to dedicate himself to satisfying the chemist's boys' eager thirst for knowledge? The wig-maker took on the task of teaching Hans Christian and Anders Sandøe to read and write German through the language of the Bible, and he fulfilled this task thoroughly and conscientiously so that his pupils not only attained a perfect command of the German language, but also acquired a knowledge of scriptural passages by heart that they never lost. In the garden of the pharmacy was a huge chestnut tree with a complex crown that offered everything that climbing boys could wish for, but this was not the place where Oldenburg's two pupils exercised their talents when school was over. Playing and socialising with other boys seemed less attractive to them than bookish activities, and in any case they were happiest in each other's company.

Latin was a language that Oldenburg had not mastered at all, so the boys had to turn to other teachers to make sure that their knowledge of Latin did not let them down in life. Nevertheless, right from their childhood they put more emphasis on learning languages that were alive rather than dead. Years later, when he was a professor, Hans Christian took a relatively relaxed view concerning the role of Latin as an academic lingua franca.

Both brothers read German books, such as *Hübner's History of the World* and *Frederick the Great's Posthumous Writings* as easily as Danish ones; nobody asked them to do so, they just consumed them voluntarily. They would ask their parents, or local citizens with a library, if they could borrow their books, or they would empty their 'piggy bank' and buy their own. They would swap the fruits of their effort. Anders was only Hans Christian's junior by sixteen months so they lived practically like twins, and when both parents slaved away day and night they were on their own and left to their mutual instruction once the indoctrination at the Oldenburgs' was over. Their independent rewading shows that they did not automatically accept Oldenburg's beatification of Lybeck nor his pietistic fundamentalism and national pride. Gradually they would develop a sceptical attitude towards authority. They did not cease to respect Oldenburg as a human being and enthusiastic teacher, but they did begin to examine his prejudices. They learnt to wonder why authorities disagree and to summon up courage to seek answers to their own questions. This development was brought about by literature such as P.A. Heiberg's *Rigsdaler-Sedlens Hændelser* ['Adventures of a Rixdollar-Note'] mocking patriotic self-complacency, and Jens Baggesen's *Labyrinten* ['The Maze'] drawing a benevolent picture of German culture without glorifying it at the expense of other nations.

When they turned twelve, Hans Christian and Anders were released from Oldenburg's infant school, and their younger siblings took their places. Both boys were asked to help their parents at the pharmacy, a job Hans Christian came to appreciate more and more, particularly working in the small lab, while Anders's interest was not equally aroused. The pharmaceutical work bound the oldest son and the father closely together and their mutual interest in chemistry established an enduring affection between them.[14]

At this stage further educational opportunities were wide open for both brothers and their interests had ramifications as complex as the crown of the chestnut tree in their back garden. Anders delved especially into matters of divinity and morality; he wrote weekly sermons that he showed to his mother, indeed he even delivered some of them, although later in life he did not find them particularly edifying.[15] At this time both brothers dreamt of following the footsteps of their forefathers and becoming students of theology. In Hans Christian's mind the post of a general-superintendent kept running alongside that of a chemist.

After infant school their reading and reflections took new directions. For a while they were instructed by a Norwegian student in chemistry and literature, and they buried themselves in more demanding works. Like Oldenburg, the Norwegian was obsessed with national pride and contempt for everything Danish, but previous experience of such prejudices had already warned them against reacting too rashly. Steffen Jørgensen, the district judge lent Anders a book on natural law by Nørregaard (a professor at the University of Copenhagen) based on Wolffian philosophy; and Jørgensen was impressed by the independent study of his borrower.[16] Hans Christian studied the aesthetics of Charles Batteux, a four-volume translation of French Enlightenment thought on poetry, music, architecture, and sculpture.[17] 'He still remembers his reflection that poetry must rank far higher among the arts than Batteux seems to be aware', he wrote much later.[18] A rather curious comment in so far as Batteux founded the entire aesthetics on poetry; but he had particularly mentioned the possibility that music (or dance) might merit a higher rank if the arts were brought together in the same performance. This bookworm must have appeared rather eccentric as a child, for as an adult he was unable to recall even one boyish prank, though he was fully aware of which branch of the arts deserved priority over the others.

A land surveyor who parcelled out fields in summer was put up at the guesthouse of the pharmacy in winter, when he gave maths classes to the two brothers. A convinced freethinker, this land surveyor was unpopular at Rudkøbing, but welcomed by the Ørsted family who recognised him for his honesty and skill. He made a lasting impression upon the boys by refuting the theology of revelation in favour of a natural religion according to Enlightenment thinking. And Jørgensen, the district judge, who had been travelling all over Western Europe and so knew the importance of speaking foreign languages, gave lessons in French to the older brother and English to the younger one. Afterwards they would swap the skills they had picked up. It is doubtful whether they gained much proficiency in either.

Now there was no more for them to learn at Rudkøbing. They had absorbed a wide range of knowledge from the few people who had something to offer. On top of that their intellectual hunger had to be satisfied by independent study and mutual instruction. The older brother was something of a prize pupil and Oldenburg's favourite. While his schoolmates were called by their christian names only, Hans Christian was called Ørsted. The admiration he received from his schoolmaster and others provoked a certain conceit, but he was also reminded that one has to deserve one's merits and live up to one's reputation.

An elderly woman who had once served at the royal court told him that he was as polite as a prince, and the boy was flattered by that compliment. He was never beaten by anybody, and his school mates were told to pay him particular respect. So, early on Hans Christian learnt to feel like a prince destined not to inherit but, after his apprenticeship and in due course, at least to merit a throne. On the one hand he could not help impressing those around him, on the other he made an effort to be polite to all, including those to whom deep down he felt superior.

The unconventional schooling with his brother at Rudkøbing and his close ties to his father had already determined Hans Christian's preferences: chemistry, theology, and aesthetics. Furthermore, the foundations of crucial features of his character had already been laid: ambition, eccentricity, independence, will power, diligence, and loyalty towards his family.

Nobody at Rudkøbing thought that the two brothers lacked any knowledge required for the University of Copenhagen entrance exam other than what they could acquire in a short time in the capital. With the approval of their father and a small allowance they left the island in spring 1794. In Copenhagen they went to see Professor Børge Riisbrigh who pointed them in the direction of appropriate tutors in Greek and Latin. They carried on with their mutual education and that autumn both passed the entrance exam with top marks.

> The laboratory is so small and so ill provided with apparatus that it cannot possibly be considered worthy of a university as rich as ours.[1]

3 | 1794–6
A University without Science

IN SPRING 1794 our two boys broke up from school in Rudkøbing to enter the University of Copenhagen. They walked along the flat country road, they passed through Roskilde to Valby Hill, and from its top they overlooked the capital lying at their feet, with its characteristic spires and towers and its hundred thousand inhabitants still squeezed in behind its narrow ramparts. Further on they walked past the Shooting Range and along the lime alley past the Memorial of Liberty—in the course of erection in commemoration of the abolition of villeinage—and finally crossed the bridge over the moat to be let into the city through Vester Port, one of the four town gates, on payment of the usual entry fee. On Vestergade near the gate they soon found themselves seated around Aunt Engelke's table in her dye-works enjoying her food.

The next day they witnessed with their own eyes what they had only heard about: Christiansborg, the royal castle, had burnt down and all that was left was a charred ruin. The following summer when the boys were away seeing their father on Langeland another great fire broke out in Copenhagen. This time the houses of citizens were destroyed in great numbers. Temporary huts were erected in front of the ruined castle to provide shelter for thousands of homeless Copenhageners. A large part of the Latin Quarter including most of the University fell victim to the flames. For years to come the city was crowded with homeless victims billeting themselves anywhere they could find.

Lodgings were no problem for the youngsters. Elers' College in Store Kannikestræde accommodated sixteen students, of which only the Ørsteds enjoyed right of priority as

Fig. 6. Vester Port, one of the four town gates of Copenhagen, through which the Ørsted brothers passed on their way to aunt Engelke Møller, whose dye-works were situated in Vestergade in the background of the picture to the right. Coloured drawing by C.W. Eckersberg (1783–1853). RL.

descendants of the founder of the college. There were eight double rooms and Hans Christian and Anders shared one of these; each room had a table, six chairs, an iron stove, a cupboard with a lock, and a wide bed shared by the roommates. Two cords of firewood a year were included. Every morning the porter would supply boiling water as well as embers for the tea-machine for three marks a month. He also lit the stove and twice a week he would clean the rooms and remove ends of tallow candles from the candlesticks.

The *inspector collegii* was formally appointed by the master, but in reality elected by the alumni. Thomas Bugge was master during the years the Ørsteds lodged there. He was Professor of astronomy and head of the cartographical project conducted by the Royal Danish Society of Sciences and Letters. Soon Hans Christian was elected inspector, and his duty was to liaise between master and students. He was keeper of the keys to the garden, the fountain, the gate, and the library and lecture hall. Every morning (at six o'clock in summer and seven o'clock in winter) he would ring the great bell in the ceiling of the hall to announce the morning prayers (*preces solennes*). The students also assembled for evening prayers, though after the great fire in 1795 homeless citizens moved into the lecture hall on the second floor and often disrupted the prayers. Under normal circumstances the students were charged four skillings, payable to the *fiscus*, the college coffer administered by the inspector. He received an increased *distributus* in return for his trouble. Hans Christian's share of the college revenue was 70 rixdollars, while ordinary students had to make do with 40. They also benefited from the so-called 'community

Fig. 7. The great fire of Copenhagen 1795. The fire broke out at Gammel Holm, site of the Royal Navy Dockyard. The first evening it leaped across Holmens Kanal to the Admiralty, through the windows of which drawings made by naval architects were thrown out. On the other side of the canal is the heartrending view of the areas on fire. Mothers and wet nurses flee with their suckling babies, pets, and utensils. Two watchmen have dropped their spiked maces and seize a looter, and at the far right a wounded man is carried away on a ladder. Drawing by C.F.F. Stanley (1769–1805), KBM.

grant', that is the current income from the University's landed property, which provided the financial support for students of that time. This amounted to four marks a week, too little to feed them, so they were lucky to be within reach of their aunt's hospitable kitchen.

The lecture hall was also the setting for debating and declamation exercises in which all had to participate once a year. In the past these exercises, conducted in Latin and ridiculed by Ludvig Holberg in his comedy *Erasmus Montanus*, had been printed in booklets of a score or so of pages, but this custom had ended. Now the students restricted themselves to defeating their opponents orally in public and celebrating their triumphs afterwards at a drinking spree.

Every evening supper was served by Engelke Møller, their widowed aunt, whose dye-works were situated, as polluting industries had to be, close to the city moat. It was here thirteen years

Fig. 8. The Ørsted brothers on their way from Elers' College to lectures at the University. Drawing by E.L. Hemmingsen (1855–1939).

earlier that Hans Christian had sailed in the wooden tub. Their college rooms and dinner table were conveniently located a few minutes walk from each other. They just had to cross Vor Frue Plads and Gammel Torv and their food was ready. 'They walked in long frocks almost reaching their heels like dressing-gowns; they clung to one another arm in arm looking like conjoined twins', as Adam Oehlenschläger, their mutual friend, described them.[2] They put their money in the same chest where it would lie unspent, since their expenses were limited as long as they had free lodgings and meals.

The college gate opened towards Store Kannikestræde. To the right they saw Round Tower and Regensen, the largest college, and to the left the dominating spire of Vor Frue Kirke,

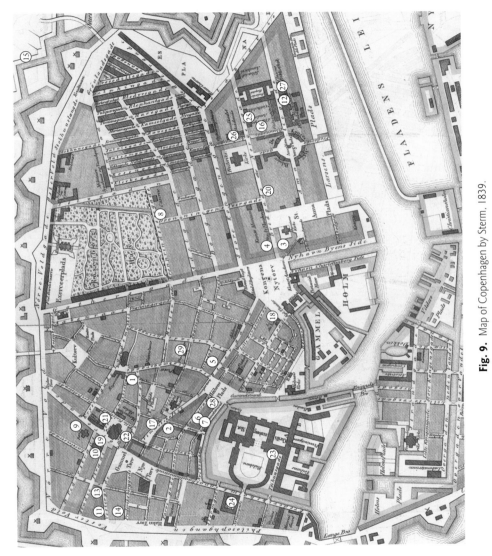

Fig. 9. Map of Copenhagen by Sterm, 1839.

HCØ'S ADDRESSES 1794–1851

1. Elers' College	Store Kannikestræde	1794–1800
2. The Lion Pharmacy	Hyskenstræde/ Vimmelskaftet	1800–1801
3. Collin's Court	Bredgade (Norgesgade) 159	1804–1806
4. Thott's Palace	Bredgade (Norgesgade) 201–202	1806–1809
5. Rubens's yard	Østergade 68	1809–1812
6.	Gammel Strand 12	1813
7.	Læderstræde 38	1813–1814
8.	Dronningens Tvergade 401B	1815–1819
9. Pingel's yard	Nørregade 35	1819–1824
10. Professor's yard	Studiestræde 106	1824–1851
Other important addresses		
11. Engelke Møller, née Ørsted, Vestergade 26		–1822
12. Søren Christian Ørsted	the pharmacy of Alm. Hospital, Amaliegade	1815–1823
13. Anders Sandøe Ørsted	Vestergade 23	1801–1812
14.	Frederiksberggade	1812–1817
8.	Dronningens Tvergade 401B	1817–1818
15.	Østerbro	1818–1824
10.	Studiestræde 106	
9.	Nørregade 35	
16. Thomasine Gyllembourg	Blancogade	
2. Sophie Probsthein	The Lion Pharmacy	
17.	Klosterstræde/ Vimmelskaftet	
18. H.C. Andersen,	Hotel du Nord, Kgs. Nytorv	
19. Adam Oehlenschläger	The Bishop's Palace, Nørregade/ Studiestræde 1820–1834	
19. Balle/Münter/Mynster	The Bishop's Palace, Nørregade/ Studiestræde	
20. Schimmelmann's palace	Bredgade (Norgesgade)	
21. Th. Bugge's professor's yard, Lille. Fiolstræde/ Store Kannikestræde		
22. B. Riisbrigh's professor's yard, Dyrkøb		
23. Videnskabernes Selskab [Royal Danish Society of Sciences and Letters]	Royal Stables, Christiansborg	

24. Skandinavisk Litterurselskab [Scandinavian Literary Society]. The Prince's Palace	
25. Chirurgisk Academi [Academy of Surgery]. Bredgade (Norgesgade)	
26. Landcadetacademiet	Academiegade
27. Det classenske Bibliotek	Amaliegade
28. Dreyer's Club	Læderstræde
29. Free Masons' Lodge	Kronprinsensgade

120 metres high. Normally, they would turn left and pass by a row of similar professorial courts with stables and cart sheds, and large gardens with old trees. The first court was inhabited by Friedrich Münter, the professor of theology, who would unpretentiously run slowly down the street, unlike Professor Moldenhawer, who would rush forward snorting with his puffy face like a turkey cock.[3] Still on the left-hand side, but further down towards Frue Plads they would meet Jacob Baden, the invalid rhetorician, leaning on his servant's arm. Before they reached Fiolstræde they would pass two more professorial courts followed by Master Bugge's house adjacent to the main building of the University, merely a modest two-storey house with a small belfry. Inside the courtyard was the medieval house of the senate and facing Nørregade the community building with its lecture halls.

At Borch's College they might turn left down Lille Kannikestræde and pass by the seventh court inhabited by Riisbrigh, the professor of philosophy. H.N. Clausen describes him as a rather old fogey with a face looking as if it were carved in wood.[4] But if they continued straight on towards Frue Plads (a cemetery at that time) they would have to pass the bishop's court. Bishop Balle is reported to have been a priceless sight when pompously stalking across the street to Vor Frue Kirke wearing his long cassock and his tricorn hat on top of his wig. It was said that Balle owned no garment other than his cassock.[5] Having passed the church the brothers would cross Gammel Torv and enter Vestergade with its many guesthouses side by side. Following this route they would miss the two last professorial courts situated in Studiestræde and Skt. Pederstræde and inhabited by Anders Gamborg, another professor of philosophy, and C.G. Kratzenstein, the professor of experimental physics, their gardens bordering each other. This is exactly the location where thirty years later Ørsted had the University extended, in the form of his Polytechnic Institute.

The daily walk through the Latin Quarter to their aunt's dinner table was a mixed experience. There were no shop windows displaying consumer goods, just a few signs indicating a guild or a commodity. There were still several cemeteries inside the ramparts of the city, but they were poorly maintained and people showed a shameful lack of respect by hanging out their washing in them. Stinking open gutters, cowpats, pig manure and horse droppings were everywhere. Walking dry-shod on the muddy pavement was nearly impossible. If a stroll was to be taken for pleasure, the only options available were along the top of the ramparts or beside the lakes outside the city where the air was tolerable.

It would be wrong to imagine student life in Copenhagen as a romance. There were no student clubs, no student choirs, no particular places to socialise for students; they resorted to alehouses and clubs for entertainment or they gathered in colleges. Another meeting place outside the lecture hall was St Petri, the church of the German-speaking congregation. Here J.G. Marezoll, a fashionable preacher siding with Kant in his controversy with Lutheran orthodox censorship in Prussia, attracted a good many students including the Ørsted brothers. They found that his sermons provided more food for thought than those of the rationalist bishop Balle.[6]

Before matriculation, however, Hans Christian and Anders had to pass the entrance exam of the University. Upon their arrival in the city, they knocked on Professor Riisbrigh's door to ask for guidance. He recommended a tutor of Greek and Latin (their weakest areas). In October they both passed the oral exam in religion, Latin, Greek, history, geography, and astronomy, and the written exam in translation from Latin, with top marks.[7] All oral exams took place on one day in the assembly hall of the Senate, where the examiners sat side by side along a big table. At the written exam the invigilation was so minimal that the students could help one other or cheat if need be.[8]

The next hurdle was the basic education all students had to go through. It had two parts. The first part, *philologicum*, was more Greek, Latin, and history; the second, *philosophicum*, consisted of philosophy, mathematics, physics, and astronomy. Each part would normally take half a year. Anders immersed himself deeply in philosophy; Hans Christian had a predilection for physics and astronomy. So, they continued their previous practice of mutual instruction and again passed everything with flying colours. Both attended Professor Riisbrigh's lectures on Kant's *Critique of Pure Reason*. In the next chapter we shall go further into the corpus of philosophical knowledge our two protagonists encountered while preparing themselves for *philosophicum*. Lecture fees were usually paid directly to the professor the day before the exam, and this practice was suspected of corruptly influencing the outcome. Rather than the usual four rixdollars, sons of noblemen would easily pay five or more, while the simple son of a public servant would to hand his *testimonium paupertatis* to the professor in return for free access.[9]

After the basic education the paths of the two brothers parted. Hans Christian took a particular interest in chemistry, a subject that could be studied together with pharmacology in the Faculty of Medicine. Anders decided to study law. However, their mutual attraction to religious, aesthetic and philosophical issues never faded, and it brought them together to talk about the books they read. So, one is fully justified in saying that they were free and independent students, philosophical minds fully devoted to their hunger for knowledge as opposed to bread-and-butter students looking upon such activity as a waste of time.

Adam Oehlenschläger found the brothers sitting in their room, just the two of them, 'as if in a dark monk's cell—serious and silent—studying!'. Adam was about to be trained as an actor at the Royal Theatre. They became friends even before he got to know their names; he noted in his diary: 'Today I made the acquaintance of two young students; there are excellent people, and we are likely to become the best of friends'. Adam, too, was served his daily dinner at Mrs Møller's where he lodged on the first floor. Only the next day did he become aware of their names.

One day, Oehlenschläger recounts, he was in the library of Elers' College which Hans Christian took care of, feeling down in the mouth. He had lost all desire to perform at the Royal Theatre and was worried about the low status of the acting profession. He felt unable to stand the rather silly (but nevertheless common) prejudice against the easy virtue ascribed to actors.

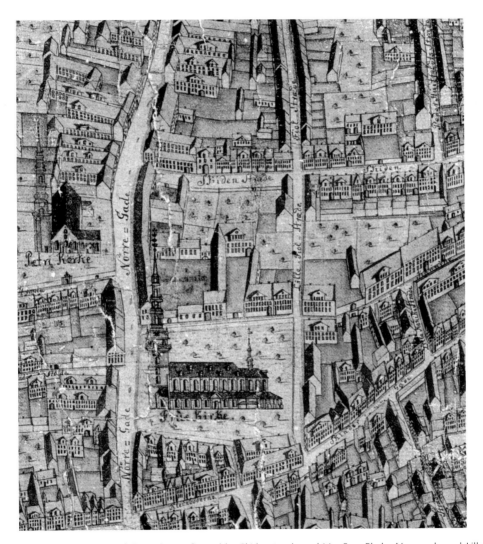

Fig. 10. The University of Copenhagen framed by Skidenstræde and Vor Frue Plads, Nørregade and Lille Fiolstræde. Fragment of Christian Gedde's elevated map of Copenhagen, 1760–1761, KBM.

Seeing the many erudite works on the shelves aroused in him a feeling of having betrayed his talent for poetry since he had left school. Hans Christian recognised this sentiment in himself. He, too, felt attracted to poetry, and the two friends had frankly shown each other their first, tender attempts as poets. In fact, Hans Christian was wrestling with his prize essay on aesthetics for the University's gold medal. However he felt it might be safer in his case to go for a bread-and-butter subject like pharmacology which would lead to a secure post as a chemist like his father, and perhaps the ability to begin his career by offering him a helping hand. Besides he could always pursue poetry in his private life *con amore*. He took his disheartened friend in hand and advised him to abandon the stage in favour of studying law. Anders could tutor him and he would then have the opportunity to become a lawyer or a judge. Adam easily grasped the point, especially as he had realised that the man he hoped would become his father-in-law, the erudite brewer and jurist Hans Heger, might be more inclined to give his daughter away to a lawyer than to an actor.

Once again the impulsive Adam was in paradise. The diligence of the two bookworms was contagious, and with their help he swotted up the examination requirements for *artium* and *philosophicum*. After Mrs Møller's dinner he would entertain the entire house performing the roles in Holberg's comedies one after the other.[10]

*

The University of Copenhagen was not a place to cultivate the sciences, but rather a national institution for educating civil servants, priests, medical doctors and jurists. The Faculty of Theology was set up according to Lutheran principles and the Faculty of Law according to the absolutism of the Danish-Norwegian monarchy. There was no Faculty of Science and hence relations with universities abroad were rather limited. The University catered for bread-and-butter students who were aiming at bourgeois safety for the graduate. Critical voices claimed that what really took place at the three faculties was an introduction to the practice of each profession. At the Faculty of Theology the undergraduates would learn to preach Christianity as the road to salvation and eternal life without having to bother about a pious or ethical way of life. At the Faculty of Medicine the undergraduates were taught to cure diseases and prolong lives with pharmaceuticals, but not to care about healthy living, and finally the student of law would acquire lawful devices to win a court case without having justice on his side. This ironical judgment had been made by Kant and taken over by critical students who dreamt of a university of reason where the Faculty of Philosophy should be ranked above the professional faculties, to provide tools for critical reflection on practice rather than adapting to prevailing circumstances. '*Sapere aude!*' was the Kantian watchword: commit yourself to your personal enlightenment and have the courage to reason for yourself![11]

It is hard to state the exact number of active students at the University, since their studies were often interrupted when they needed to work, typically as private tutors. But we know exactly how many students were matriculated each year. In 1795 when Hans Christian joined the Faculty of Medicine he only had three or four fellow students, while ten entered the Academy of Surgery. At the Faculty of Law, Anders was one of nearly seventy. The Faculty of

Theology was by far the largest with about 110 new undergraduates.[12] No science degree was available. Physics, chemistry, and botany were regarded as auxiliary disciplines of medicine, and mathematics and astronomy were only studied as part of *philosophicum*.

The University was the object of severe criticism not only because of tedious lectures and the absence of research, but also because of the economic administration and policy of appointment of staff. The course catalogue distinguished between free public lectures and lectures *privatim* taking place at a professor's residence in return for a payment, and finally lectures *privatissime* arranged for a narrow circle of fee-paying students.[13] Professors were tempted to supplement the public lectures they were obliged to offer free of charge with *privatim* lectures to make more money. There are no figures to indicate the share of income a professor could earn from fees (typically charged by his assistant), but their significance is testified to by the fact that several posts were unsalaried, meaning that the postholder's entire livelihood depended on students' fees. It was hinted that staff contrived to cut the number of their public lectures in order to force students into private ones.[14]

After the coup d'état in 1784 the way was paved for the 1788 reform which encouraged professors to write textbooks and students to read them rather than struggle with the notes they had taken down from soporific lectures read aloud from the same script year after year. Students would resort to their notes when they prepared for exams and try to sell them to younger fellow-students afterwards. Textbooks could be taken down from the shelf whenever needed, and saved lecture fees. However, this did not interfere with the professors' freedom to choose their methods of teaching. Neither did it oblige them to carry out research. No independent external examiners were introduced and marks were decided exclusively by the professors. Accusations of bribery were still on the lips of many.[15] Yet there was one important innovation. Prize essays were introduced to enable undergraduates and young graduates to draw attention to their research talent.

Most undergraduates were spending fees on tutors, that is older students or graduates who kept an eye on the exam questions posed by the professors. Normally, these questions would be the same every year. So the tutors would be on the alert at the exam board to take note of the answers required and find out if (unexpectedly) new questions turned up. This system was rather unfortunate for undergraduates like the Ørsted brothers who acclaimed and practised free and independent studies. But they were astute enough to take advantage of the system by using each other as tutors free of charge and soon they would add to their income by tutoring their fellow students.

Appointment of staff continued to be solely in the hands of the Chancellor of the University. Already prior to the 1788 reform abortive projects had been launched to pool the *corpora* of the University and pay the staff a fixed salary. At that time each of the fifteen professors was paid the surplus (his so-called *corpus*) of the landed property owned by the University. This system implied that if a new professorship was wanted it would have to be funded by existing *corpora*. Consequently, all fifteen professors (four theologians, two jurists, two medical doctors, and seven for the disciplines at the Faculty of Philosophy) resisted tooth and nail when new professorships, for instance of science or modern languages, were proposed, since these would inevitably harm their well-established interests. Therefore useful disciplines such as science, a crucial vehicle for progress according to Enlightenment thought, were refused admittance at the gate whenever anybody tried to get them into the University.

The statute of 1788 allowed *corpora* to be pooled as they gradually became available on the death of each professor. Professorships were for life. If old age prevented them from teaching they did not have to retire. They were entitled to employ a private substitute who had to make do with a fraction of the professor's salary. When *corpora* were no longer earmarked for the fifteen ordinary professorships, the sum available was more flexibly disposable, for instance to fund extraordinary posts that might in due course be converted to ordinary ones. Aesthetics, history of literature, and statistics benefited from this reform, but still there was no hope within sight for the sciences.[16]

Appointments at the Universities of Kiel and Copenhagen were not decided according to intellectual capacity, but by aristocratic bureaucrats without genuine academic knowledge. In 1795, thanks to the still existing freedom of the press, J.C. Fabricius, professor of Natural History and Economics at Kiel, published a devastating critique of the University of Copenhagen.[17] Flattery, lobbying, nepotism, and letters of recommendations meant everything; the research efforts of the applicant came second or did not count at all. Such were the circumstances the Ørsteds had to face and ponder when planning their careers.

*

Yet, for some time there had been a range of initiatives designed to bring the sciences under the aegis of the University. Government leaders of German descent such as Court Steward A.G. Moltke and Dr J.F. Struensee had pressed for the improvement of agriculture and had established two new professorships in botany and experimental physics, appointing two foreigners, G.C. Oeder and C.G. Kratzenstein, for want of qualified Danes. As expected, the Senate found the appointments provocative. Not only were sciences favoured at the expense of established disciplines, but the two professorships were even staffed by foreigners. Oeder's candidature was cleverly blocked. A fellow of Elers' College was set up against the German to demonstrate his inadequate Latin in a disputation.[18] Moltke and his candidate were defeated, and Oeder's capabilities were instead deployed as director of the botanical gardens of Frederik's Hospital. The other German, Professor Kratzenstein, had to dip into his own pocket in order to provide scientific instruments, which he did—motivated by expectations that electrical experiments would not only heal people but entertain them as well and thus attract many well-to-do spectators. Henrich Steffens for one was a guinea pig exposed to shocks from Kratzenstein's electrostatic generator.[19]

In 1759 Moltke again took the initiative to promote the sciences. Once again the University was obstinate, so he had to find leverage, for which he used The Natural Cabinet at Charlottenborg.[20] The King's Particular Coffer would pay for scientific instruments as well as for two professors appointed to teach natural history and economics respectively, both pupils of Linnaeus.[21] Three years later it became obvious that The Natural Cabinet and its staff were intended to provide leverage, when Oeder among others was assigned to work out 'A proposal to establish a fifth faculty at the University of Copenhagen to be called the Faculty of Economics'.[22] Whether the sciences were to be financed by the University's *corpora* or by the King's Particular Coffer remained an open question. Lectures and exams were to be in Danish as opposed to Latin and aimed at future civil servants and private industrialists. Once again the Senate succeeded in quashing the initiative.

The third time the matter was put on the agenda was during the short tenure of Struensee (1771–72). The Bishop of Trondheim, E. Gunnerus, famed for his interest in natural philosophy, was called in to work on a compromise. The Natural Cabinet was to amalgamate with the University. In the long run costs were to be met by the University at the expense of professorships of Oriental Languages, Greek, and Latin. It is well known that soon Struensee was broken on the wheel, so once again the Senate could breathe a sigh of relief. The threat posed by science was scattered in all directions. Oeder was downgraded to police officer in Oldenburg; Brünnich was appointed Oberhauptbergmann at the silver mines of Kongsberg, Fabricius became professor of economics at Kiel, and Gunnerus who had optimistically taken the opportunity to propose a new university in Norway, returned crestfallen to his bishopric.

Even the most powerful men were unable to create room for the sciences at the University. Surgeons were not allowed into the Faculty of Medicine; in 1785 they established a new institution for themselves, the Academy of Surgeons, in order to obtain an anatomical theatre where instruction in dissection could be offered. A laboratory for chemistry was only built in 1778 in Skidenstræde. It consisted of two rooms, each about 25 metres square, a lab and a lecture hall. There was a well and a shed for firewood, but there is no evidence that it was ever used for any significant purpose. Ørsted was unimpressed by it: in 1813 he characterised the laboratory as 'so small and so ill provided with apparatus that it cannot possibly be considered worthy of a university as rich as ours.'[23]

By 1789 a circle of natural philosophers, most of them pupils of Linnæus, had become frustrated that the regime—otherwise so keen on reforms—had totally capitulated vis-à-vis the challenge to modernise the University when adopting the 1788 statutes.[24] Around 250 of them became members of a new voluntary association, Naturhistorieselskabet [The National History Society]; they paid ten rixdollars a year for their membership, and published *Skrifter af Naturhistorieselskabet*. Its function in providing leverage for a faculty of science was mentioned in an anonymous article in *Minerva*.[25] Steffens became its first graduate.

So, in 1794 when Hans Christian entered the University, the sciences were still private hobbies for well-to-do amateurs who did not care about a safe post or a salary. The most renowned experimenter of physics in the kingdom was Court Steward Adam W. Hauch, whose collection of scientific instruments was second to none. It gained for him the kind of prestige that can only be obtained by an amateur studying for pleasure even at a cost. In 1791 Hauch was elected a member of the Royal Danish Society of Sciences and Letters. This was an honour, although he remained largely isolated because his interest in Lavoisier's chemical revolution was shared by only a few others, while far more members were busy surveying

Fig. 11. *The Clover Leaf or the Masked Jesuits* showing three distinguished civil servants. To the left Court Steward A.W. Hauch sitting on a chamber pot with the inscription 'Oxygen or Life Air' breaking wind while smelling a flask containing 'science'. In the centre the opinionated minister of state C.D.F. Reventlow, on whose hat is inscribed 'Projects' and on the paper he is handing to the genius 'you are an animal allowed to exist'. To the right the malicious surveyor C.F. Hansen with compasses and a windmill in his right hand, his left hand cutting the wings of the kneeling genius, who is being tested on a touchstone while his hand is threatened by a snake. An angel whirls copious notes. The caricature is annotated by the artist: 'To annihilate all possible charges that Jesuit-minded people might bring up against this drawing of mine I shall leave it entirely to would-be buyers to add shadows and colours according to their own taste...' Caricature by O.P. Gram, 1813, the year of the Danish-Norwegian bankruptcy, RL.

and mapping the country. Hauch was probably the only Danish chemist to have accomplished original research and written textbooks on his subject.[26] He had no connection with the University.

In 1794 a small club decided to fill a vacuum by publishing a new journal of natural philosophy called *Physikalsk-oeconomisk og medico-chirurgisk Bibliothek for Danmark og Norge*. The editors did not dare to focus on natural philosophy alone, thinking that their target group must be extended to civil servants and the medical profession in order to attract a sufficient number of subscribers. Membership of the club was limited to the authors of prize-winning essays at the University. This prerequisite reduced the number of potential members to four. Ørsted was not invited to take part and had to look for like-minded natural philosophers outside the University, for instance at the punch bowl in Dreyer's Club or in the Scandinavian Literary Society.[27]

The reform of 1788 had rendered the orthodox University a little more efficient. That was all. Its range of disciplines was unchanged. Natural philosophy had been unable to open up even the smallest crack in the defensive walls of the alma mater. Its future appeared hopeless when Ørsted took up his studies of physics and chemistry.

> Two things fill the mind with ever new and
> increasing admiration and awe . . . : the starry
> heavens above me and the moral law within me.[1]

4 | 1796–8
Two Philosophical Minds

AS INSPECTOR of Elers' College, it fell to Hans Christian's lot to buy books for the library for the tidy sum of 300 rixdollars a year. This grant stretched to around 250 books, and his choice was in no small degree determined by his and his brother's preferences. Although there are historical, medical, and theological books on his list, philosophical and aesthetical works are in the majority. First of all we notice the complete works of Kant followed by the moral philosophers Fichte and Tetens as well as Montaigne, Locke, Hume, Ferguson, and Adam Smith in German translation, and finally writers of fiction such as Lessing, Wieland, and Schiller.[2] Presumably, most of these books would appear on the shelves of our two brothers on long-term loan for the years to come. The master of the college, Professor Bugge, must have disapproved of the ostentatious priority given to philosophical and aesthetical works at the expense of the mathematical sciences. Moreover, Hans Christian neglected his series of lectures on astronomy and mechanics. Master and inspector had already started to look askance at each other.

While they were still in their teens, Oehlenschläger wrote about the Ørsted brothers that 'to all fellow students they shone like the Dioscuri and even older scholars soon noticed their extraordinary capabilities'.[3] In other words their friend compared them to Castor and Pollux, twin sons of Jupiter, twinkling in the starry heavens as the constellation Gemini. This star quality had been acquired by shutting themselves away in their monk's cell, like philosophical minds renowned for their wit and reflection. Kant's critical philosophy, which they encountered while preparing themselves for *philosophicum*, spurred on their curiosity, inspired their discussions, and constituted their frame of reference for the rest of their lives.

Even prior to their entry to the University, Christian Hornemann, a law student, had already drawn attention to Kant's critical philosophy, which had shaken the very foundations of the intellectual world. In 1784 he had studied the first edition of *Critique of Pure Reason* (1781) carefully and had become a devoted disciple of Kant, so devoted that in 1791 thanks to a recommendation from the Chancellor, the Duke of Augustenborg, he was awarded a travel grant and went off to the University of Jena. This was the place where the critical philosophy took root for the first time outside Königsberg, Kant's hometown in distant East Prussia.

Kant's philosophy was a heroic attempt to save Enlightenment confidence in human reason. On the one hand he turned against Christian Wolff's dry rationalism, which held to the view that the force of thinking alone was conducive to true knowledge. On the other hand he opposed Locke and Hume, the British empiricists, who argued that true knowledge must be based on sense-experience only. This dichotomy was resolved by Kant. Hornemann's admiration for Kant particularly hinged on his criticism of existing theodicies which resulted in a separation of religious belief and scientific knowledge. In 1793, when members of the Prussian censorship commission reproved Kant for parts of his *Religion within the Limits of Reason Alone*, their rebuke was grounded on the author's alleged atheism.[4] Hornemann's sympathy, by contrast, was motivated by the refuge Kant had created for religion. For just as science cannot prove the existence of God, it cannot disprove it either.

Hornemann was an enquiring soul who, though he realised that falling in love with a young woman, Miss Schlegel, would require him to finish his education in order to be able to provide for his future family, was however so profoundly immersed in the contemporary debate on philosophical and aesthetical issues that he prioritised the inner demands of his mind above the outer demands of his studies, money matters, and career. Hornemann never got ready for his exam; he overstretched himself and he died in 1793. Sadly, his promising series of lectures on Kantian philosophy, delivered in Copenhagen for the first time, was thus cut short, but his presentation of the critical philosophy had already attracted the attention of many students.[5] Hans Christian and Anders benefited from Hornemann's introductory lectures, which were collected posthumously by his friend J.H. Splet and published by his brother-in-law, J.F.W. Schlegel, professor of law.[6]

Preparing themselves for the *philosophicum*, both brothers attended Professor Riisbrigh's lectures on Kantian philosophy in his residence on Dyrkøb, behind Vor Frue Kirke. For thirty years Riisbrigh had taught the Wolffian rationalism which had now been swept aside by Kant. It must have been an embarrassment for him to appear like a weathervane by abjuring a philosophy for which he had previously held hundreds of students accountable at his exams.

He gained some popularity from lecturing in Danish rather than Latin. Kant, too, wrote and lectured in his native language. Many insinuated that Danish was unsuitable for philosophy, that Denmark had never produced an original philosopher, and hence, Danes had to philosophise in a foreign language. The Ørsted brothers could not disagree more. They found, by contrast, that since it is the objective of *philosophicum* to develop independent thinking, the native language must be the appropriate one simply because it is the language in which people express themselves naturally and adequately. So, to them the abolition of Latin in favour of Danish was definitely a step forward.[7]

Riisbrigh would welcome his students to Kant's philosophy by pointing immediately to its difficulty: Kant's critique had totally revolutionised epistemology and to appreciate this one had to overcome many obstacles, since Kant delves deeply into difficult problems. His terminology is obscure, the professor would continue, exceptionally using the Latin *obscuritas verborum*.[8] We know quite a lot about Riisbrigh's dispassionate interpretation of Kant's critical philosophy, because one of his students, Jonas Collin, took notes from his lectures in 1793. This source makes it obvious that Kant's 'Copernican Revolution' was set out unequivocally: true knowledge cannot be derived either from rational thinking alone (as Wolff believed), or from sense-experience alone (as Locke and Hume claimed), but only from a combination of both.

This insight was the essence of the philosophy course that Hans Christian and Anders passed with top marks, but they soon discovered that this 'combination' was more complicated than Riisbrigh seemed to be aware of, especially when applied to subjects like natural and moral philosophy. Riisbrigh's lectures had to be supplemented by an independent scrutiny of Kant's own writings to trace the difficulties; they divided the work between them and swapped the fruits of their industry. Hans Christian read his *Metaphysical Foundations of Natural Science* (1786), which resulted in his dissertation (ch. 8), while Anders studied *Groundwork of the Metaphysics of Morals* (1785), which inspired his prize essay and won him the gold medal (ch. 6).

*

Kantian epistemology is no manual of instructions, but a philosophical answer to the question: how can we know, what we (think we) know, and how can we be sure it is true?

In other words, it is not a methodology but metaphysics, in the sense that the human ability to acquire knowledge cannot in itself be made an object of investigation, since neither mind nor senses are empirical phenomena. When Kant analysed how Newton had discovered his famous laws of motion, he inferred that they had been found a priori, using mathematics, not by means of empirical observation. Newton used neither binoculars nor telescope, and he did not carry out experiments. The innocent anecdote about the apple dropping on his head is fictitious, but has a point in so far as the apple initiated a thought that came first. Proof came later on. The laws were established when the equations were solved on Newton's desk in the Master's Lodge at Trinity College Cambridge (that is, a priori), as indicated by the very title of his book *Principia Mathematica Philosophiæ Naturalis*. Mathematical principles begin in the human mind, not in nature.

Kant called the revolution in epistemology his Copernican conversion, hinting at the decisive prerequisite for Copernicus's idea of the heliocentric world picture. This Copernican prerequisite turned up when during his observations of the universe it occurred to him to divert his attention from the object to his own mind and to its function. It had been believed that the Earth was stationary at the centre of the universe and that the heavenly globe moved round, which is empirically true. Everybody could watch the sun rise in the morning and set in the evening, while nobody observed the observer who was in fact revolving around the axis of the Earth while the sun stood still. It took a new epistemology to realise that the geocentric world picture was an empirical deception.[9]

Fig. 12. Immanuel Kant (1724–1804), Professor of Philosophy at the University of Königsberg. The painting was bought by the city of Königsberg when it turned up in 1896 and was put on sale as 'das Gemälde aus dem Dresdener Kunsthandel [A painting from the Dresden Art Gallery]'. Portrait in oils c.1790 by unknown artist. *Kant-Bildnisse* 1924. RL.

Kant presented a popular outline of his Copernican conversion in several allegories. He compared the process of research with the investigation of a judge. Scientists only get the answers from nature that their questions, formulated by their own reason, ask. Answers are not arrived at when a schoolboy reels off the lesson with which his teacher has indoctrinated him. In a court case they become apparent because the judge interrogates and cross-examines the witnesses by demanding answers to his considered questions. The initiative does not emanate from nature that is passive, like the witnesses who are only allowed to reply when they are asked. Hence the researcher is governing the entire investigation in the same way as the judge is in charge of the court—in Newton's case by taking advantage of observations made by other witnesses (astronomers) and inserting them into a mathematical equation.[10]

By means of a colloquial example Kant sheds light on another aspect of his epistemological scheme. Imagine that you knock on a door on a cold winter's day. You feel the heat stream towards you before you realise that the host has lit a fire, because you expect to enter a warm room. Without watching the thermometer you take off your coat and find out that a table has been laid for six people. The hostess approaches you, now only four guests are lacking, who will they be? You understand intuitively the difference between outdoors and indoors, between cold and warmth, and that the number of plates tallies with the number of hosts and guests. How is this understanding possible? The situation does not make sense due to sense impressions alone, because they appear as a manifold of incoherent phenomena. Your understanding of what is going on depends on your mind, which by means of its innate concepts, including space and time, establishes a meaningful and coherent picture out of separate sense impressions.

What are these concepts? They are his famous categories of the understanding, guiding our mind, through which the multiplicity of impressions must pass to make sense. Without these categories the world would remain totally chaotic according to Kant. He pointed out time and again that our cognitive capability depends on two prerequisites: 'Without sense impressions no physical object would appear to us, and without concepts we would be unable to make sense of them. Thoughts without content are empty; intuition without concepts is blind.... Only if they are united can they produce knowledge.'[11] Hans Christian embraced Kant's categories of the understanding and dubbed them 'Ariadne's Clew' because to him they were the equivalent of the thread that Ariadne had presented to Theseus to enable him to escape from the labyrinth. These categories are applied unconsciously, but guided by human reason, like rules of grammar and syntax that gradually build up in the baby's mind to govern our speech although we are not aware of them as we speak.

The Ørsteds encountered yet another famous pair of concepts belonging to Kantian epistemology. They were not part of Riisbrigh's lectures but were learned about independently. They are the concepts 'phenomenon' and 'noumenon' that will appear time and again in this biography. For Kant a phenomenon is an object leaving a sense-impression and which, filtered through the relevant categories of the understanding, is cognizable. The noumenal, on the other hand, cannot be known, because it does not appear in time and space and hence leaves no sense-impression. Although we do not know anything about it, we may need the concept for purely intellectual reasons. The force of gravity serves as an example of a noumenon. It is insensible in time and space, and neither Newton nor Kant nor anybody else knows what it is. In physics it means 'something' that is beyond our knowledge, but still a necessary epistemological concept, because without the noumenon that we call gravity we would be unable to understand the reciprocal attraction of celestial bodies or the movement of tides. The concept of 'physical force' that is not represented by an object amenable to experience by the senses is a necessary epistemological concept (noumenon).

This pair of concepts was basic to Kant's dualist philosophy and influenced the natural and moral philosophy of both Ørsted brothers. Kant reflected on another dualism between the strikingly lawful determinism in nature contrasted with the not less striking freedom of the human mind. On this schism he wrote a famous passage the first sentences of which were carved unto his tombstone: 'Two things fill the mind with ever new and increasing admiration and awe, ...: The starry heavens above me and the moral law within me.'[12] None of them lend themselves as objects amenable to immediate empirical investigation. The starry heavens throw us into imaginations of unfathomable distances and infinite times. The moral law is invisibly bound to our individual consciousness and free will. The greatness of the starry heavens reduces the individual to an insignificant point in the universe. The moral law, by contrast, elevates the individual far above its physical nature and endows it with a freedom to pursue goals in life according to individual choice.[13] The starry heavens are determined by laws of nature and are the study object of theoretical philosophy, while our moral law is subject to our own free will and to practical philosophy.

In critical philosophy the universal foundation of moral philosophy can be constituted neither by laws of nature nor by any religion, but only by secular reason. Why so? Because natural laws are indifferent to morality; they do not distinguish between good and evil. The sciences can only be descriptive (how people actually behave), not prescriptive (how they ought to behave). Nor can any religion be prescriptive to mankind, because it does not emanate from universal reason, but from individual belief in revealed 'truths' such as the Ten Commandments, the Sermon on the Mount, or the Koran. Principles guiding human action must be deduced a priori, that is, before we act. Thus moral philosophy diverges basically from natural philosophy. Whereas 'homo phenomenon' is subjected to the laws of nature, 'homo noumenon' sees himself as a self-governing moral agent. This, above all, is the distinguishing feature of humanity.

Human freedom is the alpha and omega of Kantian philosophy.[14] As a noumenal being man has a potential of autonomy, because he is endowed with reason. He is in no need of authorities to tell him what to do. To Kant the essence of enlightenment is the courage to throw away the yoke of tutelage and to emancipate oneself from religious and political authorities.[15] Man's reason is his inner compass that will guide his moral steps. He has complete freedom to set up goals for his life and to motivate his vocation. To allow himself to be governed by the carrots of guardians or to be threatened by the stick of tyrants deprives man of his dignity.

Kant's moral philosophy is totally concerned with the moral agent and is formal, general, and categorical, that is, unconditional. His categorical imperative ('Act as if the maxim of your action were to become through your will a universal law!'[16]) is a formal and general principle relating to the agent, not to the ends and means of an action. If the latter had been the case it would have been hypothetical: 'If you want to retain your credibility, it is stupid to lie'. 'No!' Kant would say, 'You should never tell a lie!'. The obligation to tell the truth is unconditional. The reason is simple. If sometimes you resort to lying, nobody will be able to know when you speak the truth, and as a consequence you can never make people believe a lie either. A general maxim authorising both truths and lies would turn out to be a universal law deprived of all meaning. It would make a mockery of all communication between people. Kant's categorical imperative is unconditional and serves to test a concrete action a priori. If the maxim of the action could become—through your will—a universal law practised by everybody, make it happen! If not, reject it! Kant supplemented his categorical imperative with another principle: 'act in such a way that you treat humanity, whether in your own person or in the person of another, never simply as a means, but always at the time as an end!'.[17] This means that one should never treat other people only as means to advancing one's own interests (instrumental action). The rights and obligations of others to pursue their own moral goals must also be respected.

'Obligation' is an important word in critical philosophy. The term 'homo noumenon' implies that man is endowed with reason and free will. Of course, Kant's contention that man is endowed with free will cannot be substantiated by any empirical evidence. Free will is no phenomenon. But a man could decide to conceive of himself purely as 'homo phenomenon' that is as driven entirely by instinct. In Kant's eyes such individuals give up their dignity as human beings, reducing themselves to morally indifferent phenomena on an equal footing with animals. What Kant means by free will is man's choice to see himself *as if* he had a free will or to put it more metaphorically, to see himself as equipped with an inner compass, the magnetic needle of which points towards his self-imposed obligation to follow his maxims and defy his instinct.[18] If the human will were bound—determined by instinct—his action would be neither laudable nor responsible, because individuals cannot reasonably be blamed for actions they cannot help. Such individuals would be totally exempted from responsibility, slaves of their own instincts. But then again, they would also have abandoned their human freedom and dignity and reduced themselves to marionettes of nature that could not be made accountable for their actions.

Such—very crudely—were the challenges that the Ørsted brothers took on during their first years at university. Kant's dualism became their clue not only to their approaches to studying physics and law, but also to the frame of reference they were to share for the rest of their lives.

The Arts are the Imitation of Beautiful Nature...[1]

5 | 1796–7
Hans Christian's Gold Medals

BOTH ØRSTED brothers spent the winter months of 1795-6 on Langeland, Hans Christian giving his father a helping hand in his pharmacy. On their return to Copenhagen their roads diverged, Hans Christian studying pharmacy at the Faculty of Medicine and Anders studying law, choices corresponding to the interests they had already developed as they grew up. However it was not written in the stars that Hans Christian would become a mind reader of nature. As already mentioned, physics and chemistry were subjects that were not taught per se at the University. He admired poets and had a poetic vein himself. As an older boy he had read Jens Hvas's translation of Batteux's aesthetics (1773–4). Charles Batteux had published his *Les beaux-arts réduits à un même principe* (1746), and J.E. Schlegel had translated it into German.[2] In 1774 Batteux had written a new five-volume *Principes de la littérature* that included his first book.[3] Batteux's works became trend-setting for the aesthetical debate in Germany and Scandinavia and his central ideas were embraced by the Swiss thinker J.G. Sulzer in his four-volume encyclopaedia of aesthetics (1792–4) which Ørsted ploughed through. Batteux's overall principle conformed to the Aristotelian idea of mimesis and was expressed by Sulzer as follows:

> 'As the artist is the servant of nature and his goals are the same, he is bound to use the means of nature to achieve them. Nature is the primary and most perfect artist, and invariably she chooses the best method to serve her purpose. It is impossible to find a better method. Hence, artists must take Nature as their model... That is the true school where the artist can learn the rules of his art by imitating the universal method of Nature.'[4]

Batteux's principle, however, was far from as unequivocal as he seemed to imagine. Imitating nature is highly ambiguous. Heeding the Enlightenment cult of Newton one might think of this principle as scientific and rational, but to Batteux imitation was rather a form of idealisation. By 'nature' he meant '*la belle nature*', that is an embellishment of nature affirming the cosmic order created by God, not recalcitrant, sombre or even catastrophic nature. The principle of '*la belle nature*' was intended as a mirror, in which people could recognise themselves as individuals in accordance with or in contrast to the embellishment of nature.[5]

Early on Ørsted had been absorbed by the thought that the beauty of the arts was modelled on the beauty of nature, a thought that combined his two main interests, aesthetics and science. So he was no novice when Jacob Baden, the professor of rhetoric, set the following prize essay title for 1796: 'How can prosaic language be corrupted by moving too close to poetry; and where are the boundaries between poetic and prosaic expression?'.

The term 'aesthetic' does not appear in the title of Batteux's works. This concept comes from the German philosopher of aesthetics, A.G. Baumgarten, and it means that artistic representations of nature are not directed to our intellect (as with laws of nature), but to our emotions and taste via our senses. However, since some representations of nature address the cool brain while others appeal to the heart as the seat of emotion, they tend to be inconsistent. In other words, poetry is likely to be at variance with prose. If the boundaries between the two are blurred then confusion arises. The correspondent, the scientist, and the historian are expected to be precise while the composer and the lyric poet touch our emotions. Batteux and Sulzer provide an abundance of examples of the different genres. So when Ørsted had grasped Batteux's point, his prize essay was almost writing itself. He appreciated the ambiguity of Batteux, who represented all of nature poetically and as a consequence spoke artistically, that is elaborately, about simple nature, which he actually despised as something low, adding that 'to seem high one must walk on stilts'.[6]

Hans Christian disagreed with the view that the poet could use metre alone to define poetry because orators or correspondents appealing to the senses would also express themselves poetically although not in metric style. He was well aware, of course, that logic, mathematics, and metaphysics must be prosaic due to their purely abstract and intellectual character. He embraced Baumgarten's definition of poetry: 'A genre aiming at a sensual representation of its objects'.[7]

In short: whereas it is the aim of poetical language to titillate the senses, prosaic language serves to communicate scientific knowledge. Ørsted listed a number of texts which crossed the barrier between poetic and prosaic language and thus offended good taste. He drew attention to the flawed use of poetic language when poetry exaggerated its means of expression and became purple.[8] On the other hand, prosaic language often failed to achieve its intention if it renounced poetic expressions altogether, because it is hard to persuade reason without having moved the emotions first.[9]

The Chancellor of the University presented the gold medal to Ørsted, who stood out from the anonymous crowd of students with this modest claim to fame.

*

Ørsted's prize essay was brief and close to his sources, which is understandable in view of the fact that he was busily involved in other activities: his work as inspector, and lectures at the

University. More important, no doubt, were Ludvig Manthey's tutorials on chemistry at the Academy of Surgery. Ørsted had made chemistry his main subject, and in Professor Manthey (his senior by only nine years) he found a patron for life. Manthey used F.A.C. Gren's *Handbuch der gesamten Chemie*, the most up to date textbook at the time and preferable to the Danish translation of Macquer as well as to Tychsen's textbook of chemistry. Through marriage, Manthey had become the owner of The Lion Pharmacy that had burnt down in the great fire of 1795 and was only rebuilt in 1799, so for four years coinciding with Hans Christian's intense studies of chemistry, Manthey took it upon himself to teach chemistry at the Academy of Surgery, thereby also enhancing his own knowledge in the light of his designated post as head of the Royal Porcelain Factory in Copenhagen.[10] The relationship between Manthey and Ørsted developed into a warm and confident friendship.

At the exam on 20th May 1797 Ørsted impressed the three examining professors. The *chemico-pharmaceuticum* exam was divided into two main disciplines: chemistry and botany. Strange, therefore, that three medical professors were the examiners, while the professors of chemistry (Manthey) and botany were not. The reason was that chemistry and botany were regarded as stepchildren of the Faculty of Medicine. In addition, a state of rivalry existed between the University and the Academy of Surgery. Ørsted had acquired his knowledge at home in his father's laboratory and at the Academy of Surgery. To testify to his practical skills he presented a letter of apprenticeship issued by his father, who had passed the same exam twenty years before in front of almost the same panel; it declared his son to be a journeyman chemist by virtue of a five-year apprenticeship between the ages of twelve and seventeen, with additional practical experience in the lab during the winter of 1795–6.

Hans Christian impressed his examiners. Professor Saxtorph awarded him a *laudabilis* and the other professors a *laudabilis præ ceteris*. According to the rules *laudabilis* was the highest average of marks to appear on the diploma. Therefore, Professor Tode took steps to provide the graduate with a separate testimony that would make the holder blush and any rival blanch. Tode listed all Ørsted's academic merits, praised his knowledge on the modern anti-phlogistic chemistry as well as the previous Stahlian, and he stressed that on top of this bookish knowledge he was also well versed in practical skills in the lab. He had identified all fresh plants and explained their pharmaceutical properties correctly. Tode had examined 160 undergraduates in 28 years, he stated, but it was a long time since he had had the pleasure of meeting such a bright and at the same time such a young pharmaceutical student. He could hardly be expected to dedicate himself to the pharmaceutical profession solely, since he was endowed with an unusually fertile genius, indeed born for the sciences.[11,12]

*

So, Hans Christian was not entirely unprepared when he threw himself into writing his second prize essay on 'The Origin and Function of the Amniotic Fluid' for the Faculty of Medicine in 1797. The deadline was set at the beginning of December, and preparations for his exam prevented him from starting until June. In the course of six months he carried out a series of chemical experiments with amniotic fluid that was easily available from the maternity hospital on Amaliegade and could be analysed at the Academy of Surgery, perhaps with the help of Manthey.

No doubt Mathias Saxtorph, the professor of obstetrics, had chosen the prize essay topic. Hans Christian took '*chemia oculus medicinae alter*' ['chemistry is the second eye of medicine'] as his motto, alluding to Paracelsus and the iatrochemists of the sixteenth century. 'Use your own eyes!' Paracelsus had said, 'and burn the Galenic textbooks that lead the doctor astray.' In other words, Ørsted was determined not to rehash outdated testbooks, but to examine chemical reactions with his own eyes. This method took a sly dig at the Faculty of Medicine which still considered a chair of chemistry superfluous.

Where does the amniotic fluid originate: from the embryo or from the uterus? And what does it do? Does it nourish the embryo or is it conducive to the delivery by adding weight to the embryo and making the birth canal smoother? Is chemistry able to answer these questions 'lying in an almost impenetrable, Egyptian darkness?'. Textbooks offered several hypotheses of a speculative nature, but they often contradicted each another and Ørsted preferred natural philosophers, who relied on empirical investigations, even though he recognised that empiricists fought each other as well. He particularly took note of the works of Albrecht von Haller and J.F. Blumenbach, but hoped that his own chemical analyses might contribute to the solution.

According to the examiners, his finished report was 'well-structured, well-written and carefully worked out'. His initial chemical analyses of the elements of the amniotic fluid were particularly laudable. The outcome of his many experiments with excipient acid, salt, alcohol, and acidic quicksilver, cooling, heating, and distillation was that the amniotic fluid contained hydrogen, oxygen, carbon, and ammonium, as well as various salts, oils, phosphorus, and chalky soil. Hence, according to his analyses the fluid was 'a thin water-like solution of albumen'.[13]

The problem of the origin of the amniotic fluid was the most difficult. Nobody had found the right answer, neither Haller nor Blumenbach. If it stemmed from the embryo, there were the following possibilities: sweat, urine, phlegm and saliva. They were all rejected by the argument that the quantity of amniotic fluid decreased in proportion to the weight of the embryo during pregnancy. Moreover, the chemistry of these secretions did not match that of the amniotic fluid. If they originated in the uterus the elements of the fluid must be traceable to the cellular tissue of the uterus or to the navel string, foetal membrane, or placenta. Ørsted concluded that the amniotic fluid had several sources such as the blood via arteries in the amnion, but realised that he did not have sufficient data to reach a definitive answer. 'During our investigations experience—this faithful guardian—abandoned us, while—surrounded by a chaos of hypotheses—we did not know where to turn to.'[14] The author gained much praise for his reluctance to come up with speculative conclusions.

The amniotic fluid was useful for the protection of the embryo during pregnancy and as 'a moving force' during the birth, when the birth canal was expanded and greased to ease the liberation of the child 'from its prison'. But was the fluid nourishing, too? This was the hardest issue. The author repudiated the idea that the embryo would absorb nourishment by swallowing amniotic fluid arguing that no excrements could be traced. But he was uncertain about this as well. The essay was a methodologically sober piece of work. There were no hasty guesses, but concrete analyses, and whenever knowledge was insufficient or uncertain Ørsted concluded with reservations. He was awarded his second gold medal.

Kant was not mentioned at all. His epistemology was irrelevant to chemical analysis. Moreover, unfortunately, according to Kant, chemistry was no science and was unlikely to become one. No Danish scientist, including Ludvig Manthey, had so far taken any interest in the critical philosophy. Soon, however, when Ørsted started working on his dissertation on atomic theory, the fundamental problem of physics, he would assign a major role to Kantian metaphysics.

Ørsted's gold medals nearly suffered the same fate as that of the famous golden horns that were stolen by a goldsmith and scandalously melted down. In the 1820s his Norwegian nephew, Søren Christian Ørsted Bull, who lived with his uncle on Studiestræde at that time, seized the occasion to steal the two gold medals from a desk drawer and sell them. The theft was discovered before the treasure was melted down, and the culprit was punished by being sent to sea as a simple sailor for two years.[15] Such behaviour was not to be expected from any member of the Ørsted family.

> Act in respect of duty according to maxims
> that can become a universal law.[1]

6 | 1798
Anders's Gold Medal

HAVING PASSED *philosophicum* with Riisbrigh, Anders became a student of law and followed Anders Gamborg's lectures on moral philosophy in his professorial court in Studiestræde. To be sure, his moral philosophy touched on Kant's critical philosophy in passing, but to judge from students' notes it appears to have been an eclectic jumble of Nørregaard's natural law, which Anders had already read as a boy, and the rules of conduct of eudaemonism. Reputedly, the course was based on this maxim: 'Man should live according to nature. Consequently, all men must aspire towards perfection without harming other people, but they must also promote the happiness of other people to the best of their ability considering their individual relationship to them.'[2] This was the conventional programme of natural law at the time. For Anders, with his philosophical mind, Gamborg's series of lectures gave him material against which he could hone his arguments.

According to the maxim of natural law man must aspire towards perfection, that is, exploit his talents. This implied that he must keep himself alive (hence suicide was forbidden) and live a healthy life, be moderate in enjoyment, and work hard. Sexual desire was designed for reproduction, and as a consequence it was contrary to natural law to satisfy this desire in ways not leading to reproduction, such as masturbation. On the other hand polygamy and civil marriages did not violate natural law since they both served the purpose of reproduction. In a way veracity was an absolute obligation, Gamborg also asserted,, but he did not consider it contrary to reason 'to say something that in order to be true must be understood against the normal meaning of words, but nevertheless under particular circumstances could be thus understood as well.'[3]

Natural law referred to the 'analogy of animals', so Gamborg inferred that parents are obliged to provide for their children until they are capable of providing for themselves. This was not to say that mothers ought to breast-feed or parents bring up their children if 'the purpose of nature, that is, the well-being of children, could be otherwise achieved.' At this point Anders found that Gamborg failed completely. His maxim 'the analogy of animals' was at odds with Kant's moral philosophy in so far as animals, unlike humans, are not intelligent beings but are driven by instinct, and consequently the analogy is irrelevant for a morality of reason.

It is unclear whether Gamborg's concept of reason was rooted in obligation or religious faith. The analogy of animals, one of his hobby horses, excluded both possibilities. He gave a paper at the Scandinavian Literary Society quite seriously suggesting how to improve the quality of bird song in our forests. The idea was to let eggs from unmusical birds be hatched by musical ones, nightingales for instance. The professor had come to understand an English ornithologist to the effect that birds' voices are not innate, but acquired. So the point was to take advantage of musical parent birds and make them foster unmusical offspring. This trick would combine the useful with the agreeable, for a promenade in the forest to the accompaniment of beautiful bird song would ennoble the human mind.[4] This was the Enlightenment idea of letting the improvement of nature and the enhancement of happiness go hand in hand.

Anders also followed J.W.F. Schlegel's lectures, which he found far more profitable. Schlegel was a keen Kantian. As already mentioned, he had published the posthumous writings of his brother-in-law Christian Hornemann on Kantian epistemology, which he followed up in his own journal, *Astræa*. Anders had familiarised himself with all this. Schlegel was also the cousin of the Schlegel brothers in Jena, renowned for their journal, *Athenäum*, voicing early romanticism. Unfortunately, no records of his lectures survive. As a rule, close relations between professors and students were rare, but here was an exception. Schlegel became the patron of Anders, on a par with the Manthey-Hans Christian patronage, but there is no evidence to shed light on the way Schlegel actually supported Anders.[5] In the beginning Anders held Schlegel in high esteem calling him 'the honourable editor of *Astræa*',[6] but soon the relationship cooled, because on some points Anders was inclined to follow Fichte, while Schlegel did not tolerate the slightest deviation from Kant.

While Hans Christian was writing about the amniotic fluid, Anders pondered the problems of moral philosophy. The Faculty of Philosophy had set this prize question: '*Ostendatur nexus inter principium ethices et principius juris naturae*' ['Show the connection between the principles of ethics and the principles of natural law']. Riisbrigh and Gamborg together had initiated it and were to evaluate it. When a year later Anders was encouraged to publish his essay *Over Sammenhængen mellem Dydelærens og Retslærens Princip* ['On the Connection between the Principle of the Doctrines of Virtue and Jurisprudence'][7] he had elaborated his text considerably in the light of the criticisms put forward by his examiners.

Like Kant, Anders criticised eudaemonism as superficial, but he did not agree with Kant on his distinction between doctrines of virtue and jurisprudence. According to Kant there was a division between the duties prescribed by individual reason and fulfilled by his free will, and the obligations of jurisprudence that the citizen is punished for disobeying. Hence virtue is the moral strength by which one controls an instinctive aversion to obey the maxim of reason.

Abiding by the law is not virtuous according to Kant; it just provides a clean criminal record. Obligations of virtue are conditioned by the free will, whereas obligations of jurisprudence are unconditional. Another important difference for Kant was that obligations of virtue are infinite, because actions live up to maxims only imperfectly, whereas obligations of jurisprudence are finite: One either obeys the laws of the state, or breaks them.

Anders criticised Kant for including a large group of unconditional obligations in the doctrine of jurisprudence. In other words he was more favourably disposed towards the legislative power of the state and its motives than Kant, and a grouping together of the doctrines of virtue and jurisprudence under a superior moral philosophy was therefore less problematic for Anders.

A doctrine of virtue expresses a moral teleology for the individual way of life as a homo noumenon. But it does not answer the question as to which concrete action is necessary to reach the goal. The difference between the doctrines of jurisprudence and virtue is significant. Whereas the law orders or forbids the citizen to do something specific such as paying his taxes, not stealing, etc., nobody is entitled to set the rules and goals for somebody else's behaviour. Only the individual can do that. On the other hand, jurisprudence totally ignores the individual life-projects of citizens and applies the same yardstick to everybody.

By way of contrast, the categorical imperative, 'act according to a goal-oriented maxim that can become a universal law', takes a different direction from that which a life would take if the individual was merely following his natural instinct as a homo phenomenon. It is the prerogative of man to be guided by his own reason and free will in pursuance of his own goal by fighting the impulses in him that oppose his maxim. Kant's famous dictum, 'Enlightenment is man's emancipation from his own self-imposed tutelage', means exactly that. Man should be his own lawgiver. That is his mark of distinction.

It is obvious that Kantian moral philosophy played a crucial role in the lives of the Ørsted brothers. In addition, they were further inspired by Fichte's writings on the vocation of humans and scholars (ch. 20).

*

The following summer Anders graduated having already tutored a number of fellow students, who passed the *examen juridicum* with top marks. He had scarcely left the exam room before he took part in a competition for a position in the Faculty of Law. Delivering his trial lecture Anders revealed that subsequent to his gold medal he had come to favour some of Fichte's answers to problems Kant seemed to have left unsolved. Professor Schlegel, his patron, was one of the experts judging the candidates and he became so disappointed at his protégé's desertion of Kant that any chance of an appointment totally vanished. M.H. Bornemann won the competition. He swiftly climbed the academic ladder and was appointed professor. Anders considered him highly qualified, but the fact that Anders only came in fifth in the competition was a humiliating defeat. Part of the blame was placed on his health. A severe strain on his nerves weakened his preparations, but the real cause was his unclear or even paradoxical attitude towards jurisprudence. His prize essay had convinced Riisbrigh and Gamborg, the examiners, more than the author himself. He plunged into the writings of Fichte who in 1799 had become a hot topic in Copenhagen, partly because he had been dismissed from his professorship at Jena accused of atheism, and partly because Jonas Collin had translated and edited his book on the vocation of the scholar. Anders was already acquainted with his *Wissenschaftslehre* [theory of knowledge], and he joined the Fichtean view on jurisprudence. This cost him a career as a scholar. The University's refusal of a candidate of the first water like Anders was then and later considered one of the most embarrassing mistakes in its history.

The gold medal was a trophy and one would assume that the glorious exam he had passed (he gave the best performance since the 1788 reform) had given him satisfaction. But it did not. From the winter's day when he decided to enter for the gold medal to the trial lecture in the summer two years later, he failed to thrive. This was not only due to his overstraining himself mentally, but also because he was haunted by an increasing and painful doubt. The printed version of his prize essay ran to 470 pages, while by comparison the two prize essays of his brother were 30 and 60 pages long respectively. No doubt, Anders's workload had damaged his health.

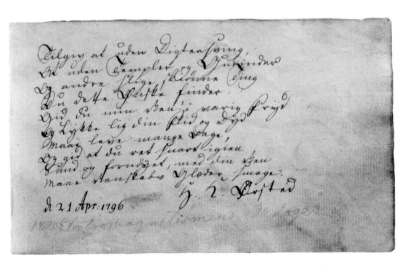

Fig. 13. The first known writing by eighteen-year-old student Hans Christian. The recipient is unknown or perhaps his brother. DTM.

Tilgiv at uden Digtersving,
At uden Templer og Gudinder
Og andre slige skiønne Ting
Du dette Ønske finder:
Gid Du min Ven! i varig Fryd
Og Lykke lig din Flid og Dyd
Maae leve mange Dage.
Og gid at du ret snart igien
Sund og fornøyet, med din Ven
Maae Venskabs Glæder smage.
d. 21 Apr:1796   H.C. Ørsted

Forgive me that without a poet's skills
And without temples and goddesses
And similar beautiful things
You find these wishes:
May you, my friend, in lasting bliss
And happiness, like your diligence and virtue,
Live for many days!
And may you pretty soon,
Healthy and cheerful,
Enjoy the pleasures of fellowship!
April 21st 1796   H.C. Ørsted.

The periods Anders spent at Langeland with his father and his younger siblings were not holidays, but just a change of workplace.[8]

Anders's diligence was legendary and tough on his health. Doctors called it nervous fever. It forced him to bed and for a while it was feared that his life was at risk. But even when he was sick he was unable to let go of a philosophical challenge. A fellow student watching at Anders's bedside argued against Fichte. The drowsy patient turned sharply and refuted his friend's objections eloquently—then fell back into apathy. The tremendous problems arising in the wake of Kant's philosophy gave Anders a headache. Adam and Hans Christian called a prominent doctor, because obviously Anders was suffering from a deep depression. Several times he found himself on the brink of suicide. 'For a long time I did not dare to walk near water or where otherwise there was a chance of precipitation', he confessed.[9]

We do not know the name of this doctor but he certainly practised Brunonianism, a cure that was fashionable at the time and was practised at Frederik's Hospital. John Brown was a Scottish doctor whose simple theory became popular in parts of the medical profession, and above all by followers of the German *Naturphilosophie*. Briefly, Brunonianism tried to explain human life-processes as a function of external stimuli of the body. Brown theorised that too many or too few stimuli were conducive to sickness. To counterbalance these external stimuli by means of drugs was the doctor's task. It was a theory of human disease that allegedly reduced previous medical wisdom to a disordered mess of empirical contingencies. However Kant was impressed, seeing in Brunonianism a real opportunity that medicine might in due course become a science, that is a system of knowledge governed by theory.[10]

Anders was treated according to Brown's theory. The doctor was certain that his patient's brain was overstrained by philosophical speculation and his body under stimulated by sedentary studying. Anders was told to take Madeira and China (opium) intended to control his fever. The cure failed. Before Brunonianism Anders had tried to overcome his spleen by rational means, because what other means were available? When reason is the distinguishing feature of man, why cannot reason redeem the pain that his obsession with it has inflicted upon him? Studying Kant's 'The power of the mind to master one's sick feelings by means of mere intentions' was laborious but for a while it seemed hopeful, 'although the effect was rather slow and imperfect'.[11] Now, Adam and Hans Christian forced him to take long daily walks and to read less. It helped him a lot to walk away from the stinking streets of the city to the park surrounding Frederiksberg Castle where old Oehlenschläger, Adam's father, lived, his house being kept by his bright and beautiful daughter, Sophie.[12] Fortunately, Anders recovered while Brunonianism were discredited. Sophie was, perhaps, his best cure.

What frustrated Anders above all was the failure of his efforts to reach the serenity he had expected or at least hoped for. Instead he had descended into an intellectual despair that enervated him and gave him migraine. Could the doctrines of virtue and jurisprudence really be contained within the bounds of one system of moral philosophy, the maxim of which was the categorical imperative? His essay had convinced Riisbrigh and Gamborg, and to that extent he had won. But he had not convinced himself. And Kant's second treatise did not help him. He turned to Fichte, believing that he could solve the problem. Anders was well aware that Fichte's theory of knowledge had not answered all his questions, but yet he hoped for solutions

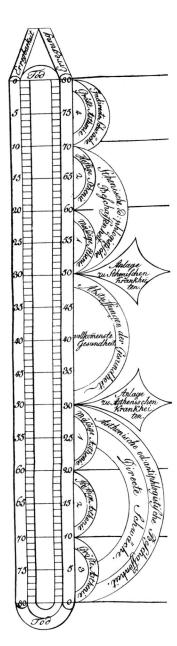

Fig. 14. Brunonianism. The theory of the British physician John Brown was systematised by his pupil Samuel Lynch and Professor C.H. Pfaff of the University of Kiel. The scale from 0 to 80 shows if the patient is sthenic (overstimulated by phlogiston) or asthenic (the opposite). Both extreme states imply death; the ideal state of health is between 30 and 50 on the scale. Overstimulated persons typically suffer from apoplexy, typhus, consumption, sleeplessness, etc. and must reduce the input of energy, e.g. by bloodletting, fast, enema or by narcotizing the muscles with opium, musk or electric charges. Understimulated persons suffer from haemorrhoids, diarrhoea, intestinal worms, paramenia, whooping cough, gout, ulcer, etc. and, by contrast, should increase their intake of energy. C.H. Pfaff 1796, RL.

in the long run, relying on Fichte's 'outstanding intellectual power'. Once again Anders was disappointed and returned to Kant. Now he had defended Kant against Fichte in his essay and Fichte against Kant in his trial lecture, but he had still not defined his own personal position.

In his memoirs Anders recounts a fierce debate taking place among a circle of friends at Vestergade while he was working on his essay. Adam Oehlenschläger had asked him if he was in agreement with Kant and Fichte that the obligation of veracity was absolutely unconditional. What if you are threatened by a violent criminal who wants you to tell him the whereabouts of his intended victim? Would you really respond that so and so is at this and that place, knowing that so and so is targeted for murder? Would it not be absurd not to tell a lie that hopefully would lead the assailant astray thus saving a human life? Anders had maintained that the obligation of veracity was unconditional, it would be contrary to reason to falsify information about so and so's whereabouts, and since the rule is absolute there is no exception to it. Adam's question was disquieting. Anders was cornered and his friends were enraged by this ethical response.[13]

Everywhere was philosophical confusion; Reinhold, Fichte, and Anders all wavered. Then Schelling appeared on the stage to reproach Fichte, although as a young man he had embraced his theory of knowledge. Everybody wished to escape from the shade of their master to become king of their own realm. Anders found it sickening. Were there no definite answers in philosophy? He decided to take a rest from his endless speculations. He resigned and diverted his energy towards reading more enjoyable works of fiction. He consumed Schiller, Goethe, and Shakespeare in A.W. Schlegel's German translation.[14] And he began his long promenades to Oehlenschläger's house at Frederiksberg. This improved his health considerably, but did not provide any philosophical serenity.

> Sense perception without concepts is blind, and
> concepts without empirical content are empty.¹

7 | 1798–1800
Editors for Kant

Having been awarded three gold medals the two Ørsted brothers had now acquired an academic status rich in the promise that they would be listened to in public debate. For four years they had thoroughly familiarised themselves with Kant's critical philosophy, and in spring 1798 they took the lead in promoting this philosophy by joining the board of editors of *Philosophisk Repertorium for den nyeste danske Litteratur* [Philosophical Repertory of Recent Danish Literature].

This weekly was the public platform of Danish Kantians, but it was not the initiative of the Ørsted brothers. The first fifteen issues were edited by others, but following an intermission of four months the Ørsteds took over. Every week an errand boy was supposed to bring an issue of sixteen pages to each of the approximately 150 subscribers, who paid three marks a quarter, but unfortunately deliveries were irregular.² In 1798 only 25 out of 52 issues appeared, and in 1799 only 18. It was hard work for the editors, who wrote the major part of the articles as the Repertory could not afford fees for contributors. The contents were intellectually demanding, and the entertainment slight, and they did not manage to attract enough subscribers: a minimum of 150 was required in order to break even. *Philosophisk Repertorium* was doomed to failure. However, Anders did review philosophical, judicial and theological literature, and wrote a few articles on such topics, while Hans Christian wrote the only scientific article they published, which spread over five issues of the journal.³

It was launched as a controversial journal emphatically attacking 'dogmatists' (rationalists of the Wolfian school), 'empiricists' (adherents of Locke and Hume), 'eclectics' (people taking a little philosophy from here and a little from there), 'eudaemonists' (people in search of happiness), certain priests belonging to the state church, arbiters of taste, and *Naturphilosophen*. The four main target groups were not necessarily Danes, but they were all opponents of Kant, the 'Genius of Reason'. First, there were professors who for professional reasons could not help but relate to critical philosophy, but lacked either the capacity or the industry to 'penetrate the architecture of his complicated construction'. They would routinely stick to traditional views insisting that 'Kant teaches us nothing new'. Secondly, there were cunning scholars pretending to respect Kant while they actually despised what they called 'his enslaved copy-cats', whom they found more rabid than their philosophical idol. Thirdly, there was a knowledgeable, even shrewd, readership of Kant's philosophy that agreed with much of what he said, but resented him for tearing down the solid foundations of empiricism. Finally, there were his superficial imitators: 'hollow places will always echo'. The programmatic intention of the journal was to reveal the first two groups and expose their prejudices. The third group would have to be repudiated with arguments, and the fourth won over for the critical philosophy through education. The editors did not hide their lights under a bushel, and altogether the programme abounded in Kantian key concepts. It resounded with an arrogance that probably rendered the cause a disservice.

Several articles touched upon political issues and probably flavoured public debate, as when they criticised the government's dismissal of civil servants, encroachment on the freedom of the press, and the banishment of Malthe Bruun, chastiser of the regime. True, the Repertory had a short life, but the Ørsteds gained their first experience as activists in the public eye. When the weekly terminated they had no difficulty in voicing their opinions elsewhere and soon they were admitted onto the editorial board of *Kjøbenhavns Lærde Efterretninger* ['The Copenhagen Learned Intelligencer'] and initiated into the select Scandinavian Literary Society which did not curtail the freedom of speech of its members.

Christian Hornemann's introductory lectures on Kant had aroused an interest in critical philosophy, particularly among philosophers of religion, morality, and law. This was the aspect of things Anders involved himself in. Elsewhere the small group of natural philosophers gathered around the Natural History Society, and the editors of *Physikalisk Bibliothek* hardly paid any attention to it (ch. 3). The editors of *Physikalisk Bibliothek* assembled at Frederik's Hospital; J.P. Mynster, the bishop of Zealand to be, has given us an account of this circle, Henrik Steffens his half-cousin in particular, and of their long pipes and exotic hubble-bubble pipe.[4] They did not master the principles of critical philosophy. Steffens wrote:

> 'My friends frequented Hornemann's lectures and were converted to philosophers. For a long time I heard them talk about space and time, about *Ding-an-sich* and forms of appearance, about categories and the absolute imperative, but so confusedly that I was definitely deterred from this philosophical school. I felt no urge to listen to Hornemann's lectures.'[5]

Steffens only heard about the theory of science when he met Schelling in person in Jena, and Schelling saw Kant not only as his teacher, but also increasingly as his adversary. Nor did J.P. Mynster convert to Kantianism:

'I was deterred not only by "the *servum pecus*" [crowd of slaves] that multiplied during those years and inanely dragged some Kantian sentences down into the filth of platitudes, but I also objected to the Kantian attempt to confine human knowledge within the narrow limits of his epistemology.'⁶

In short, for a long time Hans Christian was the only Kantian natural philosopher in Denmark.

Reading Kant's *Metaphysical Foundations of Natural Science* made Ørsted all at once aware that the prevailing atomic or corpuscular theory was fundamentally flawed. This book became a revelation to him, and *Philosophisk Repertorium* became the platform from which he would disseminate Kantian metaphysics. The occasion to do this arose when he decided to review the second edition of Court Steward A.W. Hauch's textbook of physics and chemistry *Begyndelsesgrunde til Naturlæren* [First Principles of Natural Philosophy], probably the best scientific work in Danish at the time.⁷ It was used at the University of Copenhagen and at German and Swedish universities as well. Hauch was no devotee of science for the sake of providing for his livelihood; quite the contrary, he spent a fortune cultivating his interest. He had studied in Paris and London, where he became acquainted with the leading European natural philosophers at the time, among them Lavoisier, Priestley, Cavendish, and W. Herschel, and he acquired the best scientific instruments money could provide. He had transferred Lavoisier's chemical revolution to Danish soil in 1792 and was regarded as the representative of Laplacean atomic theory.

Ørsted's critical review of Hauch's textbook was soon republished as an offprint with the new title *Grundtrækkene af Naturmetaphysikken* [Principles of the Metaphysics of Nature].⁸ Judging from the dates of publication, he set out to write *Grundtrækkene* in the winter of 1798/9.

Fig. 15. Adam Wilhelm Hauch (1755–1838), Court Steward, Royal Equerry, Director of the Royal Theatre and the Royal Orchestra, freemason, scientific writer with a large private collection of scientific instruments. DFO.

Fig. 16. A.W. Hauch's apparatus to prove that water is no element, but—according to Lavoisier—composed of two gases: oxygen and hydrogen. The cubic boxes of air are emptied and filled with oxygen and hydrogen from the two cylindrical containers. Then the two gases are conducted into the spherical container in the middle, where water is produced by combining them in the relationship 1:2. *Nye Samling af Det kgl. Danske Videnskabernes Selskab*, vol. 5, 1799. BAS.

The theme of this article was virtually presented to him. It emerged from the obvious divergence between Hauch's textbook which Professor Aasheim's course on experimental physics at the University used, and F.A.C. Gren's *Grundriss der Chemie* and *Grundriss der Naturlehre*, on which Professor Manthey's lectures at the Academy of Surgery were based. Ørsted followed the latter. Gren had worked Kant's metaphysical foundations into his textbooks and had revised his ontology of physics from atomism to Kant's so-called dynamical system. Gren's and Hauch's books were strikingly divergent. As to Lavoisier's theory on the significance of oxygen for chemistry, however, Gren had made a compromise. Kant found Lavoisier's revolutionary ideas to the effect that oxygen was a basic element of all acids and a main factor in combustion, calcification, and respiration of mammals to offer the possibility that chemistry might after all become a science proper. In spring 1798 Ørsted came forward as a spokesman for Gren's compromise and in favour of 'the celebrated Kant' with an article in *Bibliothek for Physik, Medicin og Oeconomi*.[9]

His choice of Court Steward Hauch as a debating partner was clever. Had he chosen Professor Bugge as his adversary, the controversy might have ended in a personal clash rather than a debate on substance, because they had already developed a somewhat strained relationship at Elers' College. Fortunately, Professor Bugge, his adversary at his college, had been sent to revolutionary France to join the so-called international commission of weights and measurements as the representative of his Government. Hauch, by contrast, was above intrigues at the University and as an amateur scientist neither his post nor his money was at stake. Hauch was a gentleman and his *Begyndelsesgrunde* invited debate, because to him the study of physics was not a question of being right, but of improving knowledge. The Court Steward sympathetically commented on Ørsted's review:

> 'The reason why I have followed the atomic system is that I am not adequately informed about the dynamical system nor do I consider it sufficiently elaborated to become successfully applicable except to a few aspects of natural philosophy. To my knowledge it has not yet reached the level of clarification desired, and finally I do not consider myself capable of remedying this serious deficiency of my textbook....'[10]

Hauch's comment that the dynamical system 'has not yet reached the level of clarification desired' sounds like a polite understatement. Ørsted was the first and only dynamist in Denmark, and Gren, whose textbook to some extent followed Kant, reflected an unsettled dynamism in Germany. The debate took place in an altogether unbiased and polite atmosphere. As Ørsted had expected, Hauch turned out to be 'a gentleman responding to opposition as a genuine natural philosopher and defending his view with the noble weapon that is serene reason....'[11]

*

What distinguished French atomic or corpuscular theory so definitively from Kantian metaphysics, and why did it become both the target of Ørsted's criticism and the problem he so fervently wished to solve? 'Corpus' means a body, 'corpuscle' means a small body, and 'corpuscular' designates what concerns the minute bodies that since the time of Democritus were called atoms. They were so small that they could not become any smaller. They were indivisible. The conventional wisdom of Danish natural philosophy around 1800 concerning the architecture of matter was expressed by a corpuscular theory developed by the Académie des Sciences in Paris under the leadership of Simon de Laplace on the basis of Isaac Newton's hypotheses in his *Philosophiæ Naturalis Principia Mathematica* (1687) and *Opticks* (1704). It was taken for granted, because no competing theory threatened its position.

In brief it conceived of nuclear physics as the architecture of matter and astrophysics as the structure of the universe. In the third edition of Newton's *Principia* he had set up his 'Rules of Reasoning' (methodological rules), and the analogical method was the third rule.[12] Atoms or corpuscles were too small for empirical study even under the strongest microscope. He had to find other means to defend his analogical method, and he argued that it would be contrary to reason if our Creator had constructed the microcosm according to principles different from the ones he had invented when He created the macrocosm. Following this rule of reasoning

Newton inferred that the microcosmic world consisted of atoms that were solid, stable and indivisible, had different shapes and pores of different sizes between them. These atoms were held together by attractive forces being active at a short distance. Obviously, this hypothesis was modelled on the motions of the planets in the universe. Newton imagined that the pores between the impenetrable atoms could be proportionately as large as the distances between celestial bodies, and that the attractive forces keeping the atoms together as matter resembled gravity.

Newton's notion of a vacuum between these corpuscles linked up with his theory of light. He imagined that light consisted of particular corpuscles emitted from the sun and radiating through transparent matter like glass or water that had to be porous. Corpuscular theory considered light to be on par with the intrinsic matters of heat, electricity and magnetism. These matters differed from all other matter by being imponderable and fluid to enable them to easily fill the pores between the atoms. Imponderable fluids were insensible, which was intended to explain otherwise inexplicable phenomena. For example, corpuscular theorists explained the magnetisation of a piece of iron by saying that its pores were filled with a particular imponderable fluid active at a distance like gravity and attracting iron filings. Now, everybody was quite aware that a magnet had two different poles attracting one another, while similar poles repulsed one another. At this point the analogy between macro- and microcosm broke down, because there was no force of repulsion in the universe. However, everyday observations presupposed a force of repulsion, for how else could one explain the ability of insects to walk on water, or a sewing needle to float on it, or the smell of a perfume permeating a room whenever a lady entered?

Yet Newton was cautious when he pondered the nature of atoms in his thirty-first and last query of his *Opticks*:

> 'Have not the small Particles of Bodies certain Powers, Virtues, or Forces, by which they act at a distance,... but also upon one another for producing a great Part of the Phænomena of Nature? For it's well known, that Bodies act one upon another by the Attractions of Gravity, Magnetism, and Electricity;... and make it not improbable but that there may be more attractive Powers than these. For Nature is very consonant and conformable to her self....'[13]

Newton's corpuscular theory became the paradigm of the research programme of the Académie des Sciences and from there trend-setting for Danish natural philosophy. Benjamin Franklin expanded the theory by arguing strongly that parallel to the two-fluid theory of magnetism two electrical fluids must exist. C.A. Coulomb experimented with magnetic and electric forces to measure if these forces conformed to gravity that, as Newton had discovered, varied inversely with the square of the distance. Coulomb found that both forces obey Newton's mathematical law, whereas their fluids do not share physical properties. In other words: there was a mathematical analogy between frictional electricity and magnetism, but no physical analogy.[14] The chemist C.-L. Berthollet refined atomic theory with more detailed hypotheses on chemical affinity, that is, how attractive forces could make certain substances, but not others, combine. Finally, A.L. Lavoisier produced his famous list of thirty-three elements, among them *lumière* [light-matter] *calorique* [heat-matter], but neither electricity- nor magnetic-matter were on his list. These elaborations of Newton's ideas were treated of by Simon de Laplace's mechanical-mathematical paradigm unfolded in his *Exposition du système du monde* (1796) and *Traité de mécanique céleste* (1805).

French corpuscular theory was the standard doctrine in Hauch's 'First Principles of Natural Philosophy'. He had several electrostatic generators in his exquisite Physical Cabinet at the Royal Stables of Christiansborg Castle and believed in the healing effect of these generators due to their repelling exhalation that is the discharge of the imponderable electrical fluid. Weak discharges would help patients suffering from rheumatism, gout, headache, deafness, inflammation of the eye, toothache, paralysis, as well as disorder caused by 'ladies' monthly purifications'. The benign effect of these exhalations on the growth of plants was later looked upon with some scepticism just as hatching chickens with an electrostatic generator proved more difficult than initially assumed.[15]

Hauch did not adopt the air of a smug, let alone an obstinate, partisan of corpuscular theory and referred to it with some reservation: 'all these theories show a deficiency of knowledge in this respect, and sincere admission of our ignorance does not deprive us of the hope that in the future, by means of experiments, we may improve our knowledge of the elements producing electricity'.[16] Hauch's views on magnetic fluids concluded in a similar way. He believed that the so-called animal magnetism might have an effect on certain diseases such as toothache and stomach cramp, but this magnetic effect should not be confused with the swindle committed by charlatans like Mesmer and Cagliostro, when they lured people into believing that rubbing their bodies had a healing effect, even without magnets.

To Ørsted corpuscular theory was the most problematic. Kant's critical philosophy did not approve the rule of inference by analogy, because corpuscles are not phenomena that lend themselves to empirical investigation. For sense perception without concepts is blind, and concepts without empirical content are empty. The Laplaceans ran into self-inflicted contradictions, 'apories' as Kant called them.[17] Ørsted decided to pull corpuscular theory apart and develop a dynamical alternative.

> The dynamical system assumes that substance
> is nothing other than the fundamental forces.[1]

8 | 1798–9
Doctoral Thesis on the Dynamical System

It was Kant's *Metaphysical Foundations of Natural Science*—his epistemological analysis of Newtonian physics—that compelled Ørsted to realise that the prevailing atomic theory was indefensible. 'The atomic system is shaken to its very foundations and is consequently totally unacceptable. We shall now demonstrate the foundations of the dynamical system to see if this is laid on an equally weak foundation.'[2] The task was to solve the basic problem of natural philosophy: was the physical world constructed out of substantial entities (atoms) or was the universe a product of polar forces in perpetual interplay? To start with, neither electricity nor magnetism had a significant part in the dynamical system. Both Kant's metaphysics of nature and Ørsted's thesis saw the light of day before the discovery of electrochemistry and the invention of the voltaic pile.

During the spring and summer of 1799, Ørsted began revising his *Grundtrækkene af Naturmetaphysikken* [Fundamentals of the Metaphysics of Nature] and translating it into Latin in order to submit it as his doctoral thesis.[3] The public defence took place on 5th September in the lecture hall of Elers' College. The examiners had no objections. From now on Ørsted would no longer sign himself with the anonymous 'ed.' that he had so far used for his reviews, but with the self-confident 'H.K. Ørsted, Doctor of Philosophy'.[4]

Ørsted's conclusion was that Kant's metaphysics was recondite and that 'he had not everywhere succeeded in combining intelligibility with meticulousness'.[5] Moreover, while many

German scholars had commented on other works by Kant, almost none (and in Denmark absolutely none) had dared to disentangle the elements of his *Metaphysical Foundations of Natural Sciences*. According to Ørsted, Kant's metaphysics of science was unknown in Copenhagen until he presented it in *Philosophisk Repertorium*. He now took up the task of the legendary prince Theseus who had managed to escape from the labyrinth of Knossos and save his life with the help of 'Ariadne's Clew'.[6] This was how Kant had analysed Newton and this was the way Ørsted now analysed Hauch.

All science is about physical change caused by the motions of objects. Hence Newton's three laws of motion are basic to physics. Kant never doubted their truth. They enabled us to predict the position of planets and the predictions proved correct. In his *Metaphysical Foundations* Kant analysed motion in space first and went on in the following chapter with the internal properties of matter; Ørsted followed this outline in his thesis. The first analysis is quantitative and confined to external relations: how many moving objects are countable, how much do they weigh, how fast do they move, for how long, and in what angles and orbits, etc. That is, it is expressible by numbers that can be put into a mathematical equation in order to calculate unknown quantities such as gravity. The sense-impressions emanating from matter in motion are describable by means of concepts such as distance per time units. If several objects move their orbits are describable individually. The impressive progress of physics from Galileo via Kepler to Newton owes its existence to the quantification of these mechanical phenomena and the mathematical calculations they invited. It is a precondition of this mechanical physics that matter is passive.

The second analysis turns inward to the internal properties of matter in motion asking if matter is divisible or indivisible, penetrable or impenetrable, inelastic or elastic, active or passive, etc. How could one be sure that matter is passive? If one tried to penetrate matter by contact it resisted, and this resistance could only be an indication of a force, a repulsive force impeding penetration into its space. Actually, the only thing we can know about 'it' (what we call matter or substance) is that 'it' is resisting when exposed to some external force, the pressure of a hand for instance, trying to conquer its space, that of a table for instance. Hence, the quality of 'matter' or 'substance' is a repulsive force. This is an empirical experience, in so far as it is experienced by the sense of touch of the palm of a hand. If the only activity were a repulsive force, 'matter' would expand infinitely. Consequently it must be counterbalanced by an attractive force that in turn (if it were unopposed) would make 'matter' contract into a single point. Nothing of the kind happens, therefore all 'matter' must consist of two opposed (polar) forces, somehow balancing each other.

All we can say about matter, therefore, is that it consists of polar forces in a certain proportion. We cannot observe the basic constituents of matter, the so-called atoms, and hence we know nothing about them. Whether they are solid, divisible, impenetrable, or elastic is impossible to ascertain, because these qualities defy empirical investigation. Consequently they are unfounded. Unfortunately, we can only know very little about forces. Sense impression indicates that the repulsive force acts on the surface by contact, while, apparently, the attractive forces act at a distance, as when the gravity of the moon attracts tidal water millions of kilometres away. Accordingly, corpuscular theory is pure speculation, since its concepts (solidity, impenetrability, vacuum, imponderable fluids, etc.) are empty, which is to say not founded on sense impressions. The only thing—and this is the essential point—we can know about 'matter' is that polar forces are at play. This is the fundamental cognition of the dynamical theory.

Kant's analysis was a blow to the corpuscular theory and its notion that matter consists of indivisible elements of various shapes forming a porous structure. The dynamical theory defined matter as 'something' filling space by means of its force, and since every space is infinitely divisible, it is a contradiction in terms to talk about indivisible elements (atoms). Consequently, the existence of different shapes of corpuscles is pure guesswork. It did not make sense, either, to believe in a porous structure of matter, because empty spaces are insensible, and hence we cannot even know if they exist, let alone whether they are filled with imponderable fluids. Such properties, and with them the entire corpuscular theory, must be discarded as figments of the imagination.

For Ørsted it was more rewarding to base physics on a dynamical hypothesis that reduced 'matter' to a product of polar forces. Polar forces vary in relative strength and explain the coherence of matter by assuming that the attractive force is stronger than the repulsive one. Given this, the dynamical system could explain Boyle-Mariotte's law to the effect that the volume of air decreases in proportion to the increase of an external force. By contrast, if the external pressure decreases, the volume of air increases thanks to its repulsive force.

The dynamical system opposed the mechanical materialism of the age. This is not to say, of course, that Ørsted considered humans and nature immaterial. What it does mean is that the concept of matter belongs to the categories of understanding as part of our consciousness guiding sense-impressions to enable us to make sense of our experiences.

To Kant and Ørsted, forces belong to the qualitative categories and must be understood in gradual terms of sliding scales. They are immeasurable and uncountable.[7] Consequently, they defy mathematical treatment. However Kant had always claimed mathematics to be an inescapable precondition of science, and this claim would kill the dynamical system just as it was being born, since polar forces are infinitely gradual, and inexpressible in the language of mathematics. As his analysis had brought him to realise that the corpuscular concept of matter was untenable, Kant could not avoid the conclusion that its concepts had been chosen not because they enhanced the understanding of physics, but because they enabled matter to be measured, weighed, and counted and hence mathematicised, which is exactly what the corpuscular theory required. Kant calls corpuscular concepts 'constitutive', because they fit into equations, but for a dynamical system seeking to understand physics these concepts are irrelevant.[8]

We have now arrived at the crux of the Kantian metaphysics of nature. If the object moving in space is perceived according to the quantitative category of 'substance' it can be expressed mathematically and be treated scientifically. If on the other hand this object is perceived dynamically as quality, this measurement cannot be done and there can be no science involved. Corpuscular theory of the well-known mechanical kind must fall apart for epistemological reasons. Sadly the concepts of the alternative dynamical theory are not constitutive and hence cannot produce science.

So, the outcome of *Metaphysical Foundations of Natural Science* was negative.[9] Kant ended his book with the following rather pessimistic observation on the limits of human intelligence:

> 'When the inquiring mind encourages reason to understand the sum total of all given conditions, there is nothing else to do than to turn the attention away from the physical world to our own reason; that is to say rather than explore and determine the uttermost limits of objects of nature we shall probably have to consider the outermost limits of the capacity of our own reason.'[10]

The crux is that to produce science one needs mechanical-mathematical concepts of the physical world, but the concepts developed by corpuscular theory turn out to be false. On the other hand, the dynamical theory claims to provide true concepts, but they are of no avail since they are not constitutive of science.

One serious consequence of this was that chemistry would forever remain a craft because its principles were only empirical, since the forces that might determine the combination or separation of elements could not be couched in mathematical terms.[11] For these clinching reasons, chemistry was as much excluded from the realm of science as psychology. It must have been a disappointment to Ørsted that Kant's view on chemistry was so mercilessly negative, since chemistry was the branch of science he dreamt of making his life's work. However the young doctor construed it as a personal challenge.

*

To the title of his book *Grundtrækkene af Naturmetaphysiken* Ørsted had added 'partly according a new plan'. This addition was for two reasons. First, the scope of his work differed from Kant's, who had presupposed a knowledge of the *Critique of Pure Reason*, including its scheme of forms of intuition and categories of understanding. Consequently, Kant proceeded directly to its application to Newton's laws of motion, though only using the categories that he found necessary for the analysis.[12] Ørsted on the other hand did not take this presupposition for granted and therefore followed 'Ariadne's Clew' through all parts of the maze. This thoroughness made Ørsted appear more doctrinaire than his master.[13]

The second reason why Ørsted's work 'partly' followed 'a new plan' was that he had a slight disagreement with Kant. Ørsted criticised Kant for offending against his own
a priori reasoning by tracing his main concepts from empirical observation:

> 'For the laws of nature are universal and necessary and therefore they are not empirical as Kant and many other enlightened philosophers have shown many times. So, Kant has violated his own principles with his remark that matter should be inferred from experience, since in that case the universal as well as the necessary is lost.'[14]

This comment is astonishing. Did Ørsted misunderstand Kant, who did not say that science was founded on experience alone, but on the contrary that the categories of reason are empty if they are unrelated to experience? The problem is that substance is a category, not a phenomenon. The same applies to the concept of force. It does not leave a sense-impression. Only the effect of a force does.

Ørsted said, 'Matter is nothing but the interplay of fundamental forces. Hence the quantity of matter is the same as the quantity of forces.'[15] This again is an astonishing understanding of the Kantian metaphysics of nature, because forces are exactly characterised by quality, not by quantity. In his review of Hauch's 'First Principles' Ørsted claimed, 'that the dynamical system assumes that matter is nothing but fundamental forces',[16] which must be taken to mean that the concept of force is noumenal. But if so there would be no hope of the dynamical system becoming scientific, at least not within the framework of Kantian metaphysics of nature. In his own analysis Kant had stopped at this point, stating that he had reached the outermost limit of reason and had to give in. Kant was left with his crux: the mechanical-mathematical concept of matter was empty, but scientifically productive; the dynamical system by contrast offered

Fig. 17. Adam Oehlenschläger reciting in Mrs Møller's dye-works in Vestergade *c.* 1799. The three people in front are three journeymen dyers partly turning their backs. The Ørsted brothers are sitting on each side of the reciting Adam, HCØ in profile to the left and ASØ to the right. The hostess aunt Engelke, amused, is seen at the end of the table, beneath the arm of the maid. Drawing by Carl Christian Frederik Thomsen (1847/1912).

concepts that made sense but were unable to be scientific.[17] It seems unlikely that Ørsted had found solutions to problems that Kant had had to leave unsolved.

Soon Ritter's and Volta's discoveries would provide decisive nourishment for his hope. His life's project was already articulated in 1799 and rooted in Kantian metaphysics. He stuck to it for the rest of his life and came to see his discovery of electromagnetism in 1820 as proof of its strength. But for the moment Ørsted was entirely on his own. He had to seek out foreign scientists who might help him develop his dynamical project. He realised that the battleground was not located in his own country, but on the wider European continent.

When Ørsted had defended his thesis, he could breathe more easily. The art of disputing in Latin was not his cup of tea. There are no reports of how the defence went, but when it was over it was time for a celebration, in which Oehlenschläger, of course, took part. For a long time he was sitting in silence feeling excluded from the Latin and philosophical erudition. He had no academic achievement to parade. He had given up the stage. His law studies had come to nothing, and his début as a poet was still a dream. Then, in between the toasts to young Doctor Ørsted, it was insinuated that Danish poetry had deteriorated badly since Ewald passed away. Now Adam felt taunted. He jumped up, hammered the table with his fist and shouted, 'Yes, it's true it has sunk low, but damn it all, it will rise again!' This episode indicates that Oehlenschläger was aware of his vocation as a poet and was determined to pursue it years before he became acquainted with Henrik Steffens. The company of the Ørsted brothers at the dyeworks, at Elers' College, and at Frederiksberg had awakened him. Hans Christian and Anders looked at each other. Was this exclamation a prophecy concerning all three of them? The friends were confident that they would all make great progress in their own fields and accomplish things beyond the ordinary.[18] If the Ørsteds had had the fiery temper of Adam they, too, would have jumped up to proclaim their ambitions.

> Cupid has entwined you and
> Miss P. with an indissoluble bond.[1]

9 | 1800–01
Pharmacy Manager and Fiancé

IN APRIL 1800, after five years at Elers' College, young Dr Ørsted took employment as a temporary manager of Løveapoteket [The Lion Pharmacy]. Professor Manthey had become the owner of this lucrative pharmacy by marrying Augusta, the daughter of the previous owner who had died. Løveapoteket had been destroyed by the great Copenhagen fire of 1795, but fortunately it was well insured and four years later the pharmacy, rebuilt in classical style by C.F. Harsdorff, the Architect Royal, was ready for occupation. Manthey was only eight years older than Ørsted, his protégé. During the rebuilding he had taught pharmaceutical chemistry at the Academy of Surgery as an extraordinary professor. It was here he met Ørsted who became his brightest student. In 1796 Manthey was also appointed 'artistic administrator' of the Royal Porcelain Factory in Copenhagen. To prepare himself he travelled to continental Europe on a royal grant to study porcelain making.

The patronage brought mutual benefit. Manthey needed a reliable and competent person to look after his pharmacy while he was abroad. And Ørsted needed an experienced and influential patron to promote his career. Their needs were complementary. Their plan was that Manthey would use his place on the board of the Cappel Foundation to obtain a travel grant for his protégé, who would then set out on his grand tour as soon as Manthey returned to resume the management of his pharmacy. On his journey through Germany, France, Holland, and Austria he prepared the way for his protégé's later arrival by recommending him to the scientists he met. Ørsted was not only to manage Løveapoteket, the largest pharmacy in the capital, but he would also take Manthey's place at the Academy of Surgery and deliver his lectures on chemistry. This

Fig. 18. Johan Georg Ludwig Manthey (1769–1842), owner of the Lion Pharmacy and professor at the Academy of Surgery. HCØ's friend and patron. Miniature by unknown artist. RL.

was a great opportunity for the young doctor to build up his personal finances and he was also likely to gain valuable experience as a manager, pharmacist, and teacher.

Ørsted had to manage a staff of twelve of the pharmacy. On the ground floor were the shop, laboratory, storage room, and office. An elderly head dispenser whom Manthey had inherited from his father-in-law, plus three unmarried journeymen and three apprentices lived in the basement. On the first floor above the shop was Manthey's apartment that Augusta shared with her little son. The entire household was run by 25-year-old Sophie Probsthein, assisted by a servant to carry water and firewood and two elderly maids who did the washing, ironing, and cleaning. They lived next to the kitchen on the mezzanine. They were now joined by their new manager, Dr Ørsted, whose flat was on the second floor.[2]

Soon Ørsted could send Manthey the good news that the turnover was increasing despite the fact that many Copenhageners left the city during the summer and there had been no flu epidemic. The pharmacy made substantial deliveries to Vartov Hospital, the poor-law authorities, and the Admiralty. Twice a year Ørsted had to invoice them according to the negotiated tariffs and these rather large sums could not be allowed to fall into arrears. It was also his job to monitor the stores and have them replenished with the many different herbal remedies and chemical preparations, domestic and foreign, before winter set in and ice closed the waters. Transport by sea was difficult, the British could seize entire cargoes, and insurance rates as high as ten per cent were common.[3] Stocktaking had to be done carefully, because pharmacies were

Fig. 19. The entrance of the Lion Pharmacy at the corner of Østergade and Hyskenstræde. The building was designed by C.F. Harsdorff (1735–1799), Architect Royal, erected on the site of the Copenhagen fire 1795, and inaugurated in 1799. HCØ managed the Lion Pharmacy during Ludwig Manthey's grand tour 1800–1801, during which he prepared himself to take over the post as artistic administrator of the Royal Porcelain Factory in Copenhagen. N. Høyen 1865. RL.

subject to inspection by the health authorities to make sure that there were no deficiencies that could not be accounted for. The storekeeper was often sick, and on one occasion when a consignment of goods arrived and had to be unpacked, the health authorities appeared to carry out the inspection. Luckily, Ørsted knew the chief inspector, Professor Tode, from his examination as well as from Dreyer's Club, and he obtained a postponement. Out of respect for Ørsted, Tode even offered to cancel the inspection.[4]

Part of Ørsted's job was to teach the three apprentices and to maintain discipline. In summer, the storekeeper accompanied one of the apprentices to the countryside to collect medicinal plants, and Ørsted taught him botany from fresh plants on the bench in the lab. He also taught them chemistry to prepare them for laboratory work. This instruction took place every Monday and Thursday evening. Ørsted told Manthey about his efforts in his letters:

'I try to put them on the spot by not accepting answers that only address the main point, but I also ask questions about other relevant details such as: how would you proceed if you were to actually do what you have read, what tools would you use, what temperature, what sequence of activities aiming at new results, what products will be the outcome of a given activity and in what sequence? In short: the apprentices must give me a detailed account of all activities including the causes of the end results.'[5]

The father of another apprentice, a German-speaking Jew of sixteen, complained that his boy was not taken care of according to Jewish customs and threatened to take the boy home. Ørsted treated the boys harshly. Once when he heard them playing by their bed, one of them told him that the other boy was not there, but Ørsted looked for him and found him, which led to the liar's confinement in his quarters for two Sundays and his accomplice's for one. He was not even allowed to see his parents over Christmas. He pestered Ørsted to ask Manthey to bring a portrait of Napoleon back from Paris, and the other apprentice demanded a watch for the eight rixdollars he claimed to have deposited when his apprenticeship had started. But Ørsted declined and said that he had no need for a watch at all.[6]

One of the journeymen had to be reined in, too. He was careless with his work in the lab. For instance he once mixed four grams of belladonna instead of a quarter of a gram into some powder. Fortunately (Ørsted wrote to Manthey), the quantity was so small that no damage occurred; the powder was intended for his own child and he promised not to tell anyone. Rebukes only helped for a short time. Once one of the journeymen warned Ørsted that another journeyman had a venereal disease and ought to be in hospital. His mate feared he had been infected himself, because the same maid was making both their beds. 'I assured him that venereal disease is not passed on that way, and that it must be found out if the disease was really venereal and then wait for a couple of days.'[7] However the journeyman had already alerted the apprentices about the risk of infection.

Immediately Ørsted went to see the patient, who was complaining about 'pains in his chest and colic'. The next day he grumbled about 'pains in the arse' and was offered medical help. The doctor diagnosed inflammation in his testicles and prescribed a medication. So it was a medical problem relating to a sexual organ, but it was not a venereal disease, and hospital would hardly be a good idea. The inflammation would probably have gone in eight days anyway. Manthey was content with his young protégé's handling of the matter.[8]

*

Also living in the pharmacy was the aforementioned Sophie Probsthein, who was responsible for the daily upkeep of the household. She would send a servant to the market place, or to the courtyard to fetch water and firewood, and she would cook for the staff and oversee the two maids who kept the house tidy. Our busy manager must soon have become aware of Sophie's qualities, for in November Augusta had informed her absent husband that 'Cupid has entwined him [Ørsted] and Miss P. with an indissoluble bond.'[9] This news was no surprise to Manthey, who in a letter to Ørsted confessed that he found Miss P. 'so very lovable', and that 'she is among the few people whose welfare is dear to me. Nobody but you will probably be able to make her happy, and you will hardly ever find a female who will so perfectly reciprocate your love.' Manthey made no bones about the fact that he had pondered becoming a matchmaker upon his return to Copenhagen, and was the more pleased that Ørsted had forestalled him. 'May enduring and timely happiness fall to your and your fiancée's share! And give my regards to Sophie and tell her about my pleasure and warm wishes!'[10]

Ørsted thanked him for his reassuring lines, which had given him comfort. He confided to his patron that he had pondered putting off the decision about 'such an important matter which would determine the fate of nearly all of his future' until he could discuss it with him face to face. 'Unconsciously, however, I became more and more attached to her, and although I feared nothing as much as a decision that would bind me throughout my life, I realised that Sophie's open character deprived me of every fear of making a mistake.'[11] Now, 'the unreserved approval' of his patron made him feel even more at ease.

Unfortunately, most of the copious correspondence between Sophie and Hans Christian that is known to have existed has been lost.[12] Consequently, we know hardly anything about his fiancée. The census of 1787 mentions a ten-year-old Sophie Probsthein, daughter of the first marriage of Adam Gottlieb Probsthein, a carpenter of 53, now married for a second time to Christiana, nine years his junior. Sophie had an older brother of seventeen, a painter. He did not paint doors and windows, but was studying with Bertel Thorvaldsen among others at the Royal Academy of Fine Arts. She was 26 years old, that is two years older than Hans Christian. And that is about all that is known about his fiancée. The engagement was made public; they had committed themselves for life. When seven months later he embarked on his grand tour he would leave his fiancée as part of his family. She was introduced to his siblings in Copenhagen, Anders who was soon to marry his Sophie, the sister of Adam Oehlenschläger, Jacob who was studying pharmacy, Niels Randulph, 'the Fixer' as he was called because (while studying law) he was always doing bits of business, and Tine, their sister, who was also accommodated at their aunt's dye-works on Vestergade. Sophie may also have been introduced to her prospective parents-in-law on Langeland. From now on she was attached to the Ørsted family who took it for granted that when circumstances (and particularly the pecuniary ones) were ripe Sophie and Hans Christian would celebrate their wedding feast.

*

In the spring of 1800 Doctor Ørsted was elected a member of the Scandinavian Literary Society. The following year Anders was admitted, too. The Parnassus had noticed the bright and industrious brothers. It was the goal of the Scandinavian Literary Society to unite Danish-Norwegian and Swedish-Finnish belles-lettres with philosophical, historical, and scientific works. The society was established at a meeting in 1796 presided over by Professor Jens Baggesen, the vice-provost of Regensen, the largest college in Copenhagen. A good many of the founders knew each other as freemasons.[13] They planned to publish a quarterly journal, *Skandinavisk Museum*, with articles in Danish or Swedish, intended not for scholars but for a readership of generally educated citizens. Before an article was accepted for publication it had to be read aloud at a meeting and approved by the members present. Meetings took place every fortnight, in winter at Christiansborg Castle, sharing premises with the Royal Danish Society of Sciences and Letters, and in summer at the Shooting Gallery outside the ramparts (now the

City Museum of Copenhagen). The society had a maximum of forty members, ten in each of the four classes, philosophical, historical, aesthetic and scientific.[14]

In many ways the Scandinavian Literary Society was similar to the Royal Danish Society of Sciences and Letters, with which it became a competitor. Several people were members of both institutions, the meetings took place in the same premises, and the number of members was roughly the same, as was the division of classes. The difference was the range of topics, the target group, and the level of activity, the most obvious alteration being that the philological and mathematical-mechanical classes had been dropped in favour of philosophy and aesthetics. Whereas the activities of the Royal Danish Society were focused on two projects, the cartographical survey of the kingdom and a dictionary of the Danish language, those of the Scandinavian Literary Society had a broader humanistic and outward-looking purpose. It addressed the enlightened ordinary citizen, an aim which was quaintly expressed in section 22 of its statutes: 'Academic articles that are understandable and relevant only for the erudite elite will not be approved for publication....'[15] The members were expected to give papers in a style that was likely to be appreciated by and appetising to the laity. Amongst the members who elected the two Ørsted brothers, Hans Christian (perhaps on the nomination of Manthey) on 10th May 1800,[16] and Anders one year later, we find first of all the four editors of *Physikalisk Bibliothek* and second, a number of devoted Kantians.

Before his grand tour Ørsted took part in at least twelve meetings. On 7th March 1801, Schlegel, a professor of jurisprudence, read out on behalf of a certain Mr Bardenfleth some comments on Ørsted's metaphysics of nature, but the meeting refused their inclusion in *Skandinavisk Museum* referring to section 22 of the statutes that disqualified works of a private polemic nature.

In October 1800, Jens Baggesen and his family were about to leave Copenhagen to settle down in Paris, and Ørsted and Oehlenschläger decided to arrange a farewell party in his honour in Dreyer's Club. Oehlenschläger had printed a song that Baggesen held in his hand while everybody rose to sing it. The atmosphere was passionate. The guest of honour handed the song sheet back to the author drenched in tears. Baggesen embraced and kissed him and bequeathed him his Danish lyre declaring that he did not intend to play it anymore. Oehlenschläger, who had sworn an oath at Ørsted's doctoral feast that he would resurrect Danish literature, inherited the lyre. The University had recently launched the following prize essay in aesthetics; 'Would it be favourable for Nordic belles-lettres to introduce and adopt the ancient Nordic mythology at the expense of the Greek one?' The question was grist to Oehlenschläger's mill, and his two best friends, experienced in gaining gold medals, supported him. The friendship between the three grew stronger and stronger. Oehlenschläger's essay in defence of Nordic mythology was only awarded an *accessit* meaning 'he is close to the goal'. There was only one medal to award but three essays, so Adam was left empty-handed.

His struggle to write the prize essay shows how bored he was studying law. The lasting value of his studies was part of the pretext for embarking on them, but this reason had become irrelevant because he had by now received Heger's permission to marry his daughter, and so luckily he never became a lawyer.

*

In July 1800 Ørsted wrote to Manthey in Paris that M. Saxtorph, professor of obstetrics, had suddenly passed away, and that Aasheim, professor of medicine and experimental physics, had died a few days later from typhoid fever.[17] So, the Faculty of Medicine had lost two ordinary professorships, and the two designated professors, Tode and Bang, were both decrepit. That they were designated means that their appointment had been accompanied by a promise to take over ordinary professorships as soon as the relevant *corpora* became vacant. This was an opportunity for several young applicants.

Rumours circulating among academic insiders initiated into 'the mysteries of politics' had it that the Chancellor intended to call in foreigners to take over the two posts.[18] He was sceptical about 'systems'. This was the term he used to denote what he considered ideas of sickness and health fostered by fashionable doctors from abroad and now and then taken up as their hobby horses by certain teachers at the Faculty of Medicine. To prevent such 'systems' from gaining monopoly status he wanted all disciplines to be double staffed. For the time being, however, he had no luck in finding suitable candidates from abroad.[19]

Vacancies at the University were not advertised as they are now. People aiming at an academic career would keep an eye on the fifteen ordinary professors and their state of health. Posts only became vacant by death and there was no age limit. A professor incapable of teaching would hire a substitute and pay him with part of his own salary. So, some hope was kindled in the young and ambitious Hans Christian when the rumour of fifty-year old Aasheim's life and death struggle against *Salmonella typhi* began to spread. When rumours reached Paris, Manthey was pleased on behalf of his young protégé. No death could be more heaven-sent. But it was a pity that he could not be present where decisions were being made and bring his influence to bear on them. Under the circumstances his written advice had to do. Manthey did not doubt his young manager's qualifications and that recommendations from his friends would get him the professorship, and wrote 'impatiently I await the moment when you submit your resignation. You do not have to assure me that my business will not suffer, for I know you, your heart and your talent, and that is enough.'[20]

Ørsted approached his friends in the Scandinavian Literary Society. He was annoyed that Bishop Münter supported the Chancellor's idea of calling in foreigners. Abildgaard was on his way to Røros in Norway, but one hundred per cent supportive of Ørsted, and calmed down 'his friend' (as he called him), writing that the Duke of Augustenborg went for the best-qualified candidate. He referred him to Hauch who had the ear of the Duke, 'for I know that he [Hauch] has the highest regard for your knowledge'. Abildgaard did not have the strength to take on more teaching obligations, and 'besides I'm so full of contempt for the prostituted Faculty of Medicine for academic reasons that I could not dream of joining it.'[21] Another rumour had been that Thomas Bugge, professor of mathematics and astronomy, would take over the lectures on Hauch's 'First Principles' (critically but reverentially reviewed by Ørsted), while the chemistry part would be left to Abildgaard, but this had now been disclaimed. Ørsted was not on speaking terms with Bugge, but he was sceptical, because Bugge, who had returned from Paris in the meantime, was obsessed with Laplacean mechanical physics. So, Ørsted asked, what could Bugge do regarding 'the philosophical part of physics' and electricity and magnetism, the modern part of physics his own thesis specialised in? He was advised to hand in two applications, one for a professorship of chemistry to prevent him from being rejected in favour of Bugge, and another one for a new professorship of pharmacy that insiders thought was likely to be established.[22]

Ørsted took the advice literally, attaching the finest recommendations from his friends. By the end of October it leaked out that a post as junior teacher of pharmacy, as a part of the Chancellor's overall plan for the Faculty of Medicine, was intended for Ørsted (although without a salary). Bugge was not mentioned in the Chancellor's motion for a royal resolution, which listed all of Ørsted's merits, not only his gold medals and practical pharmaceutical skills, but also his critical review of Hauch's textbook.[23] There is no evidence to substantiate the speculation that the Duke of Augustenborg worked against Ørsted's career, and an indication of this on the part of Ørsted may be due to his lack of knowledge about proceedings and perhaps to a certain paranoia. Contrary to these speculations the Duke did not take advantage of Bugge's offer, which was said to have been made only to spite Ørsted. Actually, he nominated Ørsted with an unqualified recommendation.[24] As a result undergraduates who wanted to follow private lectures in experimental physics from now on had two options: Ørsted's at the Faculty of Medicine and Bugge's at the Faculty of Philosophy.[25]

Was it half a victory that Ørsted had a foot in the door as a junior teacher? Or was it half a defeat that he could not pronounce himself a professor? Would the workload harm his management of the pharmacy? Would he have to drop his plan of a grand tour? What would Sophie say? Ørsted assured his employer that Løveapoteket would not suffer because of his lectures and he persuaded Sophie that the grand tour would be more valuable in the long run than an unpaid post. He played down the workload issue. His lectures at the Academy of Surgery were the most laborious for Gren's *Chemie* was a comprehensive textbook and were he to organise new chemical experiments he would have to prepare them.[26] On the other hand he could take advantage of this when he lectured on pharmaceutical preparations every Wednesday and Saturday from 12 to 1.[27] When lecturing privately for payment on Sundays on applied pharmacology he could recycle the experiments carried out at the Academy of Surgery. However, this optimism did not turn out to be justified; the workload turned out to be too overwhelming, and when spring approached he was so overworked that he had to cancel his lectures for a week and take to his bed. Manthey begged him to take care of his health. 'Beware of this blessing that once lost is sought again in vain.' Migraine started to torment him as previously it had his brother.

Now war was imminent. The British fleet under the command of Admiral Nelson, the terror of all European sea powers, was on its way to Copenhagen, and Ørsted fulfilled his patriotic duty by joining the Crown Prince's Corps of Students under the command of Court Steward Hauch. Once again Ørsted was sure that he would manage to do everything by keeping a stiff upper lip, but he could not find it in himself to take part in the drill, and besides there were still eight to ten lectures to deliver at the Academy of Surgery.[28] On the colours of the Crown Prince's Corps was embroidered: 'Students of Science and Letters for King and City'. However, Ørsted never became a soldier on active duty.

> We face the brightest prospects of a
> rich harvest in the field of science.[1]

10 | 1799–1801
Galvanism

ØRSTED HAD access to Manthey's well-equipped lab in the pharmacy, and during the little time his management left him, he would design and carry out experiments to develop the dynamical system he had presented in his dissertation.

At this time animal electricity, another sensational discovery, was introduced in Copenhagen by the young student of medicine Ole H. Mynster, one of the editors of *Physicalisk Bibliothek*. He had responded to the medicine prize question of 1793: 'Is electricity a decisive criterion in determining the life or death of an animal?'.[2] The idea of a specific animal electricity had been proposed by Luigi Galvani, professor of anatomy in Bologna: he believed he had discovered a new imponderable fluid in the nerves of animals. He had noticed that a dissected frog that had been hung on a brass hook to dry on an iron grid outdoors twitched even when no thunder was in the air.[3] This was a startling reaction, but several successive experiments showed that it was no accident. Galvani became firmly convinced of the existence of a natural electrical disequilibrium between the nerves and the muscles of the frog, assuming that the nerves contained the positive fluid and the muscles the negative one. His understanding rested on an analogy: the frog responded like a Leyden jar that had been charged by an electrostatic generator.[4]

All his colleagues hailed his discovery, apart from one fellow-countryman, Alessandro Volta, professor of physics at Pavia. Volta picked up on the fact that the conductors through which the electrical fluid flowed were made of two different metals. Hence, the reaction might be caused by the contact between them and have nothing to do with the frog. True, this interpretation would need a new theory of electricity in which the two metals were not only passive conductors of electrical fluids produced by friction (as by the generator), but also generating forces in themselves. Volta's new theory of contact-electricity suggested that the very touch of two different metals provoked a disequilibrium between their electrical fluids which was balanced by excess fluid from one metal flowing into the other where there was a deficit. In this case the frog's thigh was just a passive conductor. This argument made the learned republic swing over to Volta's view.

Thus the two Italians interpreted the same empirical phenomenon quite differently. Galvani the anatomist saw the world through the spectacles of vitalism. His focus was on an organism as active as a Leyden jar charged with electricity. To him it was obvious that a new imponderable fluid should be added to the existing ones: heat, light, electricity and magnetism. The new one must be a secretion produced by the organism itself. Unfortunately, this electrical secretion was reminiscent of Mesmerism or rather of the magnetic fluid that Anton Mesmer, the fashionable doctor parodied in Mozart's opera *Cosi fan Tutte*, purported to activate by stroking (or groping!) the body of a patient. And the French Académie des Sciences had exposed Mesmer as a quack and his animal magnetism as a swindle.

Volta the physicist, on the other hand, saw the frog simply as part of a physical device. But Galvani would not budge and made what he imagined would be an experimental masterstroke. Volta had placed the entire burden of proof on the contact between two different metals. So, Galvani now simply connected the nerve and the muscle of the frog directly, without a metallic link, and the frog still twitched! Volta's contact-of-metals theory had been falsified, and Italian scholars changed sides in favour of Galvani.

The course of the controversy was actually more complicated than this outline suggests. After giving it further thought, Volta hit upon this countermove: contrary to the claim that a frog produced animal electricity he now wanted to show that the electrical fluid owed nothing at all to any organism, but would equally pass through a wet piece of cloth connecting two different metals.[5] The outcome was the Voltaic pile in which the electrical fluid was produced—so he believed—by the contact between two different metals. The imponderable fluid flows from one body with a surplus of electricity via the wet conductor into another body that has a deficit of electricity. Surplus and deficit would alternate, which made the fluid circulate. How this effect was produced, Volta did not claim to know, but in honour of his colleague, who had since died, he called the phenomenon 'galvanism'. Volta's pile became a sensation and in 1800 its inventor was celebrated by the Royal Society in London as well as by Napoleon Bonaparte, First Consul, in Paris.[6]

At Jena, Alexander von Humboldt, a young mining engineer, had followed the Galvani-Volta controversy, endeavouring to contribute experiments in support of Galvani's animal electricity. A professor of botany working on the metamorphosis of plants with J.W. von Goethe in Die Naturforschende Gesellschaft in Jena advised Humboldt to seek help from a young student of exceptional talent called J.W. Ritter.[7] Although Goethe gave him financial support, Ritter was too poor to afford the entry fees for the university, and was therefore pleased to serve the well-to-do Humboldt. Ritter was now enabled to assess the Italian controversy on the basis of his own experiments. Like Gren he repudiated French corpuscular theory

Fig. 20. Napoléon Bonaparte (1769–1821), First Consul, attending a demonstration of A. Volta's (1745–1827) galvanic apparatuses in Paris after the French conquest and occupation of Northern Italy 1801. To the left Volta charges a frog's thigh with galvanism to make it twitch. In the centre an electrometer is seen and to the right Bonaparte is observing a voltaic pile with much interest. Coloured lithograph by unknown artist. Château de Versailles. RMN, Paris.

and its imponderable fluids and looked for a dynamical solution. The outcome of all this was a book that also contained a translation into German of the Galvani-Volta controversy. It was published two years before Volta's invention of the pile. Ørsted only heard about the little-known Ritter in the autumn of 1800 and immediately realised that here was significant information in support of his dynamical project.[8] Ørsted acquired and devoured Ritter's *Beweis, dass ein beständiger Galvanismus den Lebensprozess in Thierreich begleite* ['Proof that a Continuous Galvanism Accompanies Life Processes in the Animal Kingdom']. The book was like manna from heaven because, like Gren's textbooks (and indeed going beyond them) it underpinned the ideas of his project.[9] Ørsted equally appreciated Ritter's theory of light.[10]

Ritter's experiments sustained neither Galvani's organic nor Volta's inorganic contact theories, but pointed towards a completely new dynamical theory bringing galvanism, electricity, and chemical reactions together into one dynamical system, electrochemistry, of which he is the originator. Ritter's theory was in no need of substances or imponderable fluids, but claimed the following dynamical trinity:

GALVANISM

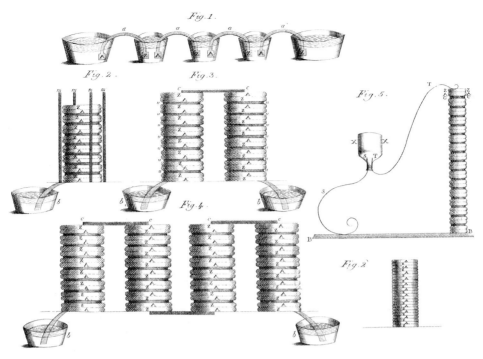

Fig. 21. Contact electricity. Volta believed that metals contain specific electrical fluids flowing from the ones with a surplus to the ones with a deficit. Thus he interpreted the electrical effect as embedded in material, but as imponderable fluids—as opposed to Galvani, who understood electricity as a particular, organic force. Like Ritter, HCØ believed that electricity is a dynamical, but noumenal force with a chemical effect. A. Volta 1816. RL.

(1) Galvanism exists when three mutually different, space-filling entities (separated, and without complete transformation of the three into one chemical entity) activate each other. Example: the voltaic pile (silver, zinc, brine-soaked cardboard).

(2) Electricity exists when two mutually different, space-filling entities activate a dynamic disequilibrium in reciprocal contact and interplay without unification. Example: silver in contact with zinc.

(3) A chemical process occurs when two space-filling entities of different qualities combine to become a new qualitative unity. Example: metal oxides, hydrogen sulphide.

According to Ritter's theory galvanism is simply an electrochemical process, where galvanism is the product of two other processes taking place at the same time, an electrical one (a disequilibrium between two metals) and a chemical one (a combination between metals and a diluted acid). Ritter concluded that since the incomplete dynamical process (the electrical one) was contained in the complete one (the chemical one) as part of

Fig. 22. Voltaic pile consisting of a wooden stand 67 centimetres high and three vertical glass rods supporting a pile of two times 60 copper and zinc discs separated by brine-soaked cardboard discs, 1802. Herlufsholm Skole. Photo by DCC.

a whole, electricity (not as it was understood in his day, frictional electricity produced by an electrostatic generator, but as it would be understood in the future) must be perceived as an electrochemical process.[11]

This theory was based on one of Ritter's extraordinary experiments. He took a glass plate and placed on it in a few drops of water two pairs of parallel metal rods of zinc and silver. He connected the first pair (but not the second) using a good conductor [see fig. 23]. After about four hours a reaction appeared: the silver rod connected with the zinc had unmistakably oxidised, while the unconnected pair was unaffected. He inferred that the dynamical process in the closed circuit was a unity, where the electrical and the chemical processes happened at the same time. It was impossible to determine if it was chemistry that produced electricity or the other way round, only that the same force must have produced both. Ritter's experiment, with a potential difference of just one volt, displayed the same phenomenon as the voltaic pile, but two years before Volta had developed it. Ritter immediately understood that an electrochemical effect had been produced.

Ørsted read Ritter's *Beweis* with its sensational experiments while he was managing the pharmacy, and soon after he received news about Volta's galvanic apparatus and was immediately fascinated. This was useful news for his project. In October he wrote to Manthey in Paris:

> When a silver disc is placed on a zinc disc, and on top of this a moist woollen cloth or map sheet and on top of this another silver disc, a zinc disc and a moist cloth and so on until they reach a certain height of say twenty of each, and they are connected to conductors, then electrical phenomena occur.[12]

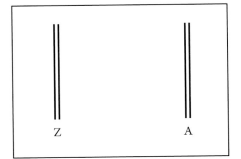 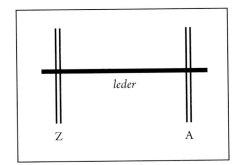

Fig. 23. Drawing of Ritter's experiment on the separation of water.

Manthey wrote back that in Paris this was not exactly recent news, because he had seen it demonstrated by J.A.C. Charles. He reported that Napoleon had invited Volta to come to Paris to show his invention to the First Consul. It had made a great stir in scientific circles.

Soon after, Ørsted received more news. In London water had been separated by means of a voltaic pile and the accumulation of hydrogen had been observed at one conductor while the other one oxidised.[13] What was in dispute was the inexplicable phenomenon observed by Nicholson and Carlisle: when a galvanic force was circulated through water, it was changed into two gases, one rose at one conductor as bubbles smelling like hydrogen, while the other conductor was oxidised. If the conductors were made of platinum the oxygen bubbled up like the hydrogen, because the platinum was unaffected by oxygen. But why did one gas bubble up at each conductor rather than both of them bubbling up in between them? And how could the idea of a chemical separation of water be compatible with the role of water as a conductor of galvanism?

This was the problem Ritter took it into his head to solve. As we shall see, this problem was rooted in fundamental scientific dilemmas of dynamism versus atomism and of empirical versus a priori research. To start with Ritter put together his own scientific equipment to test Nicholson's and Carlisle's experiment. His apparatus for separating water was designed with the idea of catching the ascending gases in two suspended tubes [fig. 24]. The two drinking glasses under the conductors were intended to catch the metal oxides deposited on conductors of base metals (zinc, tin, lead, iron, etc.). Ritter now connected the apparatus to a voltaic pile and waited until the following morning to see what had happened during the night. He found that the tube above the positive conductor had caught about one cubic inch of oxygen (O) while the other tube contained about two and a half cubic inches of hydrogen (H). When he added phosphorus to O the phosphorus ignited, the gas disappeared and H was released, ignited, and exploded as expected. When the conductor was made of a base metal he found calcified metal that was crisp when he bit it.[14]

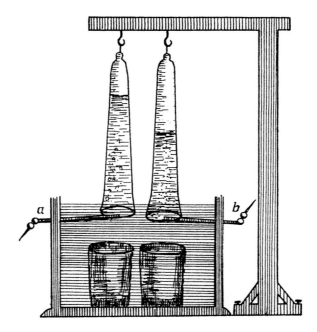

Fig. 24. Ritter's experiment on the separation of water.

Of course, only the effect (the ascending gases) was observable, while the cause (why the water was separated) could not be determined empirically. Ritter was uncomfortable about deductions based on reason alone. According to Nicholson and Carlisle molecules of water were thought to be a combination of two kinds of atoms that were split by galvanism so that the O-atoms bubbled up at one place and the H-atoms at the other. How then could the water at the same time conduct a flow of imponderable fluids in a galvanic chain? According to the first premiss, water was a combined entity separable into two kinds of atoms, but according to the other premiss water was an entity conducting one single force. These premisses contradicted one another, and hence both could not be true. But if he could isolate the two processes from each other, the problem might be solved.

Ritter then made a new apparatus for another experiment in which water was not both the object of separation and a galvanic conductor at the same time. By trial and error he found that concentrated sulphuric acid was a perfect conductor of galvanism while not developing gases in a galvanic chain [fig. 25]. He now filled the bottom of a V-tube with concentrated sulphuric acid and through a funnel he topped up the tube at each end with distilled water. He determined by means of a litmus test that the distilled water did not mix with the sulphuric acid. He then closed the two ends with corks, put gold wires halfway down into the water and connected the apparatus to the zinc and silver poles of the voltaic pile. What now happened was that H bubbled up at the golden wire connected to the zinc pole and O at the other pole. If the galvanic flow was turned around the opposite result was obtained.

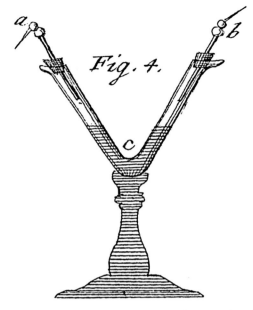

Fig. 25. Ritter's subsequent experiment on the separation of water.

What atomists conceived as a separation of water molecules turned out in Ritter's new experiment to be a galvanic effect: O was positively galvanised water and H negatively galvanised water. His point was not that the gases were something other than O and H. No, his point was that these two gases could not be the constituents of water separated by galvanism, but that the development of the two gases happened in two independent galvanic processes (separated by the sulphuric acid) so that water plus positive galvanism equals O, and water plus negative galvanism equals H. In other words: galvanism was the polar force and water remained an element as it had been since antiquity.[15]

Ritter's experiment dealt a serious blow to Lavoisier's understanding of chemistry. Like Ørsted, Hauch also took an intense interest in the voltaic pile, the separation of water described by the Englishmen, and Ritter's counter-experiment. From September until December Hauch repeated these crucial experiments by means of specially produced tubes in order to prove or disprove the theory of modern chemistry that water consisted of two gases, oxygen and hydrogen. In January 1801, he demonstrated a series of galvanic experiments to members of the Royal Danish Society of Sciences and Letters to explore the mystery of the combination and separation of water. The Court Steward repeated the known experiments, assuming in the eyes of the spectators the role of a magician who, like Galvani and Volta, was capable of seducing an audience into accepting now this and now that solution to the riddle. Of course, Hauch agreed with Ritter that the cause-and-effect issue of galvanism was not determinable by empirical observation. For whether galvanism was thought of as a force or as an imponderable fluid, it was invisible, as Ørsted and Hauch also agreed.[16]

Being poor, Ritter had made his experiments with simple equipment, but Hauch's Physical Cabinet had a thoroughly aristocratic character. He loyally repeated Ritter's experiment in which sulphuric acid separated the two bodies of water while allowing the galvanic flow to pass [fig. 26]. Like Ritter he found that the body of water adjacent to the silver pole released an odoriferous gas to the s-shaped tube and that by ignition with a sulphur match it turned out to be H. In the other s-shaped tube at the zinc pole the gas was O, since it made phosphorus re-ignite. Now, the learned assembly had seen that O and H could not derive from the same water but came each from its own portion. Consequently, O and H were not constituents of water, but generated by galvanism. Water must be a single element.

As his counter-proof, Hauch now designed an experiment [fig. 27], in which two tubes, A and B, were placed in small jars with water and the two jars were connected with a golden wire each to its own golden needle, C and D, in the two water-jars. Now, at A, O was produced and at B, H, as one would expect, but in addition small bubbles of gas were observed at C and D. Hauch presumed that these gas bubbles consisted of H at D (as at A) and O at C (as at B). To verify this he extended the experiment [fig. 28] by placing two sets of tubes (AC and DB) each in its own water-jar, and the two short ones (C and D) were connected to a golden wire, while the two long ones (A and B) were connected to the silver and zinc poles of a voltaic pile. Some hours later an emission of gas had taken place in all four tubes. H in A and D, and O in C and B. Both water-jars had released gases, and Hauch claimed that he had rehabilitated Lavoisier's discovery that water is no element as hitherto believed, but, contrary to everyday experience, a combination of two gases—*quod erat demonstrandum!*

Finally, Hauch embarked on a thought experiment intended to explain why the same gases were produced in AD and BC respectively rather than in AC and DB. His explanation was that 'the galvanic fluid' was combined with 'a fire substance' carrying the properties of both heat and light. The heating property was the more efficient when water released H, while the lighting property was the more efficient when the water released O. This enabled the entire game of patience to come out. But his way of reasoning was pure speculation and it did not pretend to be otherwise. In no way did Hauch triumph over Ritter, whose scientific work he respected.

The upheaval created by the development of the voltaic pile coincided with Hans Christian's engagement to Sophie and her introduction to his family and friends. The liaison he had longed for throughout his years of study was interrupted by his time-consuming activities as manager and lecturer as well as his membership of the Scandinavian Literary Society. No wonder his letters to Manthey testified to overwork and headache upon headache. By April his galvanic experiments had come to an end; by then he had invented a galvanometer and a new galvanic instrument.

Astounding news about Ritter's galvanic experiments on his own body and his discovery of ultraviolet light appeared in *Physikalisk Bibliothek*.[17] The article told of the shock experienced by a person who touched the contact of a voltaic pile with damp fingers, and also about the shock that hit people who held the hand of the person touching the pile. The discomfiting surprise resembled the exciting parlour game in which a man takes a firm hold of the contacts of this now familiar generator while kissing the lips of an unsuspecting lady. Standing on the

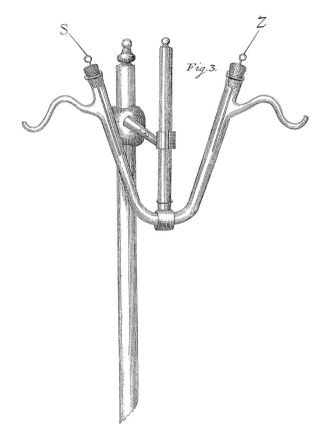

Fig. 26. Hauch's control experiment.

floor she forms an earth, and the kiss is electrified. This was probably a mixed pleasure, but nothing compared with Ritter's galvanic experiments on his own body, which approached self-torture. Firstly, he would put one contact into his mouth so that his tongue was touched. Then, taking the other contact in his hand, not only would his tongue receive a shock but his whole face would begin to shine from the gums through the now translucent cheeks. He discovered that galvanism had an acidic taste if the contact linked the zinc of the voltaic pile and the tongue via the fingers back to the silver. If the galvanism went in the opposite direction, the taste was unmistakably alkaline.

The shock was even more powerful when the contact, now with a metal button soldered onto it, was made to touch the eyeball, and the shine of the face was more conspicuous 'when the experiment took place in the dusk'. 'A very bright lightning appeared when the connection was established, accompanied by a rather strong and sudden pain in the eye.'[18] The touch of the positive contact on the eye made the objects of the room appear more distinct than normal whereas the negative one blurred them. Furthermore, galvanism evoked a range of colours. Experiments

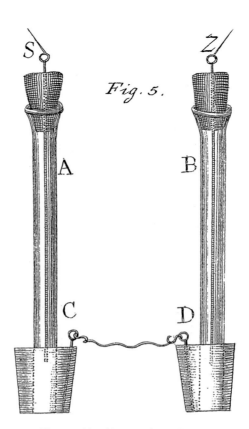

Fig. 27. Hauch's control experiment.

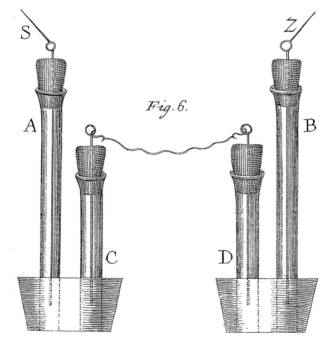

Fig. 28. Hauch's control experiment.

at dusk showed Ritter that the clear transparent glass rods that held the silver and zinc plates of the voltaic pile together first turned bluish, but when the current changed direction the otherwise colourless glass took on a distinct reddish tinge. Decades later Ørsted told his children that he had mentioned to Ritter that he was puzzled by his foolhardiness in jeopardising his sight. Ritter had retorted that since he had two eyes he could risk one of them for the benefit of science, for every time he repeated the experiment he put the contact on the same eye.[19]

The unequivocal conclusion of this series of experiments was that the polar forces of galvanism matched polar reactions in the sense organs. The phenomena were entirely empirical. It would be utterly misleading to call Ritter a speculative *Naturphilosoph*. The sense impressions were qualitative effects of galvanic forces. However, there was one inconvenient fact: nobody (or hardly anybody) had the courage to lend their bodies to the repetition of these experiments. Soon, however, Ritter had the opportunity of carrying out an experiment to corroborate his theory of the polarity of the fundamental forces.

The German-born British astronomer F.W. Herschel had succeeded in separating the sunbeams emitting light from those emitting heat (but no light) by means of a prism. He had taken the temperature with a thermometer half an inch outside the red part of the spectrum and found that it rose 6.5° in ten minutes. He concluded that there must be invisible sunbeams emitting heat and that the prism was deflecting them further outside the spectrum than the light beams inside. Herschel presented his experiment to the Royal Society in London calling these heat beams ultra-red light.[20]

Ritter was convinced of the polar nature of the fundamental forces. Like Ørsted he believed that light and heat are fundamental forces belonging to the dynamical system, as are electricity and magnetism, and that possibly all four were reducible to one and the same fundamental force. When Ritter received news of Herschel's experiment, he realised a priori that it must be possible to carry out a chemical experiment to investigate the possibility of finding the symmetrically paired phenomenon. The point of this new experiment was to show that corresponding to the ultra-red beams there must be invisible beams outside the opposite (violet) end of the spectrum. Ritter's experiment differed from Herschel's. He knew from Scheele that silver chloride changed to black at the violet end of the spectrum, because silver chloride is reduced, that is, it loses its oxygen; while at the other end of the spectrum its regular colour is restored, because it is re-oxidised.

Ritter's hypothesis was verified. In spring 1801 he presented his experiment to Die Naturforschende Gesellschaft in Jena. J.W. von Goethe, his patron, was present. The experiment had not only proven that the diffusion of light included the invisible heat rays at both ends of the colour spectrum. It had also been shown that the effect was a chemical one. This result was immediately recognised internationally to the disappointment of Goethe. For several years Ritter had eaten from Goethe's generously outstretched hand. Goethe in return had profited from the experimental talent of the poor student, a talent he needed for his own optical studies.[21] Goethe was looking for experimental results that falsified Newton's theory that white light was a product of the entire range of light beams in the colour spectrum and verified his counter-idea that colours are an interplay between light and darkness and the objective colours of physicists are a world apart from human sense impressions of colours. Sadly for Goethe,

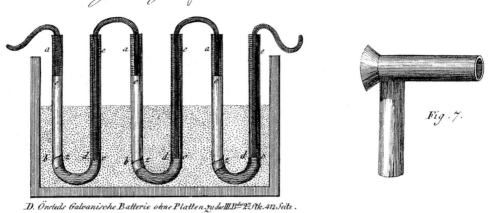

Fig. 29. HCØ's portable, galvanic battery. The U-tubes are fixed in sand, filled with acid, and connected by two metals from opposite poles of the electrochemical series, e.g. silver and zinc. This simple dynamical device created a stir when the young physicist showed it in salons and scientific societies on his grand tour 1801–04. *Magazin für den neuesten Zustände der Naturkunde*, vol. 3, RL.

Fig. 29a. HCØ as a twenty-five-year old. Drawing after G.L. Chrétien's physionotrace.

Ritter's experiment tipped the scale in Newton's favour, because the polar nature of light was now corroborated by chemistry. This amounted to treason and Goethe withdrew his support.

While this was happening at Jena, Ørsted was immersed in galvanic experiments for as long as his obligations would allow. He took a profound interest in the future of science and prophesied, 'We face the brightest prospects of a rich harvest in the field of science'. In February he witnessed a new experiment by Hauch, who had connected all the voltaic piles he could borrow in Copenhagen, amounting to six hundred silver and zinc discs. Of course, they had a colossal effect, although still less powerful than that of the electrostatic generator, according to measurements by the electrometer. But when spectators dipped their hands into the water they nevertheless got an electric shock of a tremendous strength, Ørsted noticed.

In March he invented a galvanic instrument that turned out to be more powerful than the voltaic pile.[22] By trial and error he found the construction illustrated below the most successful.[23] He packed it into his trunk for his grand tour and demonstrated it whenever he found an appreciative audience.

PART II
THE COSMOSPOLITAN

> In the evening we saw 'The Marriage of Figaro'...
> but the parts seem to be better cast at home.¹

11 | 1801–2

First Grand Tour
Tourist far away from Sophie

In the early days of the French Revolution 1789-90, Professor Manthey had been funded by the Cappel Foundation to study in Paris. Now he was on the board himself and succeeded in procuring a travel grant for his protégé as planned. Upon Manthey's return in the summer of 1801 young Dr Ørsted set out on his first grand tour.

Manthey was now ready to occupy the post of director of the Royal Porcelain Factory in Copenhagen. Moreover, as an expert in chemistry he was appointed a member of the Public Commission for the Improvement of Breweries and of a Society for the Ennobling of Craftsmen. In Manthey's mind, therefore, chemistry was tantamount to technical chemistry with a utilitarian objective. Being a freemason he associated with the elite, among them the Duke of Augustenborg, Chancellor of the University, Count E. H. Schimmelmann, Minister of Finance, and Court Steward A.W. Hauch.

Were the motive forces driving Ørsted to embark on his grand tour compatible with funding from Manthey's powerful connections? Ørsted's vocation in life obviously transcended technical chemistry. His dynamical research project went beyond everyday utility, and during Manthey's absence his ambition had been stimulated by the latest discoveries by Ritter and Volta. The difference between the motives of our ambitious young doctor and of his patron was conspicuous.

'The brightest prospects of a rich harvest in the field of science' pulled him to Europe. He did not travel to safeguard a mundane existence as a brewer, a mine engineer, a pharmacist, or a porcelain manufacturer. He also intended to enlarge his cultural horizon, win new friends, establish new networks, and immerse himself in European philosophy which, during these post-revolutionary years, was experiencing a time of transition from the Enlightenment to Romanticism. This grand tour had a price. He left his fiancée behind at the pharmacy, and for a long time their love had to find nourishment in their correspondence, if indeed their letters reached the far away object of love at all.

The Cappel Foundation's grant of 540 rixdollars for three years was awarded to Ørsted as a pharmacist for postdoctoral training.[2] On top of this grant he received 125 rixdollars for three years from Elers' Foundation administered by Professor Bugge,[3] and finally, he submitted an application to the Foundation ad usus publicos [the Foundation for Public Purposes] for a supplement to buy scientific books and instruments as well as to extend his travel to England, arguing that 'the admirable factories in that country deserve the attention of every chemist'.[4] This application was the unpredictable part of his budget, and even though Manthey pestered Schimmelmann and Reventlow, directors of the Foundation, he was kept in suspense about the outcome for more than six months. Incidentally, Professor Callisen protested that Ørsted had become the beneficiary of the Cappel grant when he already held a post.[5] He must have had the Academy of Surgery in mind, but that was a temporary job now back in Manthey's hands, and Ørsted's new post at the University was unsalaried.

Manthey was authorised to draw the grants that he transferred in instalments to various stockbrokers whom he was in the habit of using for his pharmacy transactions. However, Mr Jørgensen, the bursar of the Cappel Foundation, was very meticulous and required receipts for every single disbursement as well as documentation that the recipient attended lectures rather than frittered away his time and money. This system was inconvenient, time-consuming, and loss-making when the rate of exchange was unfavourable. Ørsted had the required testimonials issued by various lecturers, but failed to disclose that in reality they were sometimes fronts for a private study circle on Kantian metaphysics of nature or for his own untutored experiments in laboratories.

Lectures on philosophy and aesthetics also attracted him, but attending them was unjustifiable to Mr Jørgensen who kept a watchful eye that the grand tour followed the requirements of the grant to the letter.[6] For the authorities holding the purse strings, there was no end to the benefit his country would derive from his travel, but in reality Ørsted was acting in his own best interests, and Manthey consented, advising his protégé to keep a low profile as far as the more philosophical aspects of science were concerned. His books and articles helped him improve his finances and this extra income was used to balance his accounts.

*

Ørsted's itinerary is shown on the map, figure 30. Travelling on the highways was inconvenient. On 11th August he left the University of Göttingen where he had enjoyed a few days with his fellow-countrymen and Professor J. F. Blumenbach, whom he knew from his treatise on the amniotic fluid and who had demonstrated his famous collection of skulls. The journey next took him to Professor J. B. Trommsdorff, head of the pharmaceutical-chemical Institute at

FIRST GRAND TOUR

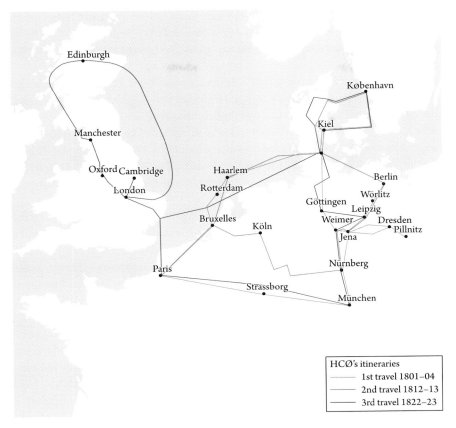

Fig. 30. Map of Europe showing HCØ's first three itineraries.

Erfurt, via Northeim, Osterode, Nordhausen, Sondershausen, and Langensalza, a distance of a hundred English miles. He had heard that the mail coach was very slow so he seized the chance of a cheap journey by horse and cart.

> 'As I realised that the driver intended to cheat me, I repeated the agreed price in the presence of some friends and received the following answer: "*Ich führe Ihnen auf den graden Weg nach Erfurt*" [I'll take you directly to Erfurt]; and when in very precise words I demanded confirmation of his promise, he said he would take me to Langensalza only. I told him to go to hell and decided to take the mail coach rather than letting this rogue lead me by the nose.'

The next day Ørsted took the mail coach from Göttingen to Nordhausen, forty English miles. It took him 28 hours or a mile and a half per hour, partly by night and along a very dangerous road. The rest of the journey from Nordhausen to Erfurt he was informed would take two days. 'So I decided to take a guide who put my luggage on a wheel barrow and brought it to Sondershausen' (a ten miles' walk). The last fourteen miles were taken by horse and cart. In this astute way he saved two overnight accommodations and a day's journey. In forty-eight hours he had covered seventy-five miles and the time he had gained was spent in erudite talks with Trommsdorff, the chemist.

One January evening he visited a dance hall in Berlin with his friend A.C. Gjerlew, a theologian who was on his grand tour to study pedagogy. His friend only enjoyed himself a little and Ørsted not at all. He assured Sophie that this was the first time in his life he had set foot in such a place, but he did not regret it because he got to know things that fully confirmed his views. Gjerlew had got into conversation with a businessman from Halle, who had a lot to tell about the misery of the place. Many of the girls had previously worked in Halle, and one of them who seemed very simple-minded, parted with information that it is generally advisory to keep quiet about.[7]

In the coach from Munich to Augsburg one of the passengers was a young Jewish girl on the way to her wedding. However, she revealed such a desire to be unfaithful to her husband-to-be (even before the wedding!) that Ørsted was perturbed. A fellow-passenger, von Hopfer, was in the habit of dealing with that kind of person and he easily obtained her promise that she would see him at Augsburg.[8] His perturbation may be read as a signal to his friends and family at home, not least to Sophie, how inappropriate it was for people engaged to be married to have affairs.

He had met von Hopfer casually at the Picture Gallery in Munich. In the evening von Hopfer had taken him to one of the well-known beer cellars where journeymen would enjoy themselves. 'I found it strange that they sang so many students' songs and so many blasphemous songs about the holy orders—in this upright and catholic country. Schiller's "*Ein freies Leben führen wir*" ["We live a free life"] and "*Frisch auf Kammeraden*" ["Cheer up comrades!"], etc. seem rather vulgar, particularly the first one, for its coarseness.'[9] A few days later Ørsted joined a party to celebrate the wine harvest and was upset to hear the grape pickers sing a song by Schiller 'written to suit the taste of brutal soldiers in the seventeenth century'. 'This testifies to a lack of delicacy and decency' he wrote to Sophie.[10] 'Nothing is more disgusting at a drinking party than bad songs and a lack of subtle entertainment. A drinking bout must be a high feast among educated people where a moderate consumption of strong drink evokes my joy of living and opens my soul to feelings of all kinds. In the beginning songs of merriness and later on songs of seriousness like Schiller's "*Ode an die Freude*" ["Ode to Joy"], that is what I wish.'[11] Schiller's ode, which later became part of Beethoven's Ninth Symphony was Ørsted's musical favourite. Giddiness, drinking to excess, and women of easy virtue were not to his taste.

*

We are comparatively well informed about Ørsted's grand tour through the letters he sent to Sophie Probsthein, which were also intended for circulation to family and friends. Manthey mentioned that he had not seen the letters to Miss Probsthein for six weeks, but a month later he had read all of them.[12] Consequently, the letters were written in a matter-of-fact tone, though occasionally this style was interrupted by personal passages with a direct form of address showing that Hans Christian had a certain recipient in mind. In one he abruptly asked Anders to deliver some books he had borrowed, and in another instance the text was interrupted by this sentence: 'Since I believe that the sixteenth is your birthday, I invited my friend W. to celebrate.'[13] His letters to Anders and Sophie were signed 'Your Christian'.

Short remarks and abbreviations of a few lines in a preserved notebook (ØC 15) inform us of his daily activities including his correspondence. As his letters are numbered in succession (from 16th February 1802 onwards) we know how many letters he received and dispatched and whether any outgoing letters are missing. So, he sent fifty-one sheets of four pages to Sophie and at least eleven to Anders, partly numbered in the notebook, though only six or seven have survived. A more thorough investigation reveals that 'Travel letters in diary form' (the title Mathilde Ørsted gave to the Ørsted Collection, ØC 80) is incomplete when compared to the short remarks found in his private notebook (ØC 15).[14] The problem is that his archive has been severely censored and part of it thrown away. As a result only 127 handwritten pages out of 250 (51 sheets to Sophie and 11 sheets to Anders) have been preserved. Of these, half have been edited by Mathilde Ørsted (marked in the margin of the preserved pages), whereas the other half of the archive is marked 'delete!' with a red pencil, while the rest has either gone missing in the mail (this can be proven in some, but far from all cases) or been thrown away later on.[15] The lacunae emerge as missing dates in 'Travel letters in diary form'. Mathilde Ørsted's criteria of selection from the preserved archival material are haphazard, and her deletions are only indicated sporadically. When occasionally Ørsted wanted to communicate more intimately he would write separate, sealed appendices as indicated in some short remarks of his notebook, but none of these love letters has been preserved.

According to his notebook his correspondence with Manthey (ØC 1–2) ran to eighty-one letters on economy, chemistry, and his career, important evidence of a productive grand tour. Some of these have gone astray in the mail.[16] Of the many letters he wrote to Anders, on philosophy in particular, only three have survived, while all confidential appendices to Sophie Probsthein, three letters to his parents in Rudkøbing, one to Aunt Engelke Møller, three to his brother Niels Randulph, and a few to other recipients, among them Adam Oehlenschläger, are lost. Furthermore, Ørsted entered in a special diary (ØC 88) scientific information concerning visits to factories, remarks on galvanic and chemical experiments, and observations, discussions, etc. with colleagues.[17]

Mathilde Ørsted's grounds for destroying letters, or deleting a vast number of passages from them, in her edition of her father's correspondence are simple. She was keen to suppress any record of Sophie Probsthein. She succeeded almost perfectly, for nowhere in the subsequent literature on Ørsted is his engagement to Sophie mentioned at all. But she did not succeed completely. Mathilde was not aware that historians might find her father's short notes (ØC 15) or her instructions 'delete!' in the margin of his travel letters in diary form (ØC 80), or his correspondence with Manthey (ØC 1–2). Sophie's name pops up now and then in the three files and Mathilde has not succeeded in deleting all traces. How his letters to Sophie have ended up in the Ørsted Collection is unknown. The fact that all the letters from Sophie are missing indicates that they may have agreed to swap letters at some point.

How did the engaged couple nurture their relationship now they were forced to live separate lives, after having lived under the same roof for a year? All we have to go on are the scant fragments that have survived. Miss Probsthein had become a member of the large Ørsted family and was invited to family events in Copenhagen where she was able to talk about Hans Christian, even if this was a poor substitute for talking with him. Of course, each of them had visions of their wedding day, setting up home together, and having a family. During the first year Hans Christian received at least four letters from Sophie (now lost) and wrote at least eight replies. Letters could go missing, and that happened to some of Sophie's.[18] Hans Christian grew impatient because of the intolerably long time it took for the next letter to arrive. In a postscript he poured out his troubles: 'Today, 12th April, it is exactly one month ago since I had a letter from S'.[19] But finally, on 23rd April he received one. And in the summer heat Manthey, his most frequent correspondent, reassured him that 'your Sophie is now very well and happy'.[20]

So, in his heart Ørsted was no doubt pleased; but he was soon forced to change his mind. In June David Probsthein, Sophie's brother, appeared at Freiberg. He, too, was on a grand tour that eventually would bring him to the community of artists in Rome. In the evening of the 23rd Hans Christian (unexpectedly on his part as it seems) bumped into his prospective brother-in-law at a party in Freiberg. Ørsted wrote:

> 'On Thursday evenings a rather numerous party of the inhabitants of that city, including ladies and foreigners, come together. Almost every time one meets Werner [the geologist] there, and he is no doubt the person holding this party together. I was there, too, last night, and found among others also Lieutenant Probsthein here. We talked about his family.'[21]

On the same day he informed Manthey about his schedule and budget for the rest of his journey. Naturally, his relationship with Sophie mattered a great deal to him. Even if his grant stretched to three years, he said that he would prefer to return home a little earlier, 'because it is undeniable that life is most agreeable on a journey like this and I appreciate it very much; but I should not let my poor Sophie wait for so long, unless it is necessary. You know how many obligations are laid upon me.'[22]

Sophie's central part in these deliberations on the very day her brother appeared in Freiburg suggests that the lieutenant had reminded his brother-in-law to-be that his prospective wife would like to see him home as soon as possible, preferably with concrete plans for their future family life together. Ørsted ended his letter to Manthey indicating his intention to return home half a year ahead of the expiry of his grant, because he could not build his marriage on debt. Even though he found Manthey's offer of lending him money generous he preferred to cover his travel expenses by the part of his grant he could save by returning earlier to Copenhagen.

The next day Ørsted paid a visit to two fellow-countrymen, Paul Steenstrup and David Probsthein. The rest of the day he spent reading Novalis's novel *Heinrich von Ofterdingen*.[23] His notes do not suggest that he had had a cordial encounter with Sophie's brother, let alone felt an attraction for his company. He saw Steenstrup for several days, but Probsthein for just a few hours. Some days later he wrote a letter of four sheets (sixteen pages) to Sophie about 'the kindred' (presumably his meeting with her brother) and enclosed 'a loose note and an address' (probably a mail address on his itinerary), to which she could reply. At this stage he had heard nothing from his fiancée for two months.

On July 10th Anders was married to his Sophie (or '*Krudttaarnet*' ['the powder keg'], which was Kamma Rahbek's pet name for her) in Frederiksberg Church. Miss Probsthein attended together with Madam Møller, 'Paalekom' (Sophie and Adam Oehlenschläger's father, unable to pronounce '*publikum*' [public] correctly), Niels Randulph ('*Commissionæren*' [the Fixer]), 'Adagio' (Adam) and Christiane Heger (his fiancée) in Paalekom's house. Christiane reported to her sister Kamma that 'all guests were moved by wholehearted joy and compassion and lost for words—except Miss Propsthein [sic] who showed no sign of emotion.' This was not exactly an amiable description of Hans Christian's fiancée who was soon to be led to the altar herself. The women of the Ørsted family were far from approving of Miss Probsthein, and Christiane did not mince her words. 'She is an absolute brute', she went on. It is hard to know what justified this harsh verdict. Was she devastated by jealousy, or did the family find her unpleasant for reasons we do not know, and did they cold-shoulder her?[24]

A love relationship that for years had to do with written greetings of love and paper kisses easily falls prey to malnutrition and gradually withers away. If Sophie had been inclined to throw herself into the arms of the Ørsted family and they had welcomed her wholeheartedly she might have found comfort and empathy. But alas, Christiane was not the only one to find Sophie unpleasant. G.J. Bull, a Norwegian student of law, staying at Madam Møller's dye-works where he had become engaged to be married to Tine, sister of the Ørsted brothers, wrote to his fiancée: 'Do you remember the first days of our relationship when you got angry because I said you beat me as hard as Sophie Probsthein? I do not understand how I could compare your sweet hands with the bony paws of that ass.'[25]

Towards the end of July Ørsted sent a letter to David Probsthein at Meissen and a good month later he had an answer.[26] The contents of these letters are unknown. He continued his series of letters to Sophie apparently unruffled. At the end of September he arrived at Salzburg, where there was a poste restante. 'I strongly expected yet another letter from a certain Sophie in Copenhagen, but the postmaster assured me that I had already received all there was. Reluctantly I had to leave empty-handed', he wrote to Sophie. She remained silent.[27]

*

After a busy winter in Berlin replete with lectures, experimenting, and social events, he plunged into tourist adventures during the summer of 1802. Sightseeing, too, was part of a young man's grand tour, even if it did not feature in the budget. His itinerary brought him into proximity with famous buildings and art collections that he might never see again. Ørsted indulged in sightseeing, especially if he had company, and he nearly always found somebody who was delighted to converse on places of interest.

His first excursion took him less than twenty miles, to Potsdam. With some friends he clubbed together to buy 'a journalière', the daily mail coach. Of course, Sanssouci and the Neues Palais, the impressive buildings in the magnificent park created in English style by Erdmannsdorff, the landscape architect,,attracted their interest. Ørsted reported home on the respect for the Prussian King so much in evidence in his rooms in Sanssouci:

> 'Nothing has been changed in his private rooms—one finds a faithfulness in the extreme, displaying the books on the table where he left them shortly before he died; the upholstery of his chairs is seen as torn apart as his favourite dogs left them; in short: everything is left exactly as it was. Of all places of a comparable beauty there is nowhere I would like to live so much as there.'[28]

From Berlin he proceeded by mail coach to Dessau, the capital of the small duchy Anhalt-Dessau by the Elbe. He followed close upon the heels of P.H. Classen, who was travelling in his own carriage with his wife, daughter, and servants. P.H. Classen was a brother of J.F. Classen, the wealthy maker of cannons, and he chaired the Classen Trust whose gargantuan fortune he administered and enlarged. Ørsted found old Mr Classen convivial, his wife reasonable and considerate, and the daughter 'a beautiful and apparently good girl'.[29] He enjoyed their company and upon his return home he was appointed a lecturer at Museion and the Classen Agricultural Institute run by the Classen Trust. They hired rooms at the Eichenkranz, a guesthouse on the borders of the Wörlitz Estate, the main attraction there, which had been established by Erdmannsdorff for Duke Friedrich Franz of Anhalt-Dessau.

Then as now, Wörlitz was a mecca for admirers of the ideas of the Enlightenment: philanthropy, religious tolerance, cosmopolitan attitudes, the unity of nature and culture, the improvement of agriculture, and English landscape architecture. Ørsted was out walking in the park at dawn, before five o'clock, and was immediately taken by the charm of the environment.

The Wörlitz Estate is composed of a series of parks separated by canals and lakes. Visitors walk through the large estate along winding paths and are intrigued by the ever-changing views, which appear as casual surprises while nevertheless, in accordance with the fashionable English landscape architecture of the time, being carefully designed. At a certain point, for instance, one is surprised all of a sudden to see three religious buildings, churches for catholic and protestant congregations and a synagogue. In the eye of the visitor the three houses of worship appear of equal importance in the physical sense, and hence in the spiritual sense as well. The duke and his architect developed the philosophical ideas for this Park of the Enlightenment on their grand tour of Europe together. Friedrich Franz imagined that his estate workers would stroll around on Sundays and see replicas of spectacular landmarks from all over Europe: from the marble urn on Rousseau's tomb on the Isle of Poplars to the first cast iron bridge at Coalbrookdale, marvellous architecture from Classical to Gothic, natural phenomena from the nocturnal eruptions of Vesuvius to exotic flowers, shrubs and trees, not to mention demonstrations of modern agricultural systems. To combine the useful with the agreeable and to bring Enlightenment ideas to the layman was the generous ambition of the duke. But let us listen to the words of our own visitor:

> 'The entire garden is intersected by canals crossed by all kinds of bridges, so that it would be difficult to suggest a kind of bridge that is not represented here. One kind I have never seen before is a suspension bridge having the property of wobbling in all directions when walked upon, as if it were rickety. This happens because a number of iron chains are hung slack across the canal, and the boards of the bridge are not joined together but each is fastened with cramp-irons to the chains. For handrails they have iron chains. Apart from

bridges there are barges, or ferries as they call them, to cross the canals. They are fastened to ropes on both sides of the canals so that one can move them to whichever side one needs by means of a windlass. One does not have to pay to get across; everybody can easily do it for himself.'

In the evening he was rowed around on the lakes in a gondola together with the Classen family. The whole party was delighted by the many surprising views that unfolded during their trip. 'More magical still is the Temple of the Night that is lit from above through coloured glass in the shape of stars…' Then follows Ørsted's description of Vesuvius which, disappointingly, did not erupt with fireworks that particular evening:

> 'Above the grotto is a mountain in the shape of a volcano that can be illuminated artistically by belching fire and emitting sparks. One can climb it and enjoy a beautiful view. Rowing on, one finds a building called Pantheon containing plaster reliefs of some Egyptian gods and goddesses. Were I to describe all the things of interest I would need several sheets of paper and you, having never seen it, would derive only moderate pleasure from the description.'

Fig. 31. 'Der Stein zu Wörlitz' [the Rock of Wörlitz] was built between 1788 and 1794 as a geographical model of the Bay of Naples and the volcano of Vesuvius on an artificial island with grottos. 'Mount Vesuvius' was brought to eruption by means of pyrotechnical illuminations. The garden kingdom of Wörlitz was inspired by English landscape architecture with its sublime views; Prince Franz of Anhalt-Dessau and his guests would sit in the castle of Wörlitz enjoying the volcanic show at the end of the artificial lake, which received its water from a diversion of the river Elbe. Aquatint by Karl Kunz (1770–1830). BPK, Berlin.

Hans Christian's account, unable to take advantage of comparisons that his fiancée could engage with, must have gone beyond Sophie's imagination. He went on:

> 'I shall tell you about yet another feature that may be coloured by my imagination, but it seems to me that animals and birds have less to fear from humans than they do elsewhere. I have a flock of swans in the canals in mind. They are so tame that they eat bread out of the hands of strangers sailing by, and often one approaches a bird quite closely without scaring it. As you see I am almost making a paradise out of this place...'[30]

From Wörlitz, Ørsted proceeded to the annual book fair in Leipzig where he met Henrik Steffens and P.A. Heiberg, who came from his exile in Paris. He also encountered Friedrich Schlegel and his wife Dorothea, and brother, August Wilhelm (who was without his wife because

Fig. 32. *Utsikt mot Pillnitz gjennom ett vindu* [View towards Pillnitz through a window]. The castle of Pillnitz with Chinese pavilions along the river Elbe south of Dresden was the summer residence of the Elector of Saxony. Here HCØ attended a performance of Mozart's opera *The Magic Flute*. Note the reflections in the window panes. Oil painting 1823 by Johan Christian Dahl (1788–1857). Museum Folkwang, Essen.

Schelling had seduced her). Schlegel had borrowed a carriage from Fichte and changed horses at mail stations when necessary. He now offered Ørsted a lift since they were going the same way along the Elbe to Dresden passing the Meissen porcelain factory in the castle on the top of the cliff.

A few days later Ørsted took a walk to the Elector's summer residence, the chinois-style castle at Pillnitz, close to the source of the Elbe. An Italian opera was about to be staged for the Elector and his court as well as for music lovers from the neighbourhood. It turned out to be difficult for Ørsted and his friends to get access, because some foreigners, English in particular, who had been admitted to a previous show, had behaved inappropriately. However, he managed to persuade the pages to open the gates. The following day he saw *The Magic Flute* for the first time.

> 'To hear German sung badly the day after one has heard Italian sung well is one of the toughest tests one's patience can be exposed to. I stood it to the end, because I have not heard this excellent piece before, and the music was so well performed that I should have been perfectly satisfied had there been no singing.'[31]

After Whitsuntide, which he blamed for wasting his time, since nothing happened then according to the normal run of things, Ørsted returned to Pillnitz on foot. He and his friends set out for a walk in the countryside in the so-called Saxon Switzerland. They were accompanied by a guide and went up and down steep cliffs and in and out of deep dales. Suddenly they found themselves in a gorge formed by

> 'a violent natural disaster. Where we entered we saw that it divided in three directions. At first we went left, then straight on, but both of these were so overgrown with trees taking advantage of the shelter from the wind that we found we were unable to move forwards.. Then we followed our guide along the third path and soon arrived at a place where the forces of nature had created the weirdest forms, looking like grotesque figures, one in the shape of a sitting animal, another one of a lion's head, and a third of an outstretched body. Then we arrived at places where the gorge widened out to double or treble its size and gave the fertile vegetation more room, and soon after that we found the narrow dale again, but so narrow that we thought it would close in on us, and we had to ascend the cliffs to get out. Finally, we met a low gate carved in stone through which we escaped from this part of the valley to enter into another one a little broader and even more beautiful. This went on for so long that we did not believe this Romantic gorge would ever come to an end.'[32]

They reached the village of Wehlen from which they could be ferried across the Elbe and begin the ascent to Königsberg which offered a splendid view of the mountains of Bohemia towards the east and Dresden towards the west. Ørsted did not often spend his time walking aimlessly in nature, but it seems that he did share a romanticist view of wild nature.

> At night he took zinc and silver discs with him to bed to make experiments on the effect of galvanism upon the eye without being interrupted.[1]

12 | 1801–2

Encountering Ritter and Winterl

WHAT FASCINATED Ørsted about Ritter above all was the originality of his experiments, which seemed capable of opening up the perspectives Kant had sketched in his metaphysics of nature. No wonder, then, that the urge to meet this pioneer of galvanism guided his itinerary directly towards Jena. No one else came close to meaning as much to Ørsted as Ritter. The first time they spent three days together, and they only had to size each other up briefly before mutual trust ensued. So shortly afterwards Ørsted returned to stay with him for two weeks. The following year they spent three whole weeks together. These three periods of being together laid the foundations of a friendship for life. Not long before his death (at the age of only thirty-three) Ritter described their friendship in these words: 'Of all my friends you have always been the closest, the most faithful, and the most honest as far as science is concerned; moreover you have always been the person who has cared for me the most'.[2] What had Ørsted done to deserve this praise? He had sent him books, letters, and money; he had promoted his scientific work in French journals; and made an attempt to win for him the Napoleon Prize for the most important galvanic discovery; finally, he had successfully recommended him to a post at the Bavarian Academy of Science.

In many ways Ritter's childhood and adolescence were reminiscent of the Ørsted brothers'. He came of provincial clergyman stock with many siblings, many of whom died young. He was apprenticed to his father, a pharmacist, at fourteen, and as a journeyman five years later he left home to matriculate at the University of Jena. For lack of capital it was out of the question for him to become a pharmacist himself, and his family was unable to support him as a student.

His parents had family and friends in the neighbourhood of Jena who would help him out with provisions. Most of them were affiliated to the Masonic order of the Evergetes, headed by I.A. Fessler (whom Ørsted was later to meet as master of the chair of the Royal York Lodge in Berlin). In addition, his father had a number of debtors in Jena, who had never settled their accounts for pharmaceuticals. Young Ritter was given a list of these bad debts and he regularly approached the debtors in the hope of squeezing some money out of them from time to time. Just as the Ørsted brothers had isolated themselves in their study, Ritter kept to himself in Jena, because his clothes were too shabby to appear in public. Ritter was a self-contained man experimenting and studying so hard that his eyes were fit to pop out of his head.[3]

But there were also significant differences between their social backgrounds and characters. The Dane was accommodated in a college; he had his daily meals in his aunt's kitchen. The German, on the other hand, often had to spend the night in damp and cold garden sheds, and he did not even have a kitchen of his own to experiment in, nor could he afford gold and platinum for experiments. If occasionally he was lucky enough to rake in a couple of silver rixdollars from his father's debts they were used as discs in his voltaic pile before they were spent on food and drink, unless, of course, this had permanently damaged them. Ritter suffered from malnutrition, and his teeth began to fall out when he was only twenty. For some time he received sums from Goethe in return for original experiments, and this money enabled him to build up a modest collection of scientific instruments. The well-to-do nobleman Novalis (pseudonym of Baron von Hardenberg) was a great admirer of Ritter's talent for experimentation and forced him to accept handouts. Behind Novalis's often cited judgement *'Ritter ist Ritter und wir sind nur Knappen'*, playing on a medieval saying that loses its meaning in literal translation [Ritter is a knight, and we are just squires], lay his fascination with an experimental genius, who (unlike many a Romantic daydreamer) knew what he was talking about. This neglected natural talent had settled a most complicated disagreement between leading international scientists Galvani and Volta and demonstrated the reality of polarity of natural forces—not by idle talk, but by scientific evidence (ch. 10).

To appreciate the admiration of the early Romanticists we must bear in mind that Ritter's experiments resolving the so-called galvanic dispute, not only in Italy but north of the Alps as well, had shown that the twitch of a frog's thigh was not due to biology as asserted by Galvani, nor to contact between two different metals as claimed by Volta, but was caused by a chemical process. Considering that the potential difference in Ritter's crucial experiment was hardly more than one volt it is impressive that he could detect any effect at all, and that in addition he could interpret this effect correctly, thus opening the door to an entirely new branch of science: electrochemistry. After two semesters of extramural studies at Jena he had explained the function of the voltaic pile before Volta had introduced his discovery to Napoleon. It was the report of this discovery, that with the support of Goethe, provided him with the peace to work at Belvedere, the Duke of Weimar's castle.

When the Duke died Ritter lost these princely research facilities. He was unable to return to Jena because his creditors there were only waiting to fleece him of his sole income, the royalties paid to him for his scientific manuscripts by C.F.E. Frommann, the publisher. Ritter had become something of a kite flyer and he was publicly blacklisted as a defaulter. Frommann, an Evergete freemason too, had become his patron.

Goethe had been away when Ritter gave his paper at Die Naturforschende Gesellschaft, and as soon as he was back in Weimar he invited him to his home and described him to Schiller as 'a sensational persona and a heavenly erudite on earth'.[4] Later, when Goethe slammed the lid of his coffers, Ritter moved with his scientific instruments to J.G. Herder's vicarage in Oberweimar. Herder was more than old enough to be his father and treated him as a son of his house.

Ritter's character was as polar as his physical theory. His temperament was manic-depressive changing suddenly between euphoria and desperation, he was often ill, despondent, and lonely, and was permanently poor. He oscillated between humility and conceit. Whenever he was following up a clue that might lead to a breakthrough in his research his powers of concentration and ability to forget himself were enormous. When carrying out galvanic experiments on his own sense organs he would show no mercy to himself and put an eye at risk—because he had two! This recklessness cost him problems with his sight, sweat, headaches, diarrhoea, and pain in the joints. Under such circumstances he would live unconcerned about his needs of tomorrow, and he would callously exploit his friends' offers to lend him more money. He would even take up a bet in a bar on whether his begging letters would be successful. If they were he would double his stake. He despised petty-bourgeois security advocates and followed the Brunonian prescription of taking opium. During his last years he would consume two to four bottles of wine a day, admitting that he ranked high on Brown's sthenic scale.

Ørsted lived for forty days with Ritter and no doubt he paid generously for his stay. They took long walks in the forest-clad hills surrounding Jena and visited beer gardens. Ørsted witnessed a situation where Ritter behaved like a madman in front of an annoying woman, putting her in a state of confusion. Ritter confided his many conquests to him, among others his relationship with a wealthy French emigrant, Mme de Gachet, who had indulged in chemical experiments on her own account, but having met Ritter took delight in their conducting experiments together. He also disclosed his intimate relationship with Dorothea Veit, Schlegel's wife, asserting that it did not harm the friendly feelings between the two men. Ørsted hinted that perhaps Ritter should watch his step concerning sexual affairs, but Ritter cursed all discretion and spoke openly about his women by name. Unfortunately, Ørsted noted, his many admirers learnt far less from his galvanic experiments than from his frivolous escapades.

The two friends discussed galvanism and did experiments together. Ørsted pumped Ritter for details of his experiments and promised to be secretive about them. He also initiated him into the Hungarian Professor Winterl's sensational chemical theories on the polar relationship between acid and base principles, which fitted nicely into their own dynamical system. Ritter soon became an enthusiast of Winterl's theories and encouraged Ørsted's plan to conduct experiments at Hermbstaedt's lab in Berlin. Ritter followed these experiments with great interest, and during the winter Ørsted kept him up to date by letter. They agreed that Ritter should come and live with Ørsted in Berlin and they also hatched plans to travel to visit Winterl in Budapest to clarify his theories. However both plans petered out as Duke Ernst of Gotha invited Ritter to his castle, Friedenstein, to polish the Duke's scientific halo. Ritter enthralled the court with his galvanic experiment and the Duke reciprocated with six louis d'or.

Fig. 33. Johann Wilhelm Ritter (1776–1810), the self-made scientist. A miniature now lost was used as a model to produce this woodcut. He wears the uniform of the Bavarian Academy of Science, Munich. *Ostwald's Klassiker der exakten Wissenschaften* No. 271, Leipzig, 1986.

At Friedenstein Ritter lived what he described as 'a life fit for the gods'—everything being free of charge. He had free access to the Duke's research library and he made seven galvanic batteries, each having three times six hundred discs as well as a still bigger one of three times one thousand discs. Ritter experimented with these batteries from early morning until late at night making notes on small slips of paper in numbers no human being could possibly manage to convert into a coherent scientific text let alone keep in order. In April 1802 the first volume of his *Beiträge zur nähern Kenntnis des Galvanismus und der Resultate seiner Untersuchung* ['Contributions to a better Knowledge of Galvanism and the Results of his Investigations'] appeared. During the period he wrote it, Ritter's life and work were particularly distinguished. He experimented during the day when light was abundant and wrote during the night by candlelight while a profusion of ideas filled his mind so that it was unable to control the chaos.[5] In addition, he had fallen in love again. Frommann, his publisher, reminded him that he had promised to have the last chapters of the second volume of his *Beiträge* ready by six in the evening of 25th July, but when the hands of the clock stood at five he had not even written a single line. Ritter wrote to Ørsted, 'Thank you very much for your Metaphysics of Nature, and congratulations on the fine foreword Mendel has written' (Ørsted 1802). And again 'Let us cheer in champagne and embrace one another'.[6]

After Ørsted's third visit in August/September 1802, Ritter wrote that for ten days after Ørsted's departure he had not left his house, because he had been constantly experimenting with galvanism and observing the stars, so he had not found time to wash. Three weeks of inspiring but confined conversation with Hans Christian cried out for immediate experimental satisfaction for both mind and body. Ørsted had been working on a lecture on Winterl's chemistry, and Ritter had aired his ideas that the classical astrological imagination of the governing of human physiology by cosmic forces was explicable by modern galvanism. To this end he sought numerical connections between the orbits of planets and comets and the pulse of earthly life, the four seasons, the phases of human life, the durability of pregnancy, and the rhythm of the circulation of the blood. He felt he was close to something sensational and would say more on the matter by return of post. However, he did not do so. Everything being in a muddle, he had mislaid Ørsted's address.

Ritter shared with other early Romanticists the state of mind in which there was a multitude of ideas and whims, more than he could carry to a conclusion and put on paper before new ideas emerged. Whereas Ørsted wrote calmly, thoughtfully, and thoroughly, Ritter like other early Romanticists was spontaneous, paradoxical, feverish, and erratic. His articles were reminiscent of Schlegel's essays and Novalis's novels in that they were fragmentary, or a brief series of telling sentences. Ritter left behind about seven hundred aphorisms. Here is one of them:

> 'We say that the force of attraction is everywhere proportionate to the quantity of matter. But what is the quantity of matter? How do we determine it? By weight? What is weight but the result of the force of attraction? So, the expression: "the ordinary force of attraction is everywhere proportionate to the quantity of matter" ought to read: "the force of attraction is everywhere identical with the force of attraction". But this is circular and explains nothing. So, how can we say "quantitative attraction"? Everything must be qualitative.'[7]

No wonder that Hans Christian allowed himself to be seduced by some of Ritter's aphorisms that in a flash of lightning brought paradoxical, but thought-provoking, formulations to bear on obscure matters. They were not ready-made solutions, but rather drew attention to problems in need of reflection.

*

Immediately on his arrival in Berlin in November 1801, Ørsted sought out Hermbstaedt, whom Manthey knew from his earlier visit and had recommended. Had money and letters arrived from home? At the post office he asked for mail from Copenhagen, letters from Anders and Adam, or good news from Manthey concerning his application for a royal grant, or what he wanted most of all, a love letter from Sophie. He also needed to find cheap accommodation. We do not know where he settled down for the winter, because the letter that informed Manthey about his address went missing, but at least he must have had a bed and a table and chairs for a study circle of three members.

He had not been able to afford new clothes for the tour, so in Hamburg he spent fifty rixdollars, the first portion of the Cappel grant, on new clothes, a set of mathematical instruments and

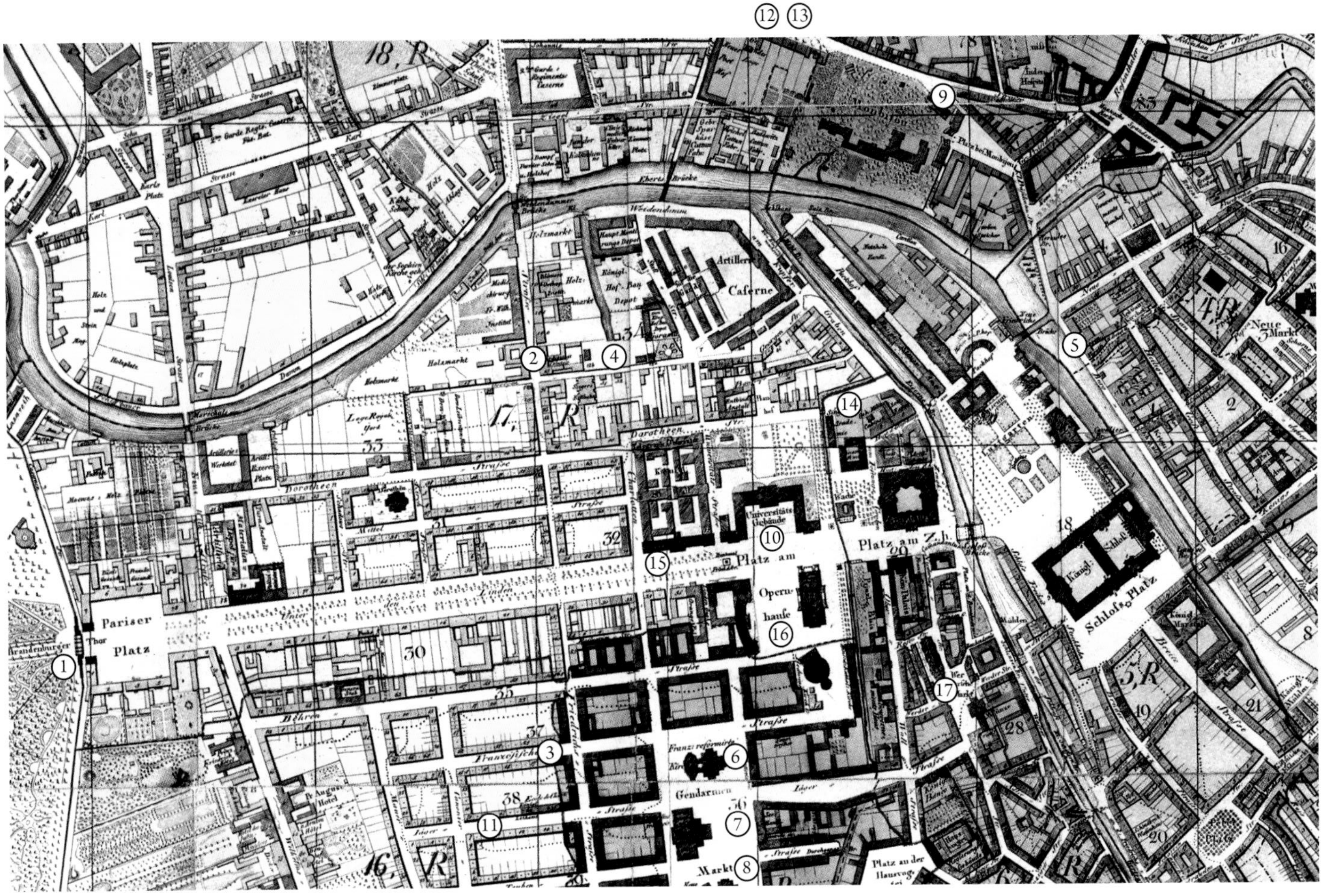

Fig. 34. Caption on p. 114

an English writing set.[8] In Berlin he took part in many social events for which he needed better clothes and he had to take advantage of Manthey's offer of a short-term loan from Hermbstaedt. His old everyday clothes were so shabby that he could not wear them in public. Taking Manthey's advice, he had a suit made consisting of a dress coat and trousers of dark grey cloth, and a short silk waistcoat.[9] This did not turn him into a dandy. What really impressed those around him was the portable galvanic battery he had invented which was illustrated in a German journal (fig. 29). This apparatus aroused a lot of curiosity and at Göttingen he had a small copy made with four u-tubes which would entertain several people on his journey. He had improved it in a number of respects so that the effect was powerful considering its small size.

There was no university in the capital of Prussia until 1810, a few years before Ørsted's next visit to Berlin. Still, in 1801/2 a good many of the early Romantic coryphaei from Jena had settled in Berlin with its public libraries and audiences, which enabled them to study and to manage on fees from private lectures. There was also Hermbstaedt's brand new chemical laboratory which he made available to Ørsted; during the winter he spent many days in Georgenstrasse where it was situated.[10] S.F. Hermbstaedt was nine years older than Manthey and seventeen years older than Ørsted. He was an influential person in Prussia as a figure of transition between the old cameralist and the modern specialised technical chemist serving the needs of pharmacies, mines, china factories, breweries, and dye-works. Hermbstaedt had married into the Rose family who owned the biggest pharmacy in Berlin, Zum Weissen Schwan, in Spandauerstrasse. First M.H. Klaproth, his patron, had married Valentin Rose's

Fig. 34. Map of Berlin, c. 1820. North is along the left.

1. Brandenburger Gate
2. Fichte, J.G. — 139, Friedrichstrasse
3. Gesellschaft Naturforschender Freunde — Französische Strasse
4. Hermbstaedt, S.F. — 43, Georgenstrasse
5. Herz, Henriette — Burgstrasse
6. The French Reformed Church
7. The Theatre — Gendarmenplatz
8. The German Lutheran Church
9. Humboldt, A. — 67, Oranienburgerstrasse
10. The University — Unter den Linden
11. Niebuhr, B.G. — Jägerstrasse
12. Schleiermacher, F.D. — Charité (to the north outside the city wall)
13. Schlegel, Fr. — Charité (shared with Schleiermacher)
14. Schlegel, A.W. lectured in Singakademie — Dorotheenstrasse
15. The Royal Academy of Science and Letters, Unter den Linden
16. The Opera — Unter den Linden
17. Weiss, C.S. — Münze am Werderschen Markt/The University (10)
18. HCØ — ?

LAB. Photo by Andreas Matschenz

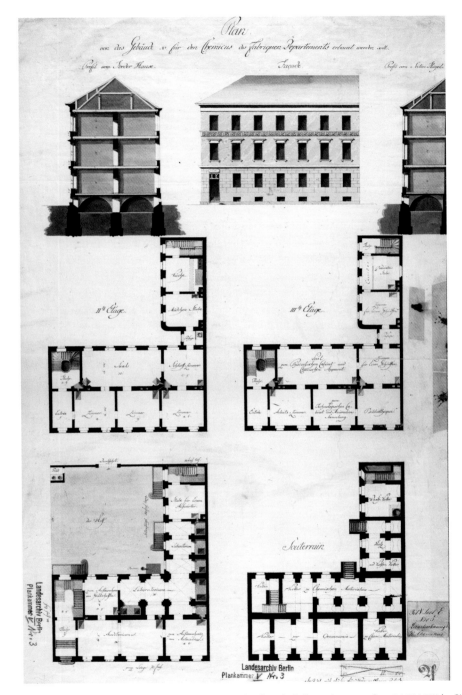

Fig. 35. The Chemical Institute of the Prussian Industrial College built for and approved on 06.11.1799 by Sigismund Friedrich Hermbstaedt (1760–1833) in 43, Georgenstrasse, Berlin. HCØ obtained access to this lab and

widow making him the guardian of her sons and her daughter Magdalena, whom Hermbstaedt then married. Ørsted became closely attached to the chemical community in Berlin professionally as well as socially.

Hermbstaedt had been appointed titular professor in 1791 and in the same year he translated and introduced Lavoisier's *Traité élémentaire de chimie*, containing the new terms 'oxygen' and 'hydrogen'. In 1800 he had become a member of the physical section of the Royal Prussian Academy of Science, with its imposing building on Unter den Linden, and he obtained a post in the government's Industrial College. However, he had serious problems when it came to suitable premises for chemical experiments. His own flat in Spandauerstrasse was useless. Fortunately, the government decided to establish a 22-metre long L-shaped building of three storeys in Georgenstrasse for the benefit of the chemists of the Industrial College.[11]

Manthey had seen Hermbstaedt's new laboratory, certainly one of the best in the German states, and encouraged Ørsted to get to work there: 'Recently the King of Prussia has ordered a magnificent building to be established and furnished with extraordinarily well equipped laboratories, a reading room, etc. to educate artisans in technical chemistry'.[12] There was a risk, however, that it would be costly to rent facilities from Hermbstaedt, and as long as Ørsted did not know whether his application to the Foundation ad usus publicos for additional money would be met, he dared not take on further liabilities, so while he was waiting he took salaried employment at the laboratory. Unfortunately, Hermbstaedt had very limited spare time to spend with his young employee.

> 'He has entrusted me with an investigation of five different kinds of alum sent to him by the Industrial College... I wish the same for Hermbstaedt that Meissner's Alcibiades wished for Phidias that he will become the idlest man in town in order to get time to work with me, because all the trust and friendship he shows me must be due to Professor Manthey's word since so far he has had no time for a scientific talk with me.'[13]

As might be expected, seeing the King's gift to Hermbstaedt soon sparked off in young Ørsted's mind a plan to establish a fully equipped chemical laboratory in Copenhagen (under his own leadership, of course). He soon had an application prepared for the Chancellor of the University. He sent it to Manthey and left it to him to decide if he should hand it in himself or let

Fig. 35. Continued
here he worked to identify Winterl's 'Andronia' and 'Thelycke' in acids and bases. On top to the left the main building in profile and façade and to the right the wing in profile. In the basement is a washroom and a room for chemical preparations, firewood, and coal. On the first floor a 60 square metre lecture hall facing the street and two laboratories facing the courtyard. In the wing is a flat for a servant. The second floor has Hermbstaedt's private flat with a dining room, a bedroom and three small rooms, while the kitchen and the maid's room are situated in the wing. On the third floor is a study and rooms for collections for technology and physics, a collection of minerals, as well as a library and a room for an amanuensis. It was a stately building with far better research facilities than Klaproth's. HCØ was lucky to obtain access to it. LAB. Photo by Andreas Matschenz.

Anders do it. The details of the application are unknown as the letter is lost, but Manthey instinctively conceived it as targeted towards technical chemistry and addressed to artisans (like Hermbstaedt's), in line with his own interests. He lifted the veil of the plan in a footnote to one of Ørsted's letters he made public.[14] Ironically, benevolent Manthey misunderstood his protégé's intentions. Ørsted's private experiments in Berlin were not concerned with technical chemistry, but aimed at proving the existence of the 'Andronia', a still fictitious element of the Hungarian chemist Winterl's grand theory. Together with a research team consisting of Rose, the pharmacist, and Richter, the analytical chemist, Ørsted and Ritter dreamt of retrieving and identifying this otherwise unknown substance in Hermbstaedt's well-equipped laboratory.

It soon became obvious to Manthey that Ørsted's plan was not aimed at technical chemistry and artisans. What he had in mind was a research laboratory that would provide the foundation for lectures targeted at an educated audience for economic gain. 'The core of the matter is', he wrote to Manthey, 'that I long to accomplish something worthy and to combine the useful with the agreeable in my own work that is to say *scientifically*. As long as I do not achieve this I do not consider myself a happy man.'[15]

*

Over Christmas, Hans Christian anticipated something of interest and paid a visit to Dr Aronson, a Jewish physician in Berlin, who had received a paraphrase of his dissertation for review. A certain Dr Mendel, who had picked up Danish in Copenhagen, had paraphrased it, and now it was going to be printed provided he could find a publisher. This proved difficult and Mendel had to pay for it himself. He dedicated the book to Lazarus Bendavid, a Jewish physicist, secretary of the Philomatic Society in Berlin, who had written on Kant's metaphysics of nature, and to another physicist, Pörschke by name, a colleague of Kant. In addition, Mendel wrote a foreword in which he described Ørsted as his famous and erudite friend, who deserved attention for his improvement of the 'architecture' of the Kantian system.[16]

Ørsted's dissertation, as we saw, had a structure different from *Grundtrækkene*, which invited comment. So in February he organised a small private study circle on Kant's *Metaphysical Foundations of Natural Science*. Apart from Ørsted there were Dr C.S. Weiss, a mineralogist recently arrived in Berlin to study chemistry, and K.J.B. Karsten, a metallurgist; all three were comparatively young (25, 22, and 20 respectively). Weiss and Ørsted focused on 'Ariadne's Clew'. They became close friends and continued their dynamical research programme for the rest of their lives.

Ørsted assumed that the study circle would have finished its work before Sophie Probstheins birthday on 16th February (he thought) 1802, which, in the absence of his fiancée, he intended to celebrate in the company of his friends, and he informed his bride-to-be of their birthday party. The study circle lasted eight evenings, and half a year later the resulting book was published under the title of *Dr Johann Christian Oersted's Ideen zu einer neuen Architektonik der Naturmetaphysik, nebst Bemerkungen über einzelne Theile derselben* ['Dr Johann Christian Oersted's Ideas of a new Architecture of the Metaphysics of Nature including Comments on some Details']. This was Ørsted's first publication in one of the major European languages (leaving aside his thesis in Latin). Now he would reach a larger readership and become known in Europe. It was widely reviewed in German journals; it was praised in Jena, but cut to pieces in Schelling's and Hegel's journal.[17] Could he expect otherwise since he had hit out at German *Naturphilosophie*?

How did this third version of Ørsted's metaphysics of nature relate to the two previous ones? *D. Oersted's Ideen* is a shortened version of his *Grundtrækkene*, combined with an appendix dealing with a theory of ether remarkably close to the ideas the elderly Kant was working on in Königsberg at the same time, though they were never prepared for publication. In Mendel's small book it became crystal clear for the first time on what grounds Ørsted was criticising Kant. One problem was Kant's incomplete application of his scheme of the categories of the understanding to the concepts of 'force' and 'motion'.[18] Ørsted wanted to pursue 'Ariadne's Clew' totally. On reflection however, this problem disappears. Kant confined his analysis to the force of gravity, while Ørsted was in need of a method that might falsify the atomic theory and promote an alternative dynamical theory.

A second problem was a lack of clarity concerning Kant's concept of force. Ørsted now saw that it made sense to see the concept of force as both cause and effect, that is to say, as force and motion at the same time. The force itself was not phenomenal, it left no sense impression: it was only noumenal. The relationship between cause and effect, or between force and motion, ought to be made understandable not only in Newton's mathematical terms, but also in physical ones.

Already in his metaphysics of nature, Kant had been critical of the mathematical solution to the mechanics of gravity. The reason was that Newton declared himself unable to explain what gravity was. The mathematical solution was to equate gravity with mass, but this did not explain anything physically, as Newton readily admitted. What Newton did accomplish was a calculation of gravity, no more. According to Ørsted the title of his immortal treatise *Philosophiae Naturalis Principia Mathematica* was a contradiction in terms, for the natural science of physics ought to present physical explanations, not mathematical ones.

How did Ørsted defend his theory in the appendix? The problem was rooted in the motion of bodies in curved lines that logically had to be caused by some combined effect of several external, attractive forces. Again logically speaking these attractive forces had to be checked by a braking effect, since otherwise the bodies would crash into the central body of attraction. This braking effect could not be a repulsive force, because it operated on the surface, and the point was that the body did not collide with any other body. Consequently, there had to be a braking effect operating at a distance. According to Ørsted this effect had to be explained by the physical presence of a substance, which he, and Kant and many physicists after him, called 'ether'. This a priori theory of ether had one advantage. Unlike Newton's law of gravity, it was physical.

Ørsted sent some copies of Mendel's book home to Anders, one of which was intended for Jørgensen, the meticulous bookkeeper in Cappel's Foundation, who was in perpetual need of appeasement by means of testimonials. Mendel's flattering foreword was most suitable in this respect. But facing the metaphysical contents of the book he felt less convinced that it would promote his career, so he asked Manthey for his opinion. Unfortunately, the answer proved his suspicions correct.

'You should not send your metaphysics of nature to anybody here for this would undeniably harm you; the rate of exchange of the newer philosophers has sunk so much in this country that everybody must make sure not to subscribe to their doctrine. Our friend Steffens is doing a great deal to aggravate matters; he elevates himself so much above other people that he is an insult to many and professes ridiculous ideas; very few people comprehend that others who, like him, talk about *a priori* knowledge are still looking for proof.'[19]

What Manthey was drawing attention to was a classic prejudice, common at the time and ever since. Only a few seemed to have taken the trouble to read Kant before they distorted his claim that metaphysics is an epistemological reflection necessary to understand experience. It is a misunderstanding to confuse metaphysics with speculation. The ether deduction is an a priori construction by our intellect, but whether it is real is still unresolved and must remain so as long as it its empirically beyond intuition. The classic blunder was and is to take an a priori deduction for a cognitive result when it is only meant to be a condition for one.

Oersted's *Ideen* was a metaphysics of nature dealing consciously with a priori cognition. To accuse him of ignoring empirical research was totally absurd.[20] A calculation of the number of days he spent at Hermbstaedt's laboratory from December 1801 to April 1802 comes to 56 days of work. This was one of the best-equipped laboratories in northern Europe at the time. Ørsted also studied books and articles in his room and, as we shall soon see, he attended about 25 meetings of various scientific societies. Moreover, he learnt to blow glass tubes and made galvanic batteries for friends and colleagues. Ørsted was definitely one of Manthey's 'others who...are still looking for proof'.

*

On 15th August 1801 Ørsted had met the fifty-year-old J.F. Westrumb at Hamelin. He was a pharmacist and a mining advisor and also had private interests in the textile industry, in dyeworks, and in bleaching works. In other words he was an all-round technical chemist and businessman, wealthy and hence envied, and, in Ørsted's opinion, greedy. From Manthey he had a letter of recommendation which procured for him the friendliest reception imaginable. They spent many hours in Westrumb's laboratory, where Ørsted was shown one of J.J. Winterl's experiments, and in his notebook Ørsted made a drawing and a description of how water and the gases ammonium and oxygen were obtained from a reduction of metal oxides, such as manganese oxide or mercury oxide.[21]

It was Westrumb who made Ørsted aware of Winterl's *Prolusiones ad chemiam in saeculi decimi noni* ['Introduction to nineteenth-century Chemistry'], which proposed an entirely new chemical system that seemed more promising than Lavoisier's antiphlogistic theory, in so far as it touched upon a reduction of incoherent empirical observations into a few dynamical principles. According to Winterl, Lavoisier's theory was about to crumble away, first of all because it lacked a theory for bases, and secondly, because its assertion that oxygen was the decisive component of all acids had turned out to be a flawed generalization. Unlike Ritter, who had established the electrochemical series of metals, Lavoisier had no theory of metals and salts. Finally, in his table of the elements, Lavoisier made use of a substance he called 'caloric', imponderable heat, which encountered increasing scepticism. When Ørsted left Hamelin, Westrumb accompanied him for two miles on foot and gave him almost thirty addresses of people belonging to his professional network.

Ørsted was attracted to Winterl's theory because it gave answers to questions Lavoisier left unsolved. Polar forces were its recurring principle. Acids and bases had polar properties, and salts, metals and earths were all basic. In addition, the acid and basic properties were linked with electricity, the bases being positive, the acids negative. Like Ørsted and Ritter he considered water a neutral element composed of positive and negative electrical forces making positively charged water basic and negatively charged water acidic. Furthermore, Winterl explained heat as a force analogous to electricity. It was a generally accepted fact that acids and bases neutralised one another and that this neutralising process increased the heat. Also in favour of Winterl's system was the fact that it had been developed independently, before the voltaic pile and Ritter's *Beweis*.

When Ørsted presented Winterl's system to his friend a month after his visit to Westrumb, Ritter immediately grasped its potential. He had discovered that the south pole of the magnet was more easily oxidized than its north pole. When a magnetised iron wire was placed in nitric acid

Fig. 36. Jacob Joseph Winterl (1732–1809), Hungarian chemist, whose rambling thesis in Latin, *Prolusiones*, kept a large proportion of European chemists busy identifying the otherwise unknown elements of 'Andronia' and 'Thelycke' in the beginning of the nineteenth century. HCØ and JWR believed they had found an ally in their struggle against the corpuscular theorists. Anonymous drawing after a lost portrait in oils. Österreichische Nationalbibliotek, Vienna.

metal oxides were generated on the south pole, and litmus paper was coloured red. In other words there were three polarities, a magnetic, a chemical and a galvanic.[22] *Prolusiones* was not a book available from a bookshop. Winterl had sent copies on his own initiative to scholars in Europe as a free gift, hoping to receive constructive criticism in return. Most scholars had probably shelved *Prolusiones* uncut. Generally, they revealed a sceptical attitude (according to Ørsted due to prejudice) as they were simply unable to understand Winterl's Latin text. Latin no longer held a monopoly among German-speaking scholars. However Bendavid, the physicist, asked Ørsted to give a written paper to the Philomatic Society in Berlin, and a committee was established to test Winterl's several hundred experiments. Valentin Rose took on the task of identifying 'Andronia', J.B. Richter was to examine 'whether it is true that alkalis can be dimmed', that is, could the basic property be weakened vis-à-vis the acidic one, while Ørsted agreed to analyse the acids. P.L. Simon was to be persuaded to examine 'deoxidised air'.[23] From the end of February to mid-March, Ørsted scrutinised Winterl's book alongside his test experiments, on deoxidised sulphuric acid among other things. Sadly, Rose had no luck in retrieving 'Andronia'.

Encouraged by Ørsted Manthey started to study *Prolusiones* and remained unimpressed.[24] Meanwhile in Jena Ritter and Ørsted were working hard to complete their manuscript for *Materialien zu einer Chemie des neunzehnten Jahrhunderts* ['Materials for a Chemistry for the Nineteenth Century']. Ørsted wanted to write a preface for the book expressing reservations, lest he be embarrassed if the examinations in progress turned out unsuccessfully.[25] Chemistry is beginning to cohere as a theoretical entity, Ørsted told his patron. Because heat and electricity appeared to be chemical effects, something dynamical, it might soon be understood how two branches of natural science were interrelated into one system.[26] But more experiments were needed. 'Everything seems to encourage the most energetic participation, *ne vitam silentio praeteriamus*' [lest we pass through our lives in silence], he wrote to Manthey.[27]

At this stage both Ørsted and Ritter corresponded with Winterl regarding their exposition of his theory. A portrait of Winterl was intended to embellish the front page of Ørsted's book, but Winterl was too modest to send him a copperplate print.[28] The manuscript concluded with a summary stating the main results in the form of a letter to a sceptical friend. This letter was a slightly revised version of Ørsted's letter to Manthey, with the language corrected by Ritter.[29] On 20th September Ørsted entered the Montag & Weiss bookstore at Regensburg to buy a couple of books as they happened to discuss Winterl's chemistry. By the end of the day the bookseller had bought Ørsted's manuscript for sixty guilders.[30] It was common knowledge that it was very difficult for an unknown, non-native speaker to find a publisher in Germany. And yet, he had now succeeded in having works published twice without having to be his own publisher, as Ritter had been obliged to do with his *Beweis*. Luckily, Ørsted could take advantage of Mendel's flattering preface while negotiating with bookseller Weiss.[31] The plan was that Ritter would publish a second volume containing experimental results, but that never appeared. Instead an abbreviated translation of *Prolusiones* saw the light of day; it was dedicated to Ørsted and had a critical preface by Ritter criticising the book for its many woolly hypotheses.[32]

> I had an interesting conversation with Fichte about genius and the mental powers that created Kant's Philosophy.[1]

13 | 1801–2
Jena: Romanticism, Salons, and Societies

THE JENA Ørsted visited was an intellectual community different from the one Steffens had seen only three or four years earlier. Then the small university town had been the focus of the early Romanticist movement. When Ørsted arrived this renowned learned republic had been scattered to the four winds.

He had many recommendations from Manthey and others to various colleagues to keep track of and in the right order, and in Jena it went wrong. Ørsted's plan was disrupted and he was wrongfooted. He actually wanted to see Voigt first, and had rehearsed his words of introduction; but it so happened that Voigt was not at home, so instead Ørsted went to see Göttling. He introduced himself, greeted him heartily from Blumenbach, and handed over his letter of recommendation to Voigt. Göttling opened the envelope and kindly ignored the mistake. Now Ørsted realised that he had made a mess of things, so he apologised sheepishly, took the letter back, and was relieved when it turned out that Blumenbach had issued such a flattering recommendation.[2]

The University of Jena, which had almost as many students as Copenhagen, was a common place of study for people from several German states, under the leadership of J.W. von Goethe. In 1785 he had appointed K.L. Reinhold, the enthusiastic Kantian, as professor of philosophy.[3] He soon became a popular lecturer attracting six to eight hundred students to the largest lecture hall, among them Jens Baggesen, with whom he was in intense correspondence.[4] When in 1794 he moved to Kiel, J.G. Fichte (recommended by Kant) took over his professorship and

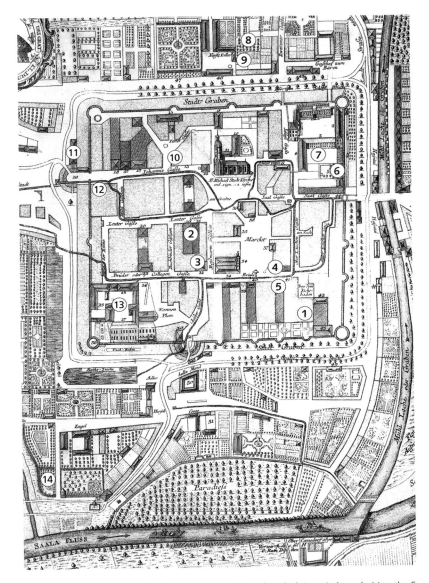

Fig. 37. Map of Jena with city wall. At the bottom the river Saale.

1. Fichte's house
2. The Schlegel brothers' house
3. Batsch's house (Die naturforschende Gesellschaft)
4. Schiller's 2nd lodgings
5. The Humboldt brothers' house
6. The University's largest lecture hall
7. Goethe's lodgings (when chairing the Senate)
8. Ritter's garden house
9. Frommann's house (the publisher)
11. Schiller's 1st lodgings
12. Schelling's lodgings
13. Reinhold's house and lecture hall
14. The University's Senate
16. Schiller's garden house

Jena um 1800, 2000.

lectured on the vocation of man in general and scholars in particular, and introduced the distinction between bread-and-butter students and philosophical minds. At that time Fichte had just published his first book anonymously. Reviewers found it so Kantian that they assumed Kant must be its author.[5] In 1795 Fichte lectured on moral philosophy in the same lecture hall as Reinhold had done (his private rooms being too small). In 1799 he was sacked following an accusation of having promoted atheism in his *Philosophische Journal*, which was confiscated. Fichte did not see himself as an atheist, quite the contrary, and made an appeal in the form of an apologia which he asked the public and authorities if they would kindly peruse before rejecting it.[6]

In Jena clashes between the authorities and the students, particularly Jacobin-minded students, were more frequent than in Copenhagen. The audacious *Burschenschaften* (students' societies) were forbidden and the students refused to accept that. The majority boycotted the University, and on behalf of the Senate Goethe felt obliged to withdraw. These events were well known in Copenhagen through Baden's *University Journal*.

> 'An Egyptian obscurity is hanging above these orders with their own legislature... They are accused of linking with Rosicrucians, Illuminati, etc.; they are the cause of all violence and duels at the University... One order hates another violently and there is perpetual war between them... Professor Fichte fulminated against these orders and organised a commission to investigate them and if possible abolish them. Inevitably he was the object of their hatred; five times they broke his windows... and in the end Fichte left Jena, concerned for his life; he always carried loaded pistols, and on his table were loaded pistols. The students smoke their pipes on their way to the lecture hall. During lectures and religious services they keep their hats on; it is considered no shame to walk around arm in arm with a prostitute in daylight; to embarrass these women while taking a stroll with them is quite common. There are daily duels, sometimes eight to ten every day, even though it is rare that anyone is killed; a stiff arm, hand or knee is a more likely outcome.'[7]

In 1789 Friedrich Schiller, whom Goethe had just met in Die Naturforschende Gesellschaft and whom he had soon afterwards appointed extraordinary professor of history, successfully lectured on the concept of 'universal history' and (in opposition to French Enlightenment thinkers) promoted his own philosophy of history, claiming that contemporary cultures grow out of past events.[8] Unfortunately, an extraordinary professorship was unsalaried, so if Schiller was to make money he needed a good many fee-paying students, and they would only pay if they received sensational lectures in return. It was also here in Jena that Schiller, supported by the Chancellor of the University of Copenhagen and by Ernst Schimmelmann, head of the Treasury, through Baggesen the writer, wrote his *Letters on the Aesthetic Education of Man* which had fallen prey to the Great Fire of Christiansborg Castle in 1794, but had been rewritten and published in Schiller's journal *Die Horen*.

Fichte and Schiller used their well-attended and compelling lectures to disseminate their political ideas of a university. Both directed harsh criticisms against the type of university that educated bread-and-butter students who, wearing blinkers, were instructed to reel off the limited examination requirements in return for a diploma securing a post and salary. Both encour-

Fig. 38. *Versuch auf den Parnass zu gelangen* [Attempt to climb the Parnassus]. Such was the tumultuous state of culture that HCØ encountered at Jena and in Berlin. The struggle between the early romanticist poets to reach the peak of recognition was brutal. On this anonymous caricature the heavily armed literary theorist A.W. Schlegel (1767–1845) has fought his way to the forefront, but is slowed down by being alone in carrying the burden. Behind him the storyteller J.L. Tieck (1773–1853) is seen on the back of the Puss in Boots. Under them wearing a black coat is the theologian F.E.D. Schleiermacher (1768–1834) absorbed in hermeneutical work. The symbol-laden poet Novalis (Friedrich von Hardenberg—1772–1801) is perambulating blindly on stilts, and Friedrich Schlegel (1772–1829) stands on his head on a lean cow his exhalations being collected by a devil flying on the back of a bat. J.G. Fichte (1762–1814) swings his whip towards all of them, while at the Parnassus, the poet August Kotzebue (1781–1819) for one (on top to the left), is offended and launches a counter-attack. The nobility, ladies and children watch in fear or indifference. *Ansichten der Literatur und Kunst unseres Zeitalters*, 1803 (re-edited Weimar, 1903). RL.

aged their students to satisfy their intellectual hunger and seek wisdom according to their interest and personal development, which would be rewarding in its own right. Like Kant and Schiller, Goethe, the Chancellor, took up a sceptical attitude towards encyclopedic lectures dishing up professional knowledge in alphabetical order. 'When we heard the encyclopedists speak or open their imposing volumes, we felt like the bobbins in a textile machine. The entire complex and incomprehensible machinery made us bow our neck and feel stupid and lost.'[9] Knowledge was not to be reproduced to the letter but to be made the object of discussion and framed within its philosophical context.

Such ideas, suggesting a different type of university, had also been advanced by Kant, who in his *Streit der Fakultäten* ['Controversy between the Faculties'] had assigned to the Faculty of

Philosophy the task of teaching students, whether of medicine, law, or divinity, to reflect on their knowledge and its use in society. Proposals to change the way students studied were promoted by Schiller when introducing his lectures on universal history and by Fichte lecturing on the vocation of the learned republic.[10] Universities should not be professional schools for bread-and-butter students, but free havens for philosophical minds to pursue a liberal education. These ideas were also discussed in Copenhagen, and we have already seen their strong impact on Ørsted.

The political reform of universities absorbed the minds of early Romanticist thinkers, the majority of whom settled in Berlin after the atheism controversy. Berlin had many attractions, but a university was not one of them. When Ørsted arrived at Jena four years after Steffens most of the reformers had left and settled in Berlin.

*

Life in Berlin and Jena was more than metaphysics of nature and chemical experiments, even for Ørsted. During the winter he would normally devote one night a week to the social life to be found in Henriette Herz's celebrated salon for intellectuals in Neue Friedrichstrasse. There he would meet other erudite gentlemen and educated ladies, there he would catch up on scientific and literary news, find an audience acclaiming his galvanic experiments, and enjoy chamber music and the singing of lieder. There he was involved in an unusual kind of entertainment established among the Jewish community of Berlin and pioneered by Henriette Herz, whose salon was the setting for a unique gathering that became a model for nineteenth-century salon culture. In Copenhagen there were only scattered examples of such social gatherings, for example at Mrs Gyllembourg's.

As a fifteen-year-old girl Henriette had been married to the Jewish physician Marcus Herz, seventeen years her senior. He was a close friend of Lessing and Moses Mendelssohn, and Lessing's play *Nathan der Weise* was modelled on Mendelssohn. He was also a disciple of Kant, who had chosen him as his respondent at his inauguration as a professor at Königsberg in 1770. Around 1780 Henriette had begun to entertain in her home, hosting informal parties for scholars and educated women. This fusion of otherwise distinct worlds of the two sexes developed into a salon culture where the hostess floated as the muse of sociability. The refreshments were simple, just tea and sandwiches; she would not know beforehand how many guests would turn up or at what time; salon life had an air of improvisation. Once one was invited to Herz's salon this meant a standing invitation. Guests of guests were welcome, too. The magnetic pull of the salon was the educated conversation, typically about music or poetry or novels. Salon culture was void of aristocratic symbols of status and orders of rank. It was bourgeois and academic, unostentatious and unceremonious. It constituted the forum for the exchange of intellectual capital, and naturally the gentlemen found an opportunity to demonstrate their chivalry to the ladies, whose charm in return opened the gates to the learned republic.[11]

Herz's salon was well established when Ørsted entered it. Perhaps one might think that for him talking with Marcus Herz was the main attraction. He was both a physician and a philosopher, and Ørsted might have heard about his correspondence with Kant, however sporadic it had become over the years. But this was not the case.[12] To Ørsted, young Henriette was the

Fig. 39. Henriette Julie Herz (1764–1847), the attractive Jewish lady, whose salon was open every Thursday for the intellectual elite of scholars and artists in Berlin. She was married to the physician Marcus Herz (1747–1803), seventeen years her senior, who in 1770 had been chosen to be Kant's respondent at his inauguration as a professor of philosophy at the University of Königsberg. HCØ was a frequent visitor in H. Herz's salon and stayed in touch with her till her death. Portrait in oils by Anna Dorothea Therbusch (1721–1782). BPK, Berlin. Photo by Jörg P. Anders.

irresistible magnet, not 'her womanish, conceited husband'.[13] She took an interest in galvanism, asking Ørsted on several occasions to demonstrate his experiments to the salon, and he was only too pleased to comply. When because of the frost he could not provide frogs for his galvanic experiments, he got hold of some mice instead, and managed to make their legs twitch. Henriette was fluent in an impressive series of foreign languages including Danish, and she sang and played the piano; Natalie from Goethe's *Wilhelm Meisters Lehrjahre* was said to be her role model as it was Sophie Ørsted's.

Friedrich Schlegel had flirted with Dorothea Veit, the daughter of Moses Mendelssohn, while she was still married to a banker whom she later left (a divorce that was referred to in Schlegel's scandalous novel *Lucinde*). Marcus Herz ordered his wife to exclude the unfaithful Dorothea from the salon, but Henriette did not let herself be bullied. It is more immoral to stay in a loveless marriage than to break out, the more so since the wife's father had arranged it, Henriette asserted. Soon Friedrich Schlegel, who admired Ritter in Jena, became one of Ørsted's friends. Later he moved to Paris with his Dorothea where they continued to see Ørsted, who was invited to contribute an article for his journal *Europa*.

An important part of the grand tour, not officially, but nevertheless indispensably for a young ambitious intellectual, was to acquire academic good manners, to learn how to get on in academia, to get one's voice heard, and to win recognition; in short, to achieve social proficiency. At the University of Landshut, Ørsted was invited to a party. He initiated a conversation with a professor of chemistry, who appeared to be reserved and professionally weak. Ørsted made efforts to guide their talk into areas he knew best in order to impress and to areas stimulating his curiosity in order to be enlightened. Ørsted learnt to oppose colleagues 'with a stroke of wit and charm to avoid offending the rules of social life'. He also discovered that flattery often gains people's confidence. A. Röschlaub, the professor of medicine, was interesting, vital, bright and witty, and 'as attractive in social life as he is acerbic in his writings. He told me that he was so well trained in writing that he put his quill away in the middle of a paragraph, when the clock struck to let him know that now it was time for him to lecture (for which he does not prepare himself), and that upon his return home would take up the quill again and carry on writing. He even assured me that once he had fallen asleep while lecturing without anybody noticing it.'[14] There was a lot about social ability to be learned in society.

*

The week had other days in it than the *jour fixe* when Mrs Herz opened her salon. So her guests would also engage in other social events in Berlin. During carnival from mid-December to the end of January the city's social distinctions were temporarily suspended, and the King would not only invite the nobility to the court at his castle, but also invite all Berliners, rich and poor, to La Redoute or to fancy dress balls in the Opera. The only condition was that everybody must wear a domino (a long carnival mantle with hood and wide sleeves) and a mask to disguise social differences. Previously, His Majesty had ordered princely guests to wear red dominos and the bourgeoisie and the common people to wear black or blue ones, and a picket fence across the Opera kept the social strata apart so that no dignitary risked being asked to dance with a social inferior.[15]

Apparently, the new King, Friedrich Wilhelm III, had given up this separation of ranks; at any rate Ørsted did not mention it. On the other hand he did report intrigues; free food and drinks were included in the royal masquerade:

> 'A masquer with a blue domino came to our table and ate with great appetite. It did not take long before the same masquer returned to guzzle with a fresh appetite. When this had repeated itself three or four times, someone followed this voracious masquer, finding that it had sneaked outside as rapidly as possible, and that the solution was that the coachmen waiting outside on their carriages had together hired a domino and a masque which they put on one after the other until they had all satisfied their appetite.'[16]

JENA: ROMANTICISM, SALONS, AND SOCIETIES

This abuse of His Majesty's gracious mercy was indecent and intolerable. At future masquerades, the King resolved, people would only get access on payment of an entrance fee, to make sure that decent citizens were not bothered by paupers with insatiable appetites.[17]

Ørsted had a similar experience at the inauguration of the new Royal Theatre on Gendarmenmarkt (between the German and the French churches in Friedrichstadt) on New Year's Day 1802.
'It was meant to be a day of celebration, but turned out to be

> a sad day for many; the crush at the entrance was so enormous that many people were brought out in a fainting fit. Ladies' shoes, trodden on and useless, were found in abundance, and for the short time I was a spectator of this drama outside the theatre I saw several ladies shoeless, dripping with sweat, and with torn clothes escape from the crowd through which they had intended to approach the entrance. One lady came out almost naked, another had her hair-piece turned around... confusion all over the place.'[18]

Ørsted had just wanted to buy a ticket, but now he postponed it. Later in the season he became a frequent theatregoer, seeing Lessing's, Iffland's, Kotzebue's, and Schiller's plays, but he did not care very much for the acoustics of the theatre or for the behaviour of its audience.

Fig. 40. The Royal Theatre on Gendarmenmarkt. It was flanked by the German Lutheran Church and the French Reformed Church—alike in design, but different in confession. The Prussian King Frederick the Great is said to have commented, 'Everybody can be saved according to his own belief'. HCØ saw several pieces by Schiller, Wieland, and Kotzebue at the theatre, where the King would arrange masquerades and concerts. Lithograph by Ludwig Eduard Lütke (1801–1850). Kupferstichkabinett, Staatliche Museen zu Berlin. BPK, Berlin. Photo by Jörg P. Anders.

'Theatregoers do not seem to be sufficiently educated for this play (*Nathan der Weise*); it was hardly possible to hear the players due to the chatting of the audience. I left one seat after another on the ground floor to obtain peace and quiet, but all in vain. When the second act was over I said in German to Dr Høyer—with whom I always speak Danish—that I found it strange that nowhere in a Berlin theatre does one find the quiet in which to listen to *Nathan der Weise*. This play is commonly praised, and the most talkative spectators left their seats or stopped their small talk.'[19]

There were voluntary associations in Berlin combining the social with the scientific that attracted Ørsted, such as Die Gesellschaft für Naturforschende Freunde [the Society for the Friends of Scientific Research], where once a month a lecture was delivered. Ørsted attended on the five occasions that he had the chance to. This society had twelve core members, real friends who were no amateurs, but professional scientists meeting successively in each other's homes, with a larger number of extraordinary members taking part in the monthly meetings. Rose and Hermbstaedt were core members.[20] Secondly, Ørsted was introduced to the recently founded Philomatic Society by its secretary, the Jewish Kantian Lazarus Bendavid, who also frequented Herz's salon. This was a small association of philosophers and scientists, Hermbstaedt being one of them. Ørsted took part in at least fourteen meetings at the Philomatic Society,[21] which had asked him to prepare a lecture on Winterl's chemical theory. He made great efforts to fulfil this assignment, finding that Winterl's theory had been repudiated for superficial reasons, 'since I do not know anybody who has criticised Winterl in a way that shows he has read him'.[22]

Bendavid also introduced Ørsted into Die Gesellschaft der Freunde der Humanität [The Society of Friends of Humanity] which was centred around Kant's philosophy. This society had been founded in 1797 by I.A. Feßler, head of the League of Evergetes, reminiscent of the Masonic orders and doing charity work for the benefit of people like Ritter. It convened on Saturday evenings in the premises of the Royal York Lodge and followed Masonic rituals. Bendavid was in charge and a member was assigned to give a paper followed by a plenum discussion of philosophical and aesthetical topics. Later on the members and their ladies had supper.[23]

Some of Ørsted's associates were Freemasons, among them the Dane Joseph Darbes, Professor of the Academy of Arts in Berlin, who made portraits of Moses Mendelssohn and Goethe. Like the brothers David and Michael Friedländer and I.A. Fessler he had joined Herz's salon. David Friedländer being an indefatigable champion of human rights for Jews, and his brother Michael having assumed the task of communicating between German and French science (he had already promoted Ritter in Delametrie's *Journal de Physique*) were both initiated Freemasons. Michael became one of Ørsted's close friends in Paris where he settled as a physician. Fessler was Grand Master of the Royal York Lodge as already mentioned. He had been a spokesman for the reform of Freemasonry, suggesting fewer ranks and less secrecy. Together with Fichte he had elaborated a reform proposal that caused the order to split. Die Gesellschaft der Freunde der Humanität can be seen as Fessler's attempt to open the gates of the lodge.

Darbes offered Ørsted Masonic hospitality on his itinerary through the German states provided, of course, that he let himself be initiated into the order. Ørsted felt unable to come to such a far-reaching decision by himself. He felt he had to ask Manthey in Copenhagen, who was a Freemason, for advice first. One evening Ørsted was introduced to a lodge although he was still not initiated. However, this lodge proved to be just a social club for both genders where members dined and played cards.[24] Soon Manthey replied that his brothers Darbes and Klaproth would be good mentors for him if he decided to join the order.[25] Some months later Darbes offered him some addresses of cheap accommodation at Freiberg on condition that he became a brother. He asked for advice, which turned out to be not to join, because it was too expensive to be worth his while. Subsequently, Manthey wrote, as a friend, not as a brother, that if Freemasonry lived up to Ørsted's expectations it might turn out to be a useful connection, but then the initiation should not be motivated by economic considerations. Ørsted would do well without that kind of connection, Manthey advised. Should he, however, be convinced by more elevated reasons in the future to join the brotherhood, 'you shall be received with open arms'.[26]

*

Ørsted made sure not to miss any opportunity to rub shoulders with the intellectual elite of Berlin. At Elers' College he had followed Anders's interest in Fichte's moral philosophy. Now, though the Jena circle had been dissolved, an opportunity to meet him occurred because Fichte and his friend, A.W. Schlegel, had moved to the Prussian capital, and Fichte was offering private lectures on his theory of knowledge.[27] In January 1802 Ørsted knocked on Fichte's door, and his wife opened it; Ørsted said he would like to subscribe to Fichte's series of lectures and pay the fee, twenty-two rixdollars. This was a lot of money, and it was not included in his budget, but 'he would hate to lose this opportunity to get to know the philosophical system of this great genius'. Fichte's wife said he could have the subscription at half-price because he was a genuine student, and neither a minister nor a diplomat nor any other kind of dabbler, and her husband had set the price with such philistines in mind.[28] Ørsted made a mental note of this practice and adopted it in Copenhagen later.

To hear Fichte was tempting for two reasons. Ørsted wanted insight into a philosophy that was so important to Anders. Just recently a new book had been published bearing the long and captivating title of *Sonnenklarer Bericht an das grössere Publikum, über das eigentliche Wesen der neuesten Philosophie. Ein Versuch, die Leser zum Verstehen zu zwingen* ['A crystal clear account for the common audience on the real character of the latest philosophy. An attempt to make the reader understand'].[29] He was equally excited 'to watch this rare talent present his thoughts'. In spite of his queer appearance, his body being so over-proportioned in relation to his short legs that he could hardly raise his face above the lectern, Fichte was said to have an unusual, almost hypnotising, aura as a speaker, a conspicuous body language, and a capacity to encourage the audience to put questions which he would answer directly without trying to avoid them. Here Ørsted might pick up clever techniques useful for his own lectures, even though he had already learnt that it did not take much to shine in Copenhagen.

After five lectures he told Anders by letter that he had plenty of reasons to envy him, for in reality Fichte was a master of the art of captivating his audience.[30] As far as his theory of science was concerned even the most abstract theses were presented 'with fire, life, and clarity'. Hans Christian was well aware, of course, that Anders was familiar with Fichte's philosophy, and he saw directly that it had only a limited relevance to his own scientific research, owing to the simple fact that Fichte's main interest was to combine Kant's theoretical with practical reason, while he had little to offer when it came to the study of nature. Nature is not free but determined, whereas Fichte's interest was in the problem of freedom. Freedom was not a theoretical, but a practical affair. It was a question of concrete action, where circumstances acted against one's intentions in life. According to Fichte an understanding of these opposing forces is not achieved through the kind of theoretical shadow boxing that Kant set up the rules for. *Das Ding-an-sich* [the thing-in-itself] is unrecognisable anyway, so why should one waste one's life by entering his metaphysical maze, however much truth it might contain?

To Fichte philosophy is not an academic matter of appropriating admission to the objective world, which might just as well be given up before one starts. No, philosophy is a way of life developed by the subject as it experiences the forces opposing its urge of freedom. Kant had begun with the objective world (the heavenly stars above) and proceeded to the subjective life (the moral law within) to end up wrestling with the issue of how to bridge the gap between theoretical and practical reason. Fichte (who never succeeded in gaining an academic degree) preferred to move in the opposite direction. He began with the self that construes its life and pursues its goals in absolute freedom. An atomic physicist lives in a world different from that of a musician or a missionary. One world is no more valid than the other. Different problems arise according to the particular way of life the subject has chosen to live. Hence Fichte's lectures started with the advice to find one's vocation in life. Only then would one learn that living one's life is a question of overcoming the opposition that realism (the non-self) constitutes vis-à-vis idealism (the self).[31]

After the lectures followed a time for conversation that offered an opportunity to ask Fichte questions, and Ørsted mentioned Schelling's system of transcendental idealism, questioning whether Schelling had any theory of knowledge that in any way addressed the opposition between realism and idealism. As far as Ørsted could see, Schelling's *Naturphilosophie* was one-sidedly idealistic. Fichte agreed, adding that Schelling's system paid no attention to this conflict and consequently did not relate to it.

Ørsted's second question concerned Schelling's *Naturphilosophie* as it appeared in his *Zeitschrift für spekulative Physik* ['Journal of Speculative Physics'].[32] He argued that Schelling's theory of science (as opposed to Fichte's theory of knowledge) was based on logic only, that is to say it dealt with logical relations between abstract concepts. Schelling had argued in a circle, which is not conducive to synthetic insight. He admitted that he was not thoroughly acquainted with Schelling's *Naturphilosophie* because he had only read part of it. Fichte was totally in agreement, and Hans Christian concluded in a letter to Anders: 'I do not find Schelling's philosophy laudable; the reason is that, except for the parts he has borrowed from Fichte, I do not find much of significance that other clever minds have not thought out as well.'[33]

Ørsted had attended one of Schelling's lectures in Jena on 29th October, but there is no record of the impression it made on him. What we know is that in the summer of 1801 Schelling delivered public lectures on the method of academic studies, and private lectures on his philosophical system.[34] From Ørsted's notebook we know that he met Schelling once in private in Jena and later bumped into him at soirées in Halle and Leipzig.

In March Ørsted and Karsten (associates from the study-circle) met with A.N. Scherer, an empirical chemist fervently adhering to Lavoisier's system. Previously, he had worked with Göttling in Jena and with Goethe and Ritter at Belvedere. Both Karsten and Ritter had been part of the editorial board of Scherer's *Allgemeines Journal der Chemie*, but this work had now been transferred to Hermbstaedt, Klaproth and Richter.[35] Scherer had left Jena and was now managing a china factory at Potsdam. Ørsted met Scherer at a soirée. His diary reports on a painful scene such as he had never experienced with any other scientist. Scherer had assaulted him in 'vulgar language', and the incident had nearly ended in 'a actual quarrel like the ones at Gammel Strand' (the fish market in Copenhagen). According to the diary Scherer had started the row by accusing Ørsted's book of exuding *Naturphilosophie* and being vitiated by 'stupidity and unreason'. Ørsted stayed cool (he wrote home), but Scherer, in an excited state of mind, continued; 'It may be my fault; Gall insists that I do not possess the wherewithal for metaphysics.' Ørsted could not agree more. Judging from Ørsted's letter he did not defend the German *Naturphilosophen*, but presented well-known Kantian arguments against the possibility of chemistry ever becoming a science proper; he admitted that including dynamical theory in textbooks of chemistry might have unfortunate consequences, because, if imperfectly explained, it might make people believe they were capable of talking about matters they knew nothing about. Despite the harsh words, they parted amicably, but Scherer continued his attacks in the company of Karsten after Ørsted had left, unaware that they had recently finished their study-circle on Kant's metaphysics of nature, which Scherer was unable to grasp.[36]

In September Ørsted met Ignaz Döllinger, a physician, at Bamberg. The University of Bamburg and its hospital were the very centre of Brunonian medicine, which was now, upon receipt of Schelling's blessings of its cures, reckoned part of German *Naturphilosophie*.[37] Ørsted spoke with the physicians and criticised Döllinger who showed them around the hospital which was clean and tidy and well ventilated. The talk dealt with *Naturphilosophie*. 'He belongs to the class of scholars that subscribes to such new ideas without having thought them through. An appropriate description of these people is one commonly used by Professor Aasheim: They have imbibed a certain doctrine; they have not received it in the original nor invented it themselves. Unfortunately, this is the usual story.'[38]

At Salzburg he met a certain Dr Wagner, also a fervent disciple of Schelling. He taught speculative physics, 'but without knowing nature from anything but textbooks... These people are marketing lame proofs and skewed physical theories, and complaining that other people refuse to accept them.'[39] Ørsted corrected Wagner in the name of 'we, the empiricists'. Often he felt himself to be pushed into taking a contrary position. Confronted with a traditional, empirical chemist like Scherer he would interpose with philosophical arguments, and when he was facing a *Naturphilosoph* he would point out their lack of empirical knowledge. 'I like to argue about philosophical matters out of a certain disposition to contradict, out of a hatred of obstinate statements on delicate issues, particularly by people of mediocre knowledge, and finally out of a disposition to see what people are like.' He enjoyed putting people into a tight spot and seeing them 'surrender unconditionally'.[40]

Ørsted kept Schelling and his doctrine at arm's length. But he associated closely with two other members of the early Romanticist avant-garde: the Schlegel brothers. August Wilhelm was ten years his senior and a celebrated translator of Shakespeare and theorist of aesthetics, while Friedrich, five years older than Ørsted, was notorious for his scandalous novel *Lucinde*. Both had a predilection for expressing themselves on questions of aesthetics in dialogues in the style of Plato, a genre that Ørsted would throw himself into. At that time he enjoyed reading *Gemäldegespräche* ['Dialogue on paintings'] and *Gespräche über die Poesie* ['Dialogue on poetry'] printed in their journal *Athenæum*.

Ørsted visited Friedrich Schlegel six times in Jena and Leipzig, and when both settled in Berlin they often saw each other there. In January 1802 he told Anders that they had developed a confidential relationship.[41] In April of the following year Ørsted handed him his manuscript, *Übersicht über die neuesten Fortschritte in der Physik* ['Overview of the Latest Progress of Physics'] for *Europa*, his journal of aesthetics.[42] This article summarised the impressions his grand tour in Germany had made on him, including Schlegel's comment: 'Schelling's *Naturphilosophie* is opposed by the acerbic empiricism the extinction of which it was meant to be; there is but little danger that his lack of scientific knowledge will prevail, because at the same time Ritter has established an example of a physics which is pure empiricism and still has satisfied the strongest demands for scientific research by rigorous methods'.[43] This hatchet job must be read with a pinch of salt, because the criticism was not thought out by a scientist, but coloured by personal aversion to Schelling, who had run away with his sister-in-law, A.W. Schlegel's wife, Caroline.

> A disgraceful tone of flattery rules the
> journals and a despicable compliance
> with the will of those in power.¹

14 | 1802–3
Post-revolutionary Paris

Ørsted's sojourn in Paris lasted almost a year, from 18th November 1802 to 13th November 1803. He took lodgings near the centre at the Hôtel de Philadelphie in the Passage des Petits-Pères. He rented two small rooms of about twelve square metres combined, sleeping in the smaller one of just four square metres. The rent was nine rixdollars a month. However, after a few months his landlord was thrown out due to insolvency, the tenant had to move, and for the rest of the time he stayed at the Grand Hôtel de Versailles, Rue Batave, near Palais-Royal. This street does not exist anymore. It was a cul-de-sac a few hundred metres to the northeast of the Louvre. Facilities were better and little dearer. The greater space made it easier to keep things tidy, but when his friend A.C. Gjerlew moved in, Ørsted must have felt as cramped as when he shared rooms with his brother at Elers' College. When Gjerlew (a theologian on his grand tour to study pedagogy and education) left in September to go to Switzerland and Italy with Jens Baggesen, he was replaced as a lodger by Martin Lehmann (the father of the National Liberal politician Orla Lehmann). They knew each other from the dye-works in Vestergade where Martin had been a tenant; they were close friends and took long walks together around Paris. Ørsted felt obliged to buy new clothes in the French style for more than sixty rixdollars. The suit he had had made in Berlin had already worn out and was not fashionable in Paris. He had not come to the capital of France to hide himself away in a den, but to indulge in its social life and make himself known.

Every Sunday evening he visited Friedrich Schlegel, who had converted to the catholic faith, and now lived in Paris with his wife Dorothea.² They arranged recitation evenings in their

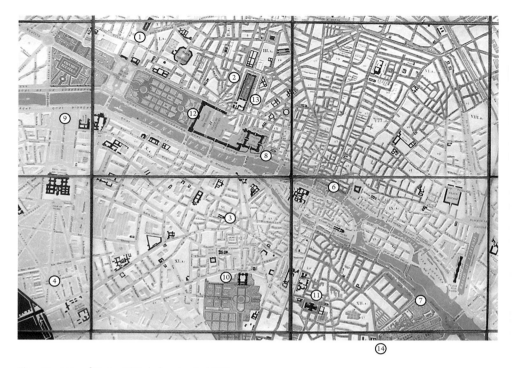

Fig. 41. Map of Paris, 1840, belonging to HCØ.

1. HCØ's lodgings in 1803, Passage des Petits-Pères.
2. HCØ's lodgings in 1803, Rue Batave (with Gjerlew and Lehmann).
3. HCØ's lodgings in 1813, Rue Colombier.
   HCØ's lodgings in 1823, Hôtel de l'Empereur (near 11).
   HCØ's lodgings in 1846, (unknown).
4. La Conservatoire des Arts et Métiers
   l'École Polytechnique
6. L'Hôtel de Ville
   L'Institut National
7. Le Jardin des Plantes
8. Le Louvre
9. Le Palais de Bourbon (9)
10. Le Palais du Luxembourg (10)
11. Le Panthéon
12. Le Palais et Jardin des Tuileries
13. Le Palais Royal
14. Société d'Arcueil, a village outside the city wall to the south. DTM.

home. Schlegel would read aloud (assuming different voices for different characters) from such works as Tieck's *Puss in Boots*, a fairy tale composed as a play in three acts. Michael Friedländer, the science journalist, had gone to Paris, too. In the beginning at least Ørsted seemed to thrive best when socialising with friends who spoke Danish or German.

The first thing was to engage a private tutor to come to Rue Batave for tea every morning at eight o'clock to teach him French for two hours at two francs a lesson. He did not start from scratch, but realised that without being able to express himself fluently he would not be taken seriously. He soon became acquainted with an ever-widening circle of scholars and was pleased to realise that he was capable of managing a conversation in French without embarrassment. After a few months, the French lessons were cut to two a week, but he kept his tutor to correct his written French.

Manthey had introduced him by letter to a certain M. Lasteyrie, a member of the Philomatic Society. Ørsted joined this society and attended its weekly meetings. Lasteyrie also organised private soirées for his republican friends every 'decade' (according to the revolutionary calendar each month was divided into three decades). At these, politics was on the agenda and 'here much was said that is considered very liberal, but would be quite ordinary at home; I have often heard far bolder statements in Dreyer's Club than anybody ventures to utter in a private party here in a so-called republic'. In reactionary Denmark citizens were at least left alone in their own homes. In post-revolutionary Paris, police spies were everywhere.[3]

As Ørsted sat in his rented room in the narrow Passage des Petits-Pères looking out of the window, the wounds caused by revolutionary vandalism met his eyes. The street name referred to an Augustinian monastery from the seventeenth century. The buildings were still there, but the monks had been thrown out and their premises and belongings confiscated soon after the outbreak of the Revolution in 1789. He could look down on the court and the gardens behind it where the baroque alleys and flowerbeds were overrun with weeds and falling into decay.[4] This was the typical fate of all catholic churches and monasteries in the capital. In Rue Batave, the old monastery site from the age of the crusades, Saint-Sépulchre, had been totally swept away in favour of the new Court Batave, a huge commercial enterprise.[5] Crossing the Seine he would catch a glimpse of the great dome on the top of the hill, Sainte Geneviève, only completed in 1790 and immediately secularised, renamed Panthéon, and dedicated '*Aux grands hommes la patrie reconnaissante*' ['In memory of the great men of our country']. A few years later the mortal remains of Voltaire and Rousseau were moved there and glorified with pomp and circumstance. The revolutionary avant-garde, the Jacobins with their ritual *bonnets rouges* [red Phrygian or 'liberty' caps], had seized several churches where from the Middle Ages onwards Franciscans and Dominicans had said their prayers at the seven canonical hours to the god whom only a few superstitious fools were expected to take seriously from now on. Another monastery, Saint-Martin des Champs, also changed its purpose. The refectory, where lazy monks had been stuffing themselves with food, became a storeroom for some of the scientific instruments and technical appliances that had been confiscated from the aristocracy. The monastery became the Conservatoire des Arts et Métiers [Conservatory for the Arts and Trades] thanks to the cooperation of Grégoire, the abbot, and J.A.C. Charles, the physicist.[6] Together they saved scales and glass flasks and tubes and other experimental equipment belonging to Lavoisier until the guillotine separated his gifted head from his body at a single stroke. His collection is still there.

*

From a scientific point of view dissatisfaction with the universities was a problem not only in Denmark but all over Europe, the exception having been in France where the sciences enjoyed exceptionally high favour. France moved to the extreme of Enlightenment fundamentalism. In September 1793 the National Convention closed all universities and colleges arguing that they were medieval catholic seminaries hatching counter-revolutionary elements. The Académie des Sciences was shut down in the name of the Revolution despite its invaluable work in establishing the metric system of weights and measurements and its contribution to the production of saltpetre so crucial to the war effort. At the same time the Académie Française, the royal foundation for arts and sciences, was abolished to the accompaniment of Marat's shouts of derision about the forty reactionary layabouts. His view seemed to be that the humanities made no useful contribution to the republic, but were only there to amuse the aristocracy. Only two years later, however, the natural sciences were resurrected as the First Class of the National Institute, but there was no room for French language and literature for another ten years.[7]

Under Napoleon (from 1799 when he was appointed First Consul, through his imperial coronation as Emperor, up to his final defeat at Waterloo in 1815) the natural sciences were particularly favoured by this increasingly absolute ruler, who distinguished himself not only as a military strategist, but also (from November 1797) as an elected member of the First Class (mathematics and physics) of the National Institute. Bonaparte's list of publications hardly entitled him to be a member. His literary achievements consisted of minor works on moral philosophy and Corsican history. His mathematical knowledge stemmed from the Military Academy, where Laplace had examined him. Bonaparte defeated his scientifically more competent rivals, thanks to the fact that a majority of the members of the First Class, headed by Laplace and Berthollet, included the First Consul in the most important scientific institution of the republic for strategic reasons. To this end the First Class was only too willing to bestow upon him the prestige emanating from being associated with the sharpest brains of the nation, even if in his letter of thanks (with more a sense of reality than genuine modesty) he indicated that he would have to be their pupil for a long time before he could hope to become their equal.[8]

Hans Christian experienced the interplay between a scientific meritocracy and a great power under an authoritarian leadership. Under the *ancien régime*, French science (apart from the exclusive Académie Royale des Sciences) had been independent and neither protected nor rewarded by royal privileges. D'Alembert, who had been the mentor of Laplace, co-edited the encyclopedia, which was a private enterprise disseminating the ideas of the Enlightenment in opposition to the absolutist regime. Lavoisier had funded his scientific instruments and research by private money earned as a tax collector, and he suffered the wrath of the revolutionary crowd in return. When he was executed in 1794, it was in his capacity as a tax collector and not as a savant. It is a pure myth that his judges ever said, 'The Revolution is in no need of scientists'. The revolutionary elite soon (and far more unequivocally than the *ancien régime*) approved the scientism of the Enlightenment, a dogmatic belief in the technical potency of science. In his 1792 report on the reorganisation of public education Condorcet had argued that the sciences enhance intelligent thinking and militate against superstition and prejudice.[9] Moreover, in contrast to the humanities, the natural sciences were characterised by unlimited progress and the accumulation of knowledge.

To these merits Bonaparte added the prestige the Empire would harvest from being associated with advanced science. Napoleon personally took part in two out of three meetings in the First Class and set the goal for France to dominate the world not only militarily, but scientifically as well. In 1798 he equipped a tremendous scientific (and military) expedition to Egypt and founded the Egyptian Institutes in Cairo and Paris. On the 18th Brumaire 1799, the day of the coup d'état, Bonaparte elected himself First Consul, and the next year he was made president of the First Class of the National Institute.[10]

*

Ørsted's first paper in French was delivered to the Philomatic Society. He presented Ritter's electrochemical discoveries to an association operating as a youth section of the National Institute, the exclusive assembly of grey-haired dignitaries. When the Académie Royale des Sciences was dissolved by decree in August 1793, the Philomatic Society assumed the role of a retreat for its members until it was resurrected as the First Class of the National Institute. The meetings of the National Institute took place in the Caryatid Hall in the Louvre, and its members attended in uniform, of which they had two versions, one for everyday use and a more extravagant one for special occasions. When Ørsted started attending the meetings (not as a member, but as a spectator introduced by Vauquelin) the members were dressed in dark green coats embroidered with light green piping, yellow waistcoats, and green breeches, i.e. *culottes* (associating them with the 'reactionary' academy) as opposed to *pantalons*, which were the emblem of progressive citizens.[11]

Ørsted preferred the Philomatic Society because here he was allowed, and even encouraged, to speak.[12] The younger members of the First Class participated in both places, whereas the older members felt aloof or did not find time, because they met in the informal Société d'Arcueil, which was confined to fewer than ten savants, constituting the absolute elite. In the Philomatic Society, Ørsted met with the chemists J.B. Biot, Charles Coulomb, A.F. Fourcroy, L.J. Gay-Lussac, L.B Guyton de Morveau, Paul Thénard, and N.L. Vauquelin, and with natural historians J.L.N.F. Cuvier, René Haüy, Alexander von Humboldt, and J.B. Lamarck, as well as technologists Jean Chaptal (who was also Minister of Domestic Affairs) and Gaspard Monge; that is to say, he met there the cream of French science apart from P.S. Laplace and C.L Berthollet, who shut themselves away in the Société d'Arcueil, although Ørsted did manage to get in touch even with them.

He also had an opportunity to see these scientists in the Athénée, which he joined immediately after his arrival even though the subscription was expensive (four louis d'or, equal to 96 francs or 24 rixdollars a year); in return he gained free access to all that went on there.

The Athénée was a social meeting place for intellectuals, including ladies. It was open every day and since it was situated near the centre of the city, Ørsted often dropped by to socialise. In the evening he would listen to popular lectures on scientific or literary subjects. He attended at least 25 times picking up useful techniques from some of those who had mastered the art of holding an audience spellbound, while others only lulled it to sleep. In particular he was impressed with Cuvier who delivered his aperçus on the history of science and emphasised the impact of science on the progress of civilisation. 'I profit a lot from this Athénée, because almost everything Paris has to offer in terms of interesting literature is available there.'[13]

At the Athénée members of both genders had an opportunity to combine the useful with the agreeable. The useful might be popular lectures on chemistry by Fourcroy or Thénard, the agreeable might be lectures on the arts. Ørsted hoped to establish a similar institution in Denmark.[14] He did not work out detailed schemes, but hit on sketchy ideas for popular education. A patriotic society ought to take up the idea and spread it to the provinces. The Royal Danish Society of Sciences and Letters as well as the Scandinavian Literary Society were fora for scholars, but nothing was available to artisans either in the capital or in the provinces. Instructing citizens in their rights and duties and rewarding people who took a critical attitude towards the government without going too far was what he had in mind: textbooks or booklets that made demands on citizens without indoctrinating them. Provincials should be educated above the elementary level of the village school. If only they were taught a little mathematics and science and were shown collections of models of modern machinery they would soon make progress. These useful measures should be combined with pleasant experiences such as concerts, art exhibitions and sea bathing, on the wonderful island of Taasinge for instance, where mineral and steam baths should be established. Ørsted found that the Paris Athénée prioritised amusement too much: further north such luxuries were indefensible. 'Everything aiming simply at the agreeable should be eliminated from the Athénée, otherwise it will get the upper hand', he noted.[15] Apparently, Ørsted's idea of disseminating science to the provinces had already been fostered in Paris in 1803.

*

He followed two private series of lectures, one by Vauquelin at thirty-six rixdollars, the other by Charles at twenty-four. Receipts kept Mr Jørgensen, the zealous bookkeeper of the Cappel Foundation, happy. The two series took place on alternate days, so Ørsted was busy every day. He was careful to take notes, and they have been preserved. Vauquelin was rumoured to be a competent analytical chemist and Ørsted was attracted to the private laboratory he shared with his disciple Fourcroy 'for the sake of his beautiful experiments'. His spoken performance, by contrast, was nothing special. 'I find no order in his lectures apart from associations of ideas...and I know nobody who outrivals him in lyrical inconsistency except Professor B.[ugge] in Copenhagen.' He had four assistants in his laboratory, which was built in the shape of an amphitheatre and equipped with a skylight enabling all participants to watch his experiments.[16]

Fig. 42. L.N. Vauquelin (1763–1829), chemist. HCØ attended his lectures, but found Charles's were better. DTM.

Ørsted appreciated Charles far more than Vauquelin. Charles was careful and diligent and had always prepared a full written text for his lectures so that they were delivered in clear sentences, although (and this is the point) he did not simply read his text out loud. In fact, people did not realise he had a written text at all. He talked vividly and freely.

> 'I find this method very good, indeed, because it provides security to a person, who does not always feel disposed to speak in public, when he has thoroughly prepared the structure and grounds of his argument. Vauquelin turns up, if not totally unprepared then only a little prepared, and so gives a worse impression, since nature has denied him what it bestowed on Charles... Either Vauquelin has not prepared his demonstration at all and therefore makes a mess of it, or he has prepared so much that the poor spectator gets to see nothing and is treated almost like the man who wanted someone to paint the Israelites' crossing of the Red Sea. The artist just painted the Red Sea and told him that the Israelis had already crossed it... Vauquelin is economical with talent and money, because he does not demonstrate any expensive experiment or any experiment requiring special skill for the four hundred louis d'or [approximately 2,400 rixdollars] a series of lectures lasting four months is calculated to yield, and he has no significant instruments... Charles has not yet let us miss any necessary experiment.'[17]

Fig. 43. J.A.C. Charles (1746–1823), chemist and balloon pilot. Châteaux de Versailles. RMN, Paris.

There is no doubt that Ørsted gleaned a lot in terms of French language skills, experimental chemistry, and pedagogy from these lectures, which he only missed when he was ill. Charles became his role model not only because he was a clever experimenter and lecturer, but also because he had had the courage to go up in the world's first hydrogen balloon. Ørsted was also impressed to hear about Biot's and Gay-Lussac's seven kilometre balloon voyage to measure the Earth's magnetism. Ørsted himself had to make do with writing poetry about ballooning.

Ørsted also had the opportunity of meeting French scientists practising in the École Polytechnique. Vauquelin had promised to introduce him to Berthollet and his assistant, Guyton de Morveau, whom the National Institute had assigned to test Winterl's theory. Guyton de Morveau was the Director of the École Polytechnique at this time and presented Ørsted with a free ticket to all the lectures there. Ørsted was particularly tempted to follow Berthollet's classes on Lavoisier's 'caloric'. The École Polytechnique was staffed by the same members of the elite that Ørsted knew from the National Institute and the Philomatic Society. Some were headhunted as bright students, placed on the lower steps of the ladder, and encouraged to climb it by assisting the older staff well, as Thénard and Guyton de Morveau had done so successfully. Ørsted did not say much about the École Polytechnique in his letters; he probably had a hunch that these things would only bore Sophie and be of little interest to Manthey, as he had been there only recently himself.[18]

The École Polytechnique was the first engineering school of any significant size in the world. It was a product of the Revolution and intended to be the technical and military vehicle of the republic. At first the institution was accommodated in the grandiose Palais-Bourbon in the Parc de Luxembourg, a royal palace confiscated by the Revolution. Access was open to young men from all over France, but they had to pass an entrance exam, which ruled out those who had an inadequate knowledge of mathematics. About four hundred of them a year passed through the eye of the needle, and since their abilities varied a great deal (some sons of peasants had been lucky enough to join the preparatory École Polymathique in Paris which Ørsted visited), they were divided into three classes. The best undergraduates studied for one year only, while the less able had to spend three years there before graduation. France needed talent, and if a bright student was poor, he was offered public funding in the name of equal opportunities—five hundred francs (around 125 rixdollars) a year or the equivalent of the salary of a sergeant.

The École Polytechnique offered a basic education, qualifying the student of engineering to pursue a number of professional occupations such as road or bridge building or mining. Since the republic found itself in an almost perpetual state of war, many students of engineering were drafted to serve as officers in the infantry or artillery, and in order to inure them to tougher times the reveille of the barracks where they lived sounded every morning at five o'clock while at the École Polytechnique they were allowed to sleep in until eight.

The École Polytechnique was endowed with an abundance of scientific instruments for reasons that are unique in the history of higher learning. Those running it could take what they needed from the property the Revolution had confiscated from the King, the Church, and the Académie des Sciences. Other equipment was taken from amateur aristocratic physicians whose private collections of scientific instruments and books had enabled their researches. The school also needed materials such as brass and mercury for laboratory exercises, and these could be supplied by the troops looting in Italy and the Rhineland.[19]

French science was based on a continuation of the Enlightenment corpuscular theory. Its leading characters, Laplace and Berthollet, represented both aspects of Newtonian science, the *Principia* on the laws of motion of celestial bodies, and the corpuscular theory of *Opticks*, assuming the validity of the laws of motion for the minutest particles. Ørsted not only read Laplace's *Exposition du système du monde*, but was particularly interested in Berthollet's *Recherches sur les lois des affinités chimiques*, while looking forward to its continuation *Essai de statique chimique*.[20] Lavoisier had nurtured a dream to be able to quantify the forces that enable one element to combine with another and to subsequently calculate the chemical affinities of substances by means of mathematics in the same way that Newton had calculated the celestial motions.[21]

Berthollet continued to hope that the laws of affinity were susceptible of discovery. The insensible gravity of celestial bodies received its analogy in the mutual attraction of atoms of substances that entered into combination with one another. Like Kant he imagined that the force of attraction was counteracted by a force of repulsion. Since a force of repulsion had no parallel in celestial bodies it might turn out to be more correct to see it as a force of cohesion in each substance. According to this way of thinking the force of cohesion would be strongest in solid matter, weak in liquids, and weaker still in gases. But what might be the chemical analogy of mass in astronomy? He imagined a measurement for affinity that would be the function of the saturation capacity of one substance—its 'force of affinity'— and the 'chemical mass' of the other substance. The saturation capacity was determined by the relationship of the quantity of acids and bases in the combining substances. He visualised that if he multiplied this proportion by the quantity of the substance in question the product would be equal to the 'chemical mass' of the substance. But Berthollet realised that this did not solve the riddle of affinity, for other properties such as the elasticity, temperature, etc. of the different substances might also have an active part in modifying the force of cohesion.[22]

Ørsted had high hopes of the shrewd Berthollet's forthcoming book. It was in the press when he visited him at Arcueil. He knew his way of reasoning from his lectures at the École Polytechnique. True, Berthollet's dawning theory of affinity with its notion of 'chemical mass' sounded materialistic and was based on mechanical philosophy, but so was Winterl's theory of 'Andronia' and 'Thelycke'. At this stage Ørsted was much closer to Ritter's theory that the causes of chemical affinity are hidden in polar electrical forces. On the other hand if Berthollet's work concluded in a discovery of quantitative laws of affinity this might somehow become adaptable to their dynamical theory. Berthollet's *Essai de statique chimique*, which Ørsted plunged into as soon as it saw the light of day, did not help to clarify his position. He found it to be more of a beginning of the formulation of a problem than a final solution to it. Even if he was critical towards the French corpuscular theory, he did not reject in advance an attempt to quantify chemistry, but considered Berthollet's investigations as 'discoveries'.[23] Consequently, he included him in his research programme on a par with Ritter and Winterl.

Ørsted must have felt a certain nervousness when, accompanied by the small, pusillanimous Michael Friedländer, he entered Berthollet's park at Arcueil. Ørsted had met Friedländer, a Freemason, in Berlin and read his *Französische Annalen*. Friedländer was an invaluable ally because he had access to the French savants. On their walk back to Paris he told Ørsted an anecdote about Berthollet from the Revolution. The army had sent a bottle of brandy to the Commission of Welfare. The drink was under suspicion of having been poisoned. Immediately it was handed over to Berthollet and a few other chemists who were ordered to report on the contents of the bottle by that evening in order that the culprits might be punished without delay. The analysis showed no sign of poisoning; the brandy had simply been mixed up with dirty water; now it had been duly filtered it was harmless. Robespierre, who suspected a conspiracy, jumped to his feet shouting that it was easy enough to hand in a report like that when the analysts were not going to drink it themselves. Immediately Berthollet called for a glass, filled it with the filtered liquid and emptied it. 'You have courage, *citoyen*!', said Robespierre. 'Less now than when I wrote the report', Berthollet answered (according to Friedländer). As a journalist he was an expert in ferreting out good anecdotes for his journal.[24]

The superb social status bestowed upon his French colleagues was Ørsted's recurrent theme in his letters to Sophie. The bourgeois ideals bringing about the Revolution had now had more than ten years to express themselves, but they were still rather confusing, whereas the appropriation of the privileges of the former aristocracy by the new meritocracy was immediately tangible. Foreigners often shook their heads when hearing the catchword 'universality', because they found it hard to name a nation that was as self-complacent as the French republic. The difference between the puritanical Nordic and German scientific institutions and miserly careers and the glamorous French equivalents was substantial.

Leading French scientists received opulent posts in the technical institutions and learned societies of the Republic, particularly after Bonaparte's ascent to power. The First Consul safeguarded his position by appointing his loyal supporters, for example Laplace, Chaptal, Monge, and Berthollet, as senators. A senator had high status, great political power, a top salary, a pension for life, and was close to the head of state. The Senate was comparable to the upper house. It appointed the highest ranks of civil servants as well as members of the legislative assembly. Senators were assigned to monitor the constitution, and they administered a number of honourable awards enabling them to acquire loyal supporters, in the same way that Napoleon who stood at the very top of the hierarchy had won their loyalty by loading them with privileges. Thus scientists filled the space left by the displaced aristocracy, the difference being that the new elite had somewhat shrunk in size but was probably shrewder and in all probability smarter. Unlike the aristocracy of the *ancien régime* the new elite did not wear wigs, but their luxurious robes and palaces were hardly distinguishable from the symbols of power that just a few years earlier had provoked the revolutionary avant-garde to drag the former rulers to the guillotine as scapegoats for their social crimes.

Under the Napoleonic Empire this elite formed an informal learned republic in possession of large villas surrounded by parks at Arcueil a few miles south of Paris. Here Berthollet had established his laboratory for the technical analysis of chemicals such as bleaching agents, dyestuffs, indigo, sugar beet, invisible ink, etc. These activities had led him deep into a debt of 150,000 francs, but Napoleon generously arranged for it to be cancelled. As a senator for life he earned 25,000 francs a year. In 1803 he had been made the senator for Montpellier, which doubled his income. The former bishop's palace at Montpellier and his villa at Arcueil were at his disposal free of charge. Each of the fifteen senators was obliged to report once a year to the Emperor what the local population thought about religion, taxation, conscription, etc., and whether the higher civil servants were fulfilling their duties. It goes without saying that these senators were loyal to Napoleon Bonaparte. They had been selected carefully. Laplace (Fig. 43a), who was president of the Senate, was paid 120,000 francs a year and had free lodgings in the Luxembourg Palace.[25] Post-revolutionary France made it possible for the son of a peasant wearing clogs and homespun clothes to embark on a scientific career and end up as a senator in a silken robe.

Compared to the scientific elite in Paris the social position of Ritter and Ørsted was miserable. Although he was a Doctor, Ørsted only held an unsalaried post as a junior teacher at the Faculty of Medicine, and Ritter's poverty and humiliating dependence on Goethe and the Duke of Gotha has already been noted (ch. 12). Ritter's reputation in Jena had grown so much that in autumn 1803 a group of students organised a petition urging him to deliver a series of lectures on galvanism. During a conversation with the court official, Ritter was reminded of the university's rule that the right to lecture was conditional on presentation of a doctoral thesis. Although the Faculty was very liberal and willing to let him off with half the normal fee and the public defence, he declined the offer and turned to Goethe for help. Goethe obtained half a year's respite to get the matter settled. Should Ritter wish to continue his lectures after the expiry of this period the University would take a benign view of his application. So, during the winter semester of 1803–4 Ritter lectured on galvanism and demonstrated the greatest experiments ever seen (as he told Goethe). His lectures did not make him richer. On the contrary, the money he had to borrow to buy equipment far exceeded the fees he collected from students.[26]

Fig. 43a Pierre-Simon de Laplace (1745–1827), mathematic physicist, who worked together with Lavoisier and instructed Napoleon at the Military Academy. He was a leading member of the Société d'Arceuil and the dominant representative of the corpuscularian school in French science. As a senator he enjoyed the privileges of Napoleonic France without losing them when the monarchy was restored after Napoleon's fall in 1815. Here he is depicted in his imperial uniform as *chancelier du Sénat*, 1838. Château de Versailles. Oil painting by Jean-Baptiste Guerin Paulin (1783–1855), RMN, Paris. Photo: Franck Raux.

> Paris is now very poor in scientific news,
> since all savants are too well paid to work.[1]

15 | 1802–03
Ritter and the Napoleon Prize

IN 1801, a year after the sensational publication of the article about his pile in the *Philosophical Transactions*, Alessandro Volta, with his personal safety guaranteed, travelled from French-occupied Italy to the National Institute in Paris where he impressed Napoleon by demonstrating his galvanic experiments at three meetings. Napoleon rewarded *citoyen* Volta with a gold medal and set up a commission of members of the First Class to be in charge of awarding a great prize of 60,000 francs to a scientist advancing galvanic science with a breakthrough of the significance of those made by Franklin and Volta. Foreigners of all nationalities would be eligible on an equal footing with French savants. Moreover, the National Institute would fund a medal to the value of 3,000 francs to be distributed annually to the scientist who had carried out the most original galvanic experiment that year. Candidates were not expected to submit their discoveries to the commission themselves. The commission had to sound out the scientific realm on its own initiative to identify candidates for the great as well as for the small Napoleon prize. The commission, consisting of Laplace, Hallé, Coulomb, Haüy, and Biot, adopted the First Consul's idea in July 1802, and the rules of competition as well as a brief historical account of the electrical science were announced in most international scientific journals.[2]

The galvanic commission of the National Institute was chaired by Biot, and Ørsted, who began to consider Ritter and himself potential candidates for the Napoleon prize, therefore took a special interest in getting to know his position. Biot was a protégé of Laplace and, like him and Berthollet, a member of the First Class. His initiation had happened on the basis of his thesis on differential equations. After Volta's demonstrations to the First Class, Biot (prompted by Laplace) looked into galvanism from a mathematical perspective. He assumed the existence of an electrical fluid that was set in floating motion by attractive and repulsive particles. Ørsted heard his lectures on electricity in the Athénée and in the Philomatic Society and met him several times at social gatherings.[3] Biot invited him to introduce Ritter's discoveries to the National Institute, and in preparation for that they carried out a series of galvanic experiments together.

Biot was a convinced atomic materialist and mathematician and despised all kinds of metaphysics. Later, when he wrote an article on Newton in *Biographie universelle*, he wriggled out of the difficulty of rationalising the religious writings of his subject by remarking that Newton had only been seriously preoccupied with theology after some psychogenic disease. Biot's interpretation of Newtonian science strictly followed the conventional corpuscular theory, including its imponderable fluids.[4]

By now Napoleon was First Consul for life and he no longer took part in the First Class. His loyal supporters in the National Institute had been appointed as powerful and exorbitantly paid senators. The way was open for the universal glory of France as the leading nation in science, epitomised by Napoleon's awarding prizes and medals with great pomp and circumstance. Ørsted, however, was more alert to the downside of what was going on:

> 'Paris is now very poor in terms of scientific news, since all savants are too busy or too well paid to work; people do nothing but financial transactions to become richer, whereas earlier they worked to earn their daily bread; people now work to achieve courtly honour, whereas earlier they sought after scientific honour. The only comfort here is that Berthollet will soon finish his chemical statics, a work that will introduce a new era.'[5]

On 4th September 1803 Ørsted revealed to Manthey what had already been on his mind for a long time. That May Ritter had already mentioned the Napoleon prize in a way indicating that he contemplated putting himself forward as a candidate together with Ørsted. This was a surprise since they had not been in touch during the winter and spring. If Ørsted had told Ritter about the Napoleon prize soon after his arrival in Paris, his friend had reacted neither promptly nor enthusiastically. By May, however, Ritter had finished his deliberations. 'The time is not ripe', he then wrote. 'Our chance will only come in 1819/20. At that time both of us will be in our forties, so, we shall live to make the discovery.'[6] The prophecy was based on astrology. According to Ritter, man like other organisms is governed by the magnetism of the earth, which depends on the ecliptic and the annual orbit of the sun through space in relation to the north-south axis. Man's greatest scientific (and artistic) achievements are subject to cosmic forces, he believed. Juxtaposition of the most epoch-making discoveries in the history of electricity and the maximum distortion of the ecliptic [JWR: 'Maximum der Schiefe der Ekliptik'] resulted in the following regular conjunctions:

MAX. DISTORTION OF THE ECLIPTIC		EPOCH-MAKING DISCOVERY IN THE HISTORY OF ELECTRICITY	
Spring 1745	The Leyden Jar	1745	Kleist
1764	The Electrophorus	1764	Wilke
Summer 1782	The Condenser	1783	Volta
Spring 1801	The Voltaic Pile	1800	Volta

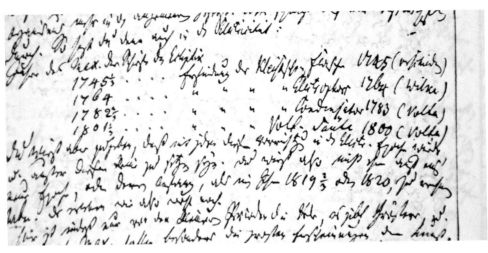

Fig. 44. J.W. Ritter's letter of 2 May 1803 to Ørsted containing the favourable horoscope for his and Ørsted's mutual and epoch-making discovery in 1819–20.

Ritter had calculated that the next favourable constellation would occur in the second third of 1819 or in 1820. Ørsted's reaction to this prediction is unknown, since Ritter, who was notoriously untidy, did not preserve his letters. What we do know from his letters to Sophie is that on 18th March he had already handed in to the Philomatic Society a report on Ritter's discovery of ultraviolet light that was received 'not without applause'.[7] On 13th May he gave a paper to the same society on Ritter's galvanic experiments. It was hardly empty boasting when he reported that 'the paper met with substantial approval', for after the meeting Biot encouraged him to request a paper on Ritter's latest discoveries, 'as he could hardly fail to receive the small prize of 3,000 livres'.[8]

Ørsted's demonstrations of Ritter's most crucial experiments to the Philomatic Society were not given to promote himself at the expense of his friend. They were part of an agreement made in Oberweimar in August–September 1802. Ritter would send his important discoveries to Paris, and Ørsted would present them to the French savants. Ørsted was only too pleased with the deal. Ørsted probably realised that promoting Ritter abroad would give him some scholarly weight and a unique opportunity to approach the savants in Paris, who were reputed to be arrogant and self-sufficient. The assignment would force him to improve his linguistic proficiency, and associating with the French learned republic would enhance his social skills, which were so crucial for a young and ambitious scientist. Manthey, working so hard in Copenhagen to further his protégé's career, fully supported his partnership with Ritter.[9] For lack of money, language skills and social competence, Ritter would never have been able to do himself what Ørsted accomplished for him.

Ørsted's conversations with Biot concluded in a flattering invitation to show Ritter's experiments to the First Class. Thus Ørsted had a tangible goal to pursue. The Napoleon prize no longer looked entirely out of reach. On 1st July he and Friedländer visited Berthollet in his villa at Arcueil. Upon his return to Rue Batave a letter from Ritter waited for him (the letter was actually addressed to Friedrich Schlegel because as usual Ritter had lost his friend's address). He could not afford the postage either and therefore he had to ask Ørsted to reimburse Schlegel for the cost of it. Ritter's horoscope about the Napoleon prize had been made independently of Biot's invitation and without any awareness that Ørsted had introduced his research to the Philomatic Society. The following day Ørsted wrote to Ritter about his plan which was to be launched there and then, not seventeen years later as suggested by his horoscope. Ritter was urged to report to Paris on his latest galvanic experiments.

However, one epoch-making experiment had already been mentioned in Ritter's prophetic letter. He had made a needle between 15 and 20 centimetres long, half of which was made of zinc, the other half of silver. The two parts were connected in the middle by a piece of brass enabling it to pivot on a piece of agate. Ritter had observed that this horizontal needle placed itself along the magnetic meridian, the zinc end pointing towards the north, the silver end towards the south. But the needle was not magnetised, and this was on purpose. Zinc and silver were situated at either end of the galvanic series (zinc: 12; silver: 0). A potential difference was thus built into the needle. Now Ritter made the needle touch a strong magnet making the zinc end the north pole, and the silver end the south pole. Since identical magnetic poles repel one another, Ritter observed that the magnetic needle reacted in a way that was 180° opposite to the galvanic one.[10]

Ritter's letter ended with a supplementary experiment that moved the contact from a voltaic pile across the zinc and silver needle: no reaction! Previously, Ritter had reported in general terms on these galvanic puzzles to his patron, the Duke of Gotha, an amateur physicist. Now Ritter added that he had ordered a tiny voltaic pile consisting of sixty pairs of discs of the smallest coins. His idea was to let this pile pivot in the same way as the galvanic needle. This miniature pile had just arrived, so for the moment Ritter was unable to give further information on the outcome.[11]

In the light of Ritter's and Ørsted's later experiments in this area, it is possible to form a reasonable idea of what the problem was that they were trying to solve. Ritter's experiment was probably intended to examine whether the galvanic needle would align itself in a certain direction and if so what the cause might be. He had hardly expected that it would align itself along the magnetic meridian. Could it be possible that the magnetic north pole attracted the zinc pole of the galvanic needle, while the Earth's magnetic south pole attracted the silver pole? In short: that magnetism affected galvanism? Everyone knew that the magnetic needle of a compass aligns itself towards the Earth's magnetic poles. In his first supplementary experiment Ritter found that the needle reacted in a diametrically opposite way when it was magnetised. The opposite reaction of the galvanic needle was astonishing. Was he picking up the scent of hitherto undiscovered galvanic forces geographically close to the magnetic ones? If so the Earth would turn out to be one huge galvanic element, a discovery that would revolutionise physics and provide further evidence in favour of the dynamical system. To investigate this question Ritter moved the conductor from a voltaic pile across the galvanised needle. No reaction! To be sure that the weak galvanism of the zinc and silver needle was not teasing him, he decided to repeat the experiment with a more powerful galvanic apparatus which would need to be constructed.

The first thing Ørsted did when he had absorbed Ritter's surprising, but rather imprecise news, was to order from 'the mechanic Le Noir's workshop' a voltaic pile with sixty pairs of big discs, an accumulator, and a zinc and silver needle identical to Ritter's, as well as a tiny voltaic pile with an agate pivot.[12] It was rather arrogant of Ørsted to label Etienne Lenoir 'a mechanic'; in fact, he was the leading instrument maker in France at the time. His workshop had produced the prototype of the new standard metre as well as a wide range of precision instruments for the National Institute, of which he was a member.[13] Ritter had asked Ørsted to borrow the one hundred francs for the equipment from the National Institute. After they had received the Napoleon prize the institute could keep the equipment and the rest of the money, he generously suggested.[14]

During the four summer months Ørsted spent thirty-seven days on galvanic experiments with his new equipment in Rue Batave. Unfortunately, he did not keep a log, but since his work up to the time of his demonstration to the galvanic commission served the purpose of winning the Napoleon prize for his friend, we must assume that he focused on testing Ritter's experiment. Or rather two experiments, since, apart from the unsettled question of the galvanic poles of the Earth, Ritter had invented an accumulator that seemed to work impeccably, although, obviously, it needed testing.[15]

Ritter responded promptly and favourably to Ørsted's letter: 'You are my true cicerone. By the end of the week you will receive the articles, to be presented to the commission when you have translated them or revised them at your own discretion. You must be careful when repeating my experiments', he wrote. If Ørsted was an optimist regarding the outcome, Ritter was a supreme optimist. He expected to gain half of the great prize of 60,000 francs, or alternatively two of the small ones (counting the two experiments/discoveries separately), or in the worst case scenario only 3,000. This small, annual prize, the equivalent of 750 rixdollars, was almost the same amount Ørsted received each year when his income from Cappel's Foundation (540), the Elers' Foundation (125) and the Foundation ad usus publicos (100) were added together.[16] 'For Heaven's sake do not entrust my share of the money to our friend [Fr.] Schlegel', Ritter added, 'He will spend it, but I need it far more than he does.'[17] Ritter did not hide his light under a bushel, and he justified his postponement of the deadline for an article, for which he had already received his fee from Frommann in Jena, by arguing that for the coming month he needed to devote himself totally to the reports to Paris, 'where people now talk about the great prize, since there can be no doubt any longer about the small one.'[18] Part of the money he intended to spend settling his debts.[19] The rest he would spend on researching the sensational galvano-magnetism. 'And please pay for the postage as usual,' Ritter added. 'You will have it all back, when the prize arrives. *Sit venia verbis.*'[20]

Ritter's papers that Ørsted used as his point of departure for his four articles in *Journal de Physique*, 1804, are missing. On 17th June he lectured in the Philomatic Society once again, this time on Ritter's magnetic experiments. Then, when his articles for the journals were finished, he went on a walking tour with his friends Gjerlew and Lehmann, visiting Rousseau's lodgings at Montmorency. They went from there to St Germain and ended up at Marly, where they saw the large steam-powered pumping facilities that had replaced Rømer's waterwheels, raising the water supply from the Seine to the fountains of Versailles. For most of July he stayed at home translating and revising Ritter's papers, and found that his French improved considerably. On 13th August he handed in the papers to Biot, the secretary of the galvanic commission. During the last two weeks in August he familiarised himself with the accumulator (the first experiment) and with the small pivoting voltaic pile aligning itself towards the Earth's galvanic poles (the second experiment). Both instruments had been unknown two months earlier both to instrument maker Lenoir and to himself.

September 4th was the day of decision at the National Institute. That same day Ørsted told Manthey: 'It is pretty certain that Ritter will be awarded the galvanic prize of 3,000 francs, and perhaps he will even receive the great one of 60,000.' He hoped to meet with Laplace. No doubt, his certainty was based on the part in Biot's history of electricity that introduced the Napoleon prize and gave prominence to the Leyden Jar as an epoch-making discovery. The accumulator ought to deserve the same acclaim, because it did for the voltaic pile exactly the same as the Leyden Jar had done for the electrostatic generator; it stored electricity.

Ritter's accumulator was built in the same way as the voltaic pile, with the critical difference that there were no zinc discs, only one copper disc between the cardboard discs. The accumulator was not galvanic per se, but when connected to a voltaic pile it collected galvanism, and retained it when disconnected. The positive and negative poles of the accumulator were opposite to those of the voltaic pile, which could be ensured by using a water separation apparatus. It turned out that the oxygen and hydrogen bubbles (or in Ritter's and Ørsted's understanding positively and negatively charged water) changed places according to whether the galvanism emanated from the accumulator or the voltaic pile.

Volta had already found out that a large voltaic pile was able to produce more galvanism at any one time than van Marum's huge electrostatic generator in Haarlem, Holland, the biggest in the world. Now Ritter showed that his accumulator had a bigger capacity than van Marum's Leyden jar in the Teyler Museum. But his most astounding observation, to French savants in particular, was that an accumulator with only a few discs had a great chemical effect, but only a weak physiological one, while an accumulator with a large number of discs produced very strong shocks, but only had a small chemical effect. Apparently, the two effects were inversely proportional. Hence, they could not be produced by the same imponderable electrical fluid as the French materialists adamantly insisted.[21] It would be a triumph for the two partners to witness the embarrassment of the galvanic commission when confronted with this observation.

As already indicated, Ørsted had more up his sleeve, a still more crucial discovery, to the effect that the Earth has two galvanic poles analogous to the well-known magnetic ones. Ritter's experiment with the zinc and silver needle had shown that it aligned itself along the magnetic meridian. Now Ritter made an experiment with his newly invented accumulator and realised that in a vertical position it was positively charged above and negatively below 'by its mere communication with the Earth'. The charge was not particularly strong; it could only be verified by the reaction of a frog. But this was no different from the first experiments transferring the electrical voltage of the pile to the accumulator. Surprisingly, the accumulator reacted in the same way as when it was connected to a pile indoors, i.e. the positive pole turned negative and vice versa. But now it was no longer connected to a voltaic pile. As a consequence the Earth must be a galvanic element per se.

Could this be corroborated if the accumulator was made to pivot horizontally? Ritter tried it out and found that his accumulator was charged when pointing in the direction north-north-east to south-south-west, but when pointing east to west the effect was gone. The electrical charge must have a source and this source had to be geographically different from the magnetic poles. The reaction of the accumulator must be caused by the Earth being analogous to a voltaic pile![22] Hence, the earth must have a galvanic meridian analogous to its magnetical one. Ritter had carefully observed bolts of lightning throughout the year at Jena and found that they followed a certain direction that had to be the same as the galvanic meridian. All living organisms, humans, animals, plants, indeed even stones and the air in the atmosphere were positively charged above and negatively below. This was confirmed by experiments on his own body.

A few days later yet another newsletter arrived from Ritter announcing that he was on his way to another sensational discovery. He had made bigger accumulators of a thousand discs, as against the two hundred previously, and repeated the experiment twenty times to achieve absolute certainty. The alignments were almost the same; they only varied between 10° and 20°. He also made experiments with iron filings and with a gold wire charged from a voltaic pile and made to pivot like the needle of a compass under a glass cupola. It turned out that the gold wire aligned itself in the direction of the galvanic poles of the Earth in the same way as the accumulator. Ørsted repeated these experiments admitting that he did not always succeed in obtaining the same results as Ritter, but that might be his own fault.

Laplace, Coulomb, Hallé, Guyton de Morveau, and Biot were assembled on the 24th Vendémiaire of the twelfth year, according to the revolutionary calendar still in use. As expected they looked at Ørsted's experiments through the lens of their own corpuscular theory. They made persistent attempts to follow Ørsted's demonstrations of the existence of a galvanic meridian, because obviously Ritter's discovery was epoch-making—if only Ørsted could verify it. But he could not. They suggested a number of possible sources of error such as impurities of the metals of the accumulator, which might be responsible for the electrical charges Ritter and Ørsted believed were emanating from galvanic poles of the Earth. The commission also tried to hang a stick of tourmaline electrified by the heat of the sun in a fine silk thread. Unfortunately, this tourmaline needle did not align itself in a distinct direction comparable to a galvanic meridian. A stick of shellac gave the same negative result.

The expected victory metamorphosed to a biting defeat. Ørsted was dumbfounded. He did not write a single syllable to anybody on the Napoleon prize. His promise to keep Manthey informed petered out. According to his diary he repeated the experiments on his own a week later. He then communicated with Ritter but we do not know what he wrote. To learn how Ritter's experiments were received by the galvanic commission we shall have to consult its minutes.[23]

The French report covers all news on galvanism for the year 1803 and the Ritter-Ørsted experiments take up four out of six pages. They attracted by far the most attention, twice as much as all other candidates for the Napoleon prize. French competitors were addressed by the revolutionary '*citoyen*', our two protagonists by the traditional '*Monsieur*'. Ørsted was considered the lowly communicator acting on behalf of Ritter who was the chief character. Nobody remarked that Ørsted was an excellent communicator—except Ritter who declared in a letter that having read Ørsted's revision of his papers he now had a much better understanding of what he had tried to say.[24] None of them had the chance to read the official minutes, because they were only printed over a hundred years later.

The galvanic commission fully acknowledged that Ritter's accumulator related to the voltaic pile as the Leyden Jar related to the electrostatic generator, that is as a condenser or in modern language a battery. They also acknowledged the importance of the accumulator's ability to show the difference between the chemical and physiological effects of galvanism, but they abstained from commenting on the conclusion drawn by our two partners, the rejection of the corpuscular theory, claiming that the same imponderable fluids flowed through the pile and the accumulator. Ritter had constructed piles with different numbers of discs:

> 'so that some gave a maximum chemical effect, while others gave a maximum physiological effect. The greatest chemical effect emanated from a pile of sixteen discs. This pile has a strong capacity, the flow of the liquid is continuous and the physiological effect is rather weak. The other apparatus has 128 pairs and the capacity is much smaller, it is re-established intermittently in sudden discharges when the resistance of the surfaces is overcome. The electricity escapes spasmodically, and the chemical effect is hardly sensible. Apparently, these differences show that the chemical effects depend above all on a strong continuous flow of the fluid while the physiological ones depend on successive and sudden discharges entailing shocks in organs.'

To Ritter and Ørsted, on the other hand, the two apparatuses showed that a flow of the same electrical fluid was an impossibility. If so it would have to be inversely proportional to itself! Hence they concluded that the dynamical theory had definitively defeated the corpuscular system.

Before Ørsted left Paris he made sure to arrange with J.C. Delamétherie, the editor of the *Journal de physique, de chimie, d'histoire naturelle et des arts* for the publication of the four papers on Ritter's discoveries that he had given in the Philomatic Society.[25] During his last two days in Paris he paid farewell calls, among others to Fourcroy whom he had heard lecture on elementary chemistry at the École Polytechnique and the Athénée. Ørsted disliked his arrogant appearance, his inclination to promote himself at the expense of his friend when he described the works of Vauquelin, his protégé, and his placing French science on a pedestal by ridiculing foreign characters from the history of science without knowing enough about them. When Ørsted came to say goodbye, Fourcroy congratulated his visitor that his grand tour had brought him to Paris where he had got to know so many scientists of excellence, far superior to any chemist from the north. Hans Christian willingly admitted that Paris offered some very impressive institutions, but the chemists of the north had one advantage: they were capable of reading Fourcroy's *Système des connaissances chimiques* in French, while Fourcroy could not read their works in the original. Finally, Fourcroy tried to discountenance Ørsted by asking him if he had acquainted himself with the École Polytechnique.[26] The guest felt he was treated like a school boy, and responded to the jibe by acknowledging that while things were admittedly lagging behind in the north, he hoped to be able to establish a similar institute upon his return. Ørsted tried to maintain a polite tone, although not one of deference, and Fourcroy finished by complimenting him on his skill in French.[27]

I cannot continue burdening Sophie
with empty promises.[1]

16 | 1802–3
The Double Game

THE MOTIVATION behind the various sources of funding for Ørsted's grand tour no doubt related to the expectation that the beneficiary would return to Denmark bringing chemical knowledge useful and profitable to the nation. This was the purpose that had moved Manthey to exert his influence on the government. But what had happened? For almost three years now Ørsted had not worried about the utility of chemistry and had even deceived his donors. He had made vigorous attempts to develop his dynamical project, and his attempt to win a great victory, epitomised in the Napoleon prize, had ended in a total fiasco. In the long run he could not conceal the fact that he was playing a double game and betraying both the confidence some benefactors had had in him, and also his faraway fiancée whom he admitted he had furnished with empty promises. Manthey tactfully reminded his protégé about how his motives appeared to have changed by asking him to look into the technical standards of a number of industries in Germany and France, and hopefully Britain, too. In addition, Ørsted ought to think about securing a livelihood for Sophie and himself.

But the beneficiary seemed to be totally immersed in his research and was pondering how to promote experimental physics and chemistry in Denmark. In 1798, P.C. Abildgaard, head of the Veterinary School, had worked out a plan: 'I am working on a project to establish a society with the purpose of carrying out experiments in physics, chemistry and physiology according to a well-considered scheme...I include you in case it materialises, which is to say if the government decides to fund the society with five hundred rixdollars annually for three years.' Abildgaard and his associates worked hard on the project. 'We benefit from our knowledge of what science accomplishes in other countries, but we do little ourselves, because as a small nation we cannot feed many scientists, who are not at the same time burdened with practical affairs.'[2] Unfortunately, the King was not interested, and Abildgaard had since died.

Now Ørsted outlined a similar plan of

> 'establishing an Institute for Physics and Chemistry upon my return with the purpose of giving instruction in the art of experimentation, rather than in technology, a matter that I believe can hardly be carried out successfully without governmental support. Nevertheless, I am also focusing on the application of science; for although I am disinclined to regard the sciences as the contemptible servants of utility and convenience, I still feel in my heart that a scientist as a good citizen ought to contribute to the prosperity of the state if he is able. He will be doubly happy when he disseminates scientific knowledge in his native country by teaching his fellow citizens. Thus, he not only elevates them to a higher level of culture, but also contributes to their insight into what they used to do purely mechanically, and in addition gives them an interest in what they do, so that they now accomplish as rational human beings what they often used to do as mere slaves.'[3]

What actually caught Ørsted's attention on the journey was Winterl's system of chemistry and Ritter's electrochemistry, and his access to Hermbstaedt's, Charles's, and Vauquelin's laboratories had been arranged with the sole intention of testing and extending their experiments. When he tried to persuade Manthey to read Winterl, his motivation was 'that here and there he offers information that is important even for technical chemistry'.[4] Not till eight months later did he receive Manthey's judgement on Winterl, which was disappointingly negative, but unfounded too.[5] Against this background he tried to summarise Winterl's theory. The result was the letter he published (after Ritter had looked over it) as the first part of his defence of Winterl, called 'Letter to a friend'.[6] This did not convince Manthey either, and he stuck to his guns. 'I have not yet read your paper on Winterl, because my bookbinder, as you know, is very slow, and I am reluctant to read anything unbound. I do not think you need to send a copy to the Duke [of Augustenborg].'[7] It became increasingly clear, and indeed embarrassing, that patron and client had completely different views of the purpose of his journey.

In his letters home, and particularly in those intended for publication, Ørsted was more calculating than truthful, as when for instance he wrote 'now and then I have had the opportunity to make some galvanic experiments, but I have not been able to extend them as I wished, because if I did the real intention of my travel would suffer.'[8] Or, as he wrote, when he had spent the entire summer in Paris trying to win the Napoleon prize: 'Now that this part of my work is almost over I can re-embark on the other part which is the more important', by which he meant visiting factories.[9]

While still in Germany he had written a letter to the Chancellor of the University requesting a loan of 2,000 rixdollars from the bursary for a collection of chemical and physical instruments. However, he sent the letter via Anders and Manthey for inspection and possible revision. Scientific instruments were absolutely necessary for 'one cannot lecture on physics and chemistry without doing experiments, and if I should lecture on speculative physics this would be ill received'.[10] The University employing him as a junior teacher had no collection of instruments after Kratzenstein's had been destroyed by fire in 1795.[11]

> 'The point is, however, that I am longing to do something upright and work hard for my own satisfaction and pleasure, that is in the scientific sense. As long as I do not do that I cannot consider myself happy. If my chemical institute materialises I suppose I could get instruments... If this plan fails... I hope it will be possible to acquire from royal coffers a grant on reasonable terms. In that case it might be useful to inform the authorities [i.e. the Chancellor] about the plan for my institute.'[12]

Anders warned his brother against running up private debts, but Hans Christian would not listen, arguing optimistically that his lectures on experimental physics would be a great success and bring in more money than he would need to discharge his loan. Manthey, too, found the idea audacious, because Steffens, who at that time was delivering his famous and much frequented lectures at Elers' College, had plans to extend them to experimental physics, and he had already received instruments from Thomas Buntzen's collection. Two series of lectures at the same time in a small capital city would probably exceed the demand, so changing the plan to technical chemistry was advisable.[13]

Of course, his marriage to Sophie was included in his plan. Since his position was unsalaried his only chance of becoming able to provide for their family was fees and royalties, but without scientific instruments his lectures would remain as speculative as those of Steffens. 'I have no fear that Steffens will take my income away, because he cannot do experiments nor does he possess sufficient meticulousness to lecture properly on experimental physics. Bugge might be a more serious obstacle although he is devoid of much necessary knowledge.' Ørsted's self-reliance was already well developed. He felt that public support for experimental equipment would get him the upper hand when advertising his lectures. It had to be in place upon his return as he intended to plunge into lecturing fired by the inspiration he had acquired on his grand tour.[14] He was developing the art of delivering compelling lectures. Charles and Berthollet were his role models, Vauquelin his bugbear. Scientific instruments were prerequisites of his dynamical project for without them it would fall apart. So he urged Manthey to press the advantage of his connections with the University Chancellor and Count Schimmelmann, Director of the Foundation ad usus publicos, as vigorously as possible.

Manthey's persistent lobbying brought mixed results. He regretted the disheartening response from the Duke. The bursary was at a low ebb, so 'the Duke has delicately declined your application', and Schimmelmann was unable to conjure up the means. Ørsted had to face the facts. 'By taking another step forward you are likely to harm your plans rather than to advance them. On the other hand the Duke reassured me that the University will be provided with equipment and that you will be able to deliver experimental lectures in due course. Moreover, he expressed a high regard and much concern for you, so I feel confident that he will find a way to let you know.'[15] In the next chapter we shall see if the Duke kept his promise.

*

Ørsted's two defeats, one by the National Institute in Paris, the other by the Chancellor in Copenhagen, did not break him. But then, in October followed a third blow that might seem to deprive him of his self-confidence. In retrospect, the absence of letters from Sophie was an ominous silence, but apparently he was so absorbed by the dream of winning the Napoleon prize that he took no news for good news. After all, her letter might have gone missing in the mail. She might have fallen ill again, or perhaps nothing worth writing about happened at the pharmacy? Her fiancé steadily persevered in sending her letters; sheet no. 44 was dispatched on 26 November 1802, and more followed: 7 December, 28 January, 7 March, 4 April, 16 May, and on 27 June two sheets nos. 50–51. In addition, he sent a special report on his walking tour with some friends around Paris on 5 and 6 August. We know these dates from his notebook. The letters were meant for circulation and written in a rather impersonal tone addressing not only Sophie, but also the wider circle of family and friends. None of the intimate letters accompanying most of his circular letters have been preserved, nor have the letters he wrote to Anders and other family members. The few places where he changed his style, because his quill was seduced by the thought of his beloved, have been censored away by Mathilde, sometimes even excised from the original. However, she has not succeeded in hiding all traces of Sophie, for example in the letter about his walk around Paris:

> '[A] I write to you from another place than Paris, but mind you, not as you might believe, from a place I shall pass through on my way home. Although it is approaching it is not yet as close as you might think. I cannot determine the day and the hour right now. At present it is only a short journey on foot around Paris. The locality I stay in now is well known to you, when I tell you that it is Montmorency, where Rousseau lived [1756-57 at Mme d'Épinay's small property 'Ermitage'], where he wrote the last parts of his [La nouvelle] *Héloïse*. I feel sure that you will appreciate a letter from me written here…
>
> [B] A person who wants to enjoy Montmorency and to recall Rousseau should stand in the morning at sunrise on the hill of Montmorency and watch luxurious, heedless Paris with its slave mentality. Then one should turn around and take a look at the modest house where one of the most profound and ingenious men once lived.
>
> [C] The scenery around Montmorency is one to which I would like to take my Sophie, should an unexpected happiness enable me to travel more often.'[16]

Like the French couple, Hans Christian and Sophie will at some future time, hopefully, enjoy this place—together.

Nevertheless, it is from the unpublished letters to Manthey (that have escaped Mathilde's censorship) that we know the fate of the engagement. Still, here too we must overcome editorial interventions. Manthey's letter of 7th May begins by mentioning Sophie's name—'in one of your letters to S. you have declared that I am the most diligent of your correspondents'—which seems to indicate a thanks and an indirect lament for Sophie's silence. In such cases Mathilde usually erases Sophie's name and adjusts the context accordingly if necessary. This solution would have been the simplest form of camouflage, but Mathilde decided to eliminate the entire letter except for the last paragraph, which is printed as if it were the end of Manthey's letter of 16th July. She gives no explanation of this unwarranted antedating by seventy days!

Here is one of Manthey's later paragraphs: 'With much pleasure have I seen the portrait you have sent and found that it resembles you very much, and this is a thing so valuable to me that I am tempted to snatch it from the person whom I would think would appreciate it less than your most dedicated friend Ludvig Manthey.' The portrait in question is the physionotrace Ørsted had commissioned from the famous Chrétien and sent home to Sophie supposing, presumably, that during their twenty-two months of separation she might easily forget what her beloved looked like.

Manthey's hint had hardly been read as a prelude to a catastrophe, but rather as an irony. It was no news that letters from Sophie were few and far between. In his reply on 12th May Ørsted listed all the factors that opposed an extension of his journey to Britain, even though the Industrial Revolution in that country had been his strongest argument when petitioning the Foundation ad usus publicos for more money. One of the grounds for dropping Britain this time was consideration for his fiancée. He felt he had to return home, 'You know why', he told Manthey. A month and a half later he reiterated this decision. 'I cannot continue burdening Sophie with empty promises. She is right to demand that I pay attention to our establishment as much as I can.'[17] He had sacrificed Britain all right, but there were other reasons besides obligations toward Sophie. Dropping Britain did not precipitate his departure as he informed Sophie from Montmorency. In reality he was totally absorbed by his dream of the Napoleon prize and the papers, instruments, and experiments involved. This was never disclosed by a single word to Sophie or anybody else for that matter.

On 10th August Hans Christian told Sophie how busy he was in Paris, and as already mentioned on 4th September he revealed to Manthey what he had been so busy to prepare.[18] He was utterly silent, however, about the fact that the Napoleon prize had eluded him. On 3rd October he was overwhelmed by a letter from Anders, who communicated the naked truth. Ørsted told Manthey: 'I have had no peace... due to a letter my brother wrote me about S.P.

Fig. 45. HCØ, physionotrace by G.L. Chrétien (1754–1811). Chrétien had invented the technology behind this art of portrayal. At first the profile of the sitter was drawn with a mechanical device by which two parallel sticks with pencils followed the contours of the head and put them on paper. Subsequently, an artist named Fouquet added the details to the silhouette, and finally the portrait was engraved and printed. DTM.

You cannot but know the contents, and you will learn my answer from Anders. I have nothing to add to what I have already told him. I have been dreadfully awakened from my slumber. I should have been warned by certain hints that I have only now come to understand.'[19]

These are the only expressions of a broken heart left by our deserted protagonist. Maybe Sophie felt that *she* had been deserted. Was her fiancé really in love with her to a degree that he was prepared to sacrifice just a few of his scientific ambitions on the altar of love? Did he really believe that he could leave her for thirty months and take it for granted that she, almost twenty-nine years old, would have the patience to wait for him that long? There is nothing to indicate that his family in Copenhagen and Rudkøbing had intimated what they might have known or thought about Sophie's situation as a grass widow. And how little did Hans Christian anticipate the obvious tensions between career and marriage that anybody but he could predict?

Anders's letter and Hans Christian's two answers have not been preserved, so we can only conjecture as to Sophie's reasons for terminating their relationship.[20] 'Travel letters in the form of a diary', as Mathilde Ørsted called her father's letters to Sophie in her vain attempt to erase her from the family history, had come to an end, the recipient was no longer interested. If Sophie had issued an ultimatum to her fiancé via Anders, perhaps as a follow up to Lieutenant Probsthein's intervention the year before, the relationship might have been saved. But we do not know what happened, only that Sophie took the initiative. Mathilde, who never married, but spent most of her life being her father's secretary both before and after his death, must have thought of the breach as an embarrassment to be hushed up. And Ørsted had to face the fact that he would have to do without the love of a woman, however distant and Platonic it had been.

How did Sophie Probsthein get on with her life? While her former fiancé, after his discovery of 1820, basked in international recognition and royal generosity, she felt quite lost in her small flat opposite the pharmacy, where she had met the young Hans Christian. Her brother, C.D. Probsthein the painter, had died recently,[21] and since she was unmarried and childless she had to support herself. She could not cope, and in January 1821 she wrote a letter begging the sculptor Bertel Thorvaldsen, her brother's former friend, who was visiting Copenhagen from Rome (and abundantly well off) to offer her a helping hand, if he so wished. Sophie (who did not spell well) set out her case:

> 'I have been offered half a lottery collection enabling me to safeguard my life against paying five hundred rixdollars, but this I am unable to do for the following reasons. True, I have been frugal and saved two thousand rixdollars but though I left this sum in the hands of a man apparently upright and affluent, recent upheavals have ruined him, so for a long time I shall have nothing in return. As I'm deprived of friends I take liberty to beg his Excellency if you out of respect for the friendship you used to have with my deceased brother and parents, would show me the favour of making a contribution to my future well-being by lending me the five hundred rixdollars which I promise to pay back by instalments of one hundred a year. Your obedient servant, Sophie Propsthayn.'[22]

Even if Thorvaldsen was reputed to be responsive to begging letters from miserable human souls, there is nothing to indicate that he reacted favourably to the letter from this solitary and destitute woman, who disappeared out of history almost unnoticed.

But nothing is so sad that no good can come of it. The loss of the loved one relieved Hans Christian of a future burden that had long weighed heavily on his shoulders, that of having responsibility for a family. In his next letter to Manthey relief shines through his sadness over Sophie's decision:

> 'Despite my attempts to familiarise myself with the applications of chemistry I do not wish to ever become responsible for a post involving duties as a practitioner in my country, because to be honest I am not suited for it nor shall I ever be. From childhood I have been absorbed by theoretical and experimental chemistry and physics, and unconsciously all my efforts are aiming in this direction. Moreover, this study has never demanded as much interest as just now, when we see a much stronger rebirth of physics than ever before in the history of science. I wish to take Berthollet as my role model, because as a genuine theoretician he has enriched his country with discoveries that are useful in industry. Against this background you will readily appreciate that I do not wish to begin my career at home with something merely technical. Even if it would nourish the hope of a happy life, it would not be a life matching my ideals.'

Perhaps it was not until now that it dawned upon him that his vocation as a scientist blocked the way to matrimonial and middle-class happiness, represented by Sophie and Manthey. Ever since Manthey had become his friend and patron he had carefully safeguarded his protégé's interests as he understood them, that is without recognising the differences between their goals in life, which Ørsted may have been inclined to suppress. The Danish monarchy had not for a long time, perhaps not since the time of Tycho Brahe, offered a post for life to an independent researcher. Now that Ørsted had played his hand and laid open his low opinion of the bread-and-butter bourgeois existence, it was crucial for him not to appear ungrateful. Hence, Ørsted expressed himself carefully, adding: 'It goes without saying that I shall not proceed in this important matter without taking the advice of my knowledgeable teacher and friend whose mature experience will often guide me.'

When Sophie refused him, his career changed track.

> 'Another way to start my career at home would be by lecturing on electricity and galvanism. I feel confident that at this moment I am stronger in these subjects than any of our fellow countrymen. If I charged a substantial fee and targeted the lectures at the higher ranks this might bring me happiness.'[23]

His defeat in the National Institute had not crushed his dedication to science, and the broken engagement had removed an obstacle to making it flourish. One cannot deny him a sense of optimism.

PART III
THE RESEARCHER AND TEACHER

> You are and remain the old Hans Christian
> Brave-Fellow, you were in your third year.[1]

17 | 1804
Alone and Abandoned in Copenhagen with a Collection of Instruments

ONLY HIS closest friends and family were informed of Hans Christian's double humiliation. His associates were kind enough not to comment on his scientific defeat, and the broken engagement does not seem to have become the object of public gossip. When Ritter was told of his friend's broken heart, he sent him these comforting words: 'I'm deeply sorry that you have received so little consolation for your broken heart, because it is you, but apart from that I find it quite understandable, yes, even natural.'[2]

There are many hints at Ørsted's relations with women, or rather at the mystery surrounding them in Ritter's many long letters. Did he have any love affairs and did they mean anything? As far as Ritter was concerned he did not take such things too seriously; in fact, he boasted of his sexual conquests. Unlike his friend, Ritter had many fleeting relationships, for instance with a French amateur chemist, Mme de Gachet, said to be not only beautiful and intelligent, but also rich; she was one of many aristocratic emigrants from the French Revolution who went into exile in the German states. Ritter shared the view of love expounded by the early Romantics of the Jena circle, as a spiritual and sexual unification of man and woman. At the time he sent Ørsted his words of consolation, he had recently got his new girlfriend, eighteen-year-old Dorothea Mönchgesang, pregnant, and having arranged for her to give birth in the reassuring environment of her native village, he married her. After the wedding he moved with wife and child to Professor Voigt's summer residence at Jena, one of the most comfortable dwellings he had experienced for a long time and a place where a few years earlier he had spent so many happy hours with Ørsted.³

A few months before his marriage to Dorothea, Ritter had set out his rather frivolous views on love: polygamy and polyandry do not lead to the extinction of the human species. He imagined this to be the typical earliest form of relationship between men and women. In the course of history it so happened that the process of civilisation, and of courtship in particular, pushed us towards monogamy, a state of affairs which eventually was enshrined in law. Polygamy, however, was perpetuated by a minority, and the human race became thoroughly confused, despite the fact that love was limited legally by matrimony. However, the laws prescribing monogamy were merely some kind of camouflage, for the moment carrying the day, while the real situation was obscured. A person who had seen through the legal façade could readily ignore the rule of monogamy and practise polygamy. Thus history could end exactly as it had begun in primeval times. Ritter wrote that, in the light of this, he might just as well not offer Ørsted his sympathy, but that would be only against this background and not out of unkindness. The more one experimented with one's love life, the better the result, yet, he added tongue in cheek, such experiments were not suitable for printing in Gilbert's Annals.⁴ In other words, Ørsted did not need to complain, because Sophie Probsthein was not the only fish in the sea. You can conduct experiments with your sex life just as well as with other forces of nature, Ritter suggested. We do not know how Ørsted responded to this sermon on libertinism, but perhaps we can allow ourselves to infer that the two friends, at this stage anyway, differed as far as their views on women were concerned. There is no evidence that Ørsted indulged in short-term affairs on his grand tour.

How did Ritter react to the collapse of their joint project in Paris? Throughout the summer of 1803 he had felt reassured that the Napoleon prize was close at hand and in his state of poverty he had already started to spend the fortune. Still, he took the fiasco very calmly. Ritter did not consider the French repudiation of the experiment validated, and saw it simply as yet another example of the crude materialism of the atomists. This was the only comment from the indomitable Ritter, and he showed no bitterness.⁵ He was used to living almost permanently in adversity and he survived on his black humour. Since neither insisted on having their experiment to show the galvanic polarity of the Earth acknowledged, and neither, as far as we know, repeated it in order to make it succeed, we may conclude that they tacitly accepted the French rejection of their experiment, but not of their idea.

While their fiasco in Paris was unknown to the public, rumours of Ørsted's attempt to promote Winterl's theory in Europe by means of his book *Materialien* soon reached Copenhagen. In a tone of schadenfreude the exiled P.A. Heiberg and young historian L. Engelstoft sent letters from Paris to their colleague Rasmus Nyerup in Copenhagen, who probably did not keep the gossip to himself. From the two almost identical letters I quote Heiberg's, which stands out as a malicious example of academic wit at its worst

> 'One of our compatriots, Doctor Ørsted, who spent last year in Paris where he is said to have found the sciences in general and his own science of chemistry in particular to be in a wretched

state, an opinion with which he will find only a few to agree, this Ørsted has recently been flogged until he bleeds by an English chemist, Chenevix. Guiton Morveau has translated this article, added some notes, and published it in *Annales de la Chimie*.⁶ The fact of the matter is that Ørsted has made himself the promoter of a certain Winterl, who is practically unknown, but according to Chenevix and Guiton a total disaster, and as a consequence they categorise Ørsted in the same way. If Winterl and Ørsted have made the blunders they are said to have done, they deserve to be locked up in each other's country's asylums... The other day I met a young German, one of Ørsted's devotees who regards him as one of the greatest geniuses in the world; this German insisted that Ørsted was right and ready to make things clear as daylight to me. I laughed and left him, for one should not mingle with fools...'⁷

Engelstoft passed on the critical review in similarly malicious terms, but did not want people in Copenhagen to know he had done so. Engelstoft, too, stuck exclusively to the review by Chenevix, stating that 'Winterl's book is complete nonsense totally written in the latest mystical terms... How will Ørsted respond? The worst thing is that no profound chemical insight nor great ingenuity is needed, only common sense, to see how nonsensical it is.'⁸

The two gossipmongers' attack on Ørsted was gratuitous, superficial, and mean. In addition to his scientific and amorous defeats he now had to endure the reputation of a lightweight. And this could not escape the attention of influential characters such as the University Chancellor, the Court Steward, or the professors. They might already have noticed these condescending remarks about *Materialien*, referred to in *Kjøbenhavns Lærde Efterretninger* [Copenhagen Learned News], which had excerpted it from a German review: 'Schelling's Naturphilosophie might be expected to stretch out its tentacles like an octopus for Winterl's chemistry. An "Andronia" or a "Thelyke", principles of acidity and alkalinity neutralising or undifferentiating themselves, provide the dualities, either opposed in "conflict" or harmonised in "indifference", concepts that Schelling's doctrine often confuses...'⁹

Ørsted did not find this parody at all funny. In the next issue he dashed off a defence to make it clear that his book on Winterl was neither a contribution to nor a development of Schelling's Naturphilosophie, but serious science deserving an empirical investigation:

'Doctor Ørsted's *Materialien* has its origins in J.J. Winterl's *Prolusiones in chemiam seculi decimi noni* published about four years ago. This book is replete with new comprehensive ideas, the result of forty years of experience and investigation. The chemists condemned it immediately, because it threatened to disrupt the noblest of their hypotheses;... This prompted Mr Ørsted to write his *Materialien* briefly sketching Winterl's ideas, drawing attention to what is already supported by known experiments, and encouraging physicists and chemists to scrutinise the new system... This year Winterl has published a new paper to which the notable Ritter has added a preface. The title is 'Darstellung der vier Bestandtheile der anorganischen Natur' ['Description of the four Constituents of Inorganic Nature']. It is dedicated to Ørsted and his *Materialien* is cited with great acclaim.'¹⁰

Ørsted also sent a reply to Chenevix, but not until a year later, in November 1805. We do not know how he defended himself, but he wrote a rejoinder for Gilbert's *Annalen der Physik*, and sent it to Oehlenschläger in Halle, where he shared a house with Steffens, asking for them both to comment on it before they sent it on to the editor.[11] However his friends did not send it. At that moment Ørsted was eager to communicate with Steffens to get feedback from him on a planned textbook of physics and chemistry, and he was disappointed when Steffens ignored the feeler he had thrown out. Instead, Steffens carried out his own attack on Chenevix in *Jenaer Allgemeine Literaturzeitung* in March 1806. Steffens took a hearty sideswipe at the haughty Briton with his prestigious 'FRS' in a farcical article denying him any insight into the profound Naturphilosophie, but without trying to repudiate his criticism. He did find space to mention Ørsted positively, but only once in passing.[12] This only became known to Ørsted a month later, when Oehlenschläger made light of his and Steffens's omission, adding 'moreover, the wretched devil is said to have drowned in the Black or Caspian Sea (wherever), and let him remain there!'.[13] Ørsted took this piece of information as an inappropriate apology for Steffens's behaviour. In addition the rumour was sheer fantasy; Chenevix died a natural death in 1830. It may also be noted the Royal Society had honoured Chenevix with the Copley medal in 1803 even though, in his rivalry with Wollaston, he had erred in alleging that palladium is not an element, but an alloy of quicksilver and platinum. Ironically, the Copley medal, which ought not to have been awarded to Chenevix, came to be the very honour bestowed on Ørsted for his discovery of electromagnetism.

*

We do not know where Ørsted lodged immediately upon his return in January 1804. The first months he probably stayed with his father at Rudkøbing, with his aunt, Mrs Møller, in Vestergade, and with his brother and sister-in-law also living in Vestergade. Of course, he could not return to his rooms at the pharmacy without bumping into Sophie. A few months into the spring he acquired his own accommodation at Collin's Court, where he shared accommodation with Buntzen's scientific instruments, to which I shall return at the end of this chapter.

During his long absence Henrik Steffens had delivered his notable introductory lectures on natural philosophy at Elers' College. These crowded lectures had gripped Adam, too, so that in order to prove his literary talent in the eyes of the reserved Steffens, he had rushed home from their famous walk in Frederiksberg Have to his den at Mrs Møller's dye-works. There he had written 'Guldhornene' [The Golden Horns] in one night, a few blocks away from the workshop where the presumptuous Heidenrich had melted the irreplaceable treasure he had stolen from the Royal Cabinet of Art.

Steffens was Schelling's favourite disciple. In 1798 he had sat at his master's feet drinking in his words. When Schelling's first writings on *Naturphilosophie* were submitted for review to *Jenaer Allgemeine Literaturzeitung*, the author received permission from the editor to suggest a reviewer, and he chose Steffens. When it became clear to the editor that Schelling and Steffens were hand in glove, the plan was foiled, and Steffens's review was published instead in Schelling's *Zeitschrift für spekulative Physik*, where, incidentally, Schelling either corrected or commented on Steffens's critical remarks in footnotes.[14] The circumstances behind the choice of reviewer were painful per se, of course. The pain was aggravated by the fact that the two journals had waged a war for and against Schelling's *Naturphilosophie*. Rumours of these rows reached Copenhagen before Steffens arrived and his adversaries took advantage of the fact to compromise him. So, Steffens was burdened with accusations of being the smart disciple of a charlatan.[15]

Had Steffens now seduced Oehlenschläger and won his friendship that used to belong to the Ørsted brothers? Ørsted had heard from Manthey about the excitement Steffens's lectures had aroused, but believed that thanks to his sound grasp of experimental physics he would be able to oust Steffens from the lecture market. His self-confidence was confirmed immediately upon his return, when he was invited to show his galvanic experiments at a soirée the Schimmelmanns hosted for the Chancellor. Ørsted soon won the sympathy of Schimmelmann thanks to his talent as an experimenter, and with his help he appropriated Buntzen's collection of instruments. Thus the government had tipped the scale in Ørsted's favour. At the same time Steffens was denied a teaching post at the University, and in distress he left his country to enter upon a professorship in Halle.[16]

Life had been merry at Mrs Møller's dye-works while Hans Christian had been away. Adam still enjoyed her food and he kept state there together with two Norwegian students, Georg Jacob Bull courting Barbara Albertine Ørsted, also known as Tine (sister of Hans Christian), and Christopher Hansteen, who had continued to study law, but found it too boring and switched to science.[17] The two younger Ørsted brothers, Jacob and Niels Randulph, had moved to Copenhagen, and were benefiting from being included in this circle. They all provided an audience for Adam when he was reciting his poems and performing his plays, *Sanct Hansaften-Spil* [Midsummer Night's Play] and *Aladdin* for instance, as soon as the ink had dried. Sometimes they would perform Holberg's comedies. Not without a certain ironical identification on Adam's part, Steffens was given the part of the conceited fool of Erasmus Montanus, while he took on the part of the simple Per Degn himself.[18] Anders was completely unfit for the stage due to his absent-mindedness and boyish clumsiness. He could never remember his lines when they were due and he was unable to coordinate them with his facial expressions and gestures. In 1802, when Anders married Sophie, Adam had written an untranslatable ironic poem hinting at the sudden improvement of the groom's personal hygiene and welcoming him into his new family.[19]

This circle of students and girl-friends also met in the flat of Anders and Sophie, 'the Assechild' [Assessor: the now obsolete title for a judge], and 'the Powder Keg'. Kamma Rahbek, Bakkehuset, poked fun at them by calling them nick-names: Hans Christian was 'the Professor with the Nose', Henrik Steffens 'the Emperor', and Adam Oehlenschläger 'Adagio'. A kind of literary salon began to develop there, although not exactly like the one Hans Christian had frequented in Henriette Herz's house in Berlin. The men had a fascination with the lady of the house while her husband was busily occupied in the courtroom or at his desk or tutoring law students. 'The Emperor' was a daily visitor and characterised 'the Assechild' as follows:

> 'He is quiet, apparently weak and compliant, silent and very modest; everybody believes they can govern this gentle man, but are surprised when they detect a resolute clarity behind his weakness. While a dispute is raging, he remains reticent; if things remain unresolved, he waits some time before expressing himself calmly, even timidly, but very much to the point; and then everyone comes to realise that the dispute has been settled.'[20]

One is tempted to conclude that the self-centred, temperamental Steffens had found his complete opposite in Anders. In summer, when life indoors within the city ramparts became too sultry, for Sophie in particular, then 'the Powder Keg' would long for her childhood's

Fig. 46. Recital of Adam Oehlenschläger's drama Hakon Jarl at a soirée in Anders and Sophie Ørsted's flat in Vestergade, 1808. The persons closest and with their backs to the viewer from the left: Christiane Heger (Adam's fiancée), Reinholdine Schønheyder, Kamma Rahbek (Christiane's sister), and in the row behind facing the viewer from the left: the actor Michael Rosing, old Oehlenschläger (Adam's and Sophie's father), standing Carl Heger (Kamma's brother), sitting HCØ and ASØ, Sophie Ørsted (Adam's sister), Albertine Ørsted (the brothers' sister, engaged to the Norwegian jurist Jakob Bull) and the brewer and jurist Hans Heger (Kamma's and Christiane's father). Standing to the right A.C. Gjerlew, HCØ's room-mate in Paris. Drawing by Carl Christian Frederik Thomsen (1847–1912). FKB.

Frederiksberg Have and Søndermarken, and 'the Assechild' would rent a house for the summer in Pile Alley in what was still the village of Frederiksberg. There he would read Goethe and Fichte aloud to her under the open sky.

The six Ørsted siblings often visited their father, Søren Christian, and their stepmother, Anna Dorothea Borring, at the pharmacy at Rudkøbing. Jacob worked there as a head dispenser without having completed his pharmaceutical studies at the University. As already mentioned, Hans Christian had helped his father during the winters of 1795 and 1796. Now, in 1804 his brother Jacob had become his father's assistant. He was not a success, not economically at any rate. Søren Christian had tried to expand his business. He was an industrious pharmacist and had tried hard to put the dilapidated pharmacy he had now owned for almost thirty years into shape. He employed many people to collect medicinal plants which he processed into pharmaceuticals for export.[21]

Twenty-six-year-old Jacob turned out to be as much a disgrace to his family as the two older brothers were a pride. In 1806 Hans Christian wrote to 'Adagio':

> 'At present I have many inconvenient undertakings. Jacob's crazy business has deprived my father of several thousand rixdollars, so he has had to sell his pharmacy. To conceal his fraud he contrived to steal all my letters to my father. It was found out, he confessed, promised to improve—then did the same thing again.'[22]

His father had lent or guaranteed him a large sum of money (a much larger one than the sum Hans Christian was granted for his grand tour) and Jacob had spent it on speculations or spent it lavishly in unknown ways. Subsequently his creditors had pestered Hans Christian, who in turn had tried to tell his father, but not given consideration to the idea that his brother might be impudent enough to steal his letters. To settle Jacob's debts, Søren Christian was forced to sell his pharmacy at Rudkøbing for 8,000 rixdollars, quite a sum in comparison with the 600 he had paid for it thirty years before. A year later their father bought the pharmacy in Storgade, Sorø, for 8,500 rixdollars and at the same time opened an annex in Ringsted. Hans Christian later told Oehlenschläger that his father had not been totally ruined by Jacob's wheeling and dealing, but it was close. Another reason for moving to Zealand was probably the desire to be nearer to his children and sister in Copenhagen. Jacob stayed back in Rudkøbing, without his post as a dispenser, and covered in shame.

Fig. 47. Søren Christian Ørsted (1750–1822), pharmacist and father of the Ørsted brothers. Pastel by Peter Copmann (1794–1850), DTM.

Anna Dorothea Borring, Søren Christian's second wife, by whom he had four daughters, developed a kind of mental illness and as a result spent a period of time in mental institutions. Who would look after the girls while the pharmacist took care of his business? At this point his sister Mrs Møller came to his rescue, for while Tine was on her way to Norway to get married to Jacob Bull, she sent Tine's best friend, her foster-child Inger Birgitte Ballum (now eighteen) to the pharmacy in Sorø to help as a maid. It was high time the neglected girls were properly looked after, but it was no easy job, for when their mother was at home it was intolerable to be near her as we know from the letters Inger Birgitte (known as Gitte) exchanged with Anders's Sophie. Even when she did an excellent piece of work, it cost the young maid so many sacrifices that often she just wanted to flee Sorø and follow Tine to Norway. Sophie ardently advised her to stay. Gitte had worked miracles on the four girls, and the chance of a relapse was considerable if she gave up now. Maybe 'the Powder Keg' had already dreamt about different plans for Gitte.[23]

Unfortunately, Jacob was not the only younger brother of the two prize-winning boys who had become embroiled in dubious financial transactions. Niels Randulph did not spend much time on academic studies either, but was always engaged in doing business deals, hence his nickname of 'the Fixer'.. Now he had also run up huge debts. He had recently turned to Count Schimmelmann for a loan, and his clever idea was to make Hans Christian his guarantor. It is difficult to refuse to help one's own brother, and Niels had probably reassured him that the loan was intended to settle old debts, which would make him solvent, etc., etc. Apparently, he had not consulted his older brother beforehand, and had put him on the spot by turning to Schimmelmann first. What if Niels was telling a pack of lies? Hans Christian did not possess the five to six hundred rixdollars that would be due in case 'the Fixer' failed. Being in an awful predicament he wrote to Schimmelmann, who had previously supported him generously:

> 'I confess that the fear of abusing your Excellency's flattering confidence on the one hand or of denying what fairness and brotherly love perhaps demands on the other, makes me uneasy; yet, I do not think I am failing any of these duties when I reassure you that my brother, Niels Randulph Ørsted, needs a sum of money partly to settle an onerous debt and partly to enhance his future happiness; but due to the present recession he has been unable to obtain such a loan anywhere despite the sufficient guarantee he is capable of offering.'[24]

Just when Jacob had drained the value of his father's property and fooled his brother, Hans Christian showed brotherliness by standing up for Niels Randulph. This was not going to be the last time 'the Fixer' caused his family trouble.

*

Soon 'Adagio' as Kamma Rahbek would nick-name her brother-in-law-to-be (Adam was engaged to be married to her sister, Christiane) followed in the footsteps of Steffens. It was scarcely without an ulterior motive that he dedicated his *Poetiske Skrifter* to Crown Prince Frederik. His petition to the Foundation ad usus publicos bore fruit, so in August 1805 he set out on his grand tour for four years, or rather, he had already set off with money he had borrowed with assurances that the petition would receive a favourable response.[25] At first he lived with the Steffens family in the house in Halle that the family shared with the theologian F.D.E. Schleiermacher. Hans Christian had assumed the role of Adam's financial agent, that is transferring grants, keeping his creditors at bay, negotiating contracts with publishers, collecting fees, promoting him as a poet to the public, and not the least paving the way for his plays at the Royal Theatre.

This turned out to be no small task. Sometimes 'Adagio' complained that Ørsted's agency taxed their friendship. There were too many letters on accountancy filling one sheet after another with prosaic reports on philistine publishers and shoemakers' bills long overdue. Could not Hans Christian write a little more poetically and personally, since Adam in his solitude abroad languished for deeper outpourings?

His lament was justified in so far that Ørsted stuck to practical matters and left literary and personal issues aside. Perhaps the jealousy Adam had aroused in him by becoming close to Steffens while relegating Hans Christian to the role of a distant and practical resource could explain the agent's attitude of formality? Nevertheless, 'Adagio' really suffered from what he perceived as his friend's cool silence, while he was living with a permanent shortage of money ('It melts in my hands'), although so poetically aloof that he did not care to pull himself together to mail the necessary authorisation for his agent to cash his grants. He also did not grasp that Hans Christian, too, was short of money and needed the thirty rixdollars he had lent the broke poet at his departure, for example to pay for his firewood for the winter.[26]

Ørsted had to use his network of contacts to promote Oehlenschläger's career as a poet and playwright at the same time as he was using it to promote his own at the University. For these endeavours money was an important tool. Both aims were subsumed under a plan to publish a joint New Year's Gift with the title 'Prometheus'. It would be an attractive book of poetry, aesthetic essays, and philosophy of nature aimed at the educated public. Hans Christian would write a preface to whet the appetite, followed by a few 'lollypops' in the form of poems by Goethe translated by 'Adagio', who would also write an essay on aesthetics which would be followed by a dialogue on mysticism by Ørsted. Sadly, he could not sell the idea to any publisher, and in his distress he offered an essay on the phrenologist Gall, the hottest conversation piece in town, but to no avail. 'The unhappy Prometheus was chained to the recondite cliff by the iron shackles of prose.'[27] Hans Christian was as disillusioned as his co-author. When it comes to the debut of young writers on the book market, publishers are beasts, he opined. They print, bind, and sell too dearly. Despite fine reviews, collections of poems do not sell. Maybe ten read and praise them when one buys them. 'Notice that about three hundred copies must be sold to indemnify the bookseller, and three hundred people who have the energy to spend three and a half rixdollars or more on poetry are not so easy to find.'[28] Both of them were quite up to date in self-pity.

In May 1806 Oehlenschläger sent home his play Hakon Jarl, and it was now up to Ørsted to pave the way for its performance at the Royal Theatre. To this end he paid a visit to Count Schimmelmann exploiting his protection. He handed over two handwritten copies of Hakon Jarl to Countess Charlotte tutoring her in Nordic mythology, which he reckoned was rather unfamiliar ground to her; for without a certain knowledge of this, the tragedy of the fall of the ancient mythology and the rise of Christianity was incomprehensible. Unfortunately, the

Countess did not grasp without prompting that the point of presenting her with two copies was that she could keep one and pass the other one to the Directors of the Royal Theatre with her recommendation. So what was Ørsted to do? To pressure the Countess might backfire. Luckily, Oehlenschläger himself had sent it to the Directors, one of whom, Knud Lyne Rahbek, could advance the matter with a remark that Goethe himself had accepted the German version of the play to be staged at Weimar.[29]

When Ørsted had been appointed professor in September 1806, it dawned on Oehlenschläger that the title of professor might open new doors to him abroad, and his agent immediately approached Schimmelmann who arranged things at the Academy of Arts. He also served his friend by obtaining from the Count an extension of his travel grant. So, 'the Professor with the Nose' proved to be a loyal agent and a valuable driving force behind 'Adagio's' artistic career.[30]

*

How was Ørsted to plan his own career? He complained to Ritter, who was on his way to Munich and to easier times. When Ørsted had talked to Franz Baader in Munich a few years earlier, he had put in a word for his German friend. He had managed to convince Baader that the way the University of Jena treated this original researcher was unworthy, and the Bavarian Academy of Science ought to offer him a well-paid post.[31] Ørsted's intervention succeeded. Now he felt uneasy about his own chances in Copenhagen and aired the possibility of an appointment in Munich, if Ritter could put in a word for him in return. Ritter responded that he could count on him, but the outcome might be a distant one and moreover the idea savoured of escapism.[32]

Ørsted decided to become a freelancer in the lecture market as Steffens had been. Now this potential rival was abroad,[33] so the only thing he needed to make it possible was a collection of scientific instruments. There were collections of scientific instruments in Copenhagen. Hauch's was by far the largest, but Becker and Manthey, the pharmacists, also had well-equipped laboratories. Manthey had access to 'a battery of two hundred pairs of discs', that is six hundred discs of copper, zinc and cardboard.[34] But these were private and not for Ørsted to use. The University did not have an experimental laboratory. Over the years Kratzenstein, the previous professor of experimental physics, had collected a considerable number of pieces of electrical apparatus for experiments and therapy. At the great fire of 1795 most of the instruments had been destroyed, and the professor died a few days later. According to his will he bequeathed not only the instruments, but also his fortune of 12,000 rixdollars 'to the advancement of the study of physics, as the cultivation of experimental physics has not received particular attention, and I must thank the noble participants of my experimental lectures for my self-made fortune'. Kratzenstein's donation later funded Ørsted's salary as a professor from 1806 onwards, 'to which was added two hundred rixdollars in consideration of the fact that this teaching post is loaded with more strain, labour, and costs than any other, and because the cultivation of this science has so far not encumbered His Majesty with any expenditure as have botany, astronomy and mineralogy, although in regard to public utility it is no less important than any of those.'[35]

Professor Aasheim, Kratzenstein's successor took over the remains of the collection so badly damaged by the fire. He supplemented it with instruments which the astronomer and industrial spy Ole Varberg bought in England, supposedly out of Aasheim's own pocket. His collection contained pieces of electrical apparatus, a hydrostatic weight, an air pump, mechanical instruments, a centrifuge, a gasometer, and a pneumatic-chemical gadget. So the University had a collection of scientific instruments, but it was Professor Bugge, his archrival, who controlled access to it, and hence it was not available to Ørsted.

In this hopeless situation Ørsted submitted his proposal for an institute of experimental chemistry to the Chancellor via his patron.[36] Manthey was very happy with the plan and pulled the strings of his academic network, but (as has already been said) in vain.[37] The Chancellor was unable to conjure up a grant and so was Schimmelmann. On the other hand they had reassured Manthey that a collection of scientific instruments was to be provided for the University and that Ørsted had been appointed to be responsible for lecturing on experimental physics and chemistry.[38] In March 1804 the Foundation ad usus publicos presented him with the much-desired collection, so the Chancellor did keep his promise. This collection had been established privately by Thomas Buntzen, a doctor of medicine. It was located in his house at Nyhavn and included a glass-grinding machine, a turning lathe, optical instruments, an electrostatic generator and a very large voltaic pile with 1050 pairs of discs.[39] Buntzen had received

Fig. 48. Collin's Court, 156, Norgesgade (today's Bredgade), from 1804–06 the site of HCØ's collection of instruments—transferred to him from the physician and physicist Thomas Buntzen (1776–1807) by means of the Foundation ad usus publicos. Oil painting by Peter Frederik Nordahl Grove (1822–1885). CMC.

loans from the Foundation ad usus publicos against the security of the collection, which was considered to be a sight no foreign scientist should omit to see while in town. The generator was the most powerful yet seen in Copenhagen, and Buntzen, who had the wildest ideas, could sometimes scare passers-by by transmitting sparks from it to the bell wire outside his entrance. The voltaic pile was dramatic, too. A frog, usually considered a resilient animal, was killed instantly when connected to the conductors, and they left conspicuous burns, while the rest of its body turned bluish.[40]

Perhaps Thomas Buntzen had had hopes of forging an academic career at the University, but instead he decided to try his luck in Russia where he died of typhoid a few years later. By entering into the service of a foreign power he forfeited his right to dispose of the collection, and the Foundation bought it for 1,200 rixdollars.[41] Adding his waived debts the value of the collection must be estimated at 2,000 rixdollars. This unforeseen turn of events was Ørsted's good fortune because it enabled the Chancellor and the Director of the Foundation to fulfil the promise they had given Manthey to support his protégée. Two years earlier Ørsted had asked for a collection of instruments at a value of two thousand rixdollars. Now, he was presented with Buntzen's collection (which was coincidentally worth exactly that sum) on condition that he insured it against fire. Moreover, he was obliged to let Henrik Steffens borrow instruments for his geognostic investigations and his brother Jacob Steffens for his instruction of the artillery. The royal resolution did not indicate that Ørsted received the collection in his capacity of a junior teacher at the University, so it must be considered a personal gesture from the Foundation to Ørsted. The instruments were to be harboured in Collin's Court in Bredgade, a few blocks from Nyhavn, where his good friend, Jonas Collin, Fichte's translator and the previous editor of *Philosophisk Repertorium*, lived. To help him pay the rent for the collection and his experiments he received three hundred rixdollars a year.[42] Now he finally had all he needed to deliver lectures on experimental physics and chemistry to the public.

> Everything 'twixt love and hate,
> To the last link must alternate...¹

18 | 1805
Rivalry and Love

TEACHING AT the University was spread over two semesters, summer from June to October and winter from December to April. November and May were reserved for examinations. Courses were announced in the course prospectus, and here we find the lectures delivered by the two rivals, Professor Bugge and Junior Teacher Ørsted.

The 64-year-old widower Thomas Bugge was Secretary of the Royal Society of Sciences and Letters representing the French scientific paradigm, the most serious obstacle to Ørsted's career and the dynamical theory he was fighting for. Bugge offered twelve lectures a week in the lecture hall of his professorial court on the corner of Vor Frue Plads and Fiolstræde (the site of today's University Library). He lived with his two sons, a daughter, a housekeeper, three maids and a servant. Part of the servant's job was to collect fees for the private lectures at the entrance to the auditorium. Four of Bugge's twelve lectures were public and free, while the remaining eight were private and for payment. Bugge's professorship was in astronomy at the Faculty of Philosophy and an important part of his post was examining students for *philosophicum*. As already mentioned, Bugge had taken over the late Aasheim's lectures on experimental physics as well as his collection of scientific instruments. From 1804 to 1806 Bugge delivered public lectures on the physics of optical instruments, on the development of electricity, magnetism and the compass, mechanics, and plane and spherical geometry. Apart from electricity the subjects accorded with his interest in astronomy and cartography. His lectures on experimental physics were private, lasted an hour, and took place every day except Sundays. They were not part of the examination requirements for *philosophicum*. In 1806 when Ørsted was appointed professor at the Faculty of Philosophy, Bugge immediately ended his lectures on experimental physics.

The 27-year-old Ørsted delivered his private lectures on experimental physics in Collin's Court which housed his collection of instruments. They took place on Thursdays, Fridays, and Saturdays, lasted two hours from four to six o'clock in order to leave plenty of time for experiments. His public lectures on pharmacology, which were compulsory for all students of medicine, took place on Wednesday afternoons. For the time being he was only a Junior Teacher at the Faculty of Medicine. His private lectures were usually on experimental physics, but sometimes he changed the syllabus to the chemistry of organic bodies, or to mechanical and dynamical physics, or chemistry, or 'the effects of internal changes in bodies that cannot be detected by weight'.[2] The variations probably reflected his work on a textbook he had started writing.

The distinction between public lectures for students aiming at an exam and hence studying a fixed syllabus on which they would be examined, and private lectures or lectures *privatissime* (for a narrow circle) was important. The first category was free of charge, the second was for payment and often aimed at people outside the University wanting to improve their competence as civil servants or industrialists or to find entertainment or general enlightenment.

In September 1804 Ørsted invited the educated public to his lectures *privatissime* on the new philosophical school and experimental physics. The new philosophy of Kant, Fichte, and Schelling set the epistemological framework for science, and the objective of physicists' experiments was to subject the philosophers' ideas to scientific control. '*Was der Geist verspricht, leistet die Natur*' ['What the spirit promises is fulfilled by nature'], a misquotation from Schiller, was cited by Ørsted in his invitation.[3] This (mis-)quote—ambiguous as it is—prompted different interpretations. Williams states that it represents the basic position of *Naturphilosophie*. If so, the point is that natural philosophy originates in the human mind and must necessarily be retrieved in nature irrespective of empirical corroboration. Empiricism is haphazard and hence deceptive. But Schiller was hardly a *Naturphilosoph* in the Schellingian/Steffensian sense. What he intended was rather a poetic way of expressing Kant's position, viz. that laws of nature are derived a priori, as in the following examples: Newton's law of gravity was worked out mathematically at his desk at Trinity College, not by empirical observation; Columbus's discovery of the new world was based on trust in a priori knowledge about the spherical shape of the Earth; the invention of the air balloon (Ørsted's favourite example, cf. ch. 48) was based on the idea of smoke ascending. No idea arising in the human brain is able to be substantiated by nature. Experiments have to be carried out, Ørsted made clear, to enable physicists to show 'empirically what philosophers put forward as phantoms of unregulated intuition'.[4] So, what Schiller tried to say, and what Ørsted wanted to stress, may be as simple as this: laws of nature are born of of human reason a priori. There is a world of difference between the two interpretations.

Nevertheless, at this stage Ørsted was advertising his lectures to the whole range of potential participants whether they subscribed to Bugge's Laplacean science or to Steffens's *Naturphilosophie*. At least one of his lectures, as we shall see, explained the forces of nature according to the speculative ideas and the analogous method epitomized by *Naturphilosophie*. Ørsted believed himself to be the only scientist in town capable of leading the public to the very frontier of scientific research by giving them insight into Volta's galvanism, Ritter's electrochemistry, Berthollet's analyses of affinity, and Winterl's theories of 'Andronia'. 'So, experiment has walked hand in hand with philosophical speculation, and promises to resolve everything into a wonderful pattern.'[5] People interested in taking part were encouraged to give notice at Doctor Ørsted's address. It is not known what the fees were, but usually people of high rank were charged twenty rixdollars, whereas people of limited means had free entry.

Carsten Hauch, the poet, attended Ørsted's lectures around 1810 and later described the particular style Ørsted used to enthral his audience: Usually he would begin quietly with simple considerations and common knowledge, possibly with definitions of concepts as prerequisites for what followed. In other words he started unexcitingly. Then he would talk about related phenomena of nature one by one, and gradually he would continue to relate them to each other and work them into a synthesis. One of his basic ideas was that synthetic insights would open the eyes of the audience to science. These phenomena were as far as possible presented concretely to the senses, not just abstractly as concepts. 'So his talking became more powerful like a flowing river growing by absorbing tributary streams, and in the end it would have the effect of an immensely strong current, so that the young at least, who were not yet wedded to preconceived ideas but susceptible to the new and extraordinary, could hardly resist him.' Hauch does not explicitly point this out, but his description becomes more cogent when it is compared with the conventional lectures at the time, which literally consisted in reading aloud from notes, often with soporific effect. Ørsted had learnt from Charles and Cuvier that a good lecturer keeps his manuscript in his pocket and seeks eye contact with his audience. Otherwise he cannot seduce them.[6]

We do not know how many listeners Bugge had, but Ørsted's private lectures were a success. Sixty people joined up for the winter semester 1805-6, and in March there were still forty left. He reported to Oehlenschläger in Paris: 'I wish to God they were better payers... I introduced these lectures by giving an overview for three hours of the differences between the states of the older and the newer sciences. These three hours were public [i.e. free]. My lecture hall could not accommodate all who wanted to get in, and I was much acclaimed.'[7] It was not twenty-year-old students who attended his private lectures, for 'nearly all my listeners are either middle-aged or older than me and hold official posts'.[8] He was disinclined to examine these civil servants, because he was afraid of insulting those who gave wrong answers.

Furthermore, he took pride in telling Oehlenschläger and Ritter: 'My lectures on chemistry are much frequented this year, and I could not accommodate them all. These lectures are also attended by five or six ladies. You will readily understand that I do not make any changes for their sake.'[9] Oehlenschläger did not respond, but Ritter was pleased with the sexual equality. Ørsted had told him the names of four ladies: Augusta, Sophie, Charlotte, and Christiane. Ritter thought that physics for women ought to be different from physics for men. He suggested that the class be divided according to sex and taught separately, with everyone coming together at the end. What he had in mind was probably the thought of sexual polar forces living separately, until they were united in love. Polarity was a crucial concept in Ritter's electrochemistry, and by taking advantage of the polarity of the sexes the masculine dominance of physics might be eased at a time when higher education was the privilege of men only. Ritter imagined that lectures should be staged as a play or an opera culminating in a wedding, where the ritual of matrimony was performed after the lecture. 'If so, they would be enabled to understand nature, and that kind of physics would be necessary for them at the moment they were to re-enter nature to be absorbed in it.'

Ritter had recently been married to Dorothea Mönchgesang and wrote time and again to Ørsted: 'Get married! Not to Augusta and not to Sophie, but take either Charlotte with the upturned eyes or Christiane with the modest, female reason.' To judge from Ørsted's letter Ritter had recommended Christiane whom he found the most charming, but Charlotte turned out to be the lady in the audience that Ørsted fell for. Were they swapping roles? Now Ritter had been faithful to his Dorothea and one child for a year, while Ørsted was already casting his eye over his female listeners.

There is no way of identifying Charlotte. Ritter can only have known her through the letters of his friend, and they must have indulged in superlatives in a way most unusual for Ørsted. The only thing we know about Charlotte is that she was twenty-two, that is seven years the junior of her suitor, which troubled him, but Ritter calmed him down and belittled the age difference. 'You will never get younger', he said, and love changes age to something different.

Fig. 49. A chemical experiment made in front of an audience. The caricatured chemist, who does not look like HCØ at all, is blowing soap bubbles to the mixed entertainment of the audience. Caricature drawing, 1805–06, by Christoffer Wilhelm Eckersberg (1783–1853). SMK.

The decisive point is 'that your Charlotte is the most amiable that could exist. And my spirit says to you in advance that she must no doubt become yours.'[10] Nature will soon unite you, not separate you. I think, Ritter wrote, that 'nature has determined her to become your sweetheart'. He would not intervene. If Ørsted would just behave properly, be honest and decent, God would manage the rest:

> 'Please give her my regards at your convenience. I mean it. To see you happy, happy as in friendship, yes, even more, that is a wish that is only extended to one close to my heart. For the concept of friendship exceeds the concept of love as far as I am concerned, because it is safe and devoid of any egoism. And what my wife is among women you are to me among men. For I cherish men while I mock women, and properly speaking I should admit to you that among all dear men, you are my dearest. This, however, must be hidden from you in order that I can always grant you manly dignity. Let me embrace you as a brother and may you live well.'[11]

For the next ten months Ritter continued to send his regards to Charlotte.[12] He steadily believed in their marriage, and when he had chosen Ørsted to be godfather to his daughter Olivia, godfather Hans Christian complained that Charlotte had not been chosen as the godmother. But Ritter defended himself: 'If you had admitted that you are really going to be married, I should certainly have asked your bride to become godmother, so put that on your own account!... You see I use all means to extract a confession from you. As long as you are mute I consider the possibility well founded. Or am I mistaken?'. Three months later Ritter went on: 'Dorothea and I think a lot about you and Charlotte. You need her. To get a good wife you must venture everything.' Unfortunately Ritter fell sick, had diarrhoea and fungal growth in his throat, lost weight, started drinking again, and in the end he succumbed to consumption. Ritter was on the verge of suffering the consequences of his debauchery.[13] Though Charlotte was no mirage, but a real lady who took pleasure in Ørsted's lectures and was adored by him, her identity is as yet undiscovered.

*

Ørsted's lecture on 'The Correspondence between the Electrical Figures and the Organic Forms' was delivered to the Scandinavian Literary Society and is undoubtedly his most speculative.[14] It was not reprinted in his literary testament, *The Soul in Nature*, so he hardly considered it worthy of posterity. It was his first appearance in his home town after his grand tour, and he placed himself in the wake of Steffens's lectures and his *Beiträge zur innern Naturgeschichte der Erde* ['Contributions to the Internal Natural History of the Earth']. In addition, there was a marked influence deriving from his conversations with Ritter on Herder's ideas of the hidden traces of God in his work of creation, in so far that he saw Lichtenberg's electrical figures as symbols interpreted in terms of bold deductions by analogy.[15] Finally, his lecture was spiced with gentle drops of Schelling's *Naturphilosophie*.[16] Bugge was a member, but absent, while Steffens, who was not, had been invited by Ørsted to join as a guest.[17] The lecture must have had a provocative effect, and Ørsted seems to have ignored Manthey's warnings as to which scientific waves it would be opportune to ride on. Above all Ørsted wanted a friendly relationship with Steffens.

The point of the lecture was to take a comprehensive view of nature as a coherent entity and to show how beneficial this point of departure was to the study of physics. Ørsted's visions were diametrically opposed to the prevailing French-oriented mathematical mechanics focusing on single phenomena. By contrast, he sought to adapt the latest research within electrochemistry (Ritter), geology (Steffens), and physiology (Treviranus) to each other in order to establish a higher synthesis. To Ørsted the dynamical forces of electrochemistry, magnetism, heat, and light kept the life processes in nature going. This appeared from the correspondence between the figures marked by the polar electrical discharges (Lichtenberg's ramifying and concentric figures) and their inherent chemical forces, reduction and combustion (release and combination of oxygen respectively) and the two basic forms of nature: the animal (animals have ramifying forms (limbs) and take in oxygen) and the vegetable (plants have concentric forms (stems) and release oxygen). The internal forces of nature are analogous to its external forms according to Ørsted. The universal electrochemical process sustaining the life of the two complementary kingdoms of nature (the animal and the vegetable) is kept going by the heat and light of the sun spreading from east to west by the rotation of the Earth on its own axis, and from north to south by the galvanic poles of the Earth (as Ørsted had demonstrated on behalf of Ritter in Paris). How does this happen? He explains that a galvanic process arises between heterogeneous types of matter below the Earth's surface, in this case between the layers of nitrogen in the southern hemisphere (coral reefs, pristine animal remnants) and the layers of carbon of the northern hemisphere (pristine forests and vegetable remnants). The Earth, in other words, is conceived on the one hand as one huge galvanic battery, around the poles of which oxygen and hydrogen bubble up to benefit animal and vegetable life, and on the other as a huge magnet. In Ørsted's own words: 'Oxygen and hydrogen, carbon and nitrogen are once again shown to be the four chemical elements, the former two corresponding to the polarity of electricity, the latter two to the polarity of magnetism, as first proven by Steffens, our great natural philosopher.'[18]

The entire lecture was a tour de force in examining the polarities of nature. 'Steffens's wonderful idea of regarding oxygen and hydrogen as representatives of east and west, and carbon and nitrogen as representatives of north and south is perfectly proven, even if it might appear paradoxical to the many who are not initiated into the latest physics.'[19] The last statement, in particular, was true. None of those present (except Steffens) can have grasped this paradox. It will hardly ease the understanding of modern readers to go deeper into the details of his way of reasoning, and even Ørsted admitted that certain points were still in need of experiments to prove his synthesis. Hence it is doubly paradoxical that the members present unanimously recommended the lecture to be included in the transactions of the society.

Ørsted's construction of this synthesis used the method of analogy, whereby a problem is solved or at least elucidated by showing that two natural phenomena correspond to one another or are analogous. This was not an unusual method in the history of science. For instance it was common to explain the physiology of plants (little known in Ørsted's day) by drawing an analogy with the physiology of animals (somewhat better known after Harvey's discovery of the circulation of the blood, around 1650). According to the method of analogy that goes back to Aristotle, plants consisted of organs corresponding to those of animals. As mammals have blood, intestines, lungs, and genitals, by analogy plants have sap, roots, leaves,

Fig. 50. HCØ's electrical figures, whose discharges are similar to Lichtenberg's. *SES* 3, Fig.1.

and flowers. Another example of the method of analogy could be Newton's conception of atoms, the minutest parts of matter, impossible to observe, which according to this method had to be analogous to the observable planetary system of the universe, for it would be utterly strange if God had applied different techniques in creating the world, because if so one technique had to be better than another, and then the world would not be perfect. Newton had explicitly suggested three methodological rules, and the third was exactly the method of analogy. So why should not Ørsted use it? Because he had been warned by Kant, his hero, whose epistemology he seems to have betrayed at this stage. In his criticism of Herder's *Älteste Urkunde*, Kant had spoken ironically about the method of analogy which was widely used by the early Romanticists, when they waved the magic wand of analogy over the phenomena of nature: 'What they do not understand, they seek to understand by means of something they do not understand at all'.[20] To build the framework of a comprehensive physics by means of the method of analogy was simply too optimistic and an unwarranted submission to Schelling's and Steffens's *Naturphilosophie*. Ørsted's attempt was never repeated and did not leave its mark on his textbook either.

*

Ørsted was probably aware that he had moved too far towards Steffens's wild speculations and in the following years he trimmed his sails. To propagate his own scheme he first had to explain the basic principles of dynamical chemistry, and then to elucidate the historical background of the controversy. This meant he also had to convince his listeners that science was not forced into a choice between prevailing extremes such as Laplacean physics and *Naturphilosophie*, but that scientists past and present shared a common focal point.

According to Ørsted the history of science revealed a series of instances where the sciences had leapt from one mistaken idea to another. Theories had been tentative and short-lived. Now, Ørsted asked, did not this view of the history of science lead to scepticism vis-à-vis human reason and ultimately to the rejection of science? It is not so, he answered, but instead he optimistically declared that scientific knowledge accumulates. There is no reason to despair, Ørsted maintained, for mistakes give rise to new questions which provide better results.

The history of chemistry had developed through the chaos of opposing theories, Ørsted continued. Previously, chemistry had been defined as the knowledge of combination or decomposition of substances without any agreement as to the causes of these changes. Since the discipline of chemistry was not regulated by governing principles it was a-theoretical, and Kant had therefore ousted it from the realm of science altogether. Nobody had suggested that electricity was included in this. Electricity was considered an independent substance (possibly two substances: a positive and a negative one), or rather a liquid without weight flowing from matter with a surplus of the electrical liquid to another matter with a deficit of the electrical liquid.

However, Ritter's discovery of electrochemistry gave corpuscular theory the deathblow. He conceived electricity not as matter but as a force, and he had found the electrochemical properties of many substances, metals in particular. Chemical combinations were no longer due to mechanical motion, but were a question of changes of properties generated by forces. Hence, the new challenge to scientists was to understand the forces of electricity, magnetism, and heat (and possibly light as well) which might very well be reducible to one fundamental force. These forces had one characteristic in common: polarity. To study the causes and effects of these forces was the task of the dynamists, who were about to replace the materialists.

Dynamism meant a radical transformation of science. Whereas previously chemistry had in misleading ways been explained by the mechanical laws of physics, natural philosophy would from now on have to be divided into a mechanical part, the theory of motion comprising physical motions due to external impact, and a dynamical part, the theory of forces dealing with the chemical phenomena that corpuscular theory had failed to explain, because it did not recognise the existence of internal, immaterial forces like electricity, magnetism, heat, and light, its proponents having wrongly assumed that these phenomena were material.

The difference between the corpuscular and the dynamical definitions of chemistry was clear. The old, mechanical view saw itself as founded on empirical observation, but discovered no general principles and no coherence. The assumption of a particular weightless fluid was not empirically founded at all and had led science astray. The new, dynamical chemistry, by contrast, was a priori, inspired by Kant, and it built on the theory of one fundamental force as the unknowable (non-empirical) cause or noumenon of its observable (empirical) effects or phenomena. Ritter's (and possibly Winterl's) successful experiments had transformed Kant's pessimism. Ørsted explained:

> 'A speculative approach to science is the opposite of that of experience, in so far as the latter supplies various objects that spur reflection and thereby are arranged in coherent chains; the former by contrast seeks first principles of the whole, sees possible constructions, preferring the basic construction of science as its definition.[21] To conduct experiments, therefore, is the physicist's true art, and if he has opened his eyes to the reconstruction of nature he takes this as his point of departure to enable him to embrace, or at least faintly see the coherence of all of nature.'[22]

Science is not directly concerned with nature, but with the way we understand nature. Nature does not change, but scientists' understanding of nature does. The history of science develops not only despite but also because of errors that turn out to have useful elements. This has happened in a progressive way as indicated by the examples. This development, according to Ørsted, did not happen at random, but was due to certain laws. He rejected contingency because all scientists throughout history had been aware of contemporary, unresolved key issues and have sought to resolve them. Consequently, scientific discoveries must be seen as answers to hitherto unsolved problems.[23] At the same time these answers invite further reconsideration, and during this process the problems that arise become more wide-ranging as a consequence of the increasing sophistication of experimental equipment and method.

Ørsted probably had his dynamical project in mind when he told his audience that revolutionary ideas of geniuses must pass through a period of maturing until they are generally accepted.[24] The history of science deals with man's hunt for true knowledge, the essence of nature and its perpetual laws. But in real empirical nature science only finds shadows of its perpetual essence, as Plato told us in his notable parable of the cave. Science does not move in a straight line from a point of departure of ignorance to its destination of complete knowledge, Ørsted remarked, but oscillates between contrasting errors. Extremes provoke opposite extremes, and new ideas only find their place after calm reflection. The human mind develops like nature in interplay between polar forces.

He saw the history of science as dialectical progress. To express this thought he chose a quotation from the peepshow-scene of Oehlenschläger's *Sanct Hansaften-Spil* ['Midsummer Night's Play'].[25]

> 'Everything 'twixt love and hate,
> To the last link must alternate ...'[26]

And since the history of science is about the development of human reason, he suggests that reason develops according to the same principles as nature, of which it is a part.

He referred the laws of nature and the scientific understanding of them to a common root: divine or universal reason. Reason pervades nature as well as ourselves, for if we did not have reason in common we should be precluded from understanding the laws of nature. In Ørsted's words:

> 'This correspondence between nature and mind is hardly coincidental. The more we progress the more perfect you will find this correspondence and the more easily you will assume with me that both natures spring from a common root... We have cast a glance at higher physics, in which the development of science itself with all its apparent contradictions belongs to the laws of nature.'

Human reason is at the same time conditioned by nature and the condition of bringing forward new insight into nature.

Later on he made it clear that the papers he gave between his first and second grand tours served the purpose of experimenting with ideas. Europe was in a state of fermentation between Enlightenment and Romanticism, and his mind was replete with the many contradictory statements and philosophical hobby horses he met on his way:

> 'I had the opportunity of getting to know many of the most excellent and varied characters, which gave me the chance of testing and correcting my own scientific views. Upon my return in 1804, I immediately started giving papers to expound the ideas I had met. For eight years I tested my views by expressing them in the most divergent ways and by absorbing objections and comments from my listeners...'[27]

So, Ørsted was very conscious of using the responses of his audience to clarify his own thoughts, not least because he had begun writing textbooks intended to disseminate his new division of physics and chemistry to a larger public.

> You are both a teacher and a researcher,
> while I am a researcher only.[1]

19 | 1804–9

Textbook Writer and Professor

SOON AFTER his return in 1804 Ørsted fostered a plan to write a textbook, but not one to be read aloud by lecturers. Printing was for a different purpose. He intended to make his textbook a piece of literature to be perused thoroughly and independently not only by the academic public, but also by amateurs. To understand Ørsted's career one must pay attention to his textbooks. A brief note in *Kjøbenhavns Lærde Efterretninger* [Copenhagen Learned News] tells us that when Ørsted lectured on 'The Correspondence between Electrical Figures and Organic Forms' he announced his plan of writing a textbook. 'The lecture is meant to be a preparation for and a part of a comprehensive work on natural science as it has developed lately thanks to an abundance of chemical, electrical, galvanic, etc. experiments.'[2] A year later the same journal mentioned his lecture 'What is Chemistry?' (misquoting the title as 'What is Theory?') and concluded:

> 'According to the author, chemistry needs to expand considerably. Everything concerning electricity, magnetism, and galvanism must be included in order that the comprehensive science of the laws of nature is divided into two main parts, viz. the investigation of laws of nature due to external changes (knowledge of motion) and those due to internal changes (knowledge of forces). Chemistry as we have known it until today will only be a part of the latter.'[3]

Ørsted's ideas for a textbook soon confronted him with almost insoluble problems. On the one hand it was meant to introduce the dynamical system and in consequence establish a new school. The textbook should explain the dynamical system and 'show the solidity of its base, the strength of the cohesion of its parts, the richness it contains, the wisdom it comprises, and the beauty it entails'.[4] On the other hand the textbook must be based on knowledge on which the scientific community is in agreement, and avoid riding the author's private hobby horses. As if this seemingly insuperable dichotomy was not enough, he soon faced yet another difficulty. Looking at the European supply of textbooks in physics, a frightening disagreement concerning the very definitions of physics and chemistry emerged. Hence it seemed impossible to fulfil all of Ørsted's objectives in one textbook.

At first his work proceeded very slowly. He had many other tasks to see to such as installing himself and his collection of instruments in Bredgade, delivering public and private lectures, giving papers to the academic elite in the Scandinavian Literary Society and the Royal Medical Society, writing articles for 'Prometheus', experimenting with Chladni's sonorous figures, organising Oehlenschläger's money affairs, etc. Just as his brother had done earlier, Hans Christian overdid things and fell sick; at times he had to stay in bed.[5]

Might Ritter help? Ørsted suggested to him that he might cooperate on the textbook project, while Ritter wanted Ørsted to join his *Taschenbuch* ['Pocketbook'] project. Ritter repeated his characterisation of his friend as both teacher and researcher, while he saw himself as a researcher only. This perception was derived from Ørsted's efforts to communicate his research in Paris. After the translation into French, a work entailing a radical revision of his papers, he had understood much better what he had wanted to say. Lacking the ability to communicate, he preferred that Ørsted wrote his textbook alone, hoping that it would be translated and published in German, as did Ørsted. However Ritter promised to help with comments and ideas, 'like the wren flying on the back of the eagle', he added playfully.[6]

If Ritter were to write a textbook, he would prefer to express himself in aphorisms rather than try to establish a complete structure. Ørsted had sent him his outline which no doubt (Ørsted's letters to Ritter have gone missing) reflected the structure of the dynamical project. But to Oehlenschläger, then on his way to Goethe in Weimar, he wrote:

> 'My textbook is beginning to gain speed. Its chemical part has acquired an entirely new structure. First I deal with all phenomena deriving from combinations and decompositions (mechanical affinity), secondly I show that in reality they depend on balances of forces (dynamical affinity), and finally I try to detect the forces contributing to all these phenomena. I have taken astronomy as my pattern, and in astronomy I deal first with the motion of celestial bodies as they seem to be in the spherical part, then as they really are in the theoretical part, and finally I look for the causes in physics. As soon as it is published I shall send you a copy if you want me to.'[7]

Obviously, Ritter's and Ørsted's two outlines did not match at all. Ritter abhorred a complete structure, meaning the building of the entire science.[8] He preferred aphorisms instead and in addition a catalogue of ideas for research projects. Ritter was keeping a diary in which he was writing down his aphorisms following the principles of Lichtenberg and Novalis. He later published it under the title of *Taschenbuch für Freunde der Natur* ['Pocketbook for Friends of Nature']. According to Ritter, the objectives of science had to be described in philosophical terms as a cosmology with existential (ethical) and cultural (aesthetic) implications. To this end Ritter recommended his friend to study Herder's *Älteste Urkunde des Menschengeschlechts* ['Earliest Document of the Human Species'], published in 1774–6, offering to get him a second-hand copy. But he also agreed to give Kant a central role, complying with what he believed to be Ørsted's intention. Ritter liked the division into a mechanical and a dynamical part, but was convinced that the order ought to be reversed so that the causes (the internal forces) came before the effects (the external motions). In addition he sent Ørsted ten sheets on the topics that ought to be included in a textbook according to his views. Among other things he suggested a dynamic explanation of the legend of Prometheus, based on Aeschylus, about the gods handing over fire, the most important physical force, to mankind.[9]

Ørsted did not comply with Ritter's advice, nor did he frame his book with Greek or Jewish accounts of the creation. He did not include any aphorisms either. He maintained his own structure, devoting only a few pages in his introduction to epistemology and philosophy of life. Instead there was room for reflections taken from the history of science.

Did Ørsted's tightrope walking succeed in keeping a balance between being open towards French corpuscular theory and German *Naturphilosophie* on the one hand and on the other holding aloof from both? The first volume of his textbook, the mechanical part or the thesis on motions susceptible to weight and measurement, was a relatively conventional description of mechanical physics. The second volume, the dynamical part or the thesis on forces untraceable by weight, was to be based on electrochemical experiments and dynamical hypotheses. The textbook was clearly dualistic, reflecting the Kantian distinction between the phenomenal and the noumenal. His conviction that 'the world is the revelation of the combined creative force and reason of the godhead' has an air of Plato about it. He describes laws of nature as the 'thoughts' of nature, or as impressions of reason, because otherwise we would not be able to detect them. As far as scientific method is concerned, Ørsted argued that empiricism and rationalism ought to walk hand in hand:

> 'Each of these scientific directions, then, needs the other. To the empiricist the idea of the Whole is to be regarded as the bright sun, which shines into the pathless chaos of experiences; and to the speculative philosopher experiences are to be regarded as guiding stars, without which he could easily lose himself in the infinite depth of reason. The further they proceed in these two opposed directions, the more they meet each other, and, like different organs of the same being, they will finally be combined into a harmonious Whole.'

Such formulations were not expressions of firm principles, but loose generalisations of an inclusive nature. There were pithy comments for both sides; to experimenters he said: 'To do experiments is to ask questions of nature, but no one can do this usefully unless he knows what to ask about'; and to speculators, as he called them: 'For mere inferences, however thorough they might be in themselves, all presume that the notion that we have formed of the object under investigation really conforms to it.' Such views echo Kant's 'Copernican revolution'. Human reason is both the prerequisite and the limitation of our cognitive capacity. Partisanship does not belong in a textbook according to Ørsted.[10]

He dedicated the 1811 edition of his textbook bearing the title *Første Indledning til den almindelige Naturlære* ['First Introduction to General Physics'] to Crown Prince Frederik, as

Oehlenschläger had done with his *Poetiske Skrifter* [Poetical Writings] in 1805. This paid off now as then. In his preface to the 1809 edition Ørsted made it clear that he did not address bread-and-butter students. 'I admit that I have not been working for the benefit of such people, who want to acquire only the knowledge necessary to get a post or earn their daily bread.'[11]

The mechanical part was printed in 1807, but the entire stock of the first edition was burnt during the bombardment that year, so the book only appeared in 1809 and was thoroughly revised only two years later, expanding from 378 pages to 608 pages.[12] Then Ørsted started writing the dynamical part, which was never published (ch. 28). The third part, with the working title 'The Theory of Higher Dynamics', scheduled to be the culmination of the dynamical project, was not written at all.[13]

*

One of Ørsted's most important research projects after his return to Copenhagen was to determine whether Winterl's chemical theory could be proven by experiment and if so whether it would advance the dynamical system. A Danish physician, H.J. Jacobsen, and a German professor of chemistry, W.G. Kastner, had achieved positive results in their search for 'Andronia', but even more decisive was a letter published in *Annalen der Physik und Chemie* by C.H. Pfaff, professor of chemistry at the University of Kiel (and a notable opponent of Schelling's *Naturphilosophie*) stating that he preferred Winterl to Lavoisier.[14]

According to Winterl, 'Andronia' was an earth in the form of a white powder and with the properties of an acid. The debate, however, identified it as the acidifying element, and in consequence its polar opposite substance, 'Thelycke', as the basifying element. Soon the debate on Winterl's work came to revolve around the question: do these two elements exist at all? Winterl willingly sent samples of both to various European academies to have them tested. He claimed that potash was replete with 'Andronia' and his adversaries refuted him by stating that potash consisted of a number of well-known elements, but 'Andronia' was not one of them.

In 1806 The Royal Danish Academy of Sciences and Letters decided to set a prize question on 'Andronia'.[15] One anonymous essay (in German) was submitted with the comment, 'The researcher often found more than he wanted to find', and Bugge, the Secretary, passed it on for evaluation in the physical class. Manthey examined it and rejected it. The experiments were incomplete and inconclusive, he reported back, and the anonymous author had agreed. The answer to the decisive question, 'what is the difference between "Andronia" and ordinary silicon' was blowing in the wind. Manthey added that the essay was less satisfactory than the experiments carried out at the same time by the German chemist C.F. Bucholz and his own protégé H.C. Ørsted, although these were not in response to the prize question.[16] The two latter analyses showed that what Winterl called 'Andronia' was nothing but ordinary silicon. 'Andronia' seemed to be a chimera. Even if Winterl's theory did not depend entirely on the existence of this substance the result was ominous for his ambitious project.[17]

At this stage Ørsted's confidence in Winterl's system was shaken. Already before Manthey's evaluation of the prize essay Ritter had related the story of the fall of the theory. Chenevix and a Spaniard called Gimbernat had sought out Winterl in Budapest. Never before had they seen a laboratory as filthy as Winterl's; the laboratory was also his larder. Winterl did not have a wife, but a couple of girls lived with him. Not one, but two, and they were not his daughters, but his laboratory assistants. During the visit Winterl had been asked about the substance 'Thelycke', in response to which he touched the breasts of Nanette, one of the girls, saying, 'this is the real "Thelycke"!'. Winterl's theory left Ritter disgusted at this swindler who had fooled truth-seeking natural philosophers. Ørsted was busy justifying himself.[18] But he was still fascinated by the theory because it so convincingly and independently supported his dynamical theory.

*

Ritter and Ørsted had not met since September 1802 in Jena, a time together they often recalled nostalgically. They never met again, although both of them longed to do so. When Steffens had left Denmark, Ørsted saw a remote chance of getting his German friend a job in Copenhagen. Often, he worried that the Chancellor might categorize him like Steffens. Such thoughts tinged him with sadness, and he pondered giving up his career in Denmark altogether and joining Ritter in Munich. Ritter was already looking forward to a reunion that turned out, however, to be unrealistic.

As an alternative, Ritter suggested a joint project involving Ørsted, Weiss, Horkel and himself that would aim at writing a kind of research manual for self-made natural philosophers, such as Ritter himself. The manual was projected as a catalogue of philosophical ideas, scientific problems, and aphorisms for inspiration. Nothing came of this either. Ørsted was up to his neck with his textbook, so Ritter was left to himself to write the manual. During the long sickness leading up to his death he only just managed to write the autobiographical introduction to his more than seven hundred aphorisms, which had striking Romanticist features in the style of Lichtenberg and Novalis. He saw his aphorisms as divine revelations that came to him unawares at happy and spiritual moments. *Fragmente aus dem Nachlasse eines jungen Physikers. Ein Taschenbuch für Freunde der Natur* ['Fragments of Posthumous Writings of a Young Physicist. A Pocketbook for Friends of Nature'] was Ritter's one-man achievement. He assumed the persona of a publisher editing the posthumous papers of a fictitious physicist, and according to Walter Benjamin the result is the most important piece of confessional writing from German Romanticism.[19] Only after Ritter's death did Ørsted receive a copy, sent to him by Gehlen via the bookseller Perthes in Hamburg. Ritter's introduction praised his Danish friend for his qualities as a human being and admired his ability as a connoisseur of the human heart in a way that must have made Ørsted blush.[20]

Ritter and Ørsted worked together as editors of Gehlen's *Journal für Physik und Chemie*, which to some extent was opposed to Gilbert's *Annalen der Physik,* the journal in which Chenevix had launched his attack on Ørsted and Winterl.[21] From 1805, Ørsted published his major articles in German translation in Gehlen's journal, and in their letters Ørsted and Ritter commented on each other's publications, loyally supporting each other. Ritter tried to have his friend elected as a corresponding member of the Royal Bavarian Academy of Science in Munich, and in order to become an ordinary member as soon as possible he submitted his *Betragtninger over Chemiens Historie* ['Reflections on the History of Chemistry']. But after the Battle of Austerlitz on 2nd December 1805, when the Austrian and Russian armies were routed by Napoleon, the academy's finances collapsed. All resources were to be pooled to resist the

French Emperor. Academy members' salaries were reduced or annulled altogether, and its yearbook could not find space for Ørsted's reflections.

Ritter fell ill frequently due to strain, drug abuse, and excessive drinking. His writing kept him up at night, his experiments needed surveillance at night, and his addiction to alcohol and opium destroyed his throat and heart. He was about to burn up inside and could hardly eat or speak. When he recovered for a short while he suffered from sleeplessness. He was broke, because his experiments cost him more than his writings brought in. Large expenses had to be paid for by issuing promissory notes. Recently, he had issued one of 250 rixdollars, but nobody would honour it, he complained to Ørsted, 'and you have no money yourself; but you have friends such as Manthey and Schimmelmann, who could help me. Please intercede with them for me! You just have to send a promissory note to Wesselhöft, printer at Jena', Ritter's landlord, a good friend, but poor. And being a helpful friend Ørsted interceded with a beautiful lady (possibly Countess Charlotte Schimmelmann) and sent a note to Wesselhöft, but nobody would cash it at Jena without security and the printer was unable to provide that. So he had to go to Leipzig to get the 192 rixdollars left when fees and exchange rates had taken their toll.[22] Meanwhile Johann Wilhelm and Dorothea and their four children had moved into some rented rooms at Maximilian Thor in Munich. Napoleon and his troops now ruled the city, and Ritter hoped for an opportunity to exhibit galvanism in front of the French Emperor in the academy as Volta had previously done in front of the First Class in Paris. Most Bavarians had to billet French soldiers, and Ritter had to billet now four, now six soldiers and officers, and besides he had to pay extraordinary war taxes, so contrary to his expectations he was now poorer as an academy member in Munich than he had been as a provisional lecturer at Jena. The mail had become unreliable, letters did not reach him, and if finally he did receive a promissory note, nobody would cash it.[23]

Then at last Ørsted's 192 rixdollars arrived, and unexpectedly spring broke out through the cold winter. 'A billion thanks, and God bless you!' Ritter wrote. Once again he was able to resume his experiments, correct his wild hypotheses, and stop himself from arguing in a circle. Ørsted was pronounced godfather to another of his children, and Ritter promised to pay back the two hundred rixdollars at midsummer or at the least at the end of September. But, of course, he was unable to keep this promise, for he was carrying old debts from his time at Jena that he needed paying back by instalments. Ritter sank deeper and deeper into an economic mire: he wrote again 'Help me! Don't be afraid, your money is safe. I'm not scared myself, if only God will let me live on for another three years.'[24] So He did, and that was not the first time that higher powers intervened exactly as Ritter had prophesied when his humour was black. More than two years later Ritter paid back an instalment of fifty rixdollars, and soon afterwards he received an advancement of 450 guilders from his publisher for *Fragmente* and promised to square the rest of his debt, or part of the rest. He forgot to mention his increasing expenses for alcohol, opium, and medicine, and now doctors' calls were unaffordable. In autumn 1809 he fell terribly ill again. At first he tried to heal himself by shocks from a Leyden jar. Nobody in Munich helped him. He had sent Dorothea and the children to live with a friend in Nuremberg, because he was unable to provide for them himself, and if they stayed on living together the whole family might be contaminated with his consumption. Living on his own he dedicated himself completely to his experiments. He promised to change his lifestyle, should he ever be well again. Ritter thought back to the forty days he had talked and experimented with Ørsted. Would those days never return, he lamented:

> 'Let peace come to Europe and let me get a post in Copenhagen, for you will never get one here in Germany. If only I were able to I should fly to you on an eagle. You have always been the most faithful, the most scientifically honest and considerate of my friends, and you have remained so. But now you must lend me six hundred guilders, not out of your own pocket, but luckily you have your good connections.'[25]

*

The first two years after his return Ørsted, too, was gloomy and complained to his friends abroad, Oehlenschläger and Ritter, about his ordeal in Copenhagen. As already indicated he toyed with the idea of going to Germany to take up a post as professor or researcher.[26] He was growing impatient and felt that the Chancellor was undermining his ambition to become a professor, and publishers refused to dance to his tune. Then at last, in September 1806, something positive happened. Gottfried Becker, pharmacist to the royal court and owner of the Elephant Pharmacy in Copenhagen, quit his extraordinary professorship due to increasing hardness of hearing. He had succeeded Kratzenstein on his death in 1795, but was wealthy enough to perform his scholarly duties without a salary, and as an extraordinary professor he had no seat in the Senate either. He was a member of the Department of Health, and Ørsted must have met him there as well as in the Scandinavian Literary Society and the Royal Medical Society, where he had given papers on eudiometry, (analysis of air into its constituent gases). When Becker resigned Ørsted saw a chance to take charge.

In 1806 Ørsted became a professor and moved from the Faculty of Medicine to that of Philosophy. From now on he would have to examine all students for *philosophicum*[27] and to teach students of medicine and pharmacology, chemistry and physics. These teaching obligations were his best chance to assert himself at the University, since as an extraordinary professor he was not part of the Senate. No doubt, he was content to see that his rival, Professor Bugge, had immediately stopped his lectures on experimental physics. Professor Ørsted was just twenty-nine years old and his appointment made his spirits rise. Now he had a future in Scandinavia after all, and had no need to go into exile like Steffens.

Finally, after a break of eighty years, the University had a professor of physics. Indirectly, the funding for this derived from Kratzenstein's will. Starting from nothing, Ørsted was now endowed with a collection of instruments and a professorship. This was a splendid basis for founding a school of disciples, an aspiration he communicated to Oehlenschläger. Kamma Rahbek had invented a new pet name for Ørsted, 'the Professor with the Nose', and it spread. Gjerlew wrote about his 'small comical nose' or about 'the small professorial soul with the snub-nose'.[28]

For the first time he was on the payroll—400 rixdollars a year. He already had 300 to cover expenses related to his collection of scientific instruments. As a professor he needed more space for a larger audience and hired premises in Thott's Palace, not far from Collin's Court.

Fig. 51. The yard behind Thott's Palace on Kongens Nytorv, 1845, to which HCØ moved his collection of scientific instruments from Collin's Court in 1806 when he was promoted as extraordinary professor. It was in fact only a question of moving the collection to the other side of the street. Oil painting by Frederik Vermehren (1823–1910). DHS.

His new lecture hall needed refurbishment to serve the double purpose of housing his instruments and his audience, and such capital expenditure was a heavy burden on him. Teachers of bookish subjects such as law and theology could hire premises between them and use them in turn, but in Thott's Palace instruments occupied half the space and he could not share the benches with anyone else. The instruments needed to be kept at a certain temperature so the rooms had to be heated permanently. Experiments needed good lighting, and moreover he needed a skilful assistant. Recurrent expenses amounted to 250 rixdollars per year: rent and insurance 110, firewood 50, lighting 40, assistant 50. The remaining 50 could not possibly pay for materials for experiments, and besides he was still carrying old debts from his grand tour.

On the other hand his private lectures provided him with an income. People of rank paid best, about 200 rixdollars a year in total, students paid less. Because he had expenses for experiments the professor of experimental physics was entitled to charge each student three rixdollars, against the humanities' two, but the number of students fell from seventy to twenty during the war with England 1807–14. As a consequence, in 1808 Professor Ørsted had a private annual income of merely 400–500 rixdollars against 600–700 in 1806.[29] So, he asked the Board of Directors for an increase of salary and did get it after a few years, but then inflation started to undermine his, and everybody else's, fixed income.

Ørsted took advantage of his promotion and soon asked the powers that be for more. How could he found a school, when physics and chemistry were not studied as independent disciplines at graduation level? In fact, he had been the only one for eighty years to devote himself totally to these disciplines and formally he was not a postgraduate of anything but pharmacology. If he were able to found a school, the first thing to do would be to establish a system of education in physics and chemistry, for only then could he keep an eye open for talent among the students he examined for *philosophicum* and encourage them to go on studying the natural sciences. It seems likely that when he was received in audience by the Crown Prince so as to thank him for his professorship, Ørsted aired these problems and pointed out the benefits that would become available to the kingdom once these things were done. In any case, soon after the death of Christian VII he submitted, by the new King's gracious orders, a 'Proposal for a Practical Institute [erased "for Chemistry and Physics"] for the Experimental Sciences'.[30] The proposal was an extension of the ideas he had sent Manthey from Paris. 'I look forward in the hope of becoming Your Majesty's tool to establish an institute that for many generations to come will have a significant impact on the prosperity of our nation.'[31]

However, Ørsted's proposal came to nothing. The establishment of a university in Norway had become the government's first priority. To start with the dual monarchy was at war with Britain, and then followed the nation's bankruptcy and the enduring agricultural crisis. But the proposal was not in vain. The King offered to pay to send Ørsted on a mission to Germany and France to investigate the significance of science for the economies of those countries. This journey was realised in 1812–13. In a number of ways the establishment of the Polytechnic Institute in 1829 was an echo of Ørsted's proposal of 1808.

> These splendid, noble, upright, purified ideas of the destiny of man.[1]
> This way of conducting a war is the most gruesome of all.[2]

20 | 1807

Fichte's Idealism and Napoleon's Wars

In October 1806 Oehlenschläger and some friends went to Weimar to visit Goethe. The Danish poet and playwright was angling for recognition from the German genius, but to his friends he made his pretext the need of good advice as to how to avoid the Napoleonic army on his way to Paris. After their victory at Austerlitz the year before, the French troops seemed invincible. In the Saale Valley, in which Weimar and Jena were only twenty miles apart, he witnessed the historic battle at Jena and Auerstedt. The Prussian army of 150,000 men was massacred by the French emperor's 200,000. Daily he heard the thunder of guns moving closer.

> 'We heard the deadly bullets whistle around our ears and windows and doors shook. Brøndsted, Koes and myself looked at each other for a moment pale and silent. Come, I said, let us go down, I think there is a cellar under the house, where the bullets cannot hit us... What did we Danes have to do with the French and the Prussians?... God knows, on April 2nd [1801 when he fought for the Students' Corps during the battle of Copenhagen] I had no fear, but here!... We lived in the most exquisite inn, being lucky to have some brave [French] hussar officers [who might protect them] also accommodated there. You should have seen them [Adam wrote to Christiane, his fiancée]. The landlord had lost his mind, and now the soldiers rushed through the gate shouting: "Du vin! Du Kirschwasser! Du l'eau de vie! [sic]". Oh, yes!... If the house had caught fire we would have lost every rag we owned and stood as beggars in a foreign country in devastating horror – yet we were in the hands of God.'[3]

After the battle, the University of Jena was closed by the occupying forces, and at the end of the month Halle, too, was taken, and another university suffered the same fate, and Steffens was out of work. Now Napoleon moved north, occupied Berlin, and seized Sanssouci and Neues Palais at Potsdam. All of Prussia from Berlin to Königsberg had been conquered by the French at a stroke. Then they marched against Russia, but while they were on their way they concluded a peace treaty with Tsar Alexander on a raft on the Njemen River near Tilsit. J.G. Fichte, holding a professorship at Erlangen after the controversy on atheism, fled from Napoleon's troops to the east and on through the Baltic territories of Russia, until he got a passage on a ship from Riga. In August 1807 he landed in Ellsinore and went to Copenhagen, where he found lodgings.

Fichte was soon sought out by Ørsted. They knew one another from Berlin, where the Danish physicist had attended Fichte's lectures on philosophy (ch. 13). Fichte was then introduced to Anders, each gaining a favourable impression of the other. He wrote to his wife: 'Filled with friendship and enthusiasm he [HCØ] called on me as soon as he knew about my arrival. He is a great man at the university here, and I reckon he will compensate for the absence of my other associates.'[4] Soon Hans Christian introduced the short, corpulent Fichte to his brother and sister-in-law, who at first did not believe her own ears. 'One evening, when Ørsted [i.e. Anders] came home telling me that Fichte was in Copenhagen, this was the first time I had not believed him...How was it possible for such a wonder to happen? I cannot describe my reaction the first time I saw the man, whom for all these years I have lived (I mean when really alive) I have prayed to God to show me in a dream', Sophie wrote to Baggesen, her admirer in Paris.[5]

Hans Christian had never seen her blush so. 'I just remember', Sophie continued, 'that I have never bowed so deeply before any mortal as before the immortal Fichte', whose works Anders had often read aloud to her. In summer the 'Asse-children' had escaped the noise and smell of Vestergade and moved to a rented house in order to be close to Frederiksberg Have and Søndermarken. There Anders would read aloud in the garden on Sundays from Fichte's works *Über das Wesen der Gelehrten* ['On the Nature of Scholars'] and *Anweisungen zum seeligen Leben* ['A Guide to Spiritual Life']. 'I can never express to you how happy I am, when I really think it over, it seems to me as if I received a revelation; I can't describe it, but these are my most blissful moments here on earth.'[6]

The 'Powder Keg' reported the encounter in Søndermarken, when her husband saw Fichte for the first time. The event remained stamped on her memory. 'I walked with Fichte. [A.S.] Ørsted hurried along and stiffened like a corpse, although with more than tenfold life in his sparkling eyes, bursting into tears. Fichte stood quietly waiting, until he came close enough to reach out his hand to him; Ørsted's voice was choked; he tried to say something. Fichte looked into his face, and having taken a close look, he clasped him to his chest. I watched them with more than earthly bliss, turned my eyes towards heaven, and the warmest tears of joy wet my cheeks.'[7]

If this report seems a bit overdramatic to a modern mind, it must be remembered that neither Anders nor Sophie ever stood a chance of meeting German writers such as Goethe and Schiller whom they admired and whose works they read aloud to each other. They did not travel as Hans Christian and Oehlenschläger did. The only coryphaeus they ever saw face to face was Fichte, hence the overwhelming excitement. What the Schimmelmann family disliked, when the day after his arrival he paid them a visit, was his neglected appearance and lax hygiene, but these were hardly noticed by the Ørsteds.[8] They were fascinated by his spiritual talk and charming nature.

Fichte also took time to see old Oehlenschläger at Frederiksberg Castle. His son Adam had learned to appreciate Fichte when he had met him on his grand tour: 'Fichte is a devil of a fellow. I shall see him soon.' Oehlenschläger had followed Fichte's lectures giving guidance on spiritual life, and he confided to Hans Christian that Fichte had mentioned him 'with warmth and respect'. By contrast Steffens was a taboo subject, 'they do not thrive together, I know that'.[9] In his memoirs Adam recounts Fichte's visit to his father:

> 'Fichte spent the whole evening with my old man chatting about various philosophical points of view, Schelling's included, in his normal naive way. He also lent him one of his books to read. The next day Kamma Rahbek came to see him and found Fichte's book in his hand. Kamma took it jokingly to read its title, but the old man interrupted her, uttering proudly: "Leave it, my child! It is Fichte. He spent the evening with me, and we talked till late; believe you me, we tackled the systems fiercely!"'.[10]

Most interesting, of course, is the relationship between Fichte and Hans Christian, who was delighted to hear about Adam's appreciation of Fichte. Ørsted read Fichte's lectures as soon as they were available in print. Like his brother he read them aloud to Sophie,

> 'the "Powder Keg", who follows him well. If it were possible to cure her of a certain hypochondria to which she is now and then susceptible and which is probably to some extent caused by her illness, but even more by a higher aspiration (not quite satisfied), Fichte's idealism might do it. At a passage in the text touching upon this higher aspiration and upon the feelings aroused in a noble being when it remains unsatisfied, she burst spontaneously into tears, because she recognised her own nature in Fichte's lecture. A woman with her soul is definitely less happy than we are, because our external conditions are far less an obstacle than hers to achieving things to which our souls aspire.'[11]

Apart from the moving empathy with his sister-in-law, the profound impact made by Fichte's lectures is worth noting. What was it that had the effect of a revelation on Sophie and that also moved the two brothers deeply when they read his works aloud to each other? It was the idealism of Fichte's comprehensive programme addressing the lifestyle of the intellectual community, a programme that inspired two generations of academics in the so-called Golden Age.[12] Hans Christian waited on Fichte during his stay in Copenhagen. 'Several points of his philosophy have been illuminated through our conversations, particularly his hostility towards *Naturphilosophie*. He sends his regards', he wrote to Adam who had arrived safely in Paris in the meantime.[13] A few days later as the English and Highland troops had disembarked at Vedbæk and other places on the coast of Zealand to proceed ominously towards the capital, he escorted Fichte out of the besieged city, before the violence of war caught up with him in our latitudes as well.

While Anders was at this stage an almost unconditional admirer of Fichte's moral philosophy, Hans Christian's attitude was more ambiguous. In Fichte's system nature and the natural

Fig. 52. Sophie Ørsted (1782–1818), née Oehlenschläger, married 1802 to ASØ. Oil painting by J.L.G. Lund (1777–1867). Bakkehusmuseet.

sciences only had an insignificant part, and his criticism of Schelling's philosophy on the identity between nature and consciousness was relentless. Ørsted always pricked up his ears when Fichte criticised Schelling. Correspondingly, Ørsted's dynamical system was irrelevant to Fichte, whose philosophy almost exclusively targeted the problem of human freedom. To him determinism in nature was a hindrance to man's freedom on a par with social limitations. But although their interests in natural and moral philosophy diverged both were preoccupied with the education of man. Ørsted's many pedagogical speeches and writings on the University and other institutions of higher learning show that Fichte's ideas were important sources of inspiration. We are concerned here with his book *Über das Wesen des Gelehrten und seine Erscheinungen im Gebiete der Freiheit* that may be translated, not literally but perhaps most meaningfully, as 'On the Vocation of Scholars and Academic Freedom'.[14]

What appealed to the Ørsteds and Sophie and a large part of the learned republic in Copenhagen was Fichte's idealism, which assumed a definite shape during the Napoleonic wars. In 1786 Kant had published a popular essay on universal history from a cosmopolitan point of view, in which history was seen as a progressive development towards the higher goal of human freedom. According to Kant, human reason more and more pervaded societies, because the rule of law increasingly triumphed over brute force, enabling people to develop their talents in freedom. Apparently, the arrow of history sped towards an upright, although hidden goal. It would be contrary to reason and 'the secret plan of nature', if human talents were wasted rather than allowed to flourish in science and art.[15]

Fichte's 'divine idea'—or the highest good, summum bonum, as Kant called it—aimed at an ideal in which society is saturated with reason and justice, and where the sciences and the arts flourish within a legal framework. The best way to achieve this end was left to the learned republic to find out. This republic was an informal association consisting of jurists, historians, philosophers, theologians, and scientists devoting their lives to the scrutiny of 'the divine idea'. Not only were they called to study the contents of this hidden idea, it was also their vocation to disseminate it and find means to its realisation. Like Kant, Fichte was indebted to Plato's dualism. The learned republic was assigned the role of a brains trust for the government, in other words it had the duty to identify 'the divine idea' and carry it into effect. To Plato, democracy was an objectionable system that had sentenced Socrates, his master, to death. The obvious

Fig. 53. J.G. Fichte (1762–1814) coming through the doorway of his admired teacher, Professor Kant in Königsberg. In 1794 the public had attributed to Kant—who was a victim of censorship as far as writings on theology were concerned—the authorship of *Versuch einer Kritik aller Offenbahrung*, 1792, which was in fact Fichte's break-through as a philosophical writer. So their ideas were closely matched. Anonymous woodcut. BPK, Berlin.

leaders of the city-state ought to be philosophers recruited from Plato's Academy, whereas peasants, soldiers, and artisans, only had the everyday chaos of sensory phenomena on their minds. They pursued only their private interests, while philosophers were endowed with talents to elevate themselves above appearances and acquire insight into lasting ideals.

Fichte divided history into five distinct periods: In the beginning man lived instinctively in a state of paradisal innocence, which was succeeded by an authoritarian state after the fall of man. The third period, which was Fichte's own, was characterised by man's self-inflicted decadence, the terror of the French Revolution, but soon people would be rising and moving towards that peak of history where they would create the most splendid art and live peacefully under just laws. The parallel to Kant was obvious. In his *Was ist Aufklärung?* ['What is Enlightenment?'] he characterised the Enlightenment as mankind's liberation from its self-inflicted subservience.

Fichte had elaborated the latter periods in three series of lectures. The contemporary history of decay was treated in *Die Grundzüge des gegenwärtigen Zeitalters* ['Main Features of the Present Time'], and the state of perfect justice setting the coping stone on human history was described in his philosophy of religion *Anweisungen zum seligen Leben oder auch Religionslehre* ['Guide to a Spiritual Life, or Philosophy of Religion']. These were the books the Ørsteds were reading when out of the blue their author materialised in Copenhagen.[16]

It was *Anweisungen* that had given Sophie a great intellectual and religious experience—particularly overwhelming perhaps, because as a woman she was excluded from the male-dominated learned republic. Fichte's philosophy of religion was anti-Lutheran in the sense that it stressed man's moral endeavours and voluntary works of charity as the road to salvation as against Luther's repeated emphasis on man's original sin and lack of free will as his fundamental limitation. No wonder that Fichte, who conferred full sovereignty on the free will, admonished his audience to carry 'the divine idea' into effect on earth. Christian ethical behaviour rather than Christian faith alone was at the core of Fichtean religious philosophy.[17]

To the Ørsted brothers Fichte's *Über das Wesen des Gelehrten* ['On the Nature of Scholars'] dealt exactly with this vocation. Here they found the clearest understanding of the role of the learned republic and universities as temples of reason, and in this work Fichte illustrated his reflections from the experiences he had garnered as a teacher at Jena. A few years later the book became the central pillar of the new university in Berlin, where for a short while Fichte was vice-chancellor (though without much success). In Copenhagen, too, his ideas had a deep and lasting influence.[18] Fichte's books were on the shelves in Ørsted's library and as already mentioned an edition from 1796 had been published in a Danish translation by Jonas Collin.[19]

The problem of dualism, according to Fichte, was that the great majority of people only noticed empirical phenomena ('what happens'), but seldom asked 'Why did x happen?', or 'What is the cause and what will be the effect?'. Clearly 'the divine idea' belonged to the noumenal level and was hidden from the common man. Even for the scholar it would remain a shadowy inkling of the goal of history, the attainment of which demanded cooperation between the capabilities of all scholars and artists articulating 'the divine idea' aesthetically. Like Plato, Fichte did not dream of a democratic constitution, but of the rule of just laws designed by the learned republic.

The scholars Fichte addressed worked to incorporate this ideal in their lives. They saw themselves as devoid of the material interests that the other classes clung to for selfish purposes. Without this vocation the scholar would have considered his life insipid and empty. Anything outside the vocation was considered a waste of time. This training of the scholar's character presupposed an inner control; no external governance would be capable of setting the course. Hence cultivating the sciences was an activity on a par with religious worship.[20] This was precisely the title Ørsted took for his speech to colleagues and students at the University's celebration of the Reformation in 1814 (ch. 29). Academic self-government should take place in absolute freedom untroubled by censorship and economic considerations. The inward-directed scholar would pay no attention to bounties or threats, but set his own course determined by his own reason and conscience.

*

The same day in August 1807 that Fichte departed, Crown Prince Frederik arrived in Copenhagen from the military headquarters at Kiel, where the army was gathered to prevent the French troops from invading the kingdom from the south and seizing the Danish-Norwegian fleet, now that Napoleon had lost his in the fateful battle at Trafalgar. The Crown Prince refused the British envoy Francis Jackson's demand to surrender the navy and ordered the capital to be defended at all costs.

As in 1801 a corps of student volunteers was rapidly improvised under the command of Court Steward A.W. Hauch and several professors became battalion commanders. An emergency hospital was established at the Freemasons' lodge.[21] The Crown Prince issued this urgent request:

> 'The danger by which the state is threatened gives me an occasion to ask students at the University to gather and join a corps under the name of the Crown Prince's Corps. I have seen this unit on the unforgettable April 2nd and taken great pleasure in witnessing its efforts for King and Country. Well now, fellow-countrymen, make the same effort this time!'[22]

In the following days the British navy of five hundred ships, meeting no opposition, landed thirty thousand soldiers and three thousand horses as well as guns and ammunition (including modern shrapnel shells) at Vedbæk and other locations on Zealand, ready to bombard Copenhagen if necessary, that is if the government did not surrender its navy.

Let us see what the British attack cost the Ørsteds and their family and close friends. Kamma and Lyne Rahbek were evicted from Bakkehuset, while the noisy Highlanders camped in Søndermarken and took possession of their home. The British generals billeted themselves in the Royal Frederiksberg Castle, but graciously left a few chambers to the decrepit King Christian VII, husband of the late English Princess Caroline Matilda. He preferred, however, to take refuge at the headquarters in Holstein. Old Oehlenschläger, too, was allowed to stay in the castle. The Crown Prince had returned to Kiel and entrusted the defence of the city to General Peyman, who according to His Majesty handled the affair so badly that he was sentenced to death after the capitulation, but on second thoughts was reprieved and dismissed without a pension.[23]

In spring that year the Ørsted brothers and Steffens had paid a visit to J.P. Mynster at his vicarage in Stevns to discuss natural philosophy and to experiment with galvanism. In August, according to the vicar,

> 'It is easy to imagine the horror spreading in the peaceful villages, the lamentation of women and children for their husbands and fathers, of mothers over their sons, who had to depart, not as usual for some bloodless manoeuvre, but to a real war. On the morning he was leaving my farm hand, who was in the territorial force, sat down on a stone in the courtyard howling....'[24]

On 29th August the territorial forces, mostly consisting of farm hands wearing clogs, met the well-trained and uniformed British soldiers under the command of Arthur Wellesley, later to become the Duke of Wellington. Mynster reports:

> 'The battle took place on a Saturday, and in the evening we saw most of our men return. Some, probably those who had followed the flag most faithfully, were taken prisoner and carried away on British ships.'[25]

Now the enemy closed in on the city of Copenhagen and on the naval harbour. On 31st August one of the voluntary corps sallied into Classens Have. The poet and naturalist Carsten Hauch, soon to become one of Ørsted's disciples, recounts:

> 'At dawn we quietly moved forward. Shooting began on the left flank. I heard the weird whistle of bullets when they hit the grass, and there was undoubtedly an abundance of them for the grass was in continuous motion. Two volunteers were wounded next to me. I saw them being carried away soaked in their own blood.'[26]

The purpose of this sally was to sweep the territory to the north of the city ramparts in order to prevent the enemy from taking shelter from the defensive gunfire behind the buildings. Classen's Agricultural Institute, The Museion in which Ørsted had taught, was set on fire for that reason, only one year after it had been built.

A few days into September Ørsted had come so much to his senses that he could send a comforting note to Oehlenschläger in Paris. Family and friends were safe. For three days and nights the city had been horribly bombarded. The tower of Vor Frue Kirke had caught fire and toppled to the ground. The navy had been stolen. Twenty ships of the line as well as ten frigates and more than forty smaller ships had been the price for getting rid of the British. Old Oehlenschläger had been forced to host the British generals at Frederiksberg Castle. Heger, Adam's father-in-law, had lost his brewery. Ørsted continued:

> 'My brother and your sister have not suffered much, I hardly at all...I am not composed enough to write you properly; a rhapsodic report will have to do...
>
> Miraculously, Lehmann has escaped death, but lost everything he owned. The three youngest of Professor Hornemann's daughters are wounded, and one of them will lose a foot. The number of wounded is inordinate. This way of conducting a war is the most gruesome of all; for the enemy bombarded the defenceless city for a period of three times twenty-four hours

Fig. 54. The sally of a corps of volunteers from Classen's Have at the end of August 1807. Museion, Classen's Agricultural Institute, is seen in flames in the background. The student Carsten Hauch was among the volunteers. Anonymous coloured copperplate engraving. Classen's Trust, Copenhagen.

> with more than twelve thousand bombs and rockets. Add to this that the British approached Denmark in the guise of friendship, until the islands were encircled; thirty thousand men ready to go on shore and their murderous arsenal operational, now you have a notion of the state of Copenhagen...'[27]

Ritter rejoiced when he received Ørsted's letter to the effect that he was left unscathed by the bombardment: 'Thank Heaven, that God preserved you and your friends and science at a time when the danger of losing everything was overwhelming. May it never return!'.[28]

During the bombardment Ørsted lived in a house close to Thott's Palace, and even though rockets hit Nyhavn, his premises were completely unharmed. Sophie Thalbitzer, a merchant's wife, living in the neighbourhood, recounts:

> 'We waded through the streets full of glass and stones... At six o'clock we reached my brother's house. Here only one bomb had fallen onto the yard next to our cart; but the gable and Buntzen's house, next to my brother's, was demolished, and all windows in Nyhavn were broken.'[29]

On the same day that Ørsted wrote to Oehlenschläger, Sophie, his sister, told Baggesen, also living in uncertainty in Paris:

'The first night...we spent in Vestergade in our apartment risking our lives. Bombs rained down and struck several yards in the neighbourhood; but the good Lord saved us. The following day we left the house and moved to Christianshavn. The following night the attack was still fiercer...and bombs struck our apartment in Vestergade. In the room we had stayed in and which we had tried to protect by putting in the loft a layer of two feet of horse manure, the ceiling was damaged and the windows blew out. Had we stayed we should have been injured or killed.'[30]

The next day they realised that this was no protection, at all. After the 'Asse-children' had saved their lives by fleeing from Vestergade, where the bombs hailed down, the 'Powder Keg' continued her report to Baggesen:

'Rumour had it that nobody would be safe in Christianshavn any longer, so we fled as we were to Amager, because the enemy would not hesitate to aim at the church spires of Christianshavn...On Amager we spent a terrible night waiting for the dawn in anxious expectation the disconsolate peace. Oh, Good God! What perpetual misery!'[31]

In Vestergade Mrs Møller and her household feared the worst. They decided to seek refuge on Amager beyond the range of British rockets. As she left her premises she handed over to her

Fig. 55. The British bombardment of Copenhagen the night between 4 and 5 September 1807 seen from north-west. CMC.

eighteen-year-old foster-daughter, Gitte Ballum, an apron with two inside pockets full of silver, money and other valuables so that she might bring them safely to Amager. The pockets were heavy, and when she found on the street outside a wheelbarrow with cloth from the dye-works that also needed saving, Gitte put the apron on the load. Then she fled with her best friend, Tine Ørsted, to Amager.

On the way Gitte and Tine had the idea of taking a look at Beyer's apartment on Gammel Torv, where Gitte's younger sister, Marie, worked as a maid. The doors were all open. The occupants had left in great haste. There was rice porridge, still lukewarm, on the table, and the girls being hungry ate the remnants before they resumed their flight. When eventually they reached their destination on Amager, Mrs Møller stood waiting impatiently for them. She immediately asked about the apron. Luckily the wheelbarrow arrived at the same moment, and Gitte exclaimed, 'Here it is!' and delivered the valuables.[32]

The result of the bombardment was devastating. A total of 181 military men had been killed and 350 wounded. The hospitals reported about 768 dead and wounded, but a larger number of civilians perished instantly in the fire or in the ruins. The total number of victims is uncertain; the city medical officer, Dr H. Callisen claimed 1,800 dead and a similar number wounded.[33] The material damage was enormous. The entire navy sailed away under British colours. The University completely burnt down along with Vor Frue Kirke, Borch's and Elers' Colleges and eight professorial yards. The University Library in the loft of Trinitatis Kirke was severely damaged, and the first fundamental work of international law, Hugo Grotius's *Defensor pacis* ['The Defender of the Peace'] from 1625 was perforated by shrapnel. Ørsted's collection of scientific instruments was undamaged, but his recently printed textbook of physics had fallen victim to the flames.[34]

Ørsted's rival, Professor Bugge, living on the corner of Vor Frue Plads and Fiolstræde, was hit still harder. He reported to his Swedish colleague, Professor Sjösten:

> 'My house was situated at the centre of the English bombs and grenades, of which on September 7th [sic] 1807 alone I received 18 bombs and grenades in my garden, house and yard. I thank God that my family and I escaped with life and limb. I have saved two thirds of my instruments and approximately 4,000 books, while one third of the instruments and 5,500 books and all my furniture and movables and a large part of my manuscripts have been destroyed by fire, so that I do not even have a chair to sit on or a dish and knife to eat with... I thank God that I have kept my good spirits as well as a desire and humour to work. One has to be philosophical enough to put up with everything, to forget everything, and to bear up under poverty as well as prosperity... Court Stewart Hauch's collection is completely unscathed.'[35]

Such was the Juggernaut of this treacherous assault. The only things Ørsted lost were the printed copies of his textbook. They belonged to the bookseller, who lost his money. From Rendsborg Ørsted's friend Gjerlew wrote: 'Your physics has burnt—No! My friend, it is only purified in the fire—you will raise a monument for yourself. I'm so pleased to know that you live entirely for the sciences. That is your proper life, but don't you sometimes play with the Graces?'.[36]

A glimpse of the world's hidden harmony...[1]

21 | 1808
Sonorous Figures

On his journey around Germany in 1801 Ørsted had met a certain Dr Panzner in Halle, who, in return for being shown his galvanic experiments, showed him his own sonorous figures, made by drawing a violin bow across the edge of a disc strewn with fine sand. In addition he presented him with a book on the phenomenon written by a certain Dr Chladni. The next day Panzner and Ørsted took a walk in the hills and on the following evenings they shared each other's company.[2] Why did the vibrations in the sand form regular, even beautiful figures? That is what they talked about, not 'en passant', but for four solid days running. Ørsted handled the book respectfully.

E.F.F. Chladni was the son of a professor of law and had grown up in the Lutheran town of Wittenberg, where his stepmother looked after him. The boy followed in his father's footsteps, studying law first at Wittenberg and later in Leipzig where he obtained a doctorate in order to qualify himself to take over his father's post. But scarcely had his father died when young Chladni dropped the planned career and embarked upon his own interests, geometry and botany. He decided to pursue his own course in life, but since his father had left him with no inheritance and he failed to make his livelihood as a tutor of law students, he tried his luck as an artist. It soon turned out that his musical talent was too modest to make a living as a piano soloist. So he decided to invent an entirely new musical instrument that he would be the only one to master. In this way he rid himself of competitors.

In 1789 his plans had assumed a firm shape. He imagined an instrument consisting of the same kind of glass tubes that are used for thermometers, which by strokes of a moistened finger would emit the tones of the scale with the intervals corresponding to the amount of lacquer smeared around the tubes. Chladni was carried away with his invention, which he called 'a euphonium'. In contrast to other instruments the euphonium kept its tone for a very long time. The only drawback was that the lacquer peeled off during transportation, but he constructed a special protective box which he could place under the seat of his cart. Now Chladni rolled happily along on the highways of Europe giving solo performances in cities as far away as St Petersburg and Copenhagen. His concerts were profitable and in between them he would carry out acoustic experiments.[3]

Of course Chladni was secretive about the construction of the internal, more specialised parts of the euphonium. He would not jeopardise his livelihood by exposing his invention to emulation. But his acoustic experiments could more readily be revealed as they might provide him with scientific fame, and this in turn was likely to attract a larger audience to his concerts. He produced sonorous figures, not of individual tones but of sounds, by running a violin bow across the edge of a glass or metal disc. The disc was supported underneath at the centre by a handle or held at the edges with two or three fingers, and strewn with a thin layer of fine sand. The violin bow produced the acoustic figure almost instantaneously as the vibrations of the disc moved the sand to the non-vibrating nodal lines of the disc. Chladni was spellbound by the regular geometrical patterns of the figures. When he divided a rectangular disc into smaller rectangles he could measure the distance between the edge of the disc and points of the nodal lines. These distances were proportional to the length of the strings of a piano. Thus the acoustic figures turned out to be visual, symmetrical representations of acoustic vibrations. Chladni interpreted the sand's movement to the nodal lines as the result of mechanical vibrations. His book was full of drawings and measurements of acoustic figures, and well-established mathematicians made great efforts to deduce the natural laws of acoustics from Chladni's sonorous figures—but in vain. Nevertheless, Chladni, like Volta before him, was invited to demonstrate his experiments to Napoleon.

The acoustic figures made a profound impression on Ørsted, and upon his return to Denmark he decided to study them more closely. He had a hunch that Chladni's mechanical observations and the mathematical interpretations of others veiled the real magic that might be hidden in dynamical physics. The soothing effect of music on the soul had always been a riddle nobody could explain. To regard Chladni figures as merely material manifestations was unimaginative and did not grasp the profound connection between physics and aesthetics, between science and art, and between body and soul. Consequently, the figures might open a new physical approach to the solution of the riddle, because they made tones visible. So, the scientist was presented with a couple of sense impressions, auditory and visual, that might put him on the empirical track of a dynamical explanation.

Ørsted hypothesized that electricity might be contributory to the production of the figures. If so, studying the process might lead to a deeper understanding of the dynamics of electricity. He subscribed to the theory of undulation according to which electrical forces are transmitted as waves. He imagined that every oscillation was the shock-effect of static electricity, like discharges of an electrostatic generator. This series of shocks produced simultaneously sound waves and visual figures.

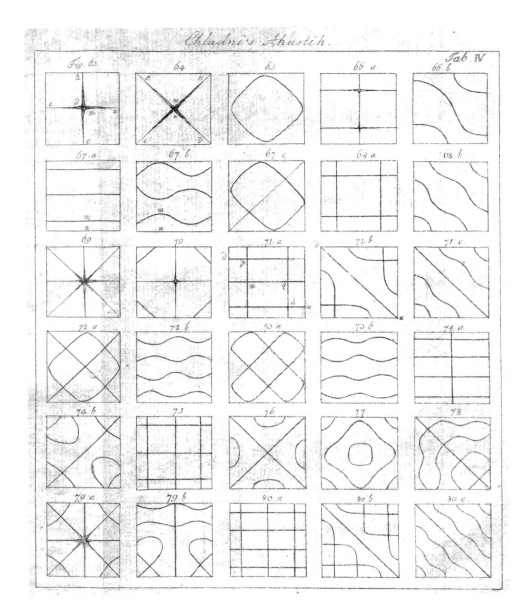

Fig. 56. Acoustic figures drawn by Chladni himself. Ørsted was presented with his book *Die Akustik*, 1802 by Dr Panzner (1777–1851) in Halle on his grand tour.

For four years from the autumn of 1804, Ørsted conducted hundreds of experiments on sonorous figures, and soon he would entertain private parties, where people were stunned as they watched his bowing making the sand dance beautifully on the plate. He adapted Chladni's method in two ways. First, whereas Chladni had copied his figures by drawing them, Ørsted copied his by pressing a sheet of paper smeared with gum Arabic onto them. The lycopodium powder he was using stuck to the paper when it was taken off the plate. The paper was then put onto a glass plate according to Lichtenberg's directions,[4] transferred to a copperplate, and finally printed. By this method (anticipating the modern photocopying machine) the hyperbolas of the figures clearly stood out.[5]

Secondly, to understand how the making of waves of electricity happened (if they were electrical at all) he had to be able to observe the motion of the sand, which was hardly possible because it happened too rapidly for the eye to follow the process. He therefore reduced the speed of the movement by replacing the comparatively coarse and heavy sand by the fine and light lycopodium powder (spores of club moss). It now turned out that the acoustic figures did not assume the form of Chladni's straight and curved lines crossing one another, but formed hyperbolas.[6] Moreover, he observed that there was not a single motion, but two separate oscillations. He called them primary and secondary motions.[7] The primary ones occurred the instant the plate was bowed. Even though it was difficult to observe the secondary ones because they happened so rapidly, Ørsted took them to be waves moving horizontally and vertically on the plate. They could not possibly have the same cause. Whereas the horizontal must be the effect of mechanical oscillations of the plate, the vertical might depend on attractions and repulsions of polar, electrical tensions between the lycopodium and the metal plate.

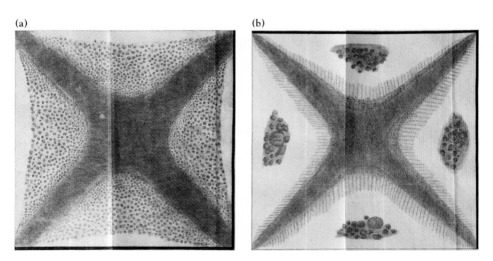

Fig. 57a, b. HCØ, 'Experiments on Acoustic Figures', KDVS Transactions 1808, Copenhagen 1810. The figures took shape gradually when the lycopodium accumulated in hyperbolic lines—a bravura showpiece HCØ loved to perform for the rest of his life. NS II (11–34) 21 and 23, SSW (264–281) 272–273.

Two sources of inspiration made Ørsted interpret the acoustic figures dynamically. One was the obvious similarly to Lichtenberg's electrical patterns. True, they had been produced by electricity directly discharged from a Leyden jar onto a resinous plate. In Lichtenberg's experiment the electricity had produced an invisible pattern on the plate, but when lycopodium was sprinkled on it, the pattern appeared. It now turned out that the negative electricity formed a concentric pattern, while the positive looked like a ramification (ch. 18). Since opposite charges attract each other, Ørsted deduced that the lycopodium acquired the charge opposite to that of the resinous plate of Lichtenberg's experiment. Might something similar be the case in the formation of Chladni's figures?

But where did the electricity come from? Whether the plate was made of glass or copper it was not an electrophorus as in the case of the Lichtenberg figures. Nevertheless, the two patterns were reminiscent of each other. Ørsted believed that the oscillations produced by the bow were accompanied by weak electrical charges, positive and negative unevenly distributed on the plate. 'No doubt the nodal lines are negative and the dust lines positive, because the lycopodium in itself is negative' and therefore is attracted by the dust lines and repulsed by the nodal lines. It is not clear what he meant by 'dust lines', and why they are charged the way Ørsted asserts. But this is the way it has to be, he imagined, 'for when he tried to shake the dust off the plate, the nodal lines would be quite free from dust, while the dust lines kept most of theirs. Consequently, a mechanical interpretation must be misleading.'[8] Ritter agreed and suggested that the waves turning upwards must be positively charged above and negatively below, while the ones turning downward were charged the other way around. When the plate was at rest, the two charges neutralised each other, but when the oscillations started the harmony was disrupted and electrical tensions occurred. So now, Ørsted tried to prove that the Chladni figures were electrical. He used Coulomb's torsion balance, a sensitive instrument capable of measuring even weak charges of electricity. Unfortunately this instrument did not show any trace of electricity on the metal plates.[9]

Another idea stemmed from Ritter's galvanic experiments on his own senses (ch. 10). A series of experiments had shown that the ear responds to shocks from the voltaic pile. The same tone was conceived differently according to whether it was exposed to positive or negative electricity. This indicated that the sound waves were not just mechanical vibrations, but either a kind of electrical charge or at least interacting with such forces on their way through space to the ear via nerves to the brain. For how else could it be explained that the ear received a tone differently when the electricity changed from the positive to the negative pole?[10] Ritter was confident that sound waves moved like light waves; Ørsted inferred that an octave of tones is analogous to the spectrum of light, and he extended this analogy by juxtaposing tones in scales and colours in the spectrum.[11] When the voltaic pile was connected to the ear the tone changed and became either higher or lower. With one pole the tone rose from G to between G sharp and A, and with the other one it fell from G to F. No other scientists ventured to repeat these hair-raising experiments on themselves, so there was great uncertainty about the interpretation of them. But Ritter had opened the door to a magical world no human being had previously entered.

Whereas Chladni believed that he had given an exhaustive explanation of his acoustic figures within the framework of Newtonian mechanics, Ørsted, keenly seconded by Ritter, was hoping, to identify a dynamical phenomenon, and that the proof of this would appear when the acoustic figures were seen in the light of Lichtenberg's and Ritter's discoveries.[12] In May 1807 Ørsted's experiments had advanced so far that he submitted his report to the Royal Society of Sciences and Letters, and Court Steward A.W. Hauch, Director of the Royal Orchestra, was appointed to chair a panel. Hauch wrote the assessment, to which the rest of the panel agreed. He did not recollect ever having seen a report that 'so beautifully and clearly had shown the order of the production of these figures. As far as he was aware nobody had ever made observations to the effect that electricity has a part in producing these acoustic figures. The concluding paragraphs may appear somewhat hypothetical, but are beautiful, nonetheless.'[13] Ørsted's report was approved for inclusion in the society's series of publications, and after some debate, he was accepted as a member of the society (ch. 24).[14]

*

His studies of acoustic figures did not bring Ørsted any closer to a breakthrough in the dynamical theory. Subsequently, he abandoned the idea that electricity was involved in producing acoustic figures. He thought instead that the vibrations of the metal plate made the air above it move. These secondary oscillations were now conceived as overtones.[15] He resumed his experiments which he had never quite abandoned, because they provided such happy entertainment at private parties. Søren Kierkegaard saw in the mature Ørsted's face a peace, harmony, and joy that always reminded him of an acoustic figure well bowed by nature.[16]

While in London in June 1823 he met (by chance, presumably, the circumstances are unknown) Charles Wheatstone, a rather shy young man producing and selling musical instruments and taking an interest in physics. Ørsted talked to him in his display rooms, where he was busy doing acoustic experiments. Ørsted demonstrated his own Chladni figures to him. In his letter home he described Wheatstone: 'He has no acquaintances among the scholars here, so I have been in a position to give him useful advice as how to make his discovery known. At this moment he is reading to me what he has written.'[17] Ørsted must have taken an interest in his experiments, because a few days later he showed them to W. Wollaston, J. Herschel, and G. Birkbeck. 'I have introduced him well here', Ørsted wrote home, and he explained the secondary oscillations to Wheatstone.[18] Meanwhile he had also demonstrated his acoustic experiments in Paris and had tried to observe the advancing motion of the lycopodiuvm by pouring a liquid onto the plate. The motions slowed down a little, but still not enough to observe them. Later, Wheatstone was appointed professor at King's College, London. Ørsted had certainly helped to advance his career.

In 1831, the year he discovered electromagnetic induction, Faraday spent six months experimenting on acoustic figures. In this connection he read Ørsted's articles from 1806.[19] Was Faraday's interest in acoustic figures prompted by the same ideas that Ørsted had entertained? Like Ørsted he was sceptical towards mere mechanical explanations, because he found them contradictory. Like Ørsted he had tried to delay the motions of the dust by pouring a liquid onto the plate. Faraday used egg white, but with little success.

Like Ørsted, too, he had come to see the secondary oscillations in the dust of the acoustic figures as produced by the air vibrations set up by the primary oscillations of the plate. This effect Faraday called acoustic induction. To prove his theory he invented a method of isolating the secondary oscillations by pieces of cardboard, an idea that had never occurred to Ørsted. The result: no secondary oscillations![20] Unlike Ørsted Faraday never entertained Ørsted's first hypothesis that electrical forces were at stake in the propagation of sonorous figures.

Faraday's deliberations seem in a number of ways to have been based on the same hope as Ørsted's; that is, that there were common features in the various ways the forces of nature were propagated. If such a reduction were possible, if sound, light, heat, electricity, and magnetism obeyed the same law of propagation, this would break new ground. Half a year after his discovery of electromagnetic induction Faraday filed a note in a sealed envelope in the Royal Society:

> 'I think also, that I see a reason for supposing that electric induction (of tension) is also performed in a similar progressive time. I am inclined to compare the diffusion of magnetic forces from a magnetic pole, to the vibrations upon the surface of disturbed water, or those of air in the phenomena of sound; i.e. I am inclined to think the vibratory theory will apply to these phenomena, as it does to sound and most probably to light. By analogy I think it may possibly apply to the phenomena of induction of electricity of tension also. These views I wish to work out experimentally: but as much of my time is engaged in the duties of my office, and as the experiments will therefore be prolonged, and may in their course be subject to the observation of others; I wish, by depositing this paper in the care of the Royal Society, to take possession as it were of a certain date, and a lone right if they are confirmed by experiments, to claim credit for the views at that date: at which time as far as I know no one is conscious of or can claim them but myself.'[21]

Thus Faraday pursued a similar dynamical physics to Ørsted's. At first Ørsted expected to find that electricity was involved in the formation of acoustic figures (as Hauch ironically noticed) but he was unable to prove it and dropped the idea. By 1823 he had come to see the secondary oscillations in the dust heaps as air vibrations caused by the oscillations of the plate. Faraday was preoccupied with the idea of an analogy between the waves of sound and light, and electricity, and magnetism. Making experiments with acoustic figures he came up with the term 'acoustic induction'. Later that year followed his epoch-making discovery of electromagnetic induction.

> What fascinates and enraptures us in the art of music
> is the deep, infinite, incomprehensible Reason.[1]

22 | 1808
The Art of Music

Ø RSTED HAD a further idea in mind as well. His experiments with sonorous figures were in fact wrestling with the classical body-soul problem. The sight of these figures had infatuated hi aesthetically. By publishing his scientific report in the physical class of the Royal Danish Society of Sciences and Letters and his Platonic dialogue on the aesthetical aspects of the acoustic figures in the Scandinavian Literary Society he succeeded in reaching a wider and more influential audience—both parts of the learned republic. His vocation was clear: science is not only a question of mechanics, cartography, and such other mundane activities as were considered useful by the waning Enlightenment. Now, in the era of Romanticism, a new dynamical physics was rising. The new era transcended the materialism of mechanical physics by bridging the gap between the sciences and the arts, and by understanding the classical problem of the body-soul relationship in a new and revolutionary way. Dynamical physics would address problems that were of interest not only to the sciences, but also to the arts and even to theologians. For the laws of nature bear closely on human consciousness. Tones penetrate the soul and evoke sublime enjoyment. Hence Ørsted finished his report on acoustic figures by pointing to their aesthetical significance:

> 'It is not the mechanical sensory stimulation which pleases us in the tone, but the mark of an invisible Reason which lies in it. So now a flow of notes floods our whole being with joy. What profundity unknown to the listener is not hidden in a single chord, what infinite arithmetic in a whole symphony! And now, joined with this, the invisible forms which appear before our soul in obscure intimations while the notes flow into the ear. In truth, we can

repeat with joy, and with pride at the nobility of our spiritual being, that what fascinates and enraptures us in the art of music and makes us forget everything while our soul soars on the flow of notes is not the mechanical stimulation of sensitive nerves. It is the deep, infinite, incomprehensible Reason of Nature which speaks to us through the flow of notes.'[2]

As a student Ørsted often had the opportunity to listen to the composer C.E.F. Weyse improvising the cadenzas of Mozart's piano concertos at the Harmonic Society. As tutor to J.L. Heiberg, Ørsted regularly bumped into Weyse who gave the boy piano lessons. In 1807, Mozart's opera *Don Giovanni* was performed for the first time in Denmark at the Royal Theatre under the conductor Edouard du Puy. Ørsted attended this performance as well as many other concertos and operas, among them *The Magic Flute* in Dresden and Copenhagen.[3] A few years later the composer Friedrich Kuhlau, on his escape from German military service, arrived in Copenhagen, where he performed several of Ludwig van Beethoven's piano concertos and symphonies. And while the Congress of Vienna was engaged in dancing in 1814–15, A.W. Hauch, Director of the Royal Orchestra and the Royal Theatre, sought out Beethoven in order to buy the scores of four of his symphonies so as to have them performed for the music-loving audience in Copenhagen.[4] Ørsted was a frequent concertgoer, and even after the bankruptcy of the state in 1813 when musical life was suddenly curtailed he maintained his membership of the Harmonic Society, although the membership fee was as much as sixteen rixdollars, four marks and eight skillings a quarter (or sixty-seven rixdollars a year), more than a maid was paid for a whole year.[5] His contribution to the theory of aesthetics came at a time when concert life flourished in Copenhagen.

Against this background Ørsted made the most of his experiments on acoustic figures and couched them in terms of a natural aesthetics. Already as a boy he had familiarised himself with Batteux's mimetic principles, in essence that good art emulates nature. To judge from the collection of books on aesthetics in his library he had delved into J.J. Rousseau's entries in the *Great French Encyclopedia* and his *Dictionnaire de Musique*, which had also inspired his great master Kant's *Kritik der Urteilskraft* ['Critique of Judgement']. Rousseau's dictionary of music mesmerised Ørsted as it did so many others, and Ørsted's visit to Montmorency was an emotional experience as is apparent from his letter to Sophie Probsthein. But every time Rousseau approached the fundamental questions (What makes music so wonderful to listen to? Why is the soul excited? What determines individual taste? Which are the sources and signs of a creative genius?), he gave up, exclaiming, '*Je ne sais quoi!*' ['I don't know']. It was this resignation that challenged Ørsted to find out if physics might become the servant of aesthetics.

Kant, who meant so much to Ørsted, was in many ways an admirer of Rousseau; according to legend, the only decoration on the walls of his study was a print of Rousseau. In his third critique, *Kritik der Urteilskraft*, he tackled some of the issues the French philosopher had come to consider insoluble. Why do individual arbiters of taste express themselves in terms of universal validity? Does the work of art (the object) or the taste of the audience (the subjects) determine its quality? What is the difference between the objects of the sciences and the arts? What provoked Ørsted in particular was the obvious contradiction between Chladni's theory that acoustic figures are produced by mechanical laws of nature, and Rousseau's and Kant's and

Fig. 58. A soirée of chamber music at the home of Chr. Waagepetersen, wine merchant by appointment to the Royal Court,.1834. The party consists of the family of the wine merchant, business associates, and friends of music. At the piano is the composer C.E.F. Weyse, fifth from the left is Professor C.W. Eckersberg, a dedicated amateur musician, and the first on the right is the composer J.P.E. Hartmann, organist at Vor Frue Kirke. The host (centre) is playing the cello while his son, L.L. Mozart Waagepetersen, sitting to his left, is playing the violin and on the wall are portraits of Friedrich Kuhlau (1782–1832), composer of *Elverhøj* [The Elf-Hill, 1828], a patriotic drama by J.L.Heiberg, and of Haydn and Beethoven. Oil painting by Nicolai Wilhelm Marstrand (1810–73), FBM.

assertion that artists are not ruled by laws and regulations, but create their works in absolute freedom. The conflict between the unconditional necessity of physical laws and the infinite freedom of the arts could hardly be any sharper.

What agitated the minds of people the most, perhaps, was the question whether the artistic process was susceptible to laws of nature in the same way as nature itself was? If so, artists had to follow certain prescriptions, and this in turn seemed to imply that creativity can be learnt. Rousseau and the Romanticists reacted fiercely against this view. Unbridled freedom was their gospel. If artists were governed by laws as nature was, they would be enslaved by them and all concertgoers would experience exactly the same thing, which would exactly match what the composer had invested in his work. But this could hardly be called artistic creativity, for art must be free as man is free. Without freedom man loses his dignity and art would lose its enchantment.

The salient feature of the arts is that they are not governed by prescriptions. A genuine work of art is created when God bestows the rules of nature on a genius without the genius being conscious of it. This is a mystical and inexplicable process. When one asks an artist how he has accomplished his work one has to put up with the answer: '*Je ne sais quoi!*'. Rousseau's belief in the freedom of the artist relied on notions as mysterious as 'genius', and 'taste'. Rousseau amused his readership with some charming characteristics of the basic concepts of aesthetics:

> 'Genius or ingenuity [he explained to a fictitious, budding artist] is not something you should roam about the country looking for. It is inside you. And if it is not, there is nothing you can do about it. You will never learn the art.'[6]

The notion of taste is equally undefinable:

> 'Of all the gifts of nature, taste is the one that feels the most and explains the least. It would not be aware what it was, if it could be defined, for taste pronounces judgement on things that cannot be judged, without putting the spectacles of reason on the nose, if I may put it that way.'[7]

Rousseau's dictionary provided only a few answers to his innumerable questions; to most of them he could only repeat his '*Je ne sais quoi!*'. Aesthetics to him was as mysterious as the lamp that brought out the hidden spirit when Aladdin rubbed it, or the tinderbox that made the dog bring the princess on its back, when the soldier struck a spark from it.

No doubt, it was this powerlessness that challenged Ørsted to scrutinise the problem from a dynamical point of view and to explore the dualistic relationship between physics and aesthetics in music. The dichotomy between body and soul, between the necessity of natural laws and the freedom of the arts, should be clarified in the aesthetics of nature. For one thing he did not accept that no rules at all governed the arts, because his experiments on Chladni's figures provided at least a visual indication that the Pythagorean theory of music and the mimetic theory of Batteux had a point. Music could not be total freedom. It seemed to incorporate a physical part that obeyed more or less well-known laws of nature, and if such laws could be better known some objectivity might be introduced into the judgement of taste.

Ørsted epitomised these unresolved issues in the title of his dialogue 'On the Cause of the Pleasure Evoked by Tones'. He read it aloud at the Scandinavian Literary Society in February 1808, when his report 'Experiments on Acoustic Figures' had already been approved for publication by the Royal Society of Sciences and Letters. The division of labour between the sciences and the humanities had already been institutionalised, although the two academic societies met in the same premises and although most people who belonged to one also belonged to the other. But it was quite unusual that the same philosophical question was addressed by one member to two overlapping audiences from two different points of view.

The dialogue was by no means a new one, but went back to Plato who, using Socrates as his spokesman and following the maieutic method, step by step released the unconscious knowledge of a slave. The dialogue did not have to unfold itself as a war of words between two incompatible views ending up with a winner and a loser. As in Plato's *Symposium* on different kinds of love, interlocutors might shed light on the topic from different perspectives. The early Romanticists breathed new life into the Platonic dialogue. A.W. Schlegel wrote his *Die Gemälde: Gespräch* [Dialogue on Paintings] in 1799, in which three people discuss paintings, particularly landscape paintings and religious motifs, on the banks of the Elbe following a visit to the Zwinger Gallery in Dresden. His brother Friedrich wrote *Gespräch über die Poesie* [Dialogue on Poetry] on a similar pattern in 1800. Schelling wrote the dialogue *Bruno: oder über das göttliche und natürliche Princip der Dinge* [Bruno: or on the Divine and Natural Principles of Things] in 1802, in which Schelling puts into the mouth of the heretic Giordano Bruno views on a range of issues within discussions on natural philosophy.[8] Ørsted was very much at home in this tradition and possessed copies of the three dialogues.[9]

In Ørsted's collected writings there are seven dialogues. Two were printed by 1810 and the genre was resumed by the end of the 1830s with another five.[10] In some of them Sophie (an uneducated, but gifted and inquisitive woman) plays the part taken by the slave in Plato's dialogues. Perhaps Ørsted had his sister-in-law in mind. The person acting as his spokesman was usually called Alfred [the peacemaker], mediating between conflicting views prevailing during the Enlightenment and early Romanticism which were voiced by other interlocutors.

Two interlocutors, Felix and Julius, represent Romanticism and deny music any objective rules. Felix is seduced by the Muses and carried away by his feelings like Rousseau, who burst into tears the first time he listened to Orfeo's lament 'Che faro senza Eurydice'. He was mystified and had no clue as to how to explain his reaction other than asserting it was totally spiritual. Felix encourages his friend: 'Go and drink in the flow of the notes and become absorbed by it. When you feel pervaded by a higher spirit and when you rejoice in unspeakable bliss you will understand the notes and not have to worry why the flow is carrying you away to heaven.'[11] Julius, too, finds neither lawfulness nor regularity. The experience of music is utterly subjective and even illusory. He founds his view on a story about a fiddler who every time he rehearses a certain passage kicks his dog. In this way he establishes a conditioned reflex that makes the dog howl at the sound of the passage in question without the fiddler touching the dog at all. Julius is so hardened in his prejudice that so-called sublime musical experiences are merely due to deception, and he does not bother to listen to Alfred's objections or take further part in the discussion. He disqualifies himself and slips out of the cast.[12]

Two other interlocutors, Valdemar and Herman, are representatives of the Enlightenment and convinced that music is objective in the sense that it exists and affects the reason and the senses independently of the subject. Valdemar is a caricature of an Epicurean who one-sidedly emphasises the sensory aspects of music. To him notes are no different from food and drink. In fact, he does not go to a concert for the sake of the music, but to see his friends, have a drink and be seen. He goes for the snobbery of high culture and does not feel that the sensual experience entails any spiritual elevation of his soul. He is a sheer philistine and a materialist through and through. Herman, too, takes it for granted that the effect of music is objective, but contrary to Valdemar he is an idealist. To him music reflects Reason with a capital 'R'. Music is the resonance of reason in the human senses. On the other hand the objective rules of music cannot possibly emanate from empirical nature, because (as maintained by Schelling's *Naturphilosophie*) it has not yet become self-conscious. Hence, mimesis, the imitation of nature, has no place in the arts, the pinnacle of human consciousness.

While the four contenders are arguing, with none of them giving way, they suddenly become aware of Alfred, who sits quietly listening to the debate while drawing figures in the sand like another Archimedes. What was he drawing? Music, of course, acoustic figures. The four open their eyes wide for now they realise that music is in fact an objective physical phenomenon and no mere imagination as Julius believes, nor a spirit floating in the air as Felix thinks, but an acoustic figure. Alfred has shown its coming into being to Felix who now relates the experiment to Herman:

> 'You should have seen it, how at the stroke of the bow the dust rises into innumerable small hills. The ones closest to the nodal lines are so small they are hardly visible; the further one gets way from there the bigger they are; Thus they are distributed among all the symmetrical parts of the figures. By a single new stroke everything is set in motion again. The hills are immediately changed to waves, and every wave seems to boil because innumerable, smaller waves are absorbed by them; but they all hurry in a symmetrical dance in prescribed ways to the large resting place. Oh, this is a motion, a life, a creation that must be seen in order to get any sense of it.'[13]

Alfred has shown that music is an objective physical phenomenon, but is it sensual as Valdemar believes or spiritual as Herman thinks? Alfred argues that it is both at the same time, that is, it is dualistic. Musical notes are sensual waves in the air perceived by the ear. They can be described by mathematics according to Pythagorean laws. Acoustic figures are visual sounds transmitted to the ear and perceived by human reason. Both Valdemar and Herman are surprised by Alfred's explanation of the dualism. The mathematical or phenomenal aspect of music is unconscious to the composer as well as to the music lover. The artistic or noumenal aspect is the feelings and ideas inherent in tones and perceived by music lovers. This explains why Valdemar and Herman perceive the same notes differently and why their tastes differ. Whereas Valdemar shrugs his shoulders haughtily, Herman's fantasy soars in the noumenal sphere.

To Valdemar the joy of music is like the sugar coating of a bitter pill. To him all sensual pleasures have equal value. But Ørsted repudiates the oft-quoted Latin phrase: '*De gustibus non est disputandum*' ['Matters of taste are not debatable'], though he readily admits that it is difficult to find criteria to rank the arts in sensory terms. He ends by ranking music the highest, because it is the most spiritual and the least material. Painting follows next. The ear and the eye are the noblest senses. Alfred (alias Ørsted) does not think that taste is haphazard: 'There is no healthy human who turns melancholic from hearing a gleeful waltz, or merry from hearing a slowly progressing chorale.' People must have a *sensus communis* that is unstated in the aesthetical debate. The problem is that discussions on the arts cannot be squeezed into fixed conceptual frames. But he admits that enjoyment of music presupposes 'a trained ear. The rules of the arts are overall the same; but the degree to which a work of art is appreciated is as different as the most cultivated European is from the wildest savage.'[14] The paradox seems to be that judgements of taste are applied to works of art as if they were universally valid, even though one is aware that there is no consensus on objective criteria of taste.

The conversation then enters into another series of questions. What is the meaning of saying that the experience of music is rationally unconscious? Why does Ørsted use the notion 'unconscious'? This is because he wants to say that even if music is following certain laws of nature, with strings vibrating in wavelengths of distinct mathematical proportions, we do not worry about these laws, indeed we are not at all conscious of the mathematics in them, when we play or listen to music. If it were necessary to know the physical laws of music, Mozart would never have managed to compose a single symphony. 'The Reason of the Pleasure evoked by Notes' is that music, in a way unconscious to us, enriches us with the beauty and reason of nature. Music develops from the laws of nature, so Ørsted's reply to Rousseau's '*Je ne sais quoi!*' is that aesthetics is the rational unconscious. Works of art can penetrate our consciousness without us knowing that it is happening.[15] Ørsted's notion of the unconscious is optimistic, because it mirrors reason and consequently beauty and freedom.[16]

Ørsted's aesthetics of nature had rescued the freedom that to the early Romanticists was so decisive for the dignity of man. Freedom became evident because taste was no longer ruled by objective laws, but was a consequence of education having the freedom and independence of man as its purpose. Aesthetical education opened the mind to deeper and nobler thoughts as shown by Schiller, too. Hence, Ørsted was a fervent advocate of music and drawing lessons for children. They were to start early with rhythmic music and dance and continue by learning to play simple instruments and train the voice. His own children were given piano and drawing lessons at home. But he also made it clear that their lives must never be wasted by forcing them into the arts against their own predispositions.[17]

Dynamical theory was crucial for the relationship between the sciences and the arts. To Kant as to Rousseau works of art were created by geniuses in total freedom. But to Ørsted the laws of nature working in unconscious ways were fundamental to the work of art and hence it became the task of the scientist to appreciate the relation between physics and aesthetics, just as he himself would do for the rest of his life. Ørsted's aesthetics of nature had developed an argument in support of the mimetic idea. The beauty of the arts was rooted in the beauty of nature.

To Kant, whereas a work of art was the creation of a genius and not just an implementation of rules, mechanical physics was simply a product of mathematical rules.[18] Hence for him Newton was not a genius, because he had discovered his laws of motion a priori by solving mathematical equations. Ørsted espoused the idea that the rules of mathematics are irrelevant for the dynamical project. Ritter had no rules at his disposal when he discovered electrochemistry. And to his regret, Ørsted found no rules available to advance his dynamical project. Ritter and Ørsted saw themselves as scientific pioneers and geniuses, to whom the Muses of Reason would graciously reveal the secrets of nature. 'Don't forget that we are artists!', Ritter reminded his friend.[19]

Without being infatuated with anything but physics and metaphysics...[1]

23 | 1808
The Royal Danish Society of Sciences and Letters

In november 1808 Ørsted was admitted to membership of the Royal Society of Sciences and Letters. He immediately took advantage of this recognition by his colleagues to take the next step in his career. He needed better premises for his collection of scientific instruments and for his lectures, but also wanted living quarters in the same house. He was keen on a tenancy of Rubens's Court in Østergade. Now he petitioned the King for the rent of four hundred rixdollars a year (four times as much as that for Thott's Palace), which he could not afford out of his salary as a professor, although this had now risen to five hundred a year. He justified his petition citing the general utility of his lectures. The Foundation ad usus publicos supported him, and in May 1809 he was in a position to announce that from now on his lectures would take place in Østergade (on Strøget on the site of today's 'Illum' department store). At the same time he received a one-off payment of four hundred rixdollars to fund expanding his collection. An audience of seventy from all trades, civil servants, the military, artists, industrialists, businessmen, etc. joined his lectures.[2] Ørsted was most grateful and told Oehlenschläger in Paris: 'As far as I am concerned I live well. Courtesy of a royal grant I have acquired large and beautiful premises in 68, Østergade. I have exquisite instruments and, thanks be to God, I can do any experiment...'[3] This tenancy enabled him to accommodate his father, who—troubled by the frequent removal of his insane wife to a mental hospital—had sold his pharmacy in Sorø and moved to Copenhagen with his daughters.

When Ørsted's report on the Chladni figures had circulated around the panel for evaluation, it was unanimously agreed it be published by the Royal Society of Sciences and Letters and their author be honoured with the society's silver medal. This was not out of the ordinary, and was strictly in accord with the regulations. However, a member of the physical class found that this gesture was less than the author was seeking. In reality he must have submitted his paper to obtain admission to ordinary membership. This view prompted a protracted debate on the criteria for qualifying for membership, resulting in a joint proposal to grant Ørsted ordinary membership by the following recommendation:

> 'Johann [sic!] Christian Ørsted, professor extraordinarius at the University of Copenhagen. The merits of this scholar are recognised inside as well as outside the country. His diligence and talents have been rewarded twice by the University. His theses '*De forma metaphysices elementaris*' and '*Materialien zu einer Chemie des 19ten Jahrhunderts*' as well as his articles for Gehlen's *Journal der Chemie* have procured foreign recognition for him. The transactions of the Scandinavian Literary Society also contain beautiful papers by this author. In addition the Royal Danish Society has awarded him its silver medal to honour his report on acoustic figures.'[4]

All members of the physical class approved the proposal. But things were not going to be that easy. Professor Bugge (a member of the mathematical class), secretary of the society, and therefore crucial in this connection issued a circular letter on errors of procedure. According to regulations the proposal had to be submitted in writing to all four classes, not only the physical one. In this case regulations had been violated, because the proposer had merely sent a messenger around to the addresses of the members of the physical class to gather their votes and signatures. At the end Bugge remarked that he had no intention of impeding Ørsted's membership, but only wished to prevent similar anomalies in the future. Twenty-five members of all classes signed Bugge's circular letter; but it did not prevent Ørsted's admission.

Apparently Ørsted had only a few enemies in the often truculent academia, in which competitors fight for posts, money and promotion. One of his few enemies was Bugge, secretary of the society since 1801. Ørsted had felt Bugge's animosity and power when he had barred him from taking over Aasheim's lectures on experimental physics by claiming them for himself. The core of this animosity is not explicit in the source material. What is immediately obvious, of course, is their opposed research interests and theoretical attitudes. Bugge was a mathematical astronomer and cartographer rooted in the Laplacean school. In his universe Ørsted's Kantian metaphysics and dynamical system was a dead end. The distrust was mutual. In Ørsted's eyes, beneath his wig Bugge was the most stubborn representative of the unreflective, technological exploitation of science. In his opinion Bugge had had an adverse effect on the research activities of the society by seizing the bulk of available resources for his own cartographical project.

The enmity was so fierce that personal animosity must have been involved. Bugge was master of Elers' College when the Ørsteds were students there, and Hans Christian became the inspector. His purchase of philosophical literature for the college library must have been a thorn in the flesh of the master. In his private letters Ørsted made no bones about his contempt for Bugge, but he was silent about actual confrontations.[5] One gets a sense of a personal enmity that could never be resolved because their scientific outlooks and personal characters were incompatible. In 1808 Ørsted was a rapidly rising star in the academic cosmos while Bugge more and more looked like an extinct comet. Yet Bugge was re-elected secretary and treasurer in 1811 for another five-year period which turned out to be his last.

*

Soon after his admission Ørsted challenged Bugge, who might well have been the only scholar of importance he had not been able to make his patron, and who turned out to be the most inconvenient obstacle to his career. Each year a prize essay was set, and Bugge, who had hardly recovered from the shock of the British bombardment that had destroyed his library and collection of instruments and for a time made him homeless, suggested a ballistic investigation of Congreve's rocket. How is its orbit affected by its decreasing load of fuel, and how can the combustion of the rocket shell be regulated so that its course towards a given target is least disturbed? What kind of bore is the best for ensuring the accuracy of the rocket?[6] Bugge's proposal is typical of his conception of the role of science in society. Mechanics were prioritised and mathematical expertise should be applied to calculations to optimise technology whether for instruments for astronomical observations, for geodesic measurements, or for civil and military machines.

As an alternative Ørsted launched a proposal to bring an unsolved problem within dynamical theory to the fore. This was a question Bugge could not possibly dream of setting, because it had already been doomed to rejection by the First Class of the National Institute in Paris. Ørsted's proposal was the core of his project:

> 'The relationship between electricity and magnetism, which is strange because of both the similarities and dissimilarities of the effects of these forces, has already for a long time attracted the attention of scientists and found keen experimenters. However, this object has gained still greater interest from discoveries recently made. Many profound ideas have been expounded without entering the field of genuine experimental physics, although many new experiments have been made without being corroborated. Hence, the society finds that successful efforts might bring greater maturity to this part of experimental physics. Against this background the society offers a gold medal for the best investigation based on evidence or at least accompanied by confirmed evidence of the relationship between electricity and magnetism.'[7]

This was a provocation that with one stroke would force a clear division of the members into supporters of Bugge or Ørsted. The latter's proposal was approved. In mid-summer 1811 an essay arrived with the title 'Memoire sur le rapport mutuel de l'électricité et du magnétisme' [Report on the relationship between electricity and magnetism].[8] Deadline was New Year's Day 1810, so it had been exceeded by six months. The sixteen pages were handed over to the physical class, who did not have to trouble to evaluate it, since it had failed on formal grounds alone. The hand-written report has been preserved in the archives of the society and it is interesting to see how this question (to which Ørsted was to succeed in finding a definitive answer a decade later) was treated by the anonymous author writing in French and working within the framework of Laplacean science. The conclusion of the essay appeared from its subtitle: 'L'électricité est un état éxalté, la magnétisme est l'état naturel du même fluide' [Electricity is a state of tension of, while magnetism is the natural state of, the same fluid].

The essay is structured as a comparative analysis of electricity and magnetism. There are no original experiments, and the assertion that they are effects of the same fluid is a mere postulate resting on analogy. Heat and cold, according to the author, are due to different intensities of the same fluid (Lavoisier's 'caloric'), it is only that our senses conceive them differently. Something similar applies to ice, water, and steam. Chemically speaking they are the same, but the physical states are different. Examples are legion: Lavoisier had discovered that water and air contain the same element (oxygen), and a bright diamond consists of the same element as black coal. Hence, electricity and magnetism are effects of the same fluid, the first one in a tense state, and the second one in its natural state. This explained their apparent differences.

The French essay referred to experiments on electricity and magnetism gleaned from scientific papers published by the Académie des Sciences and the Société Philomatique, and from entries in the *Great French Encyclopedia*. All electrical experiments referred to were made with electrostatic generators. Galvanic electricity was not mentioned at all. The theory of natural states of fluids came from the German physicist Franz Aepinus, who asserted that the magnetic fluid was accumulated at the n pole and was absent from the s pole. Consequently, a magnetic needle was in a state of tension [*exalté*], and the two poles would seek to neutralise each other and end up in 'natural' equilibrium. Two n poles would then repel one another, because this repulsive property is inherent in the magnetic fluid. Opposed poles would attract each other, because their fluids seek equilibrium. But why would two s poles repel one another, when both lacked the fluid? This Aepinus could not explain, only explain away. His theory of the 'natural' state of magnetic fluids therefore appeared self-contradictory.[9]

Even if this essay had arrived on time the physical class would undoubtedly have turned it down. It would not help anyone get any closer to advancing the dynamical project.

*

In his memoirs H.N. Clausen, professor of theology, delineated the generation gap between the mossy academic establishment and the young rebels. Bugge was portrayed rather comically:

> 'He was extremely bandy-legged, any average sized dog could easily pass between his legs. He was still riding a horse as an old man and it seemed as if his body had been designed for this...As a scholar he was probably not very profound, but sympathetic and pleasant, and his general demeanour mirrored a French elegance and polish...Once during a lecture he was called outside by a noble visitor. His absence was taken advantage of in the most juvenile ways, of course. Upon his return to the lecture hall he said: "Gentlemen! This was a French Diplomat...As he left he remarked that my stables seemed to be quite near; I blushed and said: 'Sir! Those are the students of Copenhagen.'".'[10]

The observation that the head of the cartographic survey of the kingdom had spent countless hours on horseback and that his body was marked by it does not speak against him, of course. But what Professor F.C. Sibbern told Sophie Ørsted about Bugge's treatment of the astronomer H.C. Schumacher at Altona, was worse. According to this source, Bugge had harassed Schumacher, his successor as professor and leader of the cartographic survey, demanding 'that

Schumacher was to have no say over observers and no right to use the instruments when Bugge needed them'.[11] The Norwegian scientist Christopher Hansteen, too, had been struck by Bugge's animosity towards Ørsted: 'The old poisonous rascal would rejoice to give you a kick whenever an opportunity arose', he wrote to his friend from Mrs Møller's dining table. Presumably, Hansteen belonged to the group of 'young hotspurs having unconventional ideas about mathematics' as Bugge characterised members of the new generation in the latest edition of his textbook.[12]

In December 1811 the smouldering animosity against Bugge burst into a bitter, personal clash. The occasion was trivial. A Swedish architect named Almstädt had sent to the Department of Commerce a blue dye extracted from an earth he claimed to have discovered on Zealand and which he hoped to exploit commercially. Ørsted and Holmblad, a dyer, both members of the technical committee of the Agricultural Society, were commissioned to carry out a chemical analysis, which turned out unfavourably. However, the Department of Commerce sent a sample directly to the King and Ørsted was appointed to lead a new investigation, though this only confirmed their first judgement. To cap it all, the Department sent off a third sample, this time to The Royal Society of Sciences and Letters arguing that the matter needed 'thorough and scientific knowledge'. Now Bugge seized the opportunity and carried out an investigation that came out favourably. He communicated the positive result to the Department although not on his own behalf, but on behalf of the society. Readers of his report beginning 'The Society finds that...' were meant to believe that Ørsted's negative judgement had been repudiated not by the secretary alone, but by the entire society.

Ørsted had the matter put on the agenda and presented all the details. Bugge was charged with abuse of power, and Ørsted put forward the following proposal for an amendment of the society's regulations: 'When the Society has requested a committee or an individual to examine a matter in its or his own name, the resulting judgement must in no way be communicated on behalf of the Society until it has been approved by a plenary assembly'.[13] Ørsted had good reason to feel injured, but did he have to humiliate the old 'wig', who was twice his own age (Bugge had been pronounced royal astronomer the year Ørsted was born!), in front of all the members of the society, who had just re-elected him as secretary?

Moreover Ørsted complained that the society's annual transactions were replete with irrelevant, economic trifles like the postage of letters sent or the advances paid to a certain printer, etc. The transactions ought to furnish readers with a summary of what its members had accomplished for science in terms of treatises and lectures, as happened at the National Institute in Paris. Ørsted wanted the regulations amended so that the annual transactions of the society had to be submitted before the plenary assembly for approval before it went into the press. All his proposals were aimed at tightening the regulations in order to democratise the decisions of an otherwise high-handed secretary. Ørsted's proposals were adopted, but only carried into effect after Bugge's death.[14] From now on the annual transactions appeared regularly and contained scientific news instead of trivial minutes on day-to-day economic affairs. Ørsted's own contributions in particular raised the scientific standard.

Ørsted became a member of the mathematical as well as the physical class (and Anders in 1810 of the philosophical class) and, ambitious as he was, he soon became the most active participant in the fortnightly meetings of the physical class. In 1810-11 he gave four papers on 'Investigations of the basic causes of chemical effects' that soon filled the pages of his principal work, *View of the Chemical Laws of Nature Obtained Through Recent Discoveries*, published in German in 1812 (ch. 28).[15] In 1810 he lectured on the Swedish chemist J.J. Berzelius and proposed him for foreign membership on account of his research on electrochemistry.[16]

Ørsted and Bugge also quarrelled about the recruitment of new members to the mathematical class. Bugge suggested three military engineers skilled in mathematics, fortifications, granaries, sea-maps, etc., but unable to submit scientific publications. Ørsted's opinion was not asked for and he sulked: 'Despite the fact that I have for several years taken part in the works of the mathematical class according to the regulations of this Society, I have not been informed of these elections. I abstain, however, voluntarily from using my right to protest this time and give my vote to the proposed candidates.'[17] For Ørsted's part, he put forward a number of proposals for domestic members who might become a counterweight to Bugge's supporters.[18] In addition he pointed to a number of scholars from his European network to become corresponding members: Berzelius, van Mons, Gay-Lussac, Werner, Davy, Erman, Brewster, and Brera.[19] Only one of his candidates was French, and Berthollet, Haüy, Charles, and Vauquelin were not on his list. What Ørsted's names had in common was their criticism of Lavoisier's antiphlogistic chemistry and their relevance to his dynamical project.

Ørsted attended all the meetings of the Society. Moreover, King Frederik VI ordered him to teach physics and chemistry to cadets. So twice a week he would walk to the Military Academy, unless the young officers went to his new premises in Østergade instead to watch his experiments. Overwhelmed with obligations to teach he hardly found any time for experimental research, and feeling awkward about it he excused himself in a letter to Berzelius:

> 'This winter I have delivered far more lectures than usual, as the King has ordered me to teach at several military institutions, for young officers as well as for the General Staff. Even if this is most flattering and convenient to me in a good many respects, it nevertheless has the inconvenient effect of not leaving me a single moment for myself, as the hours between lessons must be spent on preparation, and at the end of the day I feel far too exhausted to want to do anything. Hence, I have accomplished nothing this winter apart from lecturing, and cannot tell you about even one small discovery.'[20]

Where the sciences and the arts unite.[1]

24 | 1809
Family and Friends

'THE PROFESSOR with the Nose'—as we saw—now resided at Rubens's Court taking care of his father and his mentally ill stepmother, and his three stepsisters, with their twenty-year-old maid Inger Birgitte Ballum looking after the entire household. W.C. Zeise, a seventeen-year-old eccentric, had been received into the household and laboratory as a helping hand to take care of the paraphernalia of retorts and test tubes. Gradually, Ørsted, who had previously enjoyed the patronage of Manthey, was assuming the role of a patron of young students of talent. Ørsted had a life outside the laboratory and lecture hall, too. True, he was responsible for more than a dozen lectures every week, and he attended meetings at the Scandinavian Literary Society and from now on also at the Royal Danish Society of Sciences and Letters every fortnight, but this was during the winter term only, from November until April.

A good many of his friends were absent from the ruinous capital for various reasons. His room-mates from Paris, Gjerlew and Lehmann, had obtained posts in the Department of Commerce, but during the unsettled years of war with Britain (1807–14) government headquarters moved to Rendsborg in Schleswig-Holstein [now Rendsburg, in Germany]. Adam Oehlenschläger remained in Paris until 1810, so they kept in touch by means of weighty and event-filled letters. The Manthey family had acquired a summer residence in the countryside and often received visits from Ørsted, now as a friend of the family rather than as a beneficiary of Manthey's patronage. The Gyllembourg family had moved back to Copenhagen after their farmhouse by Gyrstinge Lake had burnt down. Ørsted was tutoring Johan Ludvig Heiberg, and Thomasine Gyllembourg (formerly married to the exiled P.A. Heiberg, see ch. 17) and her Swedish husband hosted a salon, where Ørsted was a frequent participant.

Hans Christian and Anders remained very close. They had been so since early childhood, at the wig-maker's school and at Elers' College. They shared their money as well as their readings of Kant, Fichte, Schiller, and Goethe. They were contributing members of the same societies and arranged recitals and staged plays for their friends. In brief, they were like the twins Castor and Pollux, as Oehlenschläger had dubbed them. But to their three younger brothers their closeness and their status as academic stars might seem like an unattainable ideal, something they could not possibly live up to, and something that may have provoked envy. Two of them, Jacob, a year and a half younger than Anders, and Niels two years his junior, are not mentioned in any book. Nor do we hear much about Barbara Albertine, who was married to a Norwegian jurist, nor of Hermann, who was two years younger than Niels. Two more siblings died very young. All these siblings have been excised from the family memoirs. Against the splendid merits of Castor and Pollux the rest of the brothers looked like black sheep.

We have already seen (ch. 17) how Jacob's fraud came very close to ruining his father. By now his father had left this prodigal son to his miserable fate at Rudkøbing while the old pharmacist himself lodged with his paragon of virtue in Copenhagen. A bad reputation had also begun to cling to the next son, Niels Randulph. He was admitted to the Faculty of Law in 1801. According to the plan he was to be tutored, as Oehlenschläger had been, by Anders. He became part of the circle of students dining at Mrs Møller's dye-works. Niels was called 'the Enterpriser' as an allusion to his many enterprises, but his 'enterprises' turned out to be gambling debts and similar endless depravities. Cunningly, Niels had taken advantage of his brother's good standing with Count Schimmelmann. This provided him with a loan of five hundred rixdollars, but it only took him a year to fritter away this sum as well. Now, he left Copenhagen and was to be found in Kiel. He had left behind a debt of more than six thousand rixdollars, a huge sum of money equal to that of Jacob's fraud. His creditors besieged Hans Christian, who had no idea what to do, for Niels had decamped with nothing but a fantastic dream that one day he would become rich. A letter from Kiel recounted optimistically that he was about to be given a secure position, but his family wanted to see it before they believed it. Anders and Sophie were well aware of his troubles, and friends tried to forestall any tendency towards compassion, indicating that Niels had known perfectly well that he would simply cheat his naive creditors and idle about in Kiel.[2]

After the Bombardment Niels Randulph changed direction of life and became second lieutenant on the isle of Lolland. On the night of 7 and 8 November 1808 a certain first lieutenant, Jacob von Hille, was assaulted in a dark forest on his ride back to his residence. He was severely wounded in the back of his head, apparently by a sabre. The following day he died in bed in the vicarage, where he was billeted, without regaining consciousness. Some of those present at his deathbed later explained to the court martial that before he died Hille, unable to speak, had indicated with his fingers a few letters that were interpreted as an M, a G, and an Ø. Most witnesses, however, were unable to confirm these indistinct gestures, let alone interpret them, and by now the first lieutenant had gone forever. Nobody had witnessed the assault itself. On this basis the three 'letters' were highly charged.

The suspicion against Niels Randulph (apart from the letter 'Ø') was founded on a series of episodes, during which he had made known his hostility toward Hille and threatened to have him removed from the battalion or to 'make him sorry'. His motivation was said to be that Hille had taken leadership of the company, of which Niels had previously been in charge, so that he was now under Hille's command. And according to another witness Niels had also threatened Hille with a duel, if Hille did not voluntarily relinquish his command. After this Niels was said to have found the supposed third man of the plot, a corporal, who also saw himself as badly treated by Hille. Niels was thought to have secured his support by 'organizing' a pair of shoes and a hat for him from the military stores.

Although Niels Randulph did not have an alibi, the court martial still had to drop the charge of murder for lack of evidence. He was sentenced for the insults and the threats the court found he had certainly issued, and in addition for insubordination and the instigation of a conspiracy against Hille. Against this background the Supreme War Commission passed sentence: 'First lieutenant Niels Randulph von Ørsted should be demoted and jailed for three months' as well as having to pay all costs of the trial.

In June 1809 Professor Ørsted intervened on behalf of his brother and petitioned King Frederik VI to modify the sentence from demotion to simple discharge from the army on the grounds that Niels intended to seek foreign military service, that is, to leave the kingdom. The King acceded to the professor's petition, ordering Niels to report himself to the billeting officer in the border town of Altona and have his papers of dismissal handed to him in return for surrendering his uniform and equipment.[3]

In October 1809 Hans Christian sent his brother (charged with murder, acquitted, yet punished and finally exiled) from Copenhagen to Riga.[4] Niels Randulph did enter into foreign military service in so far as he joined the army of the Russian tsar to fight Napoleon, Denmark's ally. In February 1810 he wrote from St Petersburg to Oehlenschläger:

> 'Tomorrow I'm off for the army in Turkey. We may never see each other again, for I shall hardly set foot on Danish soil again without being injured or killed... I remain your truest friend till death. You have now become great and famous, since we saw each other, but before we shall meet again, I hope, if not to have acquired a name, at least to have gained honour in my position... Live well my dear Oehlenschläger, do never forget me. I shall never cease to love you. Your true Randulff Ørsted.'[5]

So, Niels as a volunteer cornet in the Kiev Grenadier Regiment went to war against the Turks. In July he fought courageously for Tsar Alexander in the battle of the fortress of Schumla (deep into the Ottoman Empire). At that time he had the command of a detachment of soldiers and at great risk he had scrupulously executed the orders given, and for this brave effort the Tsar pronounced Niels a knight of the third class of the Order of St Anna.[6] In May 1811 he petitioned King Frederik VI for his gracious permission to return to his native country:

> 'Since I know that courage and manliness are recommended virtues to win Your Majesty's grace, I take liberty to submit the knightly diploma issued by my present Monarch, testifying that I am not devoid of these virtuous properties. Honour, Your Majesty, has alone been my

> motivation, and I shall do no less when serving my native country... Under the eyes of death as well as under the bliss of celebration my thoughts are always in Denmark honouring Your Royal Majesty's Grace.'[7]

Sadly, the petition was ignored. In 1812 Napoleon invaded the huge Russian Empire, but was defeated by the terrible winter's cold and on his retreat was pursued by Russian armies. At the so-called peoples' battle at Leipzig in October 1813 Hussar Niels Randulph was killed. Upon receipt of this news, Hans Christian is said to have rather callously remarked to a colleague, 'You knew Niels: I think this was probably the best thing for him to do'.[8]

*

During his grand tour in continental Europe Oehlenschläger soon had enough of Steffens's company and his unbearable know-all air. He left the Steffens family and went to Paris visiting Goethe in Weimar on the way. From there he threw out feelers to Ørsted suggesting a revival of the confidence they had enjoyed as students. The first feeler addressed Ørsted with the pet name 'brave fellow' and played with witty remarks about people born brave, but after the age of three turning into 'bad fellows' or even 'very bad fellows'. He now saw Ørsted's character composed of 'one pound of a "brave fellow", two ounces of a "bad fellow"and one tenth of a grain of a "very bad fellow"'. The two ounces of a 'bad fellow' he deserved because of various failures to meet his obligations as a commissioner, while the one tenth of a grain of a 'very bad fellow' was as negligible as a drop of homeopathic medicine. In short: Ørsted's character was still as noble as that of a three-year-old child.[9]

Their correspondence shows how the events of the previous years, when 'The Professor with the Nose' had been treated as a practical resource, while 'Adagio' had sought a close companionship with 'the Emperor', had made them drift apart. When Ørsted had been reduced to an errand boy, it was no wonder that he had become jealous of Steffens and treated Oehlenschläger more formally. But at long last 'Adagio' had seen through 'the Emperor' and as a consequence had come to miss Ørsted's honest remarks on his dramatic poetry, although he would have appreciated laudatory outpourings more. Not only had Ørsted passed these works on to people with influence on the repertoire of the Royal Theatre, but he had also been a source of feedback on works, which he saw as reflections of his own spiritual life. Was he on the right track? Was he a playwright in progress? Did his plays touch anybody's heart? Had Ørsted realised, 'how the pungency of the impetuous adolescent had matured into decency and tolerance'?[10] It caused Oehlenschläger a pang to realise that the friend of his youth, whom he had treated like his errand boy, increasingly behaved like one when sending him dispassionate, businesslike notes mixed with reproaches or even moralizing reproofs, as he had for his opportunist petitions to influential people to advance his interests on the Parnassus.

The French occupying forces had closed down the University of Halle, and Steffens's professorship no longer existed. He had not even been paid his annual salary. In his distress he returned to Denmark where Schimmelmann helped him with money. Hoping for more help, at best a professorship in Copenhagen like the one Ørsted had recently acquired, he sent Crown Prince Frederik a copy of his *Grundzüge der philosophischen Naturwissenschaft in*

Aphorismen ['Fundamental Features of Naturphilosophie in Aphorisms'], and in February 1807 he sought an audience in Kiel. Unfortunately, the King had perused the redundant *Naturphilosoph*'s book, or at least leafed through it sufficiently to grasp that its aphorisms might have disturbing side effects. The Crown Prince struck a friendly tone: 'You have a clever mind. We would like to use your skills!'. But then followed an awkward condition: 'But We cannot let you lecture. You will drive my subjects insane.'[11] Not only did the Crown Prince wish to defend his subjects against the French army and the British navy, he also cared so much about them that he did not wish them to be seduced by German *Naturphilosophie* that would impair their brains and drive the students crazy. This outcome had been more than sufficiently proven by his lectures in 1802-3.[12] To Steffens himself the seduction of his young listeners with rhetorical tricks and soaring ideas was his life's blood. An academic ivory tower was not his natural environment. He was proud enough to let His Majesty know that if those were the terms, then he was no longer interested. Every hope of a comeback had been snuffed out.

A professorship without teaching obligations sounds like a temptation that would hardly be refused by today's academics, but Steffens was different. Could he not just teach another science, the Crown Prince is said to have suggested. 'No, I wish to speak the truth. Truth is my religion, and I shall live and die for it. I am too proud and well-known in Europe not to promote and execute the desire that nature bestowed on me.'[13] The Chancellor considered Steffens to be out of his mind, and ordered the Vice-Chancellor to bar him from all lecture halls and not to allow notices to be put up advertising his lectures, even those at his private address. And to be on the safe side the chief of police was requested to keep an eye on his movements in Copenhagen. The audience had convinced the Crown Prince that 'Steffens's brain had been strangely deranged to the disfavour of his public appearance...'[14] Anders found these steps far too harsh, because Steffens did have a doctorate, after all, from the University of Kiel, that provided him with a *jus docendi*, so according to regulations he was entitled to deliver all the lectures he might want.

Having broken with Steffens, a decision he was too cowardly to communicate to him face to face, and having adopted a no less antagonistic attitude towards Schelling's school of *Naturphilosophie*, Oehlenschläger embraced Hans Christian, his 'brave fellow', recognising that he would need his firm character and sound judgement. He found that Steffens had become far too didactic, arrogant, and intolerant. Moreover, his personality was vain and conceited:

> 'I can't tell if his knowledge of the inner nature of the earth is profound, but I do know that his aesthetical judgement is most contemptible. For that man becomes contemptible, who unblushingly and repeatedly shouts that Schiller is a wretched poet, and who without having read *Ars poetica* claims Horace to be just a toper and Virgil a valetudinarian. Verily, verily, I say unto thee: Virgil and Horace...and Schiller will be read and adored long after Steffens has sunk into '*die innere Naturgeschichte der Erde*' [the inner natural history of the Earth], to quote the title of his book.'[15]

Oehlenschläger's note 'A Characteristic of Steffens' was his passionate confrontation not only with 'the Emperor' and German *Naturphilosophie* and most of the rest of the early Romanticist movement, but also with himself and the seduction he now realised he had been lured into.[16]

'The Characteristic' was not the outcome of a flash of impulsive anger, but a well thought through criticism based on several years of coexistence. The note had long sat in his desk drawer. At first it was intended for another friend, then for Kamma Rahbek, but it ended up in a letter to Ørsted. It described how Oehlenschläger had come to read Steffens like a book. His outpourings were like the fizz of champagne. Sadly, he preferred the piquant and paradoxical to the solid and mature. He looked at the world through two kinds of spectacles, one coloured by Schelling, the other by Tieck. He did not burn for the truth, but for pig-headedness. He seemed to have cocooned himself inside something resembling a drug-fuelled mentality, emitting a flow of paradoxes that certainly titillated the fantasy, but soon turned out to be simply Schellingian jargon of the kind parodied by Oehlenschläger in his handwritten joke magazine *Honde-Posten*. The jokes partly derive from bad spelling, which is hard to hit in translation.

> 'Difrence between dogg and cad (after *Naturphilosophie*): The dorg is antique; the cad is romantique. The dorg is heathen, the cadd is Christian; the dorg is necessitie, the cad is freedom, the dorg is centre, the cad is peripherie. The dorg is the old, the cadd is the new Testament. The dorg is centripetal, the cad is centrifucal. The dogg is magnetism, the cad is electricity. The dogg is the masculeen, the cad the phemineen prencipel. The dorg is the Attican Apollo; the cad the Medicæic Venus. The dogg is cad, the cadd is dorg.[17]

Worst of all, in Oehlenschläger's eyes, was Steffens's contempt for the precise observation of actual phenomena and the lack of respect for the efforts of the truth-seeking empiricist, as well as for precise chronology:

> 'A man studying mining science with him at Freiberg, a man of good will and sympathetic towards Steffens, assured me that he had only dealt with mining science for a couple of weeks, and that Werner smiled, when talking about his knowledge... I don't understand how you can avoid having noticed the shallowness of his physics and chemistry. At Halle towards the end he started lecturing on experimental physics. His young students never settled down in his lectures because they were constantly laughing at the platitudes and nonsense he uttered.'[18]
>
> As far as his accounts of the world are concerned, I am always led to think of an anecdote... He also wanted to give an account of the history of Denmark, of which he claimed to have a clear memory from his *examen artium*; he talked about Gorm the Old's father, Harold Bluetooth, at great length and with a wealth of detail. I took the liberty of interrupting him, when he paused for breath, to remind him that Harold was Gorm's son, not his father. This made him blush but also enraged him, so that he asked indignantly: "What does that signify, if anything, and does it express any idea?".'[19]

'The Emperor' had been completely divested of his academic gown. True, he could be witty at times, until one wearied of his paradox-mongering. However, his Schellingian *Naturphilosophie* was amoral, and in the field of aesthetics he was a mere cheat.

> 'He was rambling on about the arts from morning until night, but was totally devoid of an artistic mind. What is aesthetic sense other than a higher ability to carefully observe beauty in

objects without any personal interest? And how is careful attention and quiet observation of the unselfish possible for over-imaginative dreamers, who merely look to find, and don't find by looking.'[20]

'Adagio' was completely disillusioned at finding his former hero so hollow. Steffens had nothing to offer in the field of science, of which he was a professor. As we shall soon see, Ørsted took up this devastating criticism in his *Dialogue on Mysticism* (ch. 25) stating that 'the aesthetic sense is related to a capacity to understand the hidden reason in nature, that is, a sense of nature'.[21] When categorising mystics he distinguished between those who do not find by looking (because they do not observe without bias, but only see what they want to see), and those who look to find, the detached researcher looking for the unknown.[22]

Ørsted's reaction was twofold. On the one hand he was happy about Oehlenschläger's long letter, 'a true proof of friendship', on the other he was a little condescending, characterising his friend as something of a renegade betraying the new school. He was most grateful for his initiative to renew their friendship, for being called 'brave fellow' again, as their relationship, after their Prometheus project had petered out, had threatened to degenerate into some kind of businesslike routine. But wasn't his friend going too far in totally turning his back on his former enthusiasm for Romanticism? It is one thing to change one's mind, but quite another to adapt to the bigotry accompanying this conversion by confessing to his conversion and submitting a petition to a pigheaded member of the power-elite:

> 'It is unworthy of you to confess to him. When I see my friend absentmindedly walk towards a ditch, I shout, 'Look out!' and if I think that he has not heard me I yell, 'Hell, there is a ditch!' . . . You would have been a fool, if the party you had supported with more or less reasonable intensity for several years had not distinguished itself by something other than by boasting and hot air.'[23]

Thirty-year-old Ørsted reminded his friend that when he had returned from his grand tour, Oehlenschläger had opined that Fichte and Ritter were dwarfed by Steffens and Schlegel, while Ørsted had seen things the other way round. Now the fluttering and reckless poet had fallen into the opposite ditch. Ørsted totally agreed that Steffens and Schelling often behaved like superficial braggarts, straying into untenable constructions. As to Steffens, Ørsted exclaimed, 'God save me from ever becoming a Steffensian, as He has protected me against it so far.'[24] But even if Steffens's lectures did not contain convincing and profound philosophy, they did offer 'lightning flashes of guidance'.

Baggesen, whom Oehlenschläger at that time fêted in Paris, however, was seen as nastier still, a lickspittle without integrity. 'I am sure he is now ready to kiss your arse if you would only print a few words in praise of him in return.'[25] This scorn carried great weight, for he had now been witnessing Baggesen's shameless courtship of his sister-in-law for more than six months, while he sponged on the hospitality of the 'Asse-children', turning the frail Sophie's head, and opening the floodgates of frivolous gossip in Copenhagen. For almost a year now, Ørsted had been enraged by Baggesen's enchantment of 'the Powder Keg'. The poet was on a visit to Copenhagen leaving behind in Paris his French wife, Fanny, who could not stand the climate of the north. At first Baggesen was treated in friendly fashion and offered the couch for the night. He then asked for a second night, then yet another, and as the couch suited him well he settled down in Vestergade and stayed there for nine months, a little longer, perhaps, than the hospitable host and his spouse had anticipated.

Yet, Ørsted did not find Oehlenschläger's parody in *Honde-Posten* all that funny and he begged not to be classified within any distinct philosophical school. He insisted on being considered an independent scholar in no need of parroting others. His aim was to develop his own philosophy in Danish, without borrowing Latin words or Germanisms from Schelling.

Given these reservations, what were the basic ideas in *Naturphilosophie* that he sympathised with? First of all it was Schelling's great merit to have founded *Naturphilosophie*. He had not thought through the details, which he often found unattractive, but just the basic idea of seeing nature as an organised whole. 'The most beautiful and the most apt thing he has written is that nature is nothing but the revelation of the godhead. Others have said that before him, but Schelling has said it now, when only a few religious people believed it, but no philosopher realised it.'[26] Secondly, he deserved recognition for conceiving the spiritual products of man, his science, and works of art, as conscious products of nature. 'They are the highest potencies of nature and must find their analogies at all levels of nature, even if the similarity is not immediately recognised.'[27] He referred 'Adagio' to his elaboration of this view in his *Reflections on the History of Chemistry*.[28]

Ørsted let his reply to Oehlenschläger mature by leaving it in his drawer for six days before posting it, to make sure that it was not just a whim. Their very friendship was at stake, so he took no chances. He took up his letter, re-read it thoroughly, regretted nothing, but scribbled down in a hurry some loose remarks on the relationship between science and the arts: 'Actually, we want the same thing, a totality where science, the arts, and religion are joined to a higher entity, a philosophical system. But we must approach this common goal along different roads, you as a poet, I as a scientist. Hopefully, we shall meet in the end, embracing each other.'[29]

Aesthetics has no well-defined concepts, Ørsted suggested, it works from feelings and ideas under the guidance of a higher instinct. Science, by contrast, carries the torch of reason and collects solid knowledge, until a point is reached where nature is no longer studied cognitively, but is looked upon aesthetically. Now matter becomes form. When nature is observed for its beauty the scientist needs the artist. On the other hand one cannot reach true knowledge of nature by meditation. Speculation without evidence is a chimera. There must be a mutual divine reason that helps us construct knowledge as well as beauty. The two friends therefore agreed that just as good science cannot rest on speculation, so good art is impossible with empirical awareness and mimetic principles. Or in Ørsted's own terms: 'You are so right, when you say that the observations of a real poet are worth much more than all the a priori supporters' systems.'[30] They had found each other in union with Goethe's lines, 'Fortzupflanzen die Welt, sind alle vernünft'gen Discourse unvermögend; durch sie kommt auch kein Kunstwerk hervor.'[31] [No discourse of reason can propagate the world, nor produce a work of art.]

Now Ørsted made a second confession. Good works of art are not produced by artists alone, nor, of course, by scientists alone, but perhaps by both together. An intimate knowledge of the idea of beauty, natural beauty as well as artistic beauty, is necessary and this can only be provided by a coherent study of both nature and the arts. So here the two friends (both practitioners) should be ready to deliver, each in his field. Philosophers must take on the task of deriving the general principles of art through independent studies of anthropology, history, and languages, find the unity in them, and relate the synthesis to divine reason. At this point Ørsted referred to Schelling's *Vorlesungen über die Methode des academischen Studiums* [Lectures on the Methodology of Academic Studies]. What appealed to Ørsted in Schelling's methodology was its objective idealism, its point of departure not in subjectivity, but in the laws of nature containing beauty as well as divine reason. 'Here are many precious pearls.'[32]

The second confession was dashed off in a hurry, for the mail coach was waiting and the professor had three hundred essays for *philosophicum* to mark on his desk. 'I hope you will remember from our being together how I dislike becoming involved in rows, and how well I keep up a friendship with all kinds of people irrespective of their opinions, as long as they are worth it.'[33] Their correspondence mirrored the true happiness and enjoyment they had in rediscovering each other after a crisis that, given Oehlenschläger's temper, could have ended in a total breakdown. All the same, Ørsted worried about 'Adagio's' sudden change from one extreme to the opposite. This is why he encouraged him to read his *Reflections on the History of Chemistry*. Here he would learn about the effects of drastic swings of the pendulum in the history of science, and would also realise that in the long run staying calm is a better strategy than spinning around like a weather-cock.

*

Upon his return to Copenhagen Oehlenschläger at long last married his Christiane, and they joined the Gyllembourg salon in Blancogade every second Thursday. Johan Ludvig was now sixteen and ready to enter the University. Ørsted initiated him into the first principles of mathematics, Weyse gave him piano lessons, and private tutors taught him Latin, Greek and French. Ørsted was more than a tutor, he was a friend of the family, and as long as he remained a bachelor, he was a fixture every Christmas Eve in Mrs Gyllembourg's home.

In 1811, when they celebrated Christmas Eve together at the Gyllembourg's, there was also present one of Ørsted's students, the young Swedish Count Gustav Taube, who admired his professor not least because of his experiments on acoustic figures. Also, with Count Taube was his sister Anna, whom Johan Ludvig was infatuated with and tried to impress. Christmas was celebrated according to Swedish customs with glazed ham and the exchange of anonymous presents. The recipient of the present was not informed, but had to guess the identity of the giver, who might indicate her or his identity by wrapping up the gift in an extraordinary way or by writing a funny verse. This custom prolonged the distribution of presents, of course. A present for Count Taube was accompanied by a poem by Ørsted, in which he praised music and described an acoustic figure as 'a glimpse of the hidden harmony of the world'. So after some lines it dawned upon Taube (and C.E.F. Weyse the composer, also a guest, who was the tutor of music) that the present had some affinity to music. The remaining lines compared the mystifying Swedish custom with an initiation into Freemasonry and the apprentice's entry into the temple of Isis at Sais in Novalis's fragment. But the veil to be lifted in this case was

Ørsted's textbook that also served to hide a second present, a copper plate, on which acoustic figures, the scientist's visualisation of music, could be produced. The allegory was obvious: study dynamical physics and you will be initiated into the mystery of music![34]

At this stage Ørsted was about to enter a Masonic lodge. The poem described the blindfolded and helpless young man approaching the wise master of the lodge who will in turn unveil the mysteries. The ritual comprises questions to be answered, stones to be turned, knots to be untied, slips of paper to be read, and symbols to be interpreted. The ritual in Ørsted's poem is not unlike opening the door into the world of music to which physics keeps the key.

In March 1812 there was a special event and Hans Christian had been invited to bring along his father, the old pharmacist. Johan Ludvig performed *Don Juan* in his marionette theatre in honour of the beautiful Anna Taube's twenty-fifth birthday. He hoped to win her. The somewhat younger Johan Ludvig had adapted and versified Molière's original play and Weyse had helped him write the music for piano. The marionette show of *Don Juan* became a turning point in young Heiberg's life. Weyse recited the prologue and the show made tears run down Oehlenschläger's cheeks.[35] After the final fall of the curtain he leapt to his feet embracing Johan Ludvig prophesying a bright future for the budding poet. Could Heiberg wish for more from his top quality audience? As a real Don Juan he imagined he had advanced his chances of seducing Donna Anna, but alas, she was already a faithfully married woman.

> You belong to the new school of philosophy,
> so I am in no doubt that dusk must be holy to you,
> for light is no friend of mysticism...[1]

25 | 1810

Dialogue on Mysticism
Ritter's Death

I**N HIS** 'Dialogue on Mysticism', Ørsted settled his dispute with German *Naturphilosophie*, with which some of his known and unknown opponents identified him. The dialogue went through a process of writing and rewriting for five years, before he read it aloud to the Scandinavian Literary Society in December 1809. The timing was carefully chosen. Oehlenschläger had just returned from Paris, so now Denmark's celebrated poet was among the prominent listeners as a special guest. The first drafts had been intended to be the philosophical part of their planned New Year's Gift, 'Prometheus'. Perhaps they had not completely given up hope of finding a publisher, because the dialogue was the only paper he gave in the society that was not published in its periodical. The manuscript only appeared in print posthumously in *Samlede og Efterladte Skrifter* [Collected and Posthumous Works], with a few amendments.

This dialogue was in fact Ørsted's first attempt at the genre; his dialogue 'On the Cause of the Pleasure evoked by Tones' was a later offshoot of this one. The target audience here is obviously the layman, and the interlocutors express themselves in colloquial language avoiding as far as possible the use of loanwords from 'the new school'. As in Hans Christian Andersen's fairy tale 'The Bell', the dialogue is staged deep in a forest, where a group of young botanists are philosophising while lying on their backs staring dreamily through the tops of beech trees at the blue sky or sitting gazing between their upright trunks at the blue sea. The main character is Ernst, Ørsted's mouthpiece. This name is probably an error due to carelessness, because Ernst reminds his companions, Julius and Alexander, of their previous discussion in his dialogue on tones, although there his name is Alfred, and the name Alexander does not appear. In his preliminary drafts the interlocutors were called Antonio, Ludovico, and Francisco, and these Italian names suggest that Schelling's dialogue *Bruno* may have been in his mind as a model.[2]

Julius and Alexander find Ernst reading in an unusual kind of a study, a clearing in a beech forest that he calls his temple of nature. The attitude is devotional, and Ernst does not mind admitting that he is overwhelmed by the beauty and purposive organisation of nature. Ernst is depicted as a romanticist who, like Kant, is awestruck by the sight of the starry heaven visible through the foliage after sunset. Julius on the other hand is not moved at all and exclaims, 'You really are a mystic!' thus including Ernst in 'the new school' without being able to define what this school stands for. The jibe is completely repudiated by Ernst, who considers 'the new school' a many-headed monster at odds with itself and avid for sensation. 'Indeed, some of them try to inflate their own insignificance into genius of first rank, while others swear to believe whatever their masters say.'[3] Ørsted was no doubt hinting at Schelling and Steffens respectively. Kindly but firmly Ernst begs to be regarded as neither the embodiment of *Naturphilosophie* nor an appendage to Schelling.

Ørsted touched here upon the problem of categorisation that unfortunately was common among debaters. He felt victimised by this unworthy form of debating. 'The new school' was the category used by some of his associates with which to label him. Exactly what this category meant was unclear, 'because it does not like definitions'. Even so it was obvious that it was a category for metaphysicians such as Kantians, Schellingians, Steffensians, possibly adherents of Fichte, but definitely Winterlians, Ritterians as well as other romantics and mystics!

In line with his own understanding of his dynamical project, Ørsted pursued an independent course. In his petition to the King in 1817 he looked back on his career, explaining that his lectures and articles up to then were meant as experimental balloons sent up to obtain feedback as he looked for an independent place for himself in the romanticist upheaval following the French Revolution and the Napoleonic wars.[4] One of these balloons, 'On the Correspondence between Electrical Figures and Organic Forms', which he never reprinted and probably later regretted, embraced ideas from Steffens's and Schelling's *Naturphilosophie*. And before that he had compromised himself by taking Winterl's speculations seriously and thus confirming accusations of being something of a mystic. He had partly brought it on himself.

During this turbulent period Ørsted had wrestled with his 'Dialogue on Mysticism', in the final version of which he pronounced judgement on his experimental balloons as well as on the accusations he had been exposed to. He refused to be categorised, considering it a mistake to attribute to somebody the entire philosophy of a party just because that person agrees with it at some points. He declared himself an unbiased, independent, and broadminded researcher prepared to heed any serious contribution to dynamical science or to the aesthetics of nature. 'Is it not possible to respect a scholar, and to do research with him, without accepting all his views and without becoming his adherent?' he let Ernst exclaim.[5] He tried to live up to the ideal he had advocated in his works on the history of science, that many views that could now be repudiated by better knowledge, still, when they were put forward, had lifted a corner of the veil formerly hiding the truth.

The term 'mystic' was used in the press campaign against him occasioned by his support of Winterl. It was meant as an insult, of course. In his dialogue he defended himself calmly against the accusation by distinguishing between mystics and enthusiasts. A mystic is a person who is preoccupied with things he does not understand, with the intention of getting to understand them. An enthusiast, by contrast, is a person who is either an adherent or an opponent of something he does not understand and does not want to understand lest he should risk having to change his view of things. A mystic in the first sense is no insult to Ørsted, but a mark of honour.

So what is a mystic? First of all, of course, a mystic is no scientist, the difference being that a scientist studies phenomena of nature in time and space using his reason and senses. But human reason and senses are inadequate to provide full insight into nature, and in accepting this he cannot help reflecting on what is beyond the scope of reason and the senses, because there is more to human existence than objects of science. The dialogue on mysticism addresses the metaphysical issues of human existence such as ideas about nature as an entity, as well as moral, aesthetical, and religious philosophy. Or to put it in Kantian terms, the interlocutors of the dialogue discuss the noumenal realms, which have no direct bearing on scientific research.

What is this 'new school'? Ørsted asks, but he is at a loss to define it. And he is not the only one. Historians of science dealing with Ørsted seem to be focusing mainly on influences conducive to his discovery of electromagnetism, and the majority of them point to *Naturphilosophie*, but no-one among them, as far as I know, has tried to define what it is or believes. Kenneth Caneva, for one, does not 'believe that [a definition] is historiographically useful'. Nor would it be 'a trivial matter to come up with a rough characterization of what is to count as a sign of *Naturphilosophie*'. For want of a definition it must be hard to ascertain similarities and dissimilarities between Schelling and his allies or critics, a point to which I shall return soon. But as far as Ørsted is concerned he does not explain in detail his opposition to *Naturphilosophie* himself, because—as explained in his 'Reflections on the History on Chemistry'—there is something to be gained from listening even to extremists (ch. 18). His dialogue concentrated on his own view, not on his differences with others.

Dialogue on Mysticism is staged in three acts. In the first he introduces the botanists to Kantian dualism 'that there is a higher world [the noumenal] than the sensual [phenomenal]', and in consequence, 'ought there not to be a reason in nature that is beyond your reason?'.[6] Following Kant, we cannot have true theoretical knowledge about nature in itself, because nature-in-itself is infinite and can only be fully understood from God's omniscient position, while our intellectual faculty is finite and operating from one particular perspective only. Nevertheless, natural philosophers and religious thinkers have always tried to expand our knowledge about the infinite.

Alchemists tried to understand the microcosm by means of symbols depicting gold pasigraphically with the sign of the sun and other elements with signs of the planets—but to no avail. When Kant came to realise that material concepts of atoms on the one hand would be amenable to mathematical treatment, although they are not objects in time and space, whereas the forces of matter on the other hand are perceptible experiences, but lacking concepts of constitutive principles allowing for mathematisation, he resigned himself to the recognition that this apparent antinomy is rooted in our limited faculty rather than in objects of nature.[7]

In the second act the young botany students find Ernst and his friends in the clearing, but they do not join in the conversation, which touches on a quotation from Schiller often used by Ørsted: *'Was der Geist verspricht, leistet die Natur'* [What the spirit promises, is fulfilled by nature].[8] This adage has been interpreted in various ways. The laws of nature conform to human reason, Ernst claims, but Alexander demands a proof. Ernst finds this corroborated in the laws of motion, which obey mathematical principles. And mathematics is a product of human reason and therefore also the basis for the discovery of laws of nature. Ørsted's favourite example is his sonorous figures that portray tones. They reflect the geometrical laws constructed a priori by the human mind. According to this interpretation the Schiller adage could be rephrased as 'Laws of nature are a priori', that is to say we do not walk into nature looking for laws of nature at random, we think them out first in our minds. This is what Ørsted called a thought experiment. As expected, this interpretation was immediately opposed as follows: 'But it often happens that nature does not bear out the deductions of human reason.'[9] This was readily admitted by Ørsted, of course, to remove a widespread misunderstanding. Ørsted is in complete agreement with Kantian metaphysics: '…to cognise something a priori is to cognise it from its mere possibility'.[10] To assert that laws of nature arise a priori in the human mind is not at all the same thing as saying that everything that arises in the mind turns out to be a natural law. They might be figments of the imagination as was often the case with Schelling's and Steffens's speculations.

Ernst summarizes the discussion so far: Nature is governed by perpetual laws, and the form of 'matter' is constantly changing, although often very slowly. 'Matter' is nothing but the effect of forces. This applies to the macrocosm as well as the microcosm. Laws of nature conform to reason.[11] He explains to the other interlocutors that 'what makes the great Whole one world is the eternal, unbroken chain of effects. Without it everything would collapse into uniform matter.'[12] Only in the third act does he expound his fundamental philosophical ideas. Not only is nature the outcome of dynamical activity but this activity seems to be purposive. In other words a secret plan seems to be hidden in nature. And if observations of natural activities are teleological, this indicates a direction towards a goal and by implication an author behind the secret plan. While Ernst expatiates on the forces of nature behind nature's end, Julius recalls that he once read Kant's *Kritik der Urteilskraft* [Critique of Judgement] which shows that while nature is determined by the inexorable necessity of laws, apparently this happens according to a certain purpose, leading the thought to 'a secret plan of nature, and an equally wise and mighty creator of the Whole. The author [Kant] seems to me to achieve eloquence when observing the wisdom required to plan so many things for the benefit of man and for the maintenance of the Whole. I confess that I was deeply touched when I read it. It is still incomprehensible to me, how I could forget this for so long', Julius admits.[13]

Given this secret plan of nature, embodying the teleological principle of purposiveness, end, and author (God), Ørsted has adopted the Kantian synthesis of determinism and free will or the harmony between necessity (the starry heaven above me) and freedom (the moral law within me). This synthesis Kant calls 'summum bonum', or the highest good. It is not part of theoretical knowledge, of course, but is a reflective judgement, or what Ørsted calls a spiritual intuition ['aandelig Beskuelse'].[14] This understanding is not new. Ørsted had already established this harmony in his essay on acoustic figures; these were natural phenomena incorporating aesthetical qualities pointing to the infinite harmony and beauty of nature. And just as he expressed his enthusiasm for music in his dialogue on the beauty of tones, he now let the interlocutors reverence the aesthetics of nature. Ernst saw this philosophy, expressing 'a meaning of life', as an example of mysticism in the good sense. The regulative idea of 'a summum bonum' implied resolution of the apparent contradictions between the determinism of nature and the moral freedom of man. Moral philosophy aspires to the highest virtue as the social sciences aspire to the perfect society, and as science aspires to the realisation of the *summum bonum* incorporated into the secret plan of nature. This goal, of course, cannot be achieved by the determinism of laws of nature, but requires the deliberate effort of the human will.

Ørsted's dialogue is devoid of Schellingian tropes and terminology. There is no world soul, no absolute, no identity, no potencies or indifferences, and no pantheism. Ørsted weeded out all the Schellingian terms that appeared in the first edition of 'Reflections on the History of Chemistry' from the reprint in *The Soul in Nature*.[15] This shows that at an early stage he was more attracted to *Naturphilosophie* than he ended up being. Whereas Ørsted as a dualist distinguished between theoretical knowledge and reflective judgement, such distinctions are absent from Schelling's philosophy of identity.

Schelling's aim was to eliminate the contradiction between man's freedom as a conscious and law-making subject and nature as the unconscious object determined by laws. Schelling's objective idealism, his *Naturphilosophie*, is often epitomized in the catchword: 'spirit is the invisible nature, while nature is the visible spirit', which signifies pantheism. The notion of spirit means the hidden unity of natural reason and creativity, which is one and the same as the work of creation in the pantheistic sense. Thus Schelling's objective idealism is a philosophy that seeks to identify unconscious nature with man's consciousness of it. This identification is anchored in the Absolute. Organisms of nature and man's consciousness have their mutual origin in the creation, which is seen not as an event once and for all, but as a continuous process of evolution, in which both organisms of nature and man's intelligence develop from primitive stages to higher ones. They are differentiated and they achieve potencies. During this process Schelling sacrifices moral philosophy in favour of the arts that assume the task of mediating between the necessity of nature and man's free creativity. Thus *Naturphilosophie* is a philosophical construction that is not oriented towards nature in any particular sense, and makes no claim in relation to science. Schelling's is called *Naturphilosophie* because it identifies man with *die Weltseele* [the world soul], a reductionism which in practice determines man as a natural phenomenon and therefore turns its back on moral philosophy, because he sees laws of nature as ethically indifferent. The world soul takes the place of God and constitutes good as well as evil which eliminates ethical challenges from his thinking.[16]

The possibility of overcoming the schism between the phenomenal and the noumenal brought the Ørsted brothers together. More than anybody they would share an interest in constructing a coherent philosophy making it possible to unite apparently different realms. They approached the problem of dualism each from their own side and had recurrent discussions, as we shall soon see (ch. 26).

*

Five months had passed since Ørsted had heard from Munich. Ritter had been incessantly sick, and by now he was so weak that he could not even write, but had to dictate his letters to A.F. Gehlen, who took care of him on his sick bed. Gehlen edited *Journal für Chemie und Physik* and both Ritter and Ørsted were co-editors. Gehlen was also Ritter's colleague in the Bavarian Academy of Science, and Ørsted knew him from his sojourn in Berlin in 1801–02. Tuberculosis was about to end Ritter's life. His appeal for urgent money was no longer relevant. The Academy did nothing to help him, nor did Schelling, another member of the Academy. His future was gloomy. The deathblow supervened when he had worked for sixteen consecutive days until far beyond midnight, sustained by opium. He was writing the introduction to his autobiography 'Fragments of the Posthumous Writings of a Young Physicist'. Now it was complete. 'It contains the biography of the deceased, his innermost thoughts; it is my own life. It is a most serious farce dedicated to his friends', the exhausted Ritter told Ørsted.[17] More work was waiting for him. Should the vigour of youth ever return to him, he promised to sacrifice the remainder of his life on the altarpiece of science. The world needs science, he added.

Ritter's sojourn in Munich had not turned out to be the fertile life of scientific research he had hoped for. True, he had a regular salary, members of the Academy had a uniform, and he had ample time for research. The calamity was that a certain French emperor had been occupying Bavaria ever since Johann Wilhelm and Dorothea and their four children arrived there. Some relatives had moved in, sponging on them as long as the host received his salary. He also had to billet soldiers, and a time of dearth followed in the wake of war. The illnesses he had caught while staying in summer houses at Jena racked him, too. At one time, Ritter had crossed the Alps and in Italy he had met a boy called Campetti, who was said to have magical skills; he could detect water by means of a divining rod. Ritter initiated a rhabdomantic project, the art of detecting underground water or metals by a divining rod, as well as a sidereal project claiming to do the same thing by means of pendulum swings. These esoteric projects were looked at with great scepticism by Ritter's colleagues in the Academy and with a disquieting silence by Ørsted. They did not lead anywhere. When Ritter promised to sacrifice his life to science, he meant to say that he intended to resume his galvanic experiments with frogs and mimosas, should he survive.

He did not. On 23 January 1810 he died. The day before, Ørsted's last letter had arrived. As the letter is lost we do not know what he had promised Ritter. Gehlen urged Ørsted to do the same for Ritter's wife and children as he had previously promised their breadwinner, since the widow's economic situation was hopeless. And would Ørsted please be kind enough to send Gehlen the necessary information for the obituary he was going to read aloud to the Academy: 'You knew him best!'.[18]

Ritter left no archive. Thirty-two letters from Ritter to Ørsted have been preserved, but about half that number that Ørsted wrote back have gone missing. Ritter's death mask was made and from that a bust was displayed in the Academy, but this has long since disappeared. The miniature that once existed has disappeared, too. The only picture we have of him is a wood cut said to be copied from the lost miniature. The headstone to Ritter's grave, if he ever had one, is demolished.[19]

The two friends were quite dissimilar especially as far as temperament is concerned. Ørsted was an Apollonian phlegmatic, generally cool-minded, always busy, reasonable, well-organised, proficient in languages, a sharp debater, strong in will, sociable, cosmopolitan, ambitious, strategic, a talented organiser, and somewhat of a perfectionist. Ritter, in comparison, was a Dionysian bohemian, euphoric and eccentric, spontaneous and messy, a womanizer, relentless towards himself, unstructured, unstrategic, at times hypermanic, romantic, rich in ideas, sentimental, and self-made. Obviously, two such characters can be complementary and most beneficial to one another, if they pay each other respect and loyalty. Ritter taught Ørsted to make experiments and made him realise that experiments are decisive for the progress of science. They were in perfect harmony as to the dynamical project, and both agreed with Kant that polar forces are most likely reducible to a common source.

In return Ritter learnt from Ørsted's experiments on acoustic figures. And Ørsted also introduced him to Kantian epistemology.[20] Ritter gained in Ørsted an agent to try to win the Napoleon prize for his galvanic battery and a translator and editor of his papers. Ørsted helped him obtain his membership of the Bavarian Academy and he sent him money. But there were mismatches too. Ørsted did not dare take up Ritter's electrophysiological experiments on his own body, and he distanced himself entirely from Ritter's rhabdomantic and sidereal projects.

No doubt both of them had rather negative judgements of Schelling.[21] The first time Ørsted mentioned him was in his dissertation, in which he demonstrated that Schelling's attempt to show inverse proportionality between polar forces in the formation of chemical combinations was a mathematical absurdity. On his grand tour Ørsted did not spend more than a few hours with him, as against weeks spent with Ritter. There were personal reasons for their animosity. Ritter and his admirers, Friedrich Schlegel and Novalis, detested Schelling. They found that he strutted in borrowed plumes when he tried to found his *Naturphilosophie* on Ritter's experiments while not doing any research himself.[22] Ritter and Ørsted were in complete agreement finding Schelling and Steffens quixotic. Of course they had grasped a little, but the swirl from their dizzy round dance extinguished all lighting around them. Fortunately, nature herself would correct their delusions.[23]

In summer 1805, however, their views on Schelling slightly changed. Firstly, Ørsted encouraged his friend to read Schelling, since fate had brought them together as colleagues in Munich.[24] Ritter replied that he had always given Schelling a wide berth, and that he had always been more interested to know where he diverged from him rather than where he could follow him. If Ritter were to spend time with Schelling, to whom he had had a close relationship at Jena, he would read him critically.[25] But then Ritter met with him face to face as a colleague and was soon on speaking terms with him, as Ørsted had wished. In 1807 Ritter had even come to appreciate him as a kind and honest man. But praising a person is not the same thing as being a passive disciple.

Ørsted's formation as a scholar took place in a period of European wars as well as scientific and philosophical turbulence. He was convinced that the history of science follows the swings of a pendulum. In his 'Dialogue on Mysticism' he distanced himself from extreme positions, hating to be categorised by others. To determine his position in relation to the controversies of his day we can go to no better source than his letter to Manthey describing his dynamical project:

> 'I am surprised to realise how the spirit of the present age seems to unite such divergent positions into the same goal. The advent of antiphlogistic theory took all life and force away from chemistry, and one saw nothing but matter coming and going through big or small pores without knowing where, quickly or slowly without knowing why, bringing forward new qualities without knowing how. Kant's dynamical philosophy broke the ice, but even the best minds did not know where to begin, until Ritter showed, or rather began to show, that all matter must be regarded as polarities of differences—as in water. This view can hardly be experimentally proven more clearly than I tried to do in my last letter to you. Winterl's theory conforms to this although he is not aware of the entire impact. All that Kant's clever philosophy, Winterl's comprehensive observations, Ritter's profound experiments, and Steffens's poetical enthusiasm found in nature together combines into a Whole...'.[26]

So, his merit consisted in bringing these elements together into a coherent dynamical system. Kant taught him to see the weaknesses in corpuscular theory, and through Winterl's theory and Ritter's electrochemical experiments he was prompted to realise that chemistry might become a science after all. And thanks to the *natura naturans* theory that both Kant and Schelling advanced (although differently) and to Lichtenberg's and Chladni's figures he saw an aesthetics of nature taking shape. Ørsted defied categorisation. Now that Kant and Ritter had departed and Winterl had been unmasked as a fraud, he was on his own to carry through his dynamical project.

PART IV
THE SPOUSE

It is rare to meet an upright man....¹

26 | 1811

Career and Brothers Working Together

As far back as the time of King Christian VI (1730–46) there had been no degree in physics and chemistry at the University of Copenhagen. Chairs in these subjects had been founded during the reign of Christian III (1533–1559), but abolished around 1730 in favour of an additional professorship of theology. Now the University had appointed Ørsted as an extraordinary professor of physics and chemistry under the Faculty of Philosophy. Single-handedly he had resurrected two core disciplines that had been neglected for more than seventy years. Although his ambitions were high there were still no graduate programmes of physics and chemistry and no colleagues. His textbook was only used by students qualifying for *philosophicum*, since no student in Copenhagen had the opportunity to graduate in science.

As professor of physics Ørsted desired to 'form a school'; as he told Oehlenschläger in 1806: 'I now have an opportunity to establish a school of physics in Denmark and hope to find some talented people among the many students attending my lectures.'² His recruiting fields were all the undergraduates qualifying for *philosophicum* (of which physics was a part) and undergraduates of medicine and pharmacology (of which chemistry was an auxiliary subject). He had already found his first pupil, seven years his junior, the Norwegian Christopher Hansteen, whom his brother-in-law Jacob Bull had introduced to Mrs Møller's boarding house.

Ørsted's plan for establishing a school was based on mutual friendship rather than a calculated patron-client relationship. In Hansteen's case a genuine friendship developed, where Ørsted supported the undergraduate without any selfish motives. At first Hansteen studied law like Oehlenschläger, but when Ørsted returned in 1804, the Norwegian changed from law to science, following his courses without graduating, of course, since the study of science did not lead to a degree.[3] Besides Hansteen ran out of money and had to take employment as a grammar school teacher of mathematics. In 1811 Hansteen was awarded a medal by the Royal Society of Sciences and Letters for his prize essay on the question, 'Can all magnetic phenomena of the Earth be explained by a single magnetic axis or are several required?'. Ørsted helped him to have a summary of his essay published in *Journal de Physique* and in Schweigger's *Journal für Chemie und Physik*, and some years later to have the entire essay published.[4] In 1822 Hansteen assisted Ørsted in experimenting on the compressibility of water, and Ørsted helped Hansteen by measuring the earth's magnetism in Germany, France, and Britain by means of special instruments.[5] Christopher Hansteen expressed his gratitude as follows:

> 'With true gratitude I think of the favours you have shown me, my dear Professor, and of how much impact a man's better acquaintances have on his way of thinking. Allow me to express my gratitude in these few words (which are not empty): it is rare to meet an upright man in whom neither head nor heart are flawed, and when it happens eventually one feels like shouting: *eureka*! [I have found him!].'[6]

W.C. Zeise was another protégé. At seventeen years old he came to live with Ørsted in 1806. Zeise's father was a pharmacist in Slagelse, the town in whose vicarage Søren Christian had grown up, so the two families knew each other well. Young Zeise was apprenticed to Court Pharmacist Becker in Købmagergade, began studying chemistry and pharmacology, and called on Ørsted. He was a rather eccentric young boy, something of a hypochondriac, and had a mental block whenever somebody tried to force him to do anything. Apparently, Ørsted understood his thin-skinned character and never school-mastered him, but patiently encouraged his youthful initiatives. In that way Ørsted became a kind of father figure to the adolescent. In 1809 he entered the University, passed his *philosophicum*, and was admitted to Valkendorf's College to devote himself to the study of chemistry. Whenever Ørsted demonstrated his experiments at lectures, Zeise assisted his paid amanuensis. In 1811 he wrote from Slagelse to 'his great teacher and careful supervisor' to thank him for the many

> 'glittering pieces of knowledge and basic principles (the value of which I daily recognise more clearly) that you have bestowed upon me. I realise how much I have to thank you for my scientific education and force myself to restrain my emotions and the outpourings of my heart in order not to babble profusely in front of your honour. I have now recovered my health, my humour, and my former desire to acquire knowledge.'[7]

During his illness at Slagelse he had studied Gren's *Grundriß der Chemie* intending to return to the University. In 1817 he defended his doctoral thesis on the impact of alkalis on organic matter, such as sugar. Ørsted was one of the examiners and approved his student for the doctorate. As a good patron he also managed to provide him with public funding for a grand tour, recommending him to study with his friend Berzelius in Paris.[8]

The third student recruited at an early stage by Ørsted to become his pupil was Carsten Hauch, half Norwegian and half Danish. Court Steward A.W. Hauch was his uncle, so when he arrived in Copenhagen in 1803 he lived for a time in his flat in the equerry's yard close to Christiansborg Castle where the notable Physical Cabinet was housed. Here young Carsten was shown the experiment to separate water into oxygen and hydrogen. The Court Steward (we may recall) was a supporter of Ørsted despite their disagreements, and now he voiced to his nephew his discontent that Ørsted taught that Newton and Kant were the two poles on which modern science pivoted. He found it scandalous to utter such simplifications in public lectures. In 1807 Carsten Hauch volunteered under the banners of a corps making a sally against the British; as he had not yet entered the University he was not entitled to serve in the Crown Prince's Students' Corps under the command of his uncle. 'In one of the sallies a volunteer by the name of Haug was shot; rumours spread that I had been killed... Afterwards it happened over again that I was met with an exclamation of surprise: "Are you really alive? Everybody says you are dead!".'[9]

In 1809–11 he once again attended Ørsted's lectures, which at this time dealt with issues that were soon to be published in his major work *View of the Chemical Laws of Nature Obtained Through Recent Discoveries*. However, there were still reminiscences of 'the electrical figures and the organic forms (positive in ramifications and negative in circles)', and Ørsted 'hinted that sexual difference in the vegetable and animal kingdoms depends on similar opposites, although I am not sure he was explicit about it....'[10] Carsten was thrilled with these lectures, and inspired by Ørsted's enthusiasm he decided to study physics. Ørsted lent him books from his library and delighted in finding Hauch's nephew in front of his lectern: 'Ørsted expressed his opinion that I was sufficiently prepared to write my doctoral thesis on physics'.[11]

But this was not to be, because in 1814-15 Hauch entered a period in which philosophy and poetry took a firmer grip in him. He became a sworn disciple of Oehlenschläger and an equally inveterate opponent of Baggesen. Yet in November 1820 he graduated in zoology and finally, in January 1821 Carsten Hauch defended his dissertation 'Remarks on arbitrary motion as compared with the organs related to motion'. The University lying in ruins, the defence took place at Trinitatis Church and lasted seven and a half hours.[12]

After his great discovery in 1820 Ørsted was in a position to raise funds from public coffers and he succeeded in providing a travel grant for Carsten Hauch to go to Paris taking a letter of recommendation to Cuvier.[13] Sadly, Hauch fell ill in Italy and had his foot amputated. After limping home he moved to Sorø Academy to teach science for twenty years, until Ørsted came to his rescue by using his influence at court to provide him with a professorship in literature at the University of Kiel.[14] Only after the death of his patron did Carsten Hauch find an opportunity to show his gratitude. It took the shape of the most empathetic and comprehensive life of Ørsted yet written.

The three devoted pupils expressed their gratitude, but that could hardly be sufficient to satisfy the young Ørsted's dream of forming a school.

*

We have already seen, through a number of incidents, the most dramatic divisions within the Ørsted family at large. On the one side were the two older brothers, successful, ambitious, and loyal towards their family ties, their father in particular, and on the other the three younger brothers, flawed, cheating, and irresponsible. The first pair is reminiscent of other notable pairs of brothers in the history of European culture such as the Schlegels, the Grimms, and the Humboldts. In his biography Carsten Hauch praised their extraordinary qualities: 'It will already be seldom enough that one finds one man in the chronicle of Danish history comparable to either of the Ørsted brothers, but a pair ranking as high and being so spiritually close, one would hardly believe could be found.'[15] Working in such diverse fields as science and law, how could Hans Christian and Anders share each other's thoughts and work together?

To illuminate this question we must look for the shared philosophical features in their writings from this period. In 1811 Anders gave a paper at the Scandinavian Literary Society 'On the Boundaries between Theory and Practice in Moral Philosophy'.[16] At the same time Hans Christian delivered a series of lectures in the Royal Society of Sciences and Letters in preparation of his major work (ch. 28). Their papers have one thing in common: they both contain their reflections on the methods of science and moral philosophy. There are obvious parallels. Anders challenged F. Schleiermacher's view that moral philosophy should and could be amply grounded on one absolute principle, the categorical imperative, comprising doctrines of both jurisprudence and virtue,[17] as he had attempted in his prize essay in 1797, and later regretted (ch. 6).[18]

Fig. 59. Anders Sandøe Ørsted (1779–1860). While Hans Christian was auburn, Anders was fair and as a child he had his ringlets plaited into a pigtail that was cut when he settled down in Copenhagen. After *philosophicum* he became a student of law and moral philosophy, and in 1821 he obtained a leading post in the Danish Chancellery responsible for the formulation of laws. Portrait in oils by C.W. Eckersberg (1783–1853), SMK.

Anders began his paper by setting out a range of parallels between the methods of science and moral philosophy. 'Since all objects of nature are products of the same forces, only distributed in different ways, nature cannot be classified differently from the way already described', that is by infinitely small degrees, between which no sharp boundaries can be established.[19] Similarly the boundaries between the kingdoms of nature are blurred in so far as there are soft transitions between plants and animals and between plants and minerals. 'Scientists confess that even the greatest and most conspicuous differences in nature, for instance between organic and inorganic substances, between animals and plants blend together in intermediate links so that it is impossible to determine to which category the objects should be referred.'[20] Romanticists were preoccupied with the idea that it is impossible to classify nature in absolute categories, because it constitutes one great coherent whole with gradual steps of transition. This idea had often been suggested by the physicist to the jurist.

As in science, it was impossible to establish sharp boundaries in doctrines of jurisprudence and virtue between theory and practice, between motivations and consequences of actions, 'between possibility, reality, and necessity, as well as between all forms of internal practice. If this were possible we would acquire an infinite knowledge, and if so our whole character would become infinite, that is to say each of us would become God himself.'[21] This quotation is best understood by reference to the term 'modality' in Kant's scheme of the categories of the understanding. Like Hans Christian in his thesis for the doctorate, Anders had tried to follow 'Ariadne's Clew', but both had realised by now that this method was far too rigorous.[22] Anders was now convinced that moral philosophy must be divided into two: the practice of jurisprudence implementing public law, and the doctrine of virtue grounded on the categorical imperative, although in a slightly modified form.

Anders went on: 'It is thus affirmed what one of the greatest philosophers of all time [Kant], with his usually comprehensive and profound insight into the realm of truth, has taught (although as it seems to me he has not achieved absolutely everything he has attempted to do), viz. that outside mathematics...no genuine definition is possible'.[23] Hans Christian's dynamical project was beyond mathematical and categorical reach. 'How should one by means of concepts weigh and measure objects for which we do not possess either weights or measures?'[24] Anders asked with reference to his brother's lectures, emphasising that the dynamical part of science, 'the Theory of Force', had to explore 'the internal changes in bodies that cannot be detected by weight' by methods other than the mechanical-mathematical methods applicable to 'the Theory of Motion'.[25] In other words, both brothers continued to wrestle with problems that Kant had identified without being able to solve them.

Kant's categorical imperative as transformed by Anders reads as follows: 'One should adapt one's actions according to such maxims as lend themselves to becoming ordinary laws'. In real life, however, Anders witnessed a manifold array of actions that were not easy to categorise according to the principles of this sentence. 'The most distinguished scholars of this science wish this sentence to be able to draw a sharp boundary between actions prescribed and prohibited so that (provided one has true insight into its principles and is entrusted with good will) one needs no other compass to find the right way in the trackless maze of life.'[26] To Anders the impact of the pragmatic experiences he had had in court since his prize essay was such that the categorical imperative seemed insufficient by itself as a moral guide for people through their lives. Against this background he had changed the formula of the ethical challenge to this alternative: 'We should make our entire life an impression of the eternal reason that is revealed in us'.[27]

Once again we find the input from his brother as he writes: 'Just as we were given a sense of the spirit in forms by our appreciation of beauty, and a sense of reason in life by our conscience, so we were given a sense of reason in the laws of nature. The latter makes us familiar with reason—and although there is no precision in detail—we receive a majestic sense of the whole.'[28]

Anders's moral philosophy had resolved itself into two parts, the universal obligations of jurisprudence and the individual duties of virtue. Actions guided by the categorical imperative take place in utterly diverse situations in real life, perhaps in grey areas, in which a concrete appraisal is demanded to distinguish on which side of an ethical divide a given action is situated. This consideration concerns the obligations of public laws determining prohibitions and punishments, as well as moral duties that should be followed (even when it is inconvenient) by way of conscience even if not enforcable by law. Certain types of conduct in life can be repudiated out of hand by means of the maxim. The examples Anders gave were eudaemonistic doctrines entailing a waste of talent ('forces of reason' in his terminology) that should have been used to fulfil Fichte's divine idea, or a universal duty. But even here he admitted he was treading on thin ice. To Anders, serious study had always preceded enjoyment, but he had come to realise that conviviality might be seen as a refreshment of life, enabling us to better fulfil individual duty. Duties can be incompatible and our knowledge of circumstances and consequences is often far too incomplete to set universal guidelines for moral conduct. How can one superior principle like Kant's categorical imperative decide in advance whether it would be more correct to fulfil a duty towards oneself, for example to develop one's 'forces of reason', than to take care of others in need of help, or else to follow one's vocation to do scientific research or to pass just sentences? Such duties are not susceptible to final ranking according to a priori principles, although they are part of universal and individual maxims.

The categorical imperative was the regulative idea, a necessary but not sufficient basis for moral judgement, needing to be supplemented by empirical investigation of the concrete circumstances at hand. Where the regulative idea and the empirical facts are found matching, there is the appropriate action, or as Anders put it:

> 'each individual rational being has a double ideal or model to conform to, one determined by the force of reason immanent in every finite being, the other determined by the particular form in which reason reveals itself in him. The first thing he shares with everyone else, the second one, however, gives him a particular vocation of reason shared with no-one, although by obeying it he, as the necessary link, joins the great chain of rational people who together make up the entire, eternal reason. So, in conclusion, on top of common ethics, there is a particular ethic for each individual.'[29]

Hans Christian, equally, had recognised that to transform his dynamical project into science he had to focus on the point of intersection, where the line of the regulative idea crosses the experimental one. So we find again the same methodological strategy in his *Introduction to General Physics*:

> 'Each of these scientific directions [the dynamical one referring to regulative ideas and mechanical science following constitutive rules], then, needs the other; the latter the variety and lively presence of the former, which our limited creativity cannot provide; the former the unity and comprehensive view of the latter, which can only be gained from a higher standpoint. To the empiricist, the idea of the Whole is to be regarded as a bright sun, which shines into the pathless chaos of experiences, and to the speculative philosopher experiences are to be regarded as guiding stars, without which he could easily lose himself in the infinite depths of reason. The further they proceed in these two opposite directions, the more they meet each other, and like different organs in the same being they will finally be combined into a harmonious whole.'[30]

Castor and Pollux were united with strong family ties, and in addition they had come to think along similar lines under the influence of the same epistemological concepts. The parallelism shown demonstrates the fertility of their Kant studies, for even if their division of labour had brought each to his own field, both brothers were immersed in the same epistemology and as a consequence able to support each other and develop together.[31]

> You are the first physicist I have got to know whose
> lecturing... was instructive and enlightening.[1]

27 | 1812–13
Second Journey Abroad
Berlin and Paris

O N HIS first grand tour Ørsted had wondered whether he might benefit by becoming initiated into Freemasonry thus getting access to the European network of Freemasons, which included many brothers from social establishments. Manthey had advised him to be careful not to join for opportunist motives. The Danish lodge *Zorobabel til Nordstjernen* [Z. to the north star] and the German-speaking equivalent *Friedrich zur gekrönten Hoffnung* [F. to the crowned hope] had several of his patrons as brothers. Apart from Manthey, Hauch and Schimmelmann belonged to a lodge, as did several members of the royalty such as King Frederik VI, his brother-in-law the Duke of Augustenborg, his father-in-law Prince Carl of Hessen, and his cousin Prince Christian Frederik, later to become King Christian VIII, the latter two being Grand Masters.[2] There is much to indicate that Ørsted joined to promote his career.

Manthey sponsored him, and with great speed Ørsted rose to the highest degree of the brotherhood. The initiation into *Friedrich zur gekrönten Hoffnung* as Masonic brother number 806 took place on 29 January 1812. On 16 April he was promoted to the second degree, and already two days later to the third degree. This pace was unusually forced. Every quarter he paid his fees of three rixdollars. Due to the discretion of the order it is hard to tell whether his membership brought any benefits. He never assumed any honorary office in the lodge and hardly set aside much time to take an active part in the Masonic work. Still, he was appointed an honorary member in 1850. When Prince Carl of Hessen, the Grand Master, asked him for authoritative answers to his alchemical speculations, his untenable ideas were politely, but firmly, repudiated.

Ørsted had spoken to the King about the grand tour. In fact, they had been talking about his proposal to establish an institute for experimental physics and chemistry (ch. 19), and during this audience, presumably, the King had requested a report outlining how such institutions, in countries where they had already been established, contributed to national prosperity.[3] Frederik VI granted him one thousand rixdollars for the journey and authorised the Foundation ad usus publicos to pay out the sum. Sadly, during the war the financial state of the country was so destitute that the Foundation was unable to defray the costs and had to ask the Treasury to advance the sum.[4]

The endowment was earmarked for technological and economic purposes, not for scientific or aesthetical ones. We may presume that Ørsted applied for leave on full pay on condition that he provided a substitute to fulfil his usual professorial duties.[5] He employed the Norwegian chemist Jac. Krumm, lecturer at the cathedral school in Copenhagen.[6] As Ørsted had to pay Krumm's salary, he was forced to minimise his own expenses, so he terminated his tenancy of Rubens's Court, although only the living quarters, not the rooms housing his collection of scientific instruments. This curtailment turned out to be far more troublesome than anticipated.

This relinquishment of his living quarters explains why Søren Christian, his father, who had moved into Rubens's Court together with his family the year before, now bought a property in Roskilde at fourteen thousand rixdollars.[7] He planned to establish a medicine factory, or as he called it in his application to the Department of Health: 'a chemical laboratory to produce ethereal oils, extracts, etc.' from locally collected plants, as well as a range of mineral waters for which Roskilde was renowned. The Department recommended 'this notably skilful and industrious man have his application granted'. Apparently, the enterprise never got going, which is no wonder considering the bankruptcy of the monarchy. He only stayed in Roskilde for two years.[8]

Young Gitte moved there, too, to take care of the girls. Anna Dorothea Borring had come home from the mental institution, but her mental illness had left a dark scar, and the teenage girls told Gitte one lie after another. When their mother was not in bed (where she stayed until noon) she would wander around in confusion being a nuisance to everybody, to such an extent that the tender Sophie could not stand her mother-in-law, even when she and Anders were only on a brief visit.[9]

Ørsted set off in May 1812 across the Great Belt strait from Zealand to Funen. In Odense he called on pharmacist Krüger, to whom his fourth and youngest brother Hermann was apprenticed. Krüger 'is rather decent, skilled in his trade, and he seems to be a hard man. He has promised me to keep Hermann in strict order. I have asked him to take rigorous measures, as is most necessary'.[10] Why did he not tell Anders and Sophie how their brother thrived or send his regards? Why was it only his strict employer and Hermann's need to be treated rigorously that deserved mentioning? Not even in his diary did he mention his youngest brother by name.[11] Hermann, too, was unruly in his pecuniary affairs and sometimes pawned valuables that did not belong to him; on one occasion at least Hans Christian had redeemed a watch for him.

A creditor, to whom the third prodigal son owed six hundred rixdollars, had called the old pharmacist to account for the debt, and begged him to grant him a loan at least. Søren Christian felt unable to refuse him. Anders was stuck in financial troubles himself and could not help, so Hans Christian, obviously the most well-to-do in the family, had to indemnify his father by redeeming the pawned watch and returning it when he eventually came back.[12] The flaws of the three younger brothers gained more poignancy against the background of the successes of the two older, who were close to their father in Copenhagen and Roskilde, while separated from Jacob at Rudkøbing, Niels Randulph in Russia, and Hermann in Odense. Part of an explanation for this divide may lie in the fact that the two older brothers grew up with two caring parents, whereas the three younger boys lost their mother as infants thus having to compete with their half-sisters to obtain paternal affection, while an unstable stepmother haunted their home.

On his first grand tour Hans Christian had agreed with his fiancée to send her frequent letters to circulate among family and friends. Love letters were separate, of course. Now Sophie Probsthein had disappeared from the scene, so he chose another Sophie, his sister-in-law, to be his correspondent. The deal was to let friends read his letters, but they were not to be circulated. Friends who did not visit Vestergade would receive their own letters. This plan, however, was not carried out meticulously, and hence we know far less about his second journey than about his first.

As soon as Sophie Ørsted was reassured about the endowment she wrote to Sibbern, her admiring philosopher friend: 'My brother-in-law, the professor, has been granted a journey to Paris next summer. It would be great fun if the two of you could meet'.[13] Sophie's droll hope of a rendezvous between Ørsted and Sibbern, who had become her admirer after Baggesen returned to his wife in Paris, was not fulfilled. No doubt, she would have been delighted to have the two men, of whom she was so fond, meet abroad to talk about her. 'It is wonderful to know there is a friend here and there', she wrote to Sibbern,[14] who had long before set out to call on Goethe.[15] 'You lucky man on your way to Weimar. God bless you! I don't grudge you, and don't you forget your promise to beg or to steal something from Goethe for me.'[16]

They knew each other well, Ørsted and Sibbern; Sibbern was Ørsted's junior by eight years, and a postgraduate of law who had tutored students with Anders.[17] The two brothers could not possibly be ignorant of his and Sophie's romance, for their relationship was so open and innocent that any indignation would be ill-placed. Sibbern's compass was constantly oriented towards Goethe's whereabouts, whether he was in his home in Weimar or at the spa in Carlsbad. Goethe was the precipitous rock that Sibbern feared he would be unable to climb even if he reached its foot, as he wrote Sophie. Goethe was so elevated and Sibbern so unsure of himself that he would walk along memorising the words by which he could address his icon if he was lucky enough to get near him some day. His letter to Sophie described his awe like the confusion of a flustered schoolboy. And Sophie appreciated his feelings for she equally admired this precipitous rock and felt nourished by *Wilhelm Meisters Lehrjahre* [W.M.'s prentice years] and the first volume of *Dichtung und Wahrheit* [Poetry and Truth] that Sibbern had recently sent her.[18]

In Leipzig Sibbern bought a new pair of spectacles, not eyeglasses, for as Sophie was bound to remember from their walks on the ramparts around Copenhagen his eyeglasses had always had one glass only. He saw much better with two. 'This is far more convenient. Now I can always get to see the eyes of ladies passing by and I shall have no trouble identifying the lady of the house.'[19]

Finally, his long walk brought him to Carlsbad, where he ventured to hope that the magnificent poet would condescend to utter a few remarks to the peripatetic pilgrim. At the same time he shuddered at the thought of the moment the encounter might happen. Sibbern rented a room and started walking up and down the streets of Carlsbad on the lookout for the man of majestic stature in the blue coat. 'Whenever I saw a blue coat in the distance and a tall, stately figure, I was immediately agitated. And moreover, when a few days later I actually encountered Goethe in the street and he kindly asked me "How are you?", I was reminded that once a yellow shawl [Sophie's] could make me rejoice at a distance, so that I kept staring at it even when I knew it was not the right one.'[20]

Fig. 60. F.C. Sibbern (1785–1872), Professor in philosophy, psychologist and writer. Coloured print of drawing by Chresten Købke (1810–1848), RL.

*

Fig. 61. Berthold Georg Niebuhr (1776–1831), grew up in Copenhagen, son of Carsten Niebuhr the explorer of Arabia. Later he went into Prussian service as a diplomat and a professor of Roman history at the University of Berlin, where he supervised Leopold von Ranke. Lithography after a drawing by J. Schnorr v. Karolsfeld, 1823. RL.

Ørsted was not hunting Goethe, not yet at any rate, not until he could meet him with an ego of equal proportion. Ørsted was sitting in his lodgings in Bährenstrasse in Berlin writing diligently on his new book that he intended to become his major work and win him recognition all over Europe. As opposed to his stressful life in Copenhagen the capital of Prussia offered peace and quiet to work on his manuscript. When he had arrived in June a letter from Sibbern awaited him. It contained views on intellectuals worth meeting. Sibbern spoke warmly about 'Berthold Georg Niebuhr, a marvellous man, whose company you should not miss—even the most learned people here have a great respect for his erudition'. Fichte, Schleiermacher, Weiss, and not least Henriette Herz were amiable as well and would open doors everywhere. Almost every day Sibbern had seen Mrs Herz and had taught her and some of her friends Danish in her salon, and he asked Ørsted to bring some texts for an anthology of Danish poetry that he wanted to publish (anonymously and 'for the sake of the profit'). He thought the anthology would find a ready market, because the Holsteiners studying at the recently established University had to be proficient in Danish.[21]

In summer the Berliners escaped to the countryside, and in any case Ørsted had no time to socialise because he wanted to get his manuscript ready for the press. Making contact with State councillor Niebuhr, whom he knew of from the Scandinavian Literary Society in Copenhagen, was tempting however, so he paid a call, and the visit was a great success. Ørsted introduced Niebuhr to his textbook project, told him about the devastating bombardment that had destroyed the first edition, about the new edition of the mechanical part, and about the difficult drafting of the dynamical part, which he felt unable to complete. The working title of his manuscript was very much to the point: *Versuche eines Physikers sich in seiner Wissenschaft zu orientieren* [A Physicist's Attempts to Orient Himself in his Science]. Niebuhr listened with genuine interest, recognising himself in Ørsted's outpourings. 'Go ahead with your manuscript and publish it in German! I shall find a publisher and take care of the printing', the councillor encouraged him.

B.G. Niebuhr was a son of the famous explorer of *Felix Arabia*, Carsten Niebuhr, who, back in Copenhagen, had taught his son languages while he sat writing up his diary notes for the notable account of his journey. Later Barthold had devoted his time to general education at the University of Kiel and among other things familiarised himself with Kant's philosophy, under the supervision of Professor Reinhold. Here Schimmelmann had spotted his talent and employed him as his private secretary and later as deputy in the Department of Commerce. We do not know if Niebuhr and Ørsted (who were the same age) were friends in Copenhagen, only that both had their papers published in the periodical of the Scandinavian Literary Society. At the time when Ørsted was appointed professor, however, Niebuhr left his job and went to Berlin, entering into the service of the Prussian King who soon appointed him Councillor. The real reason behind his emigration can only be guessed at; officially he could not stand the Danish winter.[22]

A month after Niebuhr's arrival in Berlin, Napoleon had crushed the Prussian army at Jena and Auerstedt, after King Friedrich Wilhelm III had demanded the French troops withdraw from the south German states of the Confederation of the Rhine, which were in alliance with France. In November Napoleon entered Berlin triumphantly, setting up his headquarters in Frederick the Great's gorgeous Neues Palais. Anticipating the French occupation, the Niebuhr family fled eastwards along the same route as Fichte. When the French Emperor and the Russian Tsar had concluded a peace treaty near Tilsit in July 1807, and Copenhagen had suffered the British Bombardment that September, prospects were dire for Danes as well as Prussians. Denmark-Norway was as defenceless vis-à-vis the British navy and the allied French-Spanish troops as was occupied Prussia. Napoleon blockaded the Continent of Europe so as to cut off British trade and industry. He was at the peak of his power until 1812, when Russia ceased to implement his Continental System, and the Grande Armée retaliated by crossing the Njemen River heading for Moscow.

Despite the French occupation of Prussia, which Sibbern dreaded to see spreading to Denmark, the Prussian Minister von Stein initiated a comprehensive reform programme, which included, among other things, the establishment of the University in Berlin. It was founded on the best ideals of the University of Jena as it had flourished under Goethe, Schiller, Reinhold, Fichte, Schelling, the Schlegel brothers, and others. But the opening of a new university in temporary premises in a wing of the Prince's Palace, while French troops kept a firm grip on the country, was not the best start to wish for. Ørsted felt deeply the suppressed atmosphere in Berlin. The occupation had almost stopped all conviviality. He saw only Fichte, Mrs Herz, and Niebuhr, who invited him for tea. Niebuhr had become one of von Stein's trusted deputies, but when Napoleon demanded von Stein be deprived of his power Niebuhr's career changed course in favour of a professorship of Roman history at the University. He threw himself into study of the source materials to write his History of Rome, which was published in two volumes in 1811–12. Niebuhr and Ørsted could not find enough words of praise for one another: 'Niebuhr is a marvellous man, kind and amiable, and erudite in the best sense. Strange to say he is at home in any science. On chemistry he speaks like a chemist. He has acquired not only the general theory, but also a lot of specific knowledge of particulars, displaying better judgement than many a chemist.'[23]

It takes little effort to realise what made them admire each other. Niebuhr, by now promoted to royal court historian, was the real founder of historical source criticism, passing the method on to his disciple Leopold von Ranke, who is usually given the credit. Niebuhr's *Römische Geschichte* was acclaimed for two reasons. The first was the manner in which he really opened up the way forward for historical writing, classifying sources as primary or secondary, and rejecting the latter. The second was the creativity and imagination he applied to producing a comprehensive and vivid account of the Romans which addressed itself to the contemporary readership. Ørsted's *Ansicht* was a similar project in the realm of science.

*

Ørsted worked hard for ten weeks on his manuscript of *Ansicht der chemischen Naturgesetze durch die neueren Entdeckungen gewonnen* [View of the Chemical Laws of Nature Obtained Through Recent Discoveries]. Only in the evenings did he seek distraction. Three times he saw Fichte, who was living in a wing of the Prince's Palace in Unter den Linden. He was received 'as cordially as I might expect; but I have not seen much of him, because due to our differences on natural science there is not a lot for us to talk about.'[24] Fichte reassured him of his happy memories of Copenhagen and sent his regards to Anders and Sophie. Ørsted had tea with Henriette Herz in the Jewish quarter and showed her his bravura display of acoustic figures in her salon. He went to the theatre at Gendarmenmarkt, the opera in Unter den Linden, and the Philomatic Society; he also attended meetings in the Masonic lodge, and with Niebuhr he joined the celebrations for the King's birthday in the Academy of Science. 'The principal speaker delivered a speech in good Latin, but his modesty prolonged the introduction so much that a colleague remarked that no introduction would be needed for another seven years.'[25]

His intensive work on *Ansicht* took up all his time, so there was no time over to converse with colleagues. Hermbstaedt, whose laboratory in Georgenstrasse he had used so much during his first stay, he now considered a coarse egoist; he held his scientific skills in low esteem and looked down on his philistine lectures on chemistry which targeted industrialists only.[26] His friend C.S. Weiss was so one-sidedly absorbed in crystallography that Ørsted found him crystallised himself. Other scientists cancelled appointments so he did not get to see their instruments. Generally, he passed a harsh judgement on the science staff of the new university (Erman, Fischer, Klaproth, and Hermbstaedt), because they lectured in scattered premises and their posts were part-time. Experimental physics and chemistry was the weakest section of the University.

As soon as he had delivered his manuscript to the printing office he left Berlin. He received a preliminary fee of one hundred Prussian rixdollars, which went directly into his travelling funds. A month later, at the end of September, he arrived in Paris where he settled for the following nine months. Very little is known about this period, his correspondence shrank and not even the brief remarks in his travel diary were elaborated into letters.

In Paris Ørsted received information from Niebuhr on the progress of *Ansicht*:

> 'First I must tell you how present you are and always will be in my memory. I cannot think of you without a melancholic feeling that fate has separated us and will continue to do so. Being together with you became to me a relationship of confidence that brought me into the realm of science, so unbelievably attractive, but to do independent scientific research I lack both goal and means; you are the first physicist I have got to know whose lecturing, to a half-educated dilettante like me who is nevertheless seriously interested, was instructive and enlightening. Others assume grand airs and do not deign to talk or they lack the desired clarity. I have made attempts to bring you closer to us in Berlin, and our mutual country [Denmark] may blame me for it. The attempt had been put in hand and did not seem to be failing. L. should have been called to Berlin to replace W. and had this been the case, I should have protested in order that the post was offered to you instead. When one casts one's vote in favour of a complete stranger it is difficult to foresee the outcome. Still, prospects seemed to be rather promising. Now the whole thing is called off, because no doubt the intriguers will impose the appointment of Sprengel by force.'[27]

Obviously, Ørsted came within an ace of having the chance to swap his post in Copenhagen for one at the University in Berlin. He must have felt his lonely soul warmed by these compliments, of the kind that no doubt he received less often in Copenhagen. Niebuhr's recognition was no idle talk, but manifested itself in action: 'Please do not apologise for leaving the work of publishing *Ansicht* to me. Seldom have I accomplished an assignment with greater pleasure, and had it been inconvenient I should have paid somebody else to do it. I think your work will have an immediate and lasting effect. As soon as it is printed I shall hasten to send you copies to Paris by separate mail.'[28] Thus encouraged, Ørsted immediately set out to translate, or rather rewrite, *Ansicht* into French. In return Niebuhr asked Ørsted to collect autographs of French scholars, dead or alive, of all fields of learning, great and small, of the celebrities themselves or of their correspondents even if they were written on simple notes. Niebuhr needed the autographs as a present for Goethe, who was a keen collector.

Was it not risky that Ørsted was on a grand tour paid for by public money granted for the specific purpose of investigating technical chemistry, while in fact he was spending it on sitting at his desk in Berlin for ten weeks working on something completely different. The King wanted a report on the utility the sciences had demonstrably brought about in great countries like France and Prussia, although for the time being not Britain due to the ongoing war. Ørsted must have been conscious of this discrepancy, for he asked Sophie not to talk about the *Ansicht* project. News about the publication was not to be leaked to the press. Should the University or the King become suspicious of foul play and start to object, he already had an arsenal of justifications ready: *Ansicht* was nothing but a slight reworking of already available papers, and the German publisher had paid one hundred Prussian rixdollars to his travel fund, which would easily compensate for any possible failings.[29] However, in Paris he spent even more time translating and rewriting his major work, while ignoring the chemical industries.

In Paris Ørsted rented rooms in Hôtel Saxe in Rue Colombier in the Latin Quarter on the left bank of the Seine, between the river and Palais de Luxembourg. A large part of his time in Paris was taken up with reworking and translating *Ansicht* into French with the help of Professor Marcel Serres. The French title was *Recherches sur l'identité des forces chimique et électriques* [Investigations of the Identity between the Chemical and Electrical Forces], which will be dealt with in the following chapter. Apart from this, Ørsted met with C.W. Eckersberg the painter, who was studying historical paintings under the supervision of J.-L. David, but would soon to travel on to Rome. Eckersberg was later to paint the portraits of both Ørsted brothers. They talked a lot together, probably about aesthetics, but nothing was written down. A small circle of Danish intellectuals met regularly, and celebrated Christmas together.[30]

*

The plan of exchanging letters with Sophie soon petered out for the simple reason that she was silent. She stopped writing to both Sibbern and Ørsted. Nevertheless, Sibbern continued steadily writing and sending her pressed flowers, but Ørsted was more prone to take offence and became increasingly reticent. He sent her four letters; two were all that she could organise herself to send him.[31]

> 'Once when I shall write a moral philosophy in terms of *Beispielen* [examples] . . . you will come under the heading 'silence'. . . . You are already suffering from a bad reputation here in Berlin due to your extreme silence. Every postal delivery day people ask me if I have received news from my country, and every postal delivery day I must say no. Then they ask me who my correspondent is. And I reply that it is my sister-in-law, whom I have chosen especially because I was sure that she would write to me more often than anybody else. Then they criticise my poor country, which is not much appreciated here, and they do not accept all I say in its defence, so they go on asking what good one can expect from a country whose best correspondent does not write? Every reproach against Denmark I can handle except this one. . . . I should also berate Anders if he were there; partly because he does not write, partly because he does not make you write by whatever good or bad means necessary. Give him my regards and tell him to demonstrate once and for all that he is master in his own house.'[32]

Even this roguish lament did not bring him any news from home. So he lost patience with his correspondent and became reticent himself. Why did Sophie and Anders fail him? It was rather sad to leave Berlin, where mail if any would normally reach him, and set out on a long journey to Paris, where mail was less safe. When finally he reached Paris at the end of September there must surely to be a pile of kind letters waiting for him. But, alas, not a single line!

Unexpectedly, Ørsted received a letter of bad news from Krumm in December. Sophie's unbearable silence had been due neither to laziness nor to negligence, but to her and her husband's long-term sickness. At first Anders had fallen ill after their return from the summer's stay at Frederiksberg to the sickly city air in Vestergade. As his condition deteriorated, he had had to stay in bed, and Sophie had called for Doctor Callisen. He had prescribed an emetic, the effect of which was negligible. Anders was unable to vomit and get rid of the disease. Then Sophie, too, started to become ill, at first having headaches, then quinsy, and by the time winter set in both spouses were bedridden.[33] It was wretched misery. At last a temporary recovery began, but they were both weak, and for several months Anders had been neither in court nor in the Department, nor had he tutored law-students. And no work meant no pay.

A spontaneous thought struck Hans Christian. What if he had lost them? In a fit of guilt at having reproached the bedridden couple for their silence, he rushed to pick up his quill and wrote: 'Take my coffer! Spend the money Abrahamson manages for me. Tell him to sell my things to provide cash.... Take the income I get this winter, spend it as if it were yours.... I can do without it. I should never be able to use it more appropriately than by safeguarding my brother's life and health.... All earthly property should be shared with friends.... So take advantage of my small abundance as if it were yours. If once you acquire an abundance, as you should because of your services to the public, well, then I shall not hesitate to receive your abundance.'[34] This generosity proceeded from some heart searching and a confession: 'I don't know what it is that somehow shatters me internally when I hear you are ill; it is as if somebody is tearing a part of myself out of me. In everyday humdrum life my love for my friends, even for you, often seems cold to me, and when this humdrum life is broken by an event, happy or unhappy, I feel it as if it concerns myself'.[35]

But Anders had already for some time felt compelled to anticipate his brother's help. They had previously shared a coffer, and Anders was quite certain that what belonged to his brother, would, in case of emergency, be at his disposal, and vice versa, of course. At first Anders had spent eight hundred rixdollars of Hans Christian's travel grant. During the winter the loan increased to two thousand rixdollars. Inflation was running high, and the amount did not exceed his defaulting salary. This was not the first time Hans Christian had to help a brother in need. But as opposed to Jacob's fraud, Niels Randulph's unpaid debts, and Hermann's unredeemed pawn tickets, nobody could blame Anders for getting into financial difficulties. Hans Christian was back in Copenhagen in mid-June 1813 happy to be united with Anders and Sophie, but devastated to discover that the University had removed his collection of instruments from his former residence at Rubens's Court.

> I think your work will have an
> immediate and lasting effect[1]

28 | 1812–13
The Major Work

IN 1997 a small book that did not officially exist popped up at a Sotheby's auction in London. The first word on the first page was '*Kraftlæren*' [The Theory of Force], but there was no front page, so it was a riddle as to who the author was, who the publisher was, who the printer was, where and when the book had been printed. The latter defect suggested that publication had never taken place. Perhaps this was the reason why the book did not exist. Still it was dispatched to the highest bidder, Hans Toftlund Nielsen, a bibliophile professor of chemistry at the University of Odense. The consignment held other books as well. They were all related to Hans Christian Ørsted and some of them written by him. Obviously, the consignment came from a collector of Ørstediana. Hans Toftlund Nielsen soon ascertained that the windfall of 224 pages was Ørsted's *Kraftlære*, which Ørsted scholars had so far taken for granted had never appeared. Yet, he held a real copy of it in his hand. On the right upper corner was a hallmark showing a coronet and the letters R.L. indicating that once the publication had been registered as the property of Count Rosenørn Lehn.

It was a paradox: a printed book known in one copy only. On the one hand it looked like an unfinished, yet printed, draft. The introduction foreshadowed three parts, but it contained merely two. The third part was missing, paragraphs of the second part contradicted each other, and the numbering of them was a mess. On the other hand the book was not only printed, it was also bound, and there was a back title, 'Ørsted's Naturlære 2 Kraftlæren' ['Ørsted's Physics 2 The Theory of Force']. Everything suggests that it had served as a proof that the author (who else?) had received from the printer's for proofreading. But there were no handwritten corrections in the margin. It is generally agreed that *Kraftlæren* must be the unfinished draft of the second volume of Ørsted's planned textbook.

The Theory of Force (edited and translated 2003) must be dated to 1812. It cannot have been written later than this year, because its bibliography, seemingly up to date, contains nothing published after then. This dating is substantiated by Ørsted's promise to J.J. Berzelius. He had probably tried to have the book ready in a hurry before his scheduled departure from Copenhagen in May. At any rate that was his intention, as appears from his letter: 'Within a short time you will receive a treatise on the spirit and study of Physics that I have published as a programme for my lectures, as well as the second part of my textbook dealing with chemistry. This will be completed before my departure'.[2] Three days later Ørsted wrote to Sibbern in Berlin: 'Please leave me your address in Berlin to enable me to send you from there what I have published since we last met, above all the second part of my textbook that I am almost tempted to think will be of some importance'.[3] We may conjecture that the printer had been instructed to typeset and print his unfinished manuscript in order that Ørsted could proofread it and complete what was still missing. As late as March the busy professor took it for granted that he could make it, but late in April he realised that he would be unable to complete *The Theory of Force* for publication before embarking on his journey. The only surviving proof ended haphazardly under the hammer at an auction almost two hundred years later.[4]

Kraftlæren had delayed Ørsted's departure by a month, and when he arrived in Berlin Sibbern had left. Reading between the lines, one senses a feeling of distress in Ørsted's next letter to Sibbern. He seemed to be unable to make up his mind what to do and he needed help. He seemed to be at a loss as to estimate the range of his dynamical project. Would it be better to write shorter articles rather than a comprehensive textbook? Would it be better to write in German in order to enter into a dialogue with a wider readership?

> I have brought my manuscript with me, possibly to translate parts of it; but things do not move for me and my project is about to suffer a blow. I am engaged in a comprehensive treatise on chemistry. I have made great efforts to express myself clearly and would like to transfer linguistic clarity to the German translation. It is destined for Schweigger's journal. I shall hardly proceed with the mathematical part on this tour, because if I find the time I want to elaborate some parts still lacking in the second volume of my physics. I often miss being able to talk with you about all these problems, but if it cannot be otherwise it will partly happen in the course of printing, at least as far as the chemical theory in general is concerned.[5]

Ørsted's textbook project had started in 1804 when he had thrown out a feeler at the Shooting Range in the form of a detailed lecture for the Scandinavian Literary Society. The textbook had protracted birth pangs, because the physical and chemical sciences were developing very fast. Ørsted's original plan was shattered time and again by new discoveries by scientists such as Rumford, Biot, Berthollet, Davy, and Berzelius.[6]

He also had difficulty in drawing a line between physics and chemistry, because the two disciplines suffered from lack of any firm definition (let alone the definition used today). French corpuscular theory, for instance, tried to squeeze chemistry into the framework of Newton's mechanical paradigm by importing atoms and imponderable fluids and by embodying forces in substances. Like Kant, Ørsted distinguished between the external motions of matter and their internal forces or between a theory of effects and a theory of causes. It was more difficult for him to determine the divide between the two realms of *The Theory of Force*, the lower and the higher, indeed so difficult that he never managed to work out his projected 'higher theory of force'. To begin with he had imagined that his lower theory of force would consist of chemistry while the higher would deal with electricity, magnetism, heat, and light, when the expected breakthrough in the field of dynamical theory had happened. But no sooner had he started writing his lower theory of force than he realised that matter changed chemically when exposed to heat, combustion, or galvanism. So the distinction between lower and higher and between cause and effect was biting its own tail.

The distinction was grounded on another criterion as well. Knowledge of the lower theory of force was supposed to be sufficient for artisans such as dyers, tanners, brewers, pharmacists, and industrialists. They needed to know the properties and reactions of chemical substances. By contrast, the higher theory of force would only attract interest from 'scholars with a higher aspiration and those who want to apply physics and chemistry to enlighten other disciplines such as physiology and geology'.[7] It appears from section 4 that Ørsted was perfectly aware of the difficulties: 'If the two parts of the subject are...merely the same investigation only starting from two different sides, then, obviously, difficulties arise when we look for answers to the question: where are their boundaries?'. Once again we see the parallel between the two brothers' deliberations on theory. Hans Christian and Anders agreed that no sharp boundaries could be established within natural and moral philosophy.

His lower theory of force contained four parts covering chemistry, heat, fire, and galvanism. In the part on chemistry he classified all known substances according to their characteristics. At a time when even experts had only vague ideas about radicals and chemical combinations, Ørsted constructed an a priori scheme of electrical affinity, that is, how various substances charged with polar electrical forces could attract each other and combine. Was affinity like sexuality, so that substances combined in pairs (as Goethe believed), or did affinity work according to rules of multiple proportions so that compounds were made not in random but always in fixed proportions (as Berzelius suggested)? In the second part, he discussed the effect of heat on various kinds of matter, water for instance which underwent a change from solid to liquid to gaseous (ice, water, and steam). In the fourth, he tried to explain how chemical substances react to galvanism, what made good and bad conductors, and why the propagation of galvanism through conductors is accompanied by heat and sometimes by light, too.

His higher theory of force was intended to contain parts on electricity, magnetism, heat, and light. Today these disciplines are studied by physicists, not by chemists. Disciplinary boundaries are artificial and subject to historical revision. Ørsted's dynamical project was interdisciplinary and its primary causes presumed to be chemical. In his day physics dealt with mechanics, which he dealt with in the first part of his textbook with no particular originality. As already mentioned its entire first edition was destroyed by fire in 1807, and the book was re-edited in 1809 and considerably enlarged in 1811. Soon after his appointment as professor he had proposed what in practical terms was the equivalent of a Faculty of Science with physics and chemistry (i.e. himself) at the centre, 'since Physics is the core discipline of the natural sciences', that is, a physics including chemistry, 'because the theory of chemistry is a part of physics that should not be detached. Most modern physicists such as Gren, Lichtenberg, and Hauch have therefore included the essentials of chemistry in their textbooks'.[8]

Ørsted expressed the idea (ch. 19) that the author of a textbook should abstain from riding his private hobby horses and should represent what was commonly agreed among scholars. This idea may well be the reason why his *Theory of Forces* ended up as it did. For was not its core structure exactly a reflection of his personal hobby horse? In addition, and against Ritter's advice, he had opened with 'The Theory of Motion', i.e. the physical effects, whereas it might have been more appropriate to begin with the chemical causes, i.e. 'The Theory of Force'. As these aims turned out to be elusive his textbook threatened to become an insurmountable frustration.

When a scientific field is in a state of flux, its terminology gets confused, and a need for new definitions arises. Ørsted tried to translate scientific terms into Danish, not only to improve his chances of communicating with his domestic colleagues, but also to disseminate scientific knowledge to the laity. In *Theory of Forces* we see for the first time his Danish terms for oxygen [*ilt*], hydrogen [*brint*], nitrogen [*kvælstof*], and base [*æsk*] in print.[9]

*

Ansicht der chemischen Naturgesetze durch die neueren Entdeckungen gewonnen [View of the Natural Laws of Chemistry Obtained from Recent Discoveries] was a publication quite different from a textbook. It was 'a thought experiment', a term invented by Ørsted using Kant's proposal for a special a priori method for studying dynamical processes. The drafts of *Ansicht* were first and foremost four papers investigating the primary reasons for chemical effects, based on lectures to the Royal Society of Sciences and Letters in 1810-11.[10] There is nothing to indicate that he received critical or constructive feedback from these papers. He must have felt like a lonely voice in the wilderness meeting either silence from sceptics incapable of opposing him, or an echo from flatterers. Consequently, he now disseminated his ideas to the learned republic of Europe, in German in *Ansicht*,[11] and in French with a title more to the point in *Recherches sur l'identité des forces chimiques et électriques* [Investigations into the Identity between the Chemical and Electrical Forces].[12] The French edition was far from a literal translation of the German version. The two editions had different targets, a German readership supposed to be familiar with Kantian metaphysics of nature, and a French one supposed to be largely ignorant of it but in the habit of reading works by the Laplacean school.

What did Ørsted mean by 'a thought experiment'? It stood for the method he pursued in *Ansicht* analysing recent discoveries within chemistry and electrochemistry that would hopefully advance his dynamical project. Ørsted's method is reminiscent of Kant's deliberations in his so-called 'transition project' which was only published after his death. The project represents a possible transition from metaphysics to physics.[13] The idea seems to be that chemistry might after all become a science in the place where regulative ideas meet. Kant expressed the strategy as follows:

> 'Reflecting judgement operates with given representations to bring them within the scope of empirical concepts covering certain objects of nature, not as a scheme, but rather technically,

> not only mechanically as a tool of the understanding and intuition. In other words such concepts will have be to construed rather artificially in accordance with the universal, yet still undetermined principle ordering nature into a system. These concepts are needed to advance our capacity to pass judgement according to certain laws (about which the categories of the understanding say nothing), adapted to the possibility of providing systematic evidence. Without this precondition we dare not hope to find our way in the present maze of the multiplicity of possible, particular laws.[14]

We do not know whether Ørsted was aware of Kant's transition project, but the method of *Ansicht* was on a par with its idea. Ritter's electrochemistry and Winterl's acid-base theory were examples of such regulative ideas that might 'advance our capacity to pass judgement according to certain laws...adapted to the possibility of providing systematic evidence, without which precondition we dare not hope to find a way in the present maze...'. So what could Ørsted do other than hand the clew back to Ariadne and go on searching for regulative ideas in recent discoveries made by Ritter and Winterl as well as Berthollet, Berzelius, Davy, and himself.[15]

The first fifty pages of *Ansicht* dealt with the invisible forces that Ørsted believed to be internal and polar fundamental forces. Their relative strength appeared in three degrees or intensities. The most intensive was the combination in the electrical spark where forces produced light and heat in wave motions. In ordinary combustion the intensity was lower, but still light and heat were produced although not explosively. Thirdly, the combination produced only heat when acids and bases were mixed. This was the lowest intensity. The experiment on water separation showed oxygen ascending from the positive pole (hence the oxygen in itself must be negative) and hydrogen ascending from the negative pole (and hence the hydrogen in itself must be positive). By now Ørsted had dropped the idea that water is an element, believing instead that the otherwise neutral combination of two polar forces in water is turned into disequilibrium when the electrical circuit is closed and the released forces attract the polar forces of the water.

Immediately after the invention of the voltaic pile the nature of galvanism was in doubt, for although it seemed to have properties in common with electricity, on the other hand it was not the effect of friction as was electricity produced by an electrostatic generator. It was Ritter, who demonstrated that galvanism is an electrochemical process between two metals in contact with brine-soaked cardboard (ch. 9). The galvanic force propagated itself through conductors, and it turned out that when these touched silver chloride smeared on a plate they formed the same figures that Lichtenberg had produced by discharging a Leyden jar.

Ørsted's theory listed all inorganic substances in a sequence from the most combustible (hydrogen) at one end and the most necessary to combustion (oxygen) at the opposite end. In between are the bases, sulphur, phosphorus, and hydrocarbon at the one end and metal oxides and most acids at the other. All substances were electrical to some degree, when the balance was disturbed from outside and provoked what Ørsted called a 'conflict' between polar forces.

The essence of his electrochemical theory was inspired by Ritter's experiment with the galvanic or chemical chain. His point was to show how the galvanic process worked as interplay between repulsive and attractive 'space-filling forces', i.e. as a chemical process. Both metals of the voltaic pile contain repulsive and attractive forces and each of them attract the opposite pole. Ørsted conceived the galvanic chain, in which forces propagate, to be of the following simple form (see fig. 62):

One important force seemed difficult to include in the scheme of electro-chemistry, viz. *magnetism*. On the one hand it was reminiscent of electricity. Polar forces of the same kind repelled each other, while different ones attracted each other. On the other hand a magnet only had an effect on one substance, iron, and this effect was rectilinear at a distance like gravity. Moreover, magnetism did not operate through conductors. No substance other than iron was responsive to magnetism. This had painfully been demonstrated in Paris in 1803 when Ørsted hung a small voltaic pile on a pivot hoping it would be attracted by—indeed by what? By the magnetic north pole like the compass? Or by some unknown galvanic pole that might be located somewhere on the globe? The negative result was the absence of any regular response. The prize question set by the Royal Society of Sciences and Letters in 1810 did not shed any light on the matter either. The French essay had merely postulated a connection between

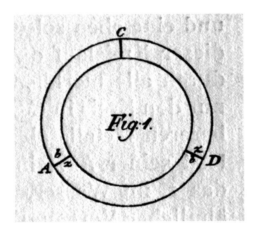

Fig. 62. The galvanic chain according to HCØ's *Ansicht*. The ring consists of a relatively combustible metal AC, a metal relatively necessary to combustion CD and water with acid AD where the two forces neutralise each other. Due to the tension between the forces the water is separated (cf. the experiment on the separation of water in ch. 7) and its combustible forces are accumulated at D and attracting forces necessary to combustion from CD, while its forces necessary to combustion are accumulated at A attracting combustible forces from AC. These accumulations of combustible forces and those necessary to combustion are indicated by b and z respectively. At the same time forces necessary to combustion are repelled from A towards C and D, and combustible ones from D towards C and A. At A and C chemical reactions are activated, by which the polar forces of the water join the opposite forces from the two metals. Now a brief equilibrium occurs, but in a split second the process begins again. In other words there is a continuous chemical process at A and D, and at the same time intermittent attractions and repulsions of the opposite forces propagate through the chain. The process can only be brought to a stop by breaching the chain at C. NS II 89, SSW 342–343, and Ole Knudsen, 1987, 64–65.

electricity and magnetism, within the mechanical paradigm to be sure, probably because the anonymous essayist had suspected that the question was a leading one and that Copenhagen would prefer a positive answer. Ritter had found a further reaction, viz. that magnetized iron oxidises more easily at the south pole than at the north pole, which pointed to a connection between magnetism and the force necessary to combustion. And every sailor had known for years that the compass needle oscillates in thunder, but that was different from proving that magnetism and electricity derived from the same fundamental force. Christopher Hansteen had shown that northern lights were centred around the magnetic poles indicating a connection between magnetism and (electrical) light. In any case Ørsted felt convinced that these recent discoveries would back up his idea of interacting fundamental forces. Magnetic forces attract and repel, they respond to chemical forces (Ritter) and to light (Hansteen).

Ørsted ended with a proposal that can be seen as a repetition of his prize question of 1810 (ch. 23). He suggested an experiment to observe the result of exerting a strong electrical effect on a magnet when electricity was in its most bound form, that is to say when it would have its strongest effect. This experiment would not be carried out without difficulties, because the electrical effect would always be diffused, in that it would influence magnetic as well as non-magnetic objects, and hence precise observations would be hard to make.[16]

According to Ørsted, all scientists have one thing in common: the conviction that nature is governed by laws that are in concordance with reason. This insight has significant consequences.

> A clearer perspective soon adds to this that there is nothing dead and rigid in nature, but that everything exists only as a result of an evolution, that this evolution proceeds according to laws, and that, therefore, the essence of every thing is based on the totality of the laws or on the unity of the laws, i.e., the higher law by which it has been created. Everything, however, must again be regarded as an active agent of a more comprehensive whole, which again belongs to a higher whole so that only the great All sets the limit of this progression. And thus the universe itself would be regarded as the totality of the evolutions, and its law would be the unity of all other laws. However, what finally gives the study of nature its ultimate meaning is the clear understanding that natural laws are identical with the laws of reason, so they are in their application like thoughts; the totality of the laws of an object, regarded as its essence, is therefore an idea of Nature, and the law or the essence of the universe is the quintessence of all ideas, identical with absolute Reason. And so we see all of nature as the manifestation of one infinite force and one infinite reason united, as the revelation of God.[17]

He summed up his reflections on the recent discoveries in chemistry into what he called a dynamical theory of atoms. What made this theory dynamical, of course, were the incessant attractions and repulsions of polar forces, combustible and necessary to combustion , when they entered a state of disequilibrium and were discharged. These forces filled a certain space or micro-field, but in themselves they filled nothing, he concluded, or almost nothing. The substance of things is merely the form in which forces appear to our intuition. He tried to forestall charges of having become an atomist by emphasising that it was true that, after the theories of Berzelius and Davy of the proportional weights of chemical compounds, he was a sort of atomist, but a particular sort. His 'atoms' had no extension, they were mathematical points. He conceived them as if they possessed conditions for possible change (the inner dynamic of motions). This idea aligned him with Gowin Knight, a British natural philosopher who in 1748 had formulated a theory of point atoms including Joseph Priestley's idea that all matter and so all atoms in the universe could be contained inside a nutshell.[18] This was a noumenal idea, of course, because these atoms could not be observed in time and space. Hence, they are not conducive to science. It differed decisively from the Laplacean one, in which corpuscles had substance and hence extension, form, and permanence.

Ansicht and *Recherches* were expressions of Ørsted's thought experiments. One cannot do research on nature without starting from a precise idea of what it is one wants to ask nature about. Science begins not with nature, but with the scientist's mind. If he does not put the appropriate question, he is not going to get the appropriate answer, of course. The good question is a priori and logical. Schiller formulated the method like this: 'What the spirit promises, is fulfilled by nature', and Ørsted paraphrased it in a number of ways, among them thus: 'The clues of nature can only make the intellect produce what is already in it'. It is self-evident that a thought experiment is not science per se, but a necessary precondition of the possibility of science (ch. 25). Therefore, new experiments or observations must be made to corroborate regulative ideas. Even having one's regulative ideas repudiated can be progress in the eyes of nature's mind reader. *Ansicht* concluded with a quotation from Goethe: 'The secret's ancient, mighty spell By insight's hand alone is freed; When revelation does succeed, The dawn of freedom comes as well'.[19]

*

Ansicht and *Recherches* targeted the scientific elite of Europe. Of course, he sent a copy home, too, to Anders and his friends in Copenhagen, 'not for your perusal (because I shall read it to you upon my return adding some explanations that I think will be required by people who are not professional chemists), but for *Ansicht*' [meaning literally: so that you can face it].[20] He also sent a copy to Berzelius with a Swede who was travelling to Stockholm, but we do not know his reaction. There were no reviews from Copenhagen. No Danish scientist ventured to print one word about Ørsted's major work. *Dansk Litteraturtidende* only referred to foreign reviews from German, French, and British sources. These translations revealed to interested Danes that Ørsted was considered on a par with foreign celebrities such as Berzelius and Davy.[21]

The man who helped Ørsted translate his major work into French was Michel Eugène Chevreul

> 'a young man of considerable practical skill and knowledge. His laboratory is only a few steps from my lodgings, so I can often visit, and sometimes do experiments there. My theoretical ideas have found his approval. He sees me in the evenings to listen to me reading the French translation of my *Ansichten* aloud to him. I have also had the opportunity of discussing my system orally with Berthollet, who responded fairly well. The next issue of *Journal de Physique* will contain a brief excerpt from my book.'[22]

Chevreul mentioned Ørsted's *Recherches* in his article on chemical substances [*corps*]. According to him Ørsted's classification of substances on a spectrum having the most combustible at one end and the most necessary to combustion at the opposite end (corresponding to Ritter's and Berzelius's electrochemical theory) was too oppositional to win general approval. Nevertheless, his system was 'one of the most philosophical imagined so far, and it cannot be denied that it has had a strong effect in restricting the validity of Lavoisier's exaggerated observations, and abolishing some of his artificial tabular limits. Consequently, it may be the embryo for the discovery of an extremely valuable, new scientific method'.[23] Ørsted had definitely found a sympathizer in Paris.

The reception in Britain was mixed. Samuel Taylor Coleridge was one of a very few people there who had a sense of Kantian philosophy. He had learnt some German to penetrate German metaphysics of nature. He read *Ansicht*, adding his comments in the margin. Coleridge was familiar with the exciting philosophical and dramaturgic environment at the University of Jena from Henry Crabb Robinson, a fellow countryman who brought his personal impressions of Schiller, Goethe, Novalis, Tieck, and the Schlegel brothers to England. Unlike Chevreul, Coleridge found Ørsted's method was insane. The most important thing in science according to the Romanticist poet was to distinguish sharply between empiricism and speculation and keep them apart. As long as these methods were running along parallel tracks, they might serve one another, but once they converged, everything went wrong. Coleridge did not write a review of *Ansicht*, just jotted spontaneous reactions in the margin as he was reading it. If Coleridge's remarks are taken at face value he did not grasp the point of a thought experiment, according to which Ørsted was exactly on the look out for the interaction between regulative ideas in preparation of an experiment. Metaphysics without experimentation was not conducive to science, and experiments without conceptual governance would just lead to empirical chaos.[24] Knowledge of Kantian epistemology was as non-existent in insular Britain as in revolutionary France.

Thomas Thomson, professor of chemistry at the University of Glasgow, is known for his support of Dalton's theory of atomic weights and the relative proportions of chemical compounds. He reviewed *Recherches* in *Annals of Philosophy*, but not until 1819. Earlier he had written a few lines about Ørsted's book while only having a second-hand knowledge of it, and on this fragile basis he found it weak and imprecise. When Ørsted became aware that his major work was suffering from such poor press coverage in Britain, he tried to alleviate the misery by sending Thomson a copy.[25] It paid off: 'It is rather surprising that a work of such originality and value should have remained unknown in this country for the past four years; for as far as I am aware nobody either in Britain or in France has paid any attention to it, disregarding the very incomplete and imprecise exposition I gave in my "Sketch of the Improvements of Chemistry of the Year 1815"'.[26] Now Thomson made it abundantly clear that his colleagues would gain inspiration from *Recherches* for their chemical research, not least due to its convincing reductions of empirical chaos to general principles. Thomson backed the idea of combustible substances and those necessary to combustion in a polar spectrum.

All things considered, the reception of *Ansicht* and *Recherches* must have been a disappointment to Ørsted. It showed how difficult it was, particularly during the Napoleonic wars, to transcend the national borders of Europe and establish some sort of contact with the centres of natural science, even when they were addressed in their native languages. There were barriers of national cultures and scientific paradigms to overcome, and as far as Britain was concerned almost insurmountable language barriers. When finally *Recherches* was reviewed in Britain, seven years after its publication, one of the reasons was the complete lack of knowledge in Europe that anything of scientific interest was happening in Denmark. As Thomson wrote:

> 'Information on chemical or physical discoveries made in Copenhagen or Kiel or any other part of the Danish kingdom, descriptions and analyses of minerals, lectures delivered in the Royal Society of Sciences and Letters in Copenhagen, obituaries of notable chemists and natural philosophers, all this would be extremely valuable. Unfortunately, I am quite ignorant of the Danish language, but since you write French without difficulty, we might use that language instead of Danish which nobody understands here or English which presumably you do not write without trouble.'

Thomson encouraged Ørsted to write articles for *Annals of Philosophy*, which was published in distant London. In return he promised that the editor would send Ørsted what he needed, possibly via the Danish Embassy.[27] Ørsted had two native languages. He had had to employ a private tutor to improve his French, and yet he needed the help of Michel Chevreul to translate *Ansicht*. As to English he could read it with some difficulty, but he was unable to write it let alone pronounce it, at least as long as Napoleon's Continental System prevented him from crossing the Channel.

When man elevates himself towards heaven
And his skull touches the stars,
His uncertain soles lose their footing...¹

29 | 1813

Controversy on Pantheism

When Sophie and Anders had recovered a little, Anders's literary interests returned. In February he reported to his brother in Paris that N.F.S. Grundtvig, theologian, historian and hymnist, had recently published his *Verdens Krønike* ['World Chronicle'] which he found to be

> 'probably the oddest work of literature recently published. This fellow's unreason and insolence and tastelessness have no limits. He takes the liberty of treating the greatest and noblest products of human reason with the greatest scorn without knowing anything about them. Everything that is not genuinely Lutheran (he stands no deviation from the right or from the left) is anathema to him, but what really binds him to Luther's teachings he does not say... He attacks nearly everybody here, and when he has forgotten someone in the text he does so rudely in the appendix. Sadly, he has praised my character, but if my health were better I should not hold back from dealing with him... myself.'²

Having read his brother's desperate lines, Hans Christian was annoyed to learn that Grundtvig's nonsense had been allowed to pass uncontradicted.

> 'Had I been in Copenhagen I should have taken on your case even if I would not be able to do it as forcefully as yourself. This man's rage against reason has for a long time made me prone to attack him. Still, it would serve him right to fall into your hands... But who says that he must be punished right away? If God gives you health in spring you might give him a thought then. He is a poisonous weed in our literature and should be uprooted. Should he become

~~powerful once, he will be one of the most detestable and malicious prosecutors. Well, enough about him.'³~~

The Ørsted brothers considered Grundtvig to be a dangerous man. His *Verdens Krønike*⁴ was an open frontal attack on the learned republic and its leading role in shaping public opinion. Their letters show that they regarded the battle to win over public opinion as a joint task. It did not matter which one of the picked up the quill, as they took it for granted that they held similar views.

Now Grundtvig had written a history of the world that he considered the beginning of a universal history in Schiller's sense, i.e. one that expressed the view that history follows a plan and that God guides this plan indirectly through peoples and individuals. Man can achieve an insight into this plan of history through the clues laid out by God. In this way mankind can learn from universal history, and the history recorded in the Bible in particular, man's *magister vitæ*.

Grundtvig (aged 31) was not only a theologian and a historian. He was an ordained priest in the Evangelical Lutheran State Church, a spiritual priest communicating with Our Lord directly in prayer, asking for guidance as to how he might become a tool for God in universal history. On his return to Copenhagen Hans Christian (aged 37) took heed of his brother's proposal and threw himself into a controversy that came to assume large proportions, as amply testified by the number and volume of the books and pamphlets that were exchanged; the tone was unusually sharp, and there can be no doubt that the controversy left all parties wounded.

The ensuing Danish pantheism controversy in many ways echoes two German debates from 1785 and 1811.[5] The first dealt with Spinoza's philosophy and the second with the alleged undermining of Evangelical Lutheran Christianity in three guises that F.H. Jacobi, Grundtvig's hero, graded according to their damaging effects: Kant's agnosticism, Fichte's atheism, and Schelling's pantheism.[6] Grundtvig was well acquainted with the arguments of this German controversy and in complete agreement with Jacobi that 'the new school' represented a religious and moral process of disintegration.[7]

To Grundtvig the second pantheism controversy in Germany was a typical reflection of the *Zeitgeist*. To him the revelation of God was in the Bible, not in nature. A philosophy equating nature with its creator was false and an echo of gnosticism, the delusion that religion, the bond between God and man, is a matter of knowledge. Features of pantheism and gnosticism were inherent in German *Naturphilosophie*, which might be spreading to Denmark. Its seductive fusion of religion and philosophy had already gained a foothold. Like Jacobi, Grundtvig felt that he was being called by God to act as his instrument in universal history. Providence governs the future, and God makes his signs known to those he has called.

With Kant and Fichte gone, Grundtvig directed his wrath against Schelling in three respects. First, *Naturphilosophie* is pantheism and therefore contradicts the gospel. Secondly, pantheism implies determinism and is thus at odds with human freedom and morality. And thirdly, *Naturphilosophie* is elitist and serves the interests of the learned republic. Ørsted saw these as attacks on 'the new school' in general and on the learned republic in particular, and he picked up the gauntlet not to defend Schelling (with whom he had differences himself), but to expose

Grundtvig to public contempt as a self-proclaimed religious fanatic. Ørsted mocked him by comparing his opponent's struggle against reason with Don Quixote's furious attack on windmills.[8]

First, let us take a look at Grundtvig's contention that *Naturphilosophie* is at odds with Christianity. Schelling's philosophy of identity enraged Grundtvig; his Christian belief forced him to keep the two levels of dualism apart. His philosophical weapon against Schelling was the principle of non-contradiction, saying that something cannot at the same time be both A and non-A. This much he had learned for *philosophicum*. If the godhead of pantheism was identical with nature and if nature developed progressively, as Schelling insisted, then this godhead could not possibly be both an acting subject and at the same time an object of natural forces, and hence changeable from an implicit being at point alpha to an explicit being at omega, indeed changeable for the better and hence at its point of departure a mixture of good and evil. Thus Grundtvig claimed to have refuted Schelling according to the principle of non-contradiction. The changing deity of pantheism could not possibly be identical with the perfect God of Christianity.

Secondly, Grundtvig contended that pantheism is deterministic and thus totally undermines moral philosophy. Had he misunderstood Schelling when he criticised the philosophy of identity for obscuring the schisms of moral philosophy and for bridging the otherwise insurmountable gap between freedom and necessity?[9] Grundtvig denied that Schelling had eliminated this schism. Closely related to moral philosophy was the classical problem of determinism as against the freedom of the will. If there were identity between the godhead and nature the issue was settled in so far that nature was determined by laws of nature expressing the will of the godhead. If so, humans would be unable to set the course of their lives and equally they would be unaccountable for their actions. A pantheist reduces himself to an organism without a will. A 'yes' to pantheism implies a 'no' to freedom according to Grundtvig.

Ørsted interpreted the text to the young chaplain on the basis of a work in which Schelling explained that determinism is not necessarily at odds with man's moral freedom:[10]

> 'The Godhead is an indissoluble harmonious unity with infinite reason. The same unity is in man, but not equally inseparable, and therefore the possibility of evil is inherent in man. From this arises the egoism of man, which in union with the principle of reason constitutes spirit. The spirit is elevated above all forces of nature and has its own will. As long as this spirit stays in harmony with the universal will, from which its essence derives, the spirit of love prevails and it is good, but when its own will asserts itself to be good, which it could only become through its identity with the universal will, then unity is disturbed, the spirit of love rules no more, and it has turned evil. Here, too, evil does not become reality ... only a fault in nature.'[11]

Evil did have an independent role in dualism as a negation of good, a fault in nature that unfortunately could become real thanks to man's freedom in history. To the priest this interpretation was nothing but professorial waffle.

Grundtvig insisted that the philosophy of identity would inevitably result in a kind of moral relativism, where good and evil were mixed into a grey area. Here vices and virtues were distributed in a continuous spectrum, not as polar opposites, because one would always be latently inherent in the other. For if nature is in continuous development, then its forces will always be in some kind of dialectic relationship in which differences are about to be exacerbated (potentiated) or equated (harmonised). Grundtvig quoted Schelling: 'Good is only a development of evil; evil is the root and good is the flower.' Good was not yet perfect, but on its way to being so. Therefore it had to contain something evil.

Ørsted found the young theologian intolerably Manichaean, too black-and-white and too haughty in his contempt for the cultural elite. But what was Ørsted's own position in this awkward situation where he was forced to choose between Schelling's determinism and Kant's freedom? He had taken on the defence of *Naturphilosophie* with certain reservations and mainly because he and Anders were enraged by Grundtvig's unrestrained attack on the newer school. He could not possibly argue against Grundtvig without countering Schelling, and if so, he would be forced to admit that the newer school evoked confusion not only between rationalism and Christianity, but also among members of the learned republic. Perhaps this was why Grundtvig subacidly remarked that

> 'certain people and Professor Ørsted among them have a lot to say about Schelling's disagreement with himself and the physical flaws of his knowledge, but I have not noticed a lot of this, and no doubt it is only noticed by those people whom Schelling kindly requests to neither praise nor criticise his system, a request he has particularly addressed to Professor Ørsted if I am not mistaken.'[12]

Obviously, Grundtvig was aware that Ørsted was actually at odds with Schelling and that as a physicist in particular he had distanced himself from *Naturphilosophie*. I am not aware of what is hidden behind Grundtvig's innuendo that Schelling had asked Ørsted to moderate his criticism of his system.

The third issue that separated the little and the great prosecutor, as they dubbed one another, was the relationship between religion and science. Grundtvig found it outrageous that *Naturphilosophie* subscribed to an all-embracing rationalism that reduced God to an object of the development process. He had no independent interest in science except when scientists transgressed the boundaries between the two by tearing Christian beliefs apart by means of their rationalist methods.[13] He feared that the learned republic might usurp God's position and seize divine power. Its seductive siren calls might readily win over the multitude of simple-minded believers who would only too late discover the emptiness of *Naturphilosophie*. Grundtvig was one of the few aware of this danger and saw himself as called from above to launch a counterattack.[14] He took on the task of protecting the faith as an affair of the heart against the rationality of the learned republic. Their merging of human reason and divine revelation was an abomination, because the gospel was no harbinger of their sky-oriented rationality, and science was no revelation from above. The noumenal part of dualism was Christian territory, whereas science belonged to the world of time and space. He accepted no merger. Christianity had no need for scientific support, in fact he found that science had nothing to offer believers as believers.

In Grundtvig's eyes the dangerousness of the newer school was its deification of nature and its naturalisation of the deity, while it abandoned the Bible to academic source criticism. Pantheism was a modern variety of ancient gnosticism, Origen's doctrine that knowledge rather than faith opens the gates of heaven. Grundtvig mercilessly attacked those who tried to change the Christian congregation from a popular assembly of believers to an academic meritocracy of scholars, or introduce a Christian ranking order: first class for the elite and second class for the common people. Passengers of the first class could cultivate their sky-oriented roots, decode holy symbols, interpret hermetic literature, and initiate each other in the subtleties of biblical criticism and Freemasonic rituals while the lost sheep wandered about losing their senses.[15]

To Ørsted the issue was more complicated, because human reason was applicable to both earthly and heavenly affairs. Nature is God's work of creation, and the forces of nature are noumenal and the thoughts of the Creator, while at the same time being the object of scientific investigation. As nature's mind-reader Ørsted had to transcend the boundary of the higher level of dualism where he met religion. The precondition of reading the thoughts of nature is that nature and the scientist are endowed with the same reason. Otherwise he would not be able to understand them. Metaphysics of nature and religion share some common ground and might thus become rivals. Hence it was important for Ørsted to draw a line between the two. Science must be autonomous. No interference from religion should be tolerated. But at the same time science pointed towards divine matters. All science can say about God is that laws of nature unequivocally point towards the existence of God. But apart from that science knows nothing about him. Inference from religious dogma to scientific research is an epistemological blunder. Whereas Grundtvig firmly believed that God's qualities had been revealed in the Bible, Ørsted (like the Freemasons) maintained a non-dogmatic and minimalist concept of God.[16]

Throughout history theologians had often misinterpreted the Bible, nourishing superstition and moving into the territory of science where they had no competence. Deterrent examples were the persecution of Bruno and Galileo by the Roman Catholic Church. Grundtvig deserved to be reminded that religion had laid violent hands on science. For when he raged against the learned republic his behaviour was as fundamentalist and intolerant as that of a medieval pope. It was the task of science to liberate people from irrational fear and to eradicate the superstition that impeded progress. Philologists and historians ought to make their critical methods available to Biblical exegesis in order to purify the Holy Scriptures from irrational prejudices. On the other hand he did not retreat from the idea of universal history that the course of history accords with a plan that will eventually be discovered to be a series of scientific laws.[17]

*

In October it fell to Ørsted to give a speech at the University on anniversary of the Reformation that every year occasioned the celebration of three things: that the Church had been reformed to Evangelical Lutheranism, that the University had become an Evangelical Lutheran Seminary, and as a welcome to a new intake of students. The professors took turns, and as it was Ørsted's turn in 1814 he decided to take the opportunity to triumph over Grundtvig. Ørsted's speech 'The Cultivation of Science Considered as Religious Worship' mirrors his arguments in the controversy on pantheism. Grundtvig (the great prosecutor) was redundant, but his opponent (the little prosecutor) was advancing at great speed. A few months later the learned republic would elect the professor to the noblest academic post in the kingdom as secretary of the Royal Society of Science and Letters. Everybody voting for him must have heard this speech. Grundtvig's career by contrast had come to a halt, or even gone into retrograde movement, as he would have put it himself. He was not even capable, having been chaplain to his father, of succeeding to his father's position on his death.

Ørsted explained how investigations into nature brought the researcher closer to things divine. True, the physical experiment was preoccupied with material things, the phenomenal, the sensual, and the changeable, but when the empirical realm had been revealed, the researcher would have to seek after causes and eternal and unchangeable forces and laws, in brief what Plato called νουσ and what Kant called 'the noumenal'. Now, the scientist will have realised that forces are polar, dependent variables that can be reduced to independent fundamental forces, and that laws of nature turn out to be expressions of an omnipresent reason that cannot be anything but God. Insight into this omnipresent reason places him in a state of worship. Scientific research leads to the extreme boundaries of human insight. On the other side of these boundaries a glimpse of the godhead comes into view.[18]

Here, on the other side, Ørsted recognised three metaphysical principles: independence, activity and harmony. The three properties of the godhead made up the trinity of his moral philosophy to the effect 'that having God in front of our eyes we should make efforts to maintain as perfectly as possible the image of God that is inherent in us'.[19] Here, too, Ørsted was diametrically opposed to Grundtvig, and in line with his brother's reformulation of Kant's categorical imperative: 'We should make our whole life an impression of the eternal reason that is revealed in us' (as we saw in ch. 26). The only difference was that this speech used religious terms ('the image of God that is inherent in us'), while Anders used secular terms ('the eternal reason that is revealed in us'). For Ørsted it was the duty of all people to follow the divine idea and live up to their moral obligations, since they had been created in the image of God. To his colleagues and freshmen listening to his speech this was a call to find the vocation corresponding to their personal talents in order to serve the learned republic.

Moreover, independence implied that God is his own cause, and when this property is included in moral philosophy it entails that humans should act as if they have free will, for if one does not govern oneself according to one's will, one surrenders to a foreign will and becomes a slave. And as little as man should 'live with the mind of a slave and degenerate into a means for the casual goals of other people'[20], nor should he reduce others to becoming instruments for his own goals. Here, too, Ørsted's reading of Kant and discussions with his brother had left obvious traces.[21]

The second property of the godhead was activity, the dynamics of *natura naturans*, that, transferred to moral philosophy, made it an obligation to use one's talents to strive to put the imprint of the spirit on everything that surrounds us, that is, not to bury them, but to develop them in the arts and sciences.[22] The third property, harmony, was synonymous with wisdom as a uniting of reason and love. This inner harmony should also be mirrored in the learned republic as reciprocal respect between the various faculties, so as to discover the divine idea that Fichte had declared to be the destiny of the intellectual.[23]

Fig. 63. Nicolai Frederik Severin Grundtvig (1783–1872) on the back of the eight-legged Sleipnir of Nordic Mythology. Caricature drawing by Constantin Hansen (1804–1880), SMK.

The prospect of material rewards motivated most people, but artists and scientists found themselves on a higher level:

> 'The love of insight that others may often have to leave behind in favour of discharging other duties is the very purpose of life for the scientist; he is destined to nourish the holy light of wisdom that will spread and radiate among other people; this is his nightly lantern that will enlighten the Earth. Woe betide him if he does not feel his vocation as the voice of the godhead!'.[24]

The young freshmen should not feel that they had entered the University because they were beguiled by the prospect of prosperity, 'the idol of dazzled mortals', but because they feel called by God in the hope of

> 'honour, I do not mean momentary honour, but that which carries a name across the waves of time to distant generations...Only the conviction that by cultivating the sciences one worships God at the same time is capable of sustaining in you the courage and the power demanded by your vocation, and this is sought for in vain in material rewards'.[25]

Ørsted's philosophy did not plagiarise German *Naturphilosophie* (as one is often led to believe if one reads the literature of his age) and he was not a moral relativist as Schelling was. Nor was he a pantheist like some Romanticists such as Oehlenschläger (at least for a while), and he was certainly not an atheist. The pantheist controversy was not to be the last time that he and Grundtvig rose up in arms against each other.

Love came from heaven above
To animate the world...[1]

30 | 1814
Love and Marriage

WHILE ØRSTED was fighting with Grundtvig in public his private life settled down in the bliss of marriage. How did Ørsted cope with love? His letters to close friends such as Ritter, Gjerlew, and Anders, to whom he confided in such matters are either silent or lost. To be sure, in his letters Ritter had teasingly hinted at his Danish friend's flirtations with his female public, but there are no indications from his own mouth that he loved any woman after Sophie Probsthein had broken off the engagement in 1803. Encouragements from friends to pursue the excitement and the sweetness of courtship were not wanting. A poem he wrote in 1809 suggests disillusionment because of Sophie's betrayal and a less committed enjoyment, or dream of enjoyment, of women's favours.

The poem may be read as an erotic progress from fear via carefree fatalism to quiet deliberation or from sensibility to reason. The two first stanzas deal with the dangers of the consequences of infatuation. Women are not necessarily angels, and if a man yields to his erotic desires and builds his happiness on women, he takes the risk of turning his life into a capricious voyage. The last stanza, however, is concerned with Platonic love pouring down upon the natural philosopher from above. A man dedicating himself to nature and aesthetics gains a firm foundation as well as warmth and peace of mind.

> Hans Christian Carefree.
> Women's favours go up and down,
> They cannot endure;
> But if you are aware of this,

> You readily let them go.
> If you found your happiness on women's favours
> You stupidly build your house on the foam of the sea;
> And soon find yourself in a state of horror in its depths.
>
> Hans Christian Careful.
> Women's favours are easy to get
> However short a time they endure;
> They may come, they may go,
> Without the slightest danger.
> I happily build on waves and streams,
> As long as I have a house that can float;
> Waves may go up and down,
> I feel secure in any case.
>
> Hans Christian Brave Fellow.
> Love came from heaven above
> To animate the world.
> The kin of the earth kneel
> In its bright temple.
> Seek with your eye to consume the flame
> That flows to you from loving glances;
> Its hot beams infuse
> Bliss into your bosom.[2]

Ørsted's reflections correspond well with his experience of life, for was not it exactly his relationship with Sophie that had thrown his emotions out onto the wild, uncontrollable waves and the cultivation of science that had brought him peace of mind? But the poem goes further than that. His life had taught him that sensibility and reason are incompatible when it comes to the erotic, but as far as scientific work is concerned they are complementary, because a man wishing to unveil the secrets of nature is driven by his commitment while harnessing them by reason. And when nature responds to inquiries the reward is not only a reasonable answer, but emotional satisfaction.

Strangely, Ørsted did not call his poem 'Women's favours', even if this is what it is about. He wrote a character sketch on each of his three nicknames: the Carefree, the Careful, and the Brave Fellow, a man fighting for a good cause.

*

Hans Christian had lived almost symbiotically with Anders and Sophie since 1804 and his observations of their marriage must have had a profound effect on him. The story that Anders managed to get ahead of Hans Christian by an hour when courting her at Frederiksberg must be denounced as pure nonsense. At that moment Hans Christian was already engaged to another Sophie. The story may have been put about because the relationship between brother-in-law and sister-in-law developed into a flirtatious confidence, and as we have seen, Hans Christian pronounced a telling characterisation of her persona and fate.

The relationship between the 'Asse-child' and the 'Powder Keg' was warm and caring, but without any sensuous giddiness. Their marriage had a pent-up character. It was childless, and this fact is probably not due to unwillingness but rather to inability. We know that they were very fond of children, Hans Christian's in particular.[3] Their childlessness may have physical causes, since both were often ill. Anders suffered from nervous fever, an imprecise diagnosis that in his case was caused by a superhuman, self-imposed stress. Only Sundays and holidays were dedicated to their marriage. Before the wedding Sophie was a patient at Frederik's Hospital in the care of Ole H. Mynster for almost two months during the autumn of 1801. Her medical record has not been preserved, but Adam, her brother, mentions the diagnosis: scarlet fever that may have been aggravated by complications and after-effects.[4] If Sophie dreaded pregnancy this may very well have been due to a psychological trauma. During her childhood she had witnessed her mother's deep depression over her younger sister, who, with a head swollen with water and the size of the rest of her frail body, was tied into her cot. For five and a half years the family was subjected to her continuous wailing. This trauma may have induced a sexual neurosis, and if so Anders was probably a prudent choice of a spouse.

There is much evidence to the effect that their love was nourished by a spiritual fellowship rather than a sexual one. Both brothers read aloud to Sophie from German philosophers and poets, and Sophie read them herself, too, in German, and in particular she immersed herself in Fichte and Goethe. Anders presented her with a piano, and she enjoyed attending to art exhibitions at Charlottenborg. Not only did she read Goethe, she sang his poetry set to Reichardt's melodies, and above her piano she set his portrait.[5] No doubt she frequented her brother-in-law's lectures.[6] Her correspondence with Sibbern and Ørsted is replete with enthusiastic descriptions of art exhibitions. She often paid visits to her female friends and they visited her frequently.

So, the Asse-children's marriage was chiefly a Platonic relationship, suffering from his overwork and her illnesses. No doubt, they loved each other deeply. Anders was lucky to be married to a woman who, unusually at that time, was as attracted to philosophy, literature, music, and art as he was. Hans Christian showed in a letter to his brother his empathy in identifying with her plight. 'A woman with her soul is really far less happy than we are, because we are far less hampered by external conditions in pursuing the aspirations of our minds.'[7] She was obliged to have her spiritual hunger satisfied by talking to people who like herself were engrossed in philosophy and art. In her day such people would mostly be men, and they could hardly avoid finding this curious and charming female attractive. And then the game was afoot. In the salon in Vestergade she would meet several such men, and Anders's kind heart did not begrudge her this male company. His pressure of work and sincere understanding of her appetite for life precluded jealousy.

Baggesen was Sophie's first philanderer. He had left his French wife in Paris in order to be put up on the couch of the 'Asse-children' for a week to finish a poem. But he ended up staying for nine months. This piece of coquetry was conducive to gossip among Copenhageners. In his diary Baggesen raved about fondling her white bosom with his poetic hands and about their heavenly dalliance; but his voluptuousness was a flight on the wings of desire. Hans Christian despised him for his improper advances and found him incapable of being infatuated with anything but his own vanity.

Sibbern's feelings for Sophie were more genuine and deeper than Baggesen's, and he knew and accepted the conditions, that his enthusiasm would be crushed at the very moment it radiated desire and lust. But as their relationship remained transparent and confined within a self-defined, but unspoken, Platonic frame, it continued to thrive. Sophie sent innumerable notes to Sibbern by messenger, for example:

> 'Thursday morning. Dear Sibbern. You have often been so kind as to offer to escort me if I wanted a walk one of the days Ørsted is busy in court. Today the weather is very nice, and I think it will serve me well to take some good exercise. If you have time I should be pleased to see you around one o'clock, but if you have business to do you really should not put it off for the sake of this, but answer me just as straightforwardly as I ask you.'[8]

So they would take a walk, and in late afternoon when Anders was on his way home from work, they would go to meet him. After dinner the three of them would stroll in the royal parks.

*

In January 1814 Søren Christian Ørsted, the old pharmacist, fell ill, and his housekeeper, 24-year-old Birgitte Ballum, sent for his son, Hans Christian. If this was a pretext for having a confidential word we do not know. The pharmacist was hardly seriously ill. But this provided the occasion when his oldest son became engaged to his housekeeper. They had lived under the same roof before, when Gitte had managed Hans Christian's household in Østergade for a couple of years. They had seen each other in Mrs Møller's dye-works where she served as a maid from 1801 after having lost her mother as a two-year old girl, then lost her father, the minister, as a three-year old, and afterwards her stepmother too, when she was twelve. In other words Birgitte, or just 'Gitte' as Hans Christian called her, had had two sets of foster-parents, before she became the maid first for Mrs Møller and then for Søren Christian's daughters. Gitte and Hans Christian were not blood relations, but with Gitte he had a long-standing relationship. She had already served his father for eight years, and now she would be married to the son for thirty-seven. It would be no exaggeration to assert that both of them knew exactly what they were getting themselves into.

The 'Powder Keg' had praised Gitte to the heavens for the way she brought up Hans Christian's half-sisters (one had died as a ten-year old in 1808). Sophie and Gitte had become confidantes, and Sophie had succeeded in encouraging her to remain at her post despite the pharmacist's unstable wife, when Gitte contemplated fleeing to the friend of her childhood, Tine, the brothers' sister, in Norway.

The day after their engagement Hans Christian wrote a letter to the bride of his choice ('My noble Miss Birgitte Ballum. To be delivered at Mr Ørsted's, Pharmacist, Roskilde') telling her that he had immediately passed on the good news to family and friends, Adam and Christiane, Mrs Møller, the Gyllembourgs, and others. Hans Christian himself, 'as half a bachelor was surprised at such a bold decision, for was he not about to surrender what one calls one's freedom?'.

Not at all, for now that Gitte had presented him with her 'yes', he felt 'a thousand agreeable fantasies painting a bright future in his soul. 'In your honest and open heart and in your fidelity and love I see the seeds of a rich family happiness'. She would be going to see him in Copenhagen, and he would see her in Roskilde, 'and in this way our frequent reunions shall soon bring us close to living together as lovers after we have for such a long time known each other as friends'.[9]

But lectures had to be delivered and papers written; still, Hans Christian had to confess that 'I cannot be for so long without seeing you now that I own you...I shall have no peace until I can call you my wife and see you sitting in front of me at our little table with Dorothea between us at one side.' The intention was to have Dorothea, Hans Christian's fifteen-year-old half-sister, accompany Gitte as the married couple's housekeeper, when she moved into his flat.

> 'Today I have no time to tell you my thoughts, I can only say that Sophie thinks that I do not have to move house to marry. Dorothea will sleep on a couch in my library and under it a drawer is put for her necessary bedclothes. As far as you and I are concerned we shall have two small beds under a canopy in the present bedroom. The library being a corner room will be my office, and you will have the beautiful room facing the street at your disposal. So far I have not ordered any furniture and the like, not even the two small beds, because the prices are expected to fall soon.'[10]

That winter's weather was awful with gales and frost. Ørsted hoped that his cold would vanish, for having a sore throat meant he was unable to lecture, and no lectures meant no fees.

> 'When the weather improves I shall no longer have to endure ten to twelve hours of piercing cold [in a horse and cart] to warm myself for five minutes in your company for morning tea...The affection you tell me that everybody bestows on me I reciprocate; it even seems to me that I love everybody more since you became the object of my highest love...I love you, I am longing to see you, and still more to have you here forever with me. I have spoken to the carpenter, and he promises that we shall not have to wait. Next week he will have the beds ready for us. The chair maker is industrious, too.'[11]

On 17 May 1814 Hans Christian was married to Gitte in Roskilde Cathedral. Curiously, their wedding day coincided with the day the Norwegians adopted their liberal constitution at Eidsvoll and which has since been celebrated as their national day. Nothing is known about wedding guests. Probably Søren Christian spent a lot of money on a big celebration in his yard in Roskilde. Sophie and Anders just managed to attend before departing for Driburg in Germany where they were to stay at a health resort. Hans Christian's cousin Albertine, who had recently been married to the well-to-do merchant Wulfsberg in Roskilde was also there, as were the three half-sisters; after the wedding they moved into the groom's flat in Copenhagen.[12] Søren Christian sold his factory and followed in the footsteps of his son to Copenhagen. During the autumn the 64-year-old pharmacist was appointed head of the pharmacy of the poor law authorities at Almindeligt Hospital, and in February 1815 he moved into the service tenancy in Amaliegade with his two younger daughters and his wife, who, for no apparent reason, seemed to have recovered her mental health.[13]

Sibbern did not hesitate to name Gitte as the friend 'the Powder Keg' appreciated the most.

> 'One might say that she was infatuated with her friend's exuberant, carefree, cheerful wit, cleverness, and healthy judgement... She had a marvellous character... and in addition, underlying her other virtues, a feeling of piety and fear of God... a woman apparently destined to spend her life without passion, yet still with a deep warmth in her heart... By the way she had no bookish education nor at that time any real aspiration for one, but she had no need of this as her daily existence was book enough for her amusement.'[14]

Sibbern's judgement was cogent and to the point and coincided with Sophie's letter of congratulation occasioned by the engagement:

> 'It was the fulfilment of a wish I have carried in my heart for a long time and that I have fervently desired more strongly than I usually allow myself to wish for something, as it does not seem right to me in this life to wish forcefully for anything. The good Lord is my witness, the joy that flowed through me when I heard these tidings, joyful because it contains a promise of so much good to come, comforting, because it proves to me what I wish to believe that good people will find each other on earth and that just things will happen. I have always had in Christian a sincere and comforting friend in life and rejoice in seeing him bound to you whom my heart loves and esteems so dearly.'[15]

But there were other views. Thomasine Gyllembourg, who knew Ørsted as a friend of the family and a tutor for her son, did not hold Gitte in high regard, at least not as far as appearance is concerned. She found her ugly.[16] So what views on love and marriage guided Hans Christian into Gitte's embrace? She was not the Muse of Providence inspiring his poetic and scientific work. Nor was she the fulfilment of an erotic fantasy; their marriage was not wrought by the sparks of passion, but by the steady flame of mature affection. There was no clash of social differences; as a virtual orphan she had been raised by two branches of the Ørsted family. She was thoroughly aware of his dedication to science, and he had long been familiar with how well she could keep a house and bring up children. These qualities were exactly what he was looking for in order to be happy.

His choice of Gitte eliminated his fear of a repetition of Miss Probsthein's infidelity. 'Women's favours go up and down...' but Ørsted only coveted a woman's favour if it was safe and did not threaten the firm foundations of his house—with a library and a laboratory. He was not after an intellectual woman who could check his manuscripts before they went to the press. Gitte was no belletrist of Sophie's calibre; Sophie's marriage, by contrast, was primarily sustained and nourished by the cultural interests of both partners. Hans Christian was not on the lookout for a partner like himself, but for a complementary character who could do what he could not and would allow him his freedom to devote himself to activities that he could do better than most people. Perhaps 'the Powder Keg' had this complementarity in mind when she considered Gitte and Hans Christian destined for one another like the affinities of her admired Goethe.

Ørsted was safe to continue his career as the extrovert and entrepreneurial cosmopolitan. She would become a clever, affectionate, and attached wife, and he would offer her comfort, security, and affluence. The attraction of a Biedermeier marriage was stronger than that of a Romanticist one. In and between the lines of their love letters it was accepted that, like his brother, he would retain the bachelor's freedom to prioritise his scientific efforts above everything else. He applied to the University for an increase of salary. 'Now, as I intend to marry shortly I shall no longer be able to live on the same income.'[17] He ordered the double bed from the carpenter. It might not be the playground of passionate Eros, but instead a place of tender embraces and the production of new Ørsteds.

The permanent spleen of Lord Byron and other romantic writers was the result of a drive towards passionate love, but as a mature man Ørsted believed he had seen through this instinctive delusion. In a notebook in which he penned aphorisms and brief fragments throughout his life he commented on Byron's 'dalliance and vile enjoyment of conquest'. People like him see love as 'the state of infatuation and take it for the main thing of human life. They are right in saying that it does not endure in matrimony; but it does not endure outside it either, when the end of desire has been reached', Ørsted claimed. 'Marriage establishes a comprehensive scope of love, confidence, and mutual happiness, which is a far more perfect way of life than infatuation. That is a state of beauty and splendour—the life of a butterfly'. But the life of a butterfly only lasts one season, then the butterfly dies. If infatuation does not develop into a higher stage of life, it remains a short-lived enjoyment, and if it is repeated with new partners, it becomes ridiculous.[18]

> She lies in his arms in the morning in bed,
> and he can make her stop crying at once...[1]

31 | 1812–15
The Prime Mover of Danish Science, and Gitte's First-Born Child

THE KING had funded Ørsted's second trip abroad in order to get a first-hand impression of the technological utility of the experimental sciences in France, the superpower, and in the Prussia she had occupied. If the absolute monarchy were to invest in the institute for the experimental sciences as Ørsted had proposed,[2] there would have to be convincing arguments in favour of the economic advantages for the belligerent Denmark-Norway. The royal coffer did not spend a thousand rixdollars on Ørsted to let him sit in a garret in Berlin writing a scholarly work on the theory of chemistry in German and subsequently shutting himself up in a hotel in Paris for six months translating it into French. Ørsted was away for fifteen months, his travel diary was only kept for half of that time and according to this half he did nothing to further the King's mission.[3] The salt-works at Schmalkalden was the only industry he visited. He saw instrument makers in Germany, buying their thermometers, barometers, Argand lamps and a microscope with an achromatic lens. He also made inquiries to Hermbstaedt about the Prussian production of gunpowder, learning that it was far beyond that of the French. There are no entries concerning visits to French industry.[4] As a matter of fact he had dedicated himself almost entirely to his scientific authorship, *Ansicht* and *Recherches*, as well as talks at universities, visits to museums, galleries, and theatres; even his conversation with J.G. Marezoll, the Kantian minister in Jena, had consumed more time than his visits to industries producing sulphuric acid, gunpowder, and so on.

Consequently, the report he submitted to the King was meagre.⁵ It only contained the usual phrases on the utility of science that could have been scribbled down in a couple of hours by any student without leaving Copenhagen. It merely stated that French factories producing saltpetre and gunpowder, textiles, calico, silk, sulphuric acid, sal ammoniac, soap, liqueurs, dye-stuffs, etc. were the fruits of the experimental sciences, chemistry in particular, the study of which had been pursued in France and Prussia in particular since 1750. A hundred years before, chemistry had been like magic to the masses, but today chemical knowledge was the property of everybody who considered himself educated. Loss-making investments happened in countries where scientific education was underdeveloped, Ørsted reported. Wherever general education had arrived, the division of labour between theory and practice was appreciated; enterprises starting on a small scale took advantage of their mistakes and ended up as large-scale successes. The most important lesson was that scholars should teach theory to artisans and keep their hands off practice. These considerations were a repetition of the arguments he had adduced previously in support of his proposal for an institute of the experimental sciences (ch. 19). Nothing new was added.

The professor's trivial remarks on the utility of science indicate that he did not have a high opinion of the intelligence of Frederik VI. They were accompanied by a proposal for a reform of the University, repeating the main lines of the proposal for the study of physics he had submitted to the Board in 1808. The Chancellor had never responded to it, so now Ørsted included it in his report to the King. He began with a historical sketch of the miserable state of the sciences in Denmark-Norway. If the sciences were to contribute to economic growth they had to be given a higher priority at the University. Chemistry had no professorship but was regarded as an auxiliary discipline for medicine, and only recently had the University acquired a chemical laboratory (in Skidenstræde), which was unworthy of the University. Ørsted was the only student in recent times, who had made physics and chemistry his core subjects. Students of physics and chemistry needed research grants and they should be more generous than those reserved for students of theology and law, because scientific instruments were costlier than Bibles and law texts. Now that the King had provided Norway with the new Frederiks University in Christiania (which implied among other things that public foundations and colleges in Copenhagen no longer had to pay for Norwegian students), finance should not be a problem. The prestige of the sciences ought to be enhanced by setting prize essay questions for students of science on an equal footing with those for students of theology and law. Hence he demanded seats in the Senate for professors of the scientific disciplines like the other disciplines, because otherwise the sciences were doomed to lose out when faculties fought about scarce resources.

As a professor in the Faculty of Philosophy, Ørsted enjoyed the privilege (that was at the same time also a heavy burden) that all students had to be examined in his discipline during their first year of study. Now he proposed a change by moving mathematics and geometry from *philosophicum* to *philologicum*.

His grounds were that mathematics is an auxiliary discipline for physics and astronomy. Things were turned upside down when mathematics were taught after instead of before these subjects. 'At present they learn trigonometry four months after they are supposed to apply it to the two subjects…If first-year students were taught mathematics first, the study of science

EXISTING SITUATION		ØRSTED'S PROPOSAL	
Philologicum	Philosophicum	Philologicum	Philosophicum
Greek	Philosophy	Greek	Philosophy
Latin	Physics	Latin	Physics
History	Astronomy	History	Astronomy
	Mathematics	**Mathematics**	

would profit tremendously.' What Ørsted did not mention was that *philosophicum* according to his plan would gain eight per cent more time, while the dead languages of *philologicum* (which were not exactly dear to him) and history would have to do with proportionally less.

This report was the King's slim pickings from Ørsted's second grand tour. First, the professor stated that the sciences are useful to the nation. Second, he demanded a higher priority for physics and chemistry at the University. Third, he required that his professorship of physics and chemistry be allotted a chair in the Senate. Finally, to hit the nail on the head, he craved an independent faculty of science, probably with himself as the dean, since his chair was 'at the centre of the sciences as all of them are enlightened by physics and chemistry'.[6] He argued that the Faculty of Philosophy was a chimera consisting of two parts, a historical-linguistic part and a mathematical-scientific part, which had little in common. 'Provided the division into faculties is to make sense, each of these parts ought to enjoy the full rights of a faculty.'[7] His dream of a faculty of science (which no other European university then possessed) was the refrain of Ørsted's university policy from 1808 until it came true in 1850.

*

On 9 January 1815, Erik Viborg, professor at the Veterinary School, circulated a letter to all members of the Royal Danish Society of Sciences and Letters that Thomas Bugge, its secretary for the past fourteen years, would have to resign as his state of health could no longer stand the workload. Consequently, Viborg called on his colleagues to meet in ten days to deliberate on measures to be taken. When thirty members assembled in the meeting room in the wing of the Court Theatre at Christiansborg Castle on the said date at seven o'clock in the evening, Bugge had already died.[8] The treasury containing cash and bonds as well as the archive of maps had been sealed up at his professorial yard by the probate court.[9] The society was in need of a new secretary.

Five candidates turned up: H.C. Ørsted, Børge Thorlacius, professor of philology, J.W. Hornemann, professor of botany, Ole H. Mynster, physician, and Gisle Thorlacius, former headmaster at the grammar school in Reykjavik. Each member was entitled to suggest one candidate, whose name was written down and circulated to the members by the society's messenger. This exercise showed that Ørsted had been nominated by 15 members (or fifty per cent), while the others received 3, 1, 1, and 10 votes respectively. At the ballot Ørsted received a majority of seventeen balls and hence the post.[10] His fee would be three hundred rixdollars a year. The society also needed a treasurer, but since the finances of the society were at a low ebb following the bankruptcy, the members were looking for a person willing to do the work on an honorary basis, and the law professor C.G.F. Schmidt-Phiseldek took on the job on those terms.[11]

The election showed from day one that Ørsted was the obvious successor. The reason why the majority chose him was probably their desire to have a dynamic secretary with a strong international network. These were the qualifications emphasised by Admiral J.B. Winterfeldt, an eighty-two-year old honorary member. Ørsted represented the informal, critical movement that had established the Scandinavian Literary Society in 1797. During the following thirty six years Ørsted was re-elected for seven five-year periods. It turned out that besides his scientific qualities Ørsted had a great talent as an organiser.

A few months later, however, Viborg became aware of an embarrassing oversight. 'Not only merited gratitude, but also general respect for the society demands an eulogy to recollect [Professor Bugge's] merits for the sciences and letters in general and for the society in particular. Consequently, this society cannot abdicate its responsibility to do so without discrediting itself in the public eye.'[12]

C. Olufsen, a professor of economics, promised to deliver such an eulogy on behalf of the society by the beginning of the following winter. His promise was never fulfilled, however, although he was often reminded. But why did not Ørsted do it himself as Bugge's successor? Viborg tried to pass an amendment to the statutes to the effect that the secretary was obliged to deliver eulogies for deceased members if no one else chose to. It is not clear whether this amendment was adopted. Perhaps Ørsted persuaded Olufsen to take on the assignment to evade an assignment so inconvenient to himself. Of course, his refusal was due to the nature of his awkward relationship with Bugge already mentioned. Only later (ch. 52) did he speak out bluntly. Bugge had to make do with being commemorated by Jonas Collin, who as one of the presidents of the Agricultural Society eulogised him there.[13] During the following years Ørsted delivered a series of commemorative speeches in a warm and respectful tone when he found it justified. Otherwise he would abstain.[14]

Probably due to its higher level of activity the Scandinavian Literary Society had outstripped its rival as a platform of scientific news. The most important reason for this was that no one could retain membership of the Scandinavian Literary Society without publishing. If no papers were contributed, the offending members were simply expelled according to the statutes. Publish or perish! But this was not the case in the rival society. True, the demand for quality *might* have been higher in the Society for Sciences and Letters, but this was hardly the case. The Scandinavian Literary Society operated by peer review, and because of the extent of the overlap between the two groups of members, it would be a mere quibble to suggest that it was easier for an article to go through the eye of the rival's needle. True, the conditions of approval differed. The Royal Danish Society accepted learned articles only, while the statutes of the Scandinavian Literary Society stipulated that 'Scientific investigations that can only be understood and considered important by scholars will not be approved by the board of editors, nor will articles containing private polemic' (§22). Their periodical was intended 'to contain philosophical, physical, historical, and aesthetical works as well as original works useful to the citizens...in the Danish or Swedish languages' (§21).[15] The Royal Danish Society had 'almost exclusively confined itself to lecturing',[16] and as the lectures were never published, the best works of the erudite community never found their way to the Royal Danish Society. As a consequence Bugge's meetings became irrelevant and boring. They reflected the practical and economic issues ensuing from the geodesic survey ordered by the government and these dominated the agenda at the expense of news on scientific research. Members with a preference for intellectual debate attended these meetings to no avail. The Royal Danish Society had become a wilderness, while its innovative rival had developed into the oasis envisioned by its founders.

As secretary, Ørsted attempted to revive the society by transferring to it the vigorous ideas initiated by the Scandinavian Society. In consequence, the relationship was reversed and the Scandinavian Literary Society declined. He removed the predominantly utilitarian activities to special committees to make room for scientific research of interest to all members. Before the beginning of the season in November a list was circulated encouraging members to give papers. In May when the activities ended for the season he worked out annual transactions [*Oversigter*] covering contributions from all four classes. They were submitted to plenum for approval before going to press. In this way the public was informed about the activities of the learned republic and the spending of taxpayers' money. The periodical that had been withering away in 1810 was revitalised and papers from the years 1809 to 1815 were printed from 1818 onwards. To attract more new ideas from abroad to the small Danish-speaking world (even smaller with the loss of Norway), several foreign scientists were elected members, mostly on Ørsted's initiative. The exchange of transactions with other European academies was reorganised and increased. It was the task of the secretary to conduct correspondence with foreign members, a task demanding proficiency in writing the three major languages. He had already mastered German in childhood, he had taken lessons in French in Paris, and he wrestled with his English by means of dictionaries. Schimmelmann, the former Treasurer, was president of the society, and he supported Ørsted's reforms; over the previous ten years Ørsted had entertained the political elite in his palace with his galvanic and acoustic experiments.

*

In June 1815, King Frederik VI returned home from the Congress of Vienna and was greeted in triumph by jubilant Copenhageners who pulled his state coach through Vester Port to Amalienborg, presumably grateful that Denmark had not been entirely wiped off the map of Europe, as if he deserved all the credit for this piece of good luck.[17] On 31 July he allowed himself be anointed in the chapel of Frederiksborg Castle; this was the event of the summer. After the miserable war, the bankruptcy of the monarchy, and the loss of Norway, the severely tried Copenhageners considered the lavish spending on royal pomp and circumstance rather too extravagant.

Anders had been awarded the Silver Cross of the Dannebrog Order in 1813 at the time when he was appointed deputy of the Danish Chancellery responsible for the judiciary.[18] Hans Christian was pronounced Knight of the Dannebrog Order two years later, probably as royal recognition of his election to the post of secretary of the Royal Danish Society of Sciences and Letters.[19] The Ørsteds now held such high ranks that both were obvious participants in the processional march preceding and the banquet following the anointing. Their wives were

Fig. 64. Frederik VI (1770–1839) is anointed after his return from the Congress of Vienna 1815, now as King of a small state in the periphery of Europe. Friederich Münter (1761–1834), Freemason, Professor of theology, and Bishop of Zealand officiated at the ceremony in Frederiksborg Slotskirke. Frederik Julius Kaas (1758–1828), ASØ's boss in the Danish Chancellery, is seen as third from the left. Copperplate print by W. Heuer, 1817. De Danske Kongers Kronologiske Samling på Rosenborg. Photo by Peter Kristiansen.

supposed to accompany them, dressed up in all their finery and looking the part, but they declined. Both were shy and Gitte had only recently given birth, so they preferred to stay in the background as spectators.[20] The Copenhageners talked about nothing but the anointing for weeks, and the two young women were getting fed up with this gossip when, thankfully, something more dramatic happened. 'The slaves of the Rasphouse prison broke out', and people found other things to chat about. This suited Sophie well, for within court circles she felt like a sparrow among hawks. She failed to remember names of all the distinguished civil servants from abroad, most of whom she found foolish, some funny, and only a few amiable. 'It was unusual for me to recollect the title of this or that gentleman or his name; I tried to avoid messing it all up, and still confess that I succeeded pretty well in failing.'[21]

As Sophie rightly said the prisoners of the rasp-house at Christianshavn revolted against the cynical treatment by the authorities on the very day that the high-ranking prison inspector attended the anointing of the King at Frederiksborg Castle. It should be noted that Sophie calls them slaves. She felt that the prisoners were treated as brutally as the black slaves in the Danish West Indies. The rebellion started when the prisoners complained that, although they were hungry, they found the food inedible. They initiated a hunger strike, and when a group of around fifty prisoners were on the verge of breaking out, the inspector called in soldiers. Fighting began, the gate of the prison was opened with a pile driver, and the wretches tried to escape in their rags, but were soon run down and surrounded by the cavalry and once again locked up.

A commission was set up to investigate conditions in the prison and pass sentences on the culprits. Out of the fifty fugitives, three were condemned to decapitation, while the rest were subjected to corporal punishment. Only later came the result of the commission's investigation, which concluded that the prisoners' complaints had been justified, and that the inspector had not complied with regulations. Against this background the death sentences were referred to the High Court, which commuted them to corporal punishment (either fifty-four or twenty-seven lashes). This turned out to be a dubious mercy, since the convicts declared that they preferred death to being locked up again.

In November the prisoners revolted again. The King was furious and immediately set up a commission. This time, however, it was not to investigate conditions in the Rasphouse but only to pass sentences. It was ordered to do so rapidly; sentences carried no right of appeal, and were to be executed at once lest the absolute monarchy risk making concessions to the convicts. This was the reason why Anders protested against the summary proceedings arguing that the prisoners should at least be allowed the right to appeal the sentences passed by the military court to High Court. But neither the King nor Frederik Julius Kaas, his cringing civil servant, head of the Danish Chancellery, would listen to any of this. Anders's protest was useless, and he loyally consented. Subsequently, the King having been obliged to second thoughts by public sentiment, demanded the Chancellery backdate an identical representation to a royal resolution lest actions of the military court seem to smack of the worst kind of absolutism. This retrospectively conferred on the military court's judgements a new legitimacy. Since Anders's protest against the lack of a right of appeal had been in vain, and since the convicts had already been executed, his potential refusal to sign the representation would change nothing, but only be a formal gesture towards alleviating his pangs of conscience. The King had his way irrespective of Ørsted's dissent.[22]

*

Gitte had been pregnant throughout the spring of 1815, and she and Hans Christian agreed that their flat was not big enough to accommodate the anticipated expanded family: Dorothea and Bodil, his half-sisters, also needed lodgings. They found a back yard flat without a view as the last house of the street obscured it. On the other hand it was convenient for Ørsted's lectures at the Military Academy, the Classen Library in Amaliegade, and his collection of scientific instruments that had moved to its old premises in Thott's Palace. Everything, including his father's new dwelling at Almindeligt Hospital, was located in the new part of Copenhagen called Frederiksstaden, within five minutes' walk.[23]

The move came late, while Gitte was near her time, and on 25 June 1815 she gave birth to their first-born whom they baptised Karen after her grandmother.[24] Her parents continued an old family tradition to recycle names that would either be a reminder of dear ancestors or honour relatives, whose responsibility and help might be needed if an accident occurred, which was likely over the course of a lifetime.

Little Karen became the lively focus of the Ørsted family. Gitte was busy refurbishing the flat, and Sophie was only too happy to look after the little girl, whom she loved as if she were her own:

> 'She looks like her father and has such nice hands, her forefingers especially pointing at her big dark blue eyes and a small snub-nose, sometimes she looks like Doctor Martin Luther's copperplate. There is a lot of fuss going on with the child, when she is asleep and I want somebody to take a look at her I wake her up right away and she smiles . . . we take her up with great pleasure. Her father loves her inexpressibly, when he is totally buried in his studies and I enter his study saying Little Karen wants your pocket watch, he leaves everything aside to look for it . . . he, who otherwise could not bear being disturbed. In the mornings she lies in his arms in bed, and he can make her stop crying at once . . .'[25]

Indeed, the professor had become a paterfamilias.

Soon after the wedding Hans Christian gave his wife a leather-bound book that, apart from seventy-six pages of hand-written drinking songs in French, German, and Danish, contained a few hundred empty pages for her household accounts.[26] This was much needed because of their scarce resources. Gitte would buy things from the grocer, the butcher, and the wine merchant and they would give her good credit, but often she had to turn to Mrs Møller, her old mistress, or to Sophie or Dorothea to borrow cash. In the beginning it was difficult to develop a routine for keeping the accounts of income and expenditure, but the time was economically chaotic, where deflation had succeeded inflation, and extra money was needed for moving house. Their household grew and they did not live puritanically. They had many mouths to feed. Apart from Hans Christian's three half-siblings they had their odd-job man who carried water and firewood, as well as two maids.

At the same time many people had to be paid for their services, the chimney sweep, the garbage collector, delivery boys bringing newspapers, the gatekeepers, and the watchmen all deserved a coin for security, not to mention cabdrivers, who expected to be tipped. Hans Christian was a member of a number of voluntary associations charging subscriptions: The Agricultural Society, the Scandinavian Literary Society, the recently formed Biblical Society, the Masonic lodge, and not least the Harmonic Society, which was by far the most expensive, but in return provided the best concerts in town. Every quarter the Ørsteds had to pay five rixdollars for poor law relief and seven rixdollars and three marks for the tax on rank and titles.

The Ørsted family ate all kinds of beef, mutton, venison, pork, and poultry (chicken, ducks, geese, turkeys), fish (eel, cod, salmon, plaice, pike, carp), potatoes, fruit, vegetables, chestnuts, cakes, white bread from Hamburg, Swiss cheese, barrels of butter, coffee and tea from the Congo. There was also wine in large quantities: at one time 136 bottles, at another time 144 bottles, later entire hogsheads containing 226 litres. They hosted many parties. Expenses for wax candles, firewood, and peat weighed heavily in the budget, but still they gave alms to the poor, not only in the form of the obligatory tax, but also voluntary amounts to beggars knocking on their door.

So, perhaps it is not strange that Gitte put a little money on the lottery, although she never won anything. Their ledger testifies to the fact that Gitte's education left much to be desired. She would spell words as they are pronounced, though it did not really matter as her spouse could read it anyway, and if she forgot something he would add it. He also taught her German to enable her to read Goethe's novels. She did not feel embarrassed at her lack of basic skills. People commenting on her character emphasise her unabashed judgements. She did not think of herself as a subordinate. Why should she? She had never had a chance to acquire bookish knowledge, and compared to her fellow sisters she had nothing to be ashamed of. Besides her spelling did improve and in due course she managed to read Goethe—in German!

*

Ørsted had always preferred living together with his instruments. He had done so on his return from his first grand tour when he moved into Collin's Court. He started his lectures in these premises although the collection was only transferred to him some months later. When two years later he was appointed professor, the room was too small, not only because his lectures were so well frequented, but also because the collection had increased. His troubles were short-lived, for his instruments were now harboured in Thott's Palace near Collin's yard, and the Foundation continued to pay the three hundred rixdollars each year for tenure and experiments.

In 1809 he had the opportunity to rent Rubens's Court. The University Board approved the contract, his subsidy was increased to four hundred rixdollars, while he had to supplement it by another two hundred, because the rent included a large private flat. He also received four to five hundred rixdollars earmarked for experiments. The following years inflation was running high, and Rubens demanded an increased rent of a thousand rixdollars. Ørsted applied for a higher subsidy shortly before he went abroad in May 1812, and took it for granted that his application would be met. He even thought he would be able to afford to improve the laboratory, and commissioned Sibbern, a master builder to put in a boiler and some tables covered in sand for holding apparatus used in experiments. Ørsted was soon to regret this decision, for while he was working on his major work in Berlin and Paris, everything went wrong at home. First, the former grant of four hundred rixdollars had run out and was not extended before his departure. Secondly, his application was misunderstood. The authorities did not increase his subsidy for the rent, only the amount for experiments from four to seven hundred rixdollars. Rubens lost his rent, and Krumm, who substituted for Ørsted, was on his way to Christiania and powerless, so the owner cancelled the contract. At this time of confusion the University bursar took the matter into his own hands, renewed the tenancy of Thott's Palace, let Ørsted's collection of scientific instruments be moved there and sent the bill for the costs of removal to Ørsted—in Paris!

Ørsted was bitter and furious. In a way he had fallen into a trap he had dug for himself, but that did not alleviate his fury. He poured out his troubles to Anders and wrote a vicious letter to the University complaining that he was made responsible for the removal of his instruments by others. He imagined that the University had made a coup by attempting to seize a collection it had never deserved to possess. He was aware that he had written in anger, and asked Anders to remove the worst insults before mailing the letter.

Now he had no home to return to. The tenancy had been cancelled, and his collection once again separated from his flat. He had no subsidy, and the builder's and smith's improvements of his laboratory were not only wasted, but also unpaid for. He was still responsible for the collection, so he had to pay for the tenancy in Thott's Palace and also for his outstanding debt to Sibbern. In addition, his travel abroad had cost him more than the grant, and helping Anders and Sophie had emptied his pockets. His debts forced him on his knees and he had to cast aside his bitterness. He begged Møsting, the new Treasurer, to pay Sibbern, which he did. Next he approached the Foundation ad usus publicos cap in hand to have the subsidy for the rent resumed and have his grant for experiments converted into part of his regular salary as a professor. Finally, he initiated a protracted row with the University bursar as to who should pay for insurance and taxes.[27] He succeeded in all respects.

So, Ørsted emerged completely victorious from his quarrels with the University. His patrons in the Foundation ad usus publicos, Schimmelmann and Møsting, phrased the outcome as follows:

> 'The King has graciously donated as a gift to the University the collection of physical and chemical instruments entrusted to Professor, Knight Ørsted's supervision and use on condition that the University continues to bear the expenses covering appropriate premises for the collection, and to take care of its maintenance and expansion as well as its being placed into the hands of Professor Ørsted for all lectures on physics and chemistry that he might take pleasure in delivering to academic citizens as well as to others.'

Officialese can be amusing. In plain English the Foundation ordered the University to pay for all expenses while allowing Ørsted to keep full sovereignty over the collection.[28]

1815 was an odd year for Ørsted. Honours flowed in his direction as he was elected secretary of the Royal Danish Society of Sciences and Letters. He was also acclaimed by the press quoting from foreign sources: 'Hans Christian Ørsted, Professor of Natural Philosophy at the High School [University] of Copenhagen, has lately been lucky enough to enhance the respect of foreigners for Danish science. Not only have his works been translated into foreign languages, but also scholars from abroad have even dedicated their works to him...Last year the Englishman, Knight Homfrede Davy's works on the elements of chemical philosophy—translated and augmented by van Mons—was dedicated to him', and a French review of *Recherches* stated: 'Of all the issues following the progress of science, only a few are more important than the inquiries Mr Ørsted has made concerning the identity of the chemical and electrical forces'.[29] Perhaps Mr Ørsted himself had been helpful in providing the newspaper with the source material for this favourable publicity.

... The academic teacher
raising the best of his students to his own level
and breathing into them true love of science
has grasped the upper link of a chain
embracing everybody.[1]

32 | 1815–17

Dynamical Research
A.S. Ørsted's Dissent

DURING THESE years Ørsted placed advertisements in the leading newspapers. This one is typical of them: 'This winter I intend to deliver a series of lectures on experimental physics in general and another one on chemistry, each three times a week from six to eight o'clock. Admission is twenty rixdollars for each. In addition, I intend to deliver a lecture on that part of astronomy that must be regarded as appropriate for every educated person. Admission is ten rixdollars. Ladies are welcome to join all lectures. Please sign up at my address. H.C. Ørsted...'.[2] In 1815 he also provided lectures for diplomats in German. They paid higher fees, and the family needed the money, and at the same time he gained a great deal of satisfaction from disseminating knowledge.

He also lectured on agricultural chemistry once a week in the Classen Library, conveniently situated close to where his father lived in Almindeligt Hospital.[3] Since 1806 Ørsted had been a member of the Agricultural Society's technical commission which evaluated innovations such as threshing machines, ploughs with cast mouldboards, and windmills with sails adjusting automatically. The Classen Trust paid for his lectures. No lists of participants have been preserved, only a few cursory notes.[4] Ørsted stressed the improvements the sciences could offer agriculture. Routine rests on traditional practice, while scientific principles, more than merely bookish knowledge, inspire experiment and reflection. The scientific farmer recognises that certain industrial crops yield a higher income than grain, but in order to profit from his knowledge he must familiarise himself with agricultural chemistry to get to know how dyestuffs are extracted from plants, how wine is produced from fruits, or how the quality of butter is improved. The soul gains and the true value for mankind rises by studying the principles guiding our practice.

In 1815 he also initiated what was to become a particular tradition. He started lecturing in Thott's Palace on the first Tuesday evening of each month to keep industrialists up to date with the latest discoveries and inventions in science and technology. These lectures attracted large audiences and continued until the summer of 1848 when the civil war broke out. Altogether 237 lectures were delivered over a period of thirty-three years. He received an annual grant of three hundred rixdollars from the Reiersen Foundation.[5] Contents changed from time to time, because it was the novelty value that mattered. His choice of subject was determined by the latest foreign journals he subscribed to and ranged from organic and inorganic chemistry through astronomy and meteorology to physics and technical inventions. Listing all the subjects would take several pages, from theories of wind and weather to measuring the speed of sound, and the use of iron chains instead of ropes for anchors. Ørsted's notes were usually quite cursory, so he probably brought along the journal in question to support his performance. He briefed his audience on discoveries and inventions, but gave them no instructions as to how they could put them into practice. He also talked about his own experiments with galvanic trough apparatuses, electromagnetism, and aluminium, and shared what he had learned from conferences in Europe.[6]

Apart from his private lectures, as a professor he was obliged to give courses in physics and chemistry to first-year students at the University and to young officers at the Military Academy. He also gave his expert opinion on enquiries from the Admiralty, such as 'Would it be correct to make the steam boiler used for the heating of tar at the Royal Navy Dockyard out of iron when the tar cauldron is made of copper?', to which the professor replied that the savings from using iron instead of copper would be greater than the drawback from the galvanic effect of rust on the iron when in contact with the copper.[7] The range of technical projects varied from experiments with a galvanic method of exploding mines to the production of wine from domestic fruits.[8]

In the summer Ørsted was not burdened by lectures, but when the winter semester started in November he was overloaded with them as from then on he would teach five hours a day.[9] A good many of the lectures were offered because he needed the fees. His creditors would pester him, and he had to spend so much time earning money to pay for his research equipment that none was left to actually do research, let alone publish the results. This was a vicious circle. As so often before he turned to the King for help through Schimmelmann and Møsting, his faithful patrons in the Foundation ad usus publicos. He let Frederik VI know that his textbook was in great demand, but that due to overwhelming pressure of work it was unfinished. The mechanical part was sold out, and the dynamical part the *Theory of Force* was still unwritten.

Ørsted accompanied his application with a brief biographical sketch of his career so far, culminating in his *Ansicht* and *Recherches*, which had made him famous abroad. 'For a period of eight years I have tested my ideas by lecturing on them in different ways and by receiving objections and comments from my audiences.'[10] He offered to write a short textbook in one year and a more comprehensive one in three years if he could graciously be awarded a grant of four times five hundred inflation-proof rixdollars. His two patrons recommended the project 'that will to no small degree contribute to the honour of Danish literature... A man of his genius, insight, and diligence needs peace... Permanent elementary instruction has a deadening effect on the spirit'.[11] However, in contrast to his dealings with Steffens, the King did not wish to take Ørsted away from teaching students and military officers. So he complied while making it clear that he 'had no intention of enabling him to reduce his teaching obligations. Quite the contrary, it is our will that he delivers all the lectures his posts direct him to do'.[12] If the short version had not been completed within the year, the rest of the grant was to be cancelled, the King decided.

It was not. The following year Ørsted poured out his troubles begging the grant to be continued even if he had not met the condition. His excuses were many. First, he had been bedridden all summer, which was true; his eye disease and other infirmities had tormented him.[13] Secondly, the Health Commission had directed him to take on extra work in negotiating the prices to be paid by the authorities to pharmacies, which had taken up as much time as writing a third of the textbook on chemistry. Thirdly, he had had to update his textbook on physics, which had sold out because of a strong demand from students. Fourthly, he was supervising a translation of Chaptal's *Chimie appliquée aux arts* ['Technical Chemistry'] and writing a preface. Fifthly, as secretary of the Royal Danish Society of Sciences and Letters he was obliged to write and edit its transactions, a volume of which was due. His workload did not decrease during the summer since he was an ordinary professor and being the youngest member of the Senate it was his duty to serve as its secretary. True, he received three hundred rixdollars for it, but this did not provide him with more time in his laboratory. He longed for 'an alleviation of the mechanical and semi-mechanical tasks imposed on me as the secretary of more institutions in order to gain time for my scientific work'.[14] The King and his patrons tempered justice with mercy. His grant continued to flow, while his textbook never saw the light of day. The assignment was transferred to W.C. Zeise, his disciple, and with it the grant, since 'there is as yet no work in Danish explaining this branch of science in its present state'.[15]

In 1817, while the University and Vor Frue Church were still bombsites, Trinitatis Church was the setting for the defence of many doctoral theses, among them those of W.C. Zeise and J.L. Heiberg. The University had not awarded a doctorate in chemistry within living memory, so Ørsted was proud when he was handed Zeise's thesis 'On the ability of alkaline bodies to change matter belonging to the organic realm'.[16] The young pharmacist had conducted experiments on the effects of alkali on sugar, at that time a comparatively new problem within the branch of organic chemistry.[17] Heiberg had defended his thesis on 'The Spanish Drama with special reference to Calderon' only a few days before Zeise.[18]

*

Ørsted's research during these years had one aim: to consolidate and elaborate the dynamical system by experiments. His first series of experiments aimed at measuring the diffusion of forces for which he used the Coulomb electrometer that his collection had at first borrowed and now bought from Manthey. The second activity concerned the development of a new galvanic apparatus that could amplify the galvanic force considerably so as to facilitate new

Fig. 65. The nave of Trinitatis Church, 1826. As long as the University was in ruins, defense of dissertations took place in the chapel of Regensen (a residential hall) or in Trinitatis Church. Coloured copperplate engraving by J.C.E. Walter (1799–1860). RL.

experiments. The third series of experiments aimed at examining the compressibility of liquids in order to repudiate the corpuscular system. These research projects each in their own way point towards the great discovery of 1820 and so we shall take a closer look at them.

Coulomb's electrometer was an ingenious piece of apparatus also called Coulomb's torsion balance. It measured the strength of magnetic and electric forces. This was before the difference between voltage (potential) and ampèrage (current) was known. Newton's force of gravity obeyed the inverse-square law, being inversely proportional to the square of the distance between the two masses. Coulomb set out to discover if other physical forces obeyed the same mathematical law, and he found it to be so as far as the forces of heat, affinity, and cohesion are concerned. At the same time he maintained the corpuscular theory in explaining the physics of these forces, involving interaction between weightless fluids. As to measurement of

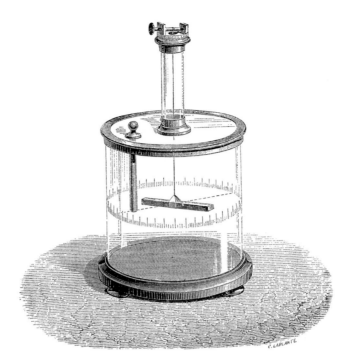

Fig. 66. In 1785 C.A. Coulomb had invented the electrometer, the so-called the torsion balance, which works in this way: A bar magnet is hung on a piece of silk thread so that it is suspended in a horizontal position; through a hole in the lid of the apparatus another bar magnet is placed vertically so that the two north poles are close to and hence repel each other. The force of repulsion can be read on a scale and the angle of repulsion can be measured. This mechanism can also register the influence of earth magnetism on the horizontal magnet, enabling the physicist to compensate for it and can see that the magnetic forces do obey the inverse-square law.

Coulomb's electrometer can furthermore be used to measuring the diffusion of static electricity. A bar of shellac (a poor conductor) is fixed to the silk string and at one end an elder-pith ball electrified by the discharge of a Leyden jar. At the other end a paper disc is placed partly to balance the elder-pith ball and partly to stabilize the movements of the bar. Through the lid another bar of shellac is lowered also carrying an electrified elder-pith ball. When both balls receive identical discharges (positive or negative) they repel each other. The degree of repulsion can be read on a scale on the cylinder wall. Coulomb found that the force by which two equally electrified balls repel each other diminishes proportionately as the square of the distance increases. If the distance between the balls is doubled, the repulsion will be four times less. And if the two balls are given opposite charges the attraction follows the same proportions. Finally, he discovered that if the balls only receive half the discharge from the Leyden jar, the reaction will be four times less. In short: Newton's law of gravity, the inverse-square law, is immutably valid for magnetism and static electricity. Poul la Cour & Jacob Appel 1897 §§215 and 286.

magnetic and electric forces they fell somewhere between showing long-distance and short-distance effects.

When working on the *Theory of Force* in 1807 Ørsted had carried out experiments with the torsion balance he had borrowed from Manthey, but without being able to make his results match the inverse-square law. A year later he told Ritter that he had paid a visit to Volta, and that Volta had equally failed to make his measurements correspond to Coulomb's. When Simon heard about this in Berlin, he set out to test Coulomb's measurements, as did Ørsted, but without reaching a final conclusion.[19] Now in 1815 he resumed these measurements.

In 1806 Ørsted had suggested that the diffusion of electricity takes place as an undulatory process like the waves of sound and light. One might find it risky to entertain any image concerning the way electricity moves in conductors, for one can hardly blink before the electricity has finished its passage through the conductor, irrespective of its length. He reckoned that the time it takes electricity to pass along a good conductor the length of a Danish mile (well over seven kilometres) was less than one second. He further believed that electricity is propagated as a series of oscillating alternations between polarity and identity, as against the corpuscular theory claiming that electricity consists of imponderable fluids flowing evenly like currents in the conductor.

If one connects a positively charged body A with a good conductor such as an iron cylinder BC, the conductor will first become polarised, becoming negative at B and positive at C. In the next phase the electrical discharges at A and B will become identical, making B positive and C negative. This polarisation and identification takes time, of course, because the process has to take place serially, and at every stage the electrical force must be both polarised and identified. Ørsted saw the empirical evidence for these successive alternations in the conductor in the thin iron wires which, when they were used as conductors of strong electrical discharges, melted in regularly alternating zones. He had made similar observations when looking at the waves which he thought that the electrical forces produced on the disc when he made acoustic figures. So, electricity was propagated not like a current but like waves.[20] Was this the reason why he had found that Coulomb's law appeared to be wrong? Ørsted must have believed that his interpretation of the propagation of electricity as a series of successive polarities had put him on the track of something crucial, for his article on this was published not only in Danish but in German, French, and English as well.[21]

However that may be, in the autumn of 1814, while he was writing his speech on the cultivation of science as religious worship, Ørsted launched a series of experiments with 'Coulomb's ingenious device'.[22] And soon after having been elected secretary in the winter of 1815 he lectured on 'The Law of the Propagation of the Electrical Forces' in the Society.[23] Did the inverse-square law apply as Coulomb thought to be the case, and if so would electricity have to be included under the same law as gravity? Or were Volta and Simon correct in claiming that 'the repulsions of electricity decrease in proportion as the distance between the reciprocally repelling bodies increases': i.e. not square but only to the first power? After a long series of measurements and mathematical calculations of varying distances and torsion powers, he found such significant deviations from the inverse-square law that he could not consider it a general law.

'My experiments did not show that the effect diminished according to any permanent law, but that it sometimes was less significant at a larger distance than a smaller one. This seems to speak against the accuracy of the experiments, and I confess that this circumstance... also evoked in me some suspicion, but upon second thoughts it enhances the reliability of the experiments. For it is well-known that the electrical forces do not propagate evenly, but that by approaching an electrical body one encounters varying zones of positive and negative electricity, and consequently the electricity of a body being placed at various different distances from another one will show now an increasing and now a decreasing effect... Hence we learn from these experiments that one cannot expect to express the law of diminishing electrical forces by any simple function of the distances, but they diminish continuously according to a ratio of their potencies, whose exponent increases as the distance increases.'[24]

This sounds very complicated, and in the Transactions, *Oversigter*, he simplified the conclusion to say that Coulomb's law applies, except over very small and very large distances. At very small distances the effect decreases in proportion as the distance increases, but at very large distances it decreases up to the third power.[25] His laboratory notes show that he put a lot of energy into long series of exact measurements of electrical phenomena and that he exploited his mathematical skills to find out if the resulting sets of figures would enable him to deduce a mathematical law. So it is simply not true to say that Ørsted was only interested in qualitative descriptions of nature.

The following winter he lectured on his theory of light in which he explained light as the reciprocal contacts of polar electrical forces in their wave motions through the conductor. 'The moment of contact produces light.' The greater the opposition, the stronger the light (and heat). Obviously this theory is not empirical, but a priori. Subsequently, in his eyes, it did not contradict the experience available concerning the chemical and heating effects of light beams, and obviously corresponded to Herschel's and Ritter's discoveries of infrared and ultraviolet light.

In the interest of his experiments with the lighting and heating effects of electrical forces in conductors, Ørsted wanted stronger galvanic devices to be put at his disposal. In 1815 he had been sent trough apparatuses made of porcelain from England.[26] In each of the forty-eight porcelain troughs filled with acid solution copper and zinc discs were sunk, and connected in pairs of Cu + Z. Porcelain was more waterproof than wood, so the acid did not flow away to be wasted. But why not simplify the device by making the trough out of copper and sinking the zinc disc into it? This would most likely be more efficient, and a coppersmith would have the skills to do it. At that time Ørsted worked with Lauritz Esmarch, a civil servant working in agricultural economics and his senior by twelve years. Ørsted must have found Esmarch easy to get on with, since he soon chose him as his travelling companion on an expedition to Bornholm. As is apparent from Gitte's ledger, as soon as the winter semester and examinations were over, large sums were spent on new instruments from Jeppe Smith and materials for experimentation such as iron filings, zinc discs, sulphuric acid, nitric acid, and alcohol.[27]

In October 1816 the new trough apparatus was ready for demonstration at one of the first monthly lectures on the latest progress of science.[28] To start with Ørsted and Esmarch had ordered forty-eight copper troughs complete with zinc discs. They were placed in two rows, interconnected, and the two troughs at one end were connected to conductors. They had also been supplied with another type of apparatus consisting of copper cylinders, at the centre of which chimneys had been soldered on for heating the liquid with glowing charcoal, and zinc cylinders of a diameter allowing them so be sunk down between the chimney and the copper cylinder.

The zinc cylinders were made of three pieces of bent zinc, since the artisan or instrument maker could not acquire rolled zinc. Only six of these cylindrical devices had been made, but their effect was considerable. Esmarch and Ørsted guided the electrical forces through mercury in a capillary tube causing it to create sparks and to evaporate and liquefy again when the circuit was broken. In only fifteen years electrochemistry had developed from Ritter's discovery using connected parallel metal bars to produce a very weak electric circuit, via the voltaic pile to Ørsted's trough apparatus that made an iron wire glow and melt. 'If an experiment was needed to show that the electric spark is just a fierce glowing of matter filling the space in which the spark appears as incandescence this experiment seems suitable.'[29] Ørsted imagined that as soon as he had acquired a sufficient number of copper cylinders he would be able to split metals!

This battery of four dozen metal troughs filled with diluted acids changed the conductors almost instantly when the circuit was closed and the electric discharges struck sparks and diffused them with a force making the cold platinum wire incandescent in the dark. Ørsted spread an atmosphere of mysticism around him as if he were an accomplice of other-worldly powers entrusting him with a magic wand. When twirling it he would enthral his audience and make

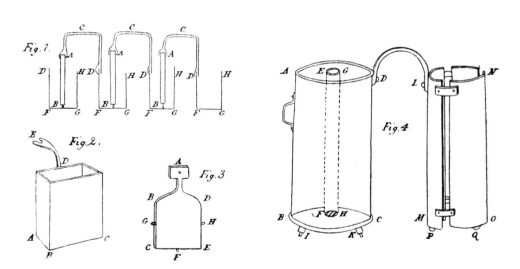

Fig. 67. HCØ's galvanic trough apparatus. Fig. 1 shows a cross-section of the trough apparatus. AB is a vertical cross-section of the square zinc discs, DFGH the copper troughs, and ACD a handle soldered onto the trough. The troughs are filled with water containing a little nitric and sulphuric acid. Fig. 2 shows a copper trough, and fig. 3 a zinc disc. HCØ discovered that the galvanic force increased by heating the acid solution and therefore constructed a cylindrical copper trough—Fig. 4—with a chimney EFGH at the centre, into which he put glowing charcoal. NS II 208.

them gape at the inconceivable effects of the invisible electric forces. They made a tremendous impact on the King as well. In the winter of 1817 Ørsted invited him to a demonstration of his galvanic trough apparatuses in Thott's Palace. He also used them to set off land mines from a distance. The King was obsessed with military matters and impressed with Ørsted's gadgets, but could not suppress a certain wonder as to how the professor could afford such wonderful devices out of his modest salary. To Ørsted this royal goodwill was taken as almost a promise of another research grant. How would he manage to have his wish fulfilled and fill his empty hands with an appropriate sum of money? He had an ability to put words together well.

> 'Your merciful Majesty', he began, 'When I enjoyed the gracious opportunity to show to Your Majesty some of my experiments, it pleased Your Majesty to remark that experimental research involves costs that generally exceed the fortune of a scientist; later on Your Majesty added that You graciously deliberated granting me a sum for the continuation of my experimental research. Despite the fact that I was already burdened by a significant debt into which I had been drawn by my scientific researches, I had decided not to bring this merciful remark to Your Majesty's mind . . .'

As is apparent from Ørsted's draft of the application, Esmarch paid part of the expenses for the big galvanic trough apparatus, but it was costly to keep going, for its forty eight containers consumed large quantities of acid that lost its strength after two to three hours. After each experiment the zinc discs had to be cleaned from oxides, and they were gradually corroding. Every experiment cost him fifteen to twenty rixdollars for acids and a similar amount for zinc discs. In addition he had to pay Zeise, his amanuensis, and Søren, his helping hand, as well as fuel for heating and lighting his lecture hall, and instruments such as chemical materials, glass retorts, u-tubes, capillary tubes, files, rasps, etc. His expenses were manifold and the discoveries that were the fruits of these experiments, gave 'no [pecuniary] advantages to the discoverers'.[30]

The third experiment dealt with the compressibility of liquids. Here, too, Manthey's instruments helped him. He had for long borrowed a device constructed by the American Jacob Perkins. The prehistory of these experiments was the Englishman John Canton's compression of water in 1761 which was later resumed by the Germans R.A. Abich and E.A.W. Zimmermann, and more recently by Perkins. By applying Boyle's law (the inverse proportionality between pressure and volume) Ørsted found that the volume of water by compression decreased by 0.00012–0.00014 at 12° C whenever the pressure increased by one atmosphere. The problem was that the measuring results of the various physicists did not correspond, which might be due to the difficulty of constructing robust enough piezometers, so that they did not expand as the pressure increased. In fact, Perkins tried to do this experiment inside a cannon. Ørsted would continue to improve his piezometers for the rest of his life in collaboration with Christopher Hansteen, amongst others.[31]

What is of particular interest here is the purpose of these experiments, which is not difficult to grasp. Measuring the reaction of various substances exposed to increasing pressures was thought by Ørsted to be the most obvious way of evaluating the corpuscular system and its

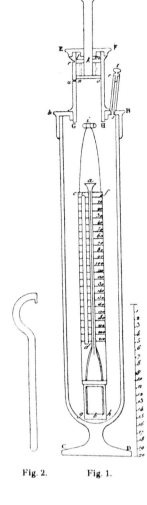

Fig. 68. HCØ's piezometer. Fig. 1 shows a cross-section of a glass cylinder with a brass hood (i) above. At GH is has an opening with a cylinder and a lid EF, in which is placed the screw IK with the sliding piston lmno. Inside the glass cylinder is a flask ab, a tube cd and a scale. Water is filled into the tube and into the flask a drop of quicksilver. When the experiment of compressibility is carried out, the brass hood is taken away, and the flask and the tube are sunk, whereupon the hood is tightened with the key (fig. 2). Now the empty part of the pump tube is filled with water through the hole rs, which is then closed. The pressure increases, and the compressibility of the water is measured by the diminishing of the volume of air. NS II 315.

rival, the dynamical theory. Experiments of compressibility must be the *experimentum crucis* to judge the controversy between Kantian dynamism and Laplacean atomism. According to the Laplaceans, matter consists of atoms with porous interspaces. If matter is compressed a sudden jump should occur when the pores are compressed as well as there being a minimal limit to compressibility when the solid atoms remain. According to Kant, by contrast, 'matter' consists of forces filling space. If this is exposed to pressure, it would diminish gradually as pressure increases, and with an infinite pressure (only existing in theory) matter would disappear into Gowin Knight's point atoms. The experiments compressing water showed no sudden jumps and consequently they did not confirm the atomic structure of matter. But this did not lead Ørsted to assert that the dynamical theory was finally proven.

The three activities show that prior to 1820 Ørsted was deeply involved in experiments aimed at determining the future of the dynamical system starting from the premises of its adversaries. These were experiments that could not easily be dismissed as 'merely qualitative'; quite the contrary, they were concerned with measurable phenomena, and the values to be found would be fed directly into mathematical equations. The experiments were intended to find out if electric forces are propagated according to the inverse-square law, and they addressed issues that might defeat corpuscular theory. Moreover, it is obvious that Ørsted was in a hurry to publish the results of his research even if they were preliminary. Maybe the reason for this haste is the same as we noticed in Faraday's career (ch. 21): Ørsted sensed that he was on his way to sensational discoveries and that he had to safeguard his priority to avoid the fate of the theses he had presented in *Ansicht*, which had subsequently been stolen by others and published in their names. This was exactly what John George Children had been guilty of when he presented Ørsted's theory of heat as his own.[32]

*

On 25 June 1817 the prisoners (or 'slaves' as Sophie called them) in the Rasphouse Prison at Christianshavn rebelled again, setting fire to the buildings. The authorities called on the military to regain control by firing cannons. Frederik VI went there in person to get an impression of the destruction that was the subject of report after report in the newspapers over the next days. This was the third insurrection in a couple of years, but this time it was even more violent than before.

Two years before, the King had immediately set up a military court to pass sentences with no right of appeal and they were carried out without delay. On that occasion Anders had only been responsible for the judiciary as a deputy in the Danish Chancellery for a little more than two years, and still he had protested against the summary proceedings demanding that the sentences passed by the military court should abide by the existing legislation and carry the right of appeal to the High Court.

This time F.J. Kaas sensed that a rapid decision to repeat the previous procedure by setting up a military court would anticipate the King's wishes and put the zealous Ørsted out of the running. But the crime was more serious this time and Kaas felt that the methods of interrogation and punishment should be tightened up. He therefore suggested a tougher mandate according to which every tenth culprit was to be shot immediately if none of them pleaded guilty. Such measures would be an unheard of escalation of martial law. True, the law did prescribe decimation of mutineers, but only among those proven guilty. And besides mutiny was by far the most serious crime covered by martial law. Ørsted argued that among the haphazardly apprehended individuals might be prisoners who had no part in the rebellion at all, indeed some of them might not even have known about the plans for it. To draw lots for the life or death of human beings could only take place at a court martial to decide

Fig. 69. The Rasphouse Prison on fire 25 July 1817 seen from the bridge crossing Christianhavn's Canal. The main building with the onion shaped spire is in flames. The square is replete with soldiers. Oil painting by Marie Bang, who is only known for this work, previously attributed to I.C. Dahl. CMC.

who among those found guilty were to be recommended for mercy. But now Kaas demanded the drawing of lots to decide who were to be executed whether the individual was found guilty or not. According to Ørsted such a procedure was a glaring example of injustice, and although Kaas was his superior he refused to sign the motion taking it for granted that Kaas, being the only one referring the matter to the King, would inform His Majesty of the reason for his refusal. Ørsted's claim that sentences ought to carry a right of appeal was rejected out of hand. The difference from 1815 was that now Anders bluntly refused to sign a motion, the unlawful consequences of which were still avoidable. And if the King implemented it, Anders would at least have made it clear that the decision had been made by the absolute monarchy alone since the deputy responsible for the judiciary disagreed.

We do not know if Kaas referred Anders's reservations to the King, but we do know that the King was furious and decided to consider his resistance as sheer insubordination. On 28 June Frederik VI called his presumptuous civil servant to account, letting him know that his attitude was completely unacceptable, and that he disapproved of Ørsted's perpetual criticism in general. As the monarch's trusted civil servant with special responsibility for the judiciary it was not up to him to peddle pernicious ideas and in particular not to interfere with the court's workings by assuming opinions damaging to the maintenance of social peace. The King ended the audience by letting the refractory Anders know that when (and not if) he submitted his request to resign it would be granted. How Anders managed to stay calm is difficult to imagine for when Frederik VI worked himself into a temper it was not wise to reply, but he did obtain the King's merciful permission to explain his conduct in writing.

Two days later his patient apology was put on the royal desk. He asked the King to realise that as it is the duty of civil servants to loyally and unconditionally subordinate their opinion to the monarch's will so they are equally obliged as councillors to heed their conviction and conscience. Otherwise they would only obey an order (as he had in 1815), although their real task is to provide advice at the royal request (as in 1817) that the King is perfectly free to accept or reject. The only thing Anders had done in the present case had been precisely to give the best piece of advice that his loyalty urged him to. Consequently, Kaas must have misunderstood the King, when he considered the royal request for the advice of the Danish Chancellery as being tantamount to a predetermined result. It would be self-contradictory to consult freely on a preordained command.

To make matters worse Anders also undertook to defend himself against the King's general reproach:

> 'As a writer as well as in my public position I have more often sided with the stricter part than with the more lenient one. In particular, I have frequently in the Danish Chancellery as well as in the High Court advised against reprieves that were recommended by a majority. For instance, I do not allow upbringing, previous behaviour, pecuniary position, etc. to have any mitigating effect as most people do, but maintain that the damage and danger of the action to society should be the predominant point of view in criminal law.[33]

Anders ignored the question of his resignation. So did the King. Anders could go on fighting for the rule of law, but he had to be careful if he wanted to be able to continue to make a difference in the Danish Chancellery, and strike a prudent balance by confining his internal opposition to matters of principle and by avoiding jeopardising his position.

The court martial took forty hours at a stretch to pass sentences on ninety-two prisoners. Seven were sentenced to death, one accused of arson and six of mutiny and conspiracy. At the place of execution (in the present free city of Christiania) their severed heads were put on stakes and their limbs and bodies on wheels. Culprits who did not plead guilty at once were locked up in powder magazines at the Citadel and interrogated again. The royal permission to force confessions by means of torture was not applied. Later another seven prisoners were sentenced to death, two were sentenced to fifty-four lashes each and one was put in solitary confinement for life. None of them was allotted a counsel for their defence, and there was no appeal. Since all the accused prisoners pleaded guilty, Kaas's suggestion of decimation was not carried out.[34]

The Rasphouse had its own physician, of course, and he had an abundance of patients among the prisoners. In 1817 Professor Howitz held the post, and Doctor Carl Otto succeeded him. Both were phrenologists, that is to say supporters of a theory of psychology established by Doctor Gall, a German physician who taught that human skills and properties are discernible by examination of the bumps of the skull. Both Howitz and Otto had opportunities to study their prisoner patients in the light of phrenology. Otto examined the skulls of the executed prisoners, identifying the lumps that according to Gall's theory proved their criminal properties; their characters showed directly from their skulls. Man had become an object of science and the alliance between criminology and phrenology had reduced man to a homo phenomenon. This was soon to become the topic of a bitter public debate, in which the Ørsteds vehemently opposed the view on human nature expounded by the phrenologists (ch. 41).

> During these years his scientific research was very limited due to his day-to-day obligations.[1]

33 | 1818–19
Shadows of Death
Expedition to Bornholm

THE CONVALESCENCE of the two 'Asse-children' at Bad Driburg did not improve their health, and Sophie never really recovered. In summer they left the sickly city to move to the countryside at Frederiksberg where they improved temporarily, only to lose their regained strength during the long and dark winter months. The damage done by the scarlet fever and an inexplicable disease of the abdomen taxed Sophie's energy and humour, and often she did not have the strength to rise from her bed in the mornings. Anders was out of his mind and felt as powerless as the doctors he called upon for help. The 'Powder Keg' herself had long anticipated where this was leading. On the last day of 1817 things looked better, and Anders invited some guests around for New Year's Eve, among them Miss Mathilde Rogert: 'Since Sophie is now rather well, we intend to say a proper farewell to the old year, although it has not been very friendly to us. If Miss Thilde would like to join this small party she will be welcome and treated at least as politely as usual.'[2]

Unfortunately, the recovery was only brief, and in January things went downhill again. Miss Thilde was a great help, but Anders was devastated. Unable to cope, he left their flat to be with Hans Christian and Gitte in Dronningens Tvergade. At night Sibbern and his sister would watch beside Sophie's sickbed, and her brother, Adam Oehlenschläger, would hold her hand. She had fever, was delirious and in pain, not much of a 'Powder Keg' really. On 9 February her life ended. She was only thirty-five.

She was buried in Frederiksberg Cemetery five days later. Her brother-in-law gave the funeral speech, her husband being unable to. Hans Christian did not address the large crowd of mourners, but talked directly to Sophie reminding her of her affiliation with Frederiksberg

Kirke. Here she was brought to be baptised the first time she left home, here her mother had been buried, here she had been confirmed, here she had been married to Anders, and here her father played the organ. Hans Christian went on:

> 'Here, too, the hope of immortality shall shine on your body, here the power of faith shall strengthen every mourning heart; it shall draw the veil aside hiding the godhead from the sensuous eye, and we shall see the truth that is only called sad names here on earth. Here it is called death and everlasting sleep, in the kingdom of God it is called awakening and entry into life. Here it is called passing away, but there a return home. Here sounds the cry that you have already finished your life, that your rich soul no longer breathes and develops for something higher, but there you will be greeted in a fresh existence...Here the mourner will brood over the sombre thought that the grave has engulfed you and that you are lying in the dark bowels of the earth.—O, but you rest in the fatherly bosom of God's everlasting mercy. Praise be to God, that hope stands comforting by the grave and that from there we may behold the peace of your better world.'

So far he had talked about the hope of resurrection. But now he changed the perspective, looked down at the bereaved Anders talking to him about heavenly matters and their reflection on Earth, about the love and friendship that Sophie had felt so strongly during her last days and that in the days to come would comfort the solitary spouse.

> Praise be to God that the richness of yonder world already reveals itself in this one; for love stems from there and it fills and pleases us already here on Earth. For you, resting there with your broken heart that has loved and been loved, has not love also accompanied you throughout your journey and given you many a happy day and relieved you from many a sorrow? Did love not abide with you; does it not follow you to your grave...? Is friendship not ready to embrace your bereavement? Does friendship for you not burn in many a bosom, for widely as you are known you are also loved?[3]

Many friends had written poems to Sophie. One of them was Grundtvig, now extending the hand of friendship and forgiveness to the Ørsted brothers after their bitter strife a few years before. However, Grundtvig published his poem 'The February Lily' (addressing and comforting Anders!) anonymously.[4] Consequently, it is hard to say whether Anders knew who the poet behind it was, and there are no signs of a reconciliation having taken place between the two. Still, the poem by Grundtvig, a clergyman without an incumbency, and the funeral speech by Ørsted, a natural philosopher and a Freemason, is evidence of the dualistic worldview they shared.

Anders was said to be the kind of man who is completely helpless without a woman at his side. He terminated his tenancy of the flat. He could not stand being there, where memories of Sophie encroached on him without his being able to determine what affected him the most: the loss of the sparkling friend or the definite ending of her melancholy. Anders took refuge with Hans Christian and Gitte. Luckily, the flat in the neighbouring house was free to let. Now the brothers were neighbours again, almost as close as in their cell at Elers' College and only a short walk from their old father.[5]

*

When Zeise had achieved his doctorate Ørsted recommended him for a travel grant for two years. He went to Göttingen and later to Paris to study under the supervision of J.J. Berzelius. In his diary he sketched the great Swedish chemist as short, but robust, with a turned-up nose, blond hair, and a square face with sparkling eyes, and 'bright but not really a genius'.[6] In this way Ørsted lost his amanuensis to a Swedish colleague who came to appreciate Zeise, not only as a scientist but also as a person. So, to Ørsted the loss threatened to become permanent and he had few chances of finding a substitute, for Zeise was the only student at the University to have dedicated himself fully to chemistry.

In April 1818, however, the overworked professor employed a young Schleswiger, Johann Georg Forchhammer, as his chemical assistant at two hundred rixdollars a year.[7] Forchhammer had studied chemistry with Professor C.H. Pfaff in Kiel, but from now on he was Ørsted's amanuensis. One of his first assignments was to help set up the new galvanic trough apparatus in an open field to set off mines at a distance. The method was to connect the galvanic conductor to a thin metal wire inserted in the powder. When the circuit was closed the wire would glow and the mine explode.[8] This was a demonstration only, for now peace prevailed in the country. Ørsted must have liked his German-speaking assistant, for a few months later he took him on a mineralogical expedition to Bornholm. Ørsted needed geological expertise, but unfortunately Forchhammer was neither a mineralogist nor a geologist, but a pharmacist and a chemist, and hence not the best choice for investigating coal deposits. He had travelled to Copenhagen believing that he was to manage a gasworks that Schimmelmann wanted to build, but the plans were shelved. His arrival turned out to become a blessing for both him and Ørsted, as the latter needed an amanuensis to replace Zeise and Forchhammer needed a new job.

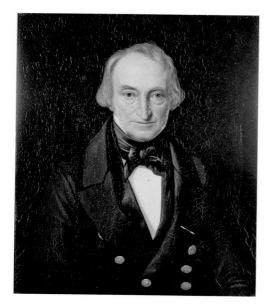

Fig. 70. W.C. Zeise (1789–1847), chemist, HCØ's amanuensis. Portrait in oils by F.F. Helsted (1809–1875), FBM.

Fig. 71. Georg Forchhammer (1794–1865), chemist, self-taught geologist, and HCØ's amanuensis. Crayon by unknown artist. RL.

In August 1818 David Coulthard, an Englishman, submitted an application to the Danish government for a monopoly of mining coal on Bornholm. It is beyond doubt that Bornholm had plenty of coal. In fact, the first thing the Agricultural Society did after its establishment in 1769 was to send an expedition to the island to investigate the possibility that Denmark-Norway might become more or less self-sufficient in coal. Møsting and Schimmelmann hoped to see the day when the country no longer depended on British coal to fuel its steam engines. After the loss of Norway, the lack of energy became a constant threat.[9] Besides, Schimmelmann needed coal for his private sugar refineries in Copenhagen. So Coulthard's application was welcome. He wanted to rationalise mining by using steam power, but he needed capital which he planned to raise through subscription, hoping that Frederik VI would subscribe to the majority of shares, worth several thousand rixdollars. The Chancellery of Agriculture was directed to ascertain if the project was sustainable. In August a commission was set up consisting of Esmarch, who had taken part in previous expeditions, and Ørsted. The mandate was to investigate the deposits of coal as well as iron ore and other raw materials in the underground rocks. Would it be advantageous to extract and use them for industry? Equally important was a survey of the geology of Bornholm.

Ørsted was fired with enthusiasm, for now he was given an opportunity to prove that his scientific knowledge was useful to the nation, as he had reassured the King time and again. Secondly, the job enabled him to get rid of old debts,[10] and thirdly he could keep his new amanuensis employed, although not with the chemistry that he knew, but with the geology that he might get to know. Ørsted had no practical experience of mining activities. His geological knowledge was confined to the little he had acquired from reading Steffens, which was of little value in the field. Forchhammer was not a member of the commission, but Ørsted obtained the King's oral permission to bring him along as a stand-in for a Norwegian specialist.

In September Ørsted, Esmarch, and Forchhammer loaded a cart with their trunks, and waved goodbye to Anders and Gitte. They landed in Rønne the next day in a terrible state, as storms and a heavy sea made them seasick. Yet they started working immediately; he wrote to Gitte:

> You should have seen me crawling up the hills at the coast, climbing fences, leaping on stones across brooks in which (let it be said for your reassurance) nobody would be able to drown; in short, you should have seen me taking a lot of exercise that is normally alien to me and you would be delighted. Only in the evening did we have lunch if I may put it that way. The food was nice, but the wine bad.[11]

The three men inspected Coulthard's test borehole. This was an ominous sight, for although it was twenty-five metres deep there was no trace of coal. But this did not dishearten them. They found seams of iron ore stretching far south of Rønne. They collected samples, numbered them, noted where they were found, and on rainy days they worked indoors writing up their descriptions. A few days later they described the coastline north of Rønne, spotting more seams of iron ore of a thickness that made them decide to dig deeper at some future time. Near Blykobberåen they discovered an abandoned coalmine and a shaft eight metres deep. 'We went down in a barrel.'[12] They also found alum slate, sufficiently rich for extracting alum, a mineral used for tawing hides, and kaolin, a fine clay used by the Royal Porcelain Factory, as well as a kind of clay that could be burned to make cement. Five weeks later the expedition returned home.

As to mining in general and Coulthard's project in particular their report smacked of over-optimism. Iron seams stretched for more than twenty kilometres. There were coal deposits in seams of varying thickness, and this was definitely not a question of brown coal, but of genuine slate coal. Samples had been sent to O. Warberg, master of the Royal Mint, and his experiments with the steam engine there had confirmed their quality. Deeper coal deposits that they had not yet reached would no doubt have a higher thermal value. The report estimated that seven million barrels of coal could be mined from a seam well over three metres thick and five million barrels from a slightly thinner seam elsewhere. 'The other somewhat smaller seams in the neighbourhood would add significantly to this output.' Systematic coal mining on Bornholm would therefore be more profitable than at the Scanian mines at Höganäs, so there was every reason to recommend Coulthard's subscription.[13]

The prospects for mining iron ore on Bornholm were equally promising. The ore contained neither sulphur nor phosphorus, it was easy to smelt, and the blast furnaces would only need a small amount of chalk for smelting, and since nature had been kind enough to locate coal in large quantities right next to them, nothing should hold them back. The position close to the coast and the low costs of transportation were benefits that the owners of many Swedish iron mines would envy them.[14]

Zeise, doing research with Berzelius in Paris, could not understand why Ørsted left his letters unanswered. Now he became aware of the reason. Ørsted told him about the expedition but remained silent about his employment of Forchhammer as his amanuensis for fear that Zeise would become nervous and jealous.[15]

*

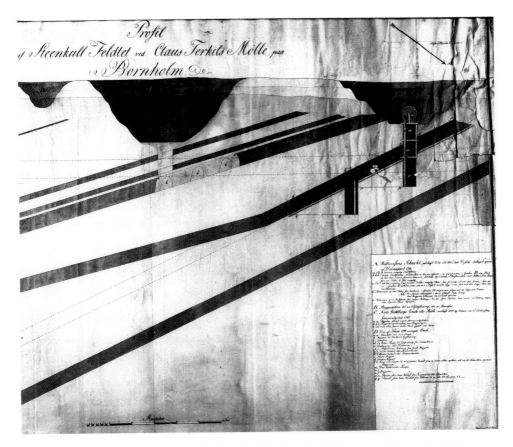

Fig. 72. Coalmine at Sorthat on Bornholm. Technical drawing. Oreby Berritzgaards Godsarkiv, LAS.

In October 1816 Gitte had given birth to their second child, a boy they christened Niels Christian, and in summer of 1818 Gitte gave birth again. This time she delivered twins, two little girls. Now Hans Christian and Gitte had had four children in four years. Karen, the firstborn, was educated at home by Mr Petersen for six rixdollars a month, and her father was longing to see if she could already read her own name (at the age of three) when he wrote it in capital letters in one of his letters to her mother.[16]

Less than a year after Sophie's funeral, Anders had become engaged to Miss Thilde. Some said that while she was alive Sophie had wanted and arranged it so. Mathilde Rogert was of the same age as Sophie, but she did not have her wit and grace. Sibbern, who had been infatuated with Sophie, took offence that Anders was getting a new Mrs Ørsted at his side so quickly. The whole winter and spring Hans Christian had planned his move to Nørregade. So, the brothers' sharing of residence came to an end and Anders found new lodgings outside the city ramparts in Østerbro. The air was better out there, and many well-to-do citizens had settled down in these rural surroundings. The distance to the city centre made a horse and cart a convenient asset. Anders and Thilde employed a farmhand to take care of their equipage.[17]

Anders's marriage to Sophie had been childless, and Thilde bore no children either. But in 1820 they took three foster-children into their home. As mentioned earlier (ch. 22) Jacob, the third brother had defrauded his father of a large sum of money. Subsequently, the swindler had tried to make a living as an innkeeper in Rudkøbing. He had married a certain Pernille Bang and became the chief fire officer of the small town, but that did not alleviate their poverty either. Their four children, three sons and a daughter born between 1812 and 1818, were neglected. Now the two older brothers intervened. Jacob and Pernille's eldest son, Søren Christian, was taken in by Hans Christian and Gitte, where he was called 'little Søren', since he bore the same name as his grandfather, a frequent guest. He was taken into the care of the family on an equal footing with their own children. Anders and Thilde received their three other children, Christian, Anders Sandøe, and Sophus in their home thus becoming 'a real family'.[18] The custom of honouring members of the family by calling newborn and helpless children after them to angle for their protection turned out to work in practice.

Anders was only allowed to keep Mathilde Rogert for four and a half years. In January 1824 he had to bury her, too. What could he do with his three foster-children? He took over Hans Christian and Gitte's flat in Nørregade, when they moved to the professorial yard in Studiestræde. Now, they were almost neighbours again, only separated by St Petri Churchyard. But Anders needed permanent help to take care of the foster-children, so he employed a Miss Bohn as his housekeeper, and she knew how to make the home a cosy place for Anders and his three nephews.

Then followed another catastrophe. '12.03.1819. My little Sophie's day of death', Gitte noted in her ledger leaving the remaining of the page empty, as if time and life had come to a standstill. The following weeks and months she was inconsolable and inactive. Hans Christian and Gitte shared only nine months with their little twin before lowering her small coffin into the grave destined to hold the rest of the family in due course.[19]

When Ørsted consoled his friends in their sorrow as they were bereaved of a dear relative, he would tell them the story of David's reaction to the death of his and Bathsheba's son, a story he remembered from his childhood Scripture lessons at the wigmaker's:

> '... While the child was yet alive, I [David] fasted and wept: for I said, Who can tell whether God will be gracious to me, that the child may live? But now he is dead, wherefore should I fast? can I bring him back again? I shall go to him, but he shall not return to me. And David comforted Bathsheba his wife, and went in unto her, and lay with her: and she bare a son, and he called his name Solomon: and the Lord loved him.'[20]

Little Sophie's parents had been under the impression that she was the more viable of the twins, as her sister Marie ailed with now this and now that. They were frightened of losing her too. The doctor prescribed the application of leeches to suck the sick blood out of her small body. They employed a woman to watch over her at night, for Gitte was exhausted, and over the following months Dr Brandis came to see the two patients time and again. Hans Christian had to take care of Karen and did so by taking her to his lectures. She walked with her father or held the assistant's hand as they walked through the city to Thott's Palace, where she would play in the garden while he lectured to the young officers.²¹ As soon as the move to Nørregade was over they celebrated the hope of a brighter future on an excursion by steamboat to Dyrehaven, the deer park by the coast north of Copenhagen, and the following month Hans Christian took his convalescing wife and their children Karen, Christian and Marie on a summer holiday to her native home on the island of Møn.²²

These journeys took place on board the paddle steamer 'Caledonia' owned by Admiral Bille. It was said that at first Frederik VI would not allow the transport of passengers onboard a vessel driven by a steam engine, as it might suddenly explode. According to a family anecdote Bille had requested Ørsted to assure His Majesty that a steamboat was a safe and expedient means of transport. In return the Ørsted family had received free tickets for and a share in 'Caledonia', so why not take the opportunity? During the voyage to Møn, the royal fear of technology came close to being justified. Something went wrong with the boiler and panic broke out. Gitte was still too weak to take care of her children, and the maid looking after Marie threw her into the arms of her mother like a bundle of rags shouting 'Take care of your child yourself, now we all perish!' Luckily it turned out to be an overreaction.²³

That entire spring Ørsted had been preoccupied with plans for the move. He was uncomfortable having his family in one place and his collection of scientific instruments somewhere else, because nearly all his lectures needed his instruments. He had found it so convenient in Rubens's Court where all his activities were united, but while he was abroad the University had

Fig. 73. The SS Caledonia. Coloured lithograph by unknown artist. CCM.

rented premises for his instruments in Thott's Palace. Now this contract was no longer renewable on good terms and Ørsted found an ideal solution in carpenter Pingel's house next to St Petri Church in Nørregade. This was a brand new building from 1813 on a site that had suffered a fire after the Bombardment. Pingel's house would provide ample space for his family as well as his instruments; indeed, it would serve as a complete professorial yard with a lecture hall. The only thing needed was the removal of a partition. If this were replaced by two columns to support the beams of the upper storey, a lecture hall holding an audience of sixty to seventy people would be established. His instruments could then be placed in the adjacent corner room. Fortunately, the University did not object to this solution, so the Board authorised Ørsted to negotiate a five-year contract with carpenter Pingel.[24]

Pingel's house had three storeys and five windows facing Nørregade. There were seven rooms on the first and second storeys. The wing had a long corridor at one side and ten windows facing the courtyard at the other. Each room had its own cast-iron stove, so the servant had a lot of work to do carrying peat and firewood let alone sawing, splitting, and stacking it. Kitchen and pantry were situated in the wing together with the maids' chambers which had a separate staircase, up which water from the trough of a well with a pump in the yard was carried in buckets, to be carried down again as waste water and thrown into the gutter. The privy emptied by the night man was in a half-timbered house in the wall separating the yard from the cemetery. Finally, behind a five-storey building at the back was a garden, also part of the tenancy, with fruit trees and a bower. Ørsted engaged a gardener.[25]

Fig. 74. Pingel's house, 35, Nørregade. When HCØ resided there (1819–1824), and later ASØ, there were only two storeys, which is to say the building went to the pilaster strip only. Photography taken after a third storey was added. DTM.

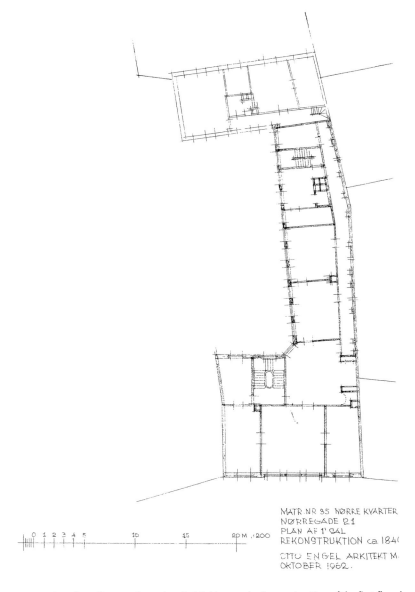

Fig. 75. Groundplan of HCØ's rented premises in 35, Nørregade. Reconstruction of the first floor by architect Otto Engel, 1962. Front, wing, and back building. DTM.

Gitte and Hans Christian agreed to use the first floor for themselves and leave the second for lectures and instruments. The lecture hall was lit by oil-burning Argand lamps as well as gas-lamps fuelled by a gas-retort in the basement.[26] Gas for lighting was a modern technology, and once a journeyman repairing the retort destroyed the mechanism with his hammer, so that the lecture hall would have been cast into darkness had Ørsted not had recourse to old-fashioned wax candles.

*

Having read Ørsted's expert report, the government recommended Coulthard's petition for royal approval in February 1819. At the same time the commission was directed to continue its investigations. So, at the end of July Ørsted, Esmarch, and Forchhammer again headed for the rocky island in the Baltic. This time their itinerary went through Scania via Kullen, Höganäs, the University of Lund, Andrarum alum works, and Simrishamn to see with their own eyes the Swedish coalmines and to discuss the quality of the raw materials with Swedish mineralogists and chemists.

A new report confirmed the optimistic result from the previous year. On 1 August they had some locals make a deep boring into the chalk south of Rønne, where Coulthards test bore had been made. Ørsted's diary shows that the commission was searching for coal and by means of the expedition's boring equipment (sent ahead) they gradually penetrated to a depth of sixty metres. A bookkeeper was hired to register the results. After three weeks water flowed into the hole faster than their pump could get rid of it and to save equipment Ørsted decided to stop boring. What did they find? Lots of fossils, but no coal! This does not seem to have worried the commission, judging from its diary.[27] On the contrary, two more promises were added: (1) 'It is hoped that borings will reveal coal further away from the sea', and (2) 'There is also hope that coal will be found under the chalk formation at other locations in Denmark'.[28] Finally, there was reason to hope for finds of copper and lead deposits that could be profitably exploited.

The conclusions were summed up in an article in *Statstidende* written by Esmarch and Ørsted and intended to be an authoritative recommendation of David Coulthard's project. The article opposed 'preconceived opinions, supported by certain popular prejudices' in circulation. The commission tried to head off some anticipated criticisms. First, it was not a question of lignite, as malicious rumours had it, but of coal, the quality of which was of the highest standard. This had been proven by professor Warberg at the Royal Mint, among others. Coulthard's coalmine would turn out to be a healthy investment, the more so because it could be connected by railway to the port of loading. 'Railway' should not be taken to mean a steam-powered train, of course, but rails of iron on which horses could easily pull coal trucks to and fro. Secondly, the commission claimed the iron ore to be of a quality comparable to that of Coalbrookdale. Forchhammer, who had studied British coal and iron works after the first expedition, was still convinced that Coulthard's enterprise was viable: 'True, at the beginning of an enterprise such as the one here recommended something must be ventured, since the investigations, despite all precautions, can only indicate probability, not certainty. But no doubt every patriot tries to provide domestic coal and iron for the nation, two kinds of raw material needed by so many branches of industry and craftsmanship'.[29]

The following year court-martial attorney Herman Schæffer launched a frontal attack on the triumvirate of the commission. In the meantime Coulthard had hardly found any investors: more than a hundred unopened copies of the subscription are still kept by the Royal Record Office.[30] Nor had any progress taken place with his projected coalmine. The only thing that had happened was that four thousand rixdollars had been advanced to Coulthard, who was still waiting for the King to buy shares for another six thousand. In addition more thousands had been paid to the commission members. Schæffer slanderously asserted that innumerable historical investigations of coal and iron deposits on Bornholm had unambiguously shown that they were mirages, and it was long ago proven that the so-called coal was lignite, and that there were no deposits of iron of any significance. Against this background it was preposterous of Ørsted to claim that the reason why enterprises had failed so far was the fact that none of their managers had been equal to the task. Why should Coulthard, no mineralogist, but a skipper and an owner of an obsolete steam engine, be successful? Schæffer was far inferior to Professor Ørsted in eloquence and structure of argument. His moral outrage was full of hot anger, as he concluded that there was neither coal nor iron worth talking about on Bornholm.[31]

People knowing Schæffer's past thought that he was motivated by a personal disappointment in so far as his own application to the Foundation ad usus publicos for a subsidy to establish a salt works on the island of Læsø had been turned down.[32] Nevertheless, Schæffer had a point and he would not stand aside unless the court acknowledged the justice of his claim, so the case went on for a long time, judgements were appealed, and the case bred further cases.

Even if Schæffer was reduced to a frustrated, litigious man, the fact remains that Coulthard's project turned out to be a fiasco and a waste of public money. In retrospect Ørsted's and Forchhammer's recommendations were blunders, but they were not fraudulent as Schæffer insinuated. Ørsted trusted his geological judgement and the sustainability of the project, as is apparent from his letters to Zeise and Gitte, none of whom he would dream of fooling: 'As to the journey [to Bornholm] it has had the luckiest outcome. Of iron ore there is enough, and it is so easy to extract... Our investigations have provided more knowledge than any before. We have found several metal seams whose richness is impossible to ascertain without further work; but some of them give the best hopes...'[33]

Schæffer's accusations wounded Ørsted's pride so much that he could not ignore them, so he brought an action for slander against him for violation of the freedom of the press. The case followed Ørsted on his travels to European cities in 1822–3 embittering his enjoyment, and in the end he carried the day. Although Schæffer called countless witnesses who were dissatisfied with the coal from Bornholm, he was fined. Ørsted's lawyer seems to have been right to emphasise that his client had made certain reservations in his report, and that it had been proven beyond doubt that there were coal and iron deposits on Bornholm even if in far smaller quantities than predicted by the commission.[34] Forchhammer was able to report that during the unusually cold winter of 1823 no less than five thousand tons of coal had been shipped from Bornholm to Copenhagen, where they were sold on competitive terms at a normal profit despite competition from Newcastle, which had dumped coal at exceptionally low prices.[35]

The only person to gain a real advantage from the expedition to Bornholm was Forchhammer. He received his baptism of fire as a geologist, and with this experience and Ørsted's protection he had the basis for his future career. Forchhammer and Zeise were the young scientists who benefited the most from Ørsted's patronage. We may assume that Forchhammer's theory on the coherent geological structure of northwest Europe was nourished by their talks on Bornholm. His theory stated: 'The Faroes and Iceland, England and Scotland, north-eastern

France and adjacent Germany in a stretch connecting to Denmark, as well as this nation including Bornholm, belong to one large basin consisting of one and the same geological formation. Comparing all the main parts of this area is bound to shed new light on the theory of planet Earth'.[36] Ørsted could muster a lot of enthusiasm when he wanted to give weight to his recommendations. But was he not on the verge of succumbing to the kind of discredited argument from analogy associated with Steffens when he endorsed such a farfetched geological theory as that put forward by his young amanuensis? Ørsted supported this purely speculative hypothesis, writing in his report, 'it gives pleasure to thoughtful people to see how an open mind discovers such a correspondence in such distant territories'.[37] Indeed, neither had ever set foot on some of them. When, on his grand tour in the summer and autumn of 1820, Forchhammer saw them for himself he had to modify his theory considerably, as he found obvious differences in the geological structures of the Orkney and Shetland islands and the Faroes.

In August 1819 Forchhammer was suddenly without accommodation when his landlady died. Ørsted was prepared to help, but not without putting the decision into the hands of Gitte:

> 'Since we let out the flat upstairs it seems appropriate to me to offer him the empty room towards the yard, provided he gets his own furniture, which he can easily borrow, I suppose. I would appreciate to hear whether you would mind this, while I am still here on Bornholm. I shall not let him have the room without your consent. I should be pleased to learn that you do not oppose my proposal; but you must tell me in earnest; I would hate to do anything if you are against it. Think it over and let me know without beating about the bush.'[38]

Forchhammer did move into one of the rooms in the wing in Nørregade, where he could do chemical experiments and write his thesis, while Zeise was in Göttingen and Paris. In April 1820 he successfully defended his works on the chemistry of manganese, and prompted by Ørsted a summary of the thesis was published in *Annals of Philosophy* by Thomas Thomson in Glasgow, who had reviewed *Recherches* so favourably.[39]

PART V
THE TRIUMPHATOR

> The year 1820 was the happiest
> in Ørsted's scientific life.[1]

34 | 1820
The Happiest Year

THIS CHAPTER does not proceed immediately to the famous discovery in July 1820. The reason is that at the beginning of the year nobody could foretell that it would become Ørsted's *annus mirabilis*, even if Ritter had predicted exactly that in his horoscope cast seventeen years earlier. Instead, I shall sketch out the events of the months prior to the discovery full of rather trivial activities ranging from pecuniary worries, helping young talents, lecturing, and writing textbooks, to administrative obligations to the Royal Danish Society and the University. Such activities confined his research to a few months in the summer, and when they finally arrived, his family would tear at him to make him leave his laboratory in the noisome capital in favour of taking them on a holiday in rural tranquillity.

The year began well for the Ørsted family. In January the Foundation ad usus publicos awarded him of a grant of one thousand rixdollars to ease his burden of debt. Originally he was granted two thousand, but after he had handed in his petition he informed Møsting, one of the directors, that since he had received one thousand from the government for his investigations on Bornholm, he only needed half of what he had asked for. The grant enabled him to pay off most of his grumbling creditors and possibly to resign his post as secretary of the Royal Danish Society to gain more time for his research. In addition he had his old grant of five hundred rixdollars a year (due to expire in 1819) extended.

The five hundred rixdollars had been granted in 1816 for writing a textbook on chemistry, on condition that the first volume was printed before the end of the year. The King did not want to hear that his ceaseless lectures for first year students had 'a deadening effect on the spirit'; 'The professor had to deliver all the lectures his public posts direct him to do'.[2] The textbook project did not advance, but the money flowed in anyway. Instead Ørsted was writing a preface to J.A.C. Chaptal's *Chemistry Applied to Craftsmanship and Industry*, which appeared in 1820. The task of writing the textbook was transferred to young Zeise, who also received a public grant. It would be regrettable, seemed to be the argument, if he were pushed into finding a job abroad, when Denmark needed his skills as a textbook writer and as a professor of chemistry (even if that post did not yet exist).³

To justify his need for the grant Ørsted had set out how much he had to pay his assistants. Forchhammer received two hundred rixdollars a year, Dyssel a hundred and forty-four, his servant one hundred, and three maids sixty rixdollars each. He owed instrument makers, the lessor of his flat, the bank for loans, booksellers, wine-merchants and his Masonic brother Carl Otto, who had lent Ørsted a considerable sum of money. The life style of the Ørsted family was still not very restricted. He also had to pay for materials for experiments such as tubes, crucibles, files, rasps, and acids, zinc discs and copper for the galvanic trough apparatuses that Frederik VI had graciously inspected.⁴

He had a range of incidental earnings, such as three hundred rixdollars from working as secretary in the Royal Danish Society, another three hundred as Dean of the Faculty of Philosophy, a similar salary from the Military Academy and almost nine hundred for lectures delivered to various government departments. But even if the professor worked every day from early morning until late at night his income did not suffice without supplementary grants from the public coffers, which were generously opened whenever he asked his patrons.

Ørsted's total annual salary was now two thousand three hundred rixdollars and Anders's just the same. This income made the brothers the best-paid academics, but both were far below the top hundred taxpayers in Copenhagen. Count Schimmelmann, for instance was taxed on the basis of an income of fifty thousand rixdollars, Søren Kierkegaard's father of eighteen thousand, Bishop Münter of almost fifteen thousand, Councillor Classen of a little more than fourteen thousand, and Anders's superior in the Chancellery of ten thousand.⁵

Ørsted was a keen supporter of ambitious young students and made efforts to advance their careers. In February he obtained another favourable response from the directors of the Foundation ad usus publicos. He had recommended his twenty-year-old pupil and friend, Henrik Gerner von Schmidten, a lieutenant in the artillery, for a travel grant to Paris. When Ørsted had started teaching physics and chemistry at the Military Academy the talent of this young aristocrat soon attracted his attention. He not only distinguished himself for his mathematical expertise (which probably surpassed Ørsted's), but also for his modesty, cordiality, spirituality and truthfulness, properties that his patron valued highly. Schmidten was just a boy when he entered the Military Academy, but already at fifteen he was made an officer. Now he had submitted his first major mathematical work *Recherches sur le calcul integral aux équations linéaires* [Investigations into Integral Calculus of Linear Equations] to the Royal Danish Society and received a favourable assessment. Curiously, nobody who had seen a portrait of Napoleon Bonaparte could escape noticing a remarkable similarity between his and Schmidten's facial features. Equipped with a bright mathematical thesis in French and features like Napoleon's, it is no wonder that Schmidten soon gained access to the famous French mathematicians in

Paris, Laplace for one, leader of the school of corpuscular theory.[6] Ørsted used his influence to provide a travel grant for him.

At this time Georg Forchhammer submitted his thesis *De mangano* for his doctorate. The subject belonged to inorganic chemistry, rather than to organic chemistry as Zeise's had done, and his interest in experiments with the chemistry of manganese derived from his studies with Pfaff at Kiel, not from his collaboration with Ørsted. Forchhammer had become his amanuensis in Zeise's absence and his assistant on the two expeditions to Bornholm, and Ørsted valued him not only as a chemist and mineralogist, but also as a person. The investigations on Bornholm switched his career onto a geological track while demoting chemistry to a side interest. The Foundation ad usus publicos had already granted him eight hundred rixdollars on Ørsted's recommendation,[7] but his journey to Scotland and the Faroes was postponed by one year because of the second expedition to Bornholm, and because Ørsted needed his amanuensis for the winter semester.

After Forchhammer had defended his thesis on 1 April he did go to England, but fortunately Doctor Zeise returned home from Paris just then and was soon working daily in Ørsted's laboratory. It was probably thanks to Ørsted that he obtained a paid job, as there was no position for him at the University. The technical committee of the Agricultural Society initiated a course of evening lectures on technical chemistry and asked Zeise to teach it. The lectures were to be accompanied by experiments and would take place in the lecture hall in Nørregade. Ørsted was chairman of the technical committee of the Agricultural Society, and he and Zeise were now sharing premises and facilities as well as a servant, J.A. Dyssel, a young student of medicine, who had been employed the year before.

Zeise placed an advertisement about the course, making it clear that he was going to talk about technical chemistry in general, and that it would be popular as well as scientific, meaning that no previous knowledge was expected. The course was not a success. The audience came with unrealistic expectations, as Ørsted had feared. He considered them artisans' prejudices, as he had made clear in 1809 in his proposal for 'A Practical Institute for Experimental Science' (ch. 19). The audience complained and soon Zeise's course was on the agenda of the Agricultural Society, whose chairman swept all criticism aside.[8] Zeise's procedure was the proper one, Ørsted argued, for it would have turned things upside down had Zeise started with particulars and left chemistry in general to the end. What Zeise had done was well in tune with the recommendations put forward by the French Minister of the Interior, J.A.C. Chaptal in his *Chimie appliquée aux arts* which had recently been published in Danish. Artisans should not have manuals or rules of thumb served to them by academics. Quite the contrary, they must learn the scientific grounds for what they do to make them understand why they work, and to enable them to intervene with the right solution when things do not work. Zeise never succeeded in developing his teaching skills.

J.A. Dyssel, a medical student, had been employed in May 1819 by Ørsted as an odd-job-man for twelve rixdollars a month.[9] The intention was that he should take care of the collection of instruments, be around to help with experiments, but also to be the professor's secretary and translate to and from foreign languages. He got himself involved in a row with military officers that evolved into a totally Kafkaesque situation. Dyssel revealed himself as a pugnacious character, writing no less than a book about the case to assert his innocence and to ridicule the officers of the company.[10] It is a mystery that Ørsted could allow himself to employ such a cantankerous individual. Only many years later did it dawn upon Ørsted how two-edged Dyssel's skills were. The fact that he did employ him indicates how limited his choice was.

As a newly appointed professor of physics and chemistry in 1806, Ørsted had decided to form a school of dynamical natural philosophy. The programme of the school would successively appear in lectures and textbooks. But when we take a look at his three first employees, Zeise, Forchhammer, and Dyssel, we can hardly call them disciples of his school. Zeise followed his own scientific path, and Ørsted, his virtual foster-father, was clever enough to give the frail youngster a free hand. Zeise's field of expertise was organic chemistry, where nobody would be likely to dwarf him. His textbook on chemistry did not follow the architecture of Ørsted's abortive *Theory of Force*; but was subdivided conventionally into organic and inorganic chemistry, neglecting dynamical sections on electricity, magnetism, light, and heat. Nor did Forchhammer become a disciple of Ørsted's in the theoretical sense; he, too, threw himself into his own expert discipline, geology, which was no integral part of the dynamical system. Dyssel abandoned the study of medicine when Ørsted took him into his laboratory and he later turned to technology. Neither Zeise, Forchhammer, and Dyssel, nor Heiberg, Hansteen, Hauch, and Schmidten can reasonably be counted as disciples of Ørsted's school. His seven protégés admired their patron; they were loyal to him, but were not reflections of him. He generously used his influence to promote their careers.

*

Throughout the winter semester, from November 1819 to May 1820, Ørsted lectured privately in Nørregade on every Monday, Wednesday, and Saturday from five to six o'clock on 'chemistry with special regard to trades and industries, the perfection of which is of particular importance to the nation'. Participants paid twenty rixdollars and the Reiersen Foundation indemnified him for possible free tickets issued to trades people. On the same days he lectured from half past six to half past seven on the physics of planet Earth, charging fifteen rixdollars. Thirdly, he offered a private series of lectures on 'electricity, galvanism, and magnetism, as well as chemical laws of nature derivable from them' accompanied by two sets of experiments for two hours, on Tuesdays and Thursdays from six to eight o'clock, also at fifteen rixdollars. Fourthly, he still delivered his monthly lectures.

In preparation for one of his lectures in April, Ørsted and Dyssel had made the new trough apparatus ready for an experiment pouring acid into the copper troughs and connecting the zinc discs. They had also prepared a long, thin platinum wire as a conductor knowing that it would emit heat and light when the circuit was closed, even if only twenty copper troughs were used. Sulphuric and nitric acids were costly and had to be used sparingly. They were effective for two to three hours only. The professor and his assistant had also procured a compass in a beautifully turned mahogany bowl with a glass hood on top to protect the magnetic needle.[11]

Ørsted's audience did not consist of casual passers-by, but of men with a sound foundation in natural philosophy. Many of them had attended his lectures previously and must be regarded as advanced amateurs well versed in his dynamical theory expounded in *Ansicht*, and familiar with the thought experiment that strong electrical forces may affect a magnet. True, this was only an idea he had put on paper during a journey when he had no chance to test it, for want of access to the necessary equipment. Now they were available in the shape of his and Esmarch's trough apparatus. So why not do the experiment? Ørsted was hesitant, probably because he feared making a fool of himself in the eyes of the audience should the experiment fail, as had happened in Paris in 1803 when he had hoped to show the National Institute that Ritter had discovered the galvanic poles of the Earth. Then suddenly it struck him that his investigations into Coulomb's law had made it doubtful whether the electric force obeyed the inverse-square law as gravity did, that is that it attracted in even, rectilinear motions inversely proportional to the distance, which he and Ritter had always taken for granted. He had now come to believe that the electrical force propagated by way of alternating shocks (ch. 32). Moreover, he had observed that heat and light (which, according to his dynamical theory derived from the same fundamental force) travelled not in a rectilinear direction but radially and transversely to the conductor. In a moment of self-confidence he decided to go ahead. While Dyssel made the final preparations he introduced the audience to the purpose of the experiment. Then he moved the platinum wire slowly around the compass at different angles, Dyssel closed the circuit between the conductor and the galvanic battery, and all of a sudden he saw the effect, the motion, and the deflection! The spectators saw it, too! The magnetic needle nodded affirmatively to Ørsted's thought experiment.

This was almost too good to be true. The effect was not very conspicuous, and he did not dare to draw definite conclusions from the reaction of the needle. Moreover, glass separated the conductor from the needle, which might explain the weak effect. The deflection might be only contingent. He hesitated to regard the outcome of the experiment as a definitive breakthrough and kept cool. Other duties were awaiting him. Easter was in late March, and he had hardly shared a moment with his family, for Forchhammer's thesis for the doctorate had been due on 1 April. If we assume that the experiment was made on Thursday April 6th, he would have to chair the Friday session of the Royal Danish Society the day after, and read out his report on the Bornholm expedition.[12] Then followed the plough competition on 12 April to which he was committed as chairman of the technical committee of the Agricultural Society.[13] This was one of the many public duties that he could not refuse. It deprived him of time to cultivate his primary interest. His five-year period as secretary of the Royal Danish Society was about to expire, and he was not quite decided whether to offer himself for re-election. Finally, a pile of students' assignments for *philosophicum* lay waiting on his desk to be marked. But his observation of the deflection of the needle certainly deserved a closer look *sine ira et studio*. In hindsight he was at a loss to comprehend why he had not simply put all obligations aside to take time to examine whether the reaction of the magnetic needle was a genuine breakthrough.

On 5 May elections were scheduled for the posts of secretary and president of the Royal Danish Society. Ørsted's first five-year term was over and he decided to stand again. Twenty-two votes were cast, seventeen in favour of Ørsted.[14] This meant re-election for another five-year term. Schimmelmann had been president for twenty-three years, succeeding A.P. Bernstorff. He, too, was re-elected and continued until his death in 1831, when Hauch succeeded him. On 2 June the last session took place, to be followed by a break until November. Now the secretary had a little more time for his research.

Fig. 76. C.F. Christensen (1805–1883), drawing of Regensen College with the linden tree around 1830. Here the idea of the Students' Association was born, and here Christian Winther's song with the refrain 'We are masters of the spiritual realm; We are the tribe that will always stand.' was sung for the first time in 1820. Here, too, and in the neighbouring Trinity Church with the Round Tower, dissertations were defended as long as the University lay in ruins. CMC.

On 5 June students were invited to a party under the linden tree in the courtyard of Regensen College—not only the alumni of the college, but all students of the University. The invitation was issued by Christian Winther, the poet, and Dyssel, Ørsted's assistant. The staff of the Faculty of Theology were afraid that the students were planning a rebellion, and insisted to Rasmus Nyerup, the college provost, that he must ensure that no drinking and singing continued beyond midnight. Almost two hundred postgraduates and undergraduates turned up, and time and again they would rise to cheer the professors they liked and shout down those they could not stand, especially the dean of the Faculty of Theology, Jens Møller. When they had finished their drinks, they threw their glasses over their heads to emphasize the seriousness of both cheers and boos. A new song was sung, written by Winther for the occasion and concluding: 'Faithful brethren! Our hymn shall rise, The stars shall witness that no one will fail. We are masters of the spiritual realm; We are the tribe that will always stand'.[15]

Although it was a quiet summer evening the noise of the party spread over the college roofs to Fiolstræde and echoed in Nørregade where the Ørsted family were having their meal. The atmosphere was convivial and the students would not permit any restrictions and certainly had no intention of stopping their merriment at midnight, when a list circulated encouraging them to enrol for membership of a students' association soon to be established. The time was ripe for stepping out of the self-imposed tutelage, as Kant had so emphatically urged in his pamphlet *Was ist Aufklärung?* Seventy-eight students signed up. Students at the University would no longer be reckoned as the professors' well-behaved disciples or as adolescents gadding about drinking and gambling. They wanted to combine into a voluntary association to voice their views self-consciously as 'masters of the spiritual realm'.

The young academic citizen who had booed at Professor Jens Møller was a student of medicine interested in natural philosophy, Henrik Krøyer by name. Now he was called to account for his presumption. But Krøyer would not be cowed. The dean had provoked the students by restricting their freedom. The first general meeting of the association was held on Sunday, 16 July in Ørsted's lecture hall in Nørregade. The room was crowded. Presumably, at least seventy-eight students, the number that had signed up a few days before, gathered within about seventy square metres. In the adjacent room were Ørsted's galvanic trough battery and the magnetic needle in the mahogany bowl. Most inconveniently, the students assembled right in the middle of the series of experiments intended to verify the still doubtful electromagnetic effect observed during the lecture in April. Ørsted must have put his lecture hall at the students' disposal a few days after the scandalous party at Regensen, without foreseeing the disturbance they would inflict on him. The University being in ruins, the students only had three choices for their first meeting: Regensen's college chapel, which was totally ruled out by the Faculty of Theology, Elers' College and Ørsted's lecture hall. It was probably Dyssel, Ørsted's assistant and one of the initiators of the students' association, who had asked his employer for a place to meet.

Now, we are not here interested in the history of the Students' Association per se, but, apart from the distraction it caused Ørsted at the greatest moment of his life, it throws light on his attitude to the students' movement. Since he had become an ordinary professor in 1817 he had gained a seat in the Senate. The Students' Association was a problem that soon appeared on its agenda. Normally, the absolute monarchy did not permit freedom of assembly to its subjects, except for the defence of the nation. According to a cabinet order of 1780 freedom of assembly was limited by a number of conditions. One was that the laws of any association must be submitted to the authorities, and another that the leader of 'all such associations or clubs must be a civil servant'.[16] The Students' Association, however, was led by five 'seniors' and according to their draft statutes they were not allowed to hold an official post. The executive of seniors sent the draft statutes to the chief of police and to the senate of the University, not to most humbly request the approval of the authorities, but as information merely.

According to the draft statutes of the Students' Association a student was defined as 'a person: (1) burning with zeal for the Great, the Good, and the Beautiful; (2) impregnated with a genuine devotion for sciences and letters; (3) animated with true love and profound respect for his peers; and (4) preferring the well-being of his peers to his own and considering everybody his friend and brother who is inspired by the same ideals'.[17] Nobody could embrace this definition of a student more wholeheartedly than Ørsted. But this was not the way the Senate looked at it. It found the Students' Association at odds with good academic order and with the Cabinet order of 1780 and would not allow it. Jens Møller suggested that students joining the association should lose their rights as members of a college. Several senators elaborated on the damage that might be caused by the association, such as on the scandalous night of recent memory at Regensen, where honourable professors were booed. One of the theologians suggested that Schelling's *Naturphilosophie* must have confused the minds of these young people.[18]

The Senate's worries were no doubt related to recent events in several German states. The German *Burschenschaften* [students' clubs] had certainly kicked over the traces on some occasions. In 1817 they had brought about an auto-da-fé at Wartburg Castle, where Luther had translated the New Testament and thrown an inkpot at the devil, and the year after another *Burschenschaft* was accused of the murder of Kotzebue, the playwright, in Mannheim. Metternich had interfered with the self-rule of Austrian universities and ordered an employment ban on student leaders. These were the dissipations or crimes that spread fear and trembling in the Senate in Copenhagen. It had received information that as vice-chancellor of the new University in Berlin, Fichte had been unable to rein in the German students and their *Burschenschaften*, who waved their black, red and yellow banners and made trouble, although they said they were only acclaiming Fichte's ideas from *Reden an die deutsche Nation* [Speeches to the German Nation].[19]

Would the Students' Association become like *Burschenschaften* in Denmark? King Frederik VI was not so nervous. He chose to look at the students as potential members of his military corps, which under the command of A.W. Hauch had so loyally volunteered to die for the country. So he did not order the chief of police to intervene. Perhaps he had already infiltrated the association just as he had his spies in place elsewhere. If so, he probably found no reason to be afraid of Danish students. But the Senate was still agitated, and saw the association as a dangerous faction within the state and an embryonic *Burschenschaft* and breeding place for debauchery, intrigues, and duels. Did not students show up at the Royal Theatre to boo the performances? And was not it utterly misleading when puerile students called themselves 'seniors'? These self-proclaimed 'masters of the spiritual realm' might be about to use their hypocritical definition of students as a cloak for an organised effort to dismantle the authority of the University.

A few professors in the Senate advised their colleagues to stay calm and to see the positive side of the new association, though they remained a minority. Was it not praiseworthy, they pointed out, that the club intends to ban card playing and 'full of worried presentiment—has asked the wardens of colleges to keep an eye on their members'?[20] The debate was cut short by a shrewd professor of law who dryly remarked that the students had not actually asked for the approval of the Senate, but were only keeping it informed. If anybody were to prohibit the establishment, it would have to be the police, and this had not happened. In the end the Senate acquiesced. The five seniors thanked it for its help, and in return changed the paragraph excluding civil servants from the leadership.

Ørsted kept a low profile as long as he belonged to the minority. He even made sure he was absent from the meeting when the ban of the Students' Association was imminent. A month and a half later he joined the tolerant minority's view, which ended up victorious, because the Board of the University and the Vice-Chancellor supported it.[21] The whole matter showed that Ørsted was marginalised in the Senate, where theologians and jurists prevailed. He did not have the strong position at the University that he enjoyed in the Royal Danish Society, where scientists had the upper hand. When Ørsted returned home from his third journey abroad in the autumn of 1823 he joined the Students' Association, along with Zeise and Sibbern. 'We, the five seniors, cannot hold back our request that you will honour our club with your presence as often as your scientific work will allow.'[22]

Finally, after the founding of the Students' Association in the lecture hall adjacent to his collection of instruments, he found a period of time to resume the experiments with his trough apparatus and compass.

*According to the great scheme of things
he had already in his earliest writings assumed
that magnetism and electricity are generated by the same forces.*[1]

35 | 1820

A Discovery by Chance?

ON SUNDAY, 9 July, Ørsted hired a cab to pick up his old father in Amaliegade. The professor appeared to be in an extraordinarily good mood, and the old pharmacist could not guess why his son would suddenly take him on an outing along the coast to the Royal Deer Park to have a drink.[2] He was bursting with excitement, for he wanted to tell his father about a series of experiments with his galvanic trough apparatus and a magnetic needle. These experiments had convinced him of what he had for so long assumed, of the interaction between electrical forces and a magnetic needle (not any needle, but a needle that was magnetised). The magnetic needle did nod approvingly each time he closed the galvanic circuit. He rejoiced at the deflection like Tycho Brahe, when back in 1572 he had observed his *stella nova* and so had received the answer to the question he had put for a long time: '*Quid si sic?*' Think if this is so!. It was no coincidence that the discovery was celebrated in Dyrehavsbakken, an amusement park in the Royal Deer Park, because this was the place where Oehlenschläger's old man with his peepshow displayed the interplay between polar forces: 'Everything 'twixt love and hate, To the last link must alternate...'.[3]

Actually, it was now already three months since he had seen the compass needle deflect. At long last he had marked the heap of *philosophicum* assignments and found time to repeat the experiment he had made before an audience in April. The past week he had several times held a conductor of platinum wire in a number of different positions close to a compass and it turned out that the deflection observed then was no hoax as had been the case with the pivoting voltaic pile in Paris seventeen years before. Now, the experiment had been subjected to controls by Esmarch and Wleugel. The effect had been tested and he had found regularities in

A DISCOVERY BY CHANCE?

Fig. 77. Dyrehavsbakken, the amusement park of the Royal Deer Park at Klampenborg. During springtime and the weeks after midsummer Copenhageners would gather around the spring of Kirsten Piil, where entertainment was abundant. Here is a menagerie including a lion and a camel, a 'manége', an exhibition of panoramas, a dancing hall, a merry-go-round, and bars. Horses are lunged, Pierrot is capering, and the upper middle classes in top-hats as well as peasant women in folk costume are enjoying themselves while soldiers are maintaining law and order. Anonymous aquatint. CMC.

the motions of the needle. The electric conflict in the conductor made the magnetic needle deflect as if he had placed another magnet parallel to it. Different magnetic poles attract each other, whereas identical ones repel each other. Now he had shown that the negative pole of the conductor attracted the north pole of the magnet and the positive one the south pole. If he reversed the electrical poles, the magnet would deflect to the opposite side. In other words, the electrical conductor behaved like a magnet. Hence he called his discovery electromagnetism. Explaining all this to his father he expected praise and recognition—and advice. How should he publicise this sensational discovery?

The decisive breakthrough had come a week before the Students' Association held its first general meeting in Ørsted's lecture hall. Much to his surprise the magnetic needle deflected in the space where the electrical conflict created a circular field of forces around the conductor. It did not matter whether he held the wire to the right or to the left of the compass. The needle deflected to the same side. But if he reversed the direction of the electrical force, the needle deflected to the opposite side. So attraction and repulsion did not happen in rectilinear movements, but as a circular power affecting the needle at a distance. Ørsted realised that all this was more difficult to explain in words to his father than he had imagined for the language did not exist in which to visualise electromagnetic phenomena. The son did his best with circular movements and concentric lines.

This discovery was neither void of preconditions nor haphazard. He did not spontaneously tamper with a galvanic battery and a compass hoping that something might happen out of the blue. The epoch-making discovery had its beginnings back in his student years and his thesis on Kant. What Kant had described as a noumenon he had produced as an empirical effect, a visual phenomenon.

An electromagnetic effect was not a phenomenon that was known to appear in nature. Nobody had ever been on the look out for natural electromagnetism. But nevertheless forces of nature are crucially involved. Electricity exists in nature as discharges in lightning followed by thunder. Apart from Benjamin Franklin, who conducted away these lethal forces and rendered them harmless, these are totally untameable. Electricity had been known since antiquity in the form of attractive and repelling forces produced by rubbing amber (in Greek, *elektron*) with animal skins. But the electricity people knew in Ørsted's day was either friction electricity produced by mechanical generators, or contact electricity, as Volta called the phenomenon generated in the pile he built, and which he generously called 'galvanism' after his rival, but which Ritter had already (although rather unnoticed) proven to be an electrochemical phenomenon. Be that as it may, the electromagnetic effect was artificially produced and belonged inside Ørsted's laboratory.

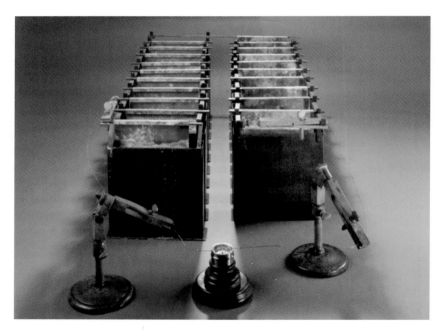

Fig. 78. HCØ's electromagnetical experiment, 1820—reconstruction with Ørsted's original instruments at DTM.

Magnetism is found in magnetic iron ore in the inner part of planet Earth and appears as polar forces attracting or repelling a compass needle. A compass is a technical instrument that in its simplest form in ancient Greece was an oblong lodestone that, attached to a floating piece of wood, turned its south pole towards the magnetic north pole and vice versa, arranged in such a way that the polar magnetic forces of the loadstone always kept it away from the edges of the water jar, on the surface of which it floated. The only observation in nature that one might call electromagnetic was when lightning struck the compass of a ship, a phenomenon many sailors had experienced. But in that case a natural phenomenon hit a special technological device, the magnetised needle. The lightning did not make anything in nature deflect, not apples for instance. When Ritter and Ørsted believed that electricity and magnetism were forces that could be understood as the same fundamental force, they imagined enormous forces of nature that might interact in northern lights or in celestial comets. But how could anybody make experiments with such cosmic hypotheses? Actually the alleged unity of electricity and magnetism had to be investigated in the laboratory as the reciprocal action between forces produced each by its own instrument: a galvanic battery and a magnetic needle.

Many people believe that this research was motivated by a dream to invent a new technological instrument, in which the two forces could be exploited for communicative purposes (such as the telegraph) or for energy production (such as the dynamo). However, there is no doubt whatsoever that such thoughts did not occur to Ørsted and were not aired between him and his father. His motivation was exclusively theoretical, like Newton's when he discovered gravity. What was the utility of electromagnetism? Absolutely none! Ørsted aimed at verifying the hypothesis he had expounded several times, for instance in *Kraftlæren* and in *Ansicht*. The overall idea was the unity of nature as the interaction of forces in perpetual transition according to unchangeable laws of nature. His dynamical project was his life and the core of the natural philosophy he and Ritter fought to enthrone in place of the mechanical world picture. If Ørsted succeeded in showing that electrical and magnetic forces interacted in his laboratory in Nørregade, they must therefore also interact in the universe, God's laboratory.

What were the prerequisites for the discovery? In hindsight we can trace a direction in Ørsted's scientific work leading to the discovery in 1820. The basic idea goes back to his thesis of Kant's metaphysics of nature. Kant was the first to challenge the mechanical, corpuscular theory and to suggest a dynamical alternative. At first Kant was at a loss to point to a method that might pave the way for a dynamical science, because 'Ariadne's Clew' would only show the way out of the labyrinth if there were phenomena analysable by the categories of the understanding. It is uncertain whether Ørsted had an indirect knowledge of Kant's so-called transition project, which suggests that dynamical physics might progress from the epistemological intersection where regulative ideas meet constitutive principles. Nevertheless, Ørsted pursued a similar method, the thought experiment. He soon realised that the dynamical system gained momentum from Ritter's *Beweis* that galvanism is an electrochemical phenomenon, and this understanding was later supported by such chemists as Davy and Berzelius. For a while he also saw Winterl's theories of the electric properties of acids and bases as contributions to the dynamical system. Together with Ritter he increasingly doubted the notion that constitutive principles are necessarily mathematical. Kant had suspected that mechanical physics was based on weighable and measurable matter, not because they were efficacious, but because they fitted in with mathematical equations. Ørsted's method of thought experimentation did not look for physical phenomena that might be expressed in mathematical language. He was looking for visual images of the effects of forces such as he had already produced in his experiments with acoustic figures.

The first time Ørsted worked with the relation between electricity and magnetism was in Paris in 1803 when he demonstrated Ritter's experiment before the National Institute. This event became the humiliation of his life. From Ritter's letters that summer it is clear that he also worked on another experiment. He made a galvanic needle consisting of silver at one end and zinc at the other and brass in the middle resting on a pivot of agate. Ritter asserted that this galvanic needle oriented itself along the magnetic meridian, the zinc acting as the north pole and the silver as a south pole. When he led the conductor from a galvanic battery across the pivot on which the needle turned, however, there was no reaction.[4] This inconclusive experiment shows that the interaction between galvanism and magnetism haunted them, but they could not find a way to verify their assumption.

Both Ritter and Ørsted imagined these forces in a cosmic context (ch. 18), where life on Earth depended on the interplay of electrical and magnetic forces moving as intersecting axes oriented east-west and north-south producing heat, light, and electrochemical processes, together constituting the cosmic order. Crucial in this notion was the Romanticist idea that nature is a unity to be understood as an interaction of polar forces deriving from a common fundamental force. If this dynamical idea should succeed in replacing the corpuscular system, it was necessary to bring the cosmic forces into the laboratory because otherwise they could not be subjected to experimentation in pairs.

The second time Ørsted worked with the relation between electricity and magnetism was shortly before Ritter's death in 1810, this time in the Royal Danish Society. His first initiative as a new member was his proposal of a prize essay on the identity or difference between electricity and magnetism (ch. 23). The anonymous French essay brought no news. It just postulated an identity founded on the assertion that both phenomena were generated by the same imponderable fluid (in electricity in tension, in magnetism in balance) as if this was the answer the Royal Danish Society wanted to hear. So, the essay contradicted the common understanding in Paris, which claimed that the molecules of electrified and magnetic bodies each contained their own fluids. How otherwise could the obvious differences between the two be explained? The electrical fluid could combine with the molecules of all substances, whereas the magnetic could only combine with iron and nickel. Whether electrified substances were negative or positive depended on a surplus or deficit of the electrical fluid. Another crucial difference was the separation of the electric fluid, which became visible in the experiment of separating water. By contrast the magnetic fluid did not separate into a north and south pole when a bar magnet was split, but each new magnet immediately re-established both poles, thus reproducing the properties of the former one. The French prize essay did not present any new experiments, neither did it refer to Coulomb's experiments twenty years earlier with the electrometer, showing that both fluids obeyed the inverse-square law thus having at least one property in common. The anonymous essay, handed in too late, was not even up to date in corpuscular terms. As Ole Knudsen the historian of science has appropriately remarked, the Laplacean theory maintained a mathematical analogy between electricity and magnetism (and gravity), while the physical analogy that Ritter and Ørsted assumed was considered impossible.[5] The prize question for the year 1810 remained unsettled until the person who set it found the answer himself.

The third time Ørsted tried to advance the issue took place in his major works, *Ansicht* and *Recherches* (ch. 28). Here he collected the properties the two forces seemed to have in common adding that the knowledge gained from a number of recent experiments suggested that the magnetic power was not confined to iron and nickel, but that all substances, although in varying degrees, could be magnetised. He now encouraged new experiments on the effect of galvanism on the magnet, but immediately faced a difficulty. For if galvanism was all pervasive it would affect magnetic as well as non-magnetic bodies, and if so how could the special electromagnetic force, if it appeared, be isolated and how could control experiments be made? Perhaps an experiment could be thought out involving a compass as well as a non-magnetic needle to find out if they reacted differently.

This idea was kept in the forefront of Ørsted's research during the years to come to the extent that his stressful life left him any leisure for scientific creativity at all. A lecture delivered to the Royal Danish Society in 1815 put forward a series of control experiments with Coulomb's electrometer testing the validity of the inverse-square law as applied to electricity (ch. 32). The occasion was that the German physicist Simon had challenged Coulomb's law: the electrical force decreases in proportion as the square of the distance increases. For seven years Ørsted had been testing this law with various electrometers, but was cautious not to rush into publicising his results, as they went against a scholar as prominent as Coulomb. Coulomb's law could only be valid if the electrical force was rectilinear and continuous as gravity, and Ørsted doubted this prerequisite. The figures Ørsted measured did not seem to correspond to the inverse-square law when the distances were either very small or very large. This might be due to the force not propagating evenly, but instead in shocks followed by pauses. In other words Ørsted questioned the mathematical analogy between gravity that was constant and rectilinear and electricity that seemed to propagate differently.[6]

The last phase on his way to the discovery of electromagnetism was his collaboration with Lauritz Esmarch on amplifying the galvanic device (ch. 32). In 1815 Ørsted acquired a new galvanic apparatus made of porcelain from England, an important step forward from the former wooden ones, from which the acid leaked. 'From England I have received a trough apparatus of porcelain that produces excellent effects. I only have forty-eight pairs of sheets with a surface of approximately ten square inches. They make coal glow as in the purest oxygen. A diluted acid and solution of salt is heated by the effect so one cannot touch it. As soon as you get to the city I shall show you the splendid effect of this device', he wrote to Manthey.[7] Soon afterwards, another innovation was effected. Esmarch and Ørsted asked themselves what the use of the porcelain was. Why not make use of the copper for the troughs? This turned out to function even better. The following year he demonstrated his new galvanic trough-apparatus to the Royal Danish Society. 'Each copper trough is connected with a loop of wire to a zinc sheet immersed into the liquid contained in the next copper trough.'[8] Another improvement was an apparatus capable of keeping the liquid hot, thus amplifying the galvanic effect even more.

This time the copper troughs were made cylindrical instead of cubic, and at the centre a chimney for charcoal was soldered on, imitating the contemporary tea urn (samovar). This apparatus made iron wires glow, even melt. 'If an experiment was needed to make a sense impression of the fact that an electrical spark is merely a strong glowing of matter filling the space in which the spark appears, this experiment seems to be suitable.'[9] This observation once again raised doubts as to whether electrical forces attracted and repelled rectilinearly or propagated otherwise around the conductor.

Now Ørsted not only celebrated the idea of the interaction between electrical and magnetic forces, he also possessed the device to test it. In June 1819 he obtained a better laboratory above his flat in Nørregade as well as an amanuensis (Forchhammer) and an assistant (Dyssel). In January 1820 he received a grant giving his overstretched economy a moment's breathing space. During a lecture on electricity and magnetism he felt inspired to do what turned out to be the crucial experiment in front of a prominent audience. He seized this opportunity in April 1820 as predicted by Ritter (ch. 15).

Three months later he sat face to face with his father in a cab proudly recounting the story of his discovery. Electromagnetism was absolutely certain, and the dynamical system was victorious. No matter how old a son gets, he is always yearning for recognition from his father. The old pharmacist listened to all the details. Ørsted and Dyssel had only set up a small galvanic apparatus, and he had hesitated to proceed, frightened of losing face in front of his audience. But as time went on the atmosphere felt safe and he found the courage to go for it. Fortunately, the outcome lived up to expectations, but the compass needle just shook a little, and he was uncertain whether the effect was merely casual. But his prominent audience opened their eyes wide when they saw the electrical force apparently passing through the glass hood and moving the needle. Nobody had ever seen that before.

His father listened attentively, I suppose. He could readily associate himself with his son's deliberations that April evening. What he failed to comprehend was that he had not become so excited by the seemingly positive outcome as to immediately throw himself upon a more systematic control experiment. Well, in hindsight the son failed to understand that too. Of course, his discovery was far more important in the history of science than any single duty he was obliged to do. But from the contemporary perspective of the government and the University, things may have looked differently. When all his obligations demanding immediate action are enumerated it is perhaps less surprising that he postponed his control experiment by three months. Until then lectures at the University took up twelve hours a week, the *philosophicum* assignments gave him weeks of work, private lectures occupied ten hours a week, and in addition he had to run courses at the Military Academy and in the Classen Institute, and attend meetings in the Royal Danish Society, the Agricultural Society, the Senate, and the Scandinavian Literary Society. Moreover, Forchhammer's thesis, Zeise's return, letters of recommendation for talented students, concerts in the Harmonic Society, private parties, evenings of playing cards, meeting the brethren in the Masonic lodge, not to forget his own family, and so on and so on—would not this suffice to render anybody breathless?

Yet, I have passed over that in June, just after the exam, he was directed to produce a report for the government on Danish weights and measurements. The occasion was a request from

Norway, where they were about to define her 'standard foot'. Christopher Hansteen, who was appointed to lead this work, suggested to go by the Danish standard, so he asked to have the étalon (calibrated measure) of the Danish Rhineland foot sent to Christiania.[10] But what was a Danish standard? And where was Ole Rømer's original étalon? A search by Schumacher (the royal astronomer and geodesist) and Ørsted revealed total confusion. The lengths of the measures varied between étalons in the care of the Municipality of Copenhagen, the Royal Archive, the artillery, and the one Bugge had used for the measurement of degrees, and instrument maker Ahl's, and so on.[11] Hence, Ørsted suggested to Hansteen that Norway wait until a Danish standard had been re-established, and the two professors put forward to their governments a scheme to link the foot to an objective and imperishable measure of nature; they used the length of a one second pendulum at forty-five degrees latitude on the meridian of Skagen (the Skaw), and then determined that the standard length of the foot should be 12/38 of this. This proposal achieved royal approval on 28 June 1820, so it was only around 1 July that Ørsted was able to resume his control experiments, for he could not tell the students to wait for their marks, or the King for his reports, while he tried to find out whether the deflection of the needle had been an April fool.

Finally, at the beginning of July he found some breathing space to work for whole days in his laboratory, which was now available except on Mondays, Wednesdays, and Saturdays from six till seven, when Zeise ran his course on technical chemistry. The time had come to use his new galvanic apparatus for the crucial control experiments. The exact date for resuming these is not known, but *Dansk Litteraturtidende* was on the street during the week following Ørsted's outing with his father to the Royal Deer Park 9th July. For that issue Ørsted had submitted his first report on the control experiments stating that Councillor Esmarch, Commodore Wleugel and Dyssel had been present as scrutineers. The first two were mentioned by name and Dyssel was on his pay roll and assigned the translation of *Experimenta* from Latin to Danish for Rahbek's journal *Hesperus*. The fair copy of Ørsted's first, undated notes on a double-folded acid-stained sheet of paper is probably in Dyssel's hand.[12] These notes correspond to the short article published in *Dansk Litteraturtidende*, in which Ørsted reported matter-of-factly that he had placed a conductor of platinum wire or of other metals or iron parallel to a magnetic needle (the positive end in the direction of the magnetic needle's north pole) and the needle's north pole deflected toward the east and consequently its south pole towards the west. He does not talk about an electric current, of course, and hence could not use the expression 'in the direction of the current'; the term 'current' would be associated with imponderable fluids, and the point of his report was exactly to prove the French corpuscular theory wrong.

In addition, Ørsted stressed the irrelevance of whether the conductor was placed a little to the right or to the left of the meridian of the needle, or if it was placed over or under it, it was only that the conductor had to be more or less parallel to it. Actually, this understatement revealed the surprising point. But he did not yet have the concepts to explain the circular form of an electromagnetic field. The surprise, of course, was the fact that the magnetic needle deflected to the same side, whether the conductor was placed to the right or to the left of the needle as long as it was tolerably parallel to it. Had the force been rectilinear, as presumed by the corpuscular theorists (analogous to gravity), the needle would have deflected to both the two sides as rectilinear attraction or repulsion. Incidentally, Ørsted refrained from any reference to an experimental setup in which the conductor was vertical to the needle, although it was tested, and produced the same effect as when it was placed horizontally.

Then he explained that he had used three different kinds of magnetic needles: (1) in a mahogany bowl with a glass hood; (2) in a glass box with a glass hood; and (3) in the open air. He had found that the three needles responded in the same way and that the glass was no obstacle to the effect. It might be wondered that Eckersberg's portrait from 1822 (ch. 37) depicts a needle in the open air, while the compass in a mahogany bowl with a glass hood exhibited in Ørsted's memorial room at Danmarks Tekniske Museum is said to be the one he used in 1820. In fact he used both. This alternative furnished him with a clue as to what was actually going on. Was it the electricity that acted on the magnetic needle while the latter was passive, or was it the other way round, or…? It was common knowledge that glass cannot conduct electricity, and consequently it must be the magnetic force that penetrated the glass. After this observation he experimented with several other substances such as wood, brass, and double-glass to block the effect. He concluded, 'they all allow the magnetic force to pass through, though not electricity, heat, and light'.[13] In other words, the phenomenon was a reciprocal effect that

Fig. 79. Diagram showing the lines of magnetic force and the direction of the magnetic field of a current-carrying wire. *The Oxford Companion to the History of Modern Science*, ed. by J. Heilbron, Oxford 2003, p. 239.

accordingly he dubbed electromagnetism, for what happened could only make sense if both forces were active simultaneously.

We do not know what advice the son asked or the father gave on their trip. Perhaps the pharmacist's heart just swelled up with pride. The professor continued his experiments the following week. His short notice in *Dansk Litteraturtidende* met with no excitement. On 21 July it is briefly referred to in the newspaper *Dagen*. As with *Ansicht*, the discovery caused no sensation; very few Danes grasped its scientific importance. Ørsted assembled some of those few to witness his further experiments on 15, 19, and 21 July, and possibly before then as well. He was now aware that the magnetic needle did not react if he placed the conductor at right angles to it. But the direction of positive and negative electricity in the conductor could be combined with magnetic north and south poles in four different ways, and if the positions were to be horizontal as well as vertical, there was plenty of work to do before all the possibilities of combination had been tried. In addition, the conductor could be bent into a kind of galvanic fork, or a simple solenoid, in which the direction of the force could be reversed. Finally, it was tempting to find out how different temperatures and chemical substances might affect the phenomenon. Ørsted called upon some of his erudite friends: the renowned Court Steward A.W. Hauch, the shrewd professor of zoology J.H. Reinhardt, his protégé the talented W.C. Zeise, and the experienced Dr L.L. Jacobson, who had examined Winterl's 'Andronia' (the adjectives are all Ørsted's), as witnesses to vouch for his discovery.

For Sunday 16 July Ørsted had been improvident enough to put his large lecture hall at the disposal of the founders of the Students' Association. So he could not work on that day, and as the decisive results had already been confirmed, he might just as well celebrate his triumph with Gitte and the children. The whole family hired a cab to drive them to the Royal Deer Park and the amusement park where for once he spent lavishly.[14]

21 July is the date of Ørsted's comprehensive report on the discovery of electromagnetism (a mere four pages written in Latin so that all European scientists might read the same text) with the title *Experimenta circa effectum conflictus electrici in acum magneticam* Experiments on the Effect of the Electrical Conflict on the Magnetic Needle.[15] Its distribution followed on 26 and 27 July. The postage to send it to forty-eight European scholars was close to ten rixdollars.[16] New to this report was that Ørsted 'being an artist with words' had by now made up concepts to express the new phenomenon. Of course 'current' was banned for designating the electrical force propagating in and about the conductor; he used the term *conflictus* or 'the electrical interplay/interaction'. He described it as moving in 'a snail-like orbit' or in 'rings' to make clear the deviation from the rectilinear attraction/repulsion (Fig. 79). He did not himself formulate the mnemonic rule of thumb which modern schoolchildren reel off in physics classes. His formula was: 'The pole of the magnetic needle, over which the negative electricity streams in, turns to the west, the needle, under which it streams in, to the east.' Here the professor made a blunder, because 'streaming in' is impermissibly associated with liquids. In the beginning the terminology was faltering and ambiguous. This is evident in his notes on the experiments. He had been muddled in reversing east to west, left to right, and then corrected himself. The mnemonic rule of thumb, unknown to Ørsted, is: place the palm of your right hand against the conductor in the direction of the force, and the magnetic needle will deflect in the direction of your thumb.

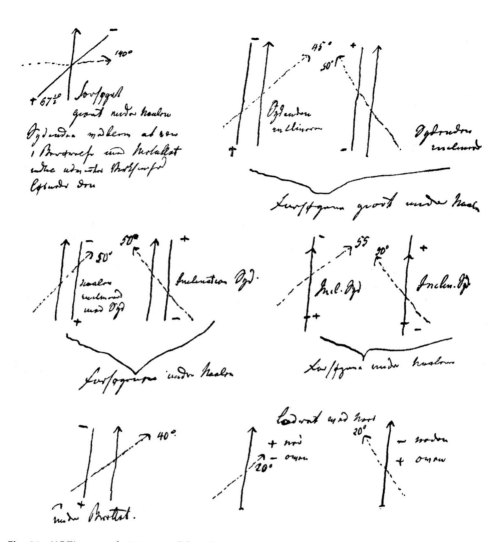

Fig. 80. HCØ's notes of 15.07.1820, ØC 23, RL.

The experiment made under the needle. The south end remained in contact with the metal until external action loosened it.
The south end deflects. The south end deflected. The Experiments made under the needle.
The deflects towards the south. Deflection south. The experiments under the needle.
Refl. south. South. The experiments under the needle.
Under the board.
Vertical at north. + at—above.—below + above.

A DISCOVERY BY CHANCE?

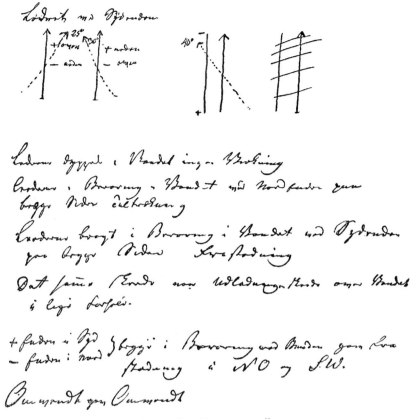

Fig. 81. Vertical at the south end. Conductors dipped in water no effect
Conductors in contact with water at the north end at both sides attraction
Conductors brought into contact in water at the south end at both sides repulsion
The same thing happened when the discharge took place above water in equal proportions
+ end in south) Both in contact with the bottom
− end in north) gave repulsion in NE and SW
Inverse on inverse.

He now sent *Experimenta* to a selection of scholars in Europe to claim his priority to the discovery. He had previously noticed that some colleagues had published some of his ideas as their own, and he would not risk a repetition. There are strong indications that it is not only empty talk when researchers behave as if their only motivation for lifelong research is not money and luxury, but the recognition and honour bestowed upon them by posterity. We have already seen Faraday register his priority to scientific discoveries through the deposit of a document in the treasury of the Royal Society. In France this procedure was customary. Lavoisier was one of the first to try to prevent a rival from taking the credit for his discovery of oxygen. This was the rationale behind the sealed letters that researchers had the right to deposit in the

EXPERIMENTA
CIRCA EFFECTUM
CONFLICTUS ELECTRICI IN ACUM MAGNETICAM.

Prima experimenta circa rem, quam illustrare aggredior, in scholis de Electricitate, Galvanismo et Magnetismo proxime-superiori hieme a me habitis instituta sunt. His experimentis monstrari videbatur, acum magneticam ope apparatus galvanici e situ moveri; idque circulo galvanico cluso, non aperto, ut frustra tentaverunt aliquot abhinc annis physici quidam celeberrimi. Cum autem hæc experimenta apparatu minus efficaci instituta essent, ideoque phænomena edita pro rei gravitate non satis luculenta viderentur, socium adscivi amicum Esmarch, regi a consiliis justitiæ, ut experimenta cum magno apparatu galvanico, a nobis conjunctim instructo, repeterentur et augerentur. Etiam vir egregius Wleugel, eques auratus ord. Dan. et apud nos præfectus rei gubernatoriæ, experimentis interfuit, nobis socius et testis. Præterea testes fuerunt horum experimentorum vir excellentissimus et a rege summis honoribus decoratus *Hauch*, cujus in rebus naturalibus scientia jam diu inclaruit, vir acutissimus Reinhardt, Historiæ naturalis Professor, vir in experimentis instituendis sagacissimus Jacobsen, Medicinæ Professor, et Chemicus experientissimus Zeise, Philosophiæ Doctor. Sæpius equidem solus experimenta circa materiam propositam institui, quæ autem ita mihi contigit detegere phænomena, in conventu horum virorum doctissimorum repetivi.

In experimentis recensendis omnia præteribo, quæ ad rationem rei inveniendam quidem conduxerunt, hac autem inventa rem amplius illustrare nequeunt; in eis igitur, quæ rei rationem perspicue demonstrant, acquiescamus.

Apparatus galvanicus, quo usus summus, constat viginti receptaculis cupreis rectangularibus, quorum et longitudo et altitudo duodecim æqualiter est pollicum, latitudo autem duos pollices et dimidium vix excedit. Qvodvis receptaculum duabus laminis cupreis instructum est ita inclinatis, ut baculum cupreum, qui laminam zinceam in aqua receptaculi proximi sustentat, portare possint. Aqua receptaculorum $\frac{1}{60}$ sui ponderis acidi sulphurici et pariter $\frac{1}{60}$ acidi nitrici continet. Pars cujusque laminæ Zinceæ in aqua submersa Qvadratum est, cujus latus circiter longitudinem 10 pollicum habet. Etiam apparatus minores adhiberi possunt, si modo filum metallicum candefacere valeant.

Fig. 82. HCØ, *Experimenta circa effectum conflictus electrici in acum magneticam*, Hafniæ, 21.07.1820. NS II 214–218. Translated by J.A. Dyssel in *Hesperus* 3, 312–21.

French Académie des Sciences and that were subsequently kept absolutely secret until the author himself decided to unveil his discovery. This procedure was particularly suitable for protecting young researchers in the middle of a series of unfinished experiments. During the forty years between 1796 and 1836 a total of 329 *plis cacheté* were deposited in the Académie des Sciences, and the number continued to rise during what has been called 'the age of science'. This system did not exist in Denmark, and Ørsted's *Experimenta* was exactly the opposite of a sealed letter, it was an author's publication on his corroborated experiment with all the risks involved in showing himself in the European limelight.

According to his mail list the recipients of *Experimenta* were

England:	Davy, Wollaston
France:	Ampère, Arago, Biot, Blainville, Boisgiraud, Fresnel, Hachette, Laplace, Petit, Savart
Germany/Austria:	Althaus, Beck, Buch, Döbereiner, Erman, Gilbert, Meier, Moll, Neef, Pfaff (Erlangen), Pfaff (Kiel), Poggendorff, Prechtl (Wien), Schweigger, Seebeck, Yelin
Italy:	Confiliachi, Ridolfi
Switzerland:	Pictet, Rive
Sweden:	Berzelius

The list contains a few more names that I cannot identify. It is not complete, however. For instance the following names are missing, although we know from other sources that they received it: Prince Christian Frederik, Forchhammer, Hansteen, Th. Thomson, Weiss, and Th. Young.[17]

Ørsted's personal reaction to the discovery was remarkably retrospective. 'According to the great scheme of things he had already in his earliest writings assumed that magnetism and electricity were produced by the same forces.'[18] He put particular emphasis on this historical justification of his previous research lest anyone belonging to his international readership should believe that the discovery was haphazard. Quite the contrary, it was the completion of his dynamical project that had preoccupied him for more than twenty years. He declared in German, French and English: 'As I have already for a long time considered the forces generated by electricity as the fundamental forces of nature, I also had to deduce the magnetic effect from them. Hence, I suggested that "the electrical force in one of the states in which it is bound, would generate an effect on the magnet as such",' quoting from his *Materialien*, 1803 and referring to *Ansicht/Recherches*, 1812-13.[19]

The day before he dispatched his *Experimenta* he once again took his family on an excursion, this time to the beach at Bellevue. Gitte and the children were longing to go on holiday and perhaps slightly upset by his recent absent-mindedness. Finally, at the end of July the family set off for Selchausdal, a manor house in the hills above Lake Tissø, where they had rented accommodation for the summer. They were now living within a convenient distance of Falkensteen, Manthey's estate. While enjoying the company of his former patron and Masonic brother, Ørsted probably pondered how the world would respond to his discovery of an undiscoverable phenomenon.

> Up to now, I had never believed
> that a Prince could think as he does.[1]

36 | 1820–21
Domestic and Foreign Reactions

IF ØRSTED had hoped for quick reactions from the recipients of *Experimenta*, he had chosen a bad time, for his European colleagues were on their summer holidays. Berzelius was guiding mineralogists around the copper mines of Falun and the iron mines Dannemora.[2] C.S. Weiss, his good friend in Berlin, was on his way to a trip along the Rhine, and Paul Erman, secretary to the Prussian Academy of Science, was taking his family to Switzerland.[3]

Immediate responses varied between difficulties in understanding the experimental setup via profound scepticism to unreserved acclaim. Pfaff in Kiel had been unable to repeat the experiment for he did not understand Ørsted's directions and kept reversing the poles of the conductor and the needle. He kindly asked Ørsted to send him a set of his marvellous copper troughs (he would pay, of course).[4] Weiss grumbled that *Experimenta* was written in such an eccentric style of Latin. 'Why have not you just written your report in plain German and French, languages you have mastery of, or just in a normal Latin?'.[5]

Others shook theirs heads. Electromagnetism? Was this one of Ritter's whimsicalities again? This was how Gilbert, the editor, reacted in Leipzig. But then he discovered that A.W. Hauch had been a witness, and Ørsted was given the benefit of the doubt. He was only convinced when he saw other scientists approving the discovery. Thomas Seebeck, who had recently moved to Berlin, scratched his head when he first encountered rumours of the discovery. Then he read the four pages and could not wait to build a big galvanic battery as quickly as possible to test the result.

Some natural philosophers were immediately sympathetic. Weiss was a case in point. Back in 1802 he had joined the study circle on Kant's metaphysics of nature and toasted Sophie Probsthein on her birthday, so to him the discovery of electromagnetism was the confirmation of an old hypothesis. 'Now you deserve the great Napoleon prize', he wrote.[6] Paul Erman, who had asked young Ørsted to present a paper to the Philomatic Society, adopted the same benevolent attitude. He did not need to wait for French approval before paying his respects. 'That Ørsted read from a magnetic needle what no electrometer had shown, whether it was because nobody had looked for it or whether it could not show it... That it was reserved to him to be the first to catch a glimpse of the transformation of electricity into some unknown "x" or to look into magnetism itself is an incomprehensible stroke of good luck. Nobody is likely to accomplish a similar thing for a thousand years to come.'[7] Schweigger, the editor from Halle, was positive towards the discovery from the very beginning; he had believed in the interdependence between electricity and magnetism for many years, and now he translated Ørsted's Latin piece into German, magnanimously adding that Ørsted's *Experimenta* was the most interesting treatise that had ever been written. Schweigger also defended Ritter's prior claim to electrochemistry against Humphry Davy. He did further work on electromagnetism inventing a multiplier that amplified the effect. His invention involved an astatic magnetic needle, a double magnet the two poles of which turned in opposite directions thus neutralising each other, and in the same way not reacting to the earth's magnetism. But it did react to the electrical conflict. In the multiplier the electrical conductor took the form of a bundle of copper wires isolated with silk filament, wound around the astatic needle. Schweigger's multiplier was a direct extension of Ørsted's discovery, registering even weak electricity such as Seebeck's thermoelectricity. Unfortunately, Ørsted did not reciprocate Schweigger's admiration.[8]

Above all Ørsted was looking forward to avenging his 1803 humiliation by the First Class of the National Institute. This was not easily done. The corpuscular theorists were prejudiced against *Experimenta* for they knew (or thought they knew) that Ørsted's discovery was impossible. The reason for this was, as Ampère wrote to a friend, 'Coulomb's hypothesis on the magnetic effect; everybody trusted this hypothesis as if it were a fact; it denied any possibility of interaction between electricity and magnetism'.[9] When Ørsted had raised his head once again, the Laplaceans would be only too pleased to cut him down to size.

The professors at the Swiss Society for Scientific Research in Geneva, M.-A. Pictet, Gaspard de la Rive, and others took on the task of examining *Experimenta* as early as August. De la Rive had a strong and suitable galvanic battery. Their control experiments did not turn out quite as they expected, since they proved Ørsted right. They decided to repeat the process, and called Arago from France to oversee the repetition of the experiment on 9 August. Now they had to give in. They translated *Experimenta* into French as an appendix to their report, giving Ørsted full credit for the discovery.[10] Pictet went on his summer holidays to Florence, where he repeated the experiment in Academia del Cimento for Marquis Ridolfi, who had already received Ørsted's piece.

In Geneva, Arago had been convinced of the validity of Ørsted's discovery and he rushed back to Paris to set up a commission to settle the matter. Besides Arago it consisted of Gay-Lussac and Laplace himself. Arago announced publicly that a demonstration of Ørsted's experiment would take place in the Louvre on 11 September. It was anticipated with much scepticism, as the members did not trust the Swiss report.[11] Arago was Ørsted's junior by nine years and had a relationship to Laplace reminiscent of Ørsted's to Bugge. Arago had long been critical of the corpuscular theory and no less critical towards the research policy advanced by Laplace in the Académie des Sciences. So the commission was rife with inner tension. Ørsted's discovery had to be seen as a frontal attack on the Laplacean paradigm and consequently as a potent weapon in Arago's struggle with the 'old wigs'. Laplace had successfully remained in power in one institution after another during the turbulent years from the *Ancien Régime*'s Académie des Sciences via the underground organisation the Société d'Arceuil and the Napoleonic First Class of the Institut National, resurrected in 1816 under Louis XVIII under the pre-revolutionary name of l'Académie Royale des Sciences. Its members did not become convinced of Ørsted's discovery until the session on the 11th, when Laplace, on behalf of the commission, authoritatively acknowledged that the electrical conductor was moving the magnetic needle, as everybody present could see for himself.

A.-M. Ampère was among the spectators attending the demonstration, and he repeated the experiment over the following days, pondering its repercussions. His criticism of it was totally different from what Coulomb's would have been had he still been alive. Ørsted was mistaken in so far as he had ignored the influence of earth magnetism on the needle, Ampère objected. This was why the deflection of the needle did not exceed forty-five degrees. And quite correctly, when he made Ørsted's experiment with an astatic needle, the effect of the electric conductor was no longer impeded by earth magnetism and the needle deflected at a right angle. He demonstrated this at the next week's session, having the idea that the astatic needle (being magnetically neutral) perhaps only reacted due to internal flows of electricity, or (put differently) perhaps a magnetism was nothing but an electrical circuit that is to say a voltaic pile. At the next session after that, on 25 September, Ampère showed that parallel electrical circuits attract and repel each other depending on the direction of the currents, reasoning that this being the case and the electrical and magnetic fluids being totally different, it must logically follow that magnetism is actually reduced to being a flow of electricity.[12] In this way corpuscular theory would survive Ørsted's discovery, or even better it would encompass and explain 'electromagnetism'. To Ørsted electromagnetism obviously disproved corpuscular theory. In Ampère's eyes it could be saved, if magnetism was considered to be a flow of electricity.

Two other members of the Académie Royale des Sciences, the physicists J.-B. Biot and F. Savart, fostered the idea of describing the electromagnetic effect in mathematical terms. By measuring the angle of the needle's deflection they found values perfectly corresponding to the inverse-square law, according to which the forces decrease inversely as the square of the distance increases. Ampère, Biot, and Savart thus maintained the Laplacean idea that electromagnetic forces are central forces acting rectilinearly like the attraction of gravity. Ørsted, of course, had quite a different view. To him the electrical conflict radiated from the conductor into the field surrounding it and moved in a circle. In this respect he was supported by Arago, demonstrating in an elegant experiment in November the circular lines of the electromagnetic field around the conductor. He wound a copper wire around a hollow glass rod inside which he put a steel rod. When he connected the copper wire to a voltaic pile the steel rod became

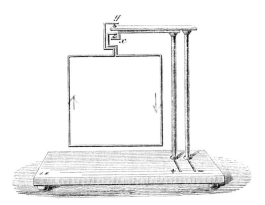

Fig. 83. A.-M. Ampère (1775–1836) disputed HCØ's electromagnetic theory by reducing magnetism to electricity. In September 1820 already Ampère carried out the following experiment: the electric force is conducted into a movable rectangle suspended in two pivots (y and x, points rotating in tiny bowls of quicksilver). When subsequently he led another electric conductor with the opposite direction of current towards the rectangle, it pivoted in accordance with the rule of thumb. From this experiment Ampère deduced his laws of electrodynamics: two parallel and unidirectional currents attract each other, while two parallel and oppositedirected currents repel each other. When in another experiment he replaced the other electric conductor with a magnet, the movable rectangle pivoted in exactly the same way. From this observation he deduced that the magnet reacted like—and actually was—an electric current. Poul la Cour & Jacob Appel 1897 §363.

magnetised, and drawing it through a disc of cardboard at a right angle, and sprinkling iron filings on the cardboard disc one could see the iron filings forming a ring in the direction of the current according to the rule of thumb. Ørsted had placed iron filings close to a galvanic conductor and observed that they reacted like magnets. He therefore saw Arago's experiment as supporting his view. Moreover, Arago's circular pattern on the cardboard disc must have reminded him of his acoustic figures making musical tones visible. An obvious division had occurred between the academy members: Arago for and Ampère against Ørsted.

H. D. de Blainville, too, despite his noble birth, opposed Laplace's doctrinaire and nepotistic leadership of the Académie Royale des Sciences. He edited the *Journal de Physique*, and Ørsted had sent him *Experimenta*, probably in the hope of having it published in his journal. However, Blainville did not dare, until he had the backing of the Academy. In November he asked Ørsted's forgiveness for this sin of omission, disclaiming the responsibility and promising (now that he had obtained confirmation of the validity of Ørsted's discovery) that in the future he would print anything that Ørsted might have the goodness to send him, as he expressed himself, from the rich vein his ingenious mind had brought to light. Blainville also informed him of the progress made by Biot's measurements of the angles of the magnetic needle's declinations, bringing hope that the research into electromagnetic phenomena could be advanced by the help of mathematics. He was now ready to approve the theories Ørsted had put forward in *Recherches*, which he had studied with the zoologist Chevreul.[13]

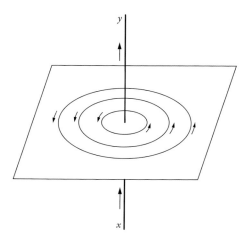

Fig. 84. D.F.J. Arago (1786–1853) substantiated HCØ's electromagnetic theory. Having repeated and corroborated HCØ's experiment in Geneva and Paris, in November 1820 Arago carried out the following experiment: He led the electric force through the vertical conductor yx penetrating a horizontal cardboard disc strewn with iron filings. Thus an electromagnetic field was established, in which every single iron filing reacts like a magnet, whose north poles point at the direction of the arrows. The circles so formed constitute the force lines of the magnetic field. Poul la Cour & Jacob Appel 1897, §362.

The editor asked Ørsted to be very discreet when talking to French scholars about their discoveries before they had been published in his journal, or alternatively to always remember to postdate references in his replies lest Blainville get into trouble for his indiscretion. It must have been unpleasant for Ørsted to find that his discovery was only published after others had elaborated on them on the basis of his experiment. And even if the letter was framed with all imaginable French phrases of politeness, it must have been an anticlimax for him to read the following lines:

> 'It will no doubt please you to learn that you have been nominated as a foreign member of the French Academy of Science together with Mr Leslie and Mr Brewster. The first has been chosen, but it is our hope that new works from your hand will increase the chances that your discovery has already given you.'[14]

Nevertheless, another three years were to pass until Ørsted was found worthy of admission into the Académie Royale des Sciences.

Apart from this correspondence Ørsted had his personal agents located in European cities. His disciples Schmidten and Forchhammer were in Paris and London respectively with all senses alert to keep their mentor informed. Strangely, Ørsted had forgotten to send *Experimenta* to Schmidten, who had very early been admitted to the sessions of the Academy in the splendid hall in the Louvre and was invited to private parties with French celebrities. Forchhammer had only received the news in Glasgow where he met with Thomson, because Ørsted's letter and piece kept failing to find their constantly moving recipient. Back in London he learned that Dr W.H. Wollaston had received information from Geneva about the approval of Ørsted's discovery. And as Wollaston had since spread the news in the most enthusiastic terms, his colleagues were by now convinced of the validity of the phenomenon. He was recognised as an authority, as he rarely allowed himself to become enthusiastic about anything.[15]

In Britain Humphry Davy started to examine Ørsted's experiment in October in the Royal Institution with his assistant Michael Faraday. Faraday's position at the institute was unclear. Originally, he had been employed by Davy as an errand boy, like Dyssel rather than Zeise, but thanks to his independent research activities and exquisite experimental skills he was soon on his way to overtaking his master. Davy was a titular professor at the Royal Institution, a private institute for scientific research and instruction, in the establishment of which Count Rumford and Thomas Young had been instrumental. Since 1820 Davy had been the President of the Royal Society, into which Faraday was only admitted as a Fellow in 1824. Both Davy and Faraday had studied *Recherches*, so Ørsted's name was familiar to both of them. Davy's expectations of *Experimenta* were not entirely positive, mainly because he was disappointed or perhaps even insulted that Ørsted did not recognize his prior claim for the electrochemical theory, but instead on page after page had given Ritter the credit, only mentioning Davy occasionally together with Berzelius. Davy saw too much metaphysical nonsense in Ritter's works even if he had to admit that sometimes he had done experiments that were both ingenious and original. But Davy could not stand to be overshadowed by anybody. So, it was from a position of slight condescension that at the beginning of October he asked Faraday to examine what in a letter to his brother he called 'some of Oersted's vague experiments'.[16]

Soon Davy, too, realised that he had to yield. On 16 November, the first session after the vacation, he reported to the Royal Society with unreserved approval. Ørsted's priority was indisputable and undoubtedly his discovery would open to the world an entirely new and exciting field of research. It was Davy who suggested honouring Ørsted with Sir Godfrey Copley's gold medal, and his proposal was immediately adopted by the research council of the society. The president, W.H. Wollaston, and the treasurer together opened the iron chest, took out a Copley medal (now there were only three left) and sent it via the Danish ambassador in London to Ørsted in Nørregade.[17]

That autumn the interest in electromagnetism flourished in scientific circles all over Europe. Faraday was overwhelmed by the flow of news on this new field of research and set out to write an article to provide an overview. His article was full of admiration for Ørsted's experiment, and having perused Ørsted's major work Faraday scored off people who maliciously hinted that the discovery was merely a stroke of luck.

> As early as 1807 he proposed to try "whether electricity the most latent had any action on the magnet". At that time no experimental proofs of the peculiar opinions he entertained were known; but his constancy in the pursuit of his subject, both reasoning and experiment, was well rewarded in the winter of 1819 by the discovery of a fact of which not a single person beside himself had the slightest suspicion; but which, when once known, instantly drew the attention of all those who were at all able to appreciate its importance and value.[18]

Fig. 85. The Copley medal, the Royal Society's token of respect sent to HCØ as soon as Humphry Davy had tested his electromagnetic experiment and obtained his Fellows' approval. The medal is named after Sir Godfrey Copley, FRS, and was awarded for the first time in 1731. At the bottom is engraved: D.JOH.CHR.OERSTED—MDCCCXX as HCØ signed himself in *Experimenta*. SNU.

In November Forchhammer received confidential news. He had attended a Royal Society meeting in Somerset House, at which Ørsted had been proposed as a foreign member along with his friend H.C. Schumacher, head of the Altona Observatory. This was a great honour, since the Royal Society only appointed a few foreign members. But for the time being they had only been nominated; the election would take place a fortnight later, so it would be prudent not to tell anybody.

Ørsted received a similar hint directly from Schumacher, who had received it himself from Thomas Young, secretary of the Board of Longitude, in which capacity he had ordered one hundred étalons from Schumacher. But Thomas Young was also secretary of the Royal Society and hence well informed about the society's marks of distinction.

> I cannot omit to congratulate you on the marvellous discovery of your countryman Dr Oersted, and of the elevation of your country to a rank in science which it has not held since the days of Tycho Brahe. Pray have the kindness to thank Prof. Oersted for his letter to me, which

I could not answer until I had procured the repetition of his experiments, and it was only upon my return to town a few days since, that I was informed of their complete success. I shall however not fail to write to him myself at some future time.[19]

In December the diploma of membership of Professor Johan Christian Oerstedt of Copenhagen was duly sealed and signed by its fellows. Still, five months passed until Young had it dispatched.

John Herschel, the astronomer, also recognised Ørsted's discovery immediately. His father, William Herschel, the discoverer of infrared light, had immigrated from Germany, and when Ørsted saw John Herschel on his first visit to England in 1823 he was pleased to be able to converse with an Englishman in German, in which language he was fluent. Herschel wrote about the discovery:

> Of all the philosophers who had speculated on this subject, none had so pertinaciously adhered on the idea of a necessary connection between the phenomena as Örsted. Baffled often he returned to the attack, and his perseverance was at length rewarded by the complete disclosure of the wonderful phenomena of electromagnetism. There is something in this which reminds us of the obstinate adherence of Columbus to his notion of the necessary existence of the New World; and the whole history of this beautiful discovery may serve to teach us reliance on those general analogies and parallels between great branches of science by which one strongly reminds us of another though no direct connection appears; as an indication not to be neglected of a community of origin.[20]

Summing up it is fair to say that the reactions to Ørsted's discovery varied a great deal and provoked controversies in learned societies. It came as a surprise to everyone. Several German scientists recognised it swiftly and without any further testing. The Swiss Academy of Science in Geneva was the first to test and acknowledge the experiment, which undermined the basic principles of corpuscular theory. The discovery split the French Académie Royale des Sciences, whereas the Royal Society of London unanimously approved it and became the first to give him a tangible token of respect. In December followed a diploma of membership from the Prussian Academy of Science in Berlin, but it was only in April 1822 that Ørsted was awarded the *small* Napoleon Prize in the form of a medal and 3,000 francs.[21]

In the meantime Ørsted had hoped that the French would recognise his discovery to be so valuable as to award the *great* Napoleon Prize of 60,000 francs that he had tried, and failed, to win for Ritter back in 1803. In a letter to King Frederik VI he grounded his hope by asserting that his discovery completed the proof that electricity is capable of generating all classes of natural effects:

> When the German physicist Ritter... believed himself to have found an experiment to prove this, the renowned Alexander von Humboldt wrote to him that he was prepared to give all his discoveries away for this one, provided it was affirmed, which did not happen. Ritter's experiments were not like mine at all. To prove the significance of my discovery it might be added that it seems to explain the magnetism of planet Earth, a phenomenon that has for thousands of years been considered one of the greatest riddles.

This assertion shows that Ørsted maintained his and Ritter's idea that the Earth functions as a galvanic battery interacting with earth magnetism. Hence, in his mind there was no doubt that his discovery had initiated 'une époque mémorable' thus fulfilling Napoleon's condition for being awarded the great prize. As a further ground he quoted a remark from Quarterly Review to the effect that his discovery 'would easily make an epoch as important as Newton's law of gravity'.[22]

But now Bonaparte was no longer Emperor of France, the monarchy had been restored, and Napoleon had been exiled. This was the depressing news from Delambre and Gay-Lussac, secretaries of the Académie Royale des Sciences, and the Danish ambassador in Paris was told the same by Alexander von Humboldt, who after his great expedition to Latin America had settled down in Paris. Many scholars were of the opinion that the discovery of electromagnetism deserved the *great* Napoleon Prize, but as a matter of fact neither the Emperor nor the *great* prize existed any more.[23] In 1803 when Ørsted had fought to win it, Ritter's discovery turned out to be an illusion. Now that his discovery made him worthy of it, the *great* Napoleon Prize turned out to be an illusion.

Prince Christian Frederik was on a grand tour with his Consort Caroline Amalie. He had dragged the Norwegian painter J.C. Dahl away from Emilie von Bloch, the young lady to whom he had recently been married in Dresden, to make him stay with him and his entourage in the castle of Quisisana overlooking the Bay of Naples. King Ferdinand of the Two Sicilies had put his country seat at the disposal of Christian Frederik. Two years later Ørsted met this distinguished landscape painter who preferred a professorship in Dresden to the one in Copenhagen that the Prince had offered him.[24]

Christian Frederik, who had been close to the Norwegian throne and taken an active part in the constitutional assembly at Eidsvoll in 1814, was an intelligent young prince who appreciated the sciences and the arts and felt strongly that the weakened Danish monarchy would only be able to assert itself internationally by means of scientific and artistic achievements. During his stay at the castle Christian Frederik received a copy of *Experimenta*, which he proudly displayed at a session in the Neapolitan Academy of Science. The Prince was pleased that Ørsted considered him capable of recognising his discovery for what it was really worth, and asked for information on 'further discoveries you might make in the physical sciences, for which you will always find unabated interest from your faithful and benevolent Christian Frederik'.[25]

From Naples the Prince proceeded to Paris, where Ørsted's protégé, the mathematician H.G. von Schmidten met him in the Collège de France. Here they attended Ampère's electromagnetic experiments for three hours. Humboldt was present, too, having a long conversation with the Prince, and Schmidten noted in his diary: 'This was not like talking to a dilettante of a princely house; but as if I was facing a professor of natural philosophy.'[26] Arago had begun investigating a possible connection between his theory of light and Ørsted's electromagnetism. This piece of news must have stunned Ørsted: imagine a Frenchman beginning to see nature as a dynamical entity the forces of which were different, but still constituted reciprocal unity. Schmidten reported that 'the Prince is totally absorbed in this and has taken pleasure in finding how much your merits are appreciated'. In Paris it occurred to Christian Frederik that

Ørsted might need new instruments or might invent some that might best be produced in Paris. If so, he was welcome to turn to Schmidten and the Prince would be happy to pay. Schmidten was fond of the royal couple: 'I have learnt to esteem both of them very much. Up to now, I had never believed that a Prince could think as he does.'[27]

*

It did not dawn at once upon Ørsted's fellow countrymen that he had made a sensational discovery, but it was rapidly rumoured once acknowledgements started to pour in from abroad. Ørsted took immediate advantage of the opportunity to advance his scientific interests. In late summer 1820 he contacted Møsting, the Treasurer, for money for a chemical laboratory in Nørregade. Møsting often invited scholars to parties at his home at Frederiksberg, Ørsted not least. 'He exposed a true zeal to find the necessary public means for the advancement of scientific projects and for travel grants to scientists.'[28]

Ørsted wanted to establish a training laboratory. Artisans, industrialists and farmers following his lectures as well as cadets, pharmacists and physicians would get an opportunity to improve their practical skills. 'Officers would learn about nitre fabrication, purification of sulphur, production of powder, etc. Physicians would find an opportunity to prepare the experiments they need to do. Pharmacists would be trained to do analyses that are not yet done in pharmacies. Industrialists and artisans would here become familiar with useful chemical experiments.'[29]

'This was the kind of institute that was graciously established thanks to His Majesty's paternal care for the diffusion of useful knowledge, in compliance with a proposal from Professor and Knight Ørsted', Zeise wrote on behalf of Ørsted in *Tidsskrift for Naturvidenskaberne* or 'Oersted's journal' as it was called in London.[30] The laboratory was located in the side wing in Nørregade, adjacent to his collection of instruments. The additional tenancy amounted to a hundred rixdollars a year. Another five to six hundred rixdollars were needed for experimental materials. Daily inspection was left in the hands of Zeise thus giving him a chance to prepare his courses, which was much needed, to judge from the complaints of his audience. Ørsted and Zeise stubbornly stuck to their principles of elevating artisans to a theoretical level and not give in to their expectations of acquiring special skills required by their trades.[31]

The professor modestly abstained from pecuniary gains, but Dr Zeise held no post at the University, and Dyssel had so far received his twelve rixdollars a month out of Ørsted's own pocket. Ørsted committed himself to submitting an annual report on the utility of the training laboratory, and on this promise his application was remunerated with one thousand five hundred rixdollars a year for a fixed term of three years. This sum took the pressure off his finances and benefited his employees. One of Ørsted's indisputable talents was gaining the ear of the authorities. Luckily Møsting, like Prince Christian Frederik, 'considered enlightenment, the sciences, and the arts the most important means to sustain the power and prestige of the state.'[32] The Prince, the Treasurer, and the professor made a powerful trinity for science. His new training laboratory was a worthy counterpart to Hermbstaedt's in Berlin (ch. 12).

Few are as happy as I...¹

37 | 1822–3
Ørsted's Triumphal Progress Germany

Ørsted's discovery did not lead to any form of public discussion of electromagnetism in Denmark. No other Danish scientist did any control experiments or elaborated the matter. No articles were written. But Ørsted did receive one letter, the only exception breaking the silence. It was sent by A.F. Tscherning, the future Minister of War, but not a member of the Royal Danish Society. However, Ørsted read it aloud to the physical class in February: 'Occasioned by Your Excellency's new discoveries concerning electricity I take pleasure in recounting an event taking place here [Frederiksværk Cannon Works] seventeen or eighteen years ago confirming your theory: a bolt of lightning struck a pole for testing gunpowder, peeling off the pole a spiral-shaped chip right down to the spot where it is supported by struts; the spiral chip stopped there and a straight chip from the strut was flung a distance of twenty to thirty metres.'² Interesting, but where was the magnetism?

Ørsted had to go abroad to be challenged, but as it was his turn as Vice-Chancellor of the University in 1821–2, he was unable to take immediate leave, and did not set out until November 1822. In return he was exempted from delivering lectures and in addition was given an extraordinary remuneration.

On his first grand tour 1801–03 he had covered long stretches on foot. The second time he had more resources and allowed himself coaches, while bargaining over the price, but now, the third time he could afford to have his own coach and change horses along the way. He could well do without walking the long distance from Hamburg to Berlin through the lonely landscapes of Mecklenburg. He had chosen the 22-year-old Gottlieb Bindesbøll as his travelling companion. He was a skilled millwright and his uncle, Jonas Collin, had commissioned him to make a grain dryer when, during the crisis in the corn market, the government decided to improve the quality of corn for export. The ambitious Bindesbøll had taken courses at the Royal Academy of Arts and frequented Ørsted's lectures on physics, and that in turn had spurred him on to study architecture. Ørsted was more than twice his age, but got on well with his fellow traveller as with so many other young students of talent. And besides, Bindesbøll could make drawings of what they saw along the way.[3]

Ørsted's aim was to call on physicists working with the theory of light. He also took on the task of measuring the earth's magnetism with a special device developed by Christopher Hansteen, to collect data needed for a worldwide map of isodynamic lines.[4] He also brought along his own instruments, enabling him to demonstrate his acoustic figures and to show his piezometer for compressing liquids, an instrument with which he had experimented with Hansteen the summer before. Finally, of course, he looked forward to basking in the admiration of colleagues wherever he went. Prince Christian Frederik had given him letters of recommendation to famous scientists, 'not to provide you with access to them', as he modestly remarked, 'for to obtain that you only need your name... My wish is only... to let these scholars be reminded of me...'. There were letters to Baron Cuvier and Sir Humphry Davy as well as to Professor Arago and J.B. Say, the political economist, 'whom you will be interested in getting to know'.[5] The tone between Ørsted and the Prince, later to become King Christian VIII, was always formal and polite in both directions, and the correct façade was always kept up, although the two entered a very close collaboration scientifically, technically, and politically, let alone in their Masonic work. Ørsted thanked him as follows: 'Your Highness, please accept my hearty and unfeigned thanks for the new proofs of your grace... The favours that a highly esteemed Prince bestows on them [the foreign scientists in question] will bring about renewed friendship for me and a new desire to assist me.'[6] In return Ørsted kept him up to date with scientific news, such as Seebeck's thermoelectricity, and he took on the role of undertaking commissions for the Prince, who was a keen collector of objects of natural history and was particularly interested in buying Haüy's large collection of minerals and crystals.

Ørsted had arranged with Gitte that friends and family should be allowed to read his ordinary letters, whereas his more intimate feelings and confidential information would be directed to her separately. Many people turned up in Nørregade to hear the latest news from Gitte's husband; Gjerlew knew from experience that his old room-mate was slow to write letters to his friends, so to get news from Berlin, Paris or London, he must ask Gitte's permission to read his letters.[7] When news of particular interest arrived, for instance on her husband's having dinner with Goethe in Weimar, Gitte could not help jumping into a cab and rushing out to see Anders and Mathilde at Østerbro to share the news.

A few love letters to Gitte ('for your own lips') have been filed in the archive with the circular letters. Frequently one does not notice the difference, for even the intimate letters are full of reports on meetings with scientists and practical things such as transferring money or bringing up their children. In the middle of writing he could pause to rebuke himself for a certain cold-heartedness or perhaps some kind of emotional shamefacedness. He caught himself disappointing his wife's yearnings for words of love. When his everyday duties extinguished a long-felt need for her he once exclaimed:

> 'I often wish that my feelings and thoughts could immediately become conscious to you and yours to me! Half an hour ago I stood glancing at the evening sky thinking about you and Karen and Christian and Maria, one after the other, and then about all of you. It was then that this wish occurred to me. A moment later a man passed by offering me an opportunity to post a letter by ship leaving in about an hour. A little something is better than nothing at all I thought to myself, even if a letter is almost nothing against all the thoughts crossing one's mind when, surrounded by quiet nature, one is yearning for those nearest and dearest to one. This is a kind of spiritual embrace that would be perfect if thoughts could meet like the eyes of two parted lovers gazing simultaneously at the moon. I often think like this, but I rarely pass it on to you, because immediately the quiet thought and the communication of it are hampered by stressful business in which they drown as it were. During the past few years I have been so overpowered by obligations that I rarely indulge in such thoughts, but I have decided to do everything I can to regain a richer inner life at the expense of the abundance of external obligations.'[8]

His plan was to travel throughout the winter semester and resume his duties in the Royal Danish Society and at the University by May, but he did not return until October. In other words he was away twice as long as planned, and even half a year was a long time to be separated from his wife and four children, the three already mentioned and Sophie, as they called their daughter born in January 1821 after the little twin who died. And Søren, their foster-son, should not be forgotten. Ørsted missed all of them. Did Søren do well at school? Was Christian able to write? Had little Sophie got her first tooth? When Gitte spoke of their father Karen would shed a tear. He sent home presents and small letters such as this: 'Dear Karen, Thanks for your letter. I am very pleased to see that your handwriting is improving so well. Your mother tells me that you have been a good girl. Let me see that you become ever more diligent so that I can be pleased with you....'[9]

In Berlin Ørsted's portrait was fashioned in a mould, from which medallions were cast in wax, gypsum or iron. He sent them to his family, but Gitte did not find them particularly lifelike. Indeed, the portrait painted by C.W. Eckersberg and delivered in a frame to Gitte at Christmas was far better. Gitte was thrilled with it. 'I can hardly believe such a good likeness...would be possible for a painter, nobody has seen it without being in perfect agreement. Your portrait has the truest and finest resemblance possible. You look so handsome that one is bound to fall in love with you whether one wants to or not. When I am out I am longing to get home to my picture.'[10]

The painting shows the red-cheeked and squint-eyed professor with his thin auburn hair in high spirits surrounded by his favourite instruments. Gitte and the children were so infatuated with the picture that it was almost as wonderful as having its sitter in his natural state in Nørregade, and seven-year-old Karen wrote to Paris: 'First of all I wish you a Happy New Year and thank you for the beautiful knitting ring and for the letter you were good enough to send me; but dear father, please do not get angry at us, when Christian and Maria and I present each of

Fig. 86. HCØ (1822), the proud discoverer of electromagnetism in his laboratory in Nørregade with his copper troughs and piezometer on the shelf in the background. His hand holds a brass plate for sonorous figures, and on the table to the right lies the bow and the compass, although not the mahogany one that is depicted in ch. 35. He wears one decoration only, the order of chivalry. His wife and children found this portrait life-like. He has a cleft chin shown by no other portrait. C.W. Eckersberg (1783–1853). Portrait in oils. DTM.

Professor Eckerberg's [sic] three little girls with our Christmas gifts from you, because their father has painted you so nicely. I shall do my best to please you this coming year in order that you may love your little Karen.'[11] Ørsted had sent gold necklaces and perfume home to the girls and a picture book to the boy from Paris. Gitte's letters were 'always a true refreshment'. The only thing he missed was 'talking to you and disentangling myself from all my businesses and throwing myself into my Gitte's arms; then I should feel a happiness that only very few would be able to enjoy'.[12]

The upbringing of the children was a recurrent theme in their correspondence. Ørsted sent instructions, and Gitte carried them out. Their own children did well at school, but Søren was a problem. He played truant, was reprimanded, and was escorted home by the caretaker. Søren did not feel at home with his foster family, and when Jacob, his poor father, came to town to attend the funeral of Mrs Møller and see the old pharmacist, Søren hurried round to the hospital in Amaliegade, where Jacob lodged until the probate case was over. The unsettled boy began to tell Gitte lies, and Ørsted's fuse was short:

> 'He must break himself of the habit of lying, otherwise he must leave our home, and I shall not spend a shilling on his education. A liar cannot have anything to do with science. The silver coin you gave him for his feigned diligence has been taken back from him, I hope, otherwise you must do it now and rather give it to some good and diligent poor child. You have forgotten to tell me if he has received a sound beating for his lies. On such an occasion he must be beaten twice, once for his laziness and a second time more harshly for his lies. Please let me know if he shows any sign of improvement...'[13]

Søren received no presents from Paris, not even greetings, but he was promised one if he improved. Later Ørsted instructed: 'Søren must write to me as soon as he has achieved his old place at school. Then he will see that I am thinking of him, too'. The beating did not help. Gitte felt unable to be strict with him and told her husband: 'When you write to me again please let Søren be greeted like the other children now that he has been punished once. I do want to avoid people seeing him as a foster-child. It is better now with him, but he has not yet got back his old place at school'.[14] He did achieve it a few months later, though. As the beech trees burst into leaf, Søren was fifth in his class at school. And the boy obtained his foster-father's permission to go on summer holidays with his biological parents in Rudkøbing.[15]

During the cold winter Gitte wanted to do something that would surprise her distant spouse. 'I have a secret for you', she writes, adding in a womanly way a secret 'you should not be told; but now I have decided I must be open. I see you are getting impatient. Listen now. I wanted to surprise you by being able to speak German when you return...'. Dr Forchhammer had offered to read with her three times a week. Gitte found she would have to pay him for his time, the more so because he was without a job. But now the inheritance from Mrs Møller, who had passed away the previous summer, would be due, and so Gitte asked her sweet-tempered husband if she might give her sister fifty rixdollars and Forchhammer the same amount. 'But you must not expect too much from me, because we are interrupted ever so often.'[16]

*

The first stop on his tour was Gottorp Castle near Schleswig, the residence of Prince Carl of Hessen, Frederik VI's father-in-law and Ørsted's superior as head of the Military Academy and General Grand Master of the Freemasons in Denmark. He was absorbed in natural philosophy and experimentation and like A.W. Hauch, another Freemason, he had a collection of scientific instruments, although not of the same quality as Hauch's. Prince Carl's interest in chemistry and physics was predominantly confined to alchemy, and at Louisenlund, another residence of his in the vicinity, he had established a laboratory for experimenting with making gold, in the company of the French General Saint Germain, who had been called on to reform the Danish army back in the 1780s. The park of Louisenlund was adorned with statues and small temples in the Egyptian style and embellished with Masonic symbols.[17]

Prince Carl was already an admirer of Ørsted whose discovery had been communicated to him from Paris. Subsequently, he had received further information from brother Manthey, and Ørsted had sent him a galvano-magnetic apparatus with a manual, enabling him to repeat his experiment. In the Prince's Masonic interpretation, electromagnetism was the very discharge of the creative force, and this being so Ørsted's experiment was the disclosure of the magic of God's power. To him the unification of electric and magnetic forces was the secret behind the productivity of nature. The element of fire was hidden in magnetism according to the cabbala, the Jewish mysticism, whose symbols Prince Carl incorporated in the Masonic doctrine. Magnetism was fire, called *Aesch*, and electricity was water, called *Majim*. The two elements were symbolised by equilateral triangles, the fire pointing upwards, the water downwards. Their unification made a Star of David, called *Schamaim*, the very driving force of nature. So the Prince believed that fire was hidden in the magnet and released by electricity thus radiating God's spark of life on the first day of creation, when the Lord said, 'Let there be light!' and there was light.

However, there was also an alternative interpretation of Prince Carl's cabbalism, in which the two equilateral triangles symbolised fire and water respectively. To Freemasons these symbols mean God (point upwards) and man (point downwards), and the combination of the two in the Star of David meaning Christ, i.e. God becoming man in Christ who was born exactly under this star of Bethlehem. In the fantasy of the General Grand Master electromagnetism, according to both interpretations, was seen as analogous to the two greatest miracles of Christianity: creation and salvation.

Prince Carl considered Ørsted a modern Prometheus. In Greek mythology Zeus had given the fire to Prometheus, who had passed it on to man. This was against the intentions of Zeus, so he took revenge by shackling Prometheus to a rock and sending an eagle to pick at his liver every day. Zeus also took revenge on man by sending Pandora to open her box thereby releasing all calamities in the world. Only hope remained. Prince Carl now instructed the discoverer of electromagnetism, whom God had called to receive the revelation of this magic force, to use it in the service of the almighty Master Builder and for the benefit of mankind. By doing so, the mercy of God would lead him to new revelations. But at the same time he alerted his brethren of the lodge never to confuse the divine powers that the Creator had bestowed on nature with the Creator himself. If they did so, materialism might easily sneak into the Christian believer, who would thus become enlightened to be sure, but not necessarily pious.[18] What Prince Carl here wanted to express was a warning against Spinoza's pantheism that was seen by many as a camouflage for materialism. A true Mason should not confuse the Master Builder with his building, or the Creator with his creation. This heresy he left to Schelling's *Naturphilosophie*.

Ørsted stayed at Schleswig for six days instead of the two he had planned, for the Prince could not get enough of the electromagnetic experiments, which he interpreted with 'a very vivid fantasy'.[19] Every day at one o'clock Ørsted went to the castle, demonstrated his experiments in the library, had tea, and stayed to a late meal. He could not completely follow the Prince's esoteric ideas, and not at all when the General Grand Master revealed that he had actually grasped very little of Ørsted's discovery. For instance, without showing any sign of embarrassment he talked about the magnetic fluid permanently streaming across planet Earth from north to south. But it would be contrary to etiquette to adopt a superior attitude towards royalty, so Ørsted exercised his diplomatic skills, not least because he was going to take advantage of Prince Carl's letters of introduction to Masonic lodges on his itinerary. In return, Ørsted was supposed to use his recently gained authority to persuade the Egyptian Institute in Paris to publish the Prince's manuscripts on hieroglyphs as a symbolic language and the key to Egyptian mythology.

Some months later a grateful and diplomatic Ørsted reported from Paris that thanks to the Prince's recommendation he had been very well received by a lodge in Paris, where he had been initiated into ancient mysteries on magnetism, among other things, by a manuscript revealing that magnetism was the creative force of nature. This was no news to him, of course, he had simply gained his insight by other methods. He had also been shown experiments with sunlight, but the outcome had been uncertain.

Ørsted aired politely, but unambiguously, his criticism of Freemasonry. If he were to make contributions to a better understanding of nature within the framework of Freemasonry, he asked his brethren to openly say what they thought they knew and about the sources and ideas that had led them to their knowledge. 'Where my spirit cannot move freely in the light, it will achieve nothing', he wrote in German to Prince Carl of Hessen, while repeating his criticism of the secretive attitude of the order. A mature spirit independently scrutinising nature is in no need of being initiated into the occult, even if it acknowledges that it has to climb an infinite number of steps to reach certainty. He emphasised that he was not grumbling, but wanted to apologise for the fact that upon his return he would only have very little to report on what had been communicated to him in the lodges on his way.[20]

Prince Carl did not grasp the deeper implications of Ørsted's criticism. The freedom of research and critical methodology, whatever that might be, was no preoccupation of the General Grand Master. To him the secret Masonic insight into the mysteries of nature was the highest wisdom imaginable, and as he saw things brother Ørsted, initiated into the third degree of the order, had been chosen to become entrusted with a secret knowledge of the most important of these mysteries: electromagnetism.

> I see from your letter of 8 March that you are not quite content with the knowledge you have acquired in the lodge in Paris. Hence, I must openly and honestly tell you my opinion. All higher knowledge is acquired piecemeal. At first the basic principles are appropriated, then the apprentice considers it and reflects, puts these thoughts on paper in note form and goes on asking. In the course of this process of learning he acquires a deeper understanding, for as a rule the explanations are given in such a way that they encourage more research. Anyway this is the objective of the Freemasonic way: step by step one is led towards the whole.[21]

However, Prince Carl needed Ørsted badly. He had given him his manuscript on Egyptian hieroglyphs to get it published by the Egyptian Institute in Paris. This had been founded by Napoleon to collect and research archaeological finds from Ancient Egypt and the region ruled by the Mamelukes, conquered during his campaign of 1798-9. The objective of this campaign had been to cut off Britain's passage to India and so eliminate her role as a colonial power in the Far East. The Egyptian Institute was partly a cloak for this and partly an attempt to glorify France by researching into the oldest civilisation of the world with its pyramids and obelisks. Many French scholars had been brought together to collect and examine the shiploads of archaeological material taken to Paris. As with the earlier expedition to Felix Arabia by Carsten Niebuhr, Berthold's father, the purpose of this scientific project was to demythologise the history of early Egypt. The results were published in *Descriptions de l'Égypte* (1809–21), but apparently the overall purpose went over the head of the Prince, who had thus, unwittingly, given Ørsted an impossible assignment.

Fig. 87. Prince Carl of Hessen (1744–1836), FVI's father-in-law, General Grand Master of the Danish-Norwegian Masonic Order. HCØ was his guest at Louisenlund for a week in 1822—being flattered more than he liked. He promised to assist his master in having his works on Egyptian mysteries published in Paris. Portrait in oils by Peter Copmann (1795–1850), 1821. DFO.

Still, Ørsted did as much as he could, calling on the leader of the Egyptian Institute, Monsieur Nicollet: 'I have spoken to Nicollet several times. He reassures me that the printing of your manuscript will cause nothing but opposition here, unless your Lordship specifies your sources, and as far as I am acquainted with the world of literature here I do not believe him to be wrong.'[22] At this stage Ørsted was familiar with Thomas Young's epoch-making work of solving the riddle of the hieroglyphs. 'He has come across the fortunate idea that proper nouns in the hieroglyphic inscriptions must necessarily be expressed by something resembling letters... In this way he was enabled to read the name "Ptolemaios" and "Berenice" on the depiction of the monument found at Rosetta.' Of course, Ørsted did not venture to write this up-to-date news to the Prince, but to his brother Anders, who was also told about J. F. Champollion, who had taken Young's method further, and in this way had been able to read the hieroglyphs forming the word 'Autocrat'. This, however forced him to bring forward the date of the famous zodiac at Dendera by two thousand years to the time of the Roman Emperor Nero.[23] This discovery completely outmanoeuvred Prince Carl's manuscript.

Of course, he was disappointed with the outcome of the matter and told Ørsted so.

> The good Nicollet knows from my letters to him the source material on the basis of which I wrote, but he has no knowledge of mysteries, picture- or symbol-language. I have written to him that I received the Egyptian books ten to twelve years ago [the relevant volumes of *Descriptions de l'Égypte*], and that before I saw the [Rosetta] stone I immediately recognised certain features... I would like to explain to the erudite community of the Egyptian Insti-

Fig. 88. Nicolas Jacques Conté, head of the Egyptian Institute in Cairo/Paris. He had lost one eye when a hydrogen balloon exploded in the military test centre at the requisitioned Château of Meudon, where experiments with special airships for use in the invasion of Britain were carried out. Conté was in charge of operations at Meudon until the less warlike subject of Egyptology was allowed to profit from his administrative skills. Lithograph by Louis-Pierre Baltard de la Fresce (1764–1846), Château de Malmaison et Bois-Préau. RMN, Paris. Photo by Gérard Blot.

tute... as clearly as possible many of the riddles they have given up interpreting, but Nicollet spins out the printing, goodness knows why. I believe that when my work is finally published, my view will at least be read as well as all the absurdities I have had to read about this important stone.

Prince Carl was sure he knew exactly what had happened in the course of Egyptian history, one of the important roots of Freemasonry.

> All knowledge was brought to Egypt by Osiris and Thoth, and there after Osiris's death it was veiled in a picture-language, and this knowledge is especially represented in Osiris's death and resurrection. Most hieroglyphic paintings in all the temples and castles show this knowledge as well as the birth of Horus. This has been the foundation of the secret religion, but this foundation, however, was only represented as fables, natural philosophy and astronomy... Speak bluntly to Nicollet about this. If he refuses to print it, ask him to return my papers, and I shall see to the matter otherwise.'[24]

In Ørsted's mind the road to insight did not pass through the occult channels of Freemasonry to higher powers unveiling the truth to especially initiated people. Modern historical research demands source criticism and open debate on the results. As far as Ørsted was concerned he had a down-to-earth view of Freemasonry as an international network of significant people who would help the initiated brother on his way and provide access to the salons and physical collections of the aristocracy and royalty.

*

From Schleswig, Ørsted and Bindesbøll went via Altona, Mecklenburg, and Brandenburg to Berlin, where the professor was offered accommodation with his old friend S.C. Weiss, professor of crystallography at the new University. The millwright had to be content with a single room in the city. Soon he met Karsten too, who had risen to the post of General Mining Engineer, and the old study circle was reassembled. Weiss congratulated him on the discovery and they were soon caught up in the old discussions on the metaphysics of forces. 'When it comes to scientific opinions it cannot be denied that Weiss is very thorough and serious in his research on the one hand; but on the other hand he is also one-sided and rather intolerant. So, our talks on several principles of science are rather tough.' They could not agree on 'the way in which the fundamental forces of matter are distributed in space', when the same fundamental forces have to explain crystallisation as well as electromagnetism. Ørsted left 'the controversy having gained new knowledge, whereas Weiss perhaps had become more careful in his judgement of the opinions of others. In any case our mutual confidence increased after these talks. In all respects Weiss and his wife treated me as a true friend.'[25]

Karsten showed him around an iron foundry, from which he supplied his family at home with Christmas presents, candlesticks and iron medallions, among them an ugly portrait of old Klaproth. But now that the famous physicist was present he had to sit for an artist named Posch in order that his likeness could be measured and modelled into a mould for making casts in plaster or iron. Ørsted medallions were sold for one rixdollar apiece.[26] Gitte used the small iron medallions as presents to Mrs Lehmann, with whose family she had celebrated Christmas. Forchhammer wanted a medallion of Klaproth, and if he had one, Zeise had to have one as well, so could Ørsted please send a few more.

He spent four mornings with Thomas Seebeck and they showed each other their experiments. Seebeck's new thermoelectrical experiments were an extension of Ørsted's electromagnetic ones. Seebeck had discovered that when he placed two bars of bismuth into separate vessels with hot and cold water and soldered them together with a copper wire so that the two soldering spots were located in different vessels, a weak electrical force was generated. If the temperature in the two vessels was kept constant, the electrical force retained the same intensity. Now, Seebeck was able to show that the force increased proportionate to the difference of temperature, although only up to a point. If the difference of temperature was increased above 550° C the intensity decreased and it reached its peak at 275° C. As Ørsted had shown interaction between electricity and magnetism, Seebeck had shown the connection between heat and electricity, henceforth called thermoelectricity.

This was more grist to Ørsted's mill, and when he arrived in Paris, he demonstrated Seebeck's experiments in the Louvre with J.B.J. Fourier. He was proud to announce to Prince Christian Frederik how they performed. They had made a metal ring by soldering together six pieces of bismuth alternating with six pieces of antimony. They had then heated every second soldering spot to generate the differences of temperature they wanted, and soon they registered an electric circuit in the ring. When they cooled down the soldering spots they had not heated, the force intensified.[27] Seebeck's experiment provided hope that the

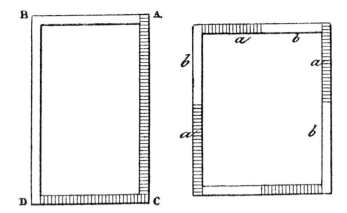

Fig. 89. In 1823 HCØ demonstrated to the Physical-Mathematical Class of the French Academy a series of Thomas Johann Seebeck's (1770–1831) experiments with thermoelectricity arising from heating/cooling the locations of the rectangle, where its two different metals—e.g. copper and bismuth—are soldered together. The electrical tension is detected by the deflection of a magnetic needle. NS II 272–282, SSW 470–477.

dynamical system would be expanded by yet another example of the unity of fundamental forces.

Ørsted spent twice the number of days he had planned to in Berlin. He wanted to see the *Gewerbeinstitut* [Mechanical Institute], the Prussian response to École Polytechnique, established on the initiative of Councillor P.C.W. von Beuth, which entertained ambitious schemes to transfer modern British technology to Prussia. The Prussian Government invested in education lest the major powers, Britain and France, should leave her behind. More engineers had to be trained, but should they fill public offices or fulfil the demands of private industry? The aim of the anglophile Beuth was that the *Gewerbeinstitut* should cover the needs of private enterprise, but it was still inadequate to provide the most highly trained engineers in great numbers. So it turned out that the *Gewerbeinstitut* offered an elitist education not adapted to the technological level of trades and crafts and hence unable to improve it.[28] Unfortunately, Ørsted did not write down any of his impressions of this mechanical institute, which was soon to be proposed as a role model for a polytechnic institute in Denmark.[29]

Ørsted also took time to walk about the city with Bindesbøll admiring the neo-classical architecture of Schinkel and others. They would stroll down Unter den Linden to the Opera House and cross over to the Gendarmen Markt with its 'two beautiful church towers on top of rather insignificant churches', in which everybody, as Frederick the Great is said to have remarked, would find salvation in his own faith. They listened to Carl Maria von Weber's *Der Freischütz* [The Marksman], and the next evening Ørsted invited his hosts, Mr and Mrs Weiss, to a concert in the new theatre. On an evening promenade he caught a glimpse of the Brandenburg Gate with the triumphal chariot on top that Napoleon in his hubris of victory had transferred to Paris, but now after Waterloo it was back on its proper place, 'an example of the lack of wisdom exercised by a conqueror taking pride in humiliating his enemy'.[30]

From Berlin their carriage took them via Wittenberg to Jena and Weimar. There Professor Döbereiner, a recipient of *Experimenta*, took the travellers to Goethe himself, with whom Ørsted immediately started conversing on electromagnetism. Goethe was well informed and quite impressed with the discovery, which, as he confessed to his diary, 'had emerged in almost dazzling light'.[31] Already on 20 October 1820 he had read the French translation of Ørsted's piece in *Bibliothèque Universelle,* prompting Döbereiner to repeat the experiment.[32] Goethe was aware that Seebeck had done the same thing in Berlin, and had also been informed about the discovery by Doctor Neef in Frankfurt (also on Ørsted's mailing list), who had proposed yet another experiment that Goethe also wanted to do. Ørsted pointed out some problems that Goethe grasped at once, and they agreed that Döbereiner and Goethe would do the experiments after Christmas. Ørsted could not conceal how happy he felt as 'a natural philosopher to associate with Goethe, whom he had for a long time admired as a poet'. 'O, yes, what is better in old age than throwing oneself into the arms of nature?', the poet exclaimed showing Ørsted the instruments he used to explain his theory of colours. Years before Ritter had refuted Goethe's optical theory while maintaining Newton's (ch. 10), which had had disastrous repercussions as his patron had withdrawn his pecuniary support and cold-shouldered him. So Ørsted had been warned only to touch on optical physics lightly. Goethe did not come to terms with Ørsted on that subject, but nevertheless invited Bindesbøll and Ørsted to a late supper after the evening's theatrical performance.

The next act belonged to Bindesbøll. Ørsted had introduced him as an architect (something of an exaggeration). His travelling companion was in the process of metamorphosis, changing from artisan to something more presentable. Goethe did not catch his name, but he did notice his title and consequently he had also invited Herr Coudray, head of Weimar's Directorate for Building.[33] He had a wide range of connections that might be helpful in promoting Bindesbøll's career, among them the architect F.C. Gau, who had familiarised himself with the architecture of Italy, Greece, and Egypt, and who had now settled in Paris, to which the two Danes were heading. At this time Goethe was immersed in a magnificent book about the cathedral of Cologne, one of the (unfinished) masterpieces of gothic architecture. The book was illustrated with a series of copperplate prints by the Boisserée brothers, and most amazing of all, they were said to have retrieved the original medieval drawings, enabling modern architects to imagine the relationship between the existing building and the originally planned cathedral. Ørsted was thrilled at the sight of these copperplate prints, and he could not find words to express his enthusiasm to Gitte. He fervently hoped to be able to afford a copy of this magnificent edition to grace their flat in Nørregade.

Ørsted communicated his enthusiasm to his wife:

> I wish you were here to be able to see all this; but since this would imply removing you two hundred Danish miles I shall have to provide you with this enjoyment second-hand. In fact, there are exquisite lithographs of some of these paintings, which I shall send to you. You will

Fig. 90. J.W. Goethe (1749–1832). Darbes was a Danish miniature painter living in Berlin, where HCØ socialised with him. In 1822 HCØ saw Goethe in Weimar and was well received. Portrait in oils by Joseph Darbes (1747–1810), Stiftung Weimarer Klassik und Kunstsammlungen, Weimar. BPK, Berlin.

take pleasure in these wonderful works, and they will embellish our home perfectly. In March you will receive eighteen such prints—bought for eighty rixdollars out of my own pocket... These prints have met with such acclaim from Thorvaldsen that he has hung up all of them in his rooms.'[34]

Gitte was delighted with the lithographs, hosting a party in order to let their friends admire them, and their friend Gjerlew (unable to come, but in need of comfort having lost not only his wife, but also his son) was allowed to borrow them for a few days.[35]

During the late evening Goethe and Ørsted talked over subjects of natural philosophy. Ørsted outlined his theory of heat and Goethe nodded approvingly. The next subject was chemistry and the affinity of matter (on which the international scientific community was far from agreement) and Goethe's theory of elective affinity in particular. The atmosphere became increasingly merry and then a proposal emerged: to arrange a congress on chemistry at Weimar to be held on the basis of papers by Ørsted accompanied by the spirit of Goethe, who would unremittingly throw in delightful thoughts to enliven the community.

> After supper we turned to his theory of light. In a way he forced the touchy subject on me. I uttered some reservations that he received in a very friendly way, adding that we might approach one another on this matter provided we spent more time together. I doubt it; but it was agreeable to me that he explained that it had not been his intention to attack the part of the theory of light concerning the refraction and reflection of light, but that he only had the

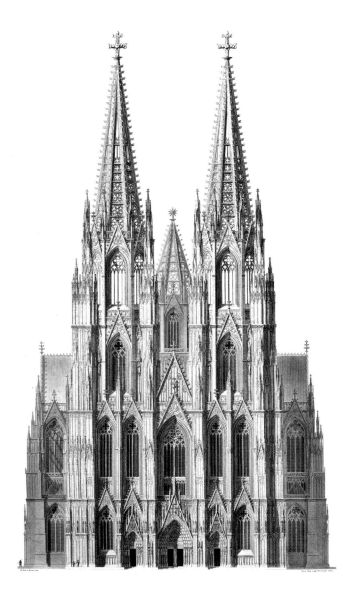

Fig. 91. Sulpiz Boisserée, copperplate of the Cathedral of Cologne. Goethe drew HCØ's attention to these prints, and together with Bindesbøll HCØ visited the exhibition arranged by the Boisserée brothers, bought the very large book, and on his return home had some of the pictures framed to adorn his sitting room. Sulpiz Boisserée 1822–31. RL.

> colours in mind. I told him that like him I did not consider our present knowledge adequate, and that I had made it one of my journey's main aims to study it more closely. We broke up around midnight very content, but without reaching agreement on everything. He expressly wished to see me on my way back, and if possible I shall be pleased to seize the opportunity.[36]

The congress never happened, nor did a second visit to Goethe, but Ørsted was asked to pass on greetings to Oehlenschläger.

Was Goethe as delighted with the visit as Ørsted? For more than a month Goethe could not get their talks out of his mind. He wrote to Boisserée, 'I would have liked to have shared his company for another day. There was a lot to be learned from him and he was a good listener. He is at such a high level of scientific and cultural erudition that I would only have needed to give the rope of the curtain a swift tug to make him appreciate my theory of colours.'[37] Really?

Ørsted and Bindesbøll arrived in their carriage in Munich on Christmas Eve. Sadly, Ritter was not around anymore. Ørsted called on Schelling without finding him at home, but the next day Schelling knocked on his door to congratulate him on his discovery. Schelling said he was no longer preoccupied with *Naturphilosophie*, the progress of which had stopped more than ten years before and would hardly advance until science had made new progress, as had happened with Ørsted's discovery, Schelling added politely.[38] The main attraction during the three weeks in Munich was Josef Fraunhofer's workshop. King Frederik had given Ørsted a free rein to purchase instruments for his collection, and Fraunhofer was ready to deliver the order as Ørsted arrived, though we do not know which instruments he bought. After the discovery of electromagnetism he did not hide his aim of eliciting a theory of light, so of course he had to pay a visit to

> Fraunhofer's wonderful workshop, in which the most perfect binoculars ever are made. Fraunhofer grinds the best optical lenses in the world and surpasses the English by far. He grinds the glass in an entirely new way. He polishes the lenses on a sophisticated machine capable of eliminating the errors of the grinding. He measures the correct rounding of the glasses with instruments that are so perfect that one can trust them to a ten thousandth of an inch. Nothing is left to chance... Fraunhofer is a very quiet and modest person with an estimable character. His face expresses wisdom and profound thinking rather than vigour. He is reticent and aloof and not seeking incidental honours unless they seek him. Although he wa brought up in workshops he has a good scientific knowledge and even reads books in Latin. By contrast Reichenbach, his master, is dedicated to public recognition and has three or four orders of chivalry... and has left the workshop to become permanent secretary of a department.'[39]

Travelling through the German states showed him that his discovery had aroused a stir in all scientific circles and that it had led to new discoveries. Schweigger's multiplier and Seebeck's thermoelectricity were the most significant. Sharing the company of Bindesbøll was a success.

Ørsted's interest in the theory of light and its relation to electromagnetism were subjects he brought forward when opportunity arose as in Fraunhofer's workshop for instance. Now the travellers set a course for Paris, 'the so-called capital of the world',[40] where he had challenged the Laplaceans twenty years before, and challenged them once again by publishing *Recherches* ten years later. By now another ten years had passed. Would the Académie Royale des Sciences recognise him this time?

> I feel a basic difference in the scientific way of
> thinking that I would not imagine to be so profound
> if I had not so often felt its living presence...[1]

38 | 1822–3
The Triumphal Progress Paris

THE TWO travellers arrived in Paris on 23 January 1823 accompanied by a countess and her seventeen-year-old daughter, to whom they had given a lift so as to have female company on their long journey. Ørsted wrote home about their charming passengers. Gitte might have become jealous had Bindesbøll not been in the carriage. 'For I thought you would not scandalise the young man. Still, you will not easily make me jealous, and whether that is due to my confidence in you or my even temper, I shall keep to myself.'[2]

Unlike 1803, when only a tiny room was affordable and had to be shared with Lehmann or Gjerlew, this time Ørsted rented two rooms separated by a hall to enable him to receive visitors in an appropriate way. The light, well-furnished rooms were located in the Hôtel de l'Empereur Joseph (map, ch. 14, fig. 41) in the Latin Quarter near the Palais de Luxembourg on the left bank of the Seine. Ørsted's lifestyle rose with his esteem. The time was ripe to make up for the humiliation of twenty years before. He carried a number of letters of introduction from Princes Carl and Christian Frederik. Also in his luggage were instruments for the demonstrations that would seal the demise of corpuscular theory and the victory of his dynamical theory: disc and bow to generate acoustic figures, piezometer, bismuth and antimony rods, copper vessels, and a compass. Evidence of French recognition of his discovery had been inexplicably slow in coming. How could he finish his many demonstrations and receptions by the month of May, when a more welcoming London awaited him?

When Ørsted had presented his discovery at the Royal Danish Society of Sciences and Letters at the first winter meeting he had touched upon the question of the propagation of fundamental forces. Did it happen in the same way when they were bound in solid matter (e.g. a conductor of platinum) as when they moved in space? Was the motion circular as in electromagnetism, undulatory as in waves of sound and light, or a zigzag as in bolts of lightning? And could the waves of light move in curved orbits? These were huge questions demanding further research. He was convinced, though, that they did not move rectilinearly like gravity. The surprising thing about the propagation of electromagnetism had turned out to be that it was not rectilinear, as Ørsted had expected. He wrote:

> How wide are the prospects opening out now as a consequence of the discovery of one single key property [electromagnetism]? The author has a feeling that he will only proceed slowly in this wide range of investigations. But he may hope that other scientists will give some of the attention they have offered to his experiments to some of the conclusions he has inferred from his discovery. If so, the many big issues seeming to him to soar in the distant twilight will emerge in the clear daylight of truth.[3]

In other words, Ørsted was aware that his discovery had opened up an entirely new world of research. The discovery was his alone, but now a great European project would arise to ascertain how forces propagate, and to find the precise relations between electricity, light, and heat. German scientists had immediately accepted Ørsted's interpretation of the new phenomenon, and two of them, Schweigger and Seebeck, had already advanced this research. In France the response was quite different. Even after the confirmation by the Geneva institute the attitude was sceptical. Only after Arago's successful repetition of the experiment on 11 September in the Louvre had the French Academy given in.[4] There was a conspicuous national difference to be seen in the pattern of reactions. 'I feel a basic difference [between Germany and France] in the scientific way of thinking, that I would not imagine to be so profound if I had not so often met its living presence', he wrote home to Gitte.

True, his protégé in Paris, Schmidten (the bright mathematician and officer resembling Napoleon) had informed his patron that Laplace himself, heading the commission awarding the Napoleon prize, had recommended Ørsted in April 1822 for the small prize of 3,000 francs. But at the same time Laplace had also remarked, 'that it [Ørsted's discovery] did not fall completely under the rubric of the prize because it is not mathematical; nevertheless it was unanimously agreed that an exception was justified due to the significance of the discovery'.[5] This suggested that there were reservations: on the one hand electromagnetism was a significant discovery, but on the other it was not mathematical. To Ørsted himself it was a matter of urgency that the French Academy recognise his theory as he had formulated and proven it in *Experimenta*. The small Napoleon prize was only half a recognition. When an unambiguous recognition failed to emerge, it must be because awarding the great prize (if it still existed) would be an indirect admission that corpuscular theory, despite its mathematical perfection, had collapsed.

Let us briefly compare the two contending systems. After Ampère's experiment (ch. 36) had shown (as he saw it) that electromagnetism is reducible to electrodynamics, thus saving the conventional corpuscular theory, the new French doctrine can be summarised in four points:

1. The forces of physics exist each as an independent imponderable fluid in electricity and magnetism, etc.
2. They are central forces acting as attraction or repulsion in rectilinear motion.
3. Forces only interact with their own kind so that the same polar forces repel each other (+ +, – –, N–N, and S–S), while opposite polar forces attract each other (+ –, – +, N–S, S–N).
4. All physical forces obey Coulomb's inverse-square law and are susceptible to mathematical treatment.

These four assumptions had now been refuted by Ørsted. His thesis on Kant had refuted the corpuscular theory with metaphysical arguments. Physical forces, according to Ørsted (and Kant) are noumenal, that is to say they cannot be observed in time and space, only their effects can. This metaphysical distinction was left unnoticed by his opponents. On the French premisses, Ørsted's counterarguments against the above four points are:

1. The idea of atoms and imponderable fluids is pure speculation (constructed to transform noumena to phenomena and adapt these to a mechanical-mathematical doctrine) and must therefore be rejected.
2. Electromagnetic forces are not central forces. They are neither attractive nor repulsive, but propagate in shocks, and consequently they differ from gravity. They do not act rectilinearly, but transversally in circular motions.
3. Electromagnetism shows that different forces interact. The electrical conductor does not act on anything, but will move a magnet as described in *Experimenta*. Ritter's electrochemistry has shown that chemical and electrical forces interact, and Seebeck's thermoelectricity has proven interaction between heat and electricity.
4. Electromagnetism does not follow Coulomb's inverse-square law and cannot be expressed in mathematical terms.

Ørsted demanded a radical choice between corpuscular theory and his dynamical system. Both could not be true, so one of them had to be false according to the law of contradiction. During the 1820s a drama concerning Ørsted's discovery was staged on the European scene. What was the discovery really about?[6] The Swede J.J. Berzelius and the Austrian J.J. Prechtl asserted that what Ørsted had discovered was not electromagnetism, but that electricity was reducible to magnetism.[7] On the other hand Ampère (ch. 36) maintained that magnetism was nothing but electricity. Both these opinions held that forces of two different elements were unable to interact. Nobody reacted so strongly to Arago's corroboration of Ørsted's experiment as Ampère, two years his senior, who immediately pushed everything else to one side to devote himself to this new field of research and to change its direction. He constructed an electric conductor, the ends of which could pivot in small bowls filled with mercury and connected them to a galvanic battery. Next to it he placed another similar construction, and he found that when the circuit was closed the two parallel electrical circuits repelled each other when the currents flowed in opposite directions, but they attracted each other when they flowed in the same direction. What did this experiment have to do with electromagnetism? There was no magnet, and that exactly was the point. It had nothing to do with Ørsted's experiment. Ampère

THE TRIUMPHAL PROGRESS: PARIS

had discovered something new: electrodynamism, the interaction between two parallel electrical circuits. Ampère now claimed that his experiment had explained what electromagnetism really was by reducing the magnetic effect to an electrical phenomenon. Ørsted, on the other hand insisted that his electromagnetic theory could explain Ampère's electrodynamics, which by the way he fully acclaimed.[8]

So Ampère believed he had proven that magnetism is a circulation of electricity within the magnet. He was a materialist sustaining the idea that each force had its own polar imponderable fluids which could not interact. When Ørsted had empirically shown that they did, this

Fig. 92. A.-M. Ampère and D.F.J. Arago discussing an experiment. Drawing by P. Fouché. *Scientific American* 1989, 73.

interaction must have been caused by the same fluids. Hence, two electrical fluids produced a magnetic effect by setting the steel particles of the magnet into motion. Ampère imagined that this effect arose in magnets when electric currents flowed along the axis of the magnet. From the electrical conductor a rectilinear effect radiated on the magnet. Why rectilinear? Because Ampère was convinced that Ørsted had misinterpreted his own experiment by ignoring earth's magnetism. If this source of error was isolated by using an astatic magnetic needle instead of the usual compass he achieved a deflection at a right angle. Thus, Ampère thought he had disproved Ørsted's theory of the circular motions of the electromagnetic field. Earth's magnetism had caused the error, he claimed. Ampère's electrodynamic theory stayed within the framework of corpuscular theory and Coulomb's inverse-square law.

Ørsted could not comprehend this theory. How could electric fluids set steel particles in motion so that they behaved like magnets? Ampère used Arago's experiment to support his view, and Ørsted had also shown that the electromagnetic field attracted iron filings. But that was hardly the same thing as saying that the electrical current attracted unmagnetised steel particles. To Ørsted, electrodynamism was explicable within the framework of his electromagnetic theory. Ørsted held that electromagnetism was a double circuit of polar forces around the conductor, one clockwise, the other anticlockwise. Consequently, when electrical forces in parallel circuits propagated in opposite directions, different forces were meeting in the field, which caused repulsion, while the same forces propagating in the same direction caused attraction.

Just as Ampère could not understand Ørsted's theory, Ørsted could not understand his. During the three months of Ørsted's stay in Paris they met several times to persuade each other, but they never succeeded. They had seen each other at parties, and on 4 and 10 February Ørsted was invited to Ampère's house together with Arago, Fresnel, Chevreul, Dulong, and others. Ørsted reported home:

> He is a very awkward disputant and does not understand how to grasp grounds or establish his own. Nevertheless, he has a profound mind. He has three significant galvanic apparatuses and his equipment to show his experiments is very complex, but what happened? Almost none of his experiments succeeded. Only when several of the spectators had left did he finally succeed in bringing about a couple of them. He is terribly confused and as awkward an experimenter as a disputer.[9]

Ampère had had a nerve-racking childhood and adolescence. His father was guillotined by the revolutionaries, his first wife, whom he loved deeply, died young in his arms, his second marriage was a disaster that was quickly annulled, and his two children gave him nothing but trouble. His relationship with 'the ultras' in the Académie went badly, and already in 1804 when a personal crisis became worse he sought spiritual guidance from people who, unlike the French so-called 'ideologues', had found more in life than mechanics and matter that can be weighed and measured.

This search led him to Kantian metaphysics, which at that time was only available in incomplete and partly misunderstood French translations. As Ampére wrote nothing systematic about his philosophical views, they are only known from some lectures he delivered at the

École Normale, and his latest biographer, James Hofmann, can only supply limited information about his views of physics and metaphysics. However, the main outline is clear. As for Ørsted, the Kantian concepts of phenomenon and noumenon became decisive for Ampère's way of thinking. According to Hofmann he used them erroneously, taking them to mean 'appearance' (phenomena) as against 'factual' (noumena).[10] An example will illustrate: apparently the Earth is flat, and apparently the sun rises in the east, but as a matter of fact the Earth is round, and the sun does not rise at all, it only appears so to us, because we find ourselves on planet Earth, which rotates around its own axis. These are the kinds of corrections that Ampère is said to have considered the objectives of science. He was not aware that to Kant the noumenal is a negative concept that can only be used as a regulative principle, but not to produce science about the factual. Noumena do not exist in time and space and cannot become objects of sense impressions. Hence his mathematical description of electrodynamics is no empirical description of the noumenal, and electrodynamics makes no progress for physics in relation to corpuscular theory.

At this point Ørsted agreed with Kant. Electromagnetism as a force was not a phenomenon, but a noumenon, for it could not be sensed in time and space per se. But its effects gave sense impressions, the deflection of the needle according to the rule of thumb. The cause of this will forever remain hidden. Consequently, he wrote explicitly: 'I repeat here what I have already said in previous works, that by electrical forces I understand nothing but the unknown cause of electrical phenomena whether they are bound to a free substance or are independent activities'.[11] Whether the force is one or the other, or even something else, is not a question for science to decide.

Two and a half months later Ampère made another attempt at converting Ørsted to his electrodynamical theory. He had invited almost all the members of the Académie who opposed Laplace as well as a few of his own supporters.

> He has thought out a very artificial theory of my discovery and was very unhappy that I still stick to my own, which is very simple... After dinner talks began and lasted for almost three hours... Even Ampère's two disciples declared that my theory was capable of explaining all phenomena. They assert that Ampère's is, too, and as his theory is nothing but the reverse of mine, in so far as he has removed the circuits I have discovered from the conductor to the magnet, I presume it will be difficult to find an altogether decisive objection to his theory.[12]

The discussion between Ampère and Ørsted spread from the laboratory to the universe. Was electromagnetism a universal phenomenon as well, and if so how did it work? Everybody knew of electrical tensions in the atmosphere that were discharged as thunder and lightning. Equally, everybody was aware that the Earth has magnetic poles, although it was uncertain how many poles there were and where exactly they were located geographically. As we remember, in 1803 Ritter had 'discovered' that planet Earth is a galvanic battery with a positive and a negative pole. Unfortunately, 'the discovery' was disclaimed the year it was made. Against this background it came as a surprise that Ampère now asserted that the freely pivoting electrical conductor oriented itself toward east and west, and so transversely to the magnetic meridian. The reason was, he said, that an electrical circuit parallel to the equator of the

Earth had to exist, primarily due to alternating layers underground and only secondarily influenced by sunlight.

Yet uncertainty about these universal phenomena was considerable. At that time Earth magnetism was a research project engaging Ørsted and Christopher Hansteen. Biot and Gay-Lussac had embarked in an air balloon, from which they intended to examine the atmosphere and earth magnetism. Alexander von Humboldt was preoccupied with similar problems. Ørsted frankly admitted that this research was skating on thin ice. His assumption was that sunlight forms an electrical orbit moving from the east towards the west. Due to earth's rotation around the sun this orbit changes throughout the seasons. North of the Arctic Circle it was freezing and pitch dark during the winter, and even if nobody had been anywhere near the South Pole, it was generally believed that the same applied there only vice-versa. He felt convinced that northern lights occurred when sunlight and consequently the electrification of the atmosphere of the earth was intensified during the summer and would be seen as long as the dark nights prevailed. Unlike Ampère and others Ørsted did not believe that the electrical orbit was a product of alternating layers in the core of the Earth, but thought that it was a surface phenomenon initiated by heating and evaporation by sunlight alone, as well as by some so far unexplained chemical effect in the east-west direction on the surface of the Earth. Planet Earth, of course, consisted of sea and land, and their different surfaces absorbed sunlight differently just as its intensity varied geographically with the seasons. The electrified surface of Earth formed a magnetic field in its core, because all substances to a higher (e.g. iron and nickel) or lower degree were susceptible to magnetisation. As the power of the sun is so enormous, the total magnetic effect will be significant and form a magnetic orbit that according to Ampère's law (and rightly so) must locate itself in the direction of 90° that is north-south. In this way Ørsted had indicated the source of the irregular reactions of the compass needle and the variations at different places on the Earth's surface. Ørsted was at that time helping his Norwegian friend Christopher Hansteen charting the isogons of Earth magnetism, i.e. lines where the declination is the same, and its isodynamic lines, i.e. lines where the intensity of magnetism is the same.[13]

In 1823 Biot and Laplace still stood on top of the scientific Parnassus of France. Biot was a chemist and had been protected by Laplace, who had assigned him research projects within galvanism and appointed him leader of the First Class galvanic commission, awarding the Napoleon prize. It was Biot who had encouraged Ørsted to demonstrate Ritter's experiments in 1803 (ch. 15). After the restoration their positions were about to be undermined by younger scientists, who looked on the old guard of the Société d'Arceuil with growing scepticism. The period when Laplace paved the way for the careers of young scholars by issuing prize questions and rewarding those who loyally based their responses on his doctrine and by reciprocating their loyalty with posts at the École Polytechnique, was running out. Indeed, an opposition of young talents against Laplace was about to organise itself, and it felt hampered by his nepotism and criticised his doctrine.

Schmidten had kept Ørsted up to date about the intrigues in the Académie Royale des Sciences. He estimated that the Académie was split into two parties which he labelled in terms taken from the political battle ground, 'ultras' and 'liberals'. The ultras were the distinguished

Fig. 93. Pierre-Simon, Marquis de Laplace (1749–1827), mathematical physicist. Signed lithograph after a portrait by an unknown artist, presented to HCØ in Paris 1823. DTM.

Marquis Laplace and the many older members who supported his view of science. The impetuous young scientists pointing out the weakness of the corpuscular theory and making unorthodox experiments were the liberals, of course. Ørsted took over these labels, obviously sympathising with the opposition.

Prominent among the liberals was Augustin Fresnel, a physicist opposing Newton's and Laplace's emission theory of light and working on the rival theory established by Huygens and Euler, who conceived light as wave motions in the ether. Fresnel won Arago, Ampère, and Petit over to his side. Fresnel and Ampère lived in the same house and followed each other's experiments. Another physicist, Joseph Fourier, developed a mathematical theory of the propagation of heat in solid matter while rejecting the existence of Lavoisier's and Laplace's 'caloric'. The publication of his controversial thesis in the periodical of the Institut National had been 'delayed' for ten years. Chemists L.-J. Gay-Lussac and L.-J. Thenard working on electrochemistry also took a critical stand against the corpuscular theory. Ørsted's friend from 1813, M.E. Chevreul, specialising in the chemistry of organic matter had publicly supported *Recherches* as soon as it had appeared (ch. 28). Some of these liberals had joined themselves into a network. Gay-Lussac and Ampère for instance had long favoured Fresnel's theory of light.

Fig. 94. Two leading members of Académie des Sciences, Paris, the mathematician P.S. de Laplace (1749–1827) and the chemist C.L. Berthollet (1748–1824). Caricature by Louis-Léopold Boilly (1761–1845). Bibliothèque de l'Institut RMN, Paris. Photo by Gérard Blot.

Some of them (Dulong, Arago, Fresnel, and Petit) knew each other from the École Polytechnique, where they had been examined in Laplace's doctrine. Others were spread around in the provinces, where they sharpened their criticism against Napoleon's old guard. It was this young opposition to which Ørsted had appealed when he challenged the elite of the Académie for the third time. This appears from his mailing list of recipients of *Experimenta,* compared with the people with whom he corresponded and whom he saw in Paris, and would in due course get elected as foreign members of the Royal Danish Society of Sciences and Letters. These liberals would no longer put up with being governed by a generation of opportunists.

This scheme only requires a few comments. Petit was part of the liberal network, but died in 1820. Victor Cousin was a philosopher and spokesman of Kantian philosophy. He was admitted into the Academy when epistemological questions were allowed on the agenda after the fall of Laplace. Ørsted met with Cousin and they talked about the turn of the tide taking place in the Academy.[14] As already indicated, Ørsted's mailing list is incomplete and must be taken with a pinch of salt. The number of times Ørsted associated with the liberals is a minimum figure derived from his letters to Gitte and Anders. The scheme shows the connection between Ørsted and those who challenged Laplace. It was predominantly the opposition that Ørsted saw, corresponded with, and had elected as foreign members of the Royal Danish Society. But it was the ultras headed by Laplace he intended to convert and hence he had to see the most influential of them as well.

THE TRIUMPHAL PROGRESS: PARIS

Fig. 95. The order of the French Legion of Honour, 'Honneur et Patrie' bestowed on HCØ in Paris 1837. DTM.

During his stay in Paris Ørsted overheard a great deal of academic gossip about scandals. Count Frédéric Cuvier, the zoologist, to whom he had been introduced by Prince Christian Frederik, was said to have sought to ingratiate himself with the new counter-revolutionary establishment by converting to Catholicism in order to achieve his goal of being appointed Vice-Chancellor of the reopened Sorbonne University. When the public came to know his ulterior motive he lost his good reputation. Soon after Ørsted met him he escorted him to a session in the Académie Royale des Sciences at the Hall of Caryatids in the Louvre. 'When I had sat down among the audience one of the functionaries stepped forward proclaiming, Monsieur Ørsted de Copenhague is here and is kindly asked to take his seat among the members, which I did.'[15]

He had supper with Laplace 'enjoying one of the most pleasant days here. His company is...exquisite...I found the conversation with Laplace very interesting. After the meal we had a long talk on French literature, with which they are very well acquainted.' Ørsted had become adroit at turning a conversation towards topics where he could lead. Now he charmed the party by airing his predilection for Pascal which flattered their patriotism. He

	LIBERALS ACCORDING TO ROBERT FOX[25]	ULTRAS ACCORDING TO ROBERT FOX	RECIPIENTS OF *EXPERIMENTA* ACCORDING TO MAILING LIST	NUMBER OF TIMES ØRSTED MET THEM IN PARIS	IN HCØ'S *CORRESPONDANCE AVEC DIVERS SAVANTS*[26]	FOREIGN MEMBERS OF THE ROYAL DANISH SOCIETY[27]
Ampére			+	6	–	–
Arago			+	4	+	+
Biot			+	3	–	–
Blainville			+	2	+	+
Boisgiraud			+	–	–	–
Fresnel			+	6	+	+
Hachette			+	2	–	–
Laplace			+	1	–	–
Petit			+	died 1820	–	–
Savart			+	–	–	–
Fourier			–	4	–	+
Dulong			–	3	–	+
Chevreul			–	8	+	+
Poisson			–	1	–	+
Gay–Lussac			–	1	–	+
Thénard			–	3	–	–
Cousin			–	2	–	+

delighted in 'the friendly way of socialising with the greatest men of Europe'.[16] Twenty years before he had felt like a sparrow among hawks, if he were admitted to their circles at all. Now he conversed with them in their own language about their own literature as their equal, whether they were liberals or ultras. Ørsted could not help feeling honoured by the old celebrity.

> He [Laplace] is a great old gentleman in France, like Goethe in Germany. Like him he is understood to have a high rank in bourgeois society being a Marquis and possessing superior posts of honour... True, their special fields differ, one a mathematician, the other a poet... Just as Goethe's poetic genius approaches science, Laplace's mathematical spirit approaches philosophy and rhetoric. Even if one is disinclined to subscribe to the scientific opinions of the former and the philosophical opinions of the latter, their great spirits shine in the territories, whose borders they have crossed... He talked a lot about my experiments on the compression of water expressing his desire to see them.'[17]

Obviously, Ørsted was charmed by Laplace. Nevertheless, his discovery of the undiscoverable electromagnetism was perhaps one of the most important contributions to his fall. The liberal opposition welcomed *Experimenta* as a gift from heaven. It precipitated the shift of the centre of power and thinking in Paris in a way that would soon dethrone Laplace.

At the beginning of March Ørsted was invited to demonstrate two different experiments to the mathematical class in the Louvre. 'I made my first attempt at addressing a big audience in French; reliable friends have told me that I did not get away with it badly.' Firstly, he showed Seebeck's thermoelectrical experiment combined with his own on electromagnetism. Ørsted proudly told Prince Christian Frederik about his demonstration. The magnetic needle had deflected as expected when he heated and cooled the soldering spots between the alternating metals of the ring functioning as a galvanic battery.[18] Three fundamental forces operating at the same time in front of the sceptics! Secondly, he demonstrated his experiments on compressibility with his own piezometer. These experiments were well suited to satisfy the Laplaceans' preference for exact measurements and comparable figures. 'This device is gaining acclaim and Pixii [a renowned French instrument maker] has already started to make them for several physicists here.' This was probably the first time since Ole Rømer's days that French instrument makers were working from a Danish prototype. 'I feel that I have slipped into praising myself and my happiness a little, or rather too much... but I hope it is all right with you, my Gitte, who I know likes to hear it.'[19]

Ørsted's discovery prompted the fall of the ultras and the already shattered corpuscular theory, but among the liberals gradually taking over the leadership of the Académie Royale des Sciences his experiments immediately caused divisions. Like Arago, some envisaged the opening of an entirely new field of research, while others like Ampère tried to force electromagnetism into the framework of the old paradigm. Ørsted did not receive the great Napoleon prize, to be awarded to a scientist whose discovery in the field of galvanism had introduced *une époque mémorable*. On the other hand his discovery had contributed to bringing about a change in the whole structure of French science. Laplace's *époque mémorable* was over.

Ørsted's sojourn in Paris ended in late April with a big public meeting held by the four royal academies together. Firstly, Joseph Fourier, secretary of the Académie Royale des Sciences (the previous First Class), lectured on recent progress in the mathematical sciences, 'on which occasion he also mentioned my works with all the acclaim desirable'.[20] Secondly, the secretary of the Academy of the Arts spoke of the sculptor A. Canova, on whose dancing girl, wearing such lovely loose drapery, Ørsted had reported in enthusiastic phrases to Sophie, his sister-in-law. Thirdly, a member of the Academy of Languages and History delivered 'an utterly one-sided paper on the one-sidedness of Egyptian art', and finally the secretary of the Academy of Aesthetics recited a poem. After the extreme specialisation and utilitarianism to which the ideas of the Enlightenment had been reduced by the revolution, Ørsted was thrilled to witness that the French elite had adopted some of the favourite ideas of Romanticism.

Ørsted took time to show Bindesbøll around in Paris. There were still conspicuous traces of the struggle between revolutionaries and clerics. Bindesbøll had to see the Conservatoire des Arts et Métiers, where J.A.C. Charles, in whose laboratory Ørsted had learnt so much, but who was now on his deathbed, had taken collections of instruments belonging to guillotined or emigrated aristocrats from revolutionary vandals into a church for safekeeping under cover of the darkness.[21] Other churches were being restored to their purpose, but polytechnic students

Fig. 96. The Hall of Caryatids in the Louvre, where the Académie des Sciences convened. After a revolutionary intermission, when most members of the Académie went underground and met informally in the Société d'Arceuil, a village a few kilometers from the city centre of Paris, the new Institut National met here again with First Consul Bonaparte heading the First Class. Copperplate print by Berthault after a drawing by Girardet. J.A. Fredericia 1918, 122.

with Jacobin sympathies tried to obstruct Jesuits from celebrating the mass by smoking them out with homemade stink bombs. 'Now the authorities have taken steps to prevent outbreaks of disturbances in the churches by sending a detachment of gendarmes to maintain order during the service', Schmidten reported to Ørsted. 'Some students have concocted a substance that produced such an enormous stench when let off that nobody could stand staying in the church.'[22]

During his first sojourn in Paris Ørsted was tempted to cross the Channel to take a closer look at the workshop of the world. Despite encouragement from Manthey he had dropped the idea, because (as he wrote) he could not keep filling Sophie Probsthein with empty promises. Now, things were different. The Royal Society was waiting for him. But if he went there he would fail to deliver his lectures at the University. There were two potential substitutes: Professor Zeise and amanuensis Forchhammer; both had offered themselves, but neither of them was a physicist. To choose one might upset the other, and Ørsted resented pushing any of his disciples away. In April he made his choice in favour of Forchhammer. He submitted his application to the Board of Directors and asked Forchhammer to explain the circumstances to Engelstoft and Mynster.[23] Zeise had to do with a letter of comfort. Ørsted reassured him that he considered both of them friends and colleagues. Being Ørsted's protégé was both advantageous and cumbersome at the same time. The choice of Forchhammer would have been an insult to Zeise if the substitution were concerned with chemistry. But Zeise lacked teaching skills and it was Forchhammer who had translated Ørsted's textbook on mechanical physics into German, and hence he was the obvious choice. Besides Forchhammer had no livelihood, and 'I take it as an obligation towards my old disciples to make as many contributions as possible towards promoting their happiness'. By now Zeise had obtained his professorship and thus had no reason to resent his patron's decision.[24]

Ørsted left Bindesbøll in Paris with the architect Gau, while he crossed the Channel on his own. Their carriage was left at Calais.

Bread-and Butter-Studies
are not really going on at this University.¹

39 | 1823
The Triumphal Progress Britain

PRINCE CHRISTIAN Frederik had asked Ørsted to hand over a small box and a letter to Humphry Davy and this was the first thing he did on his arrival in London in May 1823.² Two years earlier Davy had been elected a foreign member of the Royal Danish Society on Ørsted's initiative.³ And Ørsted had been notified about his membership of the Royal Society of London a few months before by Thomas Young,⁴ Thus the ground was well prepared for Ørsted's first visit to Britain.⁵

The voyage by steam from Calais to Dover (spelt 'Dovre', the name of a Norwegian mountain range, by Ørsted) was tranquil and he had good company, amongst others two French ladies, one of them married to P. Thénard, the chemist. They went by coach to London on well-maintained roads 'as smooth as a dance floor' thanks to turnpike money. London, already a city of more than a million inhabitants, had flagstone pavements, and he was taken aback by its gas lighting and magnificent shops. Herschel took him sightseeing in a rowing boat on the Thames explaining how the bridges were constructed on wide arches of cast-iron, so that only a few pillars impeded the traffic on the river. Had he stayed only a few years more, he would have been able to go by stagecoach through the tunnel under the Thames.

The zenith of his triumphal progress was on 8 May. The evening started at the Royal Society Club with a dinner at the Crown and Anchor Tavern in the Strand near St Clement Danes, and close to Somerset House where the Royal Society had its meeting hall. Davy was president of both club and society and had invited Ørsted. The tavern was known as the best establishment for dining in polite society. It was far from being an ordinary pub with a bar and chairs for a few

THE TRIUMPHAL PROGRESS: BRITAIN

Fig. 97. Sir Humphry Davy (1778–1829), President of the Royal Society. Oil painting 1821 by Sir Thomas Lawrence (1769–1830), NPG.

dozen people. Several notable associations frequented this establishment, associations for artists, scientists, Freemasons, and citizens of liberal not to say radical political views. Dishes were served up punctually for a couple of thousand guests and being late was considered an embarrassment.[6] Ørsted entered his name in the guest book. According to the tradition of the Society, toasts were proposed to the King, to the arts and sciences, and finally to the Royal Society.[7]

Davy left the tavern a little before the rest of the party, hurrying along the Strand to Somerset House to change into full dress, before opening the meeting. 'Strange to see that the English stand on ceremony so firmly despite loving their freedom so deeply. No doubt they stick to etiquette in public matters for the sake of freedom, and from there they have transferred the custom to other purposes.'[8] Now Ørsted followed the other Fellows a few hundred paces to the magnificent, neo-classical Somerset House, where the Royal Society was accommodated in the wing facing the Strand. Ørsted ascended the broad, winding stairs and passing through the antechamber he entered the great meeting hall with its rosette ceiling painted in light green and rose colours, with stucco medallions of Charles II, founder of the society in 1660, and George IV, protector since 1820. On the walls hung portraits of past celebrities such as Boyle, Newton, and Banks. The President, Humphry Davy, was sitting on a throne on a platform in front of the open fireplace with lit candelabra on the mantle piece. Thomas Young, secretary, and Professor Buckland, reader for the evening, were sitting at a high table, on which were placed the society's mace and two candlesticks, facing the audience of Fellows, sitting on benches along the walls like a congregation in a church.[9]

Fig. 98. Anonymous aquatint of Royal Society's assembly hall in Somerset House, The Strand, London, c1850, but as it appeared in 1823, when HCØ received its homage. The President, Humphry Davy, sits elevated on the podium on a chair hiding the fireplace. At the table in front of him sit the secretary to the left and the lecturer to the right. On the mantlepiece—hidden by the chandelier—a bust of the founder of the Royal Society, King Charles II. RS.

Now Davy stood up, took off his hat, and said a few words of welcome to the Fellows and particularly to the new foreign member, and the ceremony of admission began. 'I had to enter my name in a big book and approach to the presidential throne. The President stood up and delivered a brief speech very honourable to me, while holding my hand in his own that he had offered me as I approached him.'[10] Next Professor Buckland recited the first part of his paper proving the Biblical account of the Flood correct by finds of animal bones in some German caves.[11]

Over the following weeks Ørsted was several times invited to dinners at the Royal Society Club at the Crown and Anchor. He had to get used to the custom that 'nobody offers you anything; you just ask for what you want of the few dishes available. As to drinking it is as frugal as my medical friends could wish for. My red wine from Frölich will taste twice as good after this journey.' The Ørsted family was in the habit of buying more than a hundred bottles of red wine at a time from wine dealer Frölich in Copenhagen. On 14 June the Society was to undertake the periodic inspection of the Greenwich Observatory, and being now a Fellow Ørsted obviously joined in. In the morning he had shown John Herschel and Charles Babbage some acoustic experiments that a shy young man by the name of Charles Wheatstone had done. After lunch the three of them were rowed down the Thames in an hour and a quarter to Greenwich. On the way he saw steamships, one of them completely made of iron, and there was a crowd of vessels and workers on the river and in the docks. He was impressed by the old castle at Greenwich, 'more beautiful than any of the castles owned by the royal family', and now converted into a 'hospital or rather a refuge for old and invalid sailors'.[12]

It was not the inspection itself that was the attraction for Ørsted, but the dinner afterwards. The atmosphere was merry, toasts were proposed, and as usual a ritual accompanied them: the one proposing the toast delivered a short speech, and the recipient responded. Davy toasted Prince Christian Frederik, who had taken part in the previous year's excursion to Greenwich. Then Davy announced that John Christian Oerstedt, Secretary to the Royal Danish Society of Sciences and Letters in Copenhagen, had honoured the Society with his presence, and the two Danes definitely deserved a toast. So, Ørsted abiding by the tradition got to his feet saying:

> 'I thank you heartily for this honourable toast. I regret that I do not speak the language of this happy nation with the ease I would like to express my feelings; but I shall have to confine myself to uttering the wish that this nation, the country of Bacon and Newton, that has fostered so many distinguished men enlightening the sciences may continue to do so for the progress of science and the honour of the country.'[13]

To give a toast to a large audience in English was not an easy task for Ørsted, who had only been in the country for a month and a half and found the pronunciation of the language to be exceptionally void of rules. When he reproduced English words to Gitte he felt obliged to add phonetic notation, such as stagecoach: '*stædjekotsch*'.[14] It would take years before he was able to write a letter in almost flawless English without effort. If he could get away with it he preferred German or French when talking with Herschel and Young (German) or Davy (French). But even if the correct pronunciation was wanting, he managed to make himself understood thanks to the indulgence of his colleagues.

'Perhaps my speech sounds artificial to you', he admitted to Gitte, 'but since I had to speak unexpectedly, you may rest assured that it sounded more natural. One of my friends, Babbage, reassured me that everybody had found my speech to be the best of all... I always subtract a great deal from such compliments, concluding that it was all right...'[15] However, he rose early and took language lessons every morning from six to eight. 'I am pleased to have made that decision and to take it so seriously, because I feel that it makes me more popular here. Perhaps I take another advantage from this. I have a well-founded hope of becoming attached to one of the great encyclopedias and to be paid fifteen guineas a sheet which in our currency is one hundred and twenty rixdollars.'[16]

The row back to London took place in one of the Admiralty's boats that happened to have been built for the Danish Prince Jørgen, the one with the famous 'Prince of Denmark's March', the admiral and consort of Queen Anne, the last of the Stuarts. He was said to have worked on it himself.

*

University College London was not yet established in 1823. There were the Royal Society and some private institutions for research and training such as the Royal Institution and the London Institution, but still the renowned medieval universities of Oxford and Cambridge were the only universities in England. Of course there were also three Scottish ones, St Andrews, Edinburgh and Glasgow, to which we shall return. At the end of May Ørsted went to Cambridge. The only thing he knew beforehand he had read in some German books that rather one-sidedly described it as old-fashioned and monastery-like. His letters to Gitte offered a different impression. Of course, the colleges were old-fashioned, they were medieval, and if the buildings were not former monasteries they were at any rate designed according to such ground plans.

Oxford and Cambridge were exposed to contemporary criticism, although for reasons contrary to those applied to the French institutions of higher learning. Napoleon had eliminated classical languages and the humanities by simply closing the French universities, because according to revolutionary rhetoric they were breeding grounds for clerical tutelage and for reducing people to a state of superstition. Instead the French established new institutions to serve purely technological and utilitarian objectives. In England the study of technology and natural philosophy were supported not by public money, but exclusively by private institutions, funded by beneficiaries. Ørsted noticed that education at the universities consisted of three years cramming of classical languages and mathematics without aiming at the students' future professions. Preaching was learnt as a curate in a parish, passing sentences according to common law was learnt in the lawyer's office or in the courtroom, and treating patients in a hospital. Civil servants were trained by the administration. Or as Ørsted formulated it upon his arrival at Cambridge: 'Bread-and-butter studies are not really going on at this university'.[17] The sciences had no professorships. So in that respect the situation in England was even worse than in Copenhagen which had two, Ørsted's and Zeise's.

University tuition was generally considered boring and irrelevant to practice. What really occupied the students' time and attention were hunting, riding, rowing, brawling, drinking, and gambling. These activities were only occasionally interrupted by preparations for examinations, because students were merely expected to parrot some Latin phrases to graduate as a bachelor of arts. Of course, this is an exaggeration, but still not unlike the experience Charles Darwin remembered from his years in Cambridge between 1827 and 1830, when he spent his time hunting with a rifle for hares and a net for butterflies while his textbooks accumulated dust.[18] Then as now, those who work the hardest are not necessarily the most promising students. The fact that Trinity College had fostered Bacon in the seventeenth and Newton in the eighteenth century should not lead us to believe that the natural sciences had a strong position at Cambridge in the nineteenth century. Only in 1819, a few years before Ørsted's visit, did natural philosophy gain a platform at Cambridge, in the shape of the Cambridge Philosophical Society, a voluntary association established by a pressure group working long-term to obtain recognition for the sciences on an equal footing with mathematics and the classical languages. So there were obvious differences between the countries of Europe as far as the position of sciences at institutions of higher learning are concerned. Excluded from universities as they often were, they had to squeeze themselves in like the Natural History Society in Copenhagen, die

Naturforschende Gesellschaft in Geneva and Jena, and the Cambridge Philosophical Society; this was very different from France.

Still, Ørsted was infatuated with Cambridge. 'I visited Trinity Colledge [phonetic notation again] most often. I took great pleasure at sauntering over the spacious courtyard with green places ["lawns" was not yet part of his vocabulary] or through marvellous archways and at meeting tranquillity so much appreciated after a long period living in the noise of London... As I strolled around I thought to myself: I wish Gitte were here, but soon after I changed my mind, for had you really been here, you should no doubt have urged me to become professor at Cambridge.' And when she responded affirmatively: 'Had I been there I should have wanted to dwell there', for even if she and the children had moved outside the ramparts of Copenhagen for the summer and even if her husband was away, there was such a coming and going of people around her, that she never found a moment of peace and quiet. Hence, she would only be too pleased to emigrate to Cambridge or some other beautiful and peaceful place, if that would ensure a quieter life with her spouse.[19] But that was not the way he meant it, her busy husband commented: 'Wherever I go, I am immediately socialising, and the same applies to you, although you find it less amusing. Seize my resolution: Take pleasure in what you cannot change!'[20]

Ørsted arrived at Cambridge in the middle of examinations, finding that the University of Copenhagen was less demanding in proficiency in Greek than Cambridge, but in return the English students had spent all their time studying Greek, Latin, and mathematics. On the one hand the Cambridge curriculum was rather one-sided, but on the other hand it might benefit Danish students to acquire better foundations for their special studies.

On arrival he was invited to dine at high table at Trinity College together with Professor Whewell (mathematics), master of the college, and tutors Cumming (chemistry), Sedgwick (geology), and Henslow (botany); the latter two were Darwin's teachers. Ørsted did not mention to Gitte that these were the scholars who had founded the Cambridge Philosophical Society in 1819, where Cumming had given a paper on 'the effect of the galvanic fluid on a magnet needle' (that is to say Ørsted's discovery) as early as 1821. It was also for this society that Babbage a few years later wrote his sensational book on the decline of British science.[21] On his last evening at Trinity College Ørsted was involved in a lively discussion with Whewell on recent philosophy; we do not know the details, only that they did not agree, but still it had been interesting for both and for the rest of the party to listen to.[22]

Back in London Ørsted was busy seeing all that his hosts wanted to show him. He mentioned their names in his letters to Gitte without further details: Davy, Faraday, Wollaston, Young, South, Herschel, Babbage, Barlow, Pepys, Birkbeck, Children, Wheatstone, and Perkins. This list shows that he saw most of the people interested in research into electromagnetism on an experimental basis. This circle of scholars frequented the lecture halls of two flourishing institutions situated at opposite ends of London: the Royal Institution in Albemarle Street off Piccadilly was the older (1799). Its daily leader, Humphry Davy, was a full-time titular professor.

The Royal Institution possessed a significant collection of instruments, and its lectures were followed up by physical and chemical experiments in its combined laboratory and lecture hall.

Fig. 99. William Whewell (1794–1866), historian and philosopher of science. Lithograph by Eden Upton Eddis (1812–1901) after a painting by William Drummond, 1835. NPG.

It was not a charity. Its owners wanted useful knowledge for agriculture and industry in return for their investments. Consequently, they regarded purely scientific research as an unwarrantable luxury, unless it was also spectacular and entertaining, for they did not want to pay to be bored. Davy, who succeeded Young, was caught in the delicate dilemma between the owners' demands for useful solutions to their practical problems and his own research projects. Luckily, his invention of the safety lamp, saving the lives of thousands of miners, gave him breathing space to pursue the electrochemical experiments that excited him. In 1813 he employed Michael Faraday as an errand boy and handyman. Little was he aware that this poor and uneducated lad would exceed him in fame in less than twenty years. Faraday, too, had to learn to manoeuvre cleverly between the demands of the board and his own research projects, but fortunately he was an exceptionally gifted experimenter, and his electromagnetic experiments had an intrinsic entertainment value attracting a large audience. An anecdote has it that Sir Robert Peel, the Prime Minister, during a visit to the Royal Institution pointing at a device on its way to become the first dynamo in the world wanted to know what it was going to be used for. 'I have no idea' Faraday is said to have remarked, 'but I bet that sooner or later your Government will put a tax on it.'[23]

At the other end of London, in the City, the London Institution was situated at Finsbury Circus, another dignified building in neoclassical style 'like a castle'. It was rather similar to the

Royal Institution except that it had been founded by business capital pursuing the mutual advantages of cooperation between scientific and commercial interests. Science was poor in pecuniary terms, but rich in erudition and prestige. With the philistine commercial interest it was the other way round—what complementarity! Isolated from one another science would miss the support of capital and its own utility, and the business world would lose its fragile prestige, according to the inaugural speech. The London Institution offered series of lectures, 'attended by both sexes and whence much splendid and useful—although not profound— knowledge is disseminated'. But if the lectures were not sophisticated from a scientific point of view, they did fascinate an audience that preferred spectacular experiments and edifying physico-theological lectures.

Neither of the institutions was geared to fulfilling the needs of common mechanics, and no training in practical craftsmanship was offered. The intention was for participants to dabble in science in general, such as phrenology, physiology, zoology, and botany, as well as to combine the useful with the agreeable. To this end the London Institution kept a collection of instruments such as a galvanic battery with two thousand discs and a large galvanic trough apparatus that W.H. Pepys used for his experiments. Otherwise the laity were preoccupied with the therapeutic applications of electricity, to cure gout for instance, as was practised at hospitals in London. Peter Barlow, having a laboratory attached to the arsenal at Woolwich, worked on magnetism and on the isolation of compasses, which went out of control on board modern ships made of iron. The London Institution used gas for lighting, but there were constant leaks in the retort, threatening to poison the public, and Ørsted claimed it needed far more repairs than his own at home.

The poignant difference between the two institutions was that while the London Institution did not employ a permanent professor but used freelance lecturers of varying quality, the Royal Institution employed such eminent researchers as Davy and Faraday.[24] Ørsted was pronounced an honorary member of the latter.[25] He probably had sufficient knowledge of the internal affairs of the institution to have a hunch as to who played first fiddle. Not until autumn 1821 did the thirty-year-old Faraday have a chance to step out of Sir Humphry's shadow. He was requested by the editors of *Annals of Philosophy* to write a chronological report on the new field of research of electromagnetism resulting in the article 'A Historical Sketch of Electromagnetism', which was an illuminating introduction to the discipline Ørsted had founded. It was anonymous, meticulous, and replete with reservations when touching upon the controversy between Ørsted and Ampère. For instance the author was not sure he had understood Ørsted's theory correctly, so he wisely abstained from commenting on it. He distinguished sharply between facts by which he meant unequivocal experimental results, and theories, which were by and large rejected as figments of the imagination. Faraday was impressed by Ørsted's ability to hypothesize about his dynamical project so many years in advance, which he took as proof that his discovery was not haphazard but due to the thorough intellectual preparation of the experimenter.

Nevertheless, when Davy called Ørsted 'somewhat of a German metaphysician' and Faraday characterised the discovery as a deduction from his overall theory and not the other way around (induction) as it ought to be, these judgements were not meant to be compliments.

Theory in the British empirical tradition was generally seen as a preconceived idea leading experimentation astray. No one had any knowledge of Kantian philosophy, no one read any German, and had people been asked why, they would probably have retorted that they had hardly missed anything. For the research taking place in the Royal Institution and the London Institution followed the empirical tradition. First, it was inadmissible to construct concepts that did not refer to experimental results. Secondly, one should not hypothesize about physical phenomena that are not verifiable by experimenting. And thirdly, one must make it clear that if one still ventured to hypothesize, these must be unambiguous and capable of explaining their phenomena in mechanical terms.[26] Against this background *Recherches* contained too much nebulous talk, and it was seen as a mystery that any genuine science could be brought to life by such metaphysics. Had Ørsted begun lecturing Englishmen on Kant's epistemological doctrine including the forms of intuition and the categories of the understanding they would probably have adopted a sceptical air.

Faraday was more open-minded. He started out as a pure amateur apprenticed to his master. Actually, he was learning bookbinding, but had an unquenchable desire to read the books he was to bind for Sir Humphry. He was an unprejudiced chemist and only started to become interested in electricity and magnetism when he was asked to write the article on the history of electromagnetism already mentioned. When investigating the controversy between Ampère and Ørsted his first focus was whether the electromagnetical force propagated rectilinearly like gravity as Ampère insisted, or like a honeysuckle as Ørsted suggested. Davy and his close friend Wollaston, also a leading Fellow of the Royal Society, were both wrestling with the same issue.

Faraday experimented with an apparatus he had designed himself. Two glass vessels were filled with mercury. At the bottom of each vessel a conductor of copper was inserted and exactly perpendicular to them were placed two brass rods reaching just below the surface of the mercury. The left-hand conductor was fastened to a cylindrical magnet reaching above the mercury. As soon as the copper and brass were connected to a galvanic battery the magnet would start rotating. In the extension of the axis of the right-hand conductor he placed a fixed cylindrical magnet that also reached above the surface of the mercury. When the circuit was closed the brass rod began to rotate around the magnet. Ergo the electromagnetic force had a rotary movement like a honeysuckle. If the current went from the positive to the negative pole the rod rotated clockwise, and if the conductors were reversed it rotated anticlockwise.[27] This was definitive support for Ørsted from Faraday.

Even if Ørsted did not mention it anywhere he must have been pleased to associate with Wollaston, for it was he who back in 1804-5 had defeated Chenevix, his tormentor. Wollaston had discovered a new metal, palladium, and proved it to be a radical, whereas Chenevix having under suspicious circumstances snatched the metal and the priority insisted that palladium was merely an alloy of platinum and mercury. As already mentioned (chs 17 and 19) Chenevix had ridiculed Ørsted for his fascination with Winterl and *Naturphilosophie* and characterised him as a delirious visionary in several European journals. Ørsted's protest had been foiled by Steffens's officiousness, but Wollaston's meticulous investigation had cut Chenevix down to size and made him disappear from the arena of chemistry.[28] More support from Britain.

THE TRIUMPHAL PROGRESS: BRITAIN

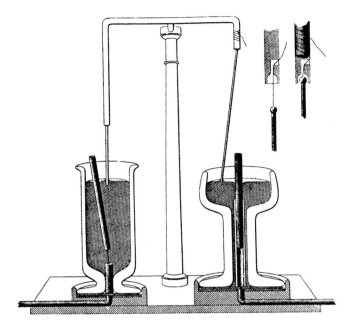

Fig. 100. Michael Faraday's (physicist, 1791–1867) instrument to prove the rotation of the electromagnetic force. The two bowls are filled with quicksilver. The electric current passes through the bottoms of the bowls and the quicksilver to the conductor above. To the left the magnet rotates around the conductor, and to the right the conductor rotates around the magnet. Consequently, the electromagnetic field can be reduced neither to electricity nor to magnetism. *Hans Christian Ørsted*, Isefjordsværket, 1987, 78.

Unfortunately, Faraday was rebuked by his employer for having published his splendid experiment on rotating electromagnetism right under the nose of Wollaston, who believed he was within an ace of reaching the same conclusion in the laboratory of the Royal Institution. This was a serious accusation of strutting in borrowed feathers. To a Sandemanian like Faraday is was a crude insinuation completely depriving him of his joy and pride in his achievement. Just at that time he had made his confession to the small Sandemanian congregation that was required prior to his initiation into the denomination, for which unconditional truthfulness was equal to the blood of Jesus Christ as the core of Christian faith.[29] The worst defamatory reproach thinkable against Faraday was the suspicion of cheating. Luckily it was the vain Davy who got angry. In a letter to Ørsted, Gjerlew had called Sir Humphry 'a conceited fool'.[30] For all we know Wollaston did not take offence at all, but Ørsted must have heard of the bad atmosphere between Davy and Faraday, for in his description of electromagnetic rotation he stressed that Wollaston had scented something similar, but that the ingenious experiment had been thought out and done by Faraday.[31] Against this background it is easier to understand that Faraday, as a consequence of the scar inflicted by the accusation, was careful to deposit a proof of his priority to his discoveries in a sealed envelope in the safe of the Royal Society as soon as he conceived the idea for anything new (chs 21 and 35).

Ørsted saw several experiments in the laboratories of these institutions. He was soon invited to visit Babbage, who showed him his calculating engine, 'a machine he had invented to make all kinds of calculations such as tables used in mathematics. At the moment the calculation is made his machine prints the complete table simultaneously. I know that this sounds like witchcraft, but in actual fact it is much simpler than one would think, although the description without a detailed drawing would hardly make it clear', he told Gitte.[32]

Before embarking on the journey by steam to Scotland, Ørsted received a letter from Gay-Lussac telling him that he had been elected to take the empty place in the physical class of the Académie Royale des Sciences. Many candidates had been proposed, but Ørsted had been elected unanimously, or almost unanimously. Out of fifty-two votes, fifty-one had been cast on his name. Such a majority rarely occurred, and he felt very honoured.[33]

In July he went first by the SS James Watt from London to Edinburgh, then as a tourist on Loch Lomond, next from Glasgow on the Clyde to Greenock, and via the islands of Arran and Man to Liverpool. Ørsted was fascinated by steam power, although it was mechanical. In Denmark he had vouched for the safety of the steam engine to King Frederik, and the royal trust in Ørsted was nicely proven in 1823, when Gitte reported that the entire royal family had ventured to go on board the SS Caledonia from Copenhagen to Kiel. The weather was bad, the Queen went to her cabin with seasickness and her physician-in-ordinary, while the King displayed his courage by walking on deck, deciding that the return voyage would also be by steam. He had finally embraced modern technology. Ørsted was acting as a kind of commissioner for Councillor L. N. Hvidt who planned to establish a network of steam routes in Denmark, and he was pleased to receive an offer via Ørsted from Jacob Perkins in London, and James Watt in Birmingham.[34]

In this respect England was miles ahead. The SS James Watt was so large that there was plenty of space for five carriages on board and yet the approximately one hundred passengers could walk around on open deck or sit down reading a novel from the ship's library, which had already become part of the fixtures. The voyage from London to Edinburgh was scheduled for sixty hours, but often, as was the case on Ørsted's tour, it was accomplished in forty-eight, equivalent to an average speed of fourteen knots. Various mechanics had worked on a gear system to link the steam engine to the paddle wheels. In the estuary of the Thames the captain had grounded the ship, and had it depended on wind power, they would all have had to stay there until the wind changed. Now, the captain just set the steam engine in reverse gear and the ship quickly freed itself by its own power.

In Edinburgh Ørsted called on David Brewster, secretary to the Scottish branch of the Royal Society. Brewster wanted to engage Ørsted as a contributor to the *Encyclopedia Britannica* and presented him with a complete copy, of the value of three to four hundred rixdollars. In return Ørsted would write an article of thirty to forty pages on electromagnetism. In the 1830 edition appeared a somewhat longer article by Ørsted under the title of 'Thermoelectricity'. The title is slightly misleading, because the article went through the history of electromagnetism from 1820 via the controversy with Ampère to Seebeck's discovery of thermoelectricity and Faraday's experiment.[35] The *Encyclopedia* was dispatched by ship to Ellsinore and Ørsted looked forward to showing his wife and children the many illustrations and reading aloud to them some of the suitable articles.[36]

THE TRIUMPHAL PROGRESS: BRITAIN

The greatest attraction in Scotland was the poet Walter Scott, who lived in Edinburgh as well as in his castle of Abbotsford by the River Tweed twenty miles to the south of the city. Ørsted and Oehlenschläger were moved by the medieval legends and romantic stories written in Gaelic by the mystic Ossian inspiring Scott's historical novels. Ørsted approached the novelist with the same awe as Sibbern, when he had swarmed around Goethe. One day he and Brewster bumped into Scott in the streets of Edinburgh, but the celebrity had rushed on after a couple of minutes. Ørsted stuttered out greetings from Oehlenschläger and Scott politely expressed his admiration for the Danish bard, to whom he ought to have written a long time ago, but sadly he had postponed it. Then Scott disappeared in the twinkling of an eye among the crowd in the High Street. Ørsted had been looking forward to filling several pages of a circular letter with an account of his meeting with Scott. Now he could hardly report what he looked like. Still, he had noticed, 'that he is big of weight and plump without being fat and limping heavily'. But his facial expression was already fading. 'If I can get a good portrait of him I shall get one for you.' Ørsted was deeply impressed with Scott's ability to earn a tremendous amount of money. 'He gets three thousand pounds sterling (fourteen thousand rixdollars) for each novel. As he writes a couple of novels a years this adds up to quite a fortune. If Oehlenschläger earned the same for his works he would soon be able to buy himself a landed estate in Denmark, should he want to.'[37]

After several unsuccessful attempts, finally one early morning some days later, and with the help of his butler, Ørsted managed to find Sir Walter at home. As Gitte noticed: 'To be in Scotland without seeing Schot [sic] would be like going to Rome without seeing the Pope.'[38] Ørsted told Oehlenschläger all the details, but although the letter has disappeared, the main

Fig. 101. Sir Walter Scott (1771–1832), Scottish poet in his study at Abbotsford. Oil painting by Sir Edwin Henry Landseer (1802–1873), c.1824. NPG.

401

thing is clear enough. Oehlenschläger had written a comedy entitled *Robinson in England*. It took place in London and dealt with Daniel Defoe and his writing of *Robinson Crusoe*. 'Adagio' was yearning to have his play translated and published and in particular to get it sold in Britain. To that end his fellow-countryman Andersen Feldborg, who was well acquainted with the British market, was commissioned to advance his case with Scott. Prospects looked rather bright, for he had already found a translator and had parts of his play *Correggio* printed in *Edinburgh Magazine*. Now Ørsted was commissioned to push the matter forward. 'Adagio' wanted an advance of two hundred pounds from Scott's publisher, but disappointingly he was asked to share the risk with the other parties involved. Seven hundred and fifty copies would be printed, and if they were all sold at full price he could count on a profit of a little more than a hundred pounds being one sixth of the total surplus. This was only half the sum he had demanded as an advance let alone his hope of an aggregate profit.[39]

Ørsted's tour of Britain took place during the summer holidays when scholars were difficult to find. Davy was fox hunting in Ireland, and Wollaston, whom he had introduced to the pleasures of angling, was away catching salmon. So Ørsted's tour rather turned into tourism, and what and whom he saw was fortuitous. He reported his impressions to Gitte and friends in copious circular letters:

> '2 July my friends had organized a trip for me. First I had lunch with Robison and Brewster who then took me to the water mains. We did it by boat drawn by a horse on a canal that sometimes is running across a road or a river. Just imagine the canal at one place crossing a sunken road so that you hear carriages driving under the canal. Here the canal passes over arches of bricks, and the canal trough (I can think of no better name for it) is cast of iron. In the same way this canal crosses a river far below at another place.'[40]

Institutions similar to the two he had visited with so much interest in London were spreading to the fast growing industrial cities in the Midlands, where engagement in the natural sciences and their utility in particular was dawning.

> 'Liverpool is also getting an establishment like the Royal Institution and the London Institution. It is aimed at education in all branches of science. Such institutions are about to spread all over England enabling more people in pursuit of scientific knowledge to get an opportunity to acquire it. The English generosity in promoting this goal is honourable. In Liverpool two hundred and twenty-four men donated one hundred pounds sterling each—a little more than a thousand rixdollars—altogether approximately four hundred thousand rixdollars [sic! Ørsted was no heavyweight at calculating]. Apart from these institutions they also have reading rooms and libraries always crowded with people. All this comes from subscriptions.'[41]

In Manchester he saw the darker sides of the industrial revolution:

> 'A huge class of workers belonging to the expanding industry is generally very crude and poor. As soon as their working hours (and they are many) are over they are crowding the streets. When there is work enough for them they are tolerable, but when circumstances occur where they have little income, they are horrible. Even now when there is plenty of work they fall into

violence and crime. I recently heard about the horrendous murder of a manufacturer committed by three such workers. In Birmingham, where I stayed recently, the workers seem to be better and living in better conditions. Immediately after the peace they were unemployed and exposed to poverty; but at that time the city paid seventy thousand pounds, that is about eight hundred thousand rixdollars, a year to sustain them. However, they did not receive this money for nothing. They had to do a lot of work that was not worth the money, but still kept them going. The improvement of education is an important contribution to their refinement. For the last fifteen years they have introduced schools according to the Bell-Lancaster method. You know that by this method large groups can be taught in a short time. This is important to industry. It has been found that among those educated in this way, no crimes are being committed and only rare offences. So, good education has provided splendid results.'[42]

Ørsted had imagined staying in England for four weeks and in Scotland for another two. At any rate this is what he told Gitte, possibly to calm her, but he stayed twice as long, that is for three and a half months.[43] This was his first time on British soil and he thoroughly enjoyed himself among the English and the Scots. He received lasting impressions pointing to a future industrial society and he acquired food for thought as to how scientific knowledge might be diffused into the population of a small, poor, and war-stricken Denmark which was lagging behind. This experience provided the inspiration for the Society for the Diffusion of Science that he founded upon his return. However, the anecdote repeated time and again in lives of Ørsted that the idea for this society was born specifically on his voyage on board the steamer bringing him across the Channel is not true. The crossing took place on his birthday 14 August and was not conducive to high-flown plans. 'The sea began to rise... Waves were violent and hit us from the side, which is most unfortunate especially for a steamboat. I have never been tossed around like that. Almost everybody was seasick. I was one of the few avoiding it... The sea splashed us terribly... Finally, we came ashore in France.'[44]

His host was awaiting him on the quay in Calais; he had taken care of his carriage that had been left longer than anticipated at the Hôtel de Bourbon. Bindesbøll was delayed on his journey from Paris, and they only arrived in Denmark on 29 August via Hamburg and Kiel. Gitte and the children welcomed him at the customhouse. The spouses realised how changed they were after almost a year of separation: both his thin and her thick hair had started to turn grey. Karen got a doll from Paris.

PART VI
THE ORGANISER

> From science, all the statesmen want are things that are useful:
> weapons, techniques, communications, and so on.
> As for scientists, what do they want of governments?
> Support, funds, independent institutions providing their
> profession with autonomy, legitimacy, and authority.[1]

40 | 1823–4
The Society for the Dissemination of Science in Denmark

WHILE ØRSTED was away, Karen and Søren had walked the long way to school inside the ramparts of the city and had lunch with Aunt Anne Marie, Gitte's sister, who was now in Copenhagen with her four children, cohabiting with a certain professor Klingberg (ch. 44). Christian, Marie and Sophie were tutored at home by Mr Petersen. Gitte had preserved berries, pickled cucumbers, and scrubbed the floors so everything would be shining for the return of her husband. For the last months she had alerted her friends to be on the lookout for a new flat, because Mr Pingel had ended their tenancy in Nørregade. An opportunity arose when Professor Nikolai Kall got seriously ill; if his health continued to go downhill his professorial yard in Studiestræde would soon be vacant. There would be space enough for Ørsted's collection of instruments.[2]

Old Kall died and during late summer the Ørsted family moved from Nørregade to Studiestræde. It was not as quiet as Cambridge, but they felt comfortable and stayed there for the next twenty-seven years. The instruments were taken to the ground floor next to the new lecture hall, and the Ørsted family installed itself on the first and second floors. In 1829 the collection of instruments was removed to the adjacent professorial yard in St Pederstræde, which had so far been occupied by Jens Møller, a professor of theology. He agreed to be bought out when the new Polytechnic Institute required all the space available in that small area.

Facing Studiestræde and looking out towards the bishop's residence were first Gitte's and Hans Christian's bedroom with its four-poster. In the middle was the four-windowed sitting room with Boisserée's copper-prints on the walls and two pier glasses and a pedestal with gilded cups symmetrically between the windows. Facing the courtyard was a three-windowed sitting room heated by a stove and with a gas chandelier, a round table, a chest of drawers and a piano, a pedestal with a tambour clock and pier glasses and small platforms in front of two of the three windows with stools enabling one to watch life in the street below. Also facing the courtyard was the children's bedroom accommodating the three younger ones not yet at school and not having any homework. Kitchen, pantry and dining room with a round table also faced the courtyard and garden. And it was here Hans Christian Andersen and the young polytechnic students would later have dinner and court the daughters of the house. The garden had a path around it and corners sheltered by windbreaks and shrubs. Besides the privies there was a two-storey building, formerly a stable and a shed for carriages. It was refurbished to become a chemical laboratory when Ørsted and Zeise left Nørregade. On the other side of a high fence was the bishop's garden belonging to Frederik Münter, Bishop of Zealand, whom Ørsted knew from the Masonic lodge. Adam and Christiane Oehlenschläger had rented a small flat there. The Ørsted brothers would visit the bishop or the poet or they would nip over to Studiestræde to have a game of ombre with their neighbours. Later in the evening the servant would lay the round table, arrange the tea machine and serve cardamom biscuits; Anders, however, preferred sandwiches.[3]

Ørsted's study was on the second floor with two windows facing the courtyard and three facing the street. In it was a desk, another table, and bookshelves from floor to ceiling for his seven thousand volumes including the *Encyclopedia Britannica* which eventually arrived from Brewster in Edinburgh. *The Great French Encyclopedia* was there already. The professor had as many square feet for his study as the one sitting room and the dining room together. Nearby were rooms for the chambermaid and for Karen and Søren. Gitte did not have a room of her own.[4]

While moving house, Ørsted was also busy providing support for his new project. Each time he had been abroad he had returned full of new plans. This time he wished to establish a new voluntary association to disseminate scientific knowledge in Denmark. The inspiration for this had obviously come to him from his visits to the Royal Institution and the London Institution, both of which he pointed to in his printed subscription list that went out in October. The very name 'The Society for the Dissemination of Science' seemed to paraphrase a translation of 'The Society for the Diffusion of Useful Knowledge' founded in 1821 by Henry Brougham. However this was rather a reading club addressed to the working class, publishing cheap editions of useful literature. Ørsted's society lacked the qualifier 'useful', and this was intentional. Ørsted had talked to Prince Christian Frederik about the objective and organisation of the society before inviting potential subscribers to the first meeting. The Prince was the head of the Board of The Royal Academy of Arts, and the arrangements they had in mind were also addressed to students at this Academy. So useful knowledge in the technical sense was not given first priority. Secondly, time and the economic situation would reveal whether the society would be able to establish its own building with laboratories, lecture halls, and permanent professors. For though Ørsted's overall idea was a voluntary association financed by private

Fig. 102. Reconstruction of HCØ's study in Studiestræde. Exhibition for the centenary in 1920 of HCØ's discovery of electromagnetism. DTM.

subscriptions, he was under no illusion that he would necessarily succeed in providing sufficient funding in a small capital city with a population only one tenth of London's, for his society to become a match for his English models.

When he compared himself to Humphry Davy there were striking similarities as well as differences. Both were heading scientific academies like Laplace in Paris and Erman in Berlin, but London had no university, and Oxford and Cambridge had neither professorships in physics and chemistry nor any collections of scientific instruments. On the other hand, London had something that few other capitals had, namely two private institutions for experimental science complete with facilities for research and training. In the London Institution the main activity was experiments with galvanism and electricity, attracting a large, paying, and self-indulgent urban audience. The Royal Institution was more the meeting place of the landed gentry who demanded agricultural chemistry and other useful knowledge, and Davy, its professor, had delivered what he was paid to do: agronomy and the safety lamp. However, it was Davy's ambition to become the Newton of electrochemistry.

It was the encounter with these two private institutions that inspired Ørsted. Davy had hardly made a secret of what he considered to be the benefit and drawback of them. The benefit, of course, was the interest in the sciences they stimulated by putting them on the public agenda as one of the goals of the liberal movement. In 1827 this was exemplified in the establishment of London University College, headed by Jeremy Bentham, the philosopher of utilitarianism, and his supporters Henry Brougham and John Bowring. The drawback that Davy felt was that the investors insisted that the research activities had to be directed towards useful projects, so as to find solutions to 'merely technical' issues. Consequently, Davy risked losing the freedom to pursue his own purely scientific research project. Ørsted no doubt would have had to lend an ear to Davy's version of the old conflict between, on the one hand, those who pay the orchestra demanding the right to decide what music is to be played, and on the other, the members of the orchestra insisting that if they are not left with freedom to play what they like, there will be no concert at all.

Ørsted's aim in founding the Society for the Dissemination of Scientific Knowledge in Denmark did not concern the training of artisans' technical skills. His motivation was to promote public interest in the sciences, for even in an absolute monarchy public opinion matters; it would not be sufficient to be a recipient of the cherished Napoleon prize if there were no pressure from below from the public or from industrialists, and no public understanding that science holds the key to national prosperity and civilizing progress. In particular, more science was needed at the University. His grand tours had always been funded in the expectation of useful knowledge by way of return. But as we have seen time and again, once he was away he changed his priorities. He had been attracted to European scientists who one way or another might contribute to the development of his dynamical scheme. In hindsight it goes without saying that Ørsted's effort in the longer run by far exceeded what the fantasy of contemporary utilitarians could imagine.

It gradually dawned upon him, however, that the tension between the expectations of his donors and his own ambitions was untenable. If he did not make a compromise he would probably lose the means of sustaining and expanding his laboratory and collection, let alone his daily bread. Nevertheless, a compromise must still be tolerable. This had not been the case with Hermbstaedt. True, he had control of a fine laboratory, but he had become a philistine in return, who had betrayed his free research in favour of solving practical problems. The only free researcher Ørsted had ever known was Ritter, but he had had to pay a heavy price, at first with his health and in the end with his life, without gaining honour let alone remuneration for his epoch-making discoveries.

Ørsted's first idea from 1802 was simply to establish a research laboratory for himself in Copenhagen. Manthey misunderstood his protégé, and the Chancellor ignored him (ch. 12). His next plan was put forward a year later and was aimed at emulating the French Athénées in Copenhagen and the provinces. In these French institutions the best French scholars gave talks on new scientific discoveries and people from the humanities entertained with papers on literature and history. To combine the useful with the amusing was the intention, but he feared that in puritanical Denmark one would have to be careful with the merely amusing or else it could get the upper hand (ch. 14).

However, he had still not abandoned the idea of getting his own research laboratory. In a detailed proposal he swept Manthey's misconceptions of a 'mere technical institution' aside, describing instead a scientific institution of physics and chemistry that would address practitioners, although not to teach them the manual trades. On the contrary, the aim was 'to elevate them to a higher level of culture...so that they now accomplish as rational human beings what they often used to do as mere slaves...'. Crucial to this institution would be experimentation and to that end the acquisition of a collection of scientific instruments was of paramount importance, even if he would have to borrow money to get it.(ch. 16). Fortunately, a year later a significant collection was put at his disposal for his private lectures (ch. 17).

His fourth plan was presented directly to the King in 1808, not to the Chancellor of the University as previously. This scheme looked like a compromise and his arguments were utilitarian. He proposed a Practical Institute for the Experimental Sciences that would have a significant impact on the prosperity of the nation for many generations to come. The very title was ambiguous. He talked about an independent institution for twenty-four pupils of two types: partly practitioners, i.e. miners, artisans, or pharmacists who needed scientific education in order to understand the scientific basis for their activities, and partly young scientists such as undergraduates, sons of well-to-do estate owners and merchants, as well as artillery cadets, needing experimental skills. Whether the two types were to be taught together or separately is not mentioned, but everybody would have to be accommodated, so the cost of a suitable building was the biggest obstacle. The plan was abandoned by the authorities, partly because of the founding of the Norwegian University in Christiania, but Ørsted did obtain a grant to go abroad to report to the King on ways the sciences contributed to the prosperity of major nations (ch. 19).

The report Ørsted submitted to the monarch in 1813 contained negligible information about the relationship between science and technology, and no concrete descriptions of factories or other industrial enterprises. His main assertion was that the practitioner would benefit from the supervision of a scientist, not the other way round. The state needed theoretical scientists for 'new theories will always sooner or later become useful to practitioners'. His line of thought was that the artisan is a specialist in narrow skills that he and nobody else needs, while the theorist is a generalist, and hence his knowledge and methodology is applicable everywhere. Having brushed aside 'the merely technical', he went on to stress his main point, that the sciences at the University must be elevated considerably in order to increase the national wealth. Interdisciplinary collaboration, with physics and chemistry at the centre, should be put in place at the University. Between the lines, Ørsted was implying that his professorship ought to constitute the core of the new Faculty of Science, standing on an equal footing with other faculties, i.e. having full representation in the Senate. The objective of his 1808 proposal was clearly reiterated in this new report (ch. 31), but sadly the monarchy was paralysed by war and national bankruptcy.

On his third grand tour in 1822–3 his visits to the London Institution and the Royal Institution in particular influenced his way of thinking. Both institutions were established through voluntary subscriptions from the citizens of London. Ørsted was impressed by the buildings available to science, but also by the interests and sacrifices of their members. He had never experienced anything like it. His earlier plans had angled for public money. This time he changed strategy. The funding was to be raised by the citizens themselves, and they would have to be motivated by the benefits they would receive from enlightenment. Hence, science should not be confined to technical chemistry and mechanics, but should be part of a general education scheme. So, Ørsted's new strategy was to pave the way for the Society for the Dissemination of Science by circumventing the state and obtaining subscriptions directly from the citizens for the sake of general education and the arts. To allay the potential worries of the regime that voluntary associations might endanger public order, Ørsted proposed that the oversight be led by Prince Christian Frederik and other members of the establishment.[5]

Prince Christian Frederik was soon won over to the idea. On 16 October 1823 Ørsted sent out invitations to subscribe to membership of the Society for the Dissemination of Scientific Knowledge in Denmark by signing up for ten rixdollars a year. The subscription list referred to the London Institution and the fact that it had received about 1.2 million rixdollars in voluntary contributions. Now for the Danish equivalent he endeavoured to collect sufficient to fund a central building for experimental science which would serve two purposes: (1) to work for the dissemination and application of science in Copenhagen; and (2) to train young men well educated in science to deliver lectures on science in the provincial towns. The central building would have a library and a collection of instruments and drawings, and the subscribers/members would elect a manager, which position Ørsted volunteered to occupy free of charge. When the invitation had been circulating for three months to little avail, Ørsted had to agitate more strongly for his plan to give it a chance. On 10 January he published his article 'Science considered as one of the basic elements of human education'.[6] Until now, his article has not been read in the context of the deadlock in which his proposal had landed.

'The grounds for disseminating science are that knowledge of science will engraft reason into the individual by insight into the laws of nature and their constancy, in relation to the transient and perishable nature around us.' Ørsted quoted the Neoplatonist motto: 'Our spirit is nourished by spirit only and seeks it everywhere'. The thinking is that the laws of nature as well as the fundamental forces are hidden from the human eye. They belong in the hidden realm of ideas, but nevertheless manifest themselves in nature and consequently in man as a part of nature. When the individual encounters science, he encounters 'a spark of the eternal reason, from which the spirit of man as well as of nature springs'. In this way enlightenment will kindle the inspiration to build one's life on the three aspects of reason: the True, the Good, and the Beautiful: on the True due to incontestable, eternal laws of reason, on the Good which we achieve when we let the moral maxim of reason guide our will and behaviour, and on the Beautiful, when we experience reason in the sensual forms in art museums, theatres, concert halls, and books.

While this general education was the primary objective, the Society for the Dissemination of Scientific Knowledge was also grounded on a secondary one which Ørsted shared with the Reform Movement in Britain, and which was behind the Society for the Diffusion of Useful Knowledge. Without science modern progress would be unthinkable. This utilitarian argument that Ørsted had set out in his previous attempts to extract funding from the government did not distinguish between science and technology. Ørsted called attention to the well-known technical benefits that astronomy, optics, metallurgy, mechanics, and chemistry had brought to commerce, shipping, and industries. In the future, British steam power would revolutionise societies more thoroughly than the political ideas of the French Revolution, he argued.

In the course of four and a half months twenty-six members subscribed, Ørsted noted at the opening meeting of the society on 27 February 1824.[7] This was a disheartening outcome, indeed. He was summing up to some of the twenty-six honourable men present, such as Prince

Christian Frederik, Count Schimmelmann, Treasurer Møsting, and Court Steward Hauch as well as some of his friends. 'I know pretty well that reality never lives up to such great expectations, that one has to accept far less due to people's lack of attention, lukewarm reactions, and sometimes even recalcitrance....'. Ørsted sulked.[8] According to his son-in-law to-be, E.A. Scharling, Ørsted had hoped for a thousand subscribers paying ten rixdollars each. A list of November 1825 shows 157 members paying the fee of ten rixdollars, while 63 paid less.[9] The economic support was thus less than a fifth of what the initiator had hoped for. Under these circumstances a central building for the society was out of the question. The Society for the Dissemination of Science in Denmark did get off the ground but as a purely lecture-based society without premises.

At the next meeting, on 26th March, Prince Christian Frederik was elected president of the society and Ørsted head of the board for life. He also became head of the scientific committee, the name of which Collin had the good fortune to get changed to the physical-technical committee. Finally, Ørsted was given sole responsibility for organising the lectures. Thus Ørsted had full control of the society and only needed backing from the Prince, who never caused him any trouble.

But soon a controversy flared up in the physical-technical committee between Collin and Ørsted. Collin did not believe in Ørsted's idea that scientific lectures would infuse artisans with insight into the laws of nature and hence 'in unpredictable ways' with a deeper understanding of their craft. A member's offer to invite apprentices of the mechanical trade into his workshop equipped with modern lathes, vices, and other tools was exactly the kind of thing the society should go for, according to Collin. This was far more useful than peripatetic scientists lecturing to middle-aged generalists and civil servants.[10] In Collin's view what artisans needed was to become more skilful artisans, and to this end scientific theorists were of little help. So technical instruction was a matter of experienced artisans training inexperienced ones. Science was largely irrelevant, and academics had no right to show off as know-alls with condescending attitudes, for they would never be on a par with the skilled artisan on the workshop floor.[11]

Ørsted did not resent the proposal in itself, he said, but the main purpose of the society according to its statutes was to disseminate science in order for artisans to better understand the laws of nature determining their crafts. To Ørsted the crucial point was that theorists had a lot to offer the practical man, whereas the practical man had nothing to offer the theorist. Ørsted tried to foster the idea that technology is derived from science, and that scientists have a mission to develop new technology and elevate its practitioners to their level; thus he espoused a hierarchical view rather than one based on a division of labour and equality. He was well acquainted with the view that theory may be damaging to practice, and to counter it he quoted James Cook, who had accompanied Joseph Banks, the president to-be of the Royal Society, in the Pacific. Cook is reported to have exclaimed: 'To Hell with all sciences!'. However, without science, Ørsted remarked dryly, Cook would never have been in a position to bring Banks safely to his destination. And against Collin's second point Ørsted argued that members of the higher classes made their living from industry and hence belonged to the target group of the society.[12]

In twenty-seven years Ørsted delivered thirty-one series of lectures for the society, from 1829 onwards in common with the Polytechnic Institute. During the same period Zeise delivered thirteen, Forchhammer seventeen, Scharling and Hummel four each, Holten three, and Ursin and Dyssel one each. The civil war between Schleswig-Holstein and Denmark and Ørsted's death in 1851 saw the end of his lectures. They took place in the lecture hall in Studiestræde. After the founding of the Polytechnic Institute in 1829 they moved to St Pederstræde and were organised under its auspices, although admission for members of the society remained free. During these twenty-seven years Ørsted delivered forty-two per cent of the lectures, Forchhammer and Zeise together forty per cent, and the last eighteen per cent were delivered by staff of the Polytechnic Institute, Scharling among them. Lectures in provincial towns provided secure jobs for a large number of polytechnic graduates. Out of eighteen lecturers touring the provinces from 1825 to 1851, twelve were postgraduates from the Polytechnic Institute, the first leaving in 1831.

M.C. Harding has classified the audience of Ørsted's first lectures on the basis of preserved lists of participants as follows:[13]

	1824–25	1825–26
Artisans and industrialists	52	40
Pupils of the Academy of Arts[14]	32	20
Officers	32	51
University students and postgraduates	23	60
Civil servants	20	34
Businessmen	9	18
Unknown	25	29
Total	193	252

It was impressive that Ørsted could assemble so many listeners to lectures in chemistry and mechanical physics on a voluntary basis when there were no diplomas in return. But did the target group of the Society for the Dissemination of Science benefit from the effort? The number of artisans and industrialists shrank from thirty-one to eighteen per cent (disregarding the unknowns), while the number of University students increased from fourteen to twenty-seven per cent, and there were less dramatic increases in the numbers of civil servants, officers, and businessmen. It was assumed then and since that the primary target group was practising artisans as Collin wished, contrary to Ørsted's intentions. Collin's view had a hard time under Ørsted's rule, and the trend was against him.

> Enlightenment is man's emancipation
> from his self-imposed tutelage.[1]

41 | 1824
The Ørsted Brothers in the Howitz Controversy

IN NOVEMBER Gitte gave birth to the youngest daughter of the family, Mathilde Elisabeth. She was named after Anders's second wife. Gitte and Hans Christian had really wanted a son, so they were slightly disappointed when they saw the baby was a girl. Karen, the oldest daughter who was nine asked if she could have the tiny infant as her child. This was readily accepted, and Karen looked after her little sister with a love and care as if she were her own. Mathilde regarded herself as her sister's first daughter. As a grown-up she became her father's secretary and editor of his papers and letters. She never married. In 1826 and 1829 Gitte gave birth to their seventh and eighth children. Both times the parents had their wishes fulfilled. The two boys were named Anders Sandøe and Albert Nicolai after their uncles.

In 1817 Anders had attracted the King's wrath by criticising the collective punishment of the rebellious prisoners (ch. 32). Two years later a young medical doctor, Franz Gothard Howitz, was appointed the physician of the Rasphouse, the prison and workhouse, which was full of extraordinary characters. When he had familiarised himself with the inmates, who were both prisoners and his patients, he began to wonder if they were prisoners because they were patients; perhaps their crimes had been committed through no fault of their own, but simply because they were mentally ill. Howitz's reflections on the ordeals of his patients made him turn against the prevailing legal thinking which he had come to consider an untenable philosophy. Instead he developed what he considered a more humane and scientific approach to justice. The penal code of the absolute monarchy took it for granted that citizens were endowed with reason and free will, so that they could be held accountable for their actions. Howitz, however, observed that the crimes for which the inmates were punished had been committed by mentally disordered people incapable of acting rationally and freely. He therefore concluded that they could not be held responsible for their crimes. According to Howitz, the legal philosophy of the time was based on Kant's moral philosophy and was upheld by the leading jurist of the country, A.S. Ørsted.

In 1819 Dr Herholdt delivered a paper in the Royal Danish Society on a very strange patient, a twenty-six-year-old woman, Rachel Hertz, the daughter of a wealthy merchant in Copenhagen.[2] During the bombardment in 1807, in the resulting terrible pandemonium, she had been punched in the abdomen. As a consequence she suffered pains, feebleness, headache, and insomnia. At first the patient was put to bed at home, later with another family, and then the symptoms disappeared, but only for a while. A year later she had convulsive fits accompanied by 'rage, madness, and noisy lunacy, often so violently that several people were unable to govern her'. Soon more new sufferings were added: vomiting blood, insensibility, and urine retention, which was helped by catheterisation of her bladder. The family doctors, Professor Callisen, and Dr Herholdt, his assistant, were unable to diagnose the disease let alone cure it.

> At that time she was only rarely conscious of her own existence and often lay so insensible that we could pinch and hit her without arousing the slightest reaction. Occasionally she had a quiet moment, in which she could take nourishment; she mostly subsisted on cold water. O, how often did we sit by her sickbed back then, Callisen and myself, with sad feelings bemoaning the impotence of the medical profession.[3]

The sufferings bound the young women to her bed for a couple of years, then suddenly all symptoms disappeared as inexplicably as they had arrived. A few years later, however, Rachel was admitted to Frederiks Hospital with severe pains in her abdomen. Herholdt examined her and found a tumour causing convulsive pain at the touch of it. He realized that death was approaching and 'doubtful about the outcome and with the consent of her family, he audaciously cut deep into the tumour right at the sensitive point, discovering a narrow, solid, alien body. When extracted by tweezers the object turned out to be a black oxidized sewing needle.'[4]

The story grew more and more fantastic.[5] Over the following years the original symptoms returned with undiminished strength, and the removal of the sewing needle brought no relief. From time to time Dr Herholdt cut out no fewer than 389 needles out of her abdomen. How they had come there was inconceivable, for there were no external signs that they had been stuck into her from the outside, and imagining that she had swallowed them was completely absurd. Miss Hertz became a good story for the newspapers; Copenhageners talked about 'the sewing needle girl', she became known abroad, and the worried Herholdt, who had never read about a similar case and hence did not know what to do, presented his case to the Royal Danish Society. They appointed an investigating committee consisting of Hauch and four professors and headed by H.C. Ørsted.

Shortly before Christmas the committee assembled in Herholdt's flat at Frederiks Hospital, where Miss Hertz had a room of her own. Herholdt suggested to seal her room and make the night nurse take an oath, but Ørsted preferred a more rigorous procedure: the committee members themselves should keep watch during the Christmas holidays.⁶ Herholdt was particularly mystified by the fact that the doctors kept cutting out one needle after another. The patient explained that many years ago she had swallowed a needle case. In addition, the doctors tapped several hogsheads (240 litres) of liquid annually from her bladder, which was far more than she consumed.

Ørsted found Herholdt's account at odds with the laws of nature and suggested to weigh the patient's intake of liquid and compare its weight with the urine delivered, or indeed, why not place her bed on a commercial weighing machine to keep a daily record. Herholdt objected arguing that it was too time consuming and costly, but Ørsted decided that the Society would pay. However the case was solved without the weighing machine, for after three days' watch a reasonable balance between intake and output of liquids occurred. This was rather lucky, Ørsted commented, as now one of the professors had refused to keep watch any longer. It was suggested that a nurse be admitted to Rachel's bed, after a body search, but the committee thought this was going too far.

One would think that 'the sewing needle maid' had been unmasked, but it was not that simple. Herholdt had previously assembled several surgeons to oversee the removal of the many needles, but now one of the professors saw the spot where a needle was to be operated on and exclaimed, 'But here is a prick of a needle!' and a colleague remarked that the head of the needle was turned towards them and hence the needle could only have been inserted from without. The following day Rachel wrote a letter to Herholdt begrudging the suspicion thrown on her; she was terribly upset and had lost her tongue. Herholdt's sympathy with Miss Hertz was unabated, and he blamed his colleagues for having doubted the sincerity of a poor patient thus provoking a serious deterioration of her state. The learned committee had to leave its assignment unresolved.

In the following years Herholdt defended his credulity. He found it heartless of people to accuse her of pulling their legs and he continued to attend her. Soon no more liquid came out of her bladder, but a gas, which was sent to Ørsted for closer examination.⁷ The gas turned out to be atmospheric air, but Herholdt would not believe that, so he sent it secretly to Professor Zeise, who said the same thing. Then Herholdt sent a new sample to Berzelius in Stockholm, but he, too, agreed that it was air.

Finally, in 1826 Herholdt, now a professor, began wavering and agreed to spy on his strange patient through a secret hole in the wall.

> Looking through the hole I discovered that she, otherwise unable to sit up, now took an upright position, and that she, having had for years a paralysed right arm, now rubbed her eyes with both hands, scratched her neck, combed her hair, and wrote her evening report for me on the blackboard. I lack words to describe my feelings at that sight. O man! I thought, what are you? Is there really a kind of madness that is not grounded in a confusion of the mind? Silent and outraged I walked away.⁸

Professor Herholdt had been fooled for nineteen years. He should have listened to Ørsted, who seven years before had already voiced his suspicions and two years before had proven the fraud. Ørsted had examined the patient as a homo phenomenon only. No matter how capricious or fraudulent she might be, her organism was by necessity subjected to the laws of nature. Miss Hertz's soul, her homo noumenon, was irrelevant—for the moment anyway.

Herholdt had taken it for granted that either a somatic disease causes a psychic effect, or a mentally ill person inflicts upon herself a somatic pain, self-mutilation or partial suicide. In both cases the patient would lose the normal use of reason. But he had never asked himself whether a person could mutilate herself and at the same time retain full exercise of her wits. Only in retrospect was he able to put the right question: 'Is there a kind of madness that is not caused by a confusion of the mind?'.

This question was outside the realm of science for there was no method to examine whether a person with a healthy mind, contrary to supposition, could behave like an insane person. States of mind belong to the noumenal world, to philosophy, not medicine, and therefore a mentally deranged person should be sent to a philosopher, not because the philosopher is a skilled therapist, but because reason is no phenomenon and therefore escapes physical investigation. If it were, an insane person might just as well go and see a watchmaker as a physician. Like cognition itself reason belongs to metaphysics.

Was Rachel Hertz endowed with reason and free will, when for nineteen years she had behaved like a mad person, yet had now revealed herself as a self-mutilating woman with the sole intention, apparently, of making herself interesting in the eyes of physicians and others? If she could not be held accountable, not being equipped with normal reason and free will, then this must be because of a somatic disease, and in that case, should not this disease be diagnosed and cured rather than punished, as Howitz's prisoners were? This point of view was presented by Howitz for the first time in 1822 to the Royal Medical Society, where H.C. Ørsted as a member of that society may have heard them. Howitz found that Miss Hertz and similar cases were mentally ill and not accountable for their actions, and were hence unsuitable for punishment. Nobody suggested punishing her, but Howitz argued that many of his prisoners who were being punished presented similar case histories.

In 1824 Howitz, who was by now a professor of forensic medicine, attempted to pave the way for judicial reform by presenting his views in A.S. Ørsted's *Juridisk Tidsskrift* [Journal of Law]. Of course, Anders realized that Howitz was touching on a serious issue: unaccountability was a well-known concept in legal circles. And a dabbler in philosophy should certainly not meddle in legal philosophy in his journal. So, they agreed that each of them would write an article for the same issue and comment on each other's views. Now both Ørsted brothers were involved in the matter, Hans Christian as a scientist, Anders as a legal philosopher. And both were well acquainted with insanity from their stepmother and the criminal behaviour of their younger brothers.

Howitz believed that man is an object of science, thus adopting an anthropological view that considered free will a metaphysical illusion. This view was grounded in a positivist notion of psychology that believed it was on its way to finding methods of deciding when a person is normal and hence accountable, and when a person is mentally ill and hence unaccountable. This method presupposed medical knowledge, and Howitz demanded that jurists left dubious cases to the medical profession. This division of competences was already part of judicial practice, but Howitz wanted it to be formally and clearly legalised. He saw his claim as a step towards a humanisation of the law, because it would lead to better protection of the insane.

Anders Ørsted had no objection to the view that man's mental activities are linked to bodily functions. Like Howitz, he considered mental illness to be a disease of the brain. It was true that science could not yet fully explain it, but the idea that the activity of the mind, the cognitive as well as the moral and the aesthetical, takes place not in the senses as believed by the empiricists but in the brain, was relatively uncontroversial. Hence, Anders shared the conviction that a criminal person who is at the same time also insane, that is suffering from a malfunction of the brain, is not accountable for his crime. His argument was based on legal philosophy. The penal code is the public instrument for upholding the rule of law by means of retribution. It is up to every citizen to decide whether to keep the law or break it, and to accept any associated punishment, but only if he is fully capable of using his reason. If he is not, retribution has no relevance, and the entire raison-d'être of the penal code is rendered null and void.[9]

As the law presupposes that every normal citizen is endowed with reason, it also presupposes his free will. Reason and freedom belong together. To declare a person insane is equal to considering his decisions predetermined. If man is predetermined, the criminal is 'born to mount the scaffold' as Anders expressed it. Considering a fellow human being as predestined to a criminal career is equivalent to depriving him of his reason, that is 'his conscience and free will' which constitutes his human dignity.[10] According to Ørsted, Howitz's fallacy was his positivist idea of grounding the law on unreliable medical indications which would subject human beings to the rule of irrationality. Anders saw Howitz as a spokesman for three dangerous tendencies. First, there was a materialism reducing the mental to the physical, secondly, a fatalism replacing free will with determinism, and thirdly, a eudaemonism imputing a pursuit of happiness to man while excluding a morality of obligations; witness Anders's own maxim: 'We should make our entire life an impression of the eternal reason that is revealed in us'.

Anders's argument with respect to Howitz's ideas is reminiscent of his brother's speech to undergraduates at the University, and an echo of dualism:

> Our soul is yearning after the eternal being from which we derive, to join this being, to understand its purposes with man, and to work cheerfully for these purposes. This exceeds by far the external world and its sensual indulgence. The force in us to live such a higher life, to pursue this higher good, and to follow the law showing us the way to this goal is the core of freedom. Since our spirit is embedded in the flesh, it is natural that we do not always abide by that law, but may let ourselves be ruled by the urges and promptings of the senses. Thus our freedom consists in choosing between these different laws into whose service we might enter.'[11]

Anders fully agreed with Howitz's view that insane people were not accountable and were unsuited to a punishment they did not stand a chance of comprehending. His insubordination in 1817 to Kaas, and indirectly to the King, had shown him to be a humane jurist. Howitz's view that his patients were punished for crimes they had not committed with their reason and free will, was self-evident to Anders. It was only when the crime itself was brought before the court that disagreement occurred about what criteria of accountability should be applied. 'In general it must be left to the judge to decide, according to the circumstances and often assisted by a physician, whether the accused has been in a state in which he is capable of understanding the relation of his act to the law', Anders wrote. Howitz argued that this decision ought to be made by the physician, since lack of accountability was a mental state with a somatic cause, and at this point the physician was the expert, the judge the amateur. The physician would have to examine the accused by confining him to 'a suitable room under the supervision of one or more competent physicians particularly experienced in the treatment of insane people', as had happened with Rachel Hertz.[12] The core of the matter, which is still topical, is therefore whether there is empirical evidence to determine whether a person is in possession of his wits or insane, and if so, whether it is up to science (the physician) or the law (the jurist) to identify it.

Anders did not believe that such scientific evidence existed. Taking Rachel Hertz as an example, it was obvious that she had managed for nineteen years to fool more than one physician, and that his brother's chemical analysis had merely unveiled her fraud, but not her mental illness. Howitz had no positive method to offer, according to Anders, when it came to somatic criteria to be incorporated into the law. Howitz's article 'On Insanity and Accountability. A Contribution to Psychology and Jurisprudence', showed that the main thing he had in mind was phrenology. He referred to J.G. Spurzheim and other disciples of F.J. Gall, asserting that certain human talents and properties are materially embedded in the brain and appear as bumps on the skull.[13] Such properties, presumably, are innate. Howitz referred to clever physicians who had examined the skulls and brains of insane people who had died finding unmistakable symptoms of cerebral disruptions. Insanity (and hence lack of accountability) is a psychological state with a somatic cause. This causal connection between brain and human characteristics was a generally accepted truth according to Howitz,[14] who referred repeatedly to Spurzheim and Gall.[15]

The jurist M. H. Bornemann, who had defeated Anders in the professorial competition years before, had given a paper on Gall's theories of the brain to the Scandinavian Literary Society in 1805, the year the German phrenologist had visited Copenhagen.[16] Like Ørsted, Bornemann criticised phrenology for its materialism and determinism. Gall's theory presupposed what it pretended to explain. These phrenological criteria (the bumps) were scientific nonsense according to the Ørsted brothers, and they should not be allowed to infringe on the penal code. More influence from doctors would not entail a humanisation of the law, but was likely to erode the rule of law.

On his first grand tour Ørsted had seen a bust of Kant and ironically noted that 'Gall's organs for shrewdness abounded', as would be expected. At that time phrenology was a hot topic. Ørsted entertained a sceptical attitude towards Gall's crude observations, but was still curious enough to attend lectures on phrenology in Berlin. Ørsted reported, 'On a skull taken down from a wheel, he [the lecturer] showed us the organ for theft, while the organ for cunning was missing, and this exactly, he suggested, was the reason why he [the criminal] had ended so sadly' [meaning that the thief had been caught, and beheaded].[17]

Howitz's successor as the physician of the Rasphouse, Dr Carl Otto, was an out-and-out phrenologist. He established his own museum exhibiting a large collection of skulls supplying

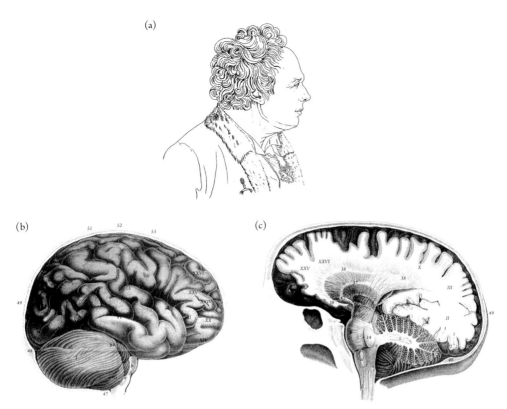

Fig. 103a,b,c. C.W. Gluck (1714–1787), *Orfeo*'s composer. b. his skull and c. his brain with the phrenologically relevant bumps holding his musical talent. Franz Joseph Gall 1809.

the scientific proof of phrenology when checked against the criminal careers of their previous owners. Otto was an enthusiast for Howitz's reform proposal and an influential spokesman for his profession, secretary of the Royal Medical Society, a Freemason, editor of *Bibliothek for Læger* [Medical Journal], and his own *Tidsskrift for Frenologi* [Journal of Phrenology]. 'Some of the best offspring of my spirit are assembled there.' Otto's lectures drew crowded houses and ended with 'a phrenological account on the skulls of the seven criminals that set fire to the Rasphouse at Christianshavn, illuminated by drawings' of the skulls. It was very important for his scientific work

> to have as many skulls of criminals as possible in order to continuously investigate the correctness of my views; not only did I appropriate the skulls of the most notorious prisoners dying in the sickrooms of the prison, before they were put into coffins to be buried in unconsecrated soil, but every time a murderer had been decapitated, I immediately established contact with the relevant people to get the skulls, in which I often succeeded, though not always. To this end I had managed, with money, to make Ravnholdt, the executioner, my friend.[18]

In 1838 Otto asked his friend Ravnholdt, offering a larger sum than usual, to procure the especially valuable skull of Worm, an especially interesting murderer, because he belonged to the upper middle class. However, the delinquent was alerted to the deal and was promised by the prison guard that they would put him into his coffin with his head on, so unfortunately the executioner missed his reward and the doctor Worm's precious head. So Otto asked two of his employees to dig it up. This was not an easy task, for plays and songs had been written about Worm who had become a kind of martyr in the capital, so wardens at the grave were on the watch for body snatchers. One night Otto's two hirelings set out for the grave armed with spades and ropes and began to dig up the coffin. They soon heard voices and steps approaching. Quickly, they broke open the lid and snatched the head, but they had no time to fill in the grave before they rushed off with their prey.

The next day the incident was on the lips of Copenhageners and the robbing of the grave was reported to Kierulf, chief of police, who was not unfamiliar with Otto's interest in skulls. However, the chief of police was a Masonic brother and friend and turned a deaf ear, calming down the enraged constables by saying that most of the body was where it ought to be, and if Worm's head was in the phrenological museum, so what? It would later appear from the museum catalogue that there was a thorough correspondence between the phrenological description of the skull and the character and crime of the delinquent.[19]

The first issue of *Maanedskrift for Litteratur* [Literary Monthly], edited by Ørsted and other members of the learned republic from 1828 onwards (ch. 46), contained an article by Schmidten, the mathematics professor and co-editor, satirically criticising Otto's *Tidsskrift for*

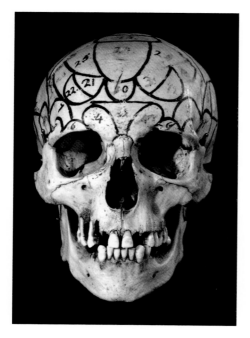

Fig. 104. Skull with marked phrenological bumps. 'Carl Ottos forbryderhoveder Frenologi og det intellektuelle miljø i København i første halvdel af 1800-tallet' [Carl Otto's Heads of Criminals and the Intellectual Fraternity in Copenhagen in the first half of the 19th Century], *Bibliotek for Læger*, 196 (2), June 2004, 132–161. Photo by Pia Bennike.

Frenologi. 'We shall not forget to mention the seven criminals executed for the uprising at Christianshavn. The entire form of their skulls and the organisation of their lives are in complete correspondence with theories for which we have to thank phrenology.'[20] No doubt Ørsted shared his protégé's judgement and his brother's criticism of phrenology.[21]

Howitz's last word pleaded for a staunch deterministic standpoint countering a Kantian moral philosophy. To Howitz the distinction between homo phenomenon and homo noumenon was metaphysical rubbish, and the categorical imperative, whether in Kant's or Ørsted's version, was idealistic nonsense. Howitz considered the ideas of human reason and free will as illusory and fought to make anthropology scientific by describing not what people ought to do but what they actually do. Man's actual behaviour is eudaemonistic, that is driven by a desire for happiness, and reason and free will are simply means to set the course towards a happy life. Contrary to what his opponents maliciously asserted, the pursuit of happiness was not necessarily selfish, because conduct conditioned by laws of nature may very well be socially utilitarian, such as the maternal love that pushes egoism aside in favour of the child. In other words, according to Howitz there was, or would soon be, a descriptive psychology that was not a critical but a positive science, understanding what man really was as opposed to what Kant and the Ørsted brothers thought he ought to be.

> He was a man of simple manners, of
> no pretensions, and not of extensive resources;
> but ingenious, and a little of a German metaphysician.[1]

42 | 1825
Aluminium
Priority and Nationalism

HUMPHRY DAVY, President of the Royal Society, who had been so generous towards Ørsted in London the previous year, suddenly arrived unannounced in Copenhagen on 22 July 1824. First he walked from the Hotel d'Angleterre passing Vor Frue Kirke, the ruined church being rebuilt (except for the spire, at which the British had aimed when bombarding the city in 1807), to Studiestræde, to find Ørsted not at home. He then sent a note by messenger saying that he would remain in the city for the following day and would be pleased to see the professor.[2] He had left England 'on the wings of hope, aided by the paddles of steam' (Davy's metaphors) on board the navy's steamship HMS Comet, testing his 'protectors'.[3] Unfortunately, the outcome was a fiasco, for in the rough weather in the North Sea the 'protectors' had been knocked off. They were zinc plates he had suggested to the Admiralty to be soldered on to the copper sheathing of the hull to inhibit 'the iron disease'.

This problem, of course, had also hit the Danish navy or rather the few ships that were left to it after the British had abducted most of it. Below the waterline the ship was a habitat for seaweeds growing like parasites on the hull thus impeding the driving force of the ship and attacking the planks, which became rotten and had to be replaced in dock.[4] To counter this problem the hull was coated with a copper sheathing fixed by copper nails. But this made a new, economic problem, for copper nails were expensive and so they were replaced by nails made of iron. But alas, another problem turned up: 'the iron disease'. The holes in the copper sheathing corroded for the simple reason that, all unawares, a galvanic element had been constructed consisting of two metals far from one another in the electrochemical series with oak soaked with salt water in between. What to do? In Britain the President of the Royal Society was asked for advice and in Denmark so was the secretary of the Royal Danish Society. Davy's solution of soldering on zinc plates (and later cast iron plates) as a positive pole generating with the copper as a negative pole an electrolytic process and forming a layer of carbonate on the copper made the corrosion stop all right, but only at the cost of resurrecting the original problem. Algae and kelp thrive well on the electrolytic carbonate, and there had been no progress. The scientific insight of the experts was unable to solve the technical problem, so shipyards had to continue fixing the copper sheathing with the expensive copper nails.

The next day Ørsted received Davy in his laboratory in Studiestræde. He proudly showed his guest his latest instrument, a thermoelectrical chain of fifty links. The Royal Danish Society had granted him one hundred rixdollars for the equipment, intended to be more efficient than the earlier versions. The problem with a thermoelectrical chain was that differences of temperatures at the spot where the two metals met had to be considerable and constant to obtain a strong 'conflict'. In existing chains differences of temperature were rapidly neutralised, and hence there was no effect. At first he had tried to solder together eight links and enclose the links in tubes letting hot and cold water alternately flow through them.[5] As 'this battery produced a big effect, and as nobody has so far made a similar one',[6] Ørsted was granted a sum to construct a still bigger one, whose effect would be still stronger, just as the effect of a voltaic pile or a trough apparatus is intensified by increasing the number of discs or troughs. Davy took a close look at Ørsted's thermoelectrical battery and was not impressed. He doubted that the effect would increase proportionately to the number of links.[7]

Davy stayed for yet another day. Ørsted had suggested that they paid a visit to Prince Christian Frederik at Sorgenfri Castle. The Prince would appreciate that, and Davy would get an opportunity to thank the Prince for his gift which Ørsted had brought to him in London a year before. Davy quivered with excitement whenever he had the chance to meet royalty. This time, however, he was disappointed. Sorgenfri did not quite match Windsor Castle, but was rather 'a villa very like an English country house of the second or third class'. Princess Caroline Amalie appeared 'quite blooming', which he did not consider an extenuating circumstance. Davy was more interested in getting permission to shoot snipe.[8]

Davy characterised Ørsted: '[He] is chiefly distinguished by his discovery of electromagnetism. He was a man of simple manners, of no pretensions and not of extensive resources; but ingenious, and a little of a German metaphysician.' Apart from communicating to Davy (in French) his admission to the Royal Danish Society (in return for his membership of the Royal Society), and his request to take care of Dr Otto in London the following year, Ørsted was never in touch with Davy again. In some formal ways they were alike: both headed the scientific societies of their nations, both had been awarded the small Napoleon prize for great discoveries, and both had a weakness for poetry. But their characters, which were opposite in many ways, were not conducive to a relationship on a par with the one Ørsted enjoyed with John Herschel, William Whewell, and Charles Wheatstone.

After his triumphal progress in Europe it must have felt a little empty to be back in Copenhagen again. There were no scientific peers to talk to. In Denmark he was the role model for young academics, Zeise, Forchhammer, Schmidten, and Dyssel. Establishment of the Society for the Dissemination of Science had languished, and the results did not come anywhere near the institutions in London that had inspired him. Ørsted's research activities after his return reflect his encounter with Davy in his laboratory in the Royal Institution, where his discoveries had been made and his fame established. In 1807 Davy had reduced and isolated four elements: potassium, sodium, boron, and chlorine. These were achievements of a kind Ørsted had not yet managed, but perhaps he could supplement the series of elements, if he could provide forces strong enough to do it either from his galvanic or from his thermoelectric battery.

Ørsted now conceived the idea of making a chemical analysis of clay soil based on the widespread assumption that this kind of soil contained a metallic element that might be reduced and isolated. So far nobody had managed to do so, but Ørsted thought out a new method, dividing the process into two phases. The first one was not so difficult, but the second one caused him a great deal of trouble. Now, in 1824 he showed Sir Humphry around in his and Zeise's new chemical laboratory demonstrating his big thermoelectric chain and apparently not being secretive. That Ørsted felt challenged by Davy is an inference on my part. If I am right that Ørsted nourished a hidden admiration for Davy (and vice versa) this explains his efforts after his third grand tour, The Society for the Dissemination of Science and his aluminium project.

Davy only stayed for two days and then left on board the HMS Comet, having sent a note to Berzelius that he would disembark at Helsingborg to see him. It must have been Ørsted who had told Davy that Berzelius was on his way to Helsingborg from Stockholm, together with a young German chemist Fr. Wöhler. Ørsted was in a hurry because the day after the visit to Sorgenfri Castle he had promised to meet Alexandre Brongniart, director of the Sèvres Porcelain Factory, whom he had met a few years earlier through a letter of introduction from Prince Christian Frederik. Brongniart and his son were travelling through the continent of Europe. The plan was that Ørsted would accompany them to visit Berzelius in Helsingborg, and from there the Frenchman and the Swede would travel around Scandinavia on a mineralogical expedition. Davy, incidentally, only showed up long after he said he would, because he had decided to spend a few days angling for salmon in the rivers Lagan and Nissan, while the HMS Comet was moored at Halmstad.

Since Davy was late, the four scientists had three or four days to share each other's company before his arrival. Brongniart was annoyed and joined Professor Nilsson from the University of Lund for a trip to the Höganäs mine. So Ørsted was left alone with Berzelius and Wöhler who were wrestling with the same kind of chemical analyses as Davy (and, indeed, Ørsted). Wöhler was 'apprenticed' to Berzelius, who had his flat and laboratory in the building of the Swedish Academy of Science in Stockholm. On 3 July Berzelius had told Ørsted that he had succeeded in reducing the elements silicon and zirconium.[9] This was a discovery he had made with Wöhler as his assistant, and was a parallel to Davy's earlier achievement. When the mineralogical expedition ended, its members parted. Brongniart went home to Sèvres, and Wöhler to Berlin. On their way they both took a break in Copenhagen, as Brongniart must absolutely see the Prince at Sorgenfri, and Wöhler must absolutely see Ørsted's laboratory. Ørsted made friends with Wöhler, Gitte invited him to dinner with Zeise and Forchhammer, and together they all visited the royal collection of minerals.[10] I take it for granted that they talked about

ALUMINIUM, PRIORITY AND NATIONALISM

Davy's method of reducing potassium and about Berzelius's reduction of silicon, but also about the fact that they had failed to produce aluminium by that method.

From the end of September 1824 to the beginning of February 1825, Ørsted trained a nineteen-year-old pharmacist, L. Schack Køster, in lecturing. He was the first to have been selected to lecture on behalf of the Society for the Dissemination of Science in the provinces. But first he needed to be approved, since he was so young and without practical experience of performing in front of an audience. At the same time he helped Ørsted in the new chemical laboratory, where Zeise experimented with his 'xanthogeneous' compounds that made the whole laboratory stink. Køster did not perform badly at the trial lecture in October, but had to practise more. He was offered an opportunity to do so at a grammar school in Østergade with its pupils as guinea pigs. Køster turned out to be a skilful experimenter and was praised for his lectures to an audience of fifty-six at the society's opening in Aarhus the following summer.[11]

Ørsted's hunt for aluminium was crowned with success thanks to his invention of a new method for reducing and isolating the aluminium in clay soil. In February he was able to

Fig. 105. Jöns Jacob Berzelius (1779–1848), Swedish chemist, secretary to the Swedish Academy of Science and like HCØ a foreign member of Académie des Sciences, Paris. Lithograph by Ambroise Tardieu, DTM.

Fig. 106. Friedrich Wöhler (1800–1882), German chemist. Lithograph by unknown artist. DTM.

present a lump of aluminium to the Royal Danish Society.[12] Having ascertained that his method worked on silicon he could then perform the crucial experiment on the clay soil. It worked, even if the output, a lump of aluminium, was not free from impurities. The new method had two phases, as he explained to his colleagues in the Royal Danish Society. First he let chlorine dry over a layer of glowing carbon and pure clay soil in a bent tube of porcelain. In this way the clay soil was reduced, i.e. separated from its oxygen, and the hydrogen combined with the chlorine making a volatile compound, that he called 'Chlorleerær' [chlorine clay metal]. This was gathered up in a vessel, while the surplus of chlorine and carbon oxide was drained off. In the next stage he heated the volatile compound with potassium amalgam, and in this way potassium chloride and 'Leeræramalgam' [clay metal amalgam] was formed, and the latter was distilled without being allowed to get into contact with air. The ensuing lump of metal, which in a letter to Hansteen Ørsted called 'chlorine argillium', looked like pewter in terms of both colour and lustre.[13] Ørsted and Køster, whose share in the successful experiment remains unclear, had now done what Davy and Berzelius had failed to accomplish.

Neither Davy nor Berzelius was informed about the sensational discovery. Ørsted wrote to Hansteen about his experiments with 'chlorine argillium', whereupon the Norwegian published it.[14] John Herschel, too, was informed, and finally and very briefly Schweigger, who was promised further details. Schweigger became curious, but despite three reminders, he failed to drag more information out of Ørsted. In other words: the discovery was not trumpeted to any great degree. Obviously, Ørsted did not know much about the properties of aluminium, and

Fig. 107. HCØ's bust designed by Mathilius Schack Elo (1887–1948) and cast by Nordisk Aluminiumindustri, 1937. In private possession.

nothing at all about its potential use. This was pure chemical research. The enormous significance this pewter-like metal would have in the twentieth century as the basic material for everything from household appliances and foil to aircraft, was undreamt of in the nineteenth. The only object made of aluminium in Ørsted's day (or rather five years after his death) was an ornamental helmet for King Frederik VII. It would have been cheaper to have made it of fourteen-carat gold.

In September 1827 Wöhler visited Ørsted again. They talked about the new metal and Ørsted explicitly gave his young German friend permission to continue the experiments, as he did not have time to go on with them himself.[15] Wöhler tried to produce aluminium according to Ørsted's method in Berlin, but since he had no success (perhaps he used too much potassium amalgam for too little chlorine clay metal) he looked for new ways.[16]

There is nothing to indicate that Ørsted felt betrayed by Wöhler. He met him at a science conference in Berlin the year after and invited him to join the Scandinavian Science Conference in Copenhagen in 1840. Considering Ørsted's relaxed attitude towards the issue of priority, especially compared to the zeal and haste he displayed in connection with his discovery of electromagnetism, it is remarkable that some of his fellow countrymen have made persistent efforts to prove that Ørsted was first, both in connection with the centenary in 1920 of his other discovery, and later.[17] There are no signs that Ørsted was the least annoyed when in 1827 Wöhler published his discovery of pure aluminium after a new method he had invented himself. Danish historians of science have added nothing to the history of aluminium but national pride. If they had taken the trouble to look into Ørsted's archive, they might have found the notes for his 118th monthly lecture dated 8 January 1828, shortly after Wöhler's publication.[18] It states: 'Perhaps Ørsted's clay metal contained a little potassium metal'. This sentence must be taken as a hint at the usual explanation given by historians of science: Ørsted's 'chlorine argillium' was impure; Wöhler's aluminium was pure.[19] In the same lecture he informed his audience about a series of properties of the aluminium Wöhler had discovered. There was no sign of envy or pique.

The precepts of Christ are so comprehensible,
that any child can understand them...[1]
The word should not just be words,
but also become flesh in every human life.[2]

43 | 1826-32
The Downfall of A.S. Ørsted
The Millennium of Christianity
The Tercentenary of the Reformation

THE 1814 pantheism controversy between Ørsted and Grundtvig had left a wounded curate in the eyes of the learned republic while the professor delivered his victory speech in Latin in Trinity Church to the effect that scientific research is a perfectly valid form of religious worship (ch. 29). Now, years later, Grundtvig once again ventured upon a battle that not only cost him his post as a priest in the Danish State Church and other injuries, but pulled A.S. Ørsted down with him as well. At the height of battle H.C. Ørsted ascended the pulpit of Trinity Church, this time as Vice-Chancellor of the University, preaching the same sermon, that Christianity and science support one another.

This time, however, Grundtvig did not clash with Ørsted, but with one of his colleagues, H. N. Clausen, professor of theology. In August 1825 Clausen published a bulky book titled *Catholicismens og Protestantismens Kirkeforfatning, Lære og Ritus*, ['The Constitution, Doctrine, and Ritual of Catholicism and Protestantism'], dedicated among others to the Bishop of Zealand, Fr. Münter.[3] Grundtvig, the curate of Vor Frelsers Kirke [Our Saviour's Church], flew into rage and in great haste distributed his response, *Kirkens Gienmæle mod Professor Theologiæ Dr H.N. Clausen* ['Retort of the Church against Professor H.N. Clausen DD']. The piece was snapped up and had to be reprinted twice. One thousand five hundred copies were rapidly sold to friends and foes. A few days later Clausen brought an action for slander against Grundtvig, invoking public protection of the freedom of expression. He was granted free legal aid. Only two months later the indictment was ready. Grundtvig was charged, and a year later sentence was passed.

The dispute was about the fundamental doctrines of the State Church. Clausen is generally considered a rationalist by church historians, that is to say an erudite theologian representing the biblical criticism of the Enlightenment and countering biblical literalism and a range of established dogmas concerning, for example, the Trinity, the damnation of unbaptised children, and eternal punishment, arguing that such dogmas have no basis in the Bible. To Clausen, a young professor and the object of the students' admiration, the State Church was out of step with the liberal ministry adopted by the majority of the clergy. The attitude had gained ground that the Holy Scriptures should be subjected to rational exegesis, in which critical reason would weed out absurdities. Hence, the clerical oath was obsolete and should be reformed. The King, the superior authority of the Lutheran Church since the Reformation, should leave it to theological experts to decide the doctrine of the church and should adjust the clerical oath accordingly. A consequence of Clausen's reform would be a constitutional change of the King's position in the State Church. The leading theologians would then become a parliamentary assembly as it were, just as the political claim of the liberal movement during the following decade demanded that the absolute monarchy should be replaced by a constitutional government in the secular realm.

Grundtvig was furious. According to Martin Luther the Bible was the word of God and it was to be taken literally and not diluted by rationalist 'scribes'. Clausen was a dangerous representative of an intellectualisation of the Christian faith, and if the State Church gave way at this point the entire confessional foundation would collapse. A clear sign of this downfall was that there was a revival movement springing in many places, where the laity came to the rescue of the common man believing with his heart, rather than with his intellect. Not that Grundtvig sympathised unconditionally with these lay groups, but they were symptomatic of the insecurity and confusion prompted by the rationalists and their critical approach to the Bible. Clausen was a dangerous heretic, a false teacher undermining popular faith.

From a legal point of view the controversy was embedded in canon law. Both Clausen and Grundtvig demanded a change, but in different directions. Clausen found that biblical criticism had significantly changed the confessional foundation and made the clerical oath, formulated way back in the age of Lutheran orthodoxy, obsolete. Hence, the section of the clerical oath was due for revision. Grundtvig wanted the opposite change, a tightening of canon law, with the state therefore imposing sanctions when it was held in contempt. Now this conflict was narrowed down to a conflict between the freedom and the bondage of the ministry. This was where A.S. Ørsted became part of the dispute partly because of his supreme responsibility for legislation, and partly as the reader of canon law at the Theological College of Education.

Anders gave a great deal of thought to the matter. Ever since 1812 he had not been able to stand Grundtvig. True, he had been praised in lauded in the *World Chronicle*, but given Grundtvig's furious attacks on named people, he would have preferred not to have been. Back then Anders's long sickness had prevented him from responding, and in his place Hans Christian had fired back comprehensively. The Ørsted brothers regarded Grundtvig as an obscurantist and an unbridled Romanticist, conceited enough to pretend that he was speaking on behalf of Our Lord. Anders was rather on the same wavelength as Clausen when it came to biblical criticism, but he disagreed with him when he demanded the adaptation of the canon law to the progress of theology.

In spring 1826, in the midst of Clausen's libel case against Grundtvig, Anders published his legal deliberations in his journal *Juridisk Tidsskrift* [Journal of Law] asking 'Does Danish Canon Law Need Radical Change?'. His reply was negative.[4] He found that the existing section on the clerical oath, unchanged since the Reformation, had shown itself to be flexible enough to contain changing notions of the confessional foundation of the State Church. A.S. Ørsted's article became fatal to his career. His thorough elaboration of the legal problem had concluded in a neither-nor. Yet, the problem was not this outcome. What defeated him was the fact that he had publicized it *sub judice*, while the case was still open and susceptible to influence. Or, closer to the truth perhaps, that somebody in the government used his article as a pretext to have the King remove him. The monarchy could not tolerate that the civil servant carrying supreme responsibly for legislation made his views on a current issue known to the public before the King had made his final decision. Anders acknowledged this point—when it was too late.

In 1817 A.S. Ørsted had gambled with his post when rebuking not only F.J. Kaas, his superior, but also His Majesty. This time he did not get off so lightly. Frantz Dahl has scrutinised the files of the Danish Chancellery and mapped the sequence of events of the summer of 1826.[5] Kaas wanted to get rid of Ørsted and to this end he exploited his mistake. The King, whose theological knowledge hardly went beyond Luther's little catechism, wrote to Kaas declaring Ørsted's article to be 'a labyrinth of ideas that has lapsed into sophistry and useless figments of an overwrought brain.'[6] When it dawned on Ørsted that his position was at stake, he talked to the King on 3 September offering to put down his pen, in other words depriving himself of the precious freedom of expression that he, the best legal mind of the country, had used so productively in his writings. The King accepted his offer, proclaiming

> 'Your article in *Juridisk Tidsskrift* on the question whether Danish canon law needs a radical change has displeased Us because We find in it a public statement of your individual opinion that We do not find becoming in your position . . . However, from your most humble letter of eighth inst. We have been satisfied to learn how you suggest gradually to confine and eventually to terminate the activities of your authorship that in several respects is hardly compatible with the management of the significant positions of trust We have graciously bestowed upon you.'[7]

But as if this was not enough, Ørsted also had to resign his lectureship at the Theological College of Education. Both resignations pained him unspeakably. He had defended the freedom of expression when it was threatened by the hardliners of the chancellery. But without voluntarily giving up his freedom of expression he would risk losing his position as the supreme civil servant for legislation. In the difficult choice between two influential careers Ørsted, at nearly fifty years old, laid down his pen to save his political post. The conscientious civil servant acknowledged his offence. The same day he informed Kaas that he intended 'to conform to the royal will concerning my authorship. I did not expect, however, that this confinement, in many respects so extremely painful to me, would be further embittered by an explicit order'.[8]

A year later Frederik VI conferred upon him the highest title but one of 'Conference Councillor' ['*Konferensraad*']. In this way the absolute monarchy flattered its troublesome, but indispensable subjects. Danish jurisprudence had lost its best brain. This was the price he had to pay to keep his post. Had he resigned it, the matter would undoubtedly have become public and people would have paid homage to its victim as a martyr. Anders did not complain and succeeded in keeping the matter a complete secret, telling only his brother. Now we know that it was Kaas who had finally managed to catch his intellectually superior colleague off guard. From now on Hans Christian would have to serve as an undercover substitute whenever Anders needed a channel of communication to the public.

On 30 October 1826 a sentence was passed in Clausen's case against Grundtvig.[9] The accused was fined one hundred rixdollars payable to the poor relief system plus the costs. But Grundtvig was not let off with a fine. He, too, had to sacrifice his pen and was subjected to advanced censorship for life.

*

On 14 May 1826 H.C. Ørsted mounted the pulpit of Trinitatis Kirke [Trinity Church], used as the ceremonial hall of the University while the new main building was under construction. On this occasion a doctor's degree was conferred on H.N. Clausen for his thesis on St Augustine. According to tradition Ørsted gave his speech in Latin. It was soon translated into Danish and printed in 1849 as part of *The Soul in Nature* under the title 'Christianity and General Education Support Each Other'. This combination of concepts is typical of Ørsted. It is found again as

Fig. 108. ASØ (1778–1860). Lithograph by Em. Bærentzen (1799–1868), after C.A. Jensen's (1792–1870) portrait in oils. RL.

Fig. 109. HCØ (1777–1851). An unsentimental, perhaps slightly idealized portrait. Etching by Erling Eckersberg (1808–1889). RL.

'Science and General Education (support each other)', which was his original proposal for the title of *The Soul in Nature*. After the pantheism controversy against Grundtvig in 1814, confident of his victory, he had talked about 'Cultivation of Science as Religious Worship' from the same pulpit. Now his message was the same as then. Divine reason is revealed in the Holy Scriptures as well as in the book of nature, but if the two contradict each other the book of nature must prevail. Modern biblical criticism has managed to eliminate the irrational features conditioned by history from the Old and New Testaments that contradicted the perpetual laws of nature. To Ørsted the two concepts, Christianity and general education, are in perfect harmony just as their respective institutions lived harmoniously together as neighbours on the same square. When the new University was completed, the bronze sculpture of an eagle above its entrance symbolically aspired towards the gilded cross on top of the tower replacing the destroyed spire of Vor Frue Kirke [Our Lady's Church]: '*coelestem adspicit lucem*'.

In his speech Ørsted made sure he stressed that this harmony was opposed to certain extreme scientific notions prompted by the godless positivism of the French Revolution which had tried to use science as a weapon to combat religion. Regrettably, history was full of examples of this attitude and its opposite: The Roman Catholic Church had attacked Galileo Galilei for defending the Copernican heliocentric world picture, and at the present time Grundtvig's repudiation of the significance of science and mathematics for human affairs was a disgrace. In Ørsted's view Greek science and the Christian Faith had survived up to his own day, not only because they both contained the same dualistic truths, but also because they had promoted and supported each other for almost two millennia. True, it was Islam rather than Christianity

Fig. 110. The University of Copenhagen, a mixture of classicist and neogothic style, built 1831–36 and designed by Peder Malling (1781–1865). Behind the shabby hoarding at the opposite corner a new service tenancy was under construction for archdeacon E. Tryde (ch 62) at the site of Th. Bugge's previous professorial yard. Note the eagle above the entrance. Etching by L.A. Winstrup (1815–1889), CMC.

that had rescued Greek science from oblivion when the Roman Empire had ended in depravity, but Christianity was not to blame for this. On the contrary, throughout the middle ages monasteries had given asylum to the sciences. Biblical studies prompted scientific disciplines, not only philology and theology, but also science and philosophy.

Ørsted used this way of reasoning to support Kant's view of history, that it seems to harbour the reasonable idea that 'man should be guided by religion to develop his full potential'. While refraining from opinions as to what God's plan might be, Kant argued that it looked as if it might be the secret plan of nature that man should fully take advantage of the capacities with which he was endowed by nature, as otherwise this endowment would seem meaningless. It is the assignment of Christians to advance the kingdom of God, and 'this kingdom we can justly call the realm of reason', and if the assignment was going to succeed, it must make use of reason (as Fichte also preached), 'the Divine Spark within us'. The eagle and the cross need each other 'in God's household'.[10]

By emphasising the way in which science and Christianity need each other's help, he stood out as a spokesman for a cooperative science. Throughout the Enlightenment the powerful Faculty of Theology had for two reasons successfully nullified one attempt after another to have the sciences admitted into the University. One was religious, rooted in a fear that science was likely to advance atheism as in France. The other one was trivially economic, in so far as more science by necessity would devour the *corpora* belonging to professorships of theology. Theologians should realize, however, that the Enlightenment had opposed the abuse of Catholic power only. Modern, dynamic physics and chemistry as promoted by Ørsted was neither atheistic nor pantheistic (ch. 29). Quite the contrary, they sided with theism, because by reflection they led to the idea of a maker of the laws of nature and hence to divine reason. Ethically, too, religion and science supported one another, for just as Christ demanded humility, the history of science makes it obvious that a researcher with finite abilities should display humility in his approach to

> 'divine reason, of which the human is but a reflection if only an infinitely small one . . . Hence it is a profound and important feeling that has led us to consider the great scientific institutions of Europe as being linked to religion, to the dissemination of which scientific endeavours will always return, even if they seem to move among earthly objects.'[11]

Finally, the sciences will open the public eye to the beauty and harmony of the laws of nature. Far from being a threat, the sciences offer insight into the three dimensions of divine reason, the True, The Good, and the Beautiful. His justification of celebrating the millennium of Danish Christianity was that it contained 'incorruptible seeds for the education of the spirit!'. At the same time as his brother was forced to give up his freedom of expression, Hans Christian reappeared on the public stage as an old hand at debating, who was tasked with advancing the cause of reason on behalf of both of them, just as in the pantheism controversy.

*

> 'Rarely in history have I met a more scandalous zeal than the one now displayed by those who, law in hand, want to force unity into the church. I have heard them called juridical Christians, and I suggest using that name; for while most of them are neither juridical nor Christians (taken literally), the name is appropriate to designate their spiritual face value. If there were such a thing as prayers of indignation it would be part of their service.'[12]

Such a sharp broadside is not expected from Ørsted's pen, but when it came to debating religious doctrines it was difficult for him to control his temper. In the 1830s he plunged into the public debate acting on his own as well as his brother's behalf. In an article in *Dansk Ugeskrift* [Danish Weekly], he chastised the priests who behaved like the Biblical scribes by exercising a kind of dogmatic terror and damnation against people who worshipped God differently.[13] In this case his impetuosity was provoked by the implication of an ecclesiastical opponent that a Freemason was hiding behind his anonymous mask. At first he thought about giving a full account of his Freemasonic creed, but decided against it.

In 1836 the University's new ceremonial hall on Vor Frue Plads [Our Lady's Square] was inaugurated and the tercentenary of the introduction of Lutheran Reformation in Denmark celebrated. That year Carl Emil Scharling, a professor of theology and the brother of Karen Ørsted's fiancé the chemist E.A. Scharling, sent a manuscript to Ørsted to have it published in

Maanedsskrift for Litteratur [Literary Monthly], which had been established on Ørsted's initiative in 1829 (ch. 46). It was a review of Bishop Tage Müller's *Udvikling af den christelige Religions Troeslærdomme* [Development of the Doctrines of the Christian Religion]. Ørsted found Scharling far too soft on the bishop and gave him some advice as to how his criticism could be sharpened. In Ørsted's eyes Scharling was too impressed by the subtleties of the bishop's exegesis and systematic theology. After all, it was only a speculative work of man intended to confirm orthodoxy, a scheme of the doctrines of the true belief serving to stamp out the infidels and heretics. But Christ did not preach infallible dogma on the Holy Trinity, on the transubstantiation of the Eucharist, on the unfathomable nature of sin and redemption, and similar matters. Religion addresses the heart and the will, not intellect and opinions. 'If one has a living faith, I imagine that a few systematic paragraphs more or less make no big difference in spiritual life.'[14]

Quite a few theologians from the University visited Ørsted in Studiestræde to debate ecclesiastical controversies in the high ceilinged sitting room. Brother Anders, living just around the corner, also took part in these gatherings. Every Sunday afternoon at four o'clock the family met together in turn in Studiestræde and Nørregade, and on Wednesdays at noon they met their erudite associates at Hans Christian's and on Thursday evenings at Uncle Anders's.[15] Among the visitors were the Professors Clausen and Martensen, and Bishop Mynster. Both brothers were looking for harmony and trying to bring about reconciliation in the State Church; they were convinced that to avoid a split with the revival movements the clergy had to watch its step very carefully.

'Christ has never taught us a system!', Ørsted asserted, and the dogmatists of the church should take care not to believe that God will pass judgement in their favour in the contest between the lay and the erudite.[16] Christianity is open to everyone, and according to the gospel the Christian belief is simple and unsophisticated. It is far more important to live like a Christian in one's vocation, which requires a noble heart rather than a sharp intellect.[17] Hence, Ørsted mocked the 'scribes' ironically, calling them juridical Christians, although finding them to be neither juridical nor Christian.[18]

For Peter Hjort's *Læsebog for Ungdommen* [Reader for Youngsters] Ørsted wrote a dialogue between a rational priest and two young people feeling very insecure as they were whirled around in the maelstrom of revivalism; Ørsted advocated tolerance. The confused interlocutors received two pieces of advice: their simple faith should only touch upon the minimum all Christians were in agreement about, and in the conduct of their lives they should make efforts to follow the inner calling they had been given by God.[19]

*

Ørsted derived his minimalist concept of God from the absence of any possible proof of the existence of God. We simply do not have the cognitive skills to ascertain that God exists, nor do they allow us to know that He does not. For Ørsted this was tantamount to saying that God or Allreason as he often called Him is the same as the laws of nature, but he did not elaborate this abstract concept so different from the common anthropocentric image of God. On another level of consciousness than the cognitive, we are invariably taken in by reflective judgement (ch. 25) when as human beings we wonder about 'the starry heaven above us' and 'the moral laws within us'. Reflections on these inconceivable metaphysical questions evoke imaginations that cannot be turned into knowledge, but nevertheless they assert themselves as imperative existentialist considerations. Ørsted formulated his considerations in this way: it is idle to discuss theological doctrines to find the truth about God, and the idler it is the more systematically these doctrines tend to be elaborated. Religious doctrines can be nothing but the work of man. Even the most erudite theologians cannot go beyond our limited cognition. Hence, it is wise to stick to the minimal creeds that people can agree on, and to abstain from speculation on dogma that only serves to divide us.

Secondly, Ørsted urged that since metaphysical speculations are pointless, the essential aspects of Christianity must be God's love of man, which he saw reflected in the truth and aesthetics of the laws of nature, and the ethical requirement that unambiguously stands out from Jesus' admonitions to his disciples: the sermon on the mount, the commandment to love God and one's neighbour, to strive towards perfection. The parallel to Kantian ethics is obvious, for the categorical imperative is when all is said and done a philosophical paraphrase of the commandment to love one's neighbour. A.S. Ørsted's alternative formulation, 'We should make our entire life an impression of the eternal reason that is revealed in us' could then be seen as an equivalent paraphrase of the commandment of perfection.[20] In the midst of ecclesiastical strife between revivalist movements and denominations within the State Church, H.C. Ørsted came forward with his view that 'in every human's life the word (of God) should not merely be words, but should also become flesh'. This was meant as an admonition to the contending parties, theologians as well as laymen that they ought to take an interest in living up to their faith, because it was in their own hands to do so, rather than squabble with one another about dogma concerning the unknowable. If ministers and laymen concentrated on the living, horizontal dimension of Christianity they would have plenty to do and would have less time to worry about their differences on its vertical dimension.

Thirdly, Ørsted, like the Enlightenment philosophers, advocated religious tolerance. Whereas systematic theologians wasted their powder on dogmatic and ritual differences that divided people, they would be better off concentrating on the core that united them. At this point he went further than Clausen, also trying to disseminate an ecumenical Christianity by focusing on what Catholics, Reformed, and Lutherans had in common. In the Park of Wörlitz and on Gendarmenplatz in Berlin Ørsted had seen temples, synagogues, and churches of different religions and denominations live in peaceful coexistence. Hence he vented his indignation in harsh words when he saw ministers of the Lutheran Church give prominence to the doctrine of Atonement and use it to degrade foreign religions such as Islam, Judaism, and Hinduism.[21] Should these other believers merely be brushed aside without making an attempt to comprehend their religious values as they appear in their eyes? Did not the Lord say: 'for the tree is known by its fruit', Ørsted, well versed in the Bible, asked.[22]

> To experiment is not to work in the dark,
> but a great art of science.[1]

44 | 1827–8
Family Life and Conferences Abroad

O N 14 August 1827 Ørsted celebrated his fiftieth birthday, and King Frederik conferred upon him the title of *Etatsraad* [titular Councillor of State]. Now the two brothers were level again. While the middle-aged Hans Christian threw himself into one engagement after another, Gitte was no less busy taking care of family life and the household in Studiestræde. Since their wedding she had given birth to seven children and she was not finished yet. During the first twelve years of marriage she had been pregnant for more than five years and another pregnancy awaited her. She was also responsible for Søren, their troublesome foster-child. He went to a grammar school for boys, Borgerdydskolen, together with Christian (11) and had become more diligent after many admonitions not to waste his talents. Karen (12) went to Miss Sibbern's school for girls. She loved drawing, flowers in particular.

When her husband was travelling, Gitte saw more of her sisters. Her older sister Bolette, married to the dyer Knud Lind, had seven children, and her younger sister Anne Marie had four, so between them the three sisters had eighteen children, quite a kindergarten. They played as cousins do and felt at home in each other's flats. Anne Marie was the sister, who as a maid for the attorney Beyer on Gammel Torv had left her rice pudding behind, which Gitte had found and eaten when looking for her on her escape from the bombarded city in 1807, with Aunt Møller's silver in her apron. Anne Marie had been widowed early and was now a lone mother of four children.[2] She was living in the vicinity of St Petri Church, close to Gitte, when Professor H.M.V. Klingberg, surgeon to the navy made a strange proposal of marriage. Both sisters wondered exactly how this proposal was to be understood and therefore consulted Ørsted, who knew Klingberg personally.

The fact of the matter was that Klingberg's sister was keeping house for him. But she was leaving to be married, and so Anne Marie could take over the job as housekeeper. If she accepted the offer and moved into Klingberg's flat she would be able to save her present rent and spend it on the education of her children instead. However, it was no easy matter for Anne Marie to cohabit. The professor assured her that in addition 'he would offer her his hand; but he also said that in his youth he had had a relationship with a bad woman with whom he had a child, and to whom he had given the promise that if he were ever to marry he would marry her' And being a man of honour Klingberg was unable to grant Anne Marie marriage without breaking his promise. What did Ørsted (then in Paris) think? Should her sister move into Klingberg's flat as a housekeeper pure and simple, or had she better hold out longer and require marriage? After all, the future of her children was at stake.

There was not time to wait for Ørsted's advice. 'My sister has accepted the offer. Yet, she has had second thoughts as to what people will say. Klingberg is considered an honest man and his reputation is good, and to always have the protection and advice of a man will greatly benefit her, particularly as the children grow up.' Ørsted was not sure about this. He had no wish to dissuade her on moral grounds, but neither would he strongly advise her to accept.[3] Reading between the lines, Gitte was suggesting that if the unfortunate event should happen that she was widowed herself she would feel compelled to do as her sister was doing. If Anne Marie was driven into the arms of Klingberg by her widowhood, even more did Gitte feel plunged into discouragement and loneliness by her grass widowhood.

> 'This winter I feel so strongly how a wretched widow is abandoned, and yet I have so many good things that she is missing. You would not believe how much I miss you, my dear Ørsted; your letters help somewhat; but they are not able to disperse the despondency that has entered into my soul this past month. The first four months passed fairly well, I did not feel it really, I kept thinking of the joy of seeing you again. This past month it has been different, I do not have the joy in me anymore, only the grief that you are absent. If it goes on like this you will only receive letters of complaint from me.'[4]

There was no marriage to Klingberg.[5] Anne Marie simply moved into his flat with her four children. How the cohabitation went we do not know, but probably it went well for in 1829 Ørsted still considered Klingberg his friend.[6] However he died in 1835, and once again Anne Marie was on her own. Soon she did get married, to Nicolai Hasle, a shipbroker, who adopted her four children, and they were given his surname. Birgitte, Peter, Theodor, and Sophie came safe and sound through the vicissitudes of life. Like Gitte's sons Anne Marie's two boys entered Borgerdydskolen.

Apart from giving birth to and looking after her children Gitte also took responsibility for the household, errand boy, servant, and maids. She had to make copious purchases, for Studiestræde was the centre of a great deal of social life. After Anders had been widowed a second time, he would turn up pretty much every evening at eight o'clock to play cards and have tea, and as long as Oehlenschläger had a tenancy in the bishop's residence he was a frequent guest as well. As time went on Ørsted's home became a regular place for social and cultural gatherings. Zeise, Forchhammer, Schmidten, and Lehmann needed no invitation. Talented students such as Martensen, a theologian, and during the 1830s polytechnic students such as Scharling and Holten were often invited to dinner; so were young poets like F. Paludan-Müller and J.L. Heiberg, who often brought his mother, Mrs Gyllembourg, with him. Finally, they had visitors from abroad, Hansteen and Berzelius from Norway and Sweden, Schumacher, the astronomer from Holstein, Doctor Brandis from Bonn, and Weiss from Berlin, to mention the most frequent guests.

Every year they celebrated Anders's birthday in Studiestræde, even though the date of 21 December was a bit awkward. The party would include not only the entire family as well as Adam Oehlenschläger, his brother-in-law, but also friends and government colleagues. Three days later the family would gather around the Christmas tree in the big dining room, and Anders with his foster-children and Miss Bohn would join them.

*

When the anxious fourteen-year old boy Hans Christian Andersen walked about knocking on the doors of the Danish cultural elite in September 1819, Ørsted was on his expedition to Bornholm. But the following year Andersen had heard the name of the celebrity and came to see him in Nørregade. The visit came off well; the lanky boy was allowed to borrow books from the professor's library and was asked to come again, for which he required no second invitation. At first Andersen had a job as a singer at the Royal Theatre, but when his voice broke his career on the stage ended, apart from walk-on parts, such as a torch bearer in 'The Marriage of Figaro', but even for that he was soon given up as hopeless by the board of directors.

He had started with the idea of showing his talent as a dancer or a singer, and when this hope was shattered he started to write for the stage, only to have his manuscripts rejected and returned.[7] So then he had knocked on the door of Collin's yard in Bredgade, where Ørsted had begun his career in 1804 lecturing alongside his scientific instruments. Jonas Collin had succeeded Hauch as director of the Royal Theatre and Andersen had hoped he could help. He also walked all the way to Bakkehuset at Frederiksberg, approaching Knud Lyne Rahbek who had the final say on plays submitted to the theatre.

The Ørsted brothers, too, were frequent guests at Bakkehuset, meeting Adam and Christiane Oehlenschläger, Kamma's sister there. Daringly, Andersen recited his Nordic tragedy *Alfsol* in five acts for these leading lights in the world of aesthetics. Ørsted and Gitte responded kindly, and so Andersen optimistically concluded that everything was well, and his confidence in himself led to *Virak i min egen Phantasie* [Homage in my own Fantasy]. Unfortunately, Rahbek turned *Alfsol* down as well, but while waiting for his verdict, Andersen had started collecting subscribers to his *Ungdoms-Forsøg* [Youthful Experiments] intending to publish them under the pseudonym of William Christian Walter—'William' referring to Shakespeare, 'Christian' to himself, and 'Walter' to Scott. They were the greatest poets of his short life.[8] He then knocked on the doors of the Royal Palace and those of the most distinguished court servants offering his *Ungdoms-Forsøg* for one rixdollar and three marks a piece. Princess Caroline Amalie subscribed to four copies, and he had the book printed.

Still, Rahbek had found a few pearls in *Alfsol*, and so, basing their decision on these, the board of directors agreed that Collin, who had also become one of the directors of the Foundation ad usus publicos, should recommend Frederik VI to grant Andersen a sum of three times four hundred rixdollars so that he could acquire education and reason, that is general grammar school knowledge. Jonas Collin probably considered the budding poet a natural talent in need of education in order to develop. Ørsted, without estimating the boy any differently, found it easy to identify with Andersen's innocent character in that he came from the provinces, as he himself and Anders had, to search for happiness in the capital city. The King agreed to Collin's proposal and together with J.P. Mynster, a member of the Board of Directors for the University and Grammar Schools, they decided that for the coming five years Andersen should subject himself to Dr Meisling's gruelling regime at Slagelse Grammar School.[9]

When Andersen went on his first Christmas holidays to Copenhagen he called on the Ørsted family in Nørregade. The professor was on his triumphant travels in Europe, but Gitte invited him to supper, presenting the boy with a two-rixdollar coin. Andersen had found a second home, one that he came to enjoy for the rest of his life.[10] The beneficiary felt that it was the promptings of providence that had thrown him into the arms of these noble people, who kept his spirits up when he had lost all faith in himself.[11]

As the years went by, Andersen would turn up on Christmas mornings and make his famous paper collages and write verses on small notes to be attached to the Christmas presents. Sometimes he would recount his fairy tales in the liveliest of ways, so that the Ørsted's children would hear the wind whistling and the owl hooting, or at other times he would dance around, flinging his arms and legs about and showing off. The children knew that if they shouted 'Faint, Andersen, faint!' he would topple over on the floor feigning unconsciousness. The atmosphere was lively and there was an air of expectation. Sometimes there would be so much tomfoolery that Andersen's pen would go off at a tangent until he realized that his verses no longer fitted the presents, and he would have to start afresh. These stanzas for Karen, however, were very much to the point:

> At 97 in Studiestræde//is Christmas joy,//stars and suns shining from the tree//and fabric for a frock//An apron, too? But who shall own it?//Dear me! She is blushing, rushing up, and curt-seying,//the pious Karen, mild and sweet//the noblest in Studiestræde![12]

This was the Swedish Christmas custom from Mrs Gyllembourg's parties that was carried on as a tradition in Studiestræde. In the evening candles would be lit everywhere, but if notable visitors provided an occasion for Ørsted to show off, the modern gas lamps in the dining room and his study would illuminate the flat.[13]

On New Year's Eve it would be Anders's turn to host a party. After his second wife Mathilde had died, he took over a part of his brother's previous tenancy in Nørregade. So between Studiestræde, Nørregade, the bishop's residence, and the professorial yards, the entire cultural elite lived within a few yards of each other. On New Year's Eve Anders's flat would be decorated with spruce branches, and there would be more presents for the lovely children. Once they played at being in the Deer Park north of Copenhagen. Miss Bohn was dressed as the women selling hot waffles, and an organ grinder played music. Perhaps the children would rehearse a

Fig. 111. HCA, paper collage of ballerinas dancing in a clearing presented to Mathilde Ørsted (1824–1906). DTM.

comedy, and Oehlenschläger would break in teasingly, revealing how clumsily Uncle Anders had been performing, when as a young poet he had staged Holberg's comedies in Aunt Engelke's dye works, with Steffens playing the part of the conceited Erasmus Montanus and Anders the unintentionally funny Per Degn.

So, both the Ørsteds had a hectic life. Often they would reach the limits of what they could manage. But there was a difference. Her activities were ceaselessly bound to their home, whereas now and then he was given the opportunity to escape from the maelstrom of obligations. This happened when he was travelling in Europe, where he was allowed to slow down a little, and where the recognition he met provided him with a fresh impetus. In spring 1827 Ørsted was invited to a congress by the medical society in Hamburg, so they decided that for once Gitte should accompany him on a journey abroad. This was to be the first and last time. The plan was that Gitte and Karen would go by the packet boat to Kiel and by stagecoach on to Hamburg, where Ørsted had rented lodgings. Eventually they would go on a holiday together.

*

Ørsted travelled ahead to Altona using his grant, and stayed with his friend, Professor H.C. Schumacher, who had been allowed to keep his residence in Altona, although he had succeeded Thomas Bugge at the observatory at Rundetaarn in Copenhagen. Schumacher and Ørsted

FAMILY LIFE AND CONFERENCES ABROAD

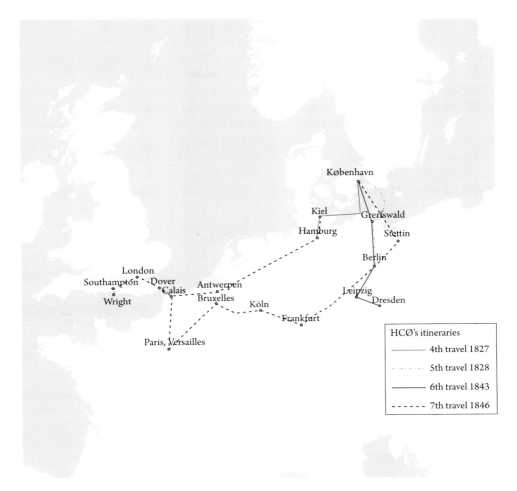

Fig. 112. Map of Europe showing HCØ's last four itineraries.

had one thing in common: both had felt inconvenienced by Bugge. It was Schumacher's task to complete the geodetical survey of the kingdom, and Schleswig, Holstein, and Lauenburg remained to be done. Moreover, Schumacher wanted to be as close as possible to his admired teacher, C.F. Gauss, the mathematician in Göttingen. He enlarged the astronomical observatory in Altona, situated on a slope declining towards the Elbe, and published the renowned journal *Astronomische Nachrichten* [Astronomical News], which gave him a central position in European astronomy.[14]

The Congress was intended to deal with earth magnetism. A worldwide project was being hatched. The idea was to make a world map showing the magnetic declination all over the globe and if possible to locate the geographical position of the magnetic poles. The map was to show the isogons of the earth, the lines connecting the points where declinations are identical. The magnetic north pole would be located where all isogons met. Such an ambitious project involving Hansteen,

445

Schumacher, and Gauss demanded international collaboration. Alexander von Humboldt, the geographer, had important contacts in Latin America, but above all it was crucial to have the British Empire included in the observations, and to this end John Herschel was a key person. Ørsted, too, was an obvious participant with his international network and talent for organisation.

The weather in Hamburg was terribly wet; Ørsted caught a cold and felt unwell. On 22 May he delivered a lecture on electromagnetism. He had spent seventy-eight rixdollars on a smart new tailcoat for the occasion. He told Gitte that he was satisfied with his performance, and the response he had received exceeded his expectations. Gitte did not enjoy meeting groups of foreigners, with whom she had to speak German, so she and Karen only arrived at the beginning of June. We do not know how long their holiday lasted nor what they did, for of course they did not need to exchange letters about it. So, it is an open question whether their only holiday abroad together was a success.

In September the following year Ørsted and other Danish and foreign scientists were invited to take part in a conference in Berlin arranged by Gesellschaft Deutscher Naturforscher und Ärzte (GDNÄ) [The Society for German Scientists and Physicians]. The steamboat took four hours to Malmö. Then from Ystad passengers spent twenty-one hours on board ship to Greifswald, and from there two days by stagecoach to Berlin. True, they had to change horses on the way, but an average speed of six kilometres an hour was hardly any faster than that at which the travellers would have been able to walk. Ørsted was accommodated with his friend Weiss in the University. He was heartily welcomed and the Weiss family was handed the presents Gitte had bought for them. Mrs Weiss had arranged for her husband's study to be Ørsted's bedroom, and upon his return he thanked them for their hospitality by sending his copperplate portrait.

The GDNÄ was founded in 1822 in Leipzig by Lorenz Oken, the biologist. The idea had originally emanated from German *Naturphilosophie*, based on the close connection between physics and metaphysics, or as the co-founder, the Romanticist painter, Dr C.G. Carus phrased it, the view of nature cannot and should not be separated from speculation. 'This is exactly why a construction of reason can only to a certain degree correspond to the empirical observation of nature, and conversely the empirical observation of nature can only to a certain degree lead to a lawfulness based on reason.'[15] The ideas of Oken and Carus had been the basis of GDNÄ's activities until the 1827 congress in Munich, but from now on the society was to be different. This year's chairman, Alexander von Humboldt, expressed the new guidelines in his welcome speech in the University's ceremonial hall as follows:

> In their true and profound feelings for the unity of nature the founding fathers of this society have linked together all branches of physical knowledge—descriptive, mensurational, and experimental. In the title of the society the terms 'scientists' and 'physicians' are almost synonymous... But although it is crucial not to break the bonds linking the exploration of organic and inorganic nature, the increasing scope and rapid development of the sciences will make it necessary to place detailed lectures on individual disciplines each in their own section, even if we keep the general scientific meetings open to the public in this hall. Only in narrower circles, only among men sharing common research interests will it be possible to have oral discussions. Without sincere exchanges of views among truth-seeking men this society would lose its principal raison d'être.[16]

From now on the GDNÄ was intended to distinguish itself from scientific academies in Europe by no longer forming a closed assembly, to which the initiated members gave papers which were subsequently printed in proceedings that nobody could find it in himself to read. Instead, researchers within the same discipline should get to know one another, meet face to face and exchange ideas, and to this end the living word was the obvious means of communication.[17] Moreover, the society aimed at establishing a platform where friendships between researchers across national borders were struck up. It was intended that the next annual meeting would to take place in Vienna, but this plan had to be shelved for a bizarre reason. The Austrian Emperor had granted funding for two Viennese scholars to enable them to come to Berlin, but as they had arrived to pick up the passports from the chief of police they had been told to stay at home and not to accept the travel grant. Clemens von Metternich, the Austrian Minister, had banned their participation. The spectre of nationalism emerged at an early stage as an obstacle to international cooperation between scientists. Despite the fact that the Viennese belonged to the German language area, Metternich had not allowed them to travel to Prussia, which was Austria's principal rival, attempting national unification.[18]

With Humboldt emphasising the importance of public backing, international cooperation, oral presentation, and personal contact, it was almost a matter of course that the programme was interspersed with social events and entertainment. The King was invited, to lend lustre to the opening, and concerts and musical evenings [*Liedertafel*] were arranged, as well as visits to the armoury and to the Egyptian Collection in Schinkel's new museum. One evening Ørsted attended a performance of Händel's 'Alexander's Feast', and he was so infatuated with the concert that for a while he put his mind to Rousseau's and Felix's musical enthusiasm in his dialogue on the tones, which diverted his mind from preparing the lecture Humboldt had asked him to deliver the next day.[19]

Being a guest of honour Ørsted was especially welcomed by Humboldt in his inaugural address, and his numbered seat was right in front of the lectern. The royal family was sitting in the gallery above, and poetry by Goethe and Schiller decorated the walls. A men's choir sang a cantata by Felix Mendelssohn-Bartholdy. Ørsted improvised his lecture to an audience of more than four hundred people including ladies. Talking without a script conformed entirely to the new spirit. The drawback for us is that the lecture does not appear in Ørsted's archive. He talked for half an hour on 'The proofs of my theory of the electromagnetic effects as opposed to Ampère's'.[20] To Gitte he reported that he was dissatisfied with his performance, but comforted himself with the fact that several listeners praised it spontaneously; King Friedrich Wilhelm III assured him that he had enjoyed his paper and that there was no need to apologize. Fortunately notes were taken, and his paper is reproduced in Oken's journal *Isis* under the title 'On the magnetism of the electrical current'.[21]

The lecture was a defence of his original theory of electromagnetism. What he had discovered in 1820 was simply that the electrical conductor is surrounded by a magnetic force making a magnetic needle respond. Ampère and other physicians interpreted the experiment differently: the molecules of the magnetic needle were surrounded by electrical circuits. Or, the magnetic effect was reduced to an electrical one. This interpretation was later challenged by Ampère himself who now asserted that identical electrical conductors repel each other, while conductors charged differently attract each other. To Ørsted, however, the decisive point was that the effect of parallel conductors on each other is not rectilinear, but transverse, which was convincingly proven by Faraday's elegant experiment. This circular motion was the new thing about his discovery that had baffled everybody, because it was not known, let alone expected, by anyone. Ampère's reduction of electromagnetism to electricity had to be erroneous, if for no other reason than the fact that electrical forces could not penetrate the glass isolating the magnetic needle. It had been established by several experiments that when a disc of glass is placed between the electrical conductor and the compass, the needle reacts at the moment the circuit is closed. Changing the positive and the negative pole makes the needle deflect to the opposite side. Consequently, electricity has an effect that most appropriately must be called magnetizing.

So, there is an electrical polarity, the positive and negative forces of which move in opposite directions in the conductor. This appears in the experiment of separating water, for instance. The electrical and magnetic circuits act simultaneously in a field around the conductor. If it is assumed that the electrical force moves along an equatorial axis east-west, then the magnetic force will move along the meridian north-south. Ørsted's apology was probably prompted by a slight embarrassment at not saying anything new, but only repeating what he had said before, including to Ampère, who had not budged.

Ørsted's discontent was real enough. He did not feel he had succeeded in moving the audience over to his side against Ampère. His dissatisfaction with the performance must have annoyed him. Why otherwise would he return to his laboratory in Studiestræde with the crucial experiment already in his mind to disprove Ampère's electrodynamical theory to the effect that electromagnetism is nothing but opposite electrical currents? In April 1829 he presented to the Royal Danish Society his *experimentum crucis*, the experiment that would finally prove to the world who was right: Ampère's mathematical theory, which according to Ørsted had been 'unable to attract many supporters outside France, and even there views are divided', or Ørsted's physical theory.[22]

What was the aim of this crucial experiment? Ørsted bent a magnetic needle so that one end pointed vertically upwards, the other down. If Ampère's theory was correct, i.e. if the needle was really the equivalent of an electrical conductor, the horizontal part would be neutral and the bent ends would neutralise one another, and there would be no effect, neither repulsion nor attraction, when an electrical conductor was placed vertically, that is parallel to the bent ends. However when Ørsted closed the electrical circuit in the conductor, the needle moved and kept moving even when he gradually moved the conductor into a horizontal position. Ørsted believed he had eliminated the possibility of an electrical circuit in the needle (as alleged by Ampère) by bending its ends into opposite directions thus nullifying the effect of possible electrical polarities. When the needle deflected he considered it proven that magnetism was not reducible to electricity.

If Ørsted had been convinced by this *experimentum crucis* himself one would presume that he would have presented it to the GDNÄ congress the following year, which actually took place in Hamburg. Instead, he lectured from the entirely new standpoint of mathematical methods in physics, a topic he had never dealt with before, and concerning a method he had never pursued. He turned the argument against Ampère, whose theory he claimed was not physical, but mathematical. Addressing an audience oscillating between the French and the

Danish theory, he regretted 'the manifold of mathematical developments making it difficult to comprehend his theory and impeding many physicists in forming their own opinion'. He did not master the integral and differential calculi Ampère used and hence he was unable to prove the French argument wrong on its own premises. The minutes do not say whether he demonstrated or even mentioned his *experimentum crucis*, so I take it that his lecture did not make a definitive impression on the audience.

As already mentioned, the GDNÄ congress of 1830 took place in Hamburg, and the largest group of participants, sixty-seven, came from the Denmark, including the Duchies. The Holsteiners did not figure independently in the minutes as would happen a few years later. Fifty-eight Prussians took part. There were twelve Swedes, but no Norwegians, and only two from France. This was due to the outbreak of the July Revolution in Paris that spread to the Netherlands, and subsequently caused Belgium to become an independent kingdom. The recreational part of the congress was an outing for two days to Heligoland, a Danish island recently ceded to Britain. Ladies participated, of course, and after a voyage of a little less than seven hours cannons and cheers from the small island in the North Sea saluted the erudite tourists. Late in the evening as those on board were dancing, the SS Willem de Eerste dragged its anchors and went adrift. A strong gale played havoc with the ship, and there were waves five metres high, but nevertheless after two days the party returned safely.[23]

Ørsted was the key speaker to the general public. As in Berlin he spoke without a script, this time on scientific method. Most confusingly, papers in his archive bear different headings.[24] Once again Oken's notes in *Isis* are helpful. According to him the lecture was on 'The difference between physical and mathematical representation, especially when they concern the same truth'.[25] According to Oken, Ørsted said something like this: 'A considerable part of physics can be described by applied mathematics, and the perspicacity and consequence by which this is done has enticed many a physicist into imposing a mathematical representation on his subject. The reason I touch upon this problem is that I want to introduce a discussion

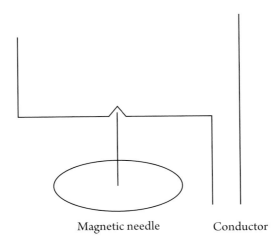

Fig. 113. HCØ's crucial experiment, reconstruction by DCC. *NS* II 479–81, *SSW* 539–541.

on method. Most scholars agree that the method of physics is the experiment, but their justification is not quite clear. When we are in the process of discovering and explaining the laws behind physical change we need to bring about these changes before our own eyes. When one continues to reflect on this matter, one realizes that experimentation is not the same thing as working in the dark, but a great scientific art. Nature is only apparently static (*natura naturata*); actually everything is changing (*natura naturans*). Hence, we do not comprehend the phenomena of nature unless we know their effects. In other words our natural knowledge is limited, and the inner essence of things is concealed from us, as Kant said. We must learn from the activities of nature.'

'Physics must learn how to represent these actions, and consequently we cannot confine our investigations to material experiments, but must venture into thought experiments dealing with actions and their consequences. Hypotheses belong to this process, where we establish relationships between factors and investigate whether they correspond to reality. But this is not the only reason. If for instance we investigate the relativity of motions, we soon realise that we move ourselves. If we move on a ship there is a triple motion, because we move in relation to the ship, and the ship in relation to the Earth, indeed a quadruple, because the Earth moves in relation to the solar system. Everything can be described mathematically, but yet it also demands a physical visualisation in a thought experiment, if we really want to understand the relativity of motions. The phenomenon must become visual to us. Nevertheless', Ørsted continued, 'mathematics and physics express the same truth.' His point was that the physical process must be described generatively, that is in becoming a new thing instead of remaining as the old one. Hence a mathematical representation like the one Laplace used in his *Traité de méchanique céleste* [Treatise on Celestial Mechanics] for instance, when describing capillary action, did not belong to physics, since the mathematics obscured the very phenomenon it was supposed to render intelligible. Laplace saw the pressure of the liquid as a negative force. But the physicist does not see a negative force in front of him. An activity cannot be negative, but must be understood by the fact that the pressure of the liquid is weaker than the positive force of the capillary tube.

Many readers of *Experimenta* were surprised that there were no mathematical equations in those four Latin pages. Paris, in particular, was disappointed, for it had always been the ambition of adherents of corpuscular theory to express the laws of physics in mathematical language. Laplace's model, Newton's *Philosophiae Naturalis Principia Mathematica* was emulated by Laplace's *Traité de méchanique céleste* which stated: 'The universe and everything it contains is subject to mathematics'. The consequence of this principle according to Schmidten was: 'If one knew all the forces of the universe, one would be able to calculate the positions of all the bodies that together constitute the universe.'[26] It was Schmidten's objective in associating with the French mathematicians 'to study everything that our present mathematics contains concerning its most important applicability to science. For physics must in its perfect state be mathematics'. He wanted to make it his life's work to enable this idea to come true, but not with the same physical concepts that the corpuscular theorists used, of course. As the dynamical system developed it would need a mathematical language of its own. Ørsted and Schmidten were looking forward to such a 'generative' mathematics.

Hence, it was a painful loss to Ørsted that Schmidten died at the age of thirty-two without having solved the problem.

Throughout the decades following Ørsted's discovery European science was split into two groups. One followed a linguistic-physical method developing new visual models and concepts suitable for thought experiments and physical experiments (Ørsted and Faraday). The other pursued a mathematical-physical method (Ampère, Hansteen, Gauss).

During Ørsted's visit to Weiss in 1828 they decided to take up the thread from their study circle on the Kantian metaphysics of nature that had united them in the dynamical project almost three decades earlier. At that time the challenge was to incorporate Winterl's chemistry and Ritter's electrochemical theory into their project. What was the state of play now after the discoveries of electromagnetism and thermoelectricity? Had the dynamical system been consolidated or did it need revision? They agreed to put their exchange of thoughts on paper, for if one commits one's thoughts to paper, one is compelled to sharpen one's formulations, and if obscurities occur, one can urge the other to express himself more accurately. It was their intention to publish their exchange of thoughts either in Copenhagen or in Berlin. Weiss was modest enough to indicate that the publication would be sold not only thanks to Ørsted's name, but also because Ørsted's style was clear while his own was obscure. Ørsted opened the debate with a ten-page note in January 1829. During the spring Weiss elaborated his response in four letters, and Ørsted concluded the discussion by commenting on them. He presented to the Royal Danish Society his 'Investigations on the internal nature of bodies with particular regard to the controversy between the atomistic and dynamical systems'; but considering it a work in progress he did not want to report any part of it for the moment.[27] Their plan was not carried through, neither was it published. It came to suffer the same fate as his *Kraftlære* [Theory of Forces], though their correspondence is preserved.

Ørsted's thought experiment began by dealing with the problem of Kant's argument that 'matter', i.e. that which fills space, exists continuously in space. Kant's argument was mathematical. Everything, matter as well as forces, is infinitely divisible, because a space, no matter how small, will always be further divisible. But does it follow from this, Ørsted asked, that matter is evenly distributed in space, as Kant seemed to think? His reason for putting this question was that water, for instance, is chemically the same whether it exists in solid form as ice, in liquid form as water, or in a gaseous state as steam. Hence, he needed an element to be chemically unchangeable but still able to change form over time.

He called his theory 'ento-strophic' to indicate that his elements (unlike the corpuscles of the atomists) had an internal force ('ento') and was able to change, ('strophe'). (The original Greek meaning of 'strophe' referred to the movements of the chorus on the stage, and later came to mean that which makes the chorus move, namely a new strophe, or stanza.) These elements are the smallest unit of matter, consisting of a nucleus surrounded by electromagnetic forces. The electrical force orbits around the nucleus inducing a transversal magnetic orbit, corresponding to an equator and a meridian. The electrical orbit is either positive or negative. Elements charged identically repel each other, and elements differently charged attract each other. Hence the elements can either be separated from each other or combine, and thus Ørsted managed to explain why matter is not necessarily evenly distributed in space, but changeable (ice, water, steam), although the elements of oxygen and hydrogen remain unchanged. His elements are dynamical, and the motions happen in orbits or oscillations. The latter are orbits, too, but have an extremely short axis.

According to Ørsted's ento-strophic theory the electromagnetic forces propagate as he had outlined in *Ansicht/Recherches* that is in perpetual stop/go rhythms. He believed that each motion of these elements is rectilinear and conforms to the inverse-square law, but the interaction of the forces of different elements is determined by a manifold of rectilinear motions in different directions. This way of reasoning makes possible the transverse motion characterising the electromagnetic field.

Ørsted believed that this theory could be extended to cover the phenomena of light and heat, or at least be the beginning of such a more comprehensive theory. Light is caused by the very fast oscillations of elements, and heat by slower ones. Different electromagnetic orbits would give different substances different specific heats. When Weiss asked Ørsted if he imagined his elements as solid or liquid, he was told that such predicates are only applicable to substances, not to individual elements. The state of a substance depends on its temperature, and since it is the electromagnetic orbits around the nuclei that generate light and heat, the question is rooted in a misunderstanding.[28]

To make his metaphysics as visually clear to Weiss as possible he suggested a metaphor. Like Newton he imagined an analogous relationship between microcosm and macrocosm. He epitomised the microcosm in this way:

> 'Imagine a stellar nebula, in which we see nothing but a point of light, disregarding the fact that it consists of innumerable suns and other celestial bodies; now, think of it as diminished in all respects and put it into your hand, while retaining the same visible size it had in its position infinitely far way; this stellar nebula will now resemble a body, so that the previous immense distances between the celestial bodies are now pores, and its internal directions of motion appear as physical and chemical properties. The compression of the body will not happen without resistance, and the reciprocal approaches between their parts will generate innumerable beams of light and heat in order to complete their to-and-fro motions between the parts more quickly, and hence also to pass through the surface more frequently and thus radiate heat and light.'[29]

These images of the inner nature of bodies were Ørsted's contribution to their exchange of ideas. If he were to elaborate on all the problems of his theory, he would have to write a whole book and still perhaps be unable to hit the weak points, he told his friend. Neither Weiss nor Ørsted managed to write works on the dynamical system. Ørsted's contribution was written at the same time that he was wrestling with plans to establish a polytechnic institution in Copenhagen, and it did not fully satisfy him.

It is the scientist who envisages great and
far-reaching innovations for the industrialist.[1]

45 | 1828–9
The Polytechnic Institute

Ø RSTED DID not launch the initial plan to establish a polytechnic school.[2] But it was his vigorous effort that transformed Professor G.F. Ursin's initiative of 1826 and made it look very like the scheme Ørsted had drafted in 1808 for 'a Practical Institute for Experimental Science' intended for two classes of pupils: practitioners and scientists.[3] Ursin's proposal was addressed to the King, that is to say the Government, on whose behalf the Board of the University transferred the scheme to Ørsted, in whose hands it underwent a root-and-branch metamorphosis. In 1829 it became known to the public as the Polytechnic Institute under Ørsted's leadership.

Ursin was a professor at the Academy of Arts teaching mathematics, geometry, and astronomy to young artists, architects, and artisans. At that time it was considered a matter of course that educating artists and artisans went hand in hand, because their work had manual skills in common. It went beyond conventional thinking to imagine that the sciences had something to offer the crafts. Dyers, brewers, and soap makers might pick up elements of organic chemistry, once this science became capable of analysing dyestuffs and processes of fermentation, but this would still belong to the future. On the other hand, the mechanical part of physics had contributed to technical innovations since Archimedes. The Industrial Revolution took off by developing new instruments, tools, and engines, but the modern inventions had little to do with traditional craftsmanship and medieval guilds. The generally accepted view was that technological progress owed a great deal to the sciences, though as we have seen, when the King sent Ørsted on a mission to leading industrial nations in Europe to deliver a report on the utility of the sciences, he returned almost empty-handed.

Professor Ursin tried to map this new technical territory in his *Magazin for Konstnere og Haandværkere* ['Magazine for Artists and Artisans']. He familiarised himself with the latest innovations from abroad, Britain in particular, and was intent on transferring them to Denmark. He had joined Ørsted's Society for the Dissemination of Science and he published Ørsted's monthly lectures in his magazine. He realized that his pupils stood no chance of acquiring the skills necessary to carry out technical modernisation at the Academy of Arts. This required mechanics who could not only handle, but also produce machine tools. This was the unsolved problem that Ursin's proposal addressed. It had probably been thought out in collaboration with Ole Winstrup, a pioneer of advanced technology who had built the first Danish steam engines and established the first iron foundry and factory for agricultural implements, Mariaslyst, at Frederiksberg. Winstrup was also a member of Ørsted's society and had pieces published in Ursin's magazine.

On 26 November 1827 Ursin submitted his humble proposal for a polytechnic school, asking that it might be discussed in depth with the committee that he imagined would be set up and of which he saw himself as an obvious member. However, the Board of the University sent it to the Senate, which passed it on to a committee to be headed by Ørsted. As he was outside the University, Ursin therefore did not become a member of it. Ørsted sketched the work of his committee in these often quoted lines: 'Professor Ursin has proposed a genuine school for artisans which in itself would be a useful institution; but by sending his proposal to the University the King has made it clear that he wants a more comprehensive institution....'[4] Making this far-reaching interpretation of the royal will Ørsted was condescendingly hinting that Ursin's proposal was amateurish and aimed too low, while at the same time establishing himself as the government's expert on technology as well as higher education. Ørsted was working towards being a shrewd politician and organiser.

Ursin's institution was designed to walk on two legs: theory and practice were to be integrated. To this end it consisted firstly of a workshop equipped with metal lathes, smithies, and planing benches, as well as a chemical laboratory. The foreman was to make the necessary equipment such as lathes and vices, which were not then available on the market. In addition, he would produce sundry other devices and models of implements that lecturers would need, and, of course, would train the pupils to master all the relevant skills. Secondly, a collection of minerals and a chemical laboratory for metallurgy and iron founding was needed to ensure the coherence with the workshop. The chemical laboratory would have its own foreman.

Then, two scientists would be appointed. One would teach pure and applied mathematics as well as mechanics, the other general and technical chemistry and metallurgy. The crucial point of Ursin's project was to integrate the theoretical sciences with the practical training in the workshop and laboratory. This integration would take place in two ways: all teachers should be doubly qualified, meaning both that the foremen should have scientific qualifications, and that the scientists, if possible, should have appropriated the relevant skills, that is be well acquainted with practice in workshops and laboratories. Practitioners and scientists should know each other well so that they could work together as peers and impart an understanding of the coherence of theory and practice to the pupils.

Finally, the pupils should learn drawing, and the drawing office would also have its own foreman teaching geometrical drawing, machine drawing, and free-hand drawing of models from the collection, as well as shading and perspective. Courses preparing applicants for admission would be included and so would language courses for pupils lacking the normal prerequisites. Lectures on technical subjects would be arranged during the winter from eight to nine by the scientific staff or by external scientists. Ursin's institute would be headed by a board consisting of the three foremen and the two scientists, as well as of two honorary scientists or practical men. From these seven people a director would be elected for no more than three years at a time. Although there is no evidence for it in the source material, we may conjecture that Ursin intended that he should be the mathematics teacher and Ole Winstrup foreman of the workshop.

The government, however, approved Ørsted's alternative proposal and so a rather different polytechnical institute emerged. His new institution was made up of three parts: (1) training in scientific experimentation (physics and chemistry); (2) lectures on scientific subjects according to textbooks; and (3) training in workshops (mechanics, technical chemistry, and drawing). There was no suggestion that the three parts might be integrated. At the Polytechnic Institute the theoretical and the practical elements were hierarchically related. The leadership was in the hands of the science staff, and the foremen of the workshops were no part of it. The entire staff was appointed in advance, to this effect:

professor at the university H.C. Ørsted	physics	director
Professor at the University W.C. Zeise	chemistry	
Professor at the University H.G. von Schmidten	mathematics	
Reader at the University G. Forchhammer	chemistry and mineralogy	
Professor at the Academy of Arts G.F. Ursin	mechanical engineering	
Professor at the Academy of Arts G.F. Hetsch	technical drawing	
Mechanic Ole J. Winstrup	workshop foreman	

The first five members of staff, Ursin, Ørsted and his three protégés, constituted the Board of the Polytechnic Institute. The three protégés already had safe jobs at the University and were also attached to the Society for the Dissemination of Science as lecturers. Ursin was the only member of the Board from outside the University, and Hetsch and Winstrup had no say at all.

Neither Ursin's nor Ørsted's proposals left any time unused. The syllabus was scheduled for eleven hours a day, six days a week over two years for both schemes, though the lessons differed. Whereas Ursin allocated six hours a day for workshop training Ørsted only allocated three, and whereas Ursin had two science lessons, Ørsted had three to four plus three for experimentation. This is the conspicuous difference. Priorities between theory and practice varied inversely, but it would be wrong to assert that Ursin's proposal lacked scientific contents.

The target group was the same in the two projects: civil servants and military officers, entrepreneurs, and mechanics. But their admission requirements differed, and so the type of students would differ correspondingly. To enter Ørsted's institution the applicant had to fulfil the same entrance requirements as for the University, i.e. *examen artium* including history, geography, Latin and Greek; artisan skills were not required. In Ursin's scheme the applicant only needed to read and write well, and master Danish and German, mathematics and geometry, as well as to have completed an apprenticeship or the equivalent, for instance training in a private workshop 'enabling him to make use of the higher practical training on an equal footing'.[5] This vague formulation was due to the fact that there were no guilds organising the modern trades needed in iron foundries, machine shops, or workshops producing scientific instruments, for the simple reason that there had been no guild established in Denmark since around 1750, and hence there was nobody to test qualifications at the close of an apprenticeship. In his proposal Ørsted compared Ursin's institution to 'a German *Gewerbeschule*' and his own as 'an École Polytechnique'. Ursin's plan aimed at an integration of three elements: technical drawing, workshop training, and science lectures, while Ørsted's scheme offered separate science lectures prioritised above workshop practice. A year later the proportion was still more lopsided.

After one year in existence Ørsted's Polytechnic Institute was subject to radical changes. It was the experience of the Board that science lectures took up so much of the students' time that they could not possibly do three hours' work in the workshop each day. Hence the Board decided to discontinue workshop training altogether in favour of the theory. This change made it hard to consider the Polytechnic Institute anything but the establishment of a Faculty of Science at the University through the back door. In order not to leave the workshops entirely empty, they were put at the disposal of young boys destined to learn a practical trade. Most of these boys were troublesome lads in need of discipline.[6] But access to the workshops, the drawing office, and the laboratory for a fee brought in a lot of money to the institute.[7] This change severed all links between theory and practice.[8]

Ørsted's new institution was accommodated at the University, it recruited most of its staff from the University, it had the same entrance requirements as the University, it had full access to the University's collection of scientific instruments, its finances were administered by the bursary of the University, and it had its prospectus in common with the University. Already the Society for the Dissemination of Science had joined hands with the University in terms of premises and staff. In his proposal Ørsted had offered financial support of six hundred rixdollars annually for the Polytechnic Institution from the Society for the Dissemination of Science, or to be quite precise he had offered to transfer the grant of the Reiersen Foundation to the Polytechnic Institute in return for free access for society members to the institute's lectures at the University. Ørsted's proposal was to recycle staff and funding to reduce the price of the Polytechnic Institute thus making it competitive and irresistible in relation to Ursin's initiative.

When Jonas Collin raised his voice in protest against an arrangement that he found rather too cunning, Ørsted defended it by referring to the benefits of rationalisation:

> 'It is obvious that it would only be a waste of time and effort if the Society for the Dissemination of Science continued to offer lectures in this city, when another institution is offering the

same thing, and even more comprehensively. True, the lectures delivered by our society have a particular character and take place when they are most convenient for manufacturers and artisans, but on the other hand when the society makes a certain contribution to the lectures of the Polytechnic Institute against favours done in return, we obtain the double satisfaction of contributing to an important institute as well as saving the means to pursue other purposes.'[9]

Since the country was short of money during the agricultural crisis of the 1820s, Ørsted's idea was to achieve as much as possible of his long cherished plans without burdening the slender coffers of the nation more than absolutely necessary. The Polytechnic Institute cost the University nothing. Here Ørsted had learnt from previous failed initiatives designed to institutionalise the sciences. The various attempts made during the Enlightenment had all failed, because the Senate, dominated by theologians, was unwilling to share the fixed number of *corpora* (the annual income of the University's landed property) with new professorships.[10] Against this background, and whatever one might think of Ørsted's understanding of what a polytechnic was or should be, one must admire the skill with which he seized the opportunity to promote his ideas. He managed to incorporate the professorial yard occupied by Jens Møller into the Polytechnic Institute for a small sum and have it redesigned into lecture halls and workshops. He also succeeded in getting his protégés, recently equipped with posts at the University, transferred to his new institute. In this way he obtained total loyalty from his staff and full control over the Polytechnic Institute. This was a great leap forward as compared to his former weak position vis-à-vis the Senate: just think of the chicanery to which he had been exposed when his collection of instruments was removed in 1812-13, and his ideological struggle against the majority in connection with the establishment of the Students' Association in 1820. Now, he was the Director of the Polytechnic Institute and in no need of support from the Senate.

When the workshop training of polytechnic students was cancelled in 1830, and the facilities left in the hands of teenage boys who were not required to follow the science lectures, all visions of uniting manual and scientific work evaporated. Ole Winstrup, the foreman, resigned for the following reasons:

> 'I ask the honourable Board of the Polytechnic Institute kindly to approve my resignation from the daily supervision of the workshop, although I do hope in the future to be of use, partly by providing practical advice, drawings, and models to the teacher of mechanical engineering [G.F. Ursin], and partly by opening my workshop [at Mariaslyst] to give access to the polytechnic students to carry out or take part in such works as cannot be made in the workshop of the institute, such as turning, boring, founding, as well as making large parts for machinery like steam engines, hydraulic presses, etc.'[11]

Apparently, Winstrup had not yet grasped that it was not the intention of the Board that the Institute's students were going to learn to build 'steam engines and hydraulic presses, etc.'. He was disappointed that the integration of science and practical workshop skills, the basic principle of Ursin's proposal, had been irrevocably abandoned. To sacrifice this principle in order to keep a bunch of badly behaved boys in order was no part of Winstrup's vocation.

When in his letter of resignation he listed the facilities to which he could provide access, he was hinting at the amateurishness of the Institute. Against this background it seems ironic that Ørsted was at the same time arranging for his postgraduates to enjoy all the privileges of guild rights without having to be tested as journeymen.[12] Ørsted's polytechnic students were educated to hold posts as civil servants in government offices, not to roam around in working clothes with tools in their dirty hands. Civil servants were educated at the University and their social status was evident from the words they spoke. The polytechnic graduates were to assume the same level of education as the jurist and the minister, far above that of the artisan.

At the same time as Winstrup felt he was wasting his time as foreman of the workshops, Professor Hetsch complained that drawing had no high rank at the Institute either. The undergraduates did not take the discipline seriously. In his thesis on the establishment of the Polytechnic Institute the historian Michael Wagner has indicated that it was rumoured or possibly even made clear that undergraduates were safe to neglect the courses held by Ursin, Hetsch, and Winstrup in late afternoons, because it was their accomplishments in the science courses run by Ørsted, Zeise, Forchhammer, and Schmidten in the mornings and early afternoons that would determine their fate at the final exam.[13] This is an assumption that evades documentation, but on the other hand is difficult to disregard because as a matter of fact the non-academic teachers felt unvalued and left within a short time. The great catastrophe happened at Ursin's examination in mechanical engineering which took place for the first time in 1831. All five students failed. They could not fail Hetsch's or Winstrup's workshop training, because this was not part of the examination requirements, but Ursin's mechanical engineering was. This was a new discipline in Denmark. There were no textbooks; there was no theory and not even a coherent description of the subject. Hence, Ursin was in a precarious position.

It was as if the exam in mechanical engineering was doomed to failure. Especially since an external examiner from the rival Military High School, Colonel F.E. von Prangen, an officer and practical man, was reputed to see red when he crossed swords with an academic. Prangen ruthlessly failed the whole lot. To a certain degree it is not only the student but also his teacher who is evaluated at an exam. So, were the teacher and his course to be blamed for the failure? Hardly! There was no better in the country. Ursin was a professor of mathematics and astronomy, an observer at Rundetaarn, and author of textbooks in drawing and mathematics, editor of *Magazin for Konstnere og Haandværkere*, the most important portal for technical news from abroad. Professor Hetsch was co-author of the textbook of drawing, renowned architect and expert of descriptive geometry. Winstrup was the owner of the biggest iron foundry and factory of agricultural implements with a branch in Germany, a millwright, and the producer of the first Danish steam engines. In addition they were dedicated, they had committed themselves to a proposal, whose technical disciplines were as important as the science disciplines. It was hardly their fault that the integration of science and practical engineering skills collapsed like a house of cards. Colonel Prangen had been a member of the commission to produce a plan for the Royal Military High school and had introduced his expert opinion with this volley: 'The exaggerated scholarly deportment is damaging to practical skills'. Now he had his satisfaction. The failed students complained, but honestly speaking, could they expect to triumph in mechanical engineering, when they had only followed the course for a quarter of the time or not at all? Ursin had to resign. He was replaced by J.A. Dyssel, Ørsted's previous amanuensis and translator of *Experimenta*. He had no success either.

Ørsted saw Ursin's proposal as an emulation of the *Gewerbeinstitut* of Berlin that the Prussian College of Commerce headed by Director Beuth had established some years earlier and that Ursin has reported on in his magazine. The aim of the *Gewerbeinstitut* was to train mechanics to work and to copy modern British machine tools, especially for metal factories, that is to say lathes, screw-cutting machines, boring machines, etc. Beuth seems to have understood that if Prussia were to catch up with the industrial lead that Britain had taken, she needed machine tools to produce other machines. The first step of this process was to acquire British machine tools, which was a delicate affair requiring export permits or, failing that, industrial espionage. In 1826 Berlin's *Gewerbeinstitut* had acquired three British lathes, probably illegally. The Germans took them apart, measured and drew the individual parts, copied them as models, and then produced their own precision lathes which were used to cut screws or teeth for cogwheels and turn rollers. They did the same thing with boring and planing machines, and others. The Prussians had understood that machine tools were an indispensable prerequisite for fulfilling the production potential of the industrial revolution: carding and spinning machines, power looms, steam engines, paper machines, printing machines, scientific instruments, locomotives, etc. In Denmark, by contrast, a two-year polytechnic education put students through scientific courses that only enabled them to administer the importation of machines. As a consequence British firms built the first railway lines in Denmark while the Prussians built their own. Technical skills were the decisive prerequisite, not theoretical knowledge.[14] Surprisingly, it had not dawned on Ørsted that the industrial revolution owed practically nothing to science, but was almost exclusively the work of skilled mechanics. Or he pretended not to know. British pioneers such as Smeaton, Rennie, Maudsley, Stevenson, Brunel, Telford, Donkin, Kay, Cartwright, Wilkinson, and others, had never been to university, and the few Danish pioneers (Bidstrup, Caspersen, Gamst, Smith, Winstrup, Nordberg, Baumgarten, Eickhoff, and Sørensen) were neither postgraduates in science nor had they been polytechnic students.[15] The Polytechnic Institute did not even have a precision lathe.

Ursin was in charge of the rebuilding of the two professorial yards.[16] The largest lecture hall, where the inauguration ceremony in the presence of Frederik VI took place, was on the first floor over the gateway in St Pederstræde. It was about one hundred square meters with four windows on each side, and there were eleven benches rising towards the back for an audience of two hundred people facing a big table for experiments and a lectern. The hall was heated by hot air circulating in brick shafts. The lighting was by gas, as in London. The gas was produced under Zeise's laboratory in the garden and channelled through hidden ducts in the floor under the lecture hall supplying one big and four small gas burners with thirty-six flames.

On the same floor were two large rooms for the physical instruments that were removed from Ørsted's professorial yard in Studiestræde. Above it was the collection of minerals and the library, and at the opposite end, above the collection of instruments there was a dark room for optical experiments. On the ground floor, opposite the gateway, was the chemical laboratory with a large forge with cast iron plates and next to it a reading room for chemistry courses where Forchhammer was based. Next to this there were two small workshops, one for carpentry work and another one for lathe turning. In the basement below, Winstrup, for a short while foreman of the workshop, was in charge of a smithy and a file-cutting establishment next to a huge stove heating the hot air circulating around the entire building. There was running water in the laboratories from a tank on the loft to which it was pumped through pipes. Running water could also have been used to flush water closets,[17] but Ursin and Ørsted did not take advantage of this recent invention, since, as is apparent from the ground plan, they built four latrines in the yard.

Zeise maintained his chemical laboratory in the garden building and had it extended to make room for 'the gas device with two retorts, two coolers and two purifying machines as well as a small and a large gas container holding about four hundred and fifty cubic feet of gas.' From this building the gas was conducted in pipes to the two professorial yards. If anyone was exposed to the risk of leaks or explosions of gas it was Zeise, but his laboratory was already stinking from xanthic acid, so that nobody could stand being there anyway. In his lecture room a stove was built according to Faraday's design in his *Manipulation of Chemistry*, 'serving as a central heating system as well as a stove with a sand bath for chemical experiments'.

At the inauguration on 5 November 1829 Frederik VI and Prince Christian Frederik and their consorts were present in the largest lecture hall together with the Board of the University, the senate, and the board and staff of the Polytechnic Institute. Before the principal speech a choir struck an appropriately emotional note by singing a cantata written for the occasion by Ørsted's lifelong friend Adam Oehlenschläger: 'Nature! From your exuberance you give Blessing to the sons of the earth; The strong researcher, the profound thinker You endow with honourable gifts. The one who descends with the torch Of science to your dark home, You adorn with your miracle, So that he can admonish spirits.'[18]

Perhaps Oehlenschläger's lines used too many Romanticist metaphors given the Enlightenment ideas Ørsted unfolded when he stepped onto the lectern, bowing humbly towards the royal guests and exclaiming: 'Great, merciful King! The institution, whose establishment we are celebrating, belongs to those evoked by Europe's recent Enlightenment.'[19] Neither aristocratic rule nor mob rule, but education should be the proper way to social mobility. The learned

Fig. 114a, b. Ground plan of the Polytechnic Institute. Studiestræde above, St Pederstræde below. Ground floor to the left, First floor to the right.

a. entry	i. drawing office	**Second floor**
b. library	k. lecture hall	a. drawing room
c. lecture hall	l. senior common room	b. amanuensis
d. entry	m. preparation	c. drawing teacher
e. study for chemistry	n. senior common room	d. amanuensis
f. study for the amanuensis	o. collection of instruments	e. physics teacher
g. analytic laboratory	p. room for optical experiments	f. meeting room for Board, collection of instruments and models
h. preparative laboratory	q. collection of instruments	

J.T. Lundbye, *Polyteknisk Læreanstalt*, Kbh. 1929, 128–129.

(a)

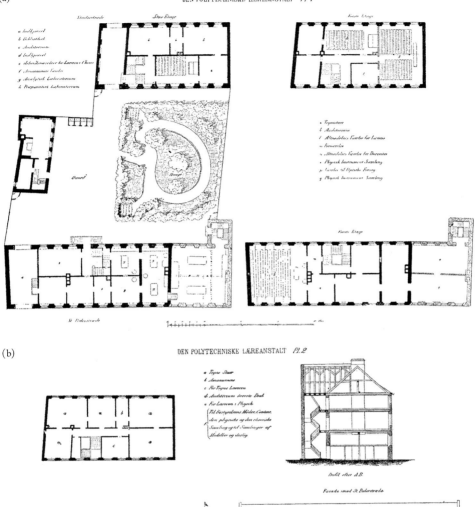

(b)

republic must be the servant of the monarchy and it should be open to any citizen of talent who is ready to use it. It is thanks to science that the arts and crafts are ennobled,... 'Reason is on the march, and as we realise that nature obeys the laws of reason, our faith in God as the creator of the laws of nature is strengthened'. Faith brought meaning and courage to our lives. Finally, experimental science had revealed and exposed the mistakes of *Naturphilosophie*. It led its Romantic adherents into idle speculations, while experiments were conducive to action. Provided his majesty was still awake at this stage of Ørsted's speech, he may have felt reminded of his conversation with Steffens back in 1807, when he extinguished his hope of getting a professorship in Denmark, because allegedly his lectures drove the minds of his subjects crazy.

In the second part of his speech Ørsted outlined his polytechnic programme of education. The aim of the institute was to provide the students with a scientific understanding of the laws of nature and make them confident with the reasons why the arts and crafts work as they do, for this knowledge would enable them to improve these activities. On the other hand, polytechnic graduates would not be specialising in specific crafts or skills. The advantage of the institute was that students interested in technology could acquire a general overview 'without having to subject themselves to the coarse treatment that barbarism, although on the wane, has left in the guild system'. The institute addressed young people who had already acquired a certain education, such as sons of well-to-do families, and who would not be put off from investing their capital in industry by any lack of scientific knowledge. It was crucial that civil servants were enabled to advise technicians and investors in order to promote their initiatives. 'Men graduating from this institute will, each in his own part of the country...form new points of departure for the dissemination of useful knowledge', Ørsted continued, probably glancing pointedly at the patron of the Society for the Dissemination of Science, Prince Christian Frederik, on the front bench. 'The national spirit will gradually take a more practical direction. The inventive spirit will assert itself more and more. The natural resources of the nation will be used more industriously. Inventions from abroad will increasingly find their way to us. And through all these efforts together and under God's blessings, prosperity will prevail, and love for our country and citizenship will be nourished and grow.'[20]

At this point Ørsted turned to the relationship between science and technology, a relationship that at the time (and even today) is unclear and unsettled and for some good reasons. Should the two activities be joined together like dancing partners, as Ursin had suggested? Or did scientific theory distract and confuse the artisan instead of supporting him, as Prangen found? Should the technician be content with rules of thumb and ignore the scientific causes, as was done in the *Gewerbeinstitut* (according to Ørsted's biased view of this institution)? If the scientist should join the artisan on the workshop floor, should not the theorist acquire practical skills as Jonas Collin advocated?

Since his first grand tour Ørsted had reflected on these issues. His present views were not spur-of-the-moment, but well thought through. They may well have been strategically motivated, which is to say that his argument was shaped by his personal interests in the matter, while his alleged care for the technological development of the nation was the price he had to pay to obtain a free scope for his research. Leaving such speculation aside I shall take Ørsted's arguments at their face value and characterise the polytechnic ideal of the science-technology relationship he defended.

Ørsted's main point was that there is a hierarchical relationship between science and technology according to which the scientist, through his cognition of the laws of nature can envisage new technological innovations, while the mechanic or artisan uses his manual and technical skills to put flesh on the ideas. In a certain sense the two competences are equal, as in Plato's organic model of society, for as little as the scientist masters the manual skills, just as little does the artisan possess the insight to invent new technology or understand the reasons why it works. However, in a different sense the relationship between them is hierarchical, because there are only a very few scientists in proportion to the multitude of practical men. Hence, the theorist should not waste his costly time in the workshop appropriating manual skills, and it would be a costly mistake to make scientists teach artisans the rules of thumb. Firstly, the complex of skills is as big as the complex of crafts, which makes the idea absurd, and secondly, the scientist makes a fool of himself in the workshop, because the practitioner will invariably wonder how a passing visitor can teach anything to someone who has been apprenticed for seven years in a trade. This was a sharp rejoinder to Collin's view.

On the other hand it would be useful to the artist, entrepreneur, and artisan to attend science lectures, as already happened at Ørsted's monthly lectures intended for such an audience, because only when they understood the basic science would they be able to improve their trade. 'In most cases it is the scientist who makes the great and visionary inventions for the tradesman; but he must in return acquire scientific insight so as to understand these inventions and appreciate their value; and he must also make a number of small extra inventions in order to entice the basic thinking of the scientist back into his workshop where so many complicated after-effects occur.'[21]

'The misunderstanding of this friendly communication between the scientist and the tradesman causes unspeakable confusion', Ørsted went on. Did he assess the situation properly when in his speech he claimed a hierarchical relationship, considered modern technical inventions as mainly science-based, and actively opposed the integration of theory and practice? Was he not aware that the incipient industry owes little or nothing to science, that inventors generally had never set foot in a University, and that Britain had become the workshop of the world without a single institution reminiscent of his new Polytechnic Institute? Add to this that his own dynamical project had led to the discovery of electromagnetism, but nobody, including Ørsted himself, could envisage its relevance to technology. To Ørsted the polytechnic bond between the scientist and the tradesman was as loose as between the metaphors of the first (already quoted) and the following last stanza of Oehlenschläger's cantata:

> Beat hammer your strong iron
> In honour of our King.
> Sail ship On the distant wave
> In honour of our King.
> Grind mill With fastest wheels
> In honour of our King.
> Wave field With your yellow wheat
> In honour of our King.

Oehlenschläger might have coordinated his poetic efforts a little better with his friend's rhetoric, for the four technologies of the cantata emphatically disavowed the relevance of science to technology.

When the Polytechnic Institute had graduated fifteen classes in fifteen years Ole J. Rawert, a factory inspector, made an appraisal of the occupations of all sixty-six of these polytechnic candidates. This showed that seventeen had become teachers (some for the Society for the Dissemination of Science) and fourteen had jobs as civil servants, while thirteen worked either for their own industries or as employees, and the last twelve had indeterminate jobs as 'practical chemists or mechanics'. What distinguished the twelve practitioners from the thirteen working in industry is not clear. But evidently Ørsted's declared aim was fulfilled in so far as well over half of the graduates became civil servants and teachers, while fewer than half had jobs in private industries. The output was modest, only four or five per year, and their contribution to the technological modernization of the nation was equally modest. Rawert cited the ruthless criticism of Professor Wilkens (to whom I shall return in ch. 58):

> 'The Polytechnic Institute is not programmed to the needs of industrialists; the courses are far too sophisticated for the students, most of whom cannot follow or understand them. Consequently, they must pick up from the courses fragments that they think will be useful to them and piece together a new course, which, however, is not coherent. From all this we may conclude that there is a need for an institution for higher education of artisans, but this need has not been satisfied by the establishment of the Polytechnic Institute.[22]

Life itself must be a work of art.
It should express the idea living in us.¹

46 | 1829–33

The Literary Critic
The Airship

In OCTOBER 1828 when Hans Christian Andersen had endured headmaster Meisling's polishing of his awkward manners and had been filled up with appropriate doses of culture and refinement, the 23-year-old budding poet moved into a garret in central Copenhagen, looking out over the tower of St Nicolas Church. Dinner was served him by turns by noble citizens, by the Ørsted family every Friday, and later also on Wednesdays.² As he writes in *The Swineherd*: 'When the kettle boiled, the bells would tinkle and play the old tune: "Oh, my dearest Augustine, All is lost, lost, lost". But that was the least of it. If anyone put his finger in the steam from this kettle he could immediately smell whatever there was for dinner in any cooking pot in town.'³ Gitte's dishes, Ørsted's edifying conversation, and the children's expectations exercised a magnetic attraction, so sometimes he would show up spontaneously. What they talked about we only know indirectly through the innumerable traces Ørsted left in Andersen's fairy tales and novels, as a maxim about the True, the Good, and the Beautiful, about 'Ariadne's Clew', about his childhood memoirs, about eccentric colleagues, about his travels and royal dinners… It must have looked rather comical to see the two talking, the young, lanky Andersen, nicknamed Little Hans Christian, at almost 185 cm, and Big Hans Christian, the middle-aged, slightly stout professor, below average height, probably no taller than 165 cm, animated and ruddy, sometimes absent-minded sometimes affable with an upturned and wandering glance, because he squinted and was very short-sighted.

His patrons' educational project could not stop until Andersen was qualified to enter the University. A coach was paid to guide him through the examination requirements. And while

Andersen was sweating over his books and as usual on Fridays was being fed in Studiestræde, he once stumbled into a young fellow, looking for all the world like Napoleon, albeit appearing shy and modest in comparison with the emperor. 'Are you heading for the exam this year?' Andersen inquired. 'Indeed!' said 'Napoleon' with a wry smile, 'I am heading there!'. 'So am I!' Andersen exclaimed, undauntedly chatting about his studies with his supposed fellow-sufferer, until Ørsted told him that his interlocutor was actually his examiner in mathematics, Professor H.G. von Schmidten. There was some embarrassment at the examination, where Andersen was supposed to prove his qualifications in mathematics to an examiner who would rather hear about the title of his next work.[4]

The *examen artium* being negotiated, Andersen was admitted to the University much to the satisfaction of his patrons. They were proved right in spotting his young talent, which, however, he preferred to manifest as a poet rather than as an academic. He found little time to delve into *philologicum* and *philosophicum*, for now 'the medley of fantasies and whims that had pursued him flew like a swarm of bees into the world' in his first published book *A Walking Tour from the Holmen Canal to the Eastern Point of Amager*, 'a humorous work, a kind of fantastic arabesque quite indicative of my entire personality'.[5] This arabesque can be seen as a juvenile rebellion against conventional ideas of education and literary norms in so far that Andersen broke with the usual, realistic framework of a narrative. Although the title was explicit that the tour ended at the eastern point of Amager, in fact the writer went on to the island of Saltholm (in the Sound), leaving the reader to his own imagination should he wish to know what happened next, as the last chapter consisted only of punctuation marks but no words.[6] This surrealistic confusion of styles and the subversion of chronology (the midnight walk only took eight to ten hours over New Year's Eve because it only took place in the mind of the poet) recommended itself to the taste of many critics.

One evening Andersen met Johan Ludvig Heiberg at the Ørsteds' dinner table. Probably the host had arranged it so, because Heiberg had given *A Walking Tour* a favourable review in *Maanedsskrift for Litteratur* [Literary Monthly], the journal Ørsted had recently founded. This was a significant compliment, and Andersen devoured it with evident relish. He handed over to Heiberg, who had recently raised the roof of the Royal Theatre with his *Elverhøi* [The Elf-Hill], some poems that he wished to be printed in Heiberg's journal, *Kjøbenhavns Flyvende Post* [The Flying Mail of Copenhagen], under the pseudonym 'h __'. One evening when the poems had been published, Andersen was on a private visit when the master of the house entered enthusiastically flashing Heiberg's journal: 'Here are two splendid poems!' he said delightedly reciting them. 'He is a fellow, this Heiberg!' Andersen was not upset by this mistake. Writing poetry as good as Heiberg's was flattering, and who would guess that 'h__' stood for Hans Christian Andersen and not for Johan Ludvig Heiberg?[7]

Andersen's second literary experiment *Love on St Nicolas Tower or what does the Pit say?* is a burlesque in Heiberg's satirical style. The pit is encouraged to vote in the form of applauding or booing at the main character, a night watchman who is oscillating between marrying his daughter to another night watchman, who is already married, or to a sailor, with whom the girl is in love. Heiberg reviewed the play in *Maanedsskrift for Litteratur* pinpointing 'its essential error, which is that it satirises what we have already overcome, namely the chivalric tragedy.'

Andersen accepted that, but the pit judged differently, acclaiming the young writer making his début for the stage, no longer as a dancer or a singer, but demonstrating his talent as a playwright. His dream of achieving his destiny in life was taking shape. This success infused him with the energy to resume his preparation for the exam. In October and November 1829 while Ørsted was busily engaged with the inauguration of the Polytechnic Institute, Andersen was examined for *philosophicum* by Ørsted. He answered all questions in physics and chemistry well, but in the end the relieved professor put another one about his own hobbyhorse:

'Tell me, please, what do you know about electromagnetism?'
'I don't know that word, Sir.' Andersen replied.
'Think about it! You have already answered everything so well, and you must know something about electromagnetism.'
'But there is nothing about it in your textbook on chemistry, Sir!'
'True', the professor said, 'but I've talked about it in my lectures.'
'I've attended them all, except one, so it must have been exactly that lecture, Sir, when you talked about it, for I know nothing whatsoever about it, I don't even know the word!'

Ørsted smiled at this bold confession, nodded his head and said, 'What a pity you didn't know, because otherwise I should have given you a "*præ*", but now you get a "*laud*" for you have answered well!'.[8]

Just imagine that despite many dinners in Ørsted's house the poet had no recollection of having ever heard the word 'electromagnetism'!

*

Towards the end of the 1820s many professors of the University and the Royal Academy of Arts began to meet in each other's homes in a way that was reminiscent of Mrs Herz's salon in Berlin. They were intended as informal monthly social events for erudite gentlemen. They would act as hosts in turns and hosts were free to invite anyone they liked in addition to the original participants. Ørsted took part, of course, and so did Oehlenschläger, though he found them boring.[9] After a few years these professorial parties petered out, but at the end of 1828 on Ørsted's initiative a smaller group of fourteen learned men decided to form an editorial board for the interdisciplinary journal of culture, *Maanedsskrift for Litteratur*. They took it upon themselves to strike 'a sober-minded and dignified tone in public debate to counterbalance the irresponsible treatment of scientific topics and the plebeian attitude that prevailed at that time'.

In handpicking the board of editors Ørsted skimmed the cream of Danish erudition in theology, law, economics, philology, literature, history, art history, mathematics, medicine, and the sciences.[10] Anders had recently been forced to put down his pen, so his participation was out of the question. It was Ørsted's idea that all articles were to be read aloud at editorial meetings for scrutiny. Ørsted explained the task of the editors as follows: if for instance a work on theology was submitted for a review, the theologians were the experts, whereas the other editors assumed the part of the educated public. They were encouraged to criticise the reviews as enlightened laity, but it was up to the theologians to decide if they would accept them.

Fig. 115. HCA at the examination table, 1829. To the left HCØ and Henrik Gerner v. Schmidten, (1799–1831) Professor in mathematics. Fresco made 1931 by Niels Larsen Stevns (1864–1941), who had been the assistant of Joachim Skovgaard decorating the restored Cathedral of Viborg. Stevns was commissioned to paint eight frescoes of decisive turning points in HCA's life for the annex of the new HCA Museum in Odense presented by the industrialist Thomas B. Thrige. Probably Jørgen Sonne's monumental frieze at Thorvaldsen's Museum in Copenhagen served as a source of inspiration. OBM.

> 'Hence, one will often find opinions in our journal that are not supported by the majority of us... (The reviewer) is obliged to listen to opposing views and friendly warning voices ; but nobody will be enslaved by any system by its being incorporated in our association. We discuss with one another in the most unrestricted open-mindedness; so much the more must our individual independence be respected.'[11]

Editorial meetings were held in the lecture hall of the Polytechnic Institute or when feasible in Ørsted's flat. Ørsted himself made twenty contributions, reviews in particular, such as of Steffens's *Polemische Blätter* [Polemic Papers] and Ursin's *Magazin for Konstnere og Haandværkere* [Magazine for Artists and Artisans], but especially of imaginative literature, as we shall soon see. Ørsted was in the chair leading the discussions of recited texts. 'This man rich in spirit and ideas and with ever-sparking thoughts spoke freely and without reservation, but he was at the same time the most amenable to any real counterargument, the most willing to take an immature thought or an ill-founded word back. Nowhere has the sense of truth, the real truth and the full truth, approached me in a nobler character than here',[12] Clausen wrote in his memoirs about the editor-in-chief.

*

In April 1829 the board of editors met at Ørsted's. Heiberg's review of Oehlenschläger's heroic epic *Hrolf Krage* was on the agenda and the reviewer read it out to his co-editors alias 'the educated public'. Heiberg's critique was serious: Oehlenschläger had no firm grip on his material and his genre. According to Heiberg's aesthetical yardstick contents and form were not coordinated, and hence *Hrolf Krage* was a mongrel.

Unfortunately, this formal blunder, Heiberg went on, was accompanied by a subjective, moral laxity. How does Oehlenschläger attempt to bring a Nordic heroic legend to life for modern people? Heiberg, the literary critic, demanded a philosophical moral, and most of all he wanted to see Hegel's philosophy expressed by historical characters. In his maturity, Oehlenschläger was being accused of a lack of principle, of being philosophically astray, and Ørsted felt unable to ward off the attack.

Heiberg's scathing review was disquieting and kept Ørsted awake the following night. First, it was worrying to witness the bitter disagreement on aesthetic criteria among reputable experts. The incident was a shock, because it clashed with Fichte's scheme of the vocation of the learned republic.[13] Secondly, the controversy forced Ørsted into a painful choice between two lifelong friendships. Ørsted had been very fond of both his talented friends even before either of them had become poets, and it was hurtful to see Oehlenschläger's work being dissected by Heiberg, particularly because his review was so convincing. First thing the next morning Ørsted went to his desk to disentangle the threads and find a compromise. He found Heiberg to be far too philosophical in his criticism, too Hegelian. Ørsted dissociated himself from Heiberg's haughty, patronising tone, which strayed from the sober attitude Ørsted expected in *Maanedsskrift for Litteratur*. His objection to Heiberg's aggressive review testifies to his own literary taste: whatever their genre, all works of literature ought to be constructive, edifying, positive. Reviews should heed these criteria and not descend to philosophical nit-picking as Heiberg had done in, as Ørsted hinted between the lines, reproaching Oehlenschläger for not being a Hegelian, but for expressing a personal view of life that the reviewer found spineless, philistine, incoherent, and sentimental.[14]

However, Heiberg had his own agenda of toppling Oehlenschläger from his pedestal in order that he might take it over and make it a platform for his own ideas. Oehlenschläger, for his part, usually called Heiberg the 'vaudeville poet', which was far from being a compliment. A year before, Heiberg had defeated Oehlenschläger in the competition set by the Royal Theatre to write a gala play for a royal wedding, with his *Elverhøi* [The Elf-Hill]. But this victory had not really displaced Oehlenschläger, at least not abroad, for the following year he was awarded the laurel as the Nordic prince of poets by Esaias Tegnér in Lund cathedral.

In the end *Hrolf Krage* was reviewed twice. Firstly, by Ørsted praising the friend of his youth for his familiarity with Nordic mythology and for his ability to breathe new life into the sources of the Nordic view of life and man. In Ørsted's eyes, Oehlenschläger had accomplished the task of presenting contemporary Scandinavia with its ancient mythology, thereby restoring her forgotten identity.[15] In fact, Heiberg did heed Ørsted's objections and softened the final version of his review. The controversy brought to light the complications of realising the purpose of

Fig. 116. Johanne Luise Heiberg (1812–1890), actress, with her husband, J.L. Heiberg (1791–1860), the playwright and literary critic, and his mother, Thomasine Gyllembourg (1773–1856), whose portrait in oils as a young woman painted by Jens Juel (1745–1802) is on the wall. Even the dog, Walter, is listening carefully to the recital. Oil painting by Nicolai Wilhelm Marstrand (1810–1873), after 1860. In private possession. Photo by FBM.

Maanedsskrift for Litteratur: to act as the platform of a united learned republic and as such become trendsetting for the public debate on cultural matters.

*

In 1833 Frederik Paludan-Müller received the best review any poet had had since Oehlenschläger's first work thirty years before. Sibbern and other reviewers were completely overcome with enthusiasm. *Danserinden* [The Danseuse] was the title of his poem, which was in philosophical rather than nationalistic vein. It was different from Oehlenschläger's style and subject matter, and Heiberg as well as Hertz found his style refined. Ørsted had introduced Paludan-Müller, the young student of law, to Heiberg at a soirée in Studiestræde, where he, Johanne Luise Heiberg and Mrs Gyllembourg had been invited, together with the Mantheys. Gitte and Hans Christian received 'The Danseuse' as a present, not for the editor, but for the hosts, not to review it, but to enjoy it. Nonetheless, Ørsted sent his laudatory and critical remarks to the author in a letter of thanks. He had also read out the poem to his wife, for he did not think that one does justice to poems by silent reading. When one reads something to oneself, one often reads for longer than one's mind stays fresh for. In company, one is likely to stop as one gets tired, and such recitals invite discussion which enhances the understanding.

'The Danseuse' was inspired by Byron's risqué and unfinished poem *Don Juan*. Byron had developed a metric form in which the final rhyming couplet of each stanza stood out as an ironic epigrammatic comment on the lyrical contents of the preceding six. This was a metre that elegantly created these shifts of tone, so that the reader's mind, becoming familiar with the rhythm, would catch both feelings, the lyrical as well as the ironical. In Byron's poem, Don Juan, the hero, is shipwrecked, but manages to make it to a Greek island. Haidee, the heroine, brings the castaway into a cave, where the inevitable is bound to happen. The two children of nature lose themselves in the delight of dalliance. Then Byron switches from the lyrical atmosphere to the irony of the two last lines: 'And thus they form a group that's quite antique,//Half naked, loving, natural, and Greek'.[16]

Paludan-Müller's hero is called Charles. He has a fall from his horse in the forests near Copenhagen, where Dione finds him badly injured and brings him to a villa nearby, nursing his wounds with her loving hands. This is where what the reader has already guessed at happens: 'Forgetting their resolve, earth, heav'n as well, Rapt in love's giddy ecstasy they fell.'[17]

Ørsted thanked him for the beautiful poem, but then the praise shifted to an unmistakably critical tone as he remarked that as the young poet was in no need of flattery, he would give a sincere assessment. He did not share the Romanticist idea of irony. He much preferred a didactic poem that without reserve elevated the reader to ideal spheres of ethics and aesthetics. He found the feelings expressed somewhat too animated. Ørsted's main point concerned 'the spirit of doubt and discontent that frequently emerges when interrupting beautiful impressions with the expression of an embittered irony.'[18]

Ørsted disliked the fact that, like Byron, Paludan-Müller had let his discontent affect those short-lived feelings of happiness. Was not the reason for the enjoyment of these feelings precisely that they are transitory? And if so, it was contradictory to lament that they were perishable. The poet should realise that the phenomenal world is not what it seems, but exists only as shadows of eternal ideas. And had he done so he could have proceeded to lead the reader towards deducing the imperishable, Plato's noumenal world. Rather than incorporating disharmonies into the poem he should have resolved them. How? Well, the juvenile love blazing up in the hearts of Charles and Dione is the transient infatuation, an illusory kind of love. The enjoyment of dalliance is short-lived, but it can lead on to true love which proceeds further to marriage where it finds its imperishable harbour.[19] In spite of his apparent conviction of Paludan-Müller's poetic calling, Ørsted's assessment of *The Danseuse* was better expressed in his passage on Byron: 'Unfortunately, there are many poets—although seldom among the greatest—who lead the inexperienced astray by imaginings contrary to nature about love and its significance in life. He should rather assert the counter force pertaining to the true conditions of life against this demoralizing influence.'[20] Byron's and Paludan-Müller's frivolous charm most certainly diverged from Ørsted's edifying ideals.

*

In December 1833 Andersen sent Ørsted a wistful letter. For almost ten years he had celebrated Christmas with the Ørsteds, cutting out paper figures for Karen, Sophie, and Mathilde, or telling stories, or gluing clippings into scrapbooks for their enjoyment.[21] Now he was sitting in his

Fig. 117. Frederik Paludan-Müller (1809–1871) riding on the back of Lord Byron (1788–1824). Caricature by Constantin Hansen (1804–1880). FBM.

room in Rome in the same street as Bertel Thorvaldsen listening to the shepherds who descended from the mountains playing their bagpipes. Instead of the Christmas tree the Scandinavian artists were decorating a laurel with candles and oranges.[22] 'This year I shall not be decorating the Christmas tree in Nørregade, not be seeing the children flock around me, not be writing verses and reading them aloud in the familiar circle, where the gas lamp burns, and...'.[23]

Andersen urgently asked his patron to speak his mind on his latest work, the dramatic poem *Agnete and the Merman*, just as he had 'previously received much wise and loving advice when I read my works to you... Please tell me your judgement; I hope my poem will please you more than what I have written so far; if you are satisfied with me, do you find a greater maturity in me? Please let me know, nothing will strengthen my courage more and make me happier. Send the letter to me c/o Café Greco a Roma'. Andersen had recited the poem to his friends in Rome and was full of expectations. He wanted to show his fellow countrymen back home that he was worthy of the royal grant that Collin and Ørsted had procured for him. Did they recognise him now? Would he become famous?

Agnete and the Merman is based on the folk ballad about a young woman divided between the Christian existence on Earth that she grew up with and the pagan realm in the sea for which she is yearning. A seductive merman emerges from the sea beguiling the girl down to the bottom of the seas and making her pregnant. However, after seven years with the merman and seven successive children Agnete is drawn back towards the land, only to find herself and her environment changed beyond recognition. She breaks down and collapses in inner confusion and turmoil, refusing the merman's entreaty to return to the sea with him. To the self-absorbed and chaste Andersen the folk ballad is no erotic drama, but an allegory of his personal rise from the bottom of society to a higher level where he also did not feel at home.[24] Of course this interpretation left the crucial points of the poem hidden, and for obvious reasons the public was disinclined to read Andersen's private feelings of rootlessness and unsatisfied yearning into the drama: from a poverty-stricken life in Odense via the educational efforts of the bourgeoisie in Copenhagen and the exotic Rome of the colony of artists to... well to what exactly? Maybe towards the dream of being hailed at the Royal Theatre?

Ørsted read the thoughts at once and criticised *Agnete and the Merman* on the basis of the same aesthetic criterion he had used to judge 'The Danseuse'. Ørsted made his point quite clear in his answer to Andersen: 'The poetic exposition of the world, with all its freedom and boldness, must be governed by the same laws that the spiritual eye finds in the real world and without which it would not be worthwhile living in it.' How could young Andersen have forgotten Ørsted's dinner talks about the Platonic ideals so quickly, that he offended them so crudely? The poet should not content himself with describing the real world as appearance but should present it as it might be, provided it reflected the ideal, noumenal world of reason revealed to him. According to Ørsted 'the forces of optimism should rule a poem, even if the vicious, or even hell itself, is its object.' By so doing a hope is cherished in the readers that their existence can change for the better.

Andersen, on the contrary, had done exactly the opposite. Agnete was drawn towards the purely sensuous world without the consciousness of the reader having been raised to a level where she would encounter the ideal laws of reason. When the Ørsted family had said farewell to Andersen in April 1833, he had written down some words of wisdom in his protégé's album:

> 'Reason in the reason = the True
> Reason in the will = the Good
> Reason in the imagination = the Beautiful
> Be reminded by this text of our conversations. Yours truly, H.C. Ørsted.'

This Kantian maxim came from his heart; it guided the life of the physicist and it followed Andersen throughout his life, becoming the very kernel in his story *The Philosophers' Stone*. But he had already forgotten them in *Agnete and the Merman*! Unwittingly, Andersen had surrendered to the imaginative universe of folk ballads and sketched a woman without higher aspirations. A piece of music should not sound like a series of dissonances. Art must reveal glimpses of the eternal reason. Ørsted ended his letter by saying 'Do not let my unchanging opinions on art weaken the friendly feelings we have shared for so long'.[25]

This cold shower was absolutely unique in the relationship between Big and Little Hans Christian. When, upon his return from his grand tour, Andersen published his first novel *The Improvisatore* and his first collection of fairy tales, *Tales, Told for Children*, Ørsted was the first to recognise the scope of his fairy tales and is renowned for his prophetic statement: '*The Improvisatore* has made you famous, but your fairy tales will make you immortal'.[26]

One Wednesday evening as Andersen was dining as usual at Gitte's table he seemed low in spirits. *Corsaren* [The Corsair] had reviewed him badly, and Andersen sighed, but to deaf ears. Ørsted's thoughts were elsewhere, only Gitte was attentive. Despondently Andersen left Studiestræde to walk home to his room in Hotel du Nord near the Royal Theatre. At half past ten Big Hans Christian knocked on his door. He was let in and said:

> 'I don't know... my wife told me that you were in such low spirits this evening, and I did not notice it. Something scientific was on my mind. She tells me that I have not been attentive to you. But you do know how fond I am of you, and I believe in you. I cannot be mistaken; it is the world that is mistaken. You'll become a poet admired by everybody.'[27]

Big Hans Christian put his arms around Little Hans Christian, who bent down to let Ørsted kiss his forehead. Andersen threw himself on his couch sobbing so loudly that his neighbour was woken up. Jonas Collin would hardly have paid such a visit. The difference between the therapeutic efforts of the two patrons was probably that whereas Collin found that Andersen needed to be provoked, Ørsted had realised that Andersen was so tender that he could not bear to be rubbed up the wrong way.

Having now met Ørsted as a critic of other people's poetry and had an impression of the aesthetic criteria on which he based his judgement, the time has come to look into his own poetic attempts to express his long nourished view that art and science are united through reason. It is the task of science 'to show the world of sense impressions as a revelation of the world of reason'. It is the task of the arts 'to depict the world of reason by means of the sensual world' or 'to represent the infinite in finite forms'. This division of labour is the same as he described to Oehlenschläger in 1807, that while science explains the noumenal in the phenomenal, the arts express the noumenal by means of the phenomenal.[28] In 1836 Ørsted finally took the manuscript of his life's poem out of the chest of drawers where it had been lying for thirty years. *Luftskibet* [The Airship] was about to soar.

'The Airship' was written at Fredensborg, where either Prince Christian Frederik (not using the castle himself) or Court Steward A.W. Hauch had offered the Ørsted family summer accommodation since 1835 in the so-called Cavalier wing.[29] They had the riding ground at one side, and on the other side the park, with Grund's famous statues of Norwegian and Faroese peasants, in the valley down to Esrom Lake. Here they relaxed every summer until Christian VIII's death in 1848.

In 1836 he was putting the finishing touches to 'The Airship'. Not many lines had survived from the first draft written when he was twenty-eight. Now he was fifty-eight:

> 'Poetic work demands a diligence no less than scientific work. The strangest thing for me is that I clearly feel that my scientific endeavours, partly to gain clarity and a spiritual overview and partly to express myself both lucidly and in accordance with the matter, have provided the same advantages in poetry as I might have gained by a longer exercise in the practice of poetry itself.'[30]

He sent copies to his brother and his girls as the poem proceeded, and gave them permission to recite it to their friends provided they rehearsed beforehand and acquired some knowledge about the ancient philosophers, which knowledge, according to the gentle poet, was a prerequisite for understanding it. Science ought to constitute an essential part of general education, like literature, and to Ørsted it was obvious to join the two branches in a poem.[31] He explained to Marie how impressed the Greeks would be, if suddenly they could come back to life and observe modern science and realise what great calamities it might have prevented in ancient Greece.[32]

Since Ørsted, unhappy with his poetic talent, had put the first version of 'The Airship' aside in 1806, the Greeks had liberated large parts of their nation from centuries-long occupation by the Ottoman Empire, to the rejoicings of Christian Europe. In the first part of the poem, the Greeks and Turks are engaged in battle. Then an air balloon rises like a golden ascending sun, not in the east, but in the western sky. Nobody has any idea what it is. Might it be a rescue from above? When it descends upon the battlefield on the island of Samos it frightens the Turks out of their wits, and to the exultation of the Greeks the enemy flees. Modern European science has saved the Christians from the tyranny of Islam. The bishop of Samos, Philon, welcomes the crew, offering them refreshments, and erects a monument of marble to commemorate the liberation.

In the second part, in Homeric hexameters, the three crew members compare the ancient civilisation with the contemporary one. Frankmann is a German physician representing the unbridled idea of progress, while Anspann, a classical archaeologist, without reservation pays homage to the greatness of ancient Greece, regarding later history as a period of decline. Kalchas, a Greek student and Ørsted's mediator and mouthpiece, encourages the disagreeing Germans to reawaken the ancient natural philosophers and to initiate them into modern science. So, the resurrected Thales realises that his hunch concerning hidden forces embedded in amber has been confirmed by the revelation of universally active electrical phenomena. Pythagoras admires Newton's laws of motion, Euclid modern integral and differential calculus, and Aristotle the anatomical and physiological discoveries of modern medicine. Anspann objects to this one-sided worship of progress alleging, like Schiller, that in the arts no progress has been made. Against this Kalchas argues that the ancient dream of flying ended in a catastrophe, when Icarus defied his father's warning and crashed; he flew too close to the sun and the wax on his wings melted. Modern airships, by contrast, land their crews safely.

The next morning the three balloon voyagers take a walk on Samos. In the short third part Ørsted describes the green forests on the spring-fresh island. At a ruined temple they find a thoughtful young man, Philon's son. Throughout the night he has racked his brains in trying to conceive the inconceivable. How did it happen that the golden bird has flown the three aliens to Samos? Kalchas delivers the answer as they gather again in the cool evening air.

He introduces the story of the two Montgolfier brothers, the pioneers who fulfilled man's ancient dream of flying, indeed of flying higher than the eagle, above the skies and across the seas to distant strands—without soaring on wings. Their father, a paper manufacturer in the valley of the Rhône, had decided that Joseph, the oldest son, was to become a priest, and Stephan, the next, a judge, while the younger, less talented brothers were to carry on with the paper mill. Soon, however, it turned out that Joseph was more than simply bookish, and he preferred roaming about in the countryside to sitting studying in his cell. The disobedient son fled his home, but the father sent out people who brought him back. Now, however, he was allowed to study chemistry, and he started his own enterprise that would renew and surpass his father's. He developed close, fraternal ties to Stephan; they tutored one another and invented new types of paper such as vellum for the printing of de luxe editions, not for the amassing of profit, but so as to use their talents for noble progress. It is hard to avoid seeing great similarities between the Montgolfier and the Ørsted brothers.

Frankmann takes over the narrative in the fifth part. The two brothers go for botanical walks in the western spurs of the Alps near their home, climbing the steep Mount Ventoux of almost two thousand metres. Then they hear about the siege of Gibraltar, immediately following the American War of Independence in 1776, when the British had been lucky enough to defend the rock against a French-Spanish attack across the isthmus and from the sea. They deplore the vain sufferings of their fellow countrymen pondering 'but what if we could soar high through the air and descend safely on the top of the rock?'. People hearing about this crazy idea shake their heads. Joseph might have lived happily as a paper manufacturer, but instead he wanders about mired in poverty. Then one day when mountaineering near Vaucluse a terrible thunderstorm takes him by surprise. Lightning strikes a peasant's cottage, but the Rhône cuts him off from going to the rescue. Joseph stands watching the sunset and the ascending smoke from the burning cottage, while the evening sky radiates the most beautiful mixture of grey, yellow and red colours. Why is the smoke rising into the air, he wonders. 'Although he had often watched the play of the smoke,//now an unknown magic seized him,//and a thought, hitherto concealed, appeared to him in clear terms.//The force lifting the smoke high above the ground,//catch it, keep it, shut it in!//Then, you will have discovered a great art//that your troubled mind often searched for.' Of course! He just has to collect smoke in a bag and the whole bag will ascend by itself. This is his thought experiment. He senses the ingenuity and already sees himself honoured and immortalised by Petrarch, the great Italian poet of the Renaissance, who also wandered about near Vaucluse at the time when popes resided in Avignon. Joseph sews a bag out of silk, finds a high-ceilinged house, turns the opening of the bag towards the floor, fills it with smoke, and watches it ascend towards the ceiling. Nature keeps the promises its spirit makes! He exults and proudly displays his experiment to some casual passers-by. Then

Fig. 118. The ascent of the hot air balloon of the Montgolfier brothers from the Tuileries, Paris, on 1 December 1783. Oil painting. Musée Carnavalet, Paris. RMN, Paris. Photo by Agence Bulloz.

he rushes home to the factory anxious to watch the faces of his father and brother when they see him send up a silken bag filled with smoke.

In the sixth part Joseph returns in high spirits to his wife, father, and brother at Annonay. The family idyll is reminiscent of Studiestræde. The incurably didactic paterfamilias is immersed in unsolved riddles of science, conveniently distancing himself from the kitchen, the washing, and the hungry children, all of which are taken care of by his wife. 'Around Montgolfier's wife the children are sitting,//each of them having something to learn.//The busy mother looks closely at them//seeing to it that none are idle,//while the husband ponders, everything else is her care;//he gladly leaves useful things to her.'

Joseph reports on the 'dragon' he has invented. The news is received with the same enthusiasm as a newborn baby. When the Assembly of the Estates of Vivarais convenes on 4 July 1783 the right moment has come to send up the first hot air balloon in history. Now the brothers build a huge balloon out of linen and paper with a diameter of forty metres and a volume of seven hundred and fifty cubic metres having a capacity to lift half a ton if need be. The first flight was unmanned. To the indescribable joy of the crowd the airship ascended to a height of two to three kilometres, and as the hot air cooled down after some time it sank to the ground. Its force had expired.

After describing the successful ascent Kalchas presents some general remarks on scientific discoveries. The first experiment with a weak, possibly accidental effect must be tested on a larger scale, which takes money, much money. Joseph sends Stephan to the King in Paris, just as Ørsted drew on the connections of his brother, when asking Frederik VI for support for his dynamical project: 'Although I acted first you are my co-inventor,//we scrutinised it together for many days....//A mutual wreath entwines our names,//names as of twins connected to the memory.' Philon rejoices at the account, spontaneously spinning a thread in Greek mythology: 'I think of Castor and Pollux as the two splendid brothers,//...One of them born to immortality, but filled with the spirit of fraternal love//he shares his destiny with his brother.'

The following year Professor J.A.C. Charles, Ørsted's admired teacher from 1802–03, had the idea of improving the balloon by filling it with the light gas, hydrogen, produced by pouring sulphuric acid over iron filings. The ascent was observed by members of the scientific elite such as Buffon, Diderot, Lavoisier, and Benjamin Franklin. Jean-Paul Marat was there as well. No scientist himself, he soon became an ultra-revolutionary politician and the political opponent of Charles, who once told Ørsted that Marat, after a row, had tried to run him through with his rapier. A speedy reaction had saved Charles's life, but now the spiteful Marat stood waiting to see blood, when the manned ascent, as he hoped, would fail. But it went well; the two voyagers rose several kilometres into the air, and landed safely.

Charles's hydrogen balloon was made of varnished silk, and as the specific weight of hydrogen is somewhat less than that of hot air, he could make do with a correspondingly smaller balloon to lift the weight of the two people standing in a wicker basket. Ørsted struggled to put into words the laws of nature exploited by aviation and to describe the proportions of the silken balloon: 'Its average is only four and a half fathoms,//a little more than a third of its height.//My airship's weight, filled with the lighter gas,//is a hundred and fifty pounds.//Add to this two voyagers and their luggage,//still the weight will not exceed six hundred.//But the

THE LITERARY CRITIC, THE AIRSHIP

air that must be displaced by the balloon//will weigh more than eight hundred pounds;//so you see, my air ball will rise easily.' Weights and measurements pose great problems, if like Ørsted one has the awkward idea of expressing them poetically.

'The Airship' has passages of great beauty, but Ørsted's urge to deliver science lectures in the genre of poetry often becomes a cocktail at odds with itself. He must absolutely pack all his insight and associations into the poem, for instance this prophecy: during the battle of Fleurus in 1794 French generals for the first time sent up a manned airship to monitor the movements of the troops from above. The airship was out of reach of the enemy's cannon. Ørsted optimistically infers that in the future it will be logically impossible to fight wars, because the airship makes all targets defenceless, and since nobody can feel safe against retaliation nobody will be so foolish as to start a war.

He sent the small, yellow printing of 'The Airship' to Heiberg and Hauch, both seeking to unite science and the arts. Hauch saw the poem as 'an expression of your poetic view that brings together a fullness, clarity, and depth of thought. "The Airship" shows that you stand near the source, from where the philosophical and poetical enthusiasms combine and emanate; it confirms the truth, that warmth and flight are just as necessary for the scientific as for the poetical genius.' Hauch was one of a few that were fully aware of Ørsted's intentions and reassured him that 'The Airship' paved the way for a future genre and by far exceeded 'the ordinary didactic poems, where mere reflection prevails'.[33]

What Ørsted intended to be a combination of two related activities, the arts and sciences, had for most critics resulted in a didactic poem, in which the didactic ruined the poetic. But only a few dared put their criticisms on paper, Bishop Mynster being one of them. Ørsted defended himself by insisting that all the historical and scientific information was the necessary raw material for the imagination, and that the sciences hold in themselves an abundance of enchantment. Precision is not necessarily prosaic. Andersen was probably the only person

Fig. 119. HCA's paper collage of an airship. OBM.

479

capable of giving an honest evaluation of 'The Airship' without hurting his mentor. He ventured to shoot it down, arguing that didactic poems are like 'mechanical puppets lacking fresh life' (ch. 59).

The next year, J.C.G. Johannsen the vicar of St Petri, the church of the German congregation, translated 'The Airship' into German, *Das Luftschiff*, and Ørsted sent it to his friend Sir John Herschel with a dedication in verse.[34] This British astronomer, mathematician, pioneer of photography, and translator of Schiller into English was the object of Ørsted's undivided admiration, because he combined the same scientific and aesthetic ideals. Ørsted's ode to research (which was essentially what his poem intended to be) now served as an appreciation of the acknowledgement Herschel had shown Ørsted by comparing his discovery of electromagnetism with Columbus's discovery of America. This was the homage that gave Ørsted a place in the history of science. Herschel thanked him for the interesting poem; he had not been aware that Ørsted also incarnated a talent for poetry.[35]

> Freedom consists in the rule of just laws,
> not in a democratic government![1]

47 | 1831–9
The Awakening of Political Life

> 'A political and public life has begun in Denmark that we never heard of before; this may be good in some sense; but I'm worried because the different views, when uttered vehemently, often separate friend from friend. The Polytechnic Institute has not been problem free; Ørsted has had to defend himself again and again. He was accused of costing the country 300,000 rixdollars. Even people outside the conflict found it too expensive for a small country like ours; so on top of all his other obligations Ørsted was up to his ears having to keep everyone informed'.[2]

Gitte wrote this to E.A. Scharling, her son-in-law to be, on his grand tour in Germany, France, and Britain in 1835.

At the time the Polytechnic Institute was established a series of crucial changes in the economic and political situation of the country occurred. After the disasters brought about by the calamitous Danish foreign policy during the Napoleonic wars, a period of recession in agriculture, the predominant commercial activity, had followed. When peace was restored in Europe, purchases of food for soldiers and feed for horses stopped. Corn prices plummeted and only began to rise again in the late 1820s. By the mid 1830s the average price of corn had doubled from its lowest point in the beginning of the 1820s, and the value of land increased correspondingly. The position of estate owners and freeholders improved, whereas tenant farmers and smallholders gained far less.

As Gitte told Scharling, a hitherto unknown political upheaval was going on. At long last King Frederik VI, under duress, had acquiesced in the part of the peace settlement he had signed up to as the Duke of Holstein with German rulers at the Congress of Vienna. The agreement concerned the introduction of consultative Assemblies of the Estates, which is to say that he consented to give representatives of the various Estates a right to be consulted on new legislation. But the absolute monarchy had shelved the agreement for fifteen years, while Prussia had carried it out long since, and the Austrian Prime Minister, Prince Clemens Metternich, time and again reminded the Danish King of his obligation. He was finally prompted to fulfil it for fear that the July revolution in 1830 in France might spread to his kingdom and the duchies.

Separatists emanating from the University of Kiel in the Duchy of Holstein had long voiced criticism of Frederik VI's absolutist rule, carried into effect by the German Chancellery in Copenhagen. The Holsteiners wanted a new constitution, by means of which a strong representation of the Estates, academics in particular, together with similar political forces in the Duchy of Schleswig, would achieve a liberal government turning its face towards the Reichstag of the German Federation in Frankfurt. These Holsteiners emphasised their inseparable bonds with the German-speaking population of Schleswig, probably more than 50% of the total. The Danish King could no longer escape the demand for a consultative Assembly of the Estates, but the establishment of it threatened to break up the United Monarchy, as if the loss of Norway had not been enough. Hesitatingly, Frederik VI set up a committee consisting of his top administrators of the German (Otto Moltke and J.P. Höpp) and the Danish Chancelleries (P.C. Stemann and A.S. Ørsted). On 28 May he signed a royal resolution ordering the establishment of a consultative Assembly of the Estates of the kingdom and of the duchies as well. Particulars concerning the electorate, meeting places of the representations, and so on were to be worked out by the committee.

The Danish King saw his monarchy and duchies as a unit, even though his sovereignty over different parts of it was not the same. Consequently, Holstein, which was a duchy in the German Federation, and Schleswig (bounded by the Eider Canal) which was not, should not be connected more closely to each other than to the Danish monarchy, so each duchy must have a representation of its own. To avoid differential treatment, the King argued, his monarchy, similarly, must have two representations, one for Zealand and the smaller islands and one for Jutland. Fearing academic agitation, *Burschenschaften*, street riots, and rule by the mob, the four Assemblies of the Estates should not convene in the university cities, but in the smaller provincial towns: Itzehoe, Schleswig, Roskilde, and Viborg.

Obviously, the introduction of the Assemblies of the Estates fundamentally changed the rules of the political game. Under the incapable kings Frederik V and Christian VII during the previous century a form of government had emerged that has fittingly been called 'absolutism tempered by public opinion'. The author of the theory, the Norwegian historian Jens Arup Seip, suggests that 'absolutism' should not be taken at face value because in reality monarchical power was never absolute, let alone despotic, but modified by opinion, taking the form during the Enlightenment of the growth of journals and voluntary associations in civil society. The rise of such new institutions prompted the monarchy to pay closer attention to public opinion. It goes without saying that it is futile to try to measure the validity of this theory. Nevertheless, it is easy to provide examples to show that the learned republic was convinced that public opinion would affect the decision-making process of the monarchy. Nor is it difficult to ascertain the King's sensibility to public opinion. Hence the freedom of the press was so important.

It was particularly important to the Ørsted brothers, whose appearances in the public arena show that they, more than most people, acted as if their division of labour was founded on Seip's theory. A.S. Ørsted's judicial articles had achieved an influence that by 1826 crossed the tolerance threshold of his superiors (ch. 43). H.C. Ørsted, since 1820 the supreme adviser on technical and scientific affairs to the monarchy, began to intervene far more comprehensively in public debate from 1828, the year he assembled the board of editors of *Maanedsskrift for Litteratur* [Literary Monthly]. In founding this journal he actually took over the function his brother had been forced to abandon. The physicist had to replace the top jurist, and there can be no doubt that they carefully planned every step together before they took it.

The way Ørsted selected the editorial board of *Maanedsskrift for Litteratur* was in full keeping with Fichte's idea of the vocation of the learned republic. Apart from J.L. Heiberg, all twelve members were his colleagues from the University, and were a comprehensive collection of the most eloquent representatives of science, moral philosophy, and the arts, or the True, the Good, and the Beautiful. Following the establishment of the constitution of the Estates the journal expanded its field to cover politics as a branch of moral philosophy.

But not only that. As the political life awakened, and the constitution of the Estates inaugurated a new epoch, reassuring people that from now on public opinion would be heeded as advisory to the government, the Ørsted brothers expected the impact of the learned republic to grow. Whereas political statements had so far been confined within the narrow limits of the freedom of the press, Pandora's box had now been opened, although the monarchy was still thin-skinned, and criticism of the paternal monarch was forbidden. Public opinion had to move skilfully and with self-restraint. Together the Ørsted brothers formed a unique team, for although A.S. Ørsted was cut off from public debate, he had his fingers on the pulse of the government, and being in the know concerning government politics he was in a position to bring his information into an intelligent dialogue with public. Moreover, in 1831 H.C. Ørsted set up another journal, *Dansk Ugeskrift* [Danish Weekly], particularly directed towards public political debate. This journal addressed knowledgeable, liberal-minded, level-headed, and well-intentioned citizens, and this target group was broader than that of *Maanedsskrift for Litteratur*. It was run by an association consisting partly of editors of his other journal and partly of liberal businessmen.

Ørsted appointed J.F. Schouw editor of *Dansk Ugeskrift*. He was a professor of botany and had been a loyal partner of the Ørsteds for some time. Originally, he had studied law with A.S. Ørsted as his tutor, and he had been recommended for a post at the Frederik University in Christiania in 1814. Nothing came of this and instead Schouw threw himself into botany, his passion, which he studied in the light of Alexander von Humboldt's plant-hunting expeditions. After his doctoral thesis he was admitted into the Royal Danish Society of Science and Letters, and the next year he became an active supporter of the establishment of the Society for the Dissemination of Science. In 1826 Ørsted appointed him archivist of the Royal Danish Society and editor of its transactions, and he kept the two posts as long as Ørsted was secretary.[3] In other words, he was a kind of straw man for Ørsted, and nobody knew that although Schouw's name appeared on the front page of *Dansk Ugeskrift* as the editor, Ørsted was the real editor. Hans Vammen, the historian, has delved into the archives of the weekly and found that hardly a single article was published without the approval of Ørsted. He demanded high standards, keeping well above the level of the gutter press. When Schouw asked H.C. Ørsted, why his brother was not part of the association, he received an evasive answer, for it was a well kept secret that A.S. Ørsted was no longer allowed to participate in public debate.[4] H.C. Ørsted was not acting on his own, but was also the mouthpiece of his brother, with Schouw the formal mouthpiece of both of them.

The Ørsted brothers were aware that they had taken on a delicate task, for it did not take much to predict that rabid opponents of the absolute monarchy would try to take over the public debate with constitutional claims that stood no chance of being taken seriously, but would only irritate the King and arouse public censorship. In this vein the young student of law, Orla Lehmann, created a role for himself on the radical wing as soon as the chance of a freer political debate in Denmark arrived. For his views to be printed in *Dansk Ugeskrift* they had to be approved by Ørsted whether they were published in Orla Lehmann's name or anonymously. Being the son of Martin Lehmann, Ørsted's close friend from his first sojourn in Paris, Orla was a frequent guest in Studiestræde. In his first year at University, Orla would be sitting in front of Ørsted's lectern preparing for *examen philosophium*. 'The personal interest he showed in me was rather a legacy from my father; but many a student will recollect the wealth of spirit and knowledge he so generously imparted to anyone who approached him with an inquisitive mind. To open one's eyes to a healthy world-view and awaken one's sense of a noble human life one could have no better tutor.'[5] A.S. Ørsted's books and articles had a greater share in his study of law than those of his university teachers.

In September 1832 Orla Lehmann asked H.C. Ørsted to take a look at a couple of reviews and an article against censorship he wished to have printed. At first he had sent them to the editor, but Schouw (alias Ørsted) found them too rabid. So did Ørsted in person. Lehmann reworked them, withdrawing formulations that might stand in the way of his intentions. Ørsted commented:

> 'You know that I am a friend of the freedom of the press, dreading any new limitation of it; but when I see the many vicious lies, distortions, and libels that not only the so-called obscurantists, but also the so-called liberals allow themselves, it seems to me that the issue of the freedom of the press cannot be debated without dealing with this directly.'[6]

H.C. Ørsted was familiar with Frederik VI's view on the matter from his brother and swiftly realised that Orla Lehmann's article, even in its revised version, would at best evoke a headshake in the Chancellery and at worst keep an undesirable sharp watch on the young hothead and the journal he wrote for. If he had nothing new to say and could not express himself to the point, he had less chance of liberalising the laws of the press than of winning the lottery. Ørsted did not wrap his advice up in polite words, but warned him, telling him plainly that he considered him 'too hot a follower of the reform party. You want to establish it in our country. Well, at first I beg you to deliberate whether it would not be better to let it simmer down abroad and see what good will be left after the many struggles... You might easily waste your best efforts either in accomplishing nothing or even in inflicting damage.'[7]

Maanedsskrift for Litteratur had already published two of Orla Lehmann's reviews of political pamphlets.[8] They had appeared under the easily recognisable pseudonym 'wa', but now the author planned to re-edit the reviews and the article on the freedom of the press separately at Reitzel's publishing house, where he was granted complete anonymity. Ørsted thought this a bad idea, because readers interested in the reviews would already have read them, and the article, Ørsted repeated, would do more harm than good. Unfortunately, there had been more recent abuses of the freedom of the press, and they had now reached a state of 'barefaced impudence and stupidity'; hence an article suggesting unrestrained freedom of the press would sound comical. 'The time has come for the friends of literature to counter this disgusting conduct; and I hope this is soon going to happen.'[9] This hope was realised by the founding of the Society for the Freedom of the Press three years later. Orla Lehmann heeded Ørsted's advice. The following year his article on freedom of the press was rewritten as a speech for the Students' Association. Their friendship survived their growing political disagreements, and Ørsted gave the young rebel letters of recommendation to his friends abroad, Arago in Paris for one. But his articles Ørsted refused; these radical views were already well known and would only exasperate the monarch.

In 1834 C.N. David, professor of economics and co-editor of *Maanedsskrift for Litteratur* organised the publication of *Fædrelandet* [The Mother Country], at first a weekly then a daily newspaper. It spoke on behalf of liberal citizens working in the public administration or in private businesses who were agitating for political reforms of the absolute monarchy during the campaign for the first election to the consultative Assemblies of the Estates. The King resented some anonymous articles in *Fædrelandet*. He was perturbed that all the thanks he received for his generous promise to introduce Assemblies of the Estates was this rampant profligacy. The paper was agitating in favour of certain political demands, as if the Assemblies of the Estates spelt dethronement of the King and popular sovereignty of the nation. Frederik VI directed the Danish Chancellery to indict David for having criticised the existing constitution for mixing legislative and executive powers. In the absolute monarchy the King was sovereign of both realms, contrary to Montesquieu's teachings on the tripartite separation of powers. It was punishable to criticise the constitution, and the deplorable fact that David's journalism had passed the censorship unnoticed was at least as bad. An example had to be made. The censor was fired and a new and tougher replacement appointed to make sure that no repetition occurred. David was immediately suspended from his professorship and summoned before the court.

Worse still, the King secretly directed his top officials P.C. Steman and A.S. Ørsted from the Danish Chancellery and Otto Moltke and J.P. Höpp from the German Chancellery to consider how the freedom of the press could be restricted further.

'As impudent writings in the newspapers and journals are becoming increasingly outspoken, We graciously order the committee to work out a report as to whether similar provisions to those adopted by the Federal Assembly in Frankfurt and applicable to Our Duchies of Holstein and Lauenburg could be prescribed for the Monarchy of Denmark as well as for the Duchy of Schleswig.'[10]

This royal initiative was a drastic one, for German conditions would be tantamount to subjecting all authors to advance censorship without the possibility of appeal. Stemann wrote in his report to the King that some of these hack journalists were 'a bunch of philosophical theorists devoid of practical knowledge of state and people, dreaming of the possibility of introducing a Platonic republic and believing themselves to be called upon to improve what they do not even understand fully'.[11] This was directly aimed at the learned republic. Stemann recommended advance censorship of each individual newspaper and journal, thus moving the absolute monarchy from the rule of law to a police state. A.S. Ørsted dissented. Restricting the freedom of the press without consulting the Assemblies of the Estates would violate the ordinance of 28 May 1831 which stated that all laws concerning the rights of the people be presented to the Estates for a hearing before being issued by the authorities. Ørsted argued that freedom of the press is a civil right (however limited) according to the ordinance of 1799, and were it to be repealed the representatives of the Estates must be given opportunity to advise the King. This form of logic was incontestable, and the King yielded at first. Possible restrictions would have to await consultation with the Estates.

In the meantime rumours of an impending restriction of the freedom of the press had leaked from the chambers of power, and signatures were being secretly collected in support of a formal petition for broadening the freedom of the press, or at the very least for the preservation of the status quo. It seems likely that A.S. Ørsted, the hard pressed government official, was prompting this appeal to the learned republic from behind the scenes. Schouw initiated and Clausen authored the petition to the King. Almost six hundred signatures, among them twenty-seven professors, eighty-two officials, and fifty-five officers, mostly from the navy, were submitted to Amalienborg on the day after the royal order. In submissive phrases the King was petitioned to reconsider the matter and to understand that a free public debate was actually in the interest of the regime, because fresh light would be shed on drafts for laws and ordinances and the risk of bad legislature reduced.

As Frederik VI received the petition H.C. Ørsted's signature did not escape his attention, and his reaction was prompt and unambiguous:

> 'We did not expect to see that several of Our dear and faithful subjects have petitioned that no change of the ordinance of the freedom of the press take place; for just as our paternal care has continuously been directed at the well-being of state and people, so nobody but We alone will be able to estimate what serves both of these the best...'[12]

These lines have subsequently been referred to as the 'We-alone-know-Rescript' in Denmark, although such words were not actually used.[13] Despite the autocratic tone nothing happened until the two Danish Assemblies of the Estates had been heard. Of course, they both opposed the monarch's plan to introduce Holsteinian conditions, that is to say transferring cases concerning the press from the courts to the police. Nevertheless, the King tightened the ordinance of the freedom of the press in line with a draft produced by the Chancellery. A.S. Ørsted succeeded in preventing the relatively independent courts from being put out of action, and ensuring that censorship was limited in time. But he had to accept with the greatest reluctance the sanction that an official violating the ordinance would lose his position.

The charges against Professor David ended up as a total fiasco for the monarchy. David was acquitted by the court, and when the government appealed, it lost again. The acquittal established a precedent and is striking evidence that A.S. Ørsted's struggle to maintain the 1799 ordinance was decisive in safeguarding the judiciary's ability to protecting a certain freedom of the press against the ad hoc interference of the monarchy.

The petition for freedom of the press and David's martyrdom resulted in a new initiative. H.C. Ørsted placed himself at the head of the voluntary Society for the Right Use of the Freedom of the Press. There can be little doubt that he was provided with all the information he needed from within the Danish Chancellery. His intention was to develop a debating culture and to counteract abuses, for even if the freedom to print all views was desirable, yet perhaps it was wrong to publish some, because they were libellous or went against respect for the feelings of others. The Society for the Right Use of the Freedom of the Press quickly acquired more than five thousand members. No other voluntary organisation had ever had as large a membership. It was not only inhabitants of the capital who wished to defend freedom of expression, there were branches in more than thirty provincial towns. The day-to-day leadership was left in the hands of an elected committee consisting of the same old crowd from Ørsted's two journals. The committee edited the society's organ, *Dansk Folkeblad* [The Danish People's Magazine], whose first editorial was written by H.C. Ørsted no doubt in collaboration with his brother.[14]

The leading article argued in favour of a golden mean between censorship barring reform schemes on the one side and unreasonable provocation on the other. Both extremes were abuses, while the well-considered argument was the proper use. The freedom of the press should make sure that laws served the common good, while preventing the state from stagnation and despotism or from 'embittering innocent citizens by false representations of their actions'.[15] Freedom of the press should at the same time guarantee the rule of law, safeguard the monarchy as tempered by public opinion, and prevent civil libels.

The aim of the Society for the Right Use of the Freedom of the Press was to bring out the truth. The truth is not the same thing as the mean. Ørsted emphasised the difference between the *juste milieu*, the liberals' derogative term for the moderates, and reason that must always prevail. To Ørsted the truth is not subject to democratic negotiation; it emanates from the laws of science and logic as well as from the maxims of moral philosophy. These are the qualitative criteria he upheld as opposed to the numbers represented by a majority. Democracy smacks of quantification.[16] The society had no political power and according to Ørsted it should not covet any. Freedom of the press ought to be used against 'the neutral sloth and giddiness of the blind mob'.[17] 'The press should be the spokesman for truth, never the slave of any party or passion.'[18] According to H.C. Ørsted 'freedom consists in the rule of just laws, not in a democratic government'.[19] A.S. Ørsted's task was to moderate the conservative forces in the government, and his brother's to moderate the destructive forces in the public debate.

Fig. 120. The triumphal procession of censorship. In front the goddess of liberty followed by newspapers and journals: *Forposten, Kjøbenhavnsposten, Corsaren, Frisindede, Fædrelandet, Dagen, Berlingske, Aftenposten*, persecuted by a cart with the inscription of the recently built town hall: 'Law should build the Country'. Caricature by Nicolai Wilhelm Marstrand (1810–1873). FBM.

Ørsted already suspected that there was a demagogue lurking inside Orla Lehmann, and he was alarmed by the increasing politicisation of social life that threatened to split old friendships asunder. He saw examples of established relationships being now dissolved into soberminded reformists and hot-headed revolutionaries or apathetic philistines. Ørsted thought that if conservatives became diehard conservatives then the reformists risked being thrown into the arms of radicals or revolutionaries and if so the almost unavoidable consequence of political (or religious) fanaticism would occur.[20] As a revolutionary physicist he had used arguments of reason to overthrow corpuscular theory, but he had witnessed too much violence in the wake of the French Revolution to wish for political extremism in Denmark.

Ørsted's political models were the Bernstorffs who had followed the golden mean. He wrote a poem simply entitled 'Bernstorfferne' ['The Bernstorffs'], which was a homage to J.H.E. Bernstorff, who had divided his estate into plots for peasants, to which he had given them legal title, and to A.P. Bernstorff, his nephew, who put the reform of the cultivation system into practice. This reform was significant, because after half a millennium of near-serfdom the peasantry was unable to figure out what to do with the freedom that was suddenly bestowed upon them; they had been so subjugated by the bailiff's whip that they might have preferred to keep the old paternalistic system, in which the nobleman, at any rate as far as the Bernstorffs were concerned, stretched out a helping hand when harvest failed or illness hit their livestock. Without the absolute monarchy taking care of raising the educational level of the peasantry and protecting their interests by new laws, the agricultural revolution would never have succeeded. The rather romanticised view of history embedded in his poem reflects his estimate of the current political situation.²¹ If the Assemblies of the Estates were to improve the lot of the common man, who had never asked for it and lacked the prerequisites for asserting himself in politics, then this common man was in need of enlightenment and this was exactly the task of the society. 'Bernstorfferne' was republished in the society's collection of historical poems and sent to the folk libraries in the provinces. Ørsted basked in Heiberg's and Hauch's admiration of his poem. His brother, too, sent his compliments: 'I find it exceedingly useful to remind people of such noble and vigorous deeds; we ought to have more encouraging poems like yours.'²²

In 1835 Ørsted reviewed 'Norway's Dawn', a polemical poem by Johan S. Welhaven in *Maanedsskrift for Litteratur*.²³ The Norwegian poet satirised his fellow contrymen's lack of vision, ambition, and education: 'Despite all the modern knowledge in print, and despite all the freedom proclaimed, the spirit that actually rules this nation is but a cloud of powder from the wigs of our ancestors'. Ørsted loved these lines. There was more patriotism in criticism that hit the bull's-eye than in flattering phrases that wrap a nation in complacency. But what had this to do with the Danes? Well, Ørsted used the poem to comment on their future political situation. Danes should keep in mind 'that the influence justly accorded to peasants, artisans, and other trades by the consultative assemblies should never outstrip higher spiritual interests'.²⁴ Ørsted fully agreed with Welhaven that the liberal Eidsvoll constitution, which had made Norway the most democratic nation in Europe, had also led to the starvation of the arts and sciences, because the people's representatives displayed an excessive interest in material things, but did not understand 'that a people is honourable, great, and splendid only to the extent that what ennobles man is brought to bloom and fruition'.²⁵

The problem was, according to Ørsted, that the majority did not comprehend the immeasurable value and impact of the sciences and the arts. A constitutional nation like Norway had a great deal to learn from an absolute monarchy like Denmark. Most people are unable to understand that in the long run research benefits everyone. Hence, it is crucial for a constitutional nation such as Denmark to become a place where the electorate chooses representatives dedicated to the sciences and the arts. As far as Ørsted's own career was concerned, he owed everything to the absolute monarchy, his travel grants, his collection of instruments, his professorship, and the Polytechnic Institute. Could he have accomplished all this had he depended on the Norwegian parliament?

During the first electoral campaign for the Assemblies of the Estates in 1835, opposing views clashed as to which Estate constitutes the core of the people and hence deserves the highest influence.²⁶ The election act targeted four groups (different from the classical Estates) of representatives in the Roskilde assembly: (1) estate owners; (2) house owners in Copenhagen; (3) house owners in the provincial towns; and (4) farmers, both owners and tenants. Actually, the Danish Government had simply copied the Prussian constitution, except for the dominance granted to Prussian *Junkers*. Members of the learned republic criticised the election act for enfranchising house owners merely, disregarding that their property might be mortgaged to the hilt. On the other hand, a professor was not eligible, unless, coincidentally, he owned a house, for instance by marriage, which was rare. The criticism focused on the probability that the electorate did not reflect reason and commonwealth. Ninety per cent of the electorate consisting of shopkeepers, distillers, brewers, artisans, merchants, manufacturers, sea captains, and the like were likely to vote for candidates representing private interests. Only a small minority of ten per cent, being officials, officers, pastors, judges, physicians, or professors were representative of reason, but without ownership they had no vote. According to their own self-image they had no private, that is material, interests, but devoted their lives to serving reason. The government defended its election act at length arguing that presumably the intellectual elite had a rhetorical edge that would favour them in the electoral campaign, which indeed turned out to be the case.[27]

Orla Lehmann urged that officials were the heart of the people, and that graduation from the University should be the only access to eligibility.[28] Others fought to favour the trades.[29] As we have already seen, the Platonic model of society was woven into Ørsted's speeches to the University, to the Polytechnic Institute, and to the Society for the Right Use of the Freedom of the Press. It is no surprise, therefore, that he repeated these ideas in a political article entitled 'No struggle for rank between the various Estates' written for *Dansk Folkeblad*. As Plato might have said, it would be detrimental to the logic of his organic metaphor to allow the peasantry to rule the state while leaving the land to be tilled by philosophers. All Estates are needed, but so is their division of labour. It is the vocation of the learned republic to enlighten the other Estates, especially now that they have a voice in politics. Plato's organic model of society and Fichte's ideas of the vocation of the erudite community were clearly orchestrated in the political views entertained by the Ørsted brothers.

In an anonymous article in *Kjøbenhavnsposten* [Copenhagen Mail], the newspaper supporting the party of reform, a shipowner argued against the Ørsteds by denying that Newton's scientific discovery had enabled him to navigate his ship across the seas.[30] Ørsted told him off: 'We cannot agree with the author when he considers the class of people working for the necessities of life the most important in the state. By contrast, it turns out that the most necessary thing, although it is the precondition of nobler existence, is not the noble thing in itself.' The hierarchical order at the Polytechnic Institute, with science above and craftsmanship below, must be recognised as the most appropriate for the entire state, Ørsted went on. '.. even in the production of material goods the spirit is the master, and if the tradesman wants to climb to a high position in society he must assert himself as an educated man. By insight he can improve his business, and by sound and good advice he can serve the commonwealth. Many tradesmen have done so in the past and we hope more will do so in the future.'[31] This is the way Ørsted's meritocracy works. The learned republic has enabled the tradesman to do his business, and science will always serve the idea of progress without being shown gratitude.

In March 1835, after the 'We-alone-know-Rescript', while A.S. Ørsted and P.C. Stemann were fighting against each other in the royal committee working on the constitution of the Estates, a group of students was organising a demonstration against the unpopular minister. They intended to deliver him a '*pereat!*' ['down with Stemann!'] and break his windows while a party was taking place in his expensive flat. The demonstrators had announced their plan with posters encouraging students to set out from the University. H.C. Ørsted noticed these posters and was informed that the Vice-Chancellor had sent a note to the chief of police asking the constables to stay calm, until he had tried to talk the youngsters out of their plans.

On that evening Ørsted and Sibbern walked out together to meet the students. Ørsted addressed the leaders, not in his capacity of professor, as he said, but as a friend of the students. He tried to exploit the prestige he knew he enjoyed. He called on them to give up their demonstration, which would accomplish nothing, and would only harm themselves as the Senate might decide to rusticate them. The students turned out to be unable to explain exactly what Stemann had done to deserve such public disapproval. Sibbern started to scold them, mentioning the name of Fichte, who had been exposed to similar treatment once in Jena. Ørsted picked up this thread and told the students an anecdote. At one of his lectures on philosophy Fichte had criticised the *Burschenschaften* for not serving any academic purpose. In the evening a student threw stones through his window. The next day Fichte began his lecture by saying: 'Last night stones were thrown through my window, but a stone is no argument. Hence, I shall proceed with my lecture.' Ørsted went on to warn the students that policemen had been posted around Stemann's house. Doubting the effect of his sermon, he ended it by saying, 'I'm afraid that what I've just said will not change your minds; but I do know that I've done my duty.' Then one of students replied, 'Yes, we'll take your advice. Long live Professor Ørsted!' The others pronounced a '*vivat*' and separated, which Ørsted immediately communicated to the chief of police. Pleased with their moderating influence, the two professors walked the few steps home to Gitte.[32]

On 1 October 1835 Roskilde was festively decorated. 'All the women were wearing their silken dresses, and the streets glittered with gold brocade caps. Tonight the whole town will be illuminated', a deputy wrote. It was an historic day for the small provincial town. The civil guards were marched up to welcome the Assembly of the Estates for Zealand and the islands. First a service was held in the cathedral, where Bishop J. P. Mynster, appointed royal deputy, preached. From there the seventy deputies walked to the Yellow Palace that had been rebuilt to make room for the session in the Queen's wing. The deputies did not sit in any planned order in the two horseshoe-shaped rows. There were no political parties. The royal commissar, A.S. Ørsted, sitting to the left of Frederik VI's bust, delivered the inaugural speech, reminding the deputies of the historical significance of the moment. To the right was the elected president of the assembly, J. F. Schouw, paying tribute to the royal commissar as a man of the people. The King was conspicuous by his absence. Afterwards dinner was served, and in the months to follow the entertainment business of Roskilde boomed.

There had been two obvious candidates for the post of royal commissar, Stemann and Ørsted. They had disagreed on most subjects when preparing the constitution of the Estates, and the King had usually adopted Stemann's line. Hence, it was a surprise when the choice of royal commissar fell on the one who had usually lost. When the choice became known in Copenhagen, many citizens put candles in their windows to express their joy and expectations. Anders received satisfaction for his many defeats in the Chancellery and stepping out of the shadows of forced silence he was once again a public figure, although his pen remained idle. The assembly elected J. F. Schouw as its president; he had been appointed by the King to represent the University. The Chancellery had suggested four professors for this post: H.C. Ørsted, J.F.W. Schlegel, F.C. Sibbern, and J.F. Schouw. At first the King had chosen Schlegel, the professor tutoring young Anders, but he declined the honour due to old age. So instead, upon the recommendation of A.S. Ørsted, the King appointed Schouw (and not Ørsted's

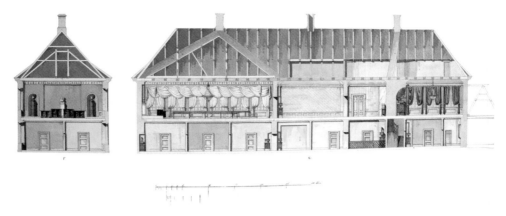

Fig. 121. The Assembly of the Estates in the Yellow Palace, Roskilde, cross section of gable and front. The assembly hall on the first floor, the flat of the royal commissar ASØ, where his nieces lent him a hand in the kitchen, on the ground floor to the left. De Rosenbergske Mapper 62, 28. Photo by DCC. PRO.

brother, which would have been unwise). Ørsted and Schouw formed a strong partnership; they knew each other well and had reason to expect loyalty of each other.

A.S. Ørsted was accommodated in the west wing of the Yellow Palace. There was plenty of room for his three nieces, or his 'lasses', as he called Karen, Marie, and Sophie. In turn they became his housekeepers during the three months of the assembly's season. Hans Christian and Gitte had always shared their seven children with Uncle Anders, who had none apart from the unstable foster children. Later they would follow him to the Viborg assembly. Anders would often invite his little nieces and nephews and their parents to the Yellow Palace at Roskilde to celebrate his brother's birthday on 14 August. On such occasions the stagecoach would take them from Fredensborg to Roskilde and then back again after a night at Hotel Prindsen. Anders employed the young jurist F.C.E. Dahlstrøm as his secretary, who soon had eyes for Sophie. She was not yet seventeen.

Unawares, Dahlstrøm had a rival, who noted his reaction to Sophie's engagement. He had not been able to pluck up courage to propose to her. He was poor, without a permanent job, and found the girl attractive, though he might also be able to manage to live by himself. Then the more courageous suitor seized the opportunity and received a positive answer. That evening the hesitating rival went to her home to take his meal and read his stories aloud to the family as usual. In front of him stood the couple engaged to be married: Sophie and Dahlstrøm!

> 'Both were so happy. I took her hand for the first time clasping it twice. My humour was splendid, or so I thought, for I did not suffer, but stayed very calm. Now I am at home, I am alone—alone! As I shall always be! By this Christmas I would have told her what would never have been good for her! Now, I shall never be married, no young girl will become dear to me anymore. From day to day I shall become more and more of a bachelor! O, yesterday, still, I was one of the young, tonight I am old. God bless you dearly beloved Sophie, you will never know how happy I might have become as a man of fortune with you at my side.'[33]

Andersen was too poor and too timid. He missed his opportunity. Had Big Hans Christian realised that Andersen melted when Sophie was near him? Was that why he had tried to get Little Hans Christian a job by supporting his application to the Royal Library? A librarian would be able to provide for a family; but the job did not materialise.

*

Right from the beginning conflict was threatening to split the Society for the Right Use of the Freedom of the Press, a conflict between liberals and moderates. Two years later it burst into flames. *Kjøbenhavnsposten* was campaigning against a leadership that it found too accommodating towards the regime. What liberal leaders like Orla Lehmann had hoped would become an offensive platform of political debate on constitutional issues was now hibernating for fear of insulting the absolute monarchy. The council of the society stood as a shield against political activities, and its leadership was only concerned about self-censorship. Therefore Lehmann and his liberal partisans suggested that a general meeting be held with the purpose of changing the objectives of the society so as to promote his political ends.[34]

Moderate members such as Schouw, Clausen, and Ørsted, on the other hand, dreaded 'any tendency to change the society into becoming a lever for political ideas, a deadly principle'.[35] H.C. Ørsted, smelling a rat at once, fulminated at the proposal to summon a general meeting to deal with the conflict. Was Lehmann planning a coup? Ørsted made the council aware that there was no certainty that a general meeting would be representative of the members of the society. In the capital alone there were one thousand two hundred members. How would everyone get a seat? Let alone the many more members in the provinces, who would be excluded. So an elected council was more suitable, and this existed already. True, it was dominated by the founders of the society, but it could be enlarged.

Secondly, Ørsted defended the leadership of which he was a part. It was abiding by the laws of the society. Its objective according to section 1 of its constitution was 'to promote the right use of the freedom of the press, to counteract abuses, and to work for popular education', which the leadership understood to mean preventing the freedom of the press from being curbed. Section 2 read: 'to print good and useful publications for all classes of citizens to ennoble the mind, to improve taste, and to disseminate knowledge', that is, work for the aesthetic and moral enlightenment of the people. These objectives transcended party-political boundaries. 'Political parties should not be promoted. The members have united, not to follow a particular party but to achieve the commonwealth on which all enlightened people of the different parties agree.'[36] Ørsted's intention was to foil the liberals' attempt to take over the society, the purpose of which, he stressed, was to safeguard the freedom of the press not to change the constitution.

Ørsted's rhetoric worked. Two thirds of the founders threatened to leave the society if Lehmann's plan was adopted. But this was a short respite. At the accession of the new King in 1839 the attitude had swung towards Lehmann's party, and at the same time a certain weakening of the support from members had occurred. Ørsted saw where things were heading. *Maanedsskrift for Litteratur* was also being whirled into the political maelstrom and the editors went off in different directions. In January 1839 Ørsted wrote to one of his young polytechnic students, who had often joined the dinner table in Studiestræde, that he had decided to withdraw from the time-wasting activities of the society and the journals in order to spend the remainder of his powers on science.

> 'Even if I have been of some use there the time spent was too long. I have begun again a more concentrated scientific life that I hope will not be without fruit. Tell this to my French friends when they ask about me. You know my many involvements that somehow provide an excuse for my having for so many years accomplished so little in physics.'[37]

PART VII
FAME AND TRIBULATIONS

> Ørsted's name is a talisman
> that opens all doors.[1]

48 | 1831–9
Technology and Industry

JUST AS the Polytechnic Institute had opened its doors a Scottish Professor J.F.W. Johnston arrived in Copenhagen to see Ørsted. He was not impressed by the University, describing its buildings in condescending terms. Even if they had not been in ruins there was no comprehensive architectural design as in Edinburgh or St. Andrews. Copenhagen had no style, the buildings no ornamentation, and the tower of the recently reconstructed Vor Frue Church no spire. Johnston saw no reason to excuse this by mentioning to his British readership that the miserable sight had been caused by British shelling. Academic standards left him cold, too. Very few studied science for its own sake; most undergraduates were bread-and-butter students aiming at simply getting their courses over and done with. The only encouraging moment was seeing Ørsted and his collection of instruments. He had willingly showed his galvanic experiments without any kind of secrecy. Modern physics uses many instruments and one understands them much better by watching them demonstrated in practice than by reading about them. Johnston visited again over the following days and every time Ørsted was busy at work, either in his laboratory or at his desk. They talked about driving forces; was research driven by money or honour? Ørsted was for honour, to which Johnston dryly remarked that the director of the Polytechnic Institute earned four to five hundred rixdollars from private lectures for two hundred *philosophicum* students, although the sons of pastors presented their *testimonium pauperitatis* to get a free place. Ørsted was the professor with the highest income.[2]

The University's buildings had now been lying in ruins for a generation, but Ørsted was lucky for he had his flat, his collection of instruments, lecture hall and workshops in his own complex, bringing all his activities together in one space. His *philosophicum* lectures took place in Studiestræde/St Pederstræde. So did his monthly lectures (having begun in Thott's Palace in 1815, see ch. 32). They were announced in the course prospectus[3] and Ursin's magazine and took place on the first Tuesday evening of every month. They were addressed to tradespeople and industrialists and were free of charge, because the Reiersen Foundation funded them. In 1830 he had reached lecture number 141, in which he reported on the GDNÄ congress in Hamburg. The audience heard about the opening sessions having the charm of being open to the laity and touching on subjects of general interest, but on the other hand also suffering from lack of debate, because scholars did not dare to oppose each other in public. Ørsted had had no scruples on that score. He criticised a German colleague for describing animal magnetism and asserting that scholars were in general agreement about it. However, Ørsted went on, speaking for himself, 'the difficulties of determining anything about animal magnetism are bigger than most people think. The obvious dishonesty of some magnetisers, the irresponsibility and lack of experimental spirit by others, and the fraudulence of patients are examples of such difficulties..., and finally it remains to be discovered whether the observed phenomena belong to the imagination or are real physical effects.'[4] Martensen, the young theologian, was on his grand tour in 1835, and in Munich he told Schelling about Ørsted's monthly lectures. The *Naturphilosoph* responded by regretting that Ørsted did not live in Munich for if he had done, he would have liked to join the audience.[5]

By establishing the Polytechnic Institute Ørsted had become the teacher of a number of students giving their minds to the natural sciences exclusively, students he could now involve in his scientific universe and hopefully recruit to his school, as he had told Oehlenschläger years before. As long as he was only teaching students for *philosophicum*, who required only elementary knowledge of physics and astronomy, the range he could choose from was small. Of course, nobody studied physics and chemistry with the prospect of getting a post in teaching let alone in research. The small staff he had gathered around him so far had not been educated at the University. Zeise was a kind of foster child, Forchhammer had studied at Kiel, Dyssel was a graduate in medicine, and Schmidten had acquired his mathematical skills at the Military Academy. This was his narrow base of recruitment as he was staffing the Polytechnic Institute. These conditions were now radically changed. He could handpick the best students from the new graduates, and equip them with letters of recommendation, travel grants, and positions. He opened the doors of his home to them, and soon bright postgraduates such as Holten, Hummel, Wilkens, and Scharling were seated at the dinner table of their patron next to Zeise, Forchhammer, Schmidten and Andersen. At long last he could form a school of his own.

A particularly familiar relationship developed with the pharmacist E.A. Scharling, who became Ørsted's new amanuensis and secretary after having submitted his prize essay on organic chemistry, done under Zeise's supervision. He graduated in 1834 and became inspector of the Polytechnic Institute, and, not least, engaged to be married to Karen, Ørsted's oldest daughter. 'Thank you for the kindness and indulgence that you have shown me lately', a grateful Scharling wrote from Hamburg:[6]

'Nobody feels more deeply than I how important the last five years have been to me; not merely have I gained much knowledge during that period, however small it still is; but my entire soul has acquired a new understanding of my obligations towards other people. Before that time I had no idea of the indulgence and the gracious kindness that make the work so easy for the serving man; in your home I found a master, a teacher, a benefactor, indeed you have pronounced me your friend!'

On the recommendation of his father-in-law to be he got a travel grant from the Foundation ad usus publicos as Ørsted's other protégés had done. He went to Göttingen, where Ørsted had recently arranged the establishment of a network of magnetic observatories with Gauss, and from there to Giessen, where Scharling had the opportunity to work with Justus Liebig.[7] His travel also brought him to Paris and London, where he carried out experiments in Faraday's laboratory.[8] Not only did he send Karen love letters he also kept his parents-in-law, who appreciated him so much, informed about his activities. In Paris he had watched children riding up and down the Champs Elysée in small carts drawn by goats, and now he thought about equipping Karen's younger brothers, Anders and Nicolai, with a similar vehicle so they could have fun in the summer in the park of Fredensborg.[9]

Unexpectedly, a lectureship at the Academy of Surgery at four hundred rixdollars a year presented itself like a windfall.[10] This was no fortune, but nevertheless a supplement to his salary at the Polytechnic Institute, and he had not lifted a finger to get it, though his father-in-law had. When the post became vacant three applicants turned up asking Ørsted to write letters of recommendation, which he did routinely and without overstating their qualities. The Academy of Surgery found all three mediocre, so in discreet ways Ørsted arranged that Forchhammer be offered the job on the quiet. However, he had not applied and to entice him to accept the offer, Ørsted found it opportune to suggest that the Academy double the salary, which it did. This was shrewd, for next Forchhammer notified his present employer that he would now only be able to teach every second semester, although both he and Ørsted already knew that such a condition could not be accepted. In consequence the position was still vacant and without qualified applicants, at a favourable salary. Now Scharling's brother, the theology professor, submitted an application on behalf of his brother, attaching a letter of recommendation by Ørsted praising the applicant to the skies. This outmanoeuvred the Academy. The job was given to Scharling, who was told to return immediately. Tongue in cheek Ørsted told young Scharling about the course of events leading to his new job, not as if it were a well thought out strategy, but as a series of contingencies that had turned up. He added that the Academy was being persuaded to accept a delay in his return, because in the meantime he had organized an extension of Scharling's travel grant for two years. It would be wrong to say that Ørsted used dirty tricks to help his son-in-law; he had just been a little smarter than everybody else. Nonetheless it smacked of nepotism.

Another manoeuvre he engineered was to transfer the Military High School to the Polytechnic Institute. The proposal was brought up in an anonymous article in *Dansk Ugeskrift* [Danish Weekly], clandestinely edited by Ørsted himself (ch. 47).[11] It was based on Ørsted's well known hierarchy of theory and practice, which applied to agriculture and forestry as well as to technology and military instruction. Theory should be acquired at the University, or, since there was no Faculty of Science, at the Polytechnic Institute. The standard of educating officers in mathematics, geometry, physics and chemistry, cartography, and modern languages would be enhanced if the Military High School were transferred to the Polytechnic Institute, where the relevant curriculum was readily available. The practice of the military craft was best acquired in real battles, and in periods of peace in the garrison, *Dansk Ugeskrift* argued.

Ørsted was very happy to keep the real author of this stunning proposal a secret. Not even his son-in-law had a clue that Ørsted was pulling the strings and that Schouw was only formally the editor of *Dansk Ugeskrift*. 'The plan seems great and promising!' Ørsted wrote to Scharling in Paris. The King had not yet taken steps to discuss the proposal, but 'I wish he would!' Ørsted went on, suggesting that the day he could call himself director of the joint institution was not around the corner.[12] He asked Scharling to be very discreet. Above all he should not talk to Captain A.F. Tscherning, who had recently been sent on indefinite leave from the Military High School by Frederik VI. The captain had settled temporarily in Paris to gather information about French military technology. Tscherning had been part of the team planning the Military High School, which was established only one year after the Polytechnic Institute, but had fallen into royal disgrace because of the democratic views Tscherning propagated in his Society of 28 May. For a short time he had taught physics and chemistry at the Military High School; he had also run a course for the Society of the Dissemination of Science at Frederiksværk.[13]

The plan of incorporating the Military High School into the Polytechnic Institute did not succeed. Ørsted may have overestimated his chances because of his close relations with Prince Carl of Hessen and Court Steward Hauch, both powerful in military matters. He had taught at the Military Academy for twenty years now, and he had impressed the King with his galvanic trough apparatus exploding mines at a distance. However, Frederik VI did not bother to ask for Ørsted's advice in this matter. His signature on the petition for freedom of the press a few months earlier had disqualified him in the King's view. What the King considered as Ørsted's disloyalty he felt might have unpredictable consequences for the military, especially if he were to join forces with the exiled Tscherning and his subversive Society of 28 May.

In 1839 Karen Ørsted was married to Scharling, he having defended his doctoral thesis, a chemical analysis of various bladder stones. Zeise and Forchhammer questioned him and awarded the degree.[14] A year later he was appointed lecturer in chemistry at the University despite the fact that he had never passed the entry exam. He therefore he needed a dispensation, which nobody opposed as the applicant was the son-in-law of the Vice-Chancellor.

Now there were two teachers of organic chemistry at the University/Polytechnic Institute: Zeise and Scharling, Ørsted's foster son and his son-in-law. The former lacked pedagogical flair and was increasingly becoming a hypochondriac, the latter an extrovert, enterprising, and already appreciated as a lecturer. Zeise could not stand the pressure from the many Polytechnic students demanding supervision in his laboratory; they got on his nerves, and he began being absent from his classes and even reported sick just before the examination period, so Scharling was paid to take over. Zeise consulted his doctor, who was also the family doctor of the Ørsteds, and the patient was exempted from duties.[15] Ørsted was put in a rather delicate position. What should he do with a sick lecturer who was not a good teacher even when he was well? On the other hand he risked being charged with nepotism if he replaced Zeise with Scharling. Rumours were already rife. Wisely, Ørsted left the decision with Zeise, but when he then asked Ørsted for advice, he found it appropriate to suggest that Zeise took care of his research at the University and for the time being let Scharling take care of the students. The drawback to this solution was that it anticipated a sharing of the chemistry laboratory, and who would then be in charge? In 1839 Zeise fell ill again, and Ørsted came up with a compromise: each of them should have his own laboratory. He applied to the Board of the University for money and got it. Zeise felt ostracised and bitter towards his old foster father for adding fuel to the rumours. Things had got out of proportion, and there was a good deal of backstage gossip about Ørsted's nepotism. In 1820 he had defended Zeise against complaints about his poor lectures (ch. 34), but at that time there had been nobody to replace him. There was now, so in these circumstances Zeise's health improved and he resumed work both as researcher and teacher. This situation could probably have continued indefinitely, but for Zeise's death in 1847.

If anything could undermine Ørsted's integrity it would be accusations of nepotism. In his commemorative speech on Zeise Ørsted delivered a balanced account of his first disciple: praise for his research, acknowledgement of his poor teaching, and regret for his hypochondria. Scharling's was not the only career he had favoured, he had also supported Zeise and Forchhammer and many others. But patronage is not quite the same thing as nepotism.

*

The government took it for granted that as Director of the Polytechnic Institute, Ørsted and his technical expertise would be at their disposal. However, one is left with the impression that he only fulfilled these obligations half-heartedly, and that the outcome was only randomly successful. In 1831 P.C. Stemann, head of the Danish Chancellery, asked the Royal Danish Society to investigate the prospects of finding drinking water beneath the ground in Copenhagen, as suggested by a French report.[16] In Paris and London clean water gushed out from underground in artesian wells. It was known that one had to bore deeply to get it, but borings had revealed regular geological structures similar to what the geologist Forchhammer had, from the evidence, expected to find beneath Copenhagen. On their first expedition to Bornholm he and Ørsted had chanced upon such a pocket of water after a few hours' boring, and it still gushed out under the name of 'Ørsted's Spring'. Now Ørsted wrote another optimistic report and on this basis the Physics Class had the attractive prospect of launching a scientific project with a double purpose. A geological investigation of the geology of Copenhagen would be undertaken and in so doing the prospects for providing clean drinking water by artesian wells would be assessed. Even if the engineering project did not extract a drop of water, the geological fieldwork would be worthwhile. Ørsted envisaged a research project that would lend international lustre to Danish science, secure the capital 'the treasure of a splendid water supply' during a siege, and forever change its sanitary level from the state of nature to modern state of the art. The society appointed a committee consisting of Ørsted, Forchhammer, A.W. Hauch and A. Schifter to carry out the first test bores. Two thousand rixdollars were granted to the project, initially.[17]

The drinking water had a bad reputation. It had a muddy taste and a bad smell. The water from the lakes surrounding the city ran through long wooden pipes joined together by lead sleeves, before users drew it from their well in buckets which they carried to the kitchen. It was not immediately suitable as drinking water, but in spite of that people drank it. Ørsted's oldest son, Christian, recounted the strangest things about the quality of the drinking water from his grammar school at Christianshavn. In the schoolyard was a pump that everybody drank from, teachers from a real glass, pupils from a communal tin mug. When Christian and his friends were sent for a glass of water for their teacher, they were supposed to filter it through a linen cloth. In this cloth a large natural history collection would appear, consisting of mosquito larvae, daphnia, nematodes, vorticella, leeches, salamanders, and tadpoles. If the teacher was unpopular, the catch was thrown back into the glass, and if he became aware of the boyish prank before drinking it he would scream, 'Are you trying to poison me?'. The water reservoirs of the city were shallow and in the hot summer sun all species of water insects throve in them, while an abundance of drowned dogs and cats floated around. Leaking gutters and overflowing toilets contaminated many of the hundreds of wells. In the worst cases the health authorities would lock up the wooden pumps.[18]

The theory of artesian well boring relied on the deposit of underground water-bearing strata of greensand. As a consequence of geological displacements these strata had an even declivity that Ørsted presumed would follow a line from a hilltop in Scania to below Copenhagen. We can recognise this way of reasoning. The two reports on coal and iron deposits on Bornholm had reinforced their theory that seams of coal stretched from Höganäs under the Baltic Sea to Bornholm. Something similar had to apply to other geological structures, Ørsted and Forchhammer concluded, 'unless some strange [historical] convulsion of nature should have disrupted the coherence so easily recognisable on both sides of the Sound'.[19] Under the greensand was a hard, sloping stratum that was impervious to the percolating rain and melting water. When the drill reached the depths of the greensand, the water would rise by high pressure according to the law of communicating vessels. They were convinced that the greensand was situated below the white chalk, and on this premiss Ørsted got the approval of the Society to initiate the boring on Nyholm, the location of the Royal Naval Dockyard. This was why Commander Schifter was in the committee. Nyholm was chosen to keep curious and (in case anything went wrong) critical spectators at bay. The Admiralty offered to supply materials, and sailors to undertake the heavy work of drilling.

The first year the muscle-powered drill reached a depth of forty-three metres. Everything went according to plan, the committee reported. Perhaps steam power ought to be used rather than sailors, which would advance the project, but also be more costly. Two thousand rixdollars was spent and in 1835 the Royal Danish Society granted another one thousand to continue the effort, and subsequently this amount appeared on the annual budget.[20] The committee kept a drilling log illustrating the geological strata with coloured drawings. The following years the drill slowly penetrated strata of white chalk and flint, in 1841 by two metres and in 1842 by eight. After eleven years the drill was down to one hundred and seventy metres below sea level and still there were no traces of greensand. In 1845 a drill stem broke, and eleven rods went missing in the shaft due to the workers' carelessness, and still no drop of water in sight. At this stage the work had cost twelve thousand rixdollars, and the impending fiasco was rumoured, to the *Schadenfreude* of the Copenhageners. A witty tailor turned to King Christian VIII promising fresh drink-

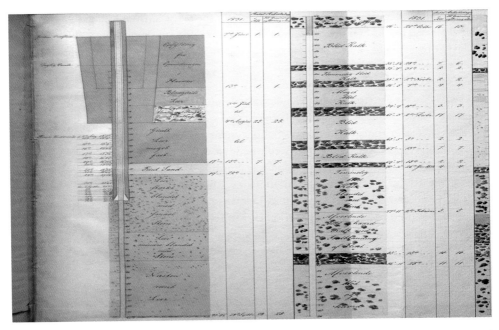

Fig. 122. Artesian well boring on Nyholm 1831–3 undertaken by the Royal Danish Society of Sciences and Letters. The table shows the total number of workdays and separately the number of days for digging and drilling alone. The various strata are described as follows: 1. strata of filling from dredging the harbour, 2. bluish-yellowish clay, 3. sand and gravel, shingles, granite, chalk, and flint, 4. yellow clay, very solid, 5. fine sand, 6. clay and pebbles, 7. flint and chalk, 8. soft chalk, 9. flint, 10. rather soft chalk, 11. flint, 12. very soft chalk, 13. hard flint, 14. soft chalk, 15. flint and hard chalk, 16. soft chalk, etc., etc. Coloured drawing in KDVS's archive.

ing water at twenty rixdollars by drilling for a fortnight elsewhere. And Christian Molbech, the historian writing in the centenary publication of the Society and having a bloodhound's sense for other peoples' weaknesses sarcastically remarked that if the Nyholm project had produced no water then at least it had proven to be an efficient drain of the coffers of the Society. In 1848 the project was discontinued. The effort had, after all, been of great scientific significance, the committee stated. Forchhammer wrote an article for *Dansk Ugeskrift*. The project had shown that the white chalk was a geological formation younger than the so-called Saltholm's chalk and hence was situated on top of it, while the greensand was absent.[21]

When Scharling returned from his grand tour in 1836 he suggested that his father-in-law should take up an idea he had encountered at the Adelaide Gallery in London and which Ørsted was to see for himself ten years later (ch. 55). The Adelaide Gallery demonstrated the latest technological innovations such as a diving bell immersed in a basin while a lecturer explained to the visitors what happened to the diver breathing through an air hose. This was both science and an entertainment at the same time. Ørsted accepted the proposal of selling tickets at one mark to the public for shows in Studiestræde.[22] Two or three ordinary workmen were employed as guides to explain the function of the latest technology, and visitors were handed a leaflet to take home as a keepsake. In 1837 the exhibition showed silver chloride plates blackened by sunlight, called daguerreotypes (or photographs as Herschel dubbed them), burning-glasses, sound experiments, bleaching agents, and Bramah's hydraulic press, all examples of spectacular science.

In April 1839 the electromagnetic telegraph was demonstrated out of doors. Ørsted had established a galvanic circuit in conductors between two wooden sheds and connected them to a magnetic needle. In principle the demonstration resembled his discovery of 1820, the only difference being the longer conductors. The needle nodded, when the circuit was closed. If the poles were shifted, the needle deflected in the opposite direction. Plus or minus, yes or no. This was the germ of a system of communication. But there was still a long way to go to make a functional telegraph. Andersen described the demonstration as magical. During the war against Britain there had only been an optical telegraph: 'quicker than the flight of birds, messages and answers from place to place and by a mute language alone, by some black shutters speaking through the air.' But now! The electromagnetical telegraph is not only more efficient, it is also harder to conceptualise: 'The naked eye sees no sign at all, the magnetic current carries the word invisibly to the remote recipient; it is the human intellect and strong natural forces that work hand in hand.'[23] Andersen wondered why there had been so few visitors. His disappointment echoed Ørsted's regrets at the dinner table, so he urged his poetic friend to turn to the press, and arranged that the royal court should honour the exhibition with its presence.[24] It helped. On the last few days of the exhibition more than a thousand people came to see it. It was a huge stride from 1820 to the establishment of the State Telegraph in 1854, and it was not Ørsted who made it.[25]

In 1834 and 1836 a committee had arranged an industrial exhibition at Charlottenburg. On a Sunday in May Ørsted took his young daughters Sophie and Mathilde by the hand to see it.[26] Why did it take place in the premises of the Academy of Arts? Would not the Polytechnic Institute have been a more obvious location, one might ask, which would enable the first generation of polytechnic students to show their technological skills and in this way to demonstrate to the public that they were worth its money? There was much gossip and scandal (and a vastly inflated impression given) about the amount of public money Ørsted had coaxed the monarch to grant out of the slender treasury. Why did not Ørsted take charge of the mutual initiative of the Agricultural Society, the Academy of Arts, and the Society for the Dissemination of Science to show that the Polytechnic Institute he had proposed as the alpha and omega of a modern nation and had now finally established, had fulfilled his promise?

The answer was self-evident if, like the eleven thousand other visitors, one walked through the ten rooms at Charlottenborg glancing at the almost seven hundred objects on display. There were ingenious locks, bricks and clinkers, textiles, homemade woollen socks, linen tablecloths, and an octagonal claw-footed parrot cage. There were chessmen, mouthpieces of pipes, and umbrella stands of ivory and ebony, spinning wheels, and sewing-tables of mahogany with inlays of rare woods, couches and rocking chairs, goldsmith's work, and samovars and chandeliers of silver, plate, and brass, chronometers and pocket watches, pianofortes, flutes and guitars, memorial crosses cast by the best iron founders of the city and statuettes of Frederik VI and Lord Byron made in biscuit ware from the Royal Porcelain Factory; in brief, an abundance of handicraft products that owed nothing to the Polytechnic Institute. The only object on display pointing towards Studiestræde was the gas lamp made a blacksmith, who had also installed these in Ørsted's lecture hall and study. Technical artefacts consist of materials worked by human hands by means of tools or machines, and this presupposes manual practice, visual skills, inventive talent, and a feeling for the demand of the market. None of these skills were taught at the Polytechnic Institute. The polytechnic education was as theoretical as the education of other civil servants.

The contradiction came out into the open in 1838, when some liberal young people, including Orla Lehmann and A.F. Tscherning, founded the Industrial Association. Their aim was not to establish industrial enterprises, but to get rid of barriers against free trade set up by the guild system. Liberal ideology was urging the reformists on. In addition, more moderate forces like Ørsted, Schouw, and Clausen joined the movement. The reformists had a political agenda rather than plans for embarking on industrial production. Whereas the interests of artisans consisted in preserving the protection traditionally granted them, members of the Industrial Association aimed at freedom of trade and saw themselves as pioneers of the modern projects including railways, telegraphs, and steamships, and Tscherning even proposed a tunnel under the Great Belt. Denmark was to be brought to the stature of Europe's modern industrial nations.

Ørsted was not one of the promoters of the Industrial Association, neither did he associate with its pioneers. He just lent it his name, which was the minimal effort to be expected from the Director of the Polytechnic Institute. At the first election to the governing body of the Association he got as many votes as Gamst and Lehmann, coming in sixth, but he declined their confidence and stayed out of it. He left it to Forchhammer, Wilkens, and Hummel to take the posts.[27] Forchhammer became chairman of the association, promoting in its periodical the significance of science.[28]

One of the first initiatives of the Industrial Association was to establish a railway on Zealand. No polytechnic graduate played any part in it, and Ørsted ignored a letter from the railway company inviting him to establish a telegraph line along the rails.

Ørsted's polytechnic project did not come up to the expectations of the public. The splendid vision that scientific education would spark off industrial modernisation was dwindling to an unrealisable myth used by Ørsted to wrench economic support from the government for his dynamical project. The undermining of the myth had started when it turned out that the optimistic promises of the underground resources of Bornholm were wildly exaggerated. Now the disillusion was amplified by the failed artesian wells. But even worse for his status as hero was the fact that he did not seem to work wholeheartedly for the realisation of his own polytechnic project. His personal engagement in the industrial exhibition at Charlottenborg and in the formation of the Industrial Association was negligible, and the same applies to his demonstration of his embryonic telegraph in Studiestræde. Rather than caring about his polytechnic project he devoted his time to writing poetry. His happiest year in 1820 had been a springboard for an equivocal polytechnic project, and the rich harvest of technological fruits that was supposed to follow in the 1830s turned out to be a disappointment; in the 1840s the criticism against Ørsted became fierce, as we shall see in ch. 58.

It is not for the sake of the school,
but for life that one seeks education.¹

49 | 1833–9
The Natural Laws of General Education

IN HIS literary testament Ørsted called electromagnetism the soul in nature. It is an invisible force universally active according to distinct laws of nature. The capacity to recognise these laws is based on the assumption that nature and the human mind are endowed with the same reason. This correspondence hinges on the existence of a ubiquitous Godhead and consequently there has to be a higher although hidden meaning for man as part of nature. Hence, it is his calling to disseminate knowledge of this omnipresent reason. The main focus of this vocation is for the teacher 'to strike divine sparks in the minds of his pupils'.² Man and nature are spiritually connected, because the electromagnetic forces as well as the laws they obey, the reason that pervades them, and the spark that mediates between them, constitute the soul in nature. To Ørsted the essence of general education is to bring to the fore this invisible 'natural law' in the minds of people.

This line of thought made no sense to Grundtvig, the theologian, historian, and hymn writer. To him human life does not obtain its meaning by reason (as infused in nature and people), no matter how spiritually this natural force is construed; human life is nourished by popular culture as embedded in historical tradition and language. The science of physics he saw as material and thus largely irrelevant to morality and cultural habits. However, he liked Ørsted's metaphor. In 1838 Grundtvig discovered his 'natural law' of human life, which had no affinity to electromagnetism and other forces of nature. That year 'the great natural law of the effect and dissemination of the real spirit dawned upon me'.[3] It was not physical forces, but *the living word* exchanged between people that empowered the community, when its members could break out of their individual isolation to come together at the folk high school for free discourse. Popular culture was in no need of improvement by academics in dignified aloofness. The two basically different views on life clashed in the debate on education in the 1830s and 1840s and set the framework for Danish educational policies in the century to follow.

In 1814 Ørsted and Grundtvig had been involved in a bitter controversy on the relationship between the philosophy of the 'new school' and Christianity. Grundtvig had accused Ørsted of diluting the Christian faith in favour of scientism and pantheism, whereas Ørsted had reproached Grundtvig for his alleged orthodoxy bordering on irrationality. This controversy culminated in Ørsted's speech at the University claiming scientific research to be akin to religious worship. This speech was an outright provocation since it was delivered from the pulpit of a church as part of a celebration of the Reformation.

Against this background Grundtvig's career had been a disappointment, since he had lost his position in the State Church and fallen victim to censorship. Both men had enjoyed long visits to Britain, and both had expressed their admiration for its tutorial system and the frank conviviality among students and staff at Trinity College, Cambridge.[4] Both endorsed the aspirations of voluntary associations to disseminate knowledge in civil society. Ørsted's colleagues in The Royal Society introduced him to the Royal Institution and the London Institution (ch. 39), while Grundtvig was invited to lectures at the Mechanics' Institutes.

Fig. 123. HCØ, 1832–33. Danish decorations awarded by FVI. Oil painting by Christian Albrecht Jensen (1792–1870). SMK.

Fig. 124. N.F.S. Grundtvig, 1843, priest, hymn-writer, historian, founder of the folk high school movement. Oil painting by Christian Albrecht Jensen (1792–1870). HS.

The Mechanics' Institute in London was a self-governing body funded by industrial philanthropists offering technical and general education to artisans and mechanics. The British government still maintained a laissez-faire attitude towards general education. The Mechanics' Institute provided not only technical instruction, but also general education in subjects such as history, literature and political economy (a much-disputed discipline ranging from profit maximisation to measures alleviating poverty and unemployment). Grundtvig plunged into the fierce debate as to who should run the Institute and what its curriculum should be. Should it be decided from below, that is by the mechanics themselves, and serve their needs, or from above by capital interests?[5] Inspired by these institutions and by the philosophy of the German historian of culture, J.G. Herder, Grundtvig had distanced himself from his former orthodoxy and discovered his 'natural law' of human life epitomized and framed by the living word in his folk high school.

On their return to Denmark, both had plans to establish new domestic institutions based on their different experiences. Ørsted, in close cooperation with Prince Christian Frederik, successor to the throne and president of the Academy of the Arts, founded The Society for the Dissemination of Scientific Knowledge offering science lectures to the laity, not only artisans and mechanics, but artists as well. Secondly, Ørsted had also succeeded in gaining an independent foothold for the sciences by founding a Polytechnic Institute headed by himself and affiliated to the University of Copenhagen. Thirdly, he acquired a leading position among the cultural elite by founding three journals which set higher standards in aesthetics, politics, liberty of the press, and general education.

Meanwhile Grundtvig, still deprived of a post in the state church or at the University, operated on a humbler level as a freelance historian and poet. He wrote beautiful hymns that the ecclesiastical authorities forbade him to print. He attracted large audiences to his lectures on historical topics, and eventually formed a voluntary association, *Danske Samfund* [The Danish Society], the forerunner of his folk high school project.

Both had responded with certain reservations to the public political awakening following the introduction of the Advisory Assemblies of the Estates in the 1830s. Whereas radical liberals aimed at putting down absolutism altogether and called for a democratic constitution in its place, Ørsted felt uncomfortable with the damaging effects a parliament was likely to inflict upon the arts and the sciences by cutting state funding. Grundtvig, on the other hand was concerned that the academic elite would browbeat the unprepared peasantry, deafen the voice of the people with know-all authority and subject the population to just another kind of tutelage. On one point they were in full agreement: both abhorred censorship.

The differences between their educational ideas were striking, so let us place them side by side. Ørsted was a follower of Kantian and Fichtean philosophy and to him the aim of education was to develop students' rational sense, that is to say to develop the whole range of their cognitive, moral, and aesthetic intellect. Learning to find the truth depended on the acquisition of the proper epistemology. Hence, science ought to have a high priority in the educational system, not only for its own sake, but because, more than other subjects, physics and chemistry were empirical and suitable for corroborating or refuting the validity of propositions.[6] Morality, according to Kant and following Plato, was also based on reason, and the maxim of duty (the categorical imperative) obliged the individual to act according to reason if need be at the expense of his happiness.[7] Beauty conformed to the perpetual laws of nature echoing divine reason. The three aspects of reason (truth, goodness, and beauty) do not attain their acme of perfection on this planet. They are eternal ideas. Man is part of nature and a creature in the image of God, although only a feeble imprint of His eternal reason, which partially reveals itself to the well educated when they seek for it. Good education is a question of awakening reason. The ideal teacher is the one who can enable a divine spark to penetrate the mind of his student. Examples abound in his seven dialogues. Teaching is a one-way process from a tutor's fund of insight to the reason of the recipient.

Exams at universities were considered a necessity, because by graduating the student acquires the key to a public position. Crudely speaking, Ørsted saw two kinds of students, the inquisitive ones, the philosophical minds, who can never learn enough because they are genuinely absorbed by intellectual curiosity, and in contrast, the so-called bread-and-butter students who see their intellectual development not as an end in itself but merely as a means of achieving their private goal in the shape of a public office that would provide a safe income for a family. The bread-and-butter student always cuts corners to achieve his goal as effortlessly as possible, asking himself what is the least knowledge that would suffice to open the door to a safe job? Hence, exams were necessary at universities to check the minimal knowledge a public servant must possess, and Ørsted worked out a scale of marking for this that outlived him by well over a century.[8]

Grundtvig's folk high school project was in no need of exams, because it did not aim at qualifying students for public posts, but aimed to equip them for their working lives as farmers, artisans, and citizens. Consequently, it would not make sense for students to confuse the ends of education with the means. A bread-and-butter student in a folk high school would be a contradiction in terms. They would join the folk high school voluntarily, step out of their isolated childhood environments to join a democratic community, and pay with their own money to be wound up like a clock and henceforth be self-motivated. At a folk high school the professor would not control the student, rather the other way round. The student's hunger for enlightenment is insatiable, and he would constantly be putting questions to his professor who would often turn out to be incapable of providing satisfactory answers. Like Ørsted's, Grundtvig's anthropology conceives man as a product of 'earth and spirit'. Both are dualists.

The difference between Ørsted's idea of a university and Grundtvig's folk high school project reflects a parallel difference in their views on the socio-economic structure of society. Ørsted endorsed the Platonist vision of a stratified society. Its members are distinguished by a range of different skills, that is to say philosophers (scientists and artists) make up the top level of the hierarchy, whereas merchants, soldiers, and peasants exist at subordinate levels, much like the parts of the human body. The brain is in command, the arms are used for fighting, the belly for enjoyment and the legs for bodily labour. Since the ultimate goal of society is to provide just laws and refined art it would be counter-productive to let the peasantry perform as lawgivers, scientists, and artists or to let philosophers defend the territory of the state. The educational system must be adapted to this concept of society. Its citizens practise an organic division of labour coinciding with their endowments, and constitute a *demos*. All they share is citizenship and obedience to common law. Ørsted did not regard this division of labour as settled once and for all. On the contrary, he subscribed to a meritocracy, where privileges acquired by birth (as by the aristocracy) or by membership of a corporation (as in a guild) should be totally replaced by intellectual merit gained via the education system.

Grundtvig resented this class society, 'where the guardians of liberty and education swallow both, whereas people are rewarded for their labour only by shadows of virtue and beauty, and are actually forced to obey, admire, and adore their haughty tyrants!'[9] He did not believe that philosophers possessed a privileged access to reason, and he suspected that the learned republic would pursue elitist interests rather than care for the commonwealth. The stratified society belonged to the past. Instead, he envisaged a more egalitarian society governed by the people themselves. Grundtvig was an ardent follower of J.G. Herder, the German philosopher, who disagreed with Kantian teachings regarding a universal reason. Like Herder, Grundtvig saw popular lifestyles rooted in specific national cultures that derived their characteristic features from their history and their language. This popular culture was an empirical fact which deserved respect and pride simply because it echoed the values and habits of the national community. Its members were not an incoherent *demos* of divided classes or Estates, but an *ethnos* bound together by its cultural heritage.[10] It was the objective of the folk high school to invigorate this tradition and to breathe self-confidence into a population that had for so long been subjected to the rule of the learned republic.

In other words Ørsted and Grundtvig both recognised that education is an instrument of power. Ørsted saw it as a political means to get rid of irrational privileges in favour of a meritocracy based on the three dimensions of reason (truth, goodness, and beauty) to enhance civilisation. Grundtvig on the other hand dreaded the idea of rule by the learned republic, and introduced his folk high school as a political instrument to pave the way for an egalitarian society founded on the values of popular culture.

*

So far I have tried to set out two opposed educational ideas in their purest form as if they were the only alternatives. But in fact during the 1830s the Advisory Assemblies of the Estates promoted their own educational interests, which mostly differed from those of Ørsted and Grundtvig. For the sake of clarity let me reduce the debate to two topics: firstly, Reform-Humanism targeting grammar schools and the University, and secondly, the idea of 'practical' education [*Realskolen*] aiming at satisfying the demands of the rising middle classes.[11] In principle, Ørsted was a supporter of Reform-Humanism, but an opponent of 'practical' education, whereas Grundtvig opposed both movements and took the outsider position of a one-man army fighting for his own project.

The Reform-Humanists cherished the culture of antiquity, prioritising the Greek arts and sciences as the core of academic education at the expense of Latin, because Greek architecture, literature, and philosophy are earlier and as a consequence possess a higher value than the Roman Empire, which they considered an inferior imitation. Another element of Reform-Humanism (and an important reason for Ørsted to support it) was its upgrading of the sciences. In an admonitory speech delivered to undergraduates he enjoined them 'never to forget that it is our spiritual being that makes us an imprint of the Godhead and that it is our study of the sciences that incessantly ignites the divine spark by showing us a reflection of our own nature and by exposing the divine imprint on nature as an omnipresent revelation'.[12] He referred to the inscription above the entrance arch to Plato's Academy in Athens: 'No access without knowledge of geometry'. His point was that scientific propositions were susceptible to the touchstone of experience, that is, testing by experiment. This argument justified his agreement with the Reform-Humanists.

In 1832 the senate of the University appointed Ørsted and J.N. Madvig, the Latin professor, to consider the status of Latin in higher education. In complete agreement, they recommended that Latin be discontinued at exams, speeches, and dissertations at the University, arguing that it was a dead language, therefore superfluous and often even ridiculous, as was the case when a student of medicine at his anatomy exam explained the shape of a bone by the sentence '*habet figuram horseshoe*'.[13] It would be wiser to spend the limited time in studying the full breadth of a subject than in learning the Latin phrases in which to express it. After all, students are brought up to think in their mother tongue. Some called Madvig 'the greatest Latinist', but when his subject lost its status as the academic lingua franca, he self-deprecatingly changed his nickname to 'the Last of the Mohicans'.[14] In 1841, when Ørsted was Vice-Chancellor, Søren Kierkegaard still had to apply for a dispensation to hand in his dissertation '*On the Concept of Irony with Constant Reference to Socrates*' in Danish. He received the permission, because nobody could deny his command of Latin. Henceforth, Latin would only be studied by those few who chose it as their primary subject.[15]

There is a striking difference between general education and competence. The movement urging 'practical' education [*Realskolen*] epitomised their campaign under the slogan 'Education for Life'.[16] They urged useful competences for the benefit of mariners, mechanics, businessmen, and the like. Their spokesmen wanted to throw the classical languages into the dustbin of oblivion and replace them with subjects that would serve the commercial interests of the middle class, such as modern languages, history, and useful sciences such as mathematics and geography, since these were the subjects that would provide useful competences for practical life.

Realskolen were neither Ørsted's nor Grundtvig's cup of tea. They did not urge competences, but advocated general education, although each advocated a different kind of general education.[17] Ørsted aimed at educating the character especially of academics, while Grundtvig dreamed of high schools empowering people to organise their lives as independent farmers or artisans, and not least as democratic citizens. Young people in need of practical skills should be apprenticed, he said. Hence, both felt unable to ride the leading hobbyhorses in the current debate.

*

Grundtvig's concept of a folk high school was at once too radical and at certain points too obscure to enable it to gain a foothold on the agenda of public debate. First he wanted the education of civil servants removed from the University and located at special institutions that he dubbed training colleges. Pastors, judges and medical doctors were not researching scholars in his sense of the word, they merely performed skills. Hence it would be a misunderstanding to educate them at a University worthy of its name as a research institute. Training to become a civil servant changed the minds of students into memory machines repeating parrot fashion what the professors taught them. To replace the existing University a dedicated research academy should be established to investigate all the historical and scientific aspects of human life in Scandinavia from time immemorial, because otherwise, when they had been eliminated from the University, these realms of research would have no place. Since research talent is scarce, a natural solution would be to cooperate with the other Scandinavian nations in establishing a joint university, an Academic Union of Scandinavia in Gothenburg, where the three nations came close to sharing a common border, where about three hundred erudite scholars could study these essential issues, with ancient Icelandic as their lingua franca.[18]

This division of the functions of the University into national training colleges and an Academic Union of Scandinavia implied a de facto elimination of the University of Copenhagen. So the only institution of higher learning in Denmark would be Sorø Academy, something between a grammar school and a university. Grundtvig's project had no need of the University of Copenhagen at all, nor of grammar schools to prepare people for it. Obviously, Sorø Academy, as it had been, would become obsolete, but its days were also numbered for more immediate reasons that had nothing to do with Grundtvig's vision.

Despite its proud history, the Academy found itself in a deep crisis. The number of students had declined and its staff was severely underemployed, which paved the way for entirely new uses of its premises and its considerable assets. Sorø Academy with its independent statutes, but under the administration of the Board of the University and Grammar Schools, became the object of a rivalry between three educational movements.[19] Grundtvig envisaged his folk high school being located there, spokesmen for *Realskolen* wanted the Academy replaced by a new institution for practical education, and Ørsted, in line with the Board, saw an opportunity to seize Sorø's treasury to spend it on expanding the University and the Polytechnic Institute.

It was Grundtvig's dream to prepare Denmark for a new era of government for the people and by the people. The justification of the high school at Sorø would be the empowerment of the untutored population by means of general education. Gothenburg was intended as the provider of its store of knowledge to Sorø, and the two institutions would be interdependent.[20] Together they would make up a pincer movement to eliminate the meritocratic system from above and below.

What confused public opinion the most about Grundtvig's project was his complete omission of any curriculum. Unlike his adversaries he deliberately abstained from setting out a syllabus. The folk high school was based on his 'natural law' of the living word. Obviously, one does not make plans for the operation of a natural law. It unfolds itself by its intrinsic driving force.[21] Nevertheless, like Ørsted he wrote dialogues, envisaging what life at his folk high school would look like once it had been established. In 1837 he dreamt of visiting Sorø in around 1950 when his great natural law of the effect and transmission of the spirit would have been operative for more than a century. A young lad showing him around would explain everything they saw to his prominent visitor.

Whereas Ørsted had applied the word 'touchstone' to designate an experiment that would decide the validity of a scientific proposition, Grundtvig saw the touchstone of Sorø folk high school as a workshop where the pupils would decide whether a technical innovation would ease their labour, or threaten to make them redundant, or reduce them to mere appendages to machines. In his dream 'it was as if I walked through the Mechanics' Institute in London where the pressure of British business and skill prevailed. Here I only saw a few very simple machines. In spite of all my admiration for British entrepreneurship it had always struck me that there death was the major player, by every new invention rewarded with thousands of victims, with skilled and industrious workers made redundant.'[22] Here at Sorø the voice of the people had succeeded in maintaining the respect for human labour and in delaying the heartless domination of machinery.

His dream went on to show him that examination would also take place at the folk high school, but it would be diametrically opposite to the control mechanism in place at the University. Why would professors fire off questions to which they knew the answers perfectly well already? However, questions were raining down from the student benches like cats and dogs . There were high jinks, and the students were taking the mickey out of the professors. Some old peasants would embarrass the professors, often forcing them to confess that they had been fooled by a notorious book or by some unfounded rumour or by their own misleading deductions... The professors would squirm like an eel.[23]

The folk high school at Sorø was going to be headed by what Grundtvig called a steward, probably a post that he himself would hold. The steward would be guided by an advisory com-

Fig. 125. H.G. Harder (1792–1873), teacher of drawing at Sorø Academy, painted its neo-classical building which was inaugurated in 1822 after the baroque complex by Laurits de Thura had burnt down in 1813. Behind the school part of the church dating back to the medieval Cistercian Monastery is visible. The Academy exemplified the historical continuity that Grundtvig dreamed of carrying on with his high school project. Vestsjællands Kunstmuseum, Sorø. Photo by Bent Ryberg.

mittee elected by the students themselves. He feared that a sudden replacement of absolutism by a 'democratic' constitution would result in a dominion by the learned republic, who were likely to take advantage and continue their dominance over the people. His high school project was intended to provide the people with a voice that would also be trained by students' participation in running the high school.

This educational idea contradicted Ørsted's meritocratic theory and practice at nearly all points. This is no coincidence. Grundtvig's dialogues and pamphlets on education were directed against Ørsted's dialogues, articles, and practice as an organiser of higher education. Ørsted's dialogues featured interlocutors who were confused one way or the other on a certain issue, until his benevolent spokesman of reason set things right and all confusion was dissolved. His scheme of a Polytechnic Institute in close alliance with the University had a detailed curriculum prioritising theoretical knowledge above workshop practice.

In fact, Ørsted made efforts to unite all institutions of higher learning under the auspices of the University. His Polytechnic Institute was the first satellite, the germ of a future Faculty of Science, which actually materialised in 1850. His next initiative to amalgamate the Military High School with his Polytechnic Institute under his own directorship failed.[24] He also recommended that a proposed Agricultural College be affiliated to the University, claiming that the integration of agricultural theory and practice was a misunderstanding. To the extent that agriculture was founded on science, young farmers should follow courses in botany, chemistry and meteorology at the Danish headquarters of science.[25] The learned republic should not be split apart, but remain united. Students demanding practical instruction must turn to landed proprietors.

His vision of a monopoly of higher education at the University presupposed extra funding, and the crisis at Sorø Academy prompted a transfer of its assets to the University. Ørsted saw no need to warn against Grundtvig's high school project finding it to be sheer lunacy. Madvig, his colleague, went so far as to comment, that 'if the peasantry is not yet ready to recognise its own incompetence, it is in need of being enlightened about it, though how this might happen I don't see at all'.[26] Ørsted believed in a sharp division of labour, but this did not entail a petrified social structure. On the contrary, Ørsted embraced social mobility, but upward mobility must depend on merit.[27] Ørsted's career shows an abundance of examples where he encouraged and helped young artisans undertake a higher education and climb the social ladder (ch. 47).

*

In December 1842 the director of Sorø Academy died, and a committee was set up to ponder the uncertain future of the institution. Ørsted represented the University with a mandate to drain the monies of the Academy's treasury. Grundtvig had no seat, but King Christian VIII allowed him to submit a report, which he did. The Deputy of Finance suggested that the Academy's fortune be spent on a much-needed expansion in terms of new buildings and higher salaries for the staff of Ørsted's Polytechnic Institute. The Advisory Assembly of Roskilde also handed in its proposal of a *Realskole*. So, the committee was up to its ears in rival ambitions and spent four years working on its recommendations.

Ørsted kept his friend Hauch informed of his plans and explored the possibilities of a post for him in Copenhagen or Kiel.[28] Therefore, Grundtvig had to suggest alternative members of staff. He would also need someone to lead the singing and a steward and a new Board that above all had to be totally independent of the Board of the University, which only longed to give his institution the kiss of death. As already mentioned, Grundtvig wanted Sorø to be governed jointly by the King's Hand and the People's Voice, but he did not have the courage to repeat his proposal that the representatives of the students should have the majority of seats, which would leave the Board virtually completely in the hands of the high school pupils themselves. He submitted his report to the King who passed it on to the committee. The Roskilde Advisory Assembly of the Estates also handed in its proposal of an institution of 'practical education'.

Finally, in October 1847 the committee came to an agreement on a compromise, to the effect that Sorø Academy must be rescued as an independent institution by harbouring a *Realhøjskole*, that is a high school for practical education. All pupils had to pass a two-year basic course in Danish language and literature. On top of this common preparation they could choose one of three branches of study: (1) history and cameralism, (2) modern foreign languages, and (3) science and mathematics. Each of these courses would give access to the University.[29]

This compromise was intended to satisfy the Academy's instinct for survival. It was intended to meet the demands of the numerous patrons of a 'practical' education and at the same time make a gesture towards Grundtvig's folk high school project. It is beyond question, however, that it would take a lot to see Grundtvig's fingerprints on this compromise. Of course, showing good will, he could recognise a few features as attempts to fulfil his educational idea, such as the two basic years, and the first branch covering history and economics. Still, this was a far cry from his utopian ideal of a genuine folk high school, though perhaps it could be seen as a gentle beginning. If Grundtvig's project was irrevocably diluted, Ørsted's plan was completely eliminated. The committee report left no hope of funding from the Academy for his Polytechnic Institute.

The *Realhøjskole*, however, never saw the light of day. A few months later the new constitutional government decided to shelve the report. Now that the country was at war funding a new educational institution was out of the question. Grundtvig complained that the official grounds for not going ahead with the school were false, since the money belonged to the Academy and not to the state, but to no avail. The Board of the University and the Grammar Schools had been abolished and its business was taken over by the new Minister of Education. Ørsted hoped that the new parliament, in which 50 out of 114 representatives were members of the learned republic, would support his plan to expand his Polytechnic Institute.[30] In October 1848 he submitted a new and enlarged application to the Ministry of Education arguing that the new regime needed engineers, agronomists, and geodesists. Ørsted offered to educate them at the Polytechnic Institute in return for funding for new buildings already designed by Bindesbøll. The Minister promised to procure 70,000 rixdollars for building costs from 'the special funds allocated to the Ministry', a political 'newspeak' term for the treasury of Sorø Academy. But first he had to consult the Minister of War, who was former captain A.F. Tscherning, in charge of the Military High School where another type of engineer was trained.[31]

As might be expected Ørsted's application met with fierce resistance. Tscherning took advantage of his membership of the government to resolve the matter in favour of the Military High School, arguing that their engineers were better equipped to handle construction and infrastructure work, mainly because they were better mathematicians. Ørsted's hopes and plans had all ended in deadlock.

In our days Scandinavians celebrate
the trio of Berzelius, Hansteen, and Ørsted.[1]

50 | 1839-47
Scandinavian Science Conferences

SOME SCANDINAVIAN scientists having participated in the annual conferences organised by the GDNÄ or the BAAS came up with the obvious idea of arranging something similar in Scandinavia. In 1838 a Norwegian physician in Christiania took an initiative together with some Swedish scientists to gather their Scandinavian colleagues at a conference in Gothenburg the following year. The idea was simple. The two Scandinavian kingdoms consisted of three nations speaking almost the same languages, which had been mastered by no other nation. Scandinavian scientists were isolated and found it hard to assert themselves on the European scene. Together, however, they would represent a tribe of six million Scandinavians whom Europe might be less likely to ignore. Within Scandinavia this cooperation would forge links between researchers and they would get more used to understanding each other's languages. Each Scandinavian country was too small to publish scientific literature in sufficient numbers, but if the whole of Scandinavia became their readership, it might stimulate scientific writers to publish more. When the divided German-speaking and English-speaking nations were able to, why should not peaceful Scandinavia do likewise?

Not everybody was convinced by this argument. Berzelius declined. He thought there were too few scientists within each discipline to form sustainable groups. He estimated that the thirty to fifty scientists that might be drummed up would at most be able to establish contact between four or five of each discipline, disregarding physicians. The noble purpose demanded a larger European cooperation, but the French and the British were so inept at speaking foreign languages that the idea was a non-starter. In addition, Swedish scientists were so poorly paid that they could not afford to travel without getting into debt, Berzelius thought. Perhaps a few might save up to take part on one occasion, but a conference every year, or every second year was quite unrealistic. So, it would be wrong for older, well-established scientists like himself and his friends Hansteen and Ørsted to encourage young scholars to join in.

Hansteen, too, declined. He had planned to see Gauss at Göttingen to work on earth magnetism, and this was his first priority. Hansteen often praised Ørsted's versatility, which he could not match having no poetic vein himself and no talent for organising things; he had to confine himself to mathematics and earth magnetism otherwise his research would get him nowhere. Hansteen's compliments smacked of ill-concealed irony. In fact, he was urging Ørsted to concentrate on the many unsolved problems of his electromagnetical theory and to stop wasting his time on poetry and Scandinavian cooperation.

It was hardly accidental that Grundtvig fostered his idea of the Academic Union of Scandinavia (ch. 49) precisely in the spring of 1839. The inference that his proposal of a joint University was a commentary on the Norwegian-Swedish idea of scientific cooperation immediately suggests itself. What induced Grundtvig to fish in troubled waters was not only the idea of Scandinavian linguistic and historical unity, which had no more fervent spokesman than himself, but also the thought of providing a strong research community. Of course, he was also attracted by the openness towards the laity and the oral communication practised by the models of GDNÄ and BAAS. Nobody could more enthusiastically approve of the idea of lecturing without a script than Grundtvig, where *the living word* (a phrase already introduced by Humboldt in 1828 and also used by Ørsted) was the acclaimed style. Sadly, the initiative excluded history and linguistics, the very subjects that had bound the Scandinavian peoples together since time immemorial.

Ørsted was disappointed at the recalcitrance of Berzelius and the lukewarm attitude of Hansteen. As far as he was concerned he was fully supportive of the project right from the beginning, even dreaming of going further and publishing joint annual transactions of the scientific societies in Copenhagen, Stockholm, and Trondheim in Danish and Swedish (at that time the Norwegians used Danish as their written language). To him the idea was an extension and a revival of the Scandinavian Literary Society, in which he and his brother had taken such an active part decades ago. Now he imagined a new society consisting of the learned republic of Scandinavia, a platform influencing public opinion and strengthening the position of science in Scandinavia as in Germany and Britain. He told Berzelius that Danish scientists, too, worked for starvation wages, but Frederik VI had promised to pay the costs of the voyage to Gothenburg for six Danish participants and this number might increase in the future. Berzelius did not give way. He had suffered a stroke and could not come, for now his doctor had prescribed bloodletting and a water cure. But could not Ørsted absent himself from the pointless meeting in Gothenburg, at least for a few days, and take the Göta Canal by boat to Stockholm where he would be sure to be received with open arms? 'Bring Oehlenschläger as well!' Ørsted thanked him for the invitation, but could not find the time, for just then he was Dean of the Faculty of Philosophy: 'I wish you a speedy recovery' he wrote.

On 13 July a number of Danes and Swedes from the south embarked on a steamship sailing from Malmø to Christiania via Gothenburg, but due to rough weather, the captain had to return to Helsingborg and the travellers continued by road. They arrived, exhausted, on time and were put up privately to save money. Twenty-one Danes took part, ten Norwegians and sixty-one Swedes. There was only one foreigner, a Prussian. This was twice the number predicted by Berzelius despite the fact that preparations had been somewhat improvised. There was as yet no society and no statutes. A committee was set up to impose some structure on the arrangement. The participants quickly agreed to emulate their German and British models by setting up open plenum meetings on subjects of general interest, and specialist meetings in three sections: (1) a physical-chemical-mathematical-geological section, (2) a natural history section for zoology and botany, and (3) a medical section with a pharmaceutical subsection. Ninety-three participants were too few to make sense of more specialised sections.

Ørsted gave a plenary lecture on 'the reciprocal utility of the artificial scientific terms of the Scandinavian languages'.[2] Lectures had to be delivered without a script to make them livelier, and this manner of extemporising seemed to be easier for Danes than for Swedes, Ørsted thought. Lecturers would have to put their speeches on paper afterwards and submit them to the publication committee, that would eventually complain about overrun deadlines. Ørsted improvised on his theme stating that he much preferred Scandinavian terms to Latin or Greek, because a language is embedding a hidden culture contributing to making the palpable truth popular. Loanwords cause misunderstandings, because those who use them are unaware of their etymology.

Ørsted was neither a language purist nor a language Romanticist in the Herderian sense, but a pragmatic rationalist. For instance he found it to be a very bad idea when some Norwegian linguists attempted to reconstruct their language by tracing and incorporating remote dialects which were otherwise obsolete. Ørsted conceived his mother tongue as the mutual Danish-Norwegian language that Norwegians and Danes had equally influenced for centuries. To replace this common language by a new 'bookish language' would only divide peoples. And the purpose of Scandinavian scientific conferences was exactly the opposite: to take advantage of the common roots and features of Swedish and Danish-Norwegian. He saw no point in artificially changing languages. Scandinavia was in no need of a lingua franca.

Two days were allocated to meetings in the sections. The problem was that the Norwegians had to return by SS Prins Carl on 20 July or else have to wait a week for the next service. So the first Scandinavian Science Conference hurried on. Ørsted talked about forces active in the capillary effect, Forchhammer on differences in the level of Danish shores, Schouw on the climate of Italy and Copenhagen, and Eschricht on the origin of intestinal worms. Reinhardt described the poison gland of a snake species from Guinea, and recited a thesis by Lund on the fauna of Brazil before the last cataclysm, on the basis of finds of petrified bones in the caves at Lagoa Santa. Sommer presented a classification of skin diseases. Norwegians talked about recent prison systems, the use of polarised light for microscopic investigations of organic bodies, and the application of sulphuric acid to the production of gypsum. The medical section dealt with the relation between a woman's pelvis and the head of a foetus and how this affected delivery, and the initiator explained how to operate on club feet and flat feet.

The idea of spicing up the scientific agenda with conviviality was also imitated from the GDNÄ and BAAS. There was an excursion to Jonsered's mechanical workshops and the day after the end of the conference the remaining scientists took the SS Polhem on the Göta Canal to the huge waterfall of Trollhättan.

The Norwegians who were to take the SS Prins Carl just managed to stay to hear Bishop Agardh's farewell address while toasting Kings Frederik and Carl Johan and their new society. The bishop standing in the ceremonial hall of the Freemasons, where the plenum meetings had taken place, interpreted that serene atmosphere: Scandinavia stretches from the fair islands with their broad beaches in the south to the barren rocks in the far north, incorporating all the beauties of nature, from flat plains of fields of grain and deep fjords where the sea settles in tranquillity to sky-high mountain peaks and fertile valleys. There is iron in the fells and strength in its peoples who worship the same God undisturbed by sectarian quarrelling. They share one mother tongue and love both freedom and order. They delve into research, and their calm serenity safeguards them from being blown over by casual squalls. Should turbulent German science so desire, it could be granted asylum in our safe haven. Nobody has seen as clearly behind the veil of Isis as Linné. And in our days Scandinavians celebrate the trio of Berzelius, Hansteen, and Ørsted.

The bishop also voiced some less formal sentiments: what the air is to the human body, science is to its spirit, and both are the freely available property of everyone. Science is disseminated through means of communication that so far have mostly been at the disposal of academies. Now it has been made easier, because thanks to modern means of transportation we have moved from stationary academies to mobile societies, such as the Scandinavian Society of Scientists facilitating cooperation between scientists (internationalisation), between the sciences (interdisciplinarity) and between science and the general public (popularisation). Thus science is brought into circulation and establishes fraternal bonds between theory and practice. States and nations are two different things. States and their borders are incessantly changing, but sciences, nations and languages are borderless and immortal. The language of science flies freely across the seas (the Kattegat) and fells (the Keel) of Scandinavia like tones in the air.

But do not the sciences extend beyond Scandinavia? he went on. Can a European society of science be established? GDNÄ and BAAS take place in German and English respectively, and non-native speakers are invariably excluded as strangers. In Scandinavian conferences all scientists are equal, and hence it is unavoidable that German and British colleagues feel left out, however welcome they are. In the long run, hopefully, cooperation between all European nations will be possible. This will secure peace in Europe. Scandinavia should be the first step towards this final goal: one European society for the promotion of science. In this way Agardh obliged the two members of the leading trio who were conspicuous by their absence, Berzelius and Hansteen. Scandinavia should not be self-sufficient, but should be open towards European cooperation. At the end everybody sang a song in Latin to the tune normally used for Horace's 'Integer vitæ, scelerisque puris...' especially written for the occasion. Grundtvig would have leapt up in protest, had he been present, but being no scientist he could not join.

The committee of statutes presented its proposal, which was adopted. The location of the next meeting had to be decided. Copenhagen won by forty votes, against sixteen for Stockholm, twelve for Gothenburg, and two for Christiania. Ørsted was elected chairman of the Copenhagen committee by sixty out of sixty-seven votes, with Schouw as his secretary.

*

In May 1840 Ørsted sent out invitations for the second Scandinavian Science Conference in Copenhagen to potential participants, among them Berzelius, Hansteen, and Steffens. Timing was an issue. The new Danish King, Christian VIII (formerly Prince Christian Frederik) was to be anointed at the end of June and would leave the capital on 10 July, but had expressed his wish to be present at some of the open meetings, and Schouw was to preside at the Assembly of the Estates at Roskilde from 15 July. The conference should also be coordinated with the steam-shipping service from Norway arriving every Friday, so all considerations would be met if the conference were planned to begin on Friday, 3 July. Scandinavian celebrities were invited to the royal ceremony to lend academic lustre to the event. This extraordinary attraction dissipated all previous hesitations, and Berzelius and Hansteen accepted to join.

Berzelius assumed that the anointing ceremony would take place in Copenhagen, but rescheduled his itinerary via Helsingborg when Ørsted told him that it would be in the church of Frederiksborg Castle. On 28 June all the bemedalled celebrities were assembled under the vaulting, among them the Ørsted brothers, Oehlenschläger, Schouw, Grundtvig, Steffens, Berzelius, Hansteen, and Andersen observing the strange ritual of Bishop Mynster anointing the last absolute monarch in Scandinavia. A royal banquet followed: 'O, it was beautiful, it was splendour', Andersen wrote.

The Copenhagen conference had more than three hundred participants, two hundred and thirty from Denmark and the Duchies, fifteen Norwegians, fifty-three Swedes, and five non-Scandinavians. The new gothic main building of the University had been completed four years earlier. The Scandinavian scientists saw the admirable sculpture of the eagle beholding the heavenly light as they entered the imposing entrance hall leading to the ceremonial hall, where plenum meetings were held. Some of the section meetings took place in the lecture rooms, but the physical-chemical section gathered at the Polytechnic Institute, the geologists in the mineralogical auditorium in the community building, and the botanists in the Botanical Gardens behind Charlottenborg.

Ørsted dealt with the same topic that Agardh had raised in Gothenburg. He defended the open plenum sessions against the criticism that they stole time from the specialist meetings. Like Humboldt in Berlin in 1828 he ranked popularity above virtuosity and also preferred themes with a unifying potential to special subjects separating the members. He also put in a word for conviviality. Making friends and having fun together does not happen at specialised sessions, but by partying and going on excursions, at which the hosts reciprocate the hospitality of guests. In addition he warned against Scandinavian self-sufficiency. Their society should be the beginning of a European network, preferably with the financial support of governments.

On Sunday 5 July the anointed King served lunch to the scientists at the Hermitage in the Royal Deer Park. As many as possible had come on the royal SS Hekla, which was a man-of-war, though as yet still without guns. One evening a dinner party was given at the Shooting

Fig. 126. King Christian VIII's anointing in Frederiksborg Church 28 June 1840. Bishop J.P. Mynster officiates at the ceremony. Only the persons mentioned in the text of this book are identified below:

H.C. Andersen, story-teller—ninth to the right of the column
J.J. Berzelius, Swedish chemist—second to the right of the column
H.N. Clausen, theologian and politician second to the right of Møsting
C.W. Eckersberg, painter—first to the right of the column
J.W. Gertner, painter—behind the back of Bishop Øllegaard with a black skullcap
N.F.S. Grundtvig, priest, historian and hymnist—fourth to the right of the column
H.L. Martensen, theologian—first to the right of Møsting
J.S. Møsting, Minister of Finance—with a golden chain below the pulpit
A. Oehlenschläger, poet—fifth from CVIII's head in the front row
A. Schifter, Vice-admiral—fourth to the right of Møsting
H. Steffens, natural philosopher—sixth to the right of the column
E.C. Tryde, archdeacon—below Eckersberg
HCØ—third to the right of the column

Etching by Johan Wilhelm Gertner (1818–1871), 1846. De Danske Kongers Kronologiske Samling på Rosenborg. Photo by Peter Kristiansen.

Range outside the city. Special songs by Andersen, Hauch, and Oehlenschläger were sung. The Students' Association assisted the Norwegian and Swedish guests, and during the weekend following the conference they were offered a voyage on board the SS Hekla to the island of Møn, where some took a look at white chalk, others at flowers, while most of them just took a walk around taking in the beautiful scenery, for instance in the park surrounding Liselund, the miniature country seat. Afterwards the estate owner of Klintholm gave a dinner party.

As a prince, Christian VIII had impressed Schmidten with his scientific insight and now he showed his generosity towards the scientists. The absolute monarchy supported science and vice versa. Immediately after his return from Møn, Ørsted reported to the King expressing the gratitude of his colleagues. 'I am happy to add that there is full agreement that the scientific value and interest of this conference are on a par with the best organised elsewhere.'[3]

The previously sceptical Berzelius found that the Copenhagen conference outdid the annual meetings of GDNÄ.[4] Christian VIII had made him a commander of Dannebrog. No doubt Ørsted had something to do with it, and now Berzelius asked him to thank the King for this unexpected and, as he modestly added, undeserved distinction. Hansteen had been made a knight of Dannebrog. Steffens had delivered a plenary lecture on his reconsideration of German *Naturphilosophie*. His paper was almost an appeal to the audience for forgiveness for his speculative misgivings. In 1824, in an emotional speech to the Students' Association he had confided to Ørsted: 'You have won the palm for which I fought.'[5] Now Steffens had been restored to favour by the new King, he had been to a royal banquet, and had been decorated. This was more than he had dreamt of.[6]

A proposal was made to award a Scandinavian prize for the best essay in response to a prize question set by the society, but this was not adopted.[7] Ørsted had obtained the King's approval for a contribution to Schouw's plan of a Scandinavian journal of science that the Swedish-Norwegian King, Carl Johan, had refused to support. Berzelius was against the idea, too, but for a different reason. In Sweden where the freedom of the press was unrestricted there was a sad tendency to politicise the sciences with verbose expositions of German *Naturphilosophie*, and this deplorable state of affairs risked spreading to Denmark, even if Danish laws hampered this kind of abuse. Ørsted agreed.[8]

*

The third Scandinavian Science Conference took place in Stockholm from 13 to 19 July 1842. Christian VIII put the SS Hekla at the disposal of the Danish delegation, numbering eighty-three scientists and three foreigners attending free. On the calm sea Captain Bille's paddle steamer could easily overtake the old-fashioned sailing ships. From the deck the passengers admired the sunset mirrored on the glassy Baltic Sea. Only the paddle wheels left small ripples. They were moved by the beauty of the archipelago outside Stockholm. After a forty-eight hour voyage SS Hekla moored in Stockholm. Ørsted was accompanied by his son Christian, his foster son Søren, a student of medicine, and Scharling, his son-in-law, and they as well as some of his colleagues from the Polytechnic Institute were all invited to dine with Berzelius and his wife. Their flat was in the noble building of the Science Academy close to the Royal Castle. Plenty of aquavit was served, but Ørsted 'took care to always have a bottle of water within

Fig. 127. HCØ welcomes Berzelius to the Scandinavian Science Conference in Copenhagen 1847. Laurits Tuxen's (1853–1927) draft proposal for the decoration of the ceremonial hall of the University 1894, which was rejected in favour of Erik Henningsen's (No. 131). DTM.

reach, otherwise the many toasts would be too much'. The Danes were accommodated at the Mint close by, and Christian went on a botanical visit to Djurgården.

More than four hundred attended the Stockholm meeting: eighty-three Danes, twenty-one Norwegians, three hundred and eight Swedes, two thirds from Stockholm and one third from the provinces, as well as twenty-four from other countries, mostly Finland. The Swedish parliament granted the necessary expenses and put Ridderhuset at their disposal for the plenary meetings.[9] Crown Prince Oskar had made great efforts to make the arrangement a successful one. He strongly encouraged his Norwegian subjects to attend and sent a special invitation to Alexander von Humboldt, who would have lent European lustre to the event had he been able to come. The Crown Prince also invited the members to take part in a voyage on Lake Mälaren to the royal castles of Gripsholm and Drottningholm. After the conference a hundred and fifty people went on an excursion to the University of Uppsala visiting the castles of Rosersberg and Skokloster on another shore of the vast lake system on the way. A choir of students welcomed the party to Uppsala, and the archbishop gave a tour of the cathedral, while the Vice-Chancellor showed them 'Linné's temple'.

On 13 July Ørsted talked to the plenum about 'the Beautiful as seen from the Scientist's Perspective'.[10] 'I was in fine fettle speaking freely... and was happy to receive great applause. After the lecture Crown Prince Oskar took my hand and said something nice to me. From all sides I had flattering comments, also from people like Professor Schouw, who normally does not flatter.'[11] Ørsted's colleagues from the Polytechnic Institute gave papers at section meetings, and he approved of them all except for Zeise's; 'despite the interesting contents he made a bad impression. Nobody understood it and most were close to falling asleep'. This information was hardly a surprise to Gitte.

King Carl XIV Johan (formerly Count Bernadotte) pronounced Ørsted Commander of the North Star, a gesture on a par with the one granted to Berzelius two years before in Copenhagen. At a royal banquet Ørsted obtained an audience with the 80-year-old King, conversing with him in French as Bernadotte had never become well versed in Swedish. Ørsted thanked him for the warm welcome the Danish delegation had received from his subjects. Carl Johan broke in: 'A constitutional king has no subjects, but makes it his honour to be the first citizen of the country and the first subject of its laws'. The King continued by urging Ørsted and his colleagues to be aware of their immense influence on public opinion and to use it to preserve the existing constitution. Then the old man began repeating himself. He was totally in favour of Scandinavia unity, of course, no country would be able to deprive another of part of its territory without European interference. So, the Scandinavian countries ought to stick together. For Sweden Denmark was a wall of defence against Germany. Similarly Sweden could be counted on to come forward as a brother-in-arms should Denmark be threatened. The King concluded by giving his Danish guests God's and his own blessing.[12]

Ørsted immediately reported on this interview to Christian VIII, who responded thus: 'I have read your report on your stay in Stockholm and your farewell audience with King Carl Johan with much interest'. He then continued his letter with the following thinly veiled criticism: 'We shall take the gracious words of the speech at their face value, but it is always very instructive to get to know such a character.'[13] Apparently, Christian VIII had never recovered from the personal defeat he had suffered in 1814, when he had been forced to give up Norway into the hands of Carl Johan, Napoleon's former field marshal.

After Stockholm the Ørsted family went on summer holidays to Roskilde to be put up in A.S. Ørsted's flat in the Yellow Palace. Christian VIII had dismissed Schouw as the royal elect representative of the University, because he had supported a proposal to amalgamate two Assemblies of the Estates thus jeopardising the fragile balance between the four parts of the kingdom. While the family enjoyed the summer at Roskilde a letter from Berzelius arrived. He was now proud and happy about the successful conference at Stockholm and totally converted to the Scandinavian Society of Science. Compared to the last BAAS meeting in Manchester he found the Stockholm conference set a new high standard. In his opinion, the English paid too much attention to amateurs.[14]

In Stockholm it was agreed to organise a conference every third year rather than every second, although this change should not be seen as a lack of interest in meeting next time in Christiania. Still, for some time it appeared as if the meeting might not go ahead; finally King Carl XIV had died, and the new Swedish-Norwegian King Oskar I was to be crowned in Nidaros Cathedral in Trondheim, which would demand the presence of a number of the scholars. Luckily, the coronation was postponed, and the fourth Scandinavian Science Conference was launched as planned from 11 to 18 July 1844. From that time on the new three-yearly rule would apply.

Christopher Hansteen ran into a series of political troubles organising the Christiania conference. He was doubtful as to whether Norway would be able to live up to the standards of the preceding conferences in Sweden and Denmark. Firstly, there was a problem for Jewish scientists. According to article 2 of the Norwegian constitution, drafted in great haste at Eidsvoll in 1814, Jews were denied admission to Norway. The reason was a suspicion that the Jewish finance market in Hamburg had conspired to drain Kongsberg Silver Works of large sums of money throughout the eighteenth century. Henrik Wergeland, the poet, considered the paragraph an embarrassment, but his proposal to repeal it had been voted down by the parliament as late as 1842.[15] A physician from Altona suggested a boycott of the conference, and a Danish professor of physiology called Norway the most intolerant nation in Europe. What could Hansteen do? He could not change the constitution single-handedly, and besides it was no business of a Dane to point the finger of scorn at Norway, because the rule was inherited from the time of the twin monarchy, and Denmark itself had seen Jew baiting as late as 1819.[16] In the end, Hansteen managed to have the ban on Jews lifted for the occasion of the conference.

The second problem bothering Hansteen had to do with a request from Grundtvig for Scandinavian students to come to Christiania to strengthen the ties between the sister nations. But however much one might be in sympathy with the idea of a Scandinavian fraternity, Hansteen and Ørsted were in complete agreement to keep religion and politics out of the Scandinavian Science Conferences. 'Even the slightest taint of a political trend may become harmful, because religious and political movements easily transmute into fanaticism, and that cannot be governed as easily as our quiet and dispassionate scientific enterprise.'[17] 'Disregarding his Scandinavian leanings', Ørsted had been careful to stay away from the Scandinavian Society in Copenhagen to keep to the straight and narrow path of science.[18] A fervent Grundtvig supporter like Frederik Barfod deplored this political indifference, but still he defended Ørsted, appreciating 'that he was in a situation of a particularly delicate nature'.[19] Both Ørsted and Hansteen were pleased to find that the Christiania conference passed off without political interruptions, 'free from the egoism that always slips into the realm of politics'.[20]

Fig. 128. Christopher Hansteen (1784–1873), Norwegian mathematician and astronomer. Lithograph by Em. Bærentzen (1799–1868), RL.

Ørsted was put up privately at Professor Stenersen's together with Bishop Agardh and his two daughters. There were fewer than two hundred participants, thirty-nine Danes, thirty-three Swedes, and ninety-nine Norwegians. From the BAAS came a few, among them the geologist Roderick Murchison. The parliament had granted a sum, but it was a mere trifle, giving Danish and Swedish participants a discount of a third on the ticket from Gothenburg to Christiania. Ørsted gave a plenary speech on the relationship between thought and imagination in the concept of nature.[21] Every evening dinner was served in the Masonic lodge of St. Olaus til den Hvide Leopard [St Olaus of the White Leopard], and afterwards there was music and dancing, and tea and ice cream for the ladies.

Hansteen had organised a cruise on board the SS Prins Carl on the fjord of Christiania. Unfortunately, the pilot was so unlucky as to run the vessel into a rock, not only once, but twice, although nobody was hurt. True, there was 'an immediate shock, particularly among the ladies', but soon good spirits returned. A salute of guns welcomed the guests, there were fireworks and music, songs were sung and toasts were proposed to the Kings and to the ladies and to Norwegian science. Murchison invited the Scandinavian scientists to join next year's BAAS conference in Southampton, and Ørsted accepted. Nevertheless, the grounding of the ship was an embarrassment to the hosts, and the pilot was prosecuted. The charge ended at High Court, whose president G.J. Bull was Ørsted's brother-in-law, married to his sister Tine, whom he had met as a student in Mrs Møller's dyeing works in Copenhagen. The pilot excused himself explaining that he was used to navigating ships with sails, not steam engines. Ørsted suggested to his brother-in-law that on behalf of the passengers he tempered justice with mercy for the poor pilot. After all no harm had been done to the distinguished passengers. However he lost his pilot's certificate and was fined and sentenced to one year's imprisonment. It might be counterproductive for a Dane to meddle in a Norwegian court case. At a royal banquet Hansteen had a chance to touch upon the matter with Oskar I, who indicated that he would appreciate a petition from Ørsted directly. Bull was jealous that his brother-in-law did not send the petition through him, for he liked to boast of his affinity. Ørsted interceded for the poor pilot to be treated leniently. Losing his certificate must suffice; there was no reason to take revenge.[22]

The fifth Scandinavian Science Conference took place in Copenhagen between 12 and 18 July 1847. The committee put every ounce of energy into the arrangement. The King granted three thousand rixdollars, and the City Council of Copenhagen was asked for a contribution. The army lent tents, and guests obtained free entry to public institutions. Ørsted notified the newspapers so as to get publicity, and they opened a reading room in the Casino Theatre, where participants could read about the conference as it happened. This time almost five hundred scientists took part, ninety-one Swedes, thirty-four Norwegians, and almost three hundred Danes including sixty from the Duchies and eight from abroad. Never before had so many Scandinavian scientists met. Berzelius and his wife were put up in Studiestræde, where Ørsted had been looking forward to showing him his new instruments, and in particular his copy of Faraday's big electromagnet. Hansteen stayed in town for another month as Ørsted's guest to share the company of his son Viggo, who had been admitted to the Polytechnic Institute. Tine and Bull were there as well, and together they called on Oehlenschläger. As in 1840 the plenary meetings took place in the ceremonial hall of the University, where once again Ørsted lectured on the Scandinavian languages.

The sessions of the by now six sections were held at the Polytechnic Institute, in the Botanical Gardens, and in the lecture halls of the University. Here are some of the topics:

Physics and Chemistry: the heat-conducting capacity of ice, the shape of the Earth, the ordinary forces of nature and their reciprocal dependence, the atomic weight of chalk, a new gasometer, an electrical duplicator, the mathematical theory of gases, the propagation of electrical currents in liquids contained in a cylindrical vessel, the construction of organ pipes.

Zoological-anatomical: two rare or new species of Nordic dolphins, the teeth of porpoises, remnants of humans and animals in the old burial mounds near Uppsala, a new species of antelope shot on Zanzibar by the lecturer himself, the internal and external ear, a misshapen foetus of a calf with two faces, gigantic octopuses stranded on Iceland in 1639 and 1791, fossil teeth of bears in Swedish bogs, a skull of a beaver from a Danish bog, the anatomy of bones in the food of a dodo, peculiarities of the size of the spine of the hedgehog, determination of vomited larvae.

Botany: classification of species of grass, Italian species of grass, the development of buds of fruit trees, demonstration of a sprout of cuscus [vetiver] from the Nicobar Islands, a comparative analysis of the vegetations of Sicily and Scandinavia, a criticism of the theory of metamorphosis.

Mineralogy and geognosy: Swedish rocks containing hornblende or augite, various strata of orthite in Sweden, newer white chalk in Denmark, coral reefs at the Nicobar Islands, geognosy in the neighbourhood of Arendal and Kragerø, the drift of ice along the west coast of Greenland, products of eruption from the Hekla volcano.

Pharmacy: tinctures of opium, Carlsbad salts, copper in preserved fruits.

Medicine: the apparatus of muscles and ligaments around the prostate and the urethra, inoculation instruments, a pelvis of a Lapp, the structure of the uterus, a demonstration of orthopaedic machines, ligation of the umbilical cord before the extraction of the foetus in connection with footling presentation and version, operation for stone, the application of ether in connection with surgery, writing apparatuses for blind people.

Copenhagen could boast of at least three new attractions likely to attract the guests. Since the former conference in 1840 Lieutenant Georg Carstensen had opened the Tivoli Gardens around the ramparts outside Vester Port. Here were tightrope dancing across the moat and trampolinists, here H.C. Lumbye performed his railway and champagne gallops, here Columbine and Harlequin and Pierrot were staged at the Pantomime Theatre, and here Gaetano Amici, an Italian master of fireworks, displayed his pyrotechnic artistry. Visitors could try swing boats and a switchback or they could just stroll in the illuminated garden enjoying the buildings in Chinese or Oriental style. A few years later Carstensen's company established a winter Tivoli dubbed Casino after another exotic location in Italy. And thirdly, the Railway Company of Zealand had just recently opened the line between the capital and Roskilde, and despite Ørsted's scruples the Scandinavian Society was allowed to host an excursion by rail to Roskilde, reaching its climax with a dinner in the courtyard in front of the Assembly of the Estates. In one week the railway carried almost thirteen thousand passengers, and during the same week Tivoli had a similar number of visitors. Here was something to show off.

On Sunday 11 July Christian VIII invited all foreign members of the conference to lunch at the Hermitage Castle, the SS Ægir brought them to Klampenborg where they also saw the hydropathic establishment. The next day the Students' Association performed Hostrup's *Genboerne* [The Neighbours Opposite] at the Court Theatre, and the day after there was sightseeing at Rosenborg Castle and the sanatorium in the Royal Gardens. On the Wednesday another sightseeing excursion was organised, this time to Christiansborg Castle and Church and in the evening a tour of Tivoli, though only a few took part since most guests had taken the train to the gala evening in Roskilde. On the Thursday a *soirée musicale* was organised in the Casino, and this was attended by a great many people because each scientist had received a couple of free tickets for ladies, and the newspapers covering the event estimated an attendance of nearly fifteen hundred people. A series of tableaux was staged, and Lumbye's orchestra struck up waltzes lasting till the early morning. On the Friday the participants were invited to the Tivoli Gardens again, this time culminating in a firework display showing the monogram of Christian VIII in red and white beside King Oskar's in blue and yellow. Finally, on Saturday evening a farewell dinner took place in the Casino Theatre, where special songs and the usual toasts were given.

The Scandinavian Society headed by Schouw and Clausen had been put in charge of Wednesday's excursion by railway to Roskilde. Six hundred participants including ladies

Fig. 129. The Tivoli Gardens, established in 1842 for the entertainment of Copenhageners on the ramparts of the city by Georg Carstensen (1812–1857). Coloured lithograph by Em. Bærentzen (1799–1868), RL.

joined up. The railway service catered for the Swedish and Norwegian guests only, and they were advised to wear their free tickets visibly. The Danes had to do with the old-fashioned stagecoach. The reserved special train arrived on a splendid summer morning in Roskilde at half past five, and the passengers were received on the platform by the local choral society: 'Welcome here, Researchers from the North, to Isefjorden!' The reception was cordial and boisterous and only a few noticed the wrong place name, but at least the lines rhymed. The whole party walked in procession under waving banners to the cathedral, where the organist and the choral society had rehearsed a specially written cantata: 'Welcome Priests from the Church of Nature!'. At half past seven in the evening tables were laid for dinner outside in the courtyard of the Yellow Palace. A rostrum had been set up surrounded by the flags of the Scandinavian nations, and a song by Hostrup was sung touching upon the somewhat frivolous metaphor of nature as the bride and the researcher as the groom. Other occasional songs used more martial metaphors of campaigns and the Scandinavian triple army, whose leaders were honoured for their bravery. Grundtvig, who had threatened to join the Christiania conference three years earlier, mounted the rostrum delivering a witty speech on the relationship between scientists and historians, whereupon the party sang the drinking song he had written for the

SCANDINAVIAN SCIENCE CONFERENCES

Fig. 130. The Casino Theatre, Carstensen's indoor venue, inaugurated in 1847, when the Scandinavian Science Conference invited ladies to a ball there, and where in March 1848 the national liberals arranged their ill-fated political meetings. The Theatre Museum, Copenhagen.

occasion, in which he praised the wine, even if the grapes were grown by the Rhine, while riding his hobbyhorse that it was not nature, but history, and especially their ancient history, that united the Scandinavian peoples. Grundtvig had to take the ordinary train at nine o'clock back to Copenhagen, but the party just managed to reciprocate with a 'Long live Grundtvig!' before he left, to the accompaniment of his own song about the mother tongue.[23]

Clausen gave a speech on behalf of the Scandinavian Society expressing his joy that three sister nations could meet 'at the end of the first railway in Scandinavia in Roskilde', as one people speaking the same basic language.[24] Then they sang Barfod's 'Good Luck to the Ancient Freehold Farm of Scandinavia!', the distinctive song of the Society, and nine cheers to the Kings sounded. Another song rejoiced in their academic freedom and the emancipation of nature from its veil of Isis by scientific research.[25] Merriment and pathos were given pride of place, and outsiders could hardly determine whether it was an assembly of slightly tipsy scientists or a crowd of enthusiastic Scandinavists. The politics that so far had been kept at bay by the trio was knocking on the door. At the farewell dinner in the Casino Theatre Ørsted expressed 'the common enjoyment of the love with which the Scandinavian Kings embraced the sciences'.[26]

Fig. 131. The party for the Scandinavian scientists, 12 July 1847, in the Yellow Palace with Roskilde Cathedral in the background. Only the persons mentioned in the text of this book are identified below:

Karl Adolf Agardh, Swedish Bishop and botanist (in the direction of HCØ's pointing right hand)
J.J. Berzelius, Swedish chemist (sitting to the left at the table in front with a grey top hat in his right hand)
H.N. Clausen, Danish theologian and politician (white haired, above Schouw and Grundtvig).
J.G. Forchhammer, Danish chemist and geologist (below Agardh, at the table in front wearing a brown jacket)
N.F.S. Grundtvig, Danish priest, hymnwriter and historian (below Clausen, to the right of Schouw)
Christopher Hansteen, Norwegian mathematician, physicist and astronomer (to the right wearing a brown tail coat and with a top hat in his right hand).
A.A. Retzius, Swedish anatomist (behind Hansteen at the same level)
E.A. Scharling, Danish chemist (to the left of Clausen wearing a black jacket and brown waistcoat with a glass in his left hand)
J.F. Schouw, Danish botanist and politician (at the table in front wearing a black skullcap)
F.C. Sibbern, Danish philosopher and psychologist (second from the left wearing a brown tail coat)
J.J.S. Steenstrup, Danish zoologist (to the left of Berzelius, his right arm crossing Sibbern)
HCØ (on the rostrum surrounded by the three Nordic flags).

Erik Ludvig Henningsen's (1855–1930) winning draft for the decoration of the ceremonial hall of the University, 1897. FBM.

The Copenhagen conference was the last before the rhythm was disrupted in 1848. The planned conference in Stockholm in 1850 had to be cancelled due to the upheavals surrounding the changing role and powers of the Danish monarchy. The Scandinavian science conferences had confirmed Ørsted's political intuition that the absolute monarchy was far more favourable than democratic parliaments towards the sciences. Together with Berzelius and Hansteen he had succeeded in separating politics and science, but only because he defined politics as socio-economic and national interests. He considered the patronage of science by the absolute monarch as totally apolitical and unconditional. The Copenhagen conference was compared by several speakers to the outburst of student Scandinavism in 1845, one of them even called for three cheers for the Scandinavian Union.[27] The Scandinavian science conferences had become a social success, and the contribution of the trio consisted in lending their names to them rather than presenting anything new scientifically. Ørsted talked exclusively about linguistic and aesthetical topics. At the same time he made his staff of the Polytechnic Institute an integral part of the Scandinavian community of science. Technology was not even allocated a modest place in a subsection as it was in the BAAS.

*Should I not be able to love this country,
without holding that of others in contempt?*[1]

51 | 1839-46
Politics and Nationalism

KING FREDERIK VI died in December 1839, and according to tradition Prince Christian Frederik, his cousin, was at once proclaimed new king, but it was the first time this happened from the balcony of Amalienborg. Being the head of three scientific institutions Ørsted was obliged to seek an audience in the royal palace to condole and congratulate. The professor thanked Christian VIII for his merciful patronage of the Royal Danish Society, of the Society for the Dissemination of Science, and of the Polytechnic Institute, adding that not only the sciences were in the King's good graces, but 'Your Majesty brings to the throne progressive, liberal ideas combined with a comprehensive view on the world matured in the school of experience'. This was a conscious hint at the monarch's political achievement in Norway during the difficult months following the peace settlement in Kiel in 1814. On that occasion he had safeguarded as much of Norway's independence as possible by joining the Eidsvoll convention, where the most liberal constitution in Europe had seen the light of day. Speaking in a hesitant voice Ørsted pronounced 'the confident hope of the people' that 'the improvement of our age will be advanced by Your Majesty's spirit uniting bold power and wise caution'. Ørsted's earnest request to take vigorous action, will to reform, and prudence in governance obviously went beyond his mandate.[2]

Christian VIII responded that he intended to continue his patronage of the three institutions, but would have less time to take part in the meetings than before and therefore asked Ørsted to keep him up to date with minutes. As far as matters of state were concerned he just referred to his open letter of the day.[3] All officials would remain in their posts and he intended to carry on the course of his predecessor in other respects as well.[4] There was a remarkable contrast between Ørsted's enthusiastic rhetoric and the Kings reserved response.

This was hardly a surprise to Ørsted who had recently disengaged himself from politics, and the open letter was no news either as it was his brother who had drafted and presented it to the council of state the day before. Among those present at the audience it was agreed that Ørsted had lost his nerve while giving his speech. In the following days people were gossiping about the episode. Some thought that he had been brave to speak bluntly about the expectations of the people, while others found that his hint at Eidsvoll was too straightforward and associated him 'with the liberal enthusiasm that is actually a kind of madness . . .'[5] Characteristically, Ørsted was seen as a liberal, even a radical by many contemporary observers, although he loathed being classified according to party politics, priding himself that both he and his brother were impartial spokesmen of reason. His public voice was never ideological, but an appeal to the rule of law and an attempt to qualify his fellow citizens for reasonable reflection. Hence, he was critical of the liberal opposition, which he believed was taking advantage of the inarticulate people for its own purposes.

> 'Genuine liberalism it seems to me consists in enabling the uneducated to reach the same spiritual level, but not in bestowing on them unprepared the power naturally and justly represented by the opinion of the educated. The liberal opposition begins by telling the uneducated that they are maltreated instead of developing in them the true sense of right and wrong.'[6]

At the beginning of his career Ørsted had asked for help from private well-wishers and government officials, but when Frederik VI was proclaimed King he went straight to the top, and not in vain. In 1820 Prince Christian Frederik realised his genius and they began a scientific cooperation giving Ørsted royal recommendations and the Prince professorial returns. In 1824 Ørsted succeeded in getting Christian Frederik and the Royal Academy of Arts to support the establishment of the Society for the Dissemination of Science. Ørsted was now frequently invited to royal banquets at Sorgenfri Castle, for instance together with Davy and later with Charles Lyell, the geologist. Once in a while he and the King would meet as Masonic brothers in the lodge. In 1826 a crisis occurred in the relationship between Frederik VI and A.S. Ørsted, and ten years later the King was upset by H.C. Ørsted's promotion of the Society for the Right Use of the Freedom of the Press. So the two brothers' relations with royalty were not always harmonious, though from 1835 onwards the Prince put a summer dwelling at the disposal of the Ørsted family at Fredensborg Castle. In 1838 Prince Christian Frederik succeeded Court Steward Hauch as president of the Royal Danish Society, and when the next year he became King, Ørsted was obliged to submit the minutes of all fortnightly meetings to him and to update his by no means inadequate knowledge of science and technology with the latest news, including the development of daguerreotypy which the Prince followed with great interest.[7]

Ørsted recounts a story illustrating the tone used at court. The episode took place at a royal banquet at Sorgenfri in 1837. The British geologist Charles Lyell and the astronomer H.C. Schumacher from Altona were guests of honour. Ørsted entertained Princess Caroline Amalie by telling her of the impressive achievements of astronomy. He had recently put the finishing touches to his dialogue on 'Christianity and Astronomy' and was on excellent form. With mathematical precision science could predict important events in space, Ørsted told his royal dinner partner, and in return the Princess teased the docile professor, facetiously wishing: 'If only these mathematicians would sometimes be wrong, for otherwise they get too conceited.' Apparently, Ørsted did not catch the joke, because he answered with an air of injured dignity on behalf of the learned republic that if the event did not occur exactly as predicted he would guarantee that somebody else would take advantage of the fact, people like Grundtvig for instance. Although that name was not mentioned, the hymnist had recently started giving historical lectures at court, and his mistrust of the prophetic cocksureness of mathematicians was well known. Ørsted realised that he had gone too far, one just does not talk to 'a lady of the highest rank' in that way, and he apologised to her adjutant. After the dinner the Princess, shedding a tear, said with charming mildness, 'I just wish that everybody could be allowed to follow their convictions'. Now Ørsted realised how badly he had blundered. He had hurt her by treating her innocent joke with professional seriousness. The episode prompted this reflection: 'I have often frankly contradicted high-ranking persons. In doing so I had no scruples, but only obeyed my urge. Perhaps I would never have become aware of my peculiar behaviour, unless somebody had drawn my attention to it.'[8]

The regular and confidential association with Christian Frederik also applied to his brother. In 1837 the Prince wrote to Ørsted that he was worried about his brother's health after he had suddenly had to leave the state council meeting:

> 'I urge you to see to it that he takes care of his health, does not overstrain himself with work, and calls upon a doctor immediately. You will not wonder that the preservation of his health is continuously on my mind, for nobody could more fervently than myself appreciate his outstanding worth as a human being and an official.'[9]

*

Christian VIII immediately promoted H.C. Ørsted to councillor of state, meaning in practice that he was now a consultant to the King on academic matters. In this capacity he sometimes ran into delicate situations, as for instance in 1840 when a group of students assembled to establish a students' union to be their own platform for university politics.[10] The University was in the melting pot. There were plans to change the length of the semester and vague thoughts were entertained of transferring the two first examinations, the *artium* and the *philosophicum-philologicum* to the grammar schools. Such changes would affect the courses, so was not it fair to let the students have their say on the matter? However, the initiators had no intention of submitting lists of members or statutes to the police or the senate as required by law, and certainly not prior to the approval by the members themselves. This insubordination provoked not only the Senate, which asked Ørsted to contact the students, but also the King, who requested a full report on the matter. Ørsted made up his mind to approach the students in person to mediate in the not-yet-established Students' Union.

Who were the initiators? They were young postgraduates mostly, being watched by the government due to their political engagements, some of them liberal leaders,[11] but also the theologian P.E. Lind, the son of Gitte's sister Anne Marie (ch. 44), in other words Ørsted's nephew. As a child he had played with Sophie and was of course part of the family. He was about to be trained as a prison chaplain.[12] Ørsted could hardly consider him a troublemaker.

At a time when the leadership of the Students' Union and the senate took up intransigent attitudes and the threat of sending down the student leaders was in the air, Ørsted mounted the rostrum and gave an almost improvised speech, although he had managed to scribble down a few notes. The atmosphere was tense. 'My honourable academic fellow-citizens!' he addressed the students, making immediate reservations. He was well aware that he was facing a sceptical audience who knew very well that he had recently been appointed a councillor of state. Hence, he was under suspicion of acting as the auxiliary arm of the government. Ørsted justified himself by stating that he had always approached the monarchy to nurture academic interests only, in the beginning to provide means for his own research and later on for young scientists in need of travel grants or posts. According to his view the students were in the same boat as himself, for under the present constitution science depended on the good will of the absolute monarchy. The only difference between students and professors was their age. Soon the students would need the same patronage that he had enjoyed, and soon he and his colleagues would give way to younger people.[13]

At last, according to the notes, Ørsted broke the Senate's rules of professional secrecy by revealing that he had voted against the threat of sending down students who had voiced their anger. Instead, he had suggested a milder response that did not make demands in the manner of an ultimatum, but only called for sober-mindedness by appealing to the reason of both parties. Students ought to recognise the academic rules and inform the Senate of their list of members and their statutes, not out of fear of being sent down, but out of love for science, just as the religious man obeys the law out of love for God, and the citizen out of love for the common weal.[14]

Everybody at the University has the same goal, Ørsted went on, the progress and dissemination of science. Hence it is important that all good proposals in favour of this goal be advanced, and if the Students' Union intended to engage constructively in the debate, it would be welcome. Ørsted talked of the purpose of the learned republic:

> 'The University should not refrain from launching and defending even the most audacious truths that may affect the whole of human society a great deal; but it is not its purpose to act directly to incorporate these truths into civil life. In the temple of knowledge truth can develop the more freely the more it is cultivated in its perfect purity. What individual members of the University find to be their calling outside its walls does not belong to the University, but will sustain their honour or shame as citizens.'[15]

Ørsted aimed at exercising a similar moderating effect when he counselled Christian VIII 'by royal orders'. The King used Ørsted as his informal deputy in the Senate, and it appears that it had been the King who, via the Vice-Chancellor and in Ørsted's absence, had directed the senate to put authoritarian measures into effect against the students. Now Ørsted tried to pull the chestnuts out of the fire. He reminded the King that at the founding of the Students' Association in 1820 the Senate had raised voices of warning, dreading that the new institution was planned as a cloak for *Burschenschaften*. Fortunately, these omens that Frederik VI had swept aside had turned out to be unfounded. Similarly, there was no reason to fear the Students' Union, unless the government by wanton carelessness provoked intransigent opposition. In other words: Ørsted supported the students' plans even if he considered that some of the initiators were political extremists.

It was Ørsted's advice that the King should support the forces seeking a calmer path towards a free constitution than 'the reform party'. The King ought to support the learned republic, 'whose liberality is kept in balance by its dedication to law and order'. If on the other hand the government took to authoritarian measures, this would only radicalise the opposition. Ørsted probably had the court case against David in mind, in which Frederik VI had pushed the learned republic aside (and lost the case) to no avail. If the King used his power to dissolve the Students' Union he risked pushing the students into organising themselves in secret *Burschenschaften* thus attracting the mob. Within a few years members of the Students' Union would be taking up appointments as priests and judges, and they would respond negatively if they met mistrust from the government.[16]

*

Like his brother, A.S. Ørsted went right to the limit of his mandate. Would the secret deal with Frederik VI from 1826 that he would not take part in public debate still be valid under his successor? Perhaps it was wiser to let this question stay unresolved. In 1841 he decided to reveal to the public some of the greatest problems posed by the reform party's demand for a free constitution.[17] In the summer of 1841 he published anonymously a review of the Norwegian jurist Ole Munch-Ræder's book on the Eidsvoll constitution, the declared liberal substitute for the absolute monarchy. Apparently, it was Ørsted's aim to problematise the constitutional issue and thus pressure the advancing opposition into putting forward constructive solutions to some of 'the knotty issues'. The constitutional issue was inextricably interwoven with social and national questions, as well as that of the succession, since King Christian and Queen Caroline Amalie were without heirs, (which problem, however, will not be dealt with here).

As the anonymous reviewer of Munch-Ræder's book Anders was formally obeying the command not to meddle in public debate; he started by pointing out the liberal party's fundamental misunderstanding when they argued as if the monarchy were an autocracy operating in strict compliance with the letter of the constitution, the lex regia. Why could not the opposition realise the difference between theory and practice? Actually, public opinion had a significant impact on the government through the Advisory Assemblies of the Estates. Ørsted emphasised that his lifelong struggle to promote the rule of law and the independence of the judiciary had made progress, and that the liberal leaders' perpetual reference to the formalities of lex regia went against their better judgement, indeed approached demagogy, and was a deliberate seduction of people who were not in possession of the facts.

Ørsted acknowledged the many positive aspects of the Norwegian constitution, and that it had been successful in Norway. But there were drawbacks as well; for instance, as his brother had pointed out time and again, the parliament's lack of understanding regarding the arts and sciences. One important difference was that Norway had a comparatively homogenous population, while Denmark had a considerable class of estate owners, whose legitimate interests were taken heed of in the Assemblies of the Estates. Had Danish estate owners in favour of the liberal wing considered their predicament if a democratic majority assumed power in Denmark? The Eidsvoll constitution could not be copied. Without a special bicameral system a democratic constitution would entail social chaos. It would take time to incorporate the peasantry in the running of state affairs. According to Ørsted the overwhelming majority of the agricultural population did not want a change of constitution, but preferred a continuous levelling out of class differences through agricultural reforms. To this end the Advisory Assemblies of the Estates had turned out to be a promising start, and the government had now gone one step further by introducing local self-rule, an initiative that would gradually prepare the rural population for active participation in politics.

The national question, perhaps, was the most burning one, smouldering as it was in the Duchies and threatening to burst into flames at any moment. The national-liberal hotspurs ought to realise that the very existence of the United Monarchy was endangered if their demand to unite the two Danish-speaking Assemblies of the Estates was endorsed. In that case the Schleswig-Holsteinian movement would be fully justified in demanding the amalgamation of their assemblies and as a consequence all hell would break loose. It was exactly to prevent a political unification between Holstein, which was a part of the German Federation, and Schleswig, which was not, that four Assemblies of the Estates had been established in 1831. Instead, the national differences found an outlet in a language struggle in which the contending parties tried to push the border between Danish and German towards the south and the north respectively. Christian VIII's rescript on the language issue of 1840 favoured Danish interests in so far as it made Danish the language of the law courts, where it was already the language of schools and churches. But the language struggle continued, and Peter Hjort-Lorenzen's provocation in 1842 in the Schleswig Assembly, when he continued to speak Danish despite the fact that the official language was German, a language he had spoken fluently so far, whipped up the excitement still further. Ørsted's bleakest suppositions had become a reality.

During the following years the national differences had sharpened between the Schleswig-Holsteinian movement and the Danish national-liberals. By then the latter had come up with the offensive Eider policy to the effect that Denmark should incorporate Schleswig into the Danish monarchy thus separating the two Duchies. A nationalist offensive was added to the demand for a new constitution, creating an explosive political cocktail that was fuelled by the popular gatherings on Skamlingsbanken, where Grundtvig gave speeches, and by the establishment of a peasant high school at Rødding. In the enlightened universe of the Ørsted brothers the phenomenon of nationalism was irrational and destructive. Bilingual as both of them were they did not accept that users of different languages were unable to live peacefully together as they had always done. The ethnic conception of the history of the nation and its language was an elementary breach with the Kantian categorical imperative, for when these cultural forces become aggressive and expansive they cannot be universalised.

*

Obviously both brothers were well known in German political and academic circles. In 1846 H.C. Ørsted and several other scientists had been invited to join a conference in Kiel by their German colleagues in GDNÄ. Ørsted arrived from a similar BAAS arrangement in Britain, where homage had been paid to him (ch. 55). In Kiel he was the guest of his friend Carsten Hauch, who had left Sorø in favour of a professorship of Danish at the University of Kiel, the second university in the United Monarchy and the focus of Schleswig-Holsteinism. The farewell dinner for the German scientists on 24 September turned out to be one of the most disagreeable Ørsted had ever been exposed to. Instead of the expected panegyrics that used to dwarf national differences, insults started to be swapped across the tables, poisoning the otherwise cosy academic atmosphere with nationalist stink bombs.[18]

In the evening Ørsted, being usually a tolerant person, wrote down in Hauch's house an account in German of his experience of this unusual festive dinner and read it aloud to his fellow countrymen. He then corrected a few details where his memory had failed him. The Danes agreed that they had been treated as if they had imposed themselves on a convention reserved for Germans, although they were invited as guests to a purely academic conference.

Fig. 132. Johannes Carsten Hauch (1790–1872) zoologist, poet, and novelist, A.W. Hauch's nephew and one of HCØ's disciples. Lithograph by Em. Bærentzen (1799–1868), RL.

According to Ørsted's account it crossed his mind during the dinner that at the Scandinavian Science Conference two years ago in Christiania he had proposed a toast to the Frederiks University on behalf of the guests. Spontaneously, he decided to do something similar and signalling to Professor Scherk, the toastmaster, he was given the floor as the first speaker:

> 'It is an honour to me to propose a toast to the University of Kiel. I am convinced that not only do I speak on behalf of everybody present, but also on behalf of many absent friends. I also feel certain that I have the support of the University of Copenhagen. I have always been pleased to notice things that favoured the works and means of the University of Kiel, and in this I do not stand on my own. This university has always benefited from distinguished teachers who have served the sciences well and done their students credit. Long live the University of Kiel!'[19]

It is hard to find anything controversial in these ritual civilities. Etiquette demanded a reciprocal toast to the guests, but instead Professor Hegewisch jumped to his feet and gave a toast to Germany. Next he effused on the German lands from the Rhine to the Baltic and the German coastline from Kiel to Mehmel bringing the great sons of these lands into memory, persons such as Kant, Fichte and Herder, to whom he added Blücher, or 'General Vorwärts' [General Forwards] as he called him. He anticipated the next conference would take place in Aachen, the historical and residential city of Charlemagne. Finally, he raised his voice wishing his dear and noble friends from the south good luck. He put a special emphasis on the words 'from the south'. This brief speech, which totally ignored the non-German participants, ended by encouraging everyone to raise their glasses and drink to Germany. Ørsted and his fellow countrymen dutifully complied with this request expecting that at least a few Germans would profess their dislike of Hegewisch's ostentatious breach of the law of hospitality. This did not happen.

Instead, Professor Forchhammer (not Ørsted's friend, but a German of the same name employed at the University of Kiel) proposed a toast to the University of Copenhagen. He explained rather circumstantially that the University of Copenhagen owed its foundation to Germany, that it had always depended on Germany, and that one would wish it to stay close to its provenance. Copenhagen was still being intellectually nourished from the south in the humanities as in the sciences. To be sure, Copenhagen was favoured by larger resources than Kiel, but if the two universities were to cooperate this must happen on the basis of equality. He totally omitted to mention that the University of Copenhagen might have made contributions to the progress of science.

Now, Ørsted was seriously worried. He began to speak on behalf of his university, extending his thanks, and acknowledging that not only the Danish, but all Scandinavian sciences and letters were more closely connected to Germany than to any other country. He hoped that such natural fraternisation might continue unmolested by political unrest. But he noted that German literature had been appropriated in accordance with the Danish character. Should anybody object to this and insist that Danes give up their identity to become German, they are demanding the impossible. Ørsted was not acclaimed for these remarks. They were not only a gesture of reconciliation, but also a disclosure of an agitated mind.

Finally, a Professor Jessen from Kiel pulled himself together to propose a toast to the Danish members of the conference, but by now the situation had got out of control. Ørsted was

Fig. 133. The Christian-Albrecht University of Kiel. Pencil drawing by Adolf Burmester of the original main building, 1850. Schleswig-Holsteinische Landesbibliothek, Kiel. Photo by Dr Jens Ahlers.

shocked. So far his public activities had been based on the assumption that the sciences and the arts were above politics, and that his fight for the supremacy of reason would carry the day in the long run as far as political affairs were concerned as well. However, the unthinkable had happened: his colleagues, otherwise so respectful, had lost their senses, poisoning the scientific atmosphere with nationalistic insults. He left Kiel with a painful impression very different from the mood he had brought back from Southampton. The hosts' festering hatred of the visiting Danes in Kiel had become so vehement that they did not even wish to say goodbye to their guests from the north.

Ørsted did not mention the fact that on 8 July Christian VIII had issued the so-called 'Open Letter' extending the hereditary right of lex regia to Schleswig, which was a direct challenge to Schleswig-Holsteinism. Ørsted and his fellow countrymen had to pay for it. Not only the leadership of the Schleswig-Holsteinian Party, but also most German lands were in a state of unrest. The Danish scientists had stirred up a hornets' nest, and whether or not they were in sympathy with the Eider policy (and Ørsted was not) they had to put up with being offended by German nationalism. Ørsted concluded: Danish and German culture should live in peaceful co-existence within the contemporary borders of the United Monarchy. To him there was no rationality in sustaining the Eider policy. The Eider Canal divided neither landscapes nor languages. To the limited extent that nature formed national character, the Kingdom of Denmark had more features in common with the Duchies to the south than with the two Scandinavian nations to the north. Applying the alternative criterion of language, it was the other way round.

In his speech in Kiel Ørsted had used the phrase 'our own character', but what was national identity according to Ørsted? Ten years earlier he had put his reflections on the Danish national character on paper. Of course, it had been unavoidable not to reflect on the question as he had met national and linguistic differences from his early childhood at the wigmaker Oldenburg's and later on his many travels. It is a common experience that a deeper understanding of one's national characteristics is only acquired after a long time abroad when one starts wondering why suddenly one feels how it is to be an alien. Few had experienced this alienation as intensely as Ørsted, who had by now spent more than five years abroad.[20] The result is no surprise. Ørsted, the cosmopolitan, applied a view of cultural relativism to the national character of Danes as well as to other peoples and regarded all forms of nationalism as conceited and irrational.

His view on Danishness was sober. He warned against a love for one's country that exalted one's own nationality at the expense of others. 'The self-sufficiency imbued by flattery is a sweet but stupefying poison.' The antidote should be fetched from a sober self-insight, from the answer to the Socratic challenge 'Know Thyself!'. National characters have not arisen independently, but have influenced each other, he claimed. He found that the Romanticists' philosophy of nature was the victim of unbridled imagination. Grundtvig believed he had found the roots of Danishness in the heroism of mythology and the age of the Vikings, but the striking fact about the Vikings was that they left Scandinavia as commercial travellers to trade with foreign cultures, and not least with Christendom, whose values they brought back. Later, for several centuries (from the Hanseatic League to the Bernstorffs) Denmark was predominantly under German influence, witness the dynasty of the House of Oldenburg, the immigration of German nobility, the Lutheran Reformation, the Scientific Renaissance, and students' grand tours. Danish language and identity were neither relics of antiquity nor medieval figments of the imagination, but something that had been brought to completion in Ørsted's own lifetime.

'Allow me to turn my attention towards the glory in which we wrap our past, as other peoples do with their pasts. It amplifies vanity into boisterous eulogies, but numbs the true and living spirit of our people.' Ørsted could hardly oppose the Romanticist philosophy of history of his contemporaries more sharply. Danes are not more heroic than other peoples, and if they were, was not that a rather beastly propensity, unworthy of modern, educated human beings? And when nationalist tricksters like Grundtvig apply understatements in praise of Danishness such as 'modesty, peacefulness, and good nature' they ought to take a critical look at themselves. Then they would discover that such properties often lapse into 'immobility, inactivity, and lethargy'. It was now, in the Advisory Assemblies of the Estates that the Danish character was flourishing, and generating optimism in the accomplishments of modern Danish literature and art, as well as in the Scandinavian Science Conferences. The crucial deeds of the present age were not related to a particular martial bravery, nor to Viking raids and chivalrous tournaments, nor to religious mysticism, but to enlightenment, education, art and science. Among the few special features Ørsted found in the Danish national character was the 'rejection of one-sidedness which is at the same time also a rejection of the extreme and the artificial that distances itself from the balance of nature'.[21] Of course, this was also very much a self-characterisation.

Thinking itself nullifies its false directions.[1]

52 | 1842–8

The Centenary of the Royal Danish Society
Magnetischer Verein
Henrik Steffens
L.A. Colding

IN HIS many positions of authority Ørsted always endeavoured to maintain a polite tone as he had been told to do by the wigmaker at Rudkøbing, and not to let factual disagreement poison personal relations. This was not only a prudent, but also a necessary precaution, for in all the initiatives he had been involved in throughout his life he had been elected or appointed to chair and govern institutions. Being at the head demanded that he enjoyed and deserved the trust and confidence of his colleagues and abided by the rules of the game, written and unwritten. He succeeded in complying with this line of conduct and never lost a position of trust. Not because he evaded conflict, but because he managed to live up to his resolution to treat his adversaries with respect and no differently from his adherents. This attitude was expressed in a letter to the hotheaded Christian Molbech, chairman of the committee to create a Danish dictionary set up by the Royal Danish Society of Sciences and Letters:

> 'I wish you would feel convinced that I am never your adversary, although it may easily happen that at some occasion I might become the opponent of an opinion of yours. But to me these two

things are sharply divided: in science as in business I do not make a difference between friends and foes. I do not think I have many of the latter, and I certainly do not count you among them.'[2]

Soon, however, the personal relationship was put to the test. In 1842 the Royal Danish Society was to celebrate its centenary, and to this end a committee of Ørsted, Schouw, Madvig, Mynster, and Ramus appointed Molbech to write the Society's jubilee publication, according to the sources available.

Molbech's working conditions were far from ideal. There were no preparatory sketches to lean on, the source material was abundant (apart from the earliest period) and time was scarce. After the committee had approved Molbech's synopsis, the publication had to be written in the course of nine to ten months. It was planned to consist of two sections, one about the institution as such, the other one about the history of the individual sciences and letters. On the day of the centenary, 25 November, he had completed the part of the first section stretching from 1742, the year of foundation, to 1815, the year Bugge died and Ørsted took over as secretary. Mynster complained to Ørsted about the manuscript's focus on scandals, jealousy, and negative criticism. True, this was ominous, but for the time being Ørsted did not want to interfere, because Molbech had not yet displayed his views on Ørsted's epoch. The centenary could be celebrated in perfect harmony, as indeed it was, in the ceremonial hall of the University where Ørsted, in the presence of King Christian VIII, the President of the Society, gave his celebratory speech. As Molbech's book was in the offing, Ørsted simply referred those interested in the history of the society to the jubilee publication, talking instead about the nature of science.[3] Sciences and letters flourish best in an ivory tower where they are elevated above skirmishes over mundane issues. Research is preoccupied with problems that only touch indirectly on social life. 'The thinker throws himself upon themes of research that are at the remotest distance from the horizon of the crowd.' The freedom of scientific research demands that it is cultivated for its own sake and not subjected to political or economic interests. In this respect science is value free. Unfortunately, both the festive mood and speech were spoiled by sudden blackouts. Irgens, the blacksmith who had made the argand lamps and was responsible for their maintenance, was booed and subsequently fined.

In the course of the following spring four chapters on Ørsted's epoch as a secretary (approximately eighty pages out of more than five hundred) were ready to be printed, an impressive achievement. But Ørsted, wishing of course to see his hard work for the Society cherished, was appalled at reading Molbech's sometimes severe critique. Ørsted found that not only was the manuscript full of errors and inaccuracies, but it also contained unfair judgements of his efforts. Molbech resented the lavish spending of scarce resources on dubious projects such as the futile borings for water, which were defended by Forchhammer. The jubilee publication also hit out at P.W. Lund's expedition to Brazil, which had been opposed by Schouw. In return, Molbech put the blame on Ørsted for being unnecessarily frugal as regards Bugge's cartographic enterprise and his own Danish dictionary both of which Ørsted was accused of having cold-shouldered.

Two points of criticism hurt Ørsted in particular. First, that Bugge's period had been painted in rosy colours and that he himself was reprimanded for his great sin of omission, that is for not giving an obituary of his predecessor. Molbech contrasted Jonas Collin's beautiful commemorative speech in the Royal Agricultural Society to the provocative silence prevailing 'in the Royal Danish Society, where Bugge had worked for half a century giving useful advice to the nation. There his memory found no spokesman, his career no description, and his merits not even the faintest appreciation'.[4] For the secretary this had been a deliberate omission (ch. 31). He simply could not stand Bugge, who in his eyes had abused his power so as to obstruct Ørsted's early career. Bugge was one of the very few examples of hostility at first sight in Ørsted's life. They had crossed swords as 'ephor' [head] and inspector respectively of Elers' College, and as to philosophy of science they were worlds apart. In Ørsted's eyes the cartographic project did not belong in the Royal Danish Society because its contribution to science was shallow, and during Ørsted's epoch it was transferred first to a special commission and later on to Schumacher in Altona. Ørsted did not withstand the temptation to assert that the only scientific merit of the cartographic project in Bugge's time was owing to Caspar Wessel's mathematical genius. 'I frankly admit that I could never have taken it upon myself to give a commemorative speech to Bugge, since the praise I could convincingly have conferred on him would easily have appeared too little to the public, and this I could not take lightly as it was known that for many years he had been hostile towards me.'[5]

Secondly, it was painful to Ørsted, that what he considered one of his merits, the *Oversigter* [Transactions] of the Society that he started writing from 1815 in a way that all papers in all disciplines were summed up in the same style, were only allotted a few ambiguous or outright erroneous comments by Molbech. He suggested that the members' papers and the editor's minutes were a mess, and that the Transactions did not come out on time. Such suggestions were both wrong and insulting, for in actual fact Ørsted had brought order to the previous chaos; from 1815 the annual transactions had been published on time and information on the Society's administrative and economic affairs had been removed, thus making the transactions more homogenous and reader-friendly.

Ørsted carefully put his objections to Molbech's manuscript on paper, had them copied and gave the King/President one copy and had others circulated among the members. He wanted them approved as corrections on behalf of the Society and published so as not to let Molbech's version of the history stand unopposed. Molbech defended his views similarly, but when the two notes were presented at a meeting to compete for the favours of the members the affair petered out. The members did not want a public polemic on Molbech's jubilee publication, but the festive committee directed him to print a supplement to this effect: 'since this historical work has been checked neither by the committee nor by other members of the Society (which perhaps no reader would easily assume [Molbech added ironically]), the author alone is responsible for the factual contents as well as the opinions expressed in this publication.'[6]

For the occasion of the centenary a medal with the portrait of King Christian VIII was made. The forty foreign members received one each in silver, while the King and Queen had one each in gold. Ørsted was nervous of being criticised for extravagance, for the medals were scheduled to cost eight hundred rixdollars. So he suggested to the treasury to cast the medal for the eighty domestic members in bronze rather than in a more extravagant metal.[7]

*

One of the relatively unnoticed projects Ørsted implemented under the auspices of the Royal Danish Society was participation in the European cooperative *Magnetischer Verein*. Its purpose was to measure the earth's magnetism globally and to draw maps of the isogons of declination, the lines connecting the places worldwide where the deflection of the magnetic needle is the same. On such maps sailors would be able to read the difference between the geographical and the magnetic poles. This cooperation between Scandinavian scientists and GDNÄ and BAAS across national borders testified to the significance of Ørsted's international network and organisational talent in a field where he was no expert himself.

We have seen (ch. 43) how Schumacher gathered some of the most distinguished German scientists, Humboldt, Gauss, and Weber at his observatory in Altona, and in 1828 Ørsted joined them. All participants were members of the Royal Danish Society, Hansteen had been awarded its gold medal for his thesis on earth magnetism, and Gauss had received the same honour for a mathematical thesis. It is obvious what made these scientists into a team. After 1820 Ørsted had resumed his interest in exploring the earth as an electromagnetic system, and no longer a galvanic one as was the case with Ritter's flawed experiment from 1803. Humboldt was a globetrotter and geographer, Gauss and Weber had the mathematical insight that Ørsted lacked, and Hansteen was already well on his way to measuring the earth's magnetism; Ørsted had been instrumental in collecting such measurements on his triumphal progress in 1822/3, and Hansteen himself had been on an expedition to the faraway taiga of Siberia.

In 1834 Ørsted joined *Magnetischer Verein* on behalf of the Royal Danish Society and received Prince Christian Frederik's backing to establish a magnetic observatory on the site of the Polytechnic Institute. Two wooden sheds were erected and some new magnetic measuring instruments invented by Gauss were installed. Instead of the usual light compass needle, Gauss had constructed a much bigger magnetic rod weighing four pounds. It was one foot long on each side of the pivot and two inches wide. This magnetic rod was less sensitive to wind and weather

Fig. 134. H.C. Schumacher (1780–1850), astronomer and cartographer. Lithograph by Ausborn after a painting by C.A. Jensen (1792–1870). RL.

and hence easier to read, among other things because, apart from the stability given by its weight, its disturbing oscillations were hampered by silken threads. Adjacent to this magnetic rod was a mirror in which one could observe a scale from six to seven metres away. To make the reading of it as precise as possible binoculars were used. Ørsted had familiarised himself with all this equipment on a study tour to Göttingen paid for by the Prince. Now he brought it home to the Polytechnic Institute which became one of a series of European magnetic observatories. Hansteen's observatory in Christiania and the Royal Society's in Kew Gardens outside London were parts of the network, too, and a new observatory was built in Milan. The idea was that all these observatories at the same time (synchronized pendulum clocks were installed) would read the declination at their location and report the results to Gauss. Ørsted was assisted in reading the instruments by his own polytechnic students, Scharling for one, and several officers.

In *Magnetischer Verein* it was agreed to measure the magnetic declination on 5–6 November 1834 and send the results to Gauss. However, these dates were haphazardly changed without notifying two observatories, Copenhagen and Milan. This omission turned out to be a blessing in disguise, for exactly on these two days there was a magnetic storm, and the two independent observations at two locations wide apart coincided. Or, as Ørsted wrote 'had so great likeness that their graphic depictions resembled each other as two drawings of the same coastline'.[8] The vocabulary is typical of Ørsted. An inveterate mathematician would have described the curve mathematically instead of using a metaphor like 'two drawings of the same coastline'.

If an electromagnetic telegraph between the two magnetic observatories had been available this omission would never have happened, and this was also a technology that had caught the

Fig. 135. K.F. Gauss (1777–1855), German mathematician and astronomer. Drawing by unknown artist. BPK, Berlin.

interest of both Gauss and Weber; Ørsted had merely demonstrated his 'telegraph' to the public at the Polytechnic Institute (ch. 50). In 1839 the Society decided to remove the magnetic observatory to the ramparts of Gyldenløves Bastion just outside the western city gate. This was Christian VIII's first contribution to the Society as King;[9] the location on the ramparts was more isolated and hence more suitable than in the courtyard of the Polytechnic Institute. Two thousand and seven hundred rixdollars were granted and so Ørsted provided the nation with a better magnetic observatory than the ones Gauss and Weber had in Göttingen and Leipzig.[10] Ørsted was not involved in the other important research projects carried out by the Royal Danish Society such as P.W. Lund's palaeontological expedition to Lagoa Santa in Brazil or the corvette Galatea's circumnavigation of the globe in 1845–7.[11]

*

Henrik Steffens had been restored to the favour of the new King of Denmark, patron of the sciences. And Prussia's new King, Friedrich Wilhelm IV had called Steffens to the University of Berlin at which he took his turn as Vice-Chancellor in 1834–5. He was the loyal subject of both monarchs. After the difficulties he had had in curbing his jealousy when Steffens had seduced young Adam and laid Copenhagen at his feet, had Ørsted now become a Steffensian, despite having sworn that this would never happen (ch. 22)? No, on the contrary, it was Steffens who had changed. He had come to recognise that Ørsted's experimental physics had produced the scientific progress of the age, whereas *Naturphilosophie* had accomplished almost nothing.[12] In his *Polemische Blätter* [Polemical Papers] (1829–35), in which he outlined the history of science from the middle ages till his own day, he had tried to make an appraisal of the relationship between philosophy and science. He regretted his earlier contempt for empirical research and acknowledged that without it the thoughts of even a totally free spirit remained empty speculation.

In 1832 Ørsted wrote a review of Steffens's *Polemische Blätter* in *Maanedsskrift for Litteratur* [Literary Monthly].[13] He translated long passages into Danish making an effort to clarify obscure German concepts. In general he found that *Naturphilosophen* expressed themselves verbosely and condescendingly to the readership so that it was justified to characterise them as intellectual 'Don Ranudos', referring to Holberg's comedy (ch. 53).[14] This criticism was not aimed at Steffens personally, but at *Naturphilosophen* en bloc. What Ørsted resented was his remaining Romanticist homage to medieval mysticism; in the spirit of the Enlightenment he rebuked Steffens that by far the best thing from the middle ages was the invention of the art of printing.[15]

The main issue brought out by Ørsted's review was prompted by the medieval controversy between 'nominalists' and 'realists' which—provided one is willing to follow Steffens in applying these concepts leniently—bears a certain resemblance to the ongoing discussion between *Naturphilosophie* (young Steffens, the disciple of Schelling) and experimental physics (HCØ). Dissatisfied with Steffens's use of these concepts Ørsted takes his definition from a German dictionary. 'Nominalists' believe that only individual things are real, whereas 'universals' are linguistic constructions without reality. 'Realists' by contrast believe that individual objects are coincidental appearances only, while only 'universals' are real or essential. However,

Fig. 136. Henrik Steffens (1776–1845), Norwegian-Danish-German *Naturphilosoph* in later life. Lithograph by unknown artist. RL.

according to Ørsted this philosophical dichotomy is unhelpful for a proper understanding of the progress of science. What earlier generations wanted was reflection.[16] Copernicus's heliocentric worldview was not born by replacing 'individuals' by 'universals', but by distinguishing appearance (the rising and setting sun) from essence (the rotation of the earth). Similarly, the modern world view on 'matter' was brought about by dynamical physics realising that the atomic theory operating with indivisible and impenetrable atoms from which 'matter' derives its form is obsolete.

The new feature of modern, dynamical science after Kant is the understanding that matter is not what everyday experience makes it seem to be. So Ørsted gave Steffens a rapid education in dynamical science. The form of 'matter' is determined by the intrinsic fundamental forces penetrating that which fills space continuously. Or in Ørsted's own words: 'Matter itself is nothing but the space filled by the fundamental forces of nature; what gives objects their unalterable characteristics, then, is the laws by which they are created'.[17] Hence, it is mistaken to perceive solid water (ice) as the essence and its aerial and fluid forms, as resulting from imponderable fluids in the pores between atoms, as derived forms. The different forms of water are explained by the internal radiation of heat, that is the different speeds of motions of the fundamental parts of water molecules. Thus modern science has radically superseded everyday experience. Physical objects—whether the motions of the planets or the forms of physical objects—are not at all what they immediately look like. And since things are not what they seem to be from their appearances, the laws determining their true essence are badly in need of becoming part of the curriculum of general education. According to everyday experience we take it as unspoken knowledge that physical objects are stable substances with a certain colour, taste, smell, etc.; but in reality our sense impressions are supplied by the various effects of repulsive, radiating, chemical, and other fundamental forces. So, Ørsted repeated to Steffens what he had previously made clear to Weiss when he compared the structure of matter with a stellar nebula, that putting the question of the decisive property of 'matter'—whether it is solid, liquid, or gaseous—makes no sense, because it depends entirely on the effects of their fundamental forces. 'Substance' is a category of the understanding. At the same time he reminded Steffens that the concept of fundamental forces is noumenal and in consequence they cannot be made objects of empirical investigation. Hence, no scientist can study them per se, one can only observe their effects.[18]

Does this mean that German *Naturphilosophie* has had no positive impact at all? No, Ørsted found that it has taught physicists to look at nature in context and drawn their attention towards comprehensive views. The principle of *natura naturans* is a powerful example of this. This presupposition stimulates thought experiments, but the method of experimental science is bound to focus on singular phenomena, one at a time. To do science presupposes epistemological consciousness, but to reduce science to philosophy is the crucial error of *Naturphilosophie*, which has not led to the discovery of a single law of nature. This was the criticism on which Steffens reflected in his speech on 'The Relationship between Naturphilosophie and Empirical Science' at the Scandinavian Science Conference in Copenhagen in 1840. Addressing an audience

> 'that has inherited the fame of the past [Tycho Brahe, Ole Rømer, Carl von Linné] and enhanced it, I wish to remove the misunderstandings that reduce the value of *Naturphilosophie* in the eyes of scientists, misunderstandings that are not only derived from the particular view of the empirical sciences, but also from the fallacies of *Naturphilosophen*. I dare not deny that they have added to the prejudices impeding the influence of this new science and the more so because I must confess to have shared them myself.'[19]

By this introductory admission the old *Naturphilosoph* apologised, confessing that his philosophy of science had hubristically thought itself capable of opening the book of nature and looking over God's shoulder. He ingratiated himself with the audience who had long ago repudiated his Schellingian programme. Steffens, an old hand at making crafty speeches, admitted that his former contempt for mechanical science and its escort, mathematics, was a mistake, but still worse he had mixed up ideas with proofs, and in this respect he found his own book *Beiträge zur innern Naturgeschichte der Erde* [Contributions to the Internal Natural History of the Earth] guilty (ch. 18).

According to Steffens, *Naturphilosophie* deserves a future as a provider of metaphysical paradigms for the sciences 'because the riddle whose solution we seek presupposes a view of the totality of nature, and the sciences are captured by the laws of the senses that will never reach an all-embracing view of nature, but will persistently explain each phenomenon by its external relationship to another phenomenon.'[20] If this was to succeed the philosophy of science would have to abandon his set of concepts that according to Kant were empty because they could not be placed in time and space. In addition the philosophy of science would have to continue Kant's work to understand the limits of human cognition.

This shows that Steffens surrendered to Ørsted's authority and yielded to his criticism. Hence, in his obituary of Steffens in 1846, Ørsted harmonised the divergent reactions he had evoked, from Frederik VI's fuming resentment to Oehlenschläger's unreserved, although temporary, admiration that soon swung to contempt. The latter's reconsideration was provoked by Steffens's sudden conversion to German *Naturphilosophie* which he had diffused at Elers' College with the lack of restraint of a recent convert accompanied by an outspoken disrespect for the empirical sciences.[21] Ørsted, too, had temporarily lost his footing and been ensnared by Steffens's seductive method. In those years of confusion several bold theories offered themselves: Winterl's 'Andronia', Steffens's *Beiträge*, and Ritter's magic wands. None of them kept their promises. By contrast experimental physics gained new insights. German *Naturphilosophie* was one extreme swing of the pendulum, French corpuscular theory the other extreme. Both turned out to be fallacies, but still left their marks. The atomists had had to recognise that their theory was metaphysical and wrong, and the experimental physicists had had to recognise the need of metaphysical beacons to orientate the focus and direction of their research. *Polemische Blätter* had opened the way to a fertile reconciliation between philosophy and science. Steffens acknowledged the errors of his youth and Ørsted's victory and so their stormy lives ended in mutual respect.[22]

*

In 1839 Ørsted received a small grant from the Royal Danish Society for his experiments on the release of heat by compressing water (ch. 32). The experiments took place with his piezometer capable of compressing liquids by the pressure of many atmospheres, and a new thermoelectric apparatus so sensitive that it registered even very small differences of temperature. As he anticipated a rather long series of experiments with innumerable measurements and calculations, Ørsted applied for funding for a research assistant and employed one of his polytechnic students, Ludvig August Colding. He had already shown 'the most undaunted diligence and the most conscientious precision'.[23] Ørsted's experiments did not really take off, but they inspired Colding's reflections and experiments which led to the discovery of the first thermodynamic law on the imperishability of forces, or as moderns physicists call it, the principle of the conservation of energy.

Like Ørsted's, Colding's discovery belongs among those that are epoch-making in the history of science, although internationally his name is overshadowed by the German Julius Robert Mayer and the British James Prescott Joule, who both made the same discovery at about the same time. None of them was aware that the other two were working on the same hypothesis. I am not concerned with priorities, but interested in the reasons why Ørsted, who assessed Colding's thesis, did not understand his experiments and conclusions. These reasons, if they can be unravelled, will advance our understanding of Ørsted's views on science and moreover explain why experimental physicists and mathematicians at the time were at odds with each other.

Colding's father knew Ørsted and asked for his advice about his son's education, as unfortunately his son could not be persuaded to carry on with the family farm. As usual Ørsted was helpful and recommended an apprenticeship with A.C. Olsen, the best carpenter in town and teacher of drawing at the Royal Academy of Arts.[24] Besides this craft Colding also learned technical drawing, as the master made his apprentice draw the Boulton and Watt steam engines at the Royal Mint in the Royal Naval Dockyard. Colding was ambitious and in 1837 he was admitted to the Polytechnic Institute, although he had difficulties passing the entrance examination in foreign languages. He joined the mechanical branch, in which Ørsted's textbook from 1811 was still part of the requirements. *Indledning til den almindelige Naturlære* [Introduction to General Science] that announced the abortive 'Theory of Force' made a tremendous impression on Colding. He also became one of Ørsted's protégés, was invited to dine with his family, and being a bright mathematician became his research assistant.

Colding now applied the knowledge he had acquired during his education, Ørsted's dynamical part of science, and began wondering about the relationship between the external pressure (a mechanical force) and the rise of temperature in the water (the force of heat). Ørsted had taught him that heat was not the imponderable 'caloric' as believed in the old textbooks, but a radiation of forces (ch. 28). He had also taught him to look at nature as a whole. The more Colding thought along those lines, the more he came to believe that the mechanical force did not vanish into the air, but by some metamorphosis was changed into a dynamical one. The correlation coefficients pointed unequivocally in that direction, but were the measurements precise, and could not pressure and heat disappear somewhere else? Despite his doubts he

Fig. 137. Ludvig August Colding (1815–1888), independent discoverer of the principle of the conservation of energy, HCØ's protégé, polytechnic student, and later director of the Water Board of Copenhagen, RL.

suggested to Ørsted that as the second Scandinavian Science Conference in Copenhagen was approaching, he should present his theory there. Ørsted found his theory interesting, but on the basis of his hard-earned experience in Paris he firmly dissuaded Colding from presenting it until he had certain evidence. However, he provided funding for Colding enabling him to continue his experiments.

The following year Colding had little time to experiment. First, he had to prepare himself for his polytechnical exam, which he passed in 1841. Then Ørsted offered him a job as lecturer at Nakskov for the Society for the Dissemination of Science. The course was advertised by drumming in the streets and was well attended. In 1843 Colding became Professor Hummel's assistant in drawing lessons at the Polytechnic Institute, and so found more time for his research. He had a thought experiment and a range of skills, he could draw and do carpentry work and mathematics. His experimental device was in part an imitation of Coulomb's sledge for studies of frictional heat. Colding's device (now at the Technical Museum, Ellsinore) looks as follows:

The aim was to measure the relation between the tractive force required to pull the sledge along the rails and the frictional heat that expanded the runners and rails. To measure the tractive force he used a dynamometer and the expansions were measured very accurately by a spherometer with a fine adjustable micrometer screw. Runners and rails were folded into lignum vitae on three sides, on the fourth they rubbed against each other, so that the expansion could only take place lengthwise. Despite the fine device he had difficulties obtaining definitive measurements, and moreover Colding was concerned whether runners and rails of different materials would give the same correlation coefficient. Results were not quite unequivocal.

In 1843 Colding submitted his thesis with the modest title 'Some Sentences on Forces'. Unknown to the author this was the second thesis in the world presenting the theory of the conservation of energy and the first that was accompanied by an experimental proof. He was confident that he had corroborated his a priori theory on the imperishability of forces. There was a mathematical correlation between mechanical motions and dynamical forces. The Royal Danish Society set up a committee of Ørsted the physicist, C. Ramus the mathematician, and J.C. Hoffmann lecturer in physics and chemistry at the Military High School, for there must be three members. The committee did not approve Colding's theory. Ørsted formulated it as follows: 'that lost mechanical forces should be changed to chemical ones (in the widest sense of the word)', the committee cannot accept.[25] He could not see for himself that a sensory change of form of the mechanical and dynamical forces had taken place, for Colding had not described any of them. As we have seen time and again, he considered the concept of force as bipolar, as when positive and negative electrical forces neutralise each other and their sum is zero. Therefore, when Colding talked about 'lost forces' he must imply that they had gone missing, because they were neutralised by a polar force, but a 'mechanical' tractive force is not the bipolar equivalent of the 'chemical' force of heat. The mechanical force could only go missing by means of an opposite (mechanical) force.

Ørsted had discussed this problem in his review of Steffens's 'Polemical Papers'. Previously, natural philosophers believed that equals unite with equals but cannot unite with opposites, but nowadays natural scientists entertained the opposite view, that is that equals abhor each

Fig. 138. L.A. Colding's apparatus for the discovery of the law of the conservation of energy. The box is loaded with cannon balls to give maximum weight. It functions either as a sledge with runners D of lignum vitae reinforced with metal, drawn on rails MM, or as a wagon on wheels driving on a beam R, which can be lifted or sunk. When the sledge is pulled forward with the handle a friction between the runner D and the rails MM occurs, which produces heat. The amount of heat corresponds to the expansion of the runner, that can be read because it is only fixed at one end, while the other is free to expand. The mechanical force required to pull the box can be read at S. To obtain a sufficiently large, i.e. measurable, expansion of the runner, the box has to be pulled several times, which is facilitated by lifting R to a higher level than the rails thus turning the sledge into a wagon running on wheels. Sinking R enables the box to be re-pulled as a sledge in order to produce more heat so that the runner is expanding still more. Colding measured the mechanical force as well as the amount of heat produced, finding them to be in direct proportion or as he said 'the mechanical equivalence of heat'. DTM.

other and unite with their opposite. Ørsted tried to reconcile this difference by arguing that one cannot talk about opposites if there is nothing in common between them. For instance it makes no sense to say that a line is the opposite of a surface or a body or that electricity is the opposite of magnetism. But it does make sense to talk about two opposite kinds of electricity

(+ and −) or two kinds of magnetism (N and S), because we perceive them as polar forces. This way of reasoning led Ørsted to conclude that only the same species have real opposites, for example with mammals and sexes. Males and females unite within their own species. In this way Steffens's two views do not contradict each other. The condition of unification is that the two parts belong to the same species, whereas the action of unification is enabled by their opposite sexes.

One may well wonder why Ørsted's concept of force prevented him from approving Colding's theory, since we have seen time and again that Ørsted followed Kant in assuming that all forces are reducible to two fundamental forces. But to understand him properly, we must remind ourselves that Ørsted made a sharp distinction between fundamental forces as a noumenal principle of which we can know nothing and their phenomenal effects, amenable to experimentation; fundamental forces merely constitute an auxiliary concept and do not exist in space and time. The dynamical view of physics has made it abundantly clear that not only polar forces interact, but also different forms of fundamental forces, as shown by the discoveries of electromagnetism, thermoelectricity, and diamagnetism. But this would not necessarily help Ørsted understand Colding's experiment, because what he tried to establish was interaction between mechanical motion (generated by an extrinsic body such as horse power) and a fundamental (intrinsic) force (in the form of heat). In Ørsted's physics, horsepower is not derived from a fundamental force and therefore it cannot interact with heat, which has a fundamental force as its source.

Two historians of science in particular, Kenneth L. Caneva and Vilhelm Marstrand, have promulgated divergent views on the relationship between Ørsted and Colding. In 1997 Caneva blamed Ørsted for operating with a changeable and confused concept of force that prevented him from appreciating Colding's discovery of the imperishability of forces of nature. His article outlining the development of Ørsted's understanding of the concept of force from his dissertation on Kant till his report on Colding's experiment it seems to me is schoolmastering and censorious. He notices that in 1805–06 Ørsted was attracted by Schelling's tripartite scheme which claimed that magnetism attracts and repels in straight lines, while electricity propagates along surfaces, and heat radiates in all directions within a given body. This observation is correct, but irrelevant, because Ørsted soon abandoned the scheme. Caneva knows that, but still he considers Ørsted a *Naturphilosoph* in general, although the only guidance that Ørsted is suggested to have borrowed from Schelling is exactly this tripartite scheme.[26] Apparently, it does not occur to Caneva that Ørsted actually takes his inspiration from Kantian epistemology, for at any rate Kant's works do not figure in his list of references, nor is his critical philosophy taken heed of anywhere in his text whatsoever. Had Caneva familiarised himself with Kantian epistemology he might not have written half a hundred pages on Ørsted's failure to come up with a precise account of his concept of fundamental forces. This concept as I have tried to show time and again is a noumenal principle from which no scientific knowledge can emanate.[27] It is purely an auxiliary concept denoting the unknowable cause of physical effects that are manifest in time and space, such as electricity, magnetism, chemistry, heat, and light. In investigating these dynamical effects Ørsted started from scratch after having rejected the explanations offered by corpuscular theory. In his abortive *Theory of Force* he has abandoned

Schelling's tripartite scheme, and in his major work *Ansicht/Recherches* he hypothesises that these effects propagate like waves. At this stage he was convinced that mechanical physics and dynamical physics belong to separate realms. This view he shared with the scientific community as a standard distinction between sensible motions (phenomena) and insensible forces (noumena).

Still, according to Caneva, Ørsted ought to have anticipated the invalidity of this distinction, which seemed to be the consequence of Colding's discovery. Following his discovery of electromagnetism he maintained the noumenal concept of fundamental forces. As is obvious from his subsequent discussion with Ampère, he did not think that his discovery had revealed the nature of fundamental forces. Against this background it is absurd to insist, as Caneva does, that 'he never succeeded in giving his ideas the kind of coherence and precision that might have made them more than just vaguely explanatory or that might have suggested specific experiments capable of testing them...His own hunches with regard to the optimal conditions [for an interaction between electricity and magnetism] were all wrong...His general conviction bore fruit while his specific ideas were barren. Here his speculations died wholly without issue.'[28] Moreover, Caneva blames Ørsted for his ambiguity as regards the concept of aether: 'What kind of matter or substance is it? What is its relation to ponderable matter? Does it have weight?'—questions that would have been meaningless to Ørsted, though perhaps not to a corpuscular theorist. But his noumenal concept helps us understand why he was unable to endorse Colding's theory.

Ørsted found the experiment unclear and could not approve of a mathematical proof of a physical change of form he did not understand: '...but whether I perhaps have misunderstood him, I dare not say with certainty', Ørsted admitted.[29] In addition the results of Colding's measurement were imprecise. Professor Ramus had reasoned similarly. One could not compare two different forces of which one is measured by a dynamometer and the other by a spherometer, and it had not dawned upon Colding to introduce the modern major term of 'energy' for the simple reason that it did not exist in the language. Nevertheless, both these examiners had to acknowledge that the thesis was erudite and thought provoking, so they encouraged Colding to go ahead on a new grant. Caneva comments with the following platitude: '...the quality of Ørsted's response owed much to the peculiarities of his own conception of force.' Yes, what else could one expect? The epoch-making theory of the conservation of energy had not yet dawned upon the scientific community. However, four years later Ørsted had overcome his difficulties in appreciating Colding's experiments and in the second report to the Royal Danish Society acknowledged 'that the forces that are lost to the obstacles against motion are not in all regards lost, but appear again in another form as heat, electricity, and the like...[Colding] assumes in general that all forces without exception, when they appear to disappear, merely go over into other forms of activity without losing anything of their true magnitude.'[30]

The heart of the matter is that if one wants a physical sensation of the mechanical force that moves the sledge and of the dynamical expansion of the rails, it is hard to imagine these forces as physically alike in a way that makes it meaningful to put them into an equation. It takes a dispensation from the demand of physical realism (ch. 28) to use the mathematical formulation of 'the mechanical equivalent of heat'. Nonetheless, every schoolboy today learns the sentence about the conservation of energy as if it expresses a straightforward lawfulness, but what has happened to the understanding? Ørsted thought that a mathematical description of a phenomenon was made at the expense of a physical understanding. And in this respect he was at one with Newton, who formulated his laws of motion mathematically without being able to explain what 'f' in the formula (standing for gravity) represented in a physical sense. This difference between a mathematical description and a physical understanding appeared most confusing to many people. Ørsted found that the mathematical description just moved the phenomenon to a higher level of abstraction, where not only he himself, but also the majority of the target group he taught would feel betrayed.

In the history of science Colding's theory and Ørsted's reservations have been explained by indicating that both were imbued with German *Naturphilosophie*.[31] Apart from the illogical statement that *Naturphilosophie* enabled one of them to understand something that at the same time it prevented the other party from accepting, it is characteristic of a large part of the literature on Ørsted, that obscure aspects of his career are usually explained (or rather explained away) by invoking *Naturphilosophie* as a wild card. It seems likely that Colding's thought experiment was informed by the dynamical idea that all forces will be reducible to one fundamental force. This paradigm had originally been suggested by Kant in his metaphysics of nature (ch. 5). He reasoned that Newton's concept of force was physically inadequate, although it seemed to work mathematically. Kant's alternative proposal was a dynamical concept of force that was unlikely to become susceptible to mathematical treatment. This was the schism that Colding (and Mayer and Joule) now overcame by the principle of the imperishableness of forces, but unfortunately, according to Ørsted, at the expense of a physical understanding.

Ørsted inferred from the insight he had acquired into the reason embedded in laws of nature that these laws must be due to a creator of reason. Ørsted's concept of 'god' was minimalist. Hence it seems to me to be a misunderstanding when in 1929 Vilhelm Marstrand, a Danish historian of science, wrote: 'Ørsted himself was of a strong religious nature and his whole concept of nature was closely interwoven into his religious attitude as is apparent in his *The Soul in Nature*...'.[32] It gets even worse when Marstrand explains Ørsted's criticism of Colding's experiment by means of the religious argument that the theory of the conservation of energy is incompatible with *The Soul in Nature*. There is a significant difference between Ørsted's and Colding's understanding of the science-religion relationship. Ørsted never used religious beliefs to elucidate scientific problems. It was always the other way round. For Ørsted the laws of nature are laws of reason created by the Allreason or God. So God is another name for the quintessence of his scientific research, but He is never used to influence scientific insight. Colding, by contrast, seems to have fostered the idea of the imperishability of forces by deduction from his belief in God's care for the everlasting life of the human soul. 'If our spirit is imperishable, then so, too, must be nature's forces.' This is deducing science from religious belief.

Colding was only able to present his theory at the fifth Scandinavian Science Conference in Copenhagen in 1847. Nothing indicates that it caused any sensation. Mayer, who had arrived at the same insight as Colding, was thrown into a struggle for priority rights and was for a time sent to a lunatic asylum. Joule harvested the fame, especially in the Anglo-Saxon tradition of the history of science, and the unit of measurement that carries his name competes to this day with the term 'calories', a reminder of Lavoisier's refuted theory of heat. As we shall soon see, Colding's polytechnic talent was used in the modernisation of the infrastructure of Copenhagen, while his great discovery ensured him membership of the Royal Danish Society.

> There is something in Ørsted's persona reminiscent of Columbus's unshakeable faith that the new world must necessarily exist.[1]

53 | 1843
Homage in Berlin

1843 was the year of two happy events for the Ørsted family. In January Marie was married to her cousin Peter Hasle, the theologian, and after the wedding she moved to the parsonage at Asnæs in northwest Zealand. A few months later Sophie was married to Dahlstrøm, the jurist. Uncle Anders no longer had 'his lasses' as his housekeepers in Roskilde and Viborg. It was also the year when Ørsted and Gitte became grandparents as Karen and Scharling produced their first-born. With their three oldest daughters well married there were only four of their children living in Studiestræde, Mathilde helping her father as a kind of secretary, Christian having difficulties in getting a post in forestry, as well as the two youngest sons, Anders Sandøe (17) and Albert Nicolai (14). Young Viggo, Christopher Hansteen's son, was also put up in Studiestræde and was working at the Polytechnic Institute.[2] In summer their parents took them to the recently opened Tivoli Gardens.

Ørsted was not only pronounced commander of the North Star Order by the Swedish King Carl XIV Johan, but the same year he also received from the Prussian King Friedrich Wilhelm IV the knight's order *pour le mérite dans les sciences et les arts*. Christian VIII basked in the honour suggesting the newly dubbed knight thank the Prussian King in person the following summer at royal expense. For the fifth time Ørsted set off for Berlin, to stay for a month and then proceed to Dresden and its environs, of which he had such pleasant memories. He was particularly fond of the art in the exhibition at the Zwinger Gallery. He wished to enjoy the marvels again in old age, while Gitte, shy of foreign parties, stayed at home as usual.

The order *pour la mérite* had the shape of a Maltese cross with spread eagle wings between the arms of the cross and like Dannebrog hearkened back to the medieval crusades. The order owed its name to Frederick the Great's admiration for everything French, which had also led him to calling his castle Sanssouci and inviting Voltaire to live there. Ørsted had previously met Friedrich Wilhelm as Crown Prince, when in 1828 he had taken part in the GDNÄ conference in Berlin and had been praised for his improvised lecture on electromagnetism. There he had also become acquainted with Alexander von Humboldt. In 1842 Humboldt was appointed the King's councillor for the civil division of his College of Heralds, and one of his first initiatives was to nominate Ørsted.³ As a geographer Humboldt was perhaps less preoccupied with physics, but Ørsted was an important link in the international project around *Magnetischer Verein* [The Magnetic Association, ch. 52] together with Hansteen, Gauss and Herschel.

Ørsted packed his best clothes and went to Berlin where as usual he stayed with the Weiss family at the University. A few days later Humboldt invited him to his chambers which were next to Sanssouci at Potsdam, because Friedrich Wilhelm wished him to be available as councillor and diplomat in foreign affairs. Ørsted travelled by steam train for the first time. The trip from Berlin to Potsdam, a distance of twenty-five kilometres, took forty-eight minutes, so the speed was modest, twenty kilometres per hour. Even if it cost twice as much to travel first class as third class, Ørsted preferred the comfort, remarking that trees whistling by just outside the windows upset the calm view of the scenery he used to enjoy from the stagecoach. The carriage rumbled and Ørsted was shaken about, he told Gitte, referring her to Andersen's fine description of touring by railway.

Humboldt, by eight years Ørsted's senior, received him as a friend and started by inquiring how he had really managed in 1820. He wondered, because after having experimented under the supervision of Ritter in Jena forty years before, he had turned to geography and reached his results by reasoning from observations, not by experiments. Ørsted had told the story of his discovery both in 1827 in Hamburg and the following year in Berlin, where Humboldt was elected president of the GDNÄ. Humboldt had heard him both times, but colleagues had been talking, and now he wanted to hear straight from the horse's mouth and in great detail how his experiment had happened. Ørsted explained once again that the discovery followed his thought experiment outlined in *Ansicht/ Recherches*, but this answer did not seem to satisfy the royal councillor entirely.

In return Humboldt started entertaining his guest with his own research, spicing it with experiences from his expedition to Latin America. After Jena he had met Georg Forster, who had been voyaging with Cook and Banks in the Pacific, and his accounts had aroused his adventurousness. He had tried to join Napoleon's expedition to Egypt, but as it turned out to be impossible to obtain a passage, his great expedition was directed towards the southwest instead, and the King of Spain had subsidized it. In Latin America he had ventured to climb the volcano of Chimborazo more than six thousand metres high; deep into the rain forest he had discovered the Casiquiare Canal connecting the Orinoco and Amazon rivers, and in Peru he found guano in inconceivable quantities, which started the European exploitation of this reservoir of fertiliser. It was here that the foundations of his investigations of climate and plant geography were laid, resulting among other things in isothermal maps of the world. To accomplish this the worldwide network of *Magnetischer Verein* was necessary. Humboldt had high hopes that the geomagnetic measurements would explain whether magnetic storms caused the latitudinal deviations of isogons that otherwise seemed inexplicable. Humboldt expressed his gratitude that Ørsted had joined this network of observatories, since through his friendship with John Herschel the project had obtained access to the British Empire, across which it would now spread its observations. Humboldt had managed to include the vast territories of the Russian Tsar and the Chinese Emperor in the network thus making it worldwide. Humboldt's stream of speech was unstoppable, Ørsted hardly got a word in.

After the expedition Humboldt had settled in Paris, where as a celebrated adventurer he triumphed in the salons. His words carried weight in the Académie des Sciences. It was to Humboldt that Frederik VI and Danish diplomacy had turned to salvage the Napoleon prize for Ørsted. Two Prussian Kings tried to entice Humboldt away from Paris by offering him the post of Court Steward, a large pension, and lodgings in the castle to make him settle down in Postdam. In 1840 the new King, the Romantic and conservative Friedrich Wilhelm IV, succeeded. In his new royal surroundings Humboldt was offered peace and quiet to write his chief work *Kosmos*, published in four volumes from 1845 onwards, where Ørsted's discovery was amply described.[4] Ørsted had found his superior in Humboldt as far as personal contacts in the international world of science were concerned. 'Acumen and wit is always at his service. He

Fig. 139. Alexander von Humboldt's (1769–1859) study in his private house in Oranienburger Strasse 67, Berlin. Watercolour by Eduard Leist, after an unknown original. BPK, Berlin.

receives information on the latest discoveries facilitated by his enormous correspondence. Thanks to his connections to governments and the most influential men he accomplishes a great deal.'[5] Humboldt invited his daunted guest to court at Sanssouci the following day at two o'clock.

On Sunday 23 July Ørsted took the train to Potsdam, took lodgings at a hotel, dressed up in his tailcoat, and put on his regalia. Humboldt fetched him by carriage. When all guests were gathered, the King arrived saying that he was pleased with the visit, whereupon Ørsted thanked him for the fine order, and passed on the compliments of Christian VIII. Now everybody went to dine. Ørsted was placed almost in front of the King with Humboldt at his side. The King talked about Bertel Thorvaldsen, whom he wanted to see in Berlin, and the Queen remarked that this would be easy now travelling had become so swift. A real conversation did not ensue except with Humboldt, but what they talked about he did not communicate to Gitte. As the King left the room he asked Ørsted to join a royal banquet once again. Humboldt's carriage took him to the hotel and from there to the railway station. Everything only took a couple of hours. At Weiss's he had a nap before joining a party at Schelling's.

Fig. 140. Neues Palais, Potsdam, c. 1820. The impressive castle was built by King Frederik the Great (1721–1786) in the park of Sanssouci after the Prussian Seven Years' War, 1756–1763. Aquatint after original by Friedrich August Calau. BPK, Berlin.

Two weeks later Ørsted was invited to a royal banquet again, this time followed by a performance at the theatre. On the short railway journey he sat in a first class compartment with noble people avoiding their clothes getting dirty. This time the banquet did not take place at the small rococo castle of Sanssouci, but at its bigger neighbour Neues Palais. Conversation was restricted to exchanges of polite phrases. Ørsted paid homage to the progress of German science in the capital of Prussia, and Friedrich Wilhelm referred the honour to his predecessor. The King reassured Ørsted that Berlin was profiting more from Ørsted's visit than the other way round, vainglorious modesty and flattery through and through. Humboldt was absent on this occasion. The banquet was soon over, and the interesting part followed in the evening in the intimate court theatre, of which Ludwig Tieck was the artistic director. The audience was entirely composed of celebrities. Euripides' tragedy *Medea* was staged and this was the most appalling Ørsted had ever seen. A mother killing her own children in revenge for her husband's adultery was at odds with human nature, he wrote to Gitte. Ørsted failed to appreciate the play's extreme psychological universe, finding that the crime of the heroine should be explained by the fact that she was not a Greek, but belonged to a barbarian and revengeful tribe. Her abominable deed could not possibly be defended on moral grounds (which was not Euripides' point at all). That Medea had been brought into a desperate situation after having broken with her family and her people in order to give her love to Jason and give birth to his children merely to be betrayed and humiliated by him went beyond Ørsted's moral universe. The myth was shocking and lacked the edification he demanded from plays.[6]

A few days earlier Ørsted had joined a party at Steffens's, where Ludwig Tieck had recited, which this virtuoso always did brilliantly. Later on the talk touched upon Ludvig Holberg whose work was a big success in those circles, with the exception of his comedy *Don Ranudo de Calibrados*, the bankrupt aristocrat. In this play Holberg overstepped a moral border according to Tieck, who found it indecent to rejoice over the poverty of other people, and asked why should Holberg ridicule a miserable Spanish nobleman for clinging to the greatness of his ancestors after having lost his fortune? The German guests were offended by the comical pillorying of this wretched aristocrat and his heraldic pride. Holberg's knack of speaking sarcastically of his protagonist was of quite a different calibre from the Romanticist irony Tieck was promoting. Ørsted, having performed Holberg's comedies with Oehlenschläger and Steffens at Mrs Møller's dyeing works forty years before, came to the defence of Holberg, the Enlightenment thinker. If it is all right to mock the pride of empty learning in *Erasmus Montanus* and the snobbery for things French in *Jean de France*, then why not also the vanity of aristocratic descent in *Don Ranudo*?

Ørsted's stay in Berlin was conveniently free from obligations. For a fortnight he was invited to private parties or to formal engagements with colleagues almost every evening. Each time he was at the centre of attention. Weiss proposed a toast for Hans Christian and Gitte, their seven children, three sons-in-law, and two grandchildren. The next Tuesday there was a festive dinner with seven colleagues in an establishment called Odeon in the Tiergarten. Ørsted was placed between Humboldt and Schelling. Weiss sitting across from Karsten reminded the guest of honour about the study circle on Kant's metaphysics of nature forty years before. Back then Volta had recently demonstrated his pile and all physicists had thrown themselves upon galvanism, but it was only in 1820 that the sensational breakthrough had occurred which later led to Seebeck's discovery of thermoelectricity. Subsequently, however, progress had come to a halt. Weiss ignored Arago, Ampère, and Faraday. Now the world was waiting for Ørsted's discovery to be applied to technology, enabling the public to travel from Berlin to Copenhagen by ships and carriages driven by electromagnetism. At the end Weiss reminded the party that the guest of honour had a brother of no less renown, who in another field had achieved the highest distinction.

Ørsted followed the ritual which required him to return the compliments. He compared Berlin as the city looked the first time he had been there in 1801 with the city he was now visiting for the fifth time, in gratitude for the magnificent knight's order. In the meantime the city had been occupied by French troops and Napoleon had humiliated Prussia by establishing his headquarters at Neues Palais. After Wellington's and Blücher's united victory at Waterloo, Berlin had been rebuilt by a distinguished architect (Schinkel) and decorated by an equally distinguished sculptor (Schadow) and painter (Cornelius). The University and museums of art and natural history had been established together with a *Gewerbeschule* and a railway linking Berlin with Potsdam. Previously, enlightenment had spread its beams to the many small principalities, but now the sunlight was gathered into one focus, Berlin, where science flourished all the more.

In several festive speeches the comparison between Ørsted and Columbus was repeated. It was Herschel who had invented it, but it was Humboldt who brought it to light in Germany for the first time.[7] Old Poul Erman, secretary to the Prussian Academy of Science, expressed it for a second time on a family excursion to Potsdam when he proposed a toast to 'Columbus Ørsted' saying that like the bold seafarer, who against all odds had stuck to his project to cross the seas to India in the opposite direction, Ørsted had been all alone in his conviction that the electrical and the magnetic forces emanated from the same fundamental force. It was only thanks to their stubbornness that the two discoverers had successfully defied conventional beliefs.[8]

On one of the first days after his arrival in Berlin he had a call from the Prussian Minister of Industry, P.C.W. von Beuth. They agreed to meet for a deeper talk later. It was Beuth who had been behind the establishment of the *Gewerbeschule* in Berlin, the institution that Ørsted had evaluated and found too light, when in 1828 he had ranked it on a par with Ursin's proposal. Now that the Polytechnic Institute had been in full swing for more than a decade he saw it again. Did he maintain the same opinion?

On 1 August Ørsted and Beuth met in the *Gerwerbeschule*. Ørsted wanted to know if it were true that the government had plans to curtail the freedom of trade. The reason for asking was that after the Napoleonic wars Prussia had abolished the guild system to pave the way for industrial development. Denmark had not reached that far, Ørsted remarked, being himself in agreement with Danish industrialists, who found that the guilds with their obsolete division of the crafts put the brakes on technical modernisation. Ørsted had succeeded in obtaining a special rule for polytechnic graduates exempting them from membership of guilds. Beuth denied the rumours adding that on the contrary, the government intended to expand the freedom of the trades geographically. Compulsory guild membership was to be repealed everywhere, entitling everybody to trade freely. The guilds would be allowed to continue, but only to take care of social affairs. Supervision of technical qualifications was to be transferred to the state and only in branches where inadequate skills might endanger the population—architects and millwrights for instance.[9] This reply was grist to Ørsted's mill.

Some days later he called on the *Gewerbeschule* again to be guided around the workshops by Professor Schubart. It had thirty workshop pupils of whom some were officers. The workshops were much larger than the ones at the Polytechnic Institute. The machines were driven by steam. Students received materials and iron that they could work and cast to make their own lathes or other kinds of tools that they were allowed to take with them when they left the school to establish their own workshops. Schubart kept a register of the workshops' supply of tools and machines. Most machines, Ørsted learned, had been smuggled out of England in defiance of the so-called tool acts.[10] This happened in the well known way that machines were bought in England and disassembled whereupon individual parts were shipped from different ports and reassembled in Prussia.[11] This form of industrial espionage had been very useful to Prussian industry, for how could they produce machine tools without possessing a prototype? Moreover, Ørsted was shown English spinning machines and looms. Ørsted's understanding of such machinery was inadequate to profit from this guided tour. Whereas the Polytechnic Institute could best be characterised as an informal Faculty of Science preparing students for official posts, the *Gewerbeschule* was aiming directly at technological practice and mechanics thus reaching out a helping hand to private entrepreneurs. Ørsted showed no sign of having realised that Beuth and Schubart had established one of the foundations for Prussia's industrialisation and her rapid rise to being a European superpower.

Ørsted added no comments to these observations, but the gap between the *Gewerbeschule* and the Polytechnic Institute was conspicuous. True, there were workshops in Copenhagen, but they were not as well equipped, the experienced foreman had quit in protest, and the polytechnic students did not show up, because their science courses occupied all their time. Ørsted did not endorse Beuth's strategy for industrial development: equality between and integration of theory and practice. His unequivocal views on the priority of science were soon to make him the target of fierce criticism in Denmark.

He called on old friends. Henriette Herz had become an old lady no longer able to keep her salon going, but at least they could remind themselves of the conviviality with the Schlegel brothers and Schleiermacher forty years earlier, when early Romanticism flourished in her drawing room. 'Perhaps she was the greatest lady in Berlin; but now she has fallen back, through old age.' She did recall the Danish lessons Sibbern had given her and Oehlenschläger's recitals.[12] Ørsted paid several visits to Steffens and his family. He also found time to see Schelling, who had become professor in Berlin in 1841 to counter the growing Hegelianism. According to his letters to Gitte they did not talk about *Naturphilosophie*, but this does not rule out, of course, that they may have done. Ørsted only mentions a chat about conscription of students. In Denmark conscription was restricted to the peasantry, while students were exempted from wasting their costly time on military service. But Prussia was different. Students had to drill although only for one year against three for less privileged classes, and they could pay others to take their place on watch duty. In this way the uneducated could earn a little and the educated could increase their popularity. Ørsted liked that, probably thinking how lucky his boys were never to have been drafted.[13]

A range of events indicate that in old age Ørsted was developing into a kind of fogey, feeling that the experience of a long life entitled him to speak with a weight defying contradiction. He felt better in the company of people who agreed with him. People sharing his views were considered knowledgeable; people who did not were waved away. In a talk with Karsten, the friend of his youth, on institutes of agriculture and forestry they soon confirmed their view that young people wishing to learn mining or road or bridge construction ought to learn practice in the field and theory at a university. It pleased him, he told Gitte, 'to find himself in agreement with a man who deserves so much trust.'[14] A few days later he talked to Mitscherlich about the same topic and once again there was agreement. His children were following courses on cameralism, and he had sent them to practise first with an estate owner, and only then to study the theoretical part at a university. 'As Mitscherlich is an eminently clever man, his example is a significant authority', Ørsted concluded.[15] In Leipzig he talked to O.L. Erdman, the chemist, who flattered Ørsted by telling him that it was *Ansicht* from 1812 that had made him choose to study chemistry, because it was only in this book he had come to understand that chemistry is a real science. Apparently, at that age Erdman had been in a phase of his life 'when a book of scientific spirit has an awakening effect'.[16]

Everywhere Ørsted was encouraged to give compliments to his renowned brother, and he did not fail, when he visited both Humboldt and the Grimm brothers, to be struck by the likeness between the close ties uniting Anders and himself and the family ties binding Alexander so closely to Wilhelm Humboldt, and Jacob so closely to Wilhelm Grimm.

From Berlin Ørsted went by train to Leipzig, where he was put up privately with Wilhelm Weber, partner in the international geomagnetic project. The memory of Niels Randulph, the black sheep of the Ørsted family, who had fallen thirty years before in the battle of Leipzig, was part of the family history that his letter to Gitte passed over in silence. He proceeded by the goods train to Dresden at a snail's pace, twenty kilometres per hour, accompanied by Weber. Here he called on the Norwegian painter J.C. Dahl, who had just then negotiated the sale of a Norwegian stave church from Vang to King Friedrich Wilhelm IV. The Norwegian congregation did not seem to have much veneration for their old wooden church, but now the symbol was re-erected outside Berlin as a manifestation of the growing German fascination for Viking art from the high north. Dahl, who was Professor at the Academy of Arts in Dresden and had earlier escorted Christian VIII on his journey to Italy, showed Ørsted around the gallery. This was a happy revisiting of the great Italian artists of the Renaissance, in particular Correggio's *The Night in Bethlehem* and Raphael's *Madonna* that had imprinted themselves on his memory. He reminded Gitte of the copy they so had often admired together in Oehlenschläger's drawing room. Now he had seen them in the original twice, though she had never had the experience. He saw in these master works 'a unity of spirituality and nature', a feeling which overwhelmed him also when he heard the music of Mozart. Dahl nodded approvingly.

They had lunch on the famous Brühl's Terrace overlooking the Elbe and in the afternoon they took a long ride by horse and carriage among the forest-clad hills stretching along the river towards Bohemia, where Ørsted had walked dreaming about his fiancée forty years before. The next day Ørsted and Dahl paid a visit to the legendary physicist and anatomist Dr C.G. Carus, who not only worked with mesmerism and magic wands, but was a practising painter as well. He was the charismatic *Naturphilosoph* at the centre of a community of artists in Saxony, including the Romanticist painters C.D. Friedrich and J.C. Dahl.

On 16 August Ørsted took the railway back to Berlin. At night the sight of flames in the opera house woke him up in his hotel room. Sparks were raining down, while on Unter den Linden he stood watching the fire that lit up the Royal Library, the University, The Academy and the Museum. He packed his suitcase to be ready to flee at a moment's notice should the fire spread. Next afternoon he took the train for Passow, and from there the following day the stagecoach to Stralsund, and that night the mail steamer to Copenhagen. The Prussian Queen was proved right, as she had conversed with Ørsted about travel times. From Dresden to Copenhagen in not quite four whole days, three by railway, one by stagecoach, and one night by steamer. Gitte and the children waited for him at the customhouse.

We could need a complete aesthetics of nature.[1]

54 | 1843–6
Aesthetics of Nature

AFTER THE great festivities surrounding Christian VIII's anointing and the second Scandinavian Science Conference the Ørsted family retired to the rural idyll at Fredensborg Castle. Here experimental physics, editing, lecturing, and politics were shelved in favour of writing aesthetical dialogues and essays that became part of Hans Christian's courses for the Society for the Dissemination of Science and *The Soul in Nature*. He dreamt of writing a comprehensive aesthetics of nature, catering for artists and scientists engaged in the cultural life of the Golden Age.

As a boy already he had read Batteux's aesthetics and acquainted himself with the idea that the arts are based on mimesis. Artists were to imitate the beautiful in nature according to an Enlightenment philosopher like Batteux. But Romanticists took a different view. A merely empirical approach to nature would provide sense impressions of the imperfect nature, which was no good, for art had to make itself independent of the real and finite world to become autonomous and construe the beauty in art in a modern, Romanticist mythology of nature that would have to come into being first. When Ørsted wrote his first dialogue on music in 1808, he had been reflecting on the relation between the beautiful in nature and the beautiful in art. At that time he had reached the conclusion that as a physicist his task was to show the fundamental significance of laws of nature for the enjoyment of art. His experiments on Chladni's acoustic figures anticipated a connection that in time might end in a physical aesthetics, in which is shown how forms and forces in nature, the physical object, influence the senses of the subject, although in a way where neither the composer nor the music lover is conscious of the natural laws active in the process. Ørsted presented a preliminary proof of this theory when in his dialogue he let his alter ego Alfred stand forward as the sage pointing out the correspondence between the visual beauty of the acoustic figure and the auditive harmony of the pure sound (ch. 22). This correspondence he called the rational subconscious, rational because the aesthetical event depends on the reason of laws of nature, and subconscious because the enjoyment is unconscious of the science in the arts.

Ørsted adored entertaining parties with his violin bow and watching the enchantment in people's eyes, as the lycopodium formed a symmetrical acoustic figure. Søren Kierkegaard and Carsten Hauch compared his happy face with an acoustic figure bowed by nature itself.[2] The experiment put him on the track of a theory that demystified the enjoyment of music. He was convinced that beauty corresponds perfectly with truth. Ørsted believed he had made a decisive and original contribution to solving Rousseau's bewilderment: *je ne sais quoi*.[3]

This correspondence between laws of nature and aesthetical experience endowed the pleasure evoked by art with a dimension of objectivity. Truth and beauty coincide. Laws of nature have objective validity, since they are no echo of man's own construction, but an event found in and corroborated by nature itself. Electromagnetism is not a law Ørsted had invented, but a phenomenon he had found. Nobody would send a researcher to Mars to find out if the inverse square law is valid there, for the universal validity of laws of nature is independent of the sense apparatuses of subjects, whether Martians or human beings, of whatever political or religious persuasion. Or as Ørsted stated: 'If our laws of reason did not exist in nature, it would be futile to try to force them upon it; if the laws of nature were not in our reason, it would be impossible for us to comprehend them.'[4]

Just as he had planned in his unfinished textbooks, he divided the impressions of beauty into visual geometrical forms according to the laws of mechanical physics on the one hand, and on the other hand dynamical physics that produced impressions of sound and light waves propagated in the air. Our aesthetic enjoyment is due to the symmetry of geometric forms that we find harmonious and sympathetic. Symmetrical forms may be circles, rectangles, polygons, hyperbolas, parabolas, equilateral triangles, etc. Asymmetrical forms, by contrast, are not aesthetic and we respond in antipathy. Ørsted put down on paper the names of his heroes Newton, Goethe, Schiller and Kant and folded the paper while the ink was still drying so that the letters came off mirror-wise and a symmetrical figure was created. The same impression is produced when one cuts folded sheets of paper with scissors. If they are cut in sheets folded both vertically and horizontally, one gets a doubly symmetrical paper collage.

Geometrical forms are often seen in nature. When the sun sets over the sea we see a circle on a straight line (the horizon), and the reflection of the sun's disc doubles the image symmetrically. Or if we throw a stone into the water we can watch the ripples propagate in circular undulations. If we throw another stone these undulations are broken in regular motions forming a different geometrical pattern. Gauss made an experiment in a flat ellipse-shaped vat full of mercury and found that he could produce undulations in the focal points of the ellipse, so that regular patterns appeared where the two undulations intersected each other. The effect was particularly beautiful, because the lustre of mercury reflected the light and in this way amplified the contrast between the lit wave tops and the shadows of their valleys.

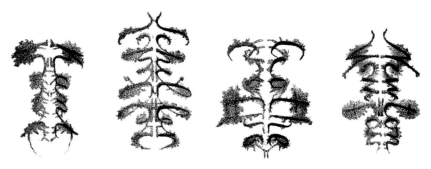

Fig. 141. HCØ's symmetrical figures formed as the names of his great examples written in ink and immediately folded to make a symmetrical mirror image. *SES* 3, 159 and Fig. 2.

Fig. 142. HCA's symmetrical ink blot made in the same way as fig. 141. Andersen has created a recognisable drawing of a person wearing a turban out of a random splotch. In around 1871 he noted, '… you make an ink blot on paper, fold it together pressing from all sides, and a figure emerges. With a little fantasy and some skill of drawing you help the image to stand out as a clear figure. If you are a genius you will get an ingenious image, or so it works with poets. A splotch is made by chance on their paper; then they squeeze it and help it out by drawing a little and the story is there.' OBM.

The dynamical part of physics, a product of Ørsted's own time, is a different category of sense impressions that are not concerned with form. Ørsted focused on the aesthetic effects transmitted by waves of light and sound. He hoped that the dynamical part of physics, still in its infancy, would in time shed more light on the aesthetics of nature than he had been able to do in his 1808 essay 'On the Cause of the Pleasure Evoked by Tones'. The hope was nourished by the progress made in the study of light. Optical analyses of white light and its polarisation in the spectrum together with the deeper insight into the function of the eye would have favourable results for the visual artists. The dependence of light and colours on changing positions of the sun in its orbit, changing angles of incidence at sunrise and sunset, the white moonlight, the light Northern nights, aurora borealis, etc.; how did these natural phenomena impact our sense of aesthetics? On the basis of Ørsted's overall theory that the same reason prevails in nature as in humans, his dynamical theory offered a new possibility of explaining why the beautiful in nature is subconsciously transformed into sheer pleasure. Hence it was particularly important for artists to understand the visual and auditory effects on our senses and minds.

Let us take a look at some specific examples of Ørsted's physical aesthetics. In his dialogue 'The Fountain', the scientist Alfred and the artist Frank are sitting in the Tuileries in Paris growing lyrical over the wonderful fountains. Frank asserts provocatively that the sciences have little to offer when it comes to explaining the experience of beauty.[5] But Alfred, alias Ørsted, opposes him, 'Just consider what optics have done by means of telescopes and microscopes. Let him watch a drop of water through the microscope, the world of life and motion presented to his eye will compel him to confess that science has revealed secrets that would otherwise remain hidden to him.'[6]

Likewise, it is beyond dispute that the theories of aesthetics of Aristotle and Winkelmann have influenced the development of art. There are mimetic rules for art, classical ideals desiring an art true to nature to the extent that laws of nature are known: note systems, harmonics, rhythm, geometry, chromatology, etc. Next, Alfred explains the curves of the jets of water as a consequence of gravity, and their geometrical form, the parabola fascinating for its symmetry. The drops reflect the light of the sun, big drops in one way, small ones in a different way, so that a variation of the visual impression occurs that becomes particularly conspicuous when the fountain is illuminated and the ambient world darkens. But there are audible effects, too, when the water jets rise and fall. Evenly flowing water gives peace of mind. In bright sunlight the colours of the rainbow stand out distinctly. The experience of beauty is conditioned by the natural laws of gravity, light, and sound, objective sense impressions. Correspondingly we experience beauty, a term that falls short of expressing the combined experience of the sublime that fills us with humility. We are confronted by forces that provide an impression of activity and a harmony that comforts and warms the soul. This combined aesthetic experience becomes the richer the better we understand its complexity and become able to read what Ørsted called the mind of nature.[7]

Ørsted took examples of the beautiful in nature from the animal kingdom. The swan is a case in point. Just imagine a person who has never seen a swan and suddenly is made aware of this peculiar bird in a hen run. He probably would not find it beautiful, but rather helpless and

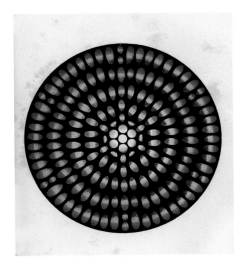

Fig. 143. HCØ's colour circle showing the polarisation of sunlight through a prism, so that light and darkness and the colour spectrum (red, yellow, orange, light and dark green, blue, and violet) illuminate objects in symmetrical forms while forming together a harmony. Being inspired by Goethe's theory of colours HCØ spoke in favour of a concord between the objective laws of nature and the collective human intuition of colours. *SES* 3, 170–206 and Fig. 3.

clumsy as it waddles along on its short legs with broad webbed feet and a filthy coat of feathers. There is a disproportion between its long neck and its lumpy body. However, a hen run is not the proper element of a swan. Floating on the still lake, one sees its reflection in the water and discovers a beautiful symmetrical figure. One might think it was a different bird altogether. Now the feathers are clean and shining white, a colour setting off the red or yellow beak and the black eyes so well. Now the swan is stable, with its neck gracefully in the air and its body sailing majestically on the water. When flapping its wings it is like a ship at full sail, when flying its wings sing in the air. Its beauty depends on its entire figure in its proper surroundings in addition to the symbols attached to it: the sublime in its posture, the forceful in its character, and the harmonic in its form.[8]

But is everything in nature beautiful? Or are there examples of ugly phenomena? Frederika Bremer, the Swedish writer, who visited Ørsted in Copenhagen and corresponded with him, was a warm lover of nature. Both were devoted to the aesthetics of nature, but Miss Bremer challenged Ørsted, because she could not go along with the view that all of nature is so reasonable and harmonic.[9] He was knocked off balance with the suggestion that the bat is an example of an ugly animal. Ørsted argued that its ugliness depends on its lack of independence. It does not look like a mammal due to its wings, and it does not look like a bird due to its fur. It is neither one thing nor the other, but a hybrid, and since it does not meet the ideals of beauty in any of the two categories we find it ugly. Hanging there in caves head down scratching its naked wings, we are unpleasantly affected by the sight of it. Fortunately, the unaesthetic features disappear when we observe it in its element. Flying around at dusk and navigating with its peculiar radar its hybrid character is hardly noticeable, and in turn we start admiring its eminent flying quality and sense of locality.[10] The more reason pervades the aesthetic intuition of the unreflected impression, the easier we drop our prejudices about it and realise its beauty. The ugly is the negation of the beautiful. The ugly is not an

Fig. 144. HCA's clip of a swan with an added rhyme mirroring HCØ's aesthetical view of the nature of a swan: 'On water the swan has the rank of beauty//On land it has an unsteady way of walking.' Astrid Stampe's Billedbog, 36, OBM.

ontological concept any more than the wicked is definable as anything but a negation of the good. According to Ørsted the ugly is the contrast of the beautiful, that which is asymmetrical, disharmonic (in tones and colours), the dependent on others, the monsters, and the lifeless, or the corrupt. The ugly and the wicked are finite qualities, whereas the beautiful and the good are eternal qualities.

Ørsted also chose examples from the vegetable kingdom: the white lily as the symbol of single-mindedness, purity and pride:

> 'It is easy for everybody to see the reason for this view: the straight stem, the beautiful, large, white flower, unfolded to full openness, whose symmetry is so easy to conceive, evokes this impression; its fragrance adds yet another pleasure.'[11]

The rose is different. Its flower is bushy as it is composed of a multitude of petals that are soft as velvet and sit on a thin pedicel full of thorns. This makes a contrast that enhances the aesthetical pleasure of the flower itself, which spreads an exciting fragrance. Its blood red colour becomes the symbol of love above all. The tiny blue violet that almost vanishes in the grass evokes quite different imaginations. Were it not for its strong fragrance it would hardly be noticed at all. But because the sweet violet is one of the first flowers of the spring it brings more happiness. Paradoxically, its attraction is its modesty. It is so unflamboyant that it is almost an excuse for itself. But it smells wonderful. It appears that Ørsted did not think that the sciences were ready to explain the entire aesthetics of nature. Only in small parts is there still room for the imagination: in fact he had to admit that the poet was a better interpreter of the symbolism of flowers than the physicist.

Once again Little Hans Christian seized a visual symbol from the dinner table in Studiestræde, which he could transplant into his autobiographical fairy tale *The Ugly Duckling*.[12] As small as the other feathered beings in the hen run, the 'duckling' was unaware that he was a cygnet. At first he was merely too big and too strange and therefore he needed a good beating. He was so disgusting that not even the dog would care to bite him. But then he grew up and became aware of the other swans and found them beautiful, they were shining white with long supple necks and splendid long wings. And seeing his own mirror image in the clear water he gained self-consciousness and was no longer clumsy, filthy, ugly, and nasty. The ugly duckling had turned into a beautiful swan.[13] In his fairy tale *The Wild Swans* princes are transformed into swans and back again, which agrees finely with the symbol of the majestic.[14] In *The Drop of Water* Kribble-Krabble held his magnifying glass to his eye, and looked at a drop of water that had been taken out of a puddle by the ditch. But what a kribbling and krabbling was there! All the thousands of little creatures hopped and sprang and tugged at one another, and ate each other up.[15] And when Andersen played with Ørsted's children he would fold paper and cut out the finest symmetrical figures. Andersen also shared Ørsted's view that although science gradually reduces the space of the imagination, this is not the same thing as saying that science is at war with poetry. His ironic fairy tale *The New Century's Goddess* takes place three generations later (that is, around 1935) when 'Time is too short and precious for the mere plays of fantasy, and, to speak seriously for once, what is poetry? These resonant outpourings of feeling and thought, they are only the offspring of nervous vibrations. Enthusiasm, joy, pain, all the movements of the organism, the wise men tell us, are but nerve vibrations. Each of us is but a stringed instrument.'[16] These examples show the great and lasting influence Big Hans Christian's aesthetics of nature had on Andersen's art. Carsten Hauch and Frederika Bremer were affected by it too.

In his first dialogue in *The Soul in Nature* Alfred explains to Sophie that there are physical and chemical and biological laws of nature that work together as thoughts of nature.[17] Nature is conceived as *natura naturans*, as an organic entity from an aesthetical perspective. Phenomena work together to fulfil a comprehensive purpose. There is nothing superfluous in nature and no devious courses. Like a great novel or quartet or painting each part is justified by its meaning for the whole. Alfred agrees with Sophie that the Sarps Force [a waterfall] is a well-chosen example of this. Both had seen it recently and admired its beauty, which Erik Paulsen had immortalised in his painting.

Parts of nature would fill man with horror. The horror scenario would be to imagine oneself engulfed in this maelstrom. The Romanticist notion of this experience was 'the sublime', which for Edmund Burke, the Irish born Whig politician and man of letters, whom Ørsted criticised, was the frightening experience of nature that makes one's hair stand on end.[18] Ørsted could not accept that the anxiety-prone sentiment should appeal to our sense of beauty, nor Burke's interpretation that the sublime consists in the intuition of healing we experience when the horror turns out not to be a concrete threat against the spectator. It is not the frightening object in itself that evokes the sense of beauty, but the grandiose and powerful. Ørsted, on the other hand, was more preoccupied with the idyllic or picturesque

Fig. 145. Erik Pauelsen (1749–1790), oil painting of the Sarps Force [waterfall], 1788. HCØ's description: 'The fundamental idea, in as far as the fundamental idea of an object of nature can be expressed at all, is a rushing river. Its waters continuously renewed, it hurls down from great heights. It obeys the same laws of falling bodies as any other physical object and consequently gains increasing speed as it falls; as water it has the property that its parts move quickly and separate lightly and the soaring parts form drops. At the ever increasing speed... a giant separation is brought about; at every clash infinite numbers of drops are splashing in different directions; a world of drops is formed full of motion... The air that must mix with the falling water makes foam, innumerable bubbles of air contained by water membranes, whose changing, uneven, and white surface is as peculiar as it is well known. The loudness of the sound every falling part makes, is determined by its height of fall... The foam and the noise of its roar testify to its destructive power, which also shows when something breakable enters its gulf. All this constitutes a coherent whole, in which every part is generated according to laws of nature, or in other words: all thoughts of nature are inseparable from the fundamental idea. It gets its unique difference from other falls from its particular position in nature, the many changes of the speed of the river flowing in, changes in the direction and strength of light, changes in the motion of air, heat, etc. So the Sarps Force stands out almost as a creature with its own character, filling our imagination with the idea of a huge, although unconscious juggernaut, a thrall of nature with an almost untameable power.' Vestsjællands Kunstmuseum, Sorø. Photo by Anders Sune Berg.

in nature. The waterfall is only part of the scenery. The river came from idyllic brooks in the mountains and would later pass quietly into a delta before it disappeared into the sea. The brook was filled with rainwater arising from the sea evaporated by the heat of the sun and condensed into clouds. The unconscious juggernaut was violent, but on the whole in Ørsted's view picturesque, and as soon as we realise that it is determined by the laws of nature, the anxiety-prone imagination is reined in.

Ørsted summed up his comprehensive experience of impressions of nature in three categories. When nature shows independence in relation to humans we experience it as sublime, and it arouses our admiration (not our fear). At other occasions nature presents itself as particularly active and evokes our enthusiasm because we understand that nature is an incessant process of creation. And thirdly we are spellbound by nature's correspondence with itself, its harmonious tranquillity filling the mind with peace and warm feelings. Ørsted is once again turning against Romanticism by rationalising the sentiments involved in experiences of nature. He makes reservations, admitting that his aesthetics of nature is in its infancy. There is still a long way to go, but one thing is certain: it is the reason of nature that addresses the human.

But if Ørsted is right in asserting that art is about the rational subconscious, why then do all people not experience the same thing when they ramble in the countryside, or walk in a gallery, or go to a concert? According to Ørsted this is due to the unequal distribution of reason as a consequence of innate varieties and differences in upbringing and education. Differences in taste depend on differences in intuition and education. Thus experience of nature depends on the scientific insight a human has acquired. Life is enriched and people are given greater enjoyment of the beautiful in nature by making them familiar with the laws of nature. And vice versa: by neglecting their science education in disciplines such as geology, zoology, botany, cartography, meteorology, physics and chemistry people remain chained to spiritual poverty. Scientific illiterates have poor prerequisites to enjoy natural and by consequence artistic beauty. Presumably, Ørsted considers his 'Two Chapters of the Physics of Beauty' as a draft for a textbook in the aesthetics of nature structured in paragraphs like *Kraftlæren*, his abortive textbook on the theory of forces. Even as a draft it was translated into German by H. Zeise, who also translated some of Andersen's works.[19] The distance from Kant's *Critique of Judgement* is obvious and shows the significance he attributed to the growth of science within the previous half century, not the least to his own contributions to dynamics. His last twelve years in life were dedicated to the task of writing a more elaborated aesthetics of nature, the unfinished result of which had a prominent place in *The Soul in Nature*, his life's testimony (chs 60–61).

*

For well over a century historians of art have argued that Danish paintings of the Golden Age have a number of features in common. Some have talked about the Eckersberg School and it has been discussed how Danish painters of the Golden Age distinguished themselves from other European schools at that time. Realism is a term often used to characterise it; realism as opposed to the symbolism so pronounced in German Romanticist art. But realism has turned out to be a multi-faceted notion. When Hannover, the art historian, used it, he meant 'mimetic', that is, the work of art was an immediate imitation of nature. Later other art historians have noticed the obvious differences between the painter's outdoor sketches and the later reworking of motif, light, and colours in the studio.[20] And it is not only lesser details artists have changed on the way from sketch to finished work of art, it is often the impression of the whole represented by the motif that has been reworked. This is documented for Eckersberg as well as for Købke and Lundbye. So 'realism' no longer means 'mimetic', but rather typical of the ideal, for the fundamental thought these painters are now seeking in the landscape in an attempt to put it onto the canvas is reminiscent of Ørsted's dualism, i.e. the relation between the immediately observable but also random and changeable phenomena of nature, and the invisible laws of nature behind them that are eternal and unchangeable and universal.

Can the investigation of Ørsted's aesthetics of nature that began with poetry, went on with music and painting, make a contribution to this debate? There are two ways to shed light on this issue. First we must examine Ørsted's direct contact with the painters of the Golden Age, and secondly we must take a look at their paintings to find out if Ørsted's aesthetics of nature have left a mark on them. As far as direct contact is concerned there is little to be found in diaries and correspondence. Historians prefer to have solid evidence to prove influence of one person on another to avoid accusations of pure speculation. Unfortunately, very few letters were exchanged between Ørsted and Danish painters of the Golden Age such as Dahl and Eckersberg and their pupils. Eckersberg's diaries do not reveal what he talked to Ørsted about in Paris at Christmas time 1812, although we know that they spent several hours in each other's company. Tangible proof stems from a few lists of participants informing us that thirty-two pupils from the Academy of Arts followed Ørsted's lectures in the Society for the Dissemination of Science in 1824–5, thirteen of them from the drawing school, among them Wilh. Bendz, Ditlev Blunch, Albert Küchler, Martinus Rørbye, Jørgen Sonne, and Chr. Holm. Twenty pupils attended his next course. In 1830–1 Constantin Hansen listened to Ørsted talking about the physics of the Earth, as we know from his letters to his sisters.

Wherever Ørsted lived in Copenhagen he was only a few minutes' walk from Charlottenborg, so why would he write letters to Dahl and Eckersberg, whom we know he counted among his friends? Ørsted was within close range, only a few steps separated Charlottenborg from Thott's Palace. Eckersberg left Copenhagen in 1811 and met Ørsted in Paris the following year, and when Dahl left Copenhagen in 1818 heading for Dresden, Eckersberg returned home and was appointed professor at the Academy of Arts, a year after Ørsted became *ordinarius* at the University. The best documentation relating to private relations between the Ørsted and Eckersberg families is supplied by little Karen asking her father's approval of giving her Christmas presents from papa in Paris to her playmates, Eckersberg's daughters. The presents were meant as thanks for the portrait Eckersberg had been commissioned to paint of her father. Now the paterfamilias was hanging on the wall soothing their grief at his long absence. 'When I'm out, I'm longing to get home to my picture', Gitte wrote, and the parents employed a private drawing teacher for little Karen.

Ørsted was close to Dahl, too, who accompanied Prince Christian Frederik, President of the Academy of Arts, to Italy. They always met when Ørsted was in Dresden, and in Copenhagen when Dahl was on his way to Norway. Together they called on Dr Carus, the painter, to discuss modern landscape painting (ch. 53). His letters to Gitte do not say what they talked about, but we know a little about the contents of Dahl's and Carus's talks. They dissociated themselves from the contemporary postcard 'art', which aimed at topological accuracy and a quick profit. Carus snorted at a public rambling in the countryside like illiterates in a library, and Dahl complained about stunted persons taking pleasure in breathing the fresh mountain air, lapping up sunshine and raindrops without the slightest reflection on the causes of these phenomena. It is the task of the landscape painter to open people's eyes to the beautiful in nature and to the laws behind it—the divine reason.[21]

AESTHETICS OF NATURE

Fig. 146. Christoffer Wilhelm Eckersberg's (1783–1853) drawing of himself, HCØ, Jens Peter Møller (1782–1854), landscape painter, and Arnold Wallick (1779–1779–1845), scene painter. Paris 1812–13. SMK.

Ørsted gave Martensen a letter of introduction to Dahl, and C.A. Jensen one to Herschel, who was going to sit for him in London. G. Bindesbøll, who followed Ørsted on his triumphal progress to the capitals of Europe in 1822–3, was a pupil at the building school of the Academy of Art and had followed Ørsted's monthly lectures. They had frequent discussions on changes of style in European architecture, for instance the breach of classicism by gothic. Later, Bindesbøll became included in the narrow circle of painters of the Golden Age in Rome. In fact, all the artists who appeared in Constantin Hansen's painting from Rome in 1837 had taken part in Ørsted's science lectures! It is therefore an idea that immediately suggests itself that the direction the compass would be pointing to, when Ørsted called upon Prince Christian Frederik to organize science courses for the laity, would be to the Academy of Arts in Charlottenborg. But yet another event makes the pieces fall into place almost by themselves. During Ørsted's triumphal progress the professors had introduced a reform of the education at Charlottenborg, which is hardly without connection to the Society for the Dissemination of Science, and the link was Prince Christian Frederik, President of the Academy as well as the Society.

Landscape painting was flourishing in Eckersberg's period. He had begun as a painter of historical themes and continued his training as such in France, but something happened in 1812 after encountering Ørsted in Paris. Napoleon's star plummeted and the hero worship faded away. In Rome Eckersberg started seriously painting landscapes in the open air. Worship of nature was the new fashion, and the reform at the Academy of Arts supported this new trend. So far training in drawing had taken place in artificial light, and consisted of copying other drawings, or drawing statues, or models (first only male ones, later female ones as well). While the history painting focused on heroes and heroines of mythology, the Bible, or royalty, and was commissioned by the royal court or the aristocracy, landscape painting opened the eyes of the spectator to the beautiful in nature addressing and educating the rising middle classes. Landscape painters taught the Danes to take an aesthetical look at nature, just as

Fig. 147. Constantin Hansen (1804–1880), 'A Party of Danish Artists in Rome', 1837. All the artists had been attending HCØ's lectures on natural philosophy. At the centre of the picture HCØ's travel companion, Gottlieb Bindesbøll (1800–1856), wearing a Turkish fez, is airing his aesthetical points of view, and obviously he enjoys the full attention of his friends. To the left Constantin Hansen himself, who together with Bindesbøll had been traveling Greece—the motherland of aesthetics—and the Ottoman Empire; hence he knew what was what. In the window Albert Küchler (1803–1886) and Ditlev Conrad Blunck (1799–1854) are listening, while Wilhelm Marstrand (1810–1873) is admiring the scenery. Martinus Rørbye (1803–1848) sits in the doorway looking deeply in the coffee grounds, while Jørgen Sonne (1801–1890) sits on the table puffing his long pipe. Bertel Thorvaldsen (1768–1844) has possibly been replaced by a dog on the chair to the right. Apart from their shared interest in landscape painting and HCØ's lectures on light, colours, and the hidden forces of nature, the seven artists were all commissioned by the Art Society in Copenhagen to paint pictures. SMK.

landscape architects and gardeners were shaping nature according to the aesthetic ideals at the time. As landscape painting advanced, perspective became crucial and Eckersberg published his textbook on linear perspective. While the old drawing lessons were black and white and concentrated on the structure of the motif, light and colours in nature and their representation on the canvas now gained significance. Landscape painting was facing new artistic challenges: a true conception of changing daylight, the theory of shadows and colours, and this in a double sense: the colours of nature depending on sunlight and the colours of pigments by means of which the painter expressed his observations. Eckersberg wanted to teach 'painting after nature', i.e. outdoors, in changing light, colours, and shadows.

AESTHETICS OF NATURE

Fig. 148. Christoffer Wilhelm Eckersberg (1783–1853), caricature drawing of the model school of the Academy of Arts, Charlottenborg, 1806 The Professor is chastising the shameful pupil, who has had difficulties in getting the awkward position of the model down on paper. Kobberstiksamlingen [Collection of Copperplates]. SMK.

The reform entailed three innovations at the Academy. First, painting lessons were now transferred to studios where daylight streamed in as a substitute for artificial light. Secondly, Eckersberg started organising excursions with his pupils outside the ramparts of the city, often in master baker Købke's carriage, and competitions were set on painting natural themes. Thirdly, and most important, what could Eckersberg teach his pupils on the light, colours, and shadows of nature? He could teach them to use their eyes, of course, but eyes only see changing empirical phenomena, not the laws of nature determining the incessant changes of light, colours and shadows. This was a branch of scientific knowledge and as far as the pigments of colours were concerned a problem of technical chemistry, subjects that had no professorships at the Academy. Who but Ørsted could supply that knowledge? And as the pupils of the Academy were not admitted to the University, they were referred to Ørsted's monthly lectures, as was Bindesbøll. After the reform Eckersberg could refer his pupils to Ørsted's public lectures under the auspices of the Society for the Dissemination of Science. Aided by Prince Christian Frederik, courses at the two places were coordinated so that every Thursday and Saturday from five to six o'clock the pupils would leave the studios in Charlottenborg and walk up Strøget to attend Ørsted's lectures.[22]

So what was the contribution of science to the art of painting? Above all Ørsted initiated the young artists into the theory of light, in the deflection of the spectrum of colours in the atmosphere. Everyday common sense cannot explain what colours actually are. Ørsted taught that colours are not properties of the things in themselves, but of the reflection by objects of the white light of the sun. The things are colourless per se, so that colours are born out of the deflections of the spectrum from different angles, red and yellow light is deflected the least, and the blue and violet the most, and the sky in itself is colourless. This is why it is able to assume all the colours of the rainbow varying throughout the day and the year. Ørsted also introduced meteorology, the theory of the rainbow and the secondary rainbow, the theory of the aurora borealis, the electromagnetic theory of thunder and lightings, the formations of clouds according to Howard, and plant geography and climate according to Humboldt and Schouw. From 1824 budding artists had to turn to the Society for the Dissemination of Science if they wanted to understand why the colours of nature are shifting from sunrise to sunset throughout the seasons, and what effects diffuse light and concentrated sources of light would have on the formation of shadows.

Ørsted's scientific contribution to the landscape painting of the Golden Age found its definitive form in the seventeen Sunday lectures on the Beautiful (which were actually delivered on Saturday evenings) from 1848 to 1849. They constitute the quintessence of Ørsted's aesthetics of nature, the sum of scientific contributions to aesthetics and his reflections on the relationship between science, aesthetics and ethics, or as he used to express it himself, the relations between truth, beauty, and goodness. They were taken down verbatim by speed-writers and later made into a fair copy.[23] They conclude that the highest reason is embedded in the laws of nature, the dynamics of which present man with his ideals of beauty.

Ørsted does not inflate himself by seeking to beguile his audience that as a scientist he is capable of handing out to them the appropriate rules of their artistic activity, for despite the progress made by modern science more laws of nature are concealed than discovered, and furthermore, neither the genius of art nor the genius of science works according to prescribed rules. Hence Ørsted warns 'the many magnificent artists and connoisseurs' he is pleased to recognise among his audience 'above all not to believe that he is aiming at prescribing laws to art'—and hey presto, a work of art emerges! Quite the opposite, he modestly encourages his listeners to object to his statements in order that they may discuss them to approach the truth together. Science is still in its infancy, and as far as the lecturer himself is concerned he is an art lover certainly, but this is far from being the same thing as a connoisseur.

The physical theory of light and colours (optics) is anything but common sense. It is about a complicated interplay between the light of the sun, its refraction from surfaces of objects of nature to our faculty of seeing, implying on its part a not yet fully understood communication between the eye and the mind. It would take us too far to go deeper into these theories here. Suffice it to say that it was precisely in Ørsted's day that important parts of optics began to be understood, and he was fully absorbed by the discoveries made by his friends Ritter, Herschel, and Fresnel. He had outlined his own hypotheses in *Ansicht/ Recherches* about the fundamental unity between the forces of electricity, magnetism, light, and heat. He had discussed Newtonian optics with Goethe and the polarisation of light with Thomas Young in London and with Fresnel in Paris, and although he did not produce original contributions of his own to optics he delivered several lectures in the Royal Danish Society, such as on morning and evening glows and on Northern lights. No Dane was as well acquainted with these matters as Ørsted.

The seventeen lectures were divided between general considerations on the aesthetics of nature and particular scientific knowledge on the nature of light and colours that are the result of the absorption or reflection of light from different surfaces, on the anatomy of the eye, on the mental effects of light and colours, on Goethe's doctrine of colours, on the waves of sound and light and their speed, on humans' intuition of colours, etc. When the lectures were over he thanked the audience for the comments 'that had given him much pleasure and brought him to ponder things more profoundly'.

Ørsted's contribution to the Danish landscape painting of the Golden Age, it seems to me, consists in encouraging the artists to express his philosophical dualism, heed the mimetic principle in their art, and thus enable the spectator to understand the eternal idea of the subject matter. This encapsulates seeing not only the subject matter as an immediate phenomenon, but also at the same time the natural laws behind it that radiate the sublime, the powerful, and the divine. We encounter this dualism in many works of art of the Golden Age, for instance in Chresten Købke's famous painting of *Frederiksborg Slot i Aftenbelysning* [F. Castle in the

Fig. 149. Chresten Købke (1810–1848). 'Frederiksborg Castle in Evening Glow'. Hans Tybjerg's thorough analysis shows that first Købke made drafts of the phenomena of nature on location as they presented themselves immediately to his senses. Then he painted his motif in his studio making the final painting stand out in noumenal ideality—corresponding to HCØ's aesthetics of nature. HS.

Evening Glow] from 1835. Hans Tybjerg's study of this masterpiece shows the dualism in the production of this painting: at first the many sketches of the subject as a phenomenon of immediate sensibility, and then the noumenal reflection added in the studio throwing a diffuse light over the castle and making it stand out clearly without shadows and sublime in heavenly light. The effect of light gives the viewer an impression of a coherent whole and seeks to inspire a sense of wonder about the inconceivable forces evoking this magnificent sight.

> He did not remember when Mr John Oerstedt for the first time
> had presented his ingenious idea about the unity of the forces,
> but in comparison the electrical telegraph was just foam on the surface.[1]

55 | 1846
Homage in Britain

ON BEHALF of the BAAS John Herschel invited Ørsted to attend the annual conference in Southampton in September 1846 as a guest of honour. He was now nearly sixty-nine and doubted whether he could stand the hardships. Hence, he invited Mathilde, his youngest daughter of twenty-one, to come along, and also Professor Forchhammer and his wife. As the tour developed it was Mrs Forchhammer who often had to stay in bed while old Ørsted coped without a problem. Gitte moved out of town with the two boys, Nikolai and Anders. She did not like the fuss that inevitably surrounded her husband everywhere he went. Her language skills were limited to a little German, not enough for conversation, while Mathilde turned out to be adroit at using her French.[2]

The four travellers embarked on the SS Geyser paddling them to Stettin, whence the train brought them to Berlin. There was a happy reunion with the Weiss family, but Humboldt, like so many other intellectuals, had left the capital during the holidays, so Ørsted could only leave his card. He enjoyed showing his daughter around, but this time Germany was only a series of Principalities to pass through swiftly, and they rushed through Erfurt, Gotha, and Fulda on the express mail to Frankfurt. Never had Ørsted paid so much turnpike money, and the collectors could not even be bothered to leave their sheds, but just handed out a rod with a small bag for the coins. From Frankfurt the small party went by steamer down the Rhine passing Lorelei and Bonn where they called on Dr Brandis, the former court physician who had prescribed Anders and Sophie a cure at Driburg and still practised a kind of mesmerism.

But the cathedral of Cologne was too fascinating for Ørsted to rush by without seeing it. He told his escorts about his previous visit when Goethe and the Boisserée brothers had helped

Fig. 150. Mathilde Elisabeth Ørsted (1824–1906), HCØ's unmarried daughter, who helped him as a secretary and published his Danish correspondence in 1870. Daguerreotype c.1846. RL.

him deepen his insight into Gothic architecture. Mathilde, not least, knew the cathedral already, for lithographs of it embellished her parents' drawing room in Studiestræde. In the meantime, however, he had adopted a more critical view of the matter. He still admired its architectonic spirit. It evoked the appropriate solemn atmosphere, but now it seemed to him that the original plans of gothic cathedrals were overambitious. They were never completed. The building of the cathedral of Cologne had now lasted six hundred years, and it was still unfinished. Like Notre Dame it lacked its spires, and the cathedral of Strasbourg still also lacked one. The construction of it exceeded the powers of the church, the state, and the people. Ørsted criticised the Romanticists' propensity to prostrate themselves before the taste of the Middle Ages. Catholic projects that could only be carried out by means of income from the sale of indulgences seemed to Ørsted to be repellant examples of irrational and unrealistic thinking of the past, and not worth emulating as the Romanticists did. The middle ages had surrendered to fantasies running riot because they were incapable of taming them with science and technology. Unfortunately, modern thinking had not yet succeeded in establishing its own artistic expression, except in the form of cast iron bridges and steam engines. But that would come, he confided to his diary.[3]

*

After this confession to rationality the four embarked on the train for Aachen and Brussels and on to Paris. They were accommodated in a hotel near Palais Royal in a quarter where Ørsted felt at home. Here, too, scholars were on holiday, so without a bad conscience he could spend the time as a tourist guide to Mathilde and the Forchhammers. First they walked around in the Jardin des Plantes where a zoological garden was set up and they saw wild animals from exotic lands. Afterwards they looked in the windows of many luxury shops at Palais Royal which attracted Mathilde and Mrs Forchhammer. Ørsted was invited to join a meeting of the Scandinavian Society, but declined. He did not like the prospect of being caught up in political discussions. He explained to Haxthausen from the Danish Embassy that unfortunately, like the Society for the Right Use of the Freedom of the Press, the Scandinavian Society in Copenhagen, despite the wishes of many members, had turned into a political debating society, and hence he turned his back. In return, the friendly diplomat organised four tickets for a royal concert in the Tuilerie Gardens.

The Gardens were illuminated for the outdoors concert. The Danes had seats right next to where the Citizen King Louis Philippe was enthroned on a balcony with his family. While the orchestra played the first piece, a shot sounded, and the horrified audience saw the King point in a certain direction. The music went on. It was difficult to see and hear what was happening. Was the shot part of the show or was it an attempt at assassinating the King? The evening culminated with a roar of fireworks that almost deafened the crowd's '*Vive le Roi!*'. Only the next day did the newspapers report that a shot had been fired, but it missed, and the King's gestures signalled that he was unharmed. The criminal was apprehended at once, a worker by the name of Charles Henry. The police found cash in his pocket. Was it his thirty pieces of silver?[4]

The travelling companions stayed for almost a month in Paris. The departure was delayed because the Danish diplomats attempted to obtain an audience with the King.[5] But this was not the only reason. Mathilde and Mrs Forchhammer had joined the tour to experience sightseeing in Paris. A voyage down the Seine to St. Cloud and Versailles was indispensable. They saw the art collection in the castle, the big fountains in the park, and they enjoyed the views. In the Opéra Comique they attended a performance of the *Queen's Musketeers* that was supposed to be funny, but they found it boring.

The Danish party dined with Monsieur Guizot, the prime minister, but as nearly all guests wished to talk to the host, Ørsted only managed to express a few formal remarks. A few days later they attended a party at Monsieur Salvandy's, the minister of education. At the dinner table Ørsted was placed close to Victor Hugo, but they exchanged only the usual phrases of politeness. The diplomats also arranged an invitation from the prefect (mayor) of Paris, Count Rambuteau, to dine at the City Hall. They entered into a fine conversation and a few days later the small Danish delegation were given a guided tour of the renaissance palace by Count Rambuteau himself and were also invited to dine in his private home. To him Ørsted was not just a name, but a man of whose fame he knew the cause.

Finally, the day came when the busy Citizen King had found time to receive Forchhammer and Ørsted. Louis Philippe was well prepared and soon initiated a conversation about the Roskilde Assembly of the Estates and how brilliantly his Danish guest was leading it. Ørsted discretely indicated to the King that perhaps he mixed him up with his brother, the renowned jurist. The King was slightly perturbed and changed the subject. Instead he started to talk about his vacation in Denmark and Norway interlarding his story with a few Danish words. The empty talk made Ørsted feel a fool to have wasted so much time waiting for this nonsensical situation, but he was comforted by the thought that at least Mathilde had had the tourist experience of a lifetime.

*

The next day they left Paris by stagecoach, the wheels of which could be taken off enabling the coach to be loaded onto a train; the travellers got their wheels back at the terminal station of the railway, the town of Arras more than half way to Calais. There they became the victims of two competing shipping companies' persuasive advertisements. A Frenchman offered a passage by a steamer bigger and faster than that of the British rival promising that the Danes would arrive at Dover first, although the British steamer would be leaving half an hour earlier. Ørsted did not trust him and embarked on the British ship. When they sat down in the train leaving Dover a few hours later, they still could not spot the French ship in the Channel. The railway took them to London at fifty kilometres an hour. When more than twenty years before he had travelled between the two cities for the first time, the means of transportation had been reminiscent of that used in the Roman Empire. Now steam power had reduced the time drastically. They had arrived in the world's leading industrial nation where everyone was busy, and everything went quickly, but all the same Ørsted found peace and order. He accommodated himself and his companions in the Hotel Sablonière on Leicester Square, lodgings he had recommended to Andersen who had been happy to stay there. The sun would shine right onto my bed, he wrote.[6] But to Reventlow, the Danish ambassador, this was not fashionable enough. It was bordering Soho, the quarter of prostitutes, he said. But to Ørsted this was good enough. Newton had lived in Leicester Square, too.

They had two weeks at their disposal as tourists in London, before the BAAS opened the conference in Southampton. First they went to the Danish embassy at Westminster to get cash, then they took a walk in St James's Park, where the swans were black, then along Piccadilly and Regent Street, where Mathilde was mirrored in the tempting display windows, through Hyde Park to the Polytechnic Institution as far out as Regent's Park. The Polytechnic Institution had not existed when Ørsted was in London twenty-three years earlier. He had never seen anything like it. It turned out to be a popular exhibition set up to make money on experiments and inventions. Ørsted had always been keenly alive to the entertainment value of science and took in new impressions everywhere, because to obtain funding for the sciences they must catch the public interest. He had been enchanted by Ritter's spectacular experiments and throughout his life his own galvanic experiments and acoustic figures had cast a spell. But here at the Polytechnic Institution the showmanship had gone too far. Entertainment triumphed to the detriment of the scientific spirit, and Ørsted considered the exhibition a sacrilege.

The British commercialisation of science sold tickets. In fact the Polytechnic Institution was an imitation of Adelaide Gallery that had created the concept and carried it into practice in 1832. In 1836 these shows had made such an impression on Scharling that he had persuaded the Society for the Dissemination of Science to transfer the idea to Studiestræde (ch. 48). Coincidentally, it was an old acquaintance of Ørsted's, the American Jacob Perkins, with whom he had discussed steam engines in 1823, who had set up the establishment. The main attraction

Fig. 151. Adelaide Gallery, London. Drawing by Thomas Kiernan, The John Johnson Collection. Bodleian Library, Oxford University.

was a steam driven machine gun he tried to sell to the British army arguing that if the Duke of Wellington had had this weapon at this disposal he would have beaten Napoleon at Waterloo in an hour instead of wasting a whole day. When it turned out that the Adelaide Gallery was far from able to satisfy the enormous demand for entertainment, it acquired a competitor.[7]

Among the curiosities in the Polytechnic Institution was a pneumatic telegraph, a model of the tunnel under the Thames, and an 'electric' steam engine igniting powder that did not explode, however, at any rate not on the day Ørsted was there. He felt duped:

> 'It is hard to think of a worse mixture than the one we are witnessing here. Almost all the time one heard loud music and a numbing noise from machines and human voices. The lighting was strong, but situated so stupidly that at several places one could not see anything. After the spectators had walked around for a while a lecture on the electromagnetic telegraph was announced. Now everybody rushed to a gallery, but there was not enough room for everybody and of those who stayed, a quarter could neither see nor hear.'[8]

In his note book Ørsted expressed his indignation as follows: 'In the guidebook *New Picture of London* the Adelaide Gallery is said to be intended to display the sciences and their applications; but since it did not attract enough people, the investors resorted to simple treachery against science. This strong expression for the degradation of science is highly appropriate.'[9] When a few years later the great exhibition in the Crystal Palace became all the rage, far exceeding the expectations of the organisers, this was partly due to the preparatory work done by these two places of entertainment prompting a public yearning for sensations. Crystal Palace opened only a few months after Ørsted's death.

The next day the four walked down the Strand passing by the Crown and Anchor and Somerset House, where the Royal Society in 1823 had celebrated Ørsted. Davy had died long ago, but Charles Babbage was still in full vigour with his lifelong project to build a calculating engine. Ørsted knocked on his door and was immediately asked to lunch the next Sunday and when Babbage learned that Forchhammer and the ladies were with him they were invited as well. Who else should be invited? Ørsted suggested some other European scientists heading for the BAAS conference and told Babbage that just now he was on his way to the Colosseum. Babbage offered to escort them.

The Colosseum was another new tourist attraction. It displayed a panorama painting of London of no less than four thousand square metres (bigger than sixty-five times sixty metres) showing the view from the top of St Paul's Cathedral. In the middle of this huge picture there was a lookout turret, and the higher one went the wider the view. This sounds impressive, but Ørsted was disappointed and descended to find out if there was anything else to look at. There was. On the floor copies of English sculptures were on display accompanied by organ music; everywhere were tables with refreshments. Through a window a waterfall could be seen, the water of which fell thirteen metres to drive a mill. So it looked, but Forchhammer quickly realised that it was an optical illusion. The entire exhibition was organised to ensnare the senses by means of optical tricks. A subterranean walk led them to a stalactite cave in which a glass cylinder was displayed containing a male body incessantly changing form like Proteus, the Greek god of the seas. There were magical light effects and magic mirrors making the big look small and vice versa. Everything was a search for sensations. Ørsted found that the atmosphere of this public entertainment undermined the reason of the Enlightenment. A pedagogical opportunity had been sacrificed in favour of profit making. Once again Ørsted felt betrayed by the mummery 'that must have cost incredible sums, but nonetheless survives by the sale of tickets'.[10]

Ørsted, Forchhammer, and the ladies had seen elephants, giraffes, lions, and tigers before, but only in pictures, never in flesh and blood. True, there were occasional displays of wild animals in the Deer Park north of Copenhagen. Ørsted's colleague, the professor of zoology, Reinhardt, had obtained the police's permission to show chained kangaroos, gnus, and tigers, but no one knew how dangerous they might become, and when a predatory animal escaped, Frederik VI forbade them.[11] The last time the two scientists had been in London there had been no zoo, but in 1828 a consortium had been given a corner of Regent's Park where wild beasts and birds were caged in to amuse the public. Copenhagen had not come so far. Only in 1859 a few birds of prey and a seal in a tub were exhibited in the Danish equivalent of Regent's Park at Frederiksberg. In the Royal Chambers of Art there were stuffed animals, but no exotic beasts. Now for the first time they saw real live lions, tigers, panthers, hyenas, bison, elephants, giraffes and a single camel, even a rhinoceros, of which Mathilde did not know the name in Danish.[12] There was also a clever galvanic invention. The garden had some rare plants which were a temptation to snails that liked to eat them. To protect the plants the area was surrounded by a zinc plate ten centimetres high with a copper edge on top. When the snails approached the delicious plants, their slime came to touch both metals, and a galvanic shock put an end to the greedy molluscs.[13]

At the British Museum they saw the remnants of prehistoric bones of a mastodon associated with P.W. Lund's finds at Lagoa Santa in Brazil, and Forchhammer had an opportunity to make great play with his knowledge of palaeontology. There they also saw Egyptian mummies, including 'the wig of a lady abounding with curls arranged with such taste, that one would believe it was made at the present time.'[14] One day Mathilde went with her father to the British Museum to see the Rosetta stone, and Ørsted told the story about his friend Thomas Young and how he had deciphered the hieroglyphs by comparing them with the two other inscriptions of the same contents, but in different languages and letters. They also saw the solitary caryatid and the Parthenon frieze that Lord Elgin had cut out from the Acropolis.[15]

Ørsted recalled the great experience he had had in 1823, when he, Herschel, and Babbage had sailed down the Thames with the Royal Society to inspect the Greenwich Observatory (ch. 39). Now he suggested to Forchhammer and the ladies that they repeat this tour. They took the steamer from Charing Cross, went under the cast iron bridges past Somerset House, above the tunnel that was not there the last time, and past the Tower of London. There was such life on the river, a forest of masts and a multitude of steamers they had to constantly circumnavigate. At Greenwich it was not so much the astronomical observatory that caught Ørsted's attention as the meteorological and magnetic one that was a link in the worldwide chain of which his Polytechnic Institute was another. Just as they were watching a straw-electrometer, there was a rumble of thunder and the sky clouded over. The electrometers worked beautifully. In passing he was introduced to Sir William Hamilton from Dublin, entirely unknown to him. As it dawned upon the Irishman whom he was facing his face lit up, and he told Ørsted that he had just borrowed *Recherches* from a friend to read at home. This was the second time on the tour he had the satisfaction of meeting a stranger who was reading his major work.[16]

Then the Sunday arrived, when they had been invited to lunch (or 'lundge' as Ørsted spelt it in his diary). Babbage turned monomaniac when he met people who ought to show interest in his calculating engine. Grundtvig as well as Ørsted had had to put up with him earlier. The machine Ørsted had been shown last time was still not completed and probably never would be. In the meantime Babbage had thought out an entirely new principle. The government had granted seventeen thousand pounds for his first project, but since the new principle was both better and simpler, it would be a waste of time to finish the first one. Babbage could not convince the government, but worked on the new version, uncertain as to whether his project could be financed and completed. His four guests found his account terribly boring and they were relieved when they were released four to five hours later. Babbage presented each of his guests with a part of the first, unfinished machine asking them to cherish it as a relic.[17]

Fig. 152. HCØ at sixty-nine, obviously unhappy in the situation. To obtain a fine portrait Ørsted had to put up with being powdered on his face to hide the old and grey parts of his skin. During exposure for twenty-five seconds he had to sit as quiet as a mouse. He was hardly as good at sitting naturally and relaxed as the vainer HCA (ch. 58). A. Claudet, London. DTM.

One day Ørsted passed by London's leading daguerreotypist, Antoine Claudet. They went inside, and Claudet asked permission to take a picture of the celebrity. Ørsted was handed some books to symbolise that here was an erudite sitter who had spontaneously been interrupted in his studies. He soon took four portraits presenting Ørsted with one of them. It is exposed and fixed on a silver plate, and Ørsted looks terrified as if he feared a bodily attack by the camera obscura. 'No doubt, this is the most alike of the daguerreotypes of me. A little stiffness and alienation is unavoidable', he remarked.[18] We notice his not particularly well-sitting toupee, his squinting left eye, and his receding chin, but the angle has the effect of making his nose (which Kamma Rahbek found so remarkable) less conspicuous. The following year John Herschel told him how it was possible to make colour photographs, and Ørsted immediately passed on the news to his audience while encouraging J.A. Jerichau, one of his polytechnic students, to take up the research.[19]

Now the British members of the BAAS began to arrive in London to go on to the conference at Southampton, and dinner parties were arranged by Leonard Horner, the President of the Geographical Society, Edward Sabine, who worked with Herschel in the earth magnetic mapping project, as well as Lyell and Murchison, the geologists. Ørsted was also asked to visit Whewell in Cambridge after the conference, but this was prevented by the GDNÄ conference at Kiel. He did not succeed in meeting Faraday either. He was told that Faraday would call on him at the hotel on Leicester Square a certain Monday, but then Ørsted had gone to Greenwich.[20] Afterwards he had tried to pay him a visit at the Royal Institution, but only managed to have his assistant show him the big horseshoe-shaped electromagnet that Ørsted subsequently copied for the Polytechnic Institute.[21] They could not avoid seeing each other at Southampton, for Faraday was chairing the chemical section.[22] Ørsted's meticulous diary does not mention him at all during the conference. But he probably made the appointment that Ørsted would see him at the Royal Institution before he left England, for the first day after his return to London Ørsted showed up for two hours in Faraday's laboratory to see some of his diamagnetic experiments.[23] Faraday did not show his bravura piece consisting in placing a coal box replete with iron filings under the big electromagnet. When the circuit was closed iron sparks ascended like fireworks. This orgy of light was best enjoyed in total darkness. Even the serious Faraday deigned to perform as a showman. He had learnt it from Davy, for the Royal Institution depended on a public in a strong financial position that would be attracted to it like iron filings to a magnet.[24]

*

On 10 September the four travelling companions went by train to Southampton where private accommodation had been arranged with George Jones and his family in Portswood House. He was the mayor of Southampton and utterly delighted that the BAAS had chosen his town for the conference that year. The BAAS meant prestige and a royal visit, and both brought business and publicity.

Fig. 153. Michael Faraday (1791–1867) lecturing in the Royal Institution, London, c.1856. Coloured lithograph by Alexander Blaikley (1816–1903). Bridgeman Art Library, London.

Fig. 154. Michael Faraday's great electromagnet, which HCØ copied for his collection of instruments at the Polytechnic Institute, Copenhagen. *NS* II 552–555. DTM.

The BAAS had been founded in 1831 taking the Prussian GDNÄ as its model, but was not a direct imitation. It was prompted by the dissatisfaction of a number of liberal members with the conservative line dominating the Royal Society. To divide the members into Whigs and Tories might make some believe that the same line of separation applied to the Royal Society as to the House of Commons. So it was, but of course the disagreement concerned the policy regarding science conducted by the leadership of the Royal Society. The conservatives were blamed for what Babbage called the decline of British science, indeed his book *Reflections on the Decline of Science in England* was precisely an attack against the right wing. Its high esteem of the amateur was determined by the aristocratic attitude that if voluntary activities such as science and sports were changed into professional and salaried work these activities would no longer be cultivated for being pleasurable or noble in themselves. This attitude had died out in Denmark with Court Steward A.W. Hauch.

But in the Royal Society this schism was acute. Firstly, practising researchers were a minority among the members; allegedly, the majority were there out of snobbery in order to boast of

Fig. 155. Faraday's box—another present containing various materials such as glass pearls, types of wood, and metals to examine diamagnetism. Diamagnetism is the name of the effect made when both magnetic poles repel certain substances. The effect also arises when magnets influence polarised light, since the magnetism turns the plane of the polarisation. HCØ demonstrated these experiments in the KDVS, *Oversigter* [Transactions] 1847 (*NS* II 550–552), 1848 (*NS* II 568–574), and 1849 (*NS* II 574–581). DTM.

the title FRS—Fellow of the Royal Society. Many of these members were general practitioners of medicine abusing the title to charge their patients exorbitant fees. Secondly, the society was claimed to be run by an aristocratic clique, as opposed to the whiggish conviction that it ought to be meritocratic, and thus run by distinguished researchers. The Tory clique in power had never done a stroke of work to make the House of Commons grant public money to advance science and technology. The Royal Society was too proud to receive public support believing that it would spoil its honourable idea of the amateur. It was this arrogance, Babbage wrote, that caused England to come staggering up behind France, the arch enemy, and Prussia, its ally. Science was scarcely cultivated at Oxford and Cambridge. The few existing research institutions such as the Royal Institution or Mechanics' Institutes were established and financed by industrial money. Finally, the Tories were reproached for not have disseminated science and technology to the people and the provinces.

The BAAS was founded to remedy these failings: to enhance research, to raise public money and to arouse the interest of the public. Among the initiators was John Herschel, who had gone to South Africa to study the southern hemisphere in disappointment at having been rejected as presidential candidate by a small majority in favour of the Duke of Sussex. Babbage, whose project aimed at applying mathematics to rationalising industrial production, advocated a utilitarian goal for science, leading the way with his calculating engine. Whewell fought to introduce scientific subjects at Cambridge. The Whigs' reform policies promulgated in 1832 were quite down-to-earth and utilitarian, and the founders of the BAAS found that scientific research needed proselytes among industrialists in the Midlands. This meant they had to become convinced that a strengthening of the sciences would provide progress for industry and people. This implied increased professionalisation and division of labour, leading to BAAS conferences being divided into sections for the disciplines as the Prussians had done. On one point the BAAS diverged from the GDNÄ; learning from bad experience in the Royal Society there would be no section for physicians. They were allowed to remain members, but their numbers would be reduced by a new set of statutes. The BAAS was no mere substitute for the Royal Society, but a supplement with new purposes.

In 1837 Britain acquired a new sovereign, Queen Victoria, and the Royal Society a new President, the Marquis of Northampton, who was not an active researcher, but was clever enough to pursue the middle course, supportive of the BAAS's purpose and allowing the society to live on in its exclusivity. It might be fair to say that while the BAAS became the liberal House of Commons of the sciences, the Royal Society remained the conservative House of Lords. The new President saw to it that Ross's arctic expeditions and Sabine's earth magnetic observatories in the colonies were paid for by the state, as the Danish Galatea expedition had been. In this new climate Herschel decided to return from South Africa to take an active part in the BAAS.[25]

Although Ørsted must have known about these conflicts and the political atmosphere he was likely to meet in Southampton, he did not give vent to his own judgement in his diary or letters. Did he find parallels with the Danish scene, and how did he respond to the British politicisation of science? Ørsted still had the ability to organise the First Chamber of the sciences, the Royal Danish Society, the honorary lodge of selected professors, as well as their Second Chamber, the Society for the Dissemination of Science addressing the laity and the provinces. Both institutions enjoyed the protection of the King, and the British problem of physicians did not exist in any of them, because in Denmark they were organised separately in the Royal Medical Society. Ørsted's entrepreneurial initiatives as regards the Polytechnic Institution and the Scandinavian Science Conferences had contrived to ensure that nothing new that took place in Europe would be missing in his own country.

Of course, Ørsted indulged himself in the warm reception he was offered in Southampton. Murchison, organiser of the conference, welcomed Ørsted and Forchhammer introducing them in a way that could have made them feel like high priests in the holy temple of science. Murchison alluded to the pride and joy that must have filled Ørsted when he landed in Britain seeing that his great discovery had been followed up by Faraday, so that one was now aware that diamagnetic forces (a weak magnetism provoked by strong electrical tensions) were generated ubiquitously in nature. Then he thanked the King of Denmark, the knowledgeable patron of science, for sending Forchhammer as well, the maker of the first geological maps of Denmark. Murchison thanked them for the hospitality he had encountered in Odin's realm two years before and reassured the two Scandinavians that the hearty feelings were mutual.[26] Next the Foreign Secretary, Lord Palmerston, and the President of the Royal Society, the Marquis of Northampton spoke. Prince Albert was an honorary member of the BAAS and present for the first time. Afterwards Ørsted was introduced to the Prince, and it seemed to him that the Prince was 'somewhat confused about substance for a conversation'.[27]

A few days later there was a festive dinner and Ørsted was given the seat of honour between Herschel and Murchison. His conversation with Herschel was not particularly successful, because the old astronomer was hard of hearing and had slurred speech for want of several teeth, Ørsted thought. The dinner started chaotically, for someone had forgotten the wine, which only arrived after a couple of dishes, and then toasts were proposed for Queen Victoria and Prince Albert. And then for Ørsted, rising and asking for indulgence, although as he said it did not take much eloquence to give thanks for the touching hospitality that had met all foreign scientists. He wished the British society good luck in the advancement of science for the benefit of humanity. All must assume this task and solve problems through international cooperation. He stretched out his hand in the spirit of brotherhood. Ørsted's speech was no more turgid than any other, and no doubt sincere.[28]

There were seven sections: for mathematics and physics, chemistry, geology and physical geography, botany and zoology, physiology, statistics, and finally mechanics. Ørsted had brought a not particularly sensational paper on a problem with mercury that after several years of storage in a hermetically sealed glass received a dark membrane that one would immediately believe was mercuric oxide, but if so where did the oxygen come from, for the glass was airproof? The paper was read aloud by a native speaker in Faraday's section for chemistry. Ørsted took long walks with Herschel airing his idea of giving another paper, which would be more relevant to the theme of this annual meeting, the worldwide project on meteorological and earth magnetic observations. Ørsted proposed that the BAAS arranged a series of experiments to decide if the falling of bodies from great heights declined towards the east and possibly towards the south in relation to a vertical line of falling. Various uncorroborated historical experiments, among others in mine shafts, seemed to suggest that Galileo's laws of falling bodies are invalid due to the rotation of the earth, but perhaps the electricity induced by earth magnetism in the falling body could explain the aberration. To clarify these ideas Ørsted proposed experiments in very tall vacuum cylinders. He was aware that they would be costly, and he therefore suggested they be undertaken as a common European project. Herschel encouraged Ørsted to proceed with his paper so he put down his ideas on paper in Portswood House.[29] His colleagues applauded the idea, while others were sceptical.

A whole day was reserved for an outing to the Isle of Wight for the ladies. Mrs Forchhammer and Mathilde accompanied the three daughters of the Scot, Leonard Horner, who chaired the geological section. They were Susan, Joanna, and Leonora, and they were well-educated young women interested in botany, geography, and history. They were skilled in languages, German in particular, which was rare in Britain then. They had picked up German naturally during a long sojourn in Germany with their family. On the Isle of Wight Mathilde made friends with these sisters. They conversed in German, exchanged presents, visited each other in Bloomsbury and later in Copenhagen, and a few years later Joanna Horner took on the assignment of translating the German version of *The Soul in Nature* into English.

At their departure the mayor thanked the BAAS, because the society had chosen to hold its annual meeting at Southampton. Never before had the city received so many noble visitors

Fig. 156. Leonard Horner (1785–1864) chemist and mineralogist, FRS and FGS, and his wife. Leonard Horner studied in Bonn during the 1830s, where his daughters Leonora and Joanna learnt German. DTM.

(hear, hear!), and he was sure the conference would have a lasting effect. When he looked around in the audience he spotted a Herschel who had discovered a new planet, and an Ørsted whose discovery had made his name soar all over the globe. Southampton was really honoured. Herschel replied that even if Britain was strong in science it was just one of many nations competing to bring civilisation forward by disseminating science to the whole world (hear, hear!). England was taken to be an island, and so it was geographically, but he repudiated this island status as far as the connection to European science and civilisation was concerned (hear, hear!). A bridge had been built spanning the former gap between Great Britain and the Continent. The human spirit has conquered the physical substance (hear, hear!). He did not remember when Mr John Oerstedt for the first time had presented his ingenious idea about the unity of the forces, but in comparison the electrical telegraph was just foam on the surface (hear, hear!). In electromagnetism Mr Oerstedt had discovered a principle that would develop incessantly in new directions, fascinate the eye and delight the heart...[30]

This was a moment of emotions. Ørsted had never met such an unreserved acclaim in such a friendly atmosphere. What to say? He was quite overwhelmed by feelings of gratitude and apologised for not being able to find adequate words in English. He had tried to learn English by reading the greatest poets of the country (Shakespeare, Walter Scott, and Oliver Goldsmith were in his library), but he had only been in England for a short time and was lacking the skills necessary to address the present, noble audience. He was personally touched, but more proud on behalf of the great scientists of Britain, and he had no doubt that England's leading position in the world was due to the sciences and the encouragement and dissemination they enjoyed in this society. Hence it was a great honour to be among its best cultivators in this temple of learning (cheers! cheers!).[31]

The four travelling companions had been away for two and a half months. At first ten days through Prussia and other German states at high speed, then four weeks in Paris and two in London followed by a week in Southampton, the summit of the journey, for Ørsted to find at the end the anticlimax in Kiel already mentioned (ch. 51).

> Teaching young people to understand
> the laws of nature in the mathematical way,
> will make the large majority get used to a one-sided view
> that will make them lose the intended harmonious development.[1]

56 | 1840–50
Polytechnic Criticism

IN 1844, at long last, Ørsted came out with his revised version of *Naturlærens Mechaniske Deel* 'The Mechanical Part of Physics'. He had promised it to Frederik VI as early as 1815 and had received grants to write it (ch. 31), but did not complete it until now, almost thirty years later.[2] It was an updated version of his textbook from 1809–11, it had more than three hundred pages divided into paragraphs and cost two rixdollars and forty-eight shillings. An anonymous reviewer in *Kjøbenhavnsposten* pointed out that Ørsted's language was almost void of loanwords and hence understandable for the laity. This sounded rather harmless, but was meant to be ironical. The critic thought a textbook on physics should be based on mathematics and took to parading his expertise by flinging the finest loanwords about in his review.[3] A month later a more detailed review followed. Ørsted's formula for the centrifugal force was wrong, as it had been also in the 1809–11 edition. Ørsted was no longer handled with kidgloves; not only was he accused of being a bad practitioner, obviously he was also a poor theoretician and an irresponsible teacher, for when he examined polytechnic students according to the wrong formulas of his textbook, then they must by consequence have graduated on false premisses. Had they been examined at Cambridge or in Paris (or at Irkutsk, the reviewer sarcastically added) they would have been bound to fail. If they practised on this flawed basis their technology would break down.[4]

In Christiania Christopher Hansteen was reading the Danish newspapers. He was a proficient mathematician and had to agree with the reviewer, although he found the gibe indecent.[5] He told Ørsted how centrifugal forces should be formulated in mathematical language, and

Ørsted asked C.V. Holten, the physics reader at the Polytechnic Institute, to come forward in the press, not in *Kjøbenhavnsposten*, which had compromised itself by giving space to the rude review, but in *Fædrelandet*. However, Holten's rejoinder did not halt the criticism. Apparently, the anonymous reviewer was involved in some captious campaign against Ørsted, even if he gave assurances in elegant turns of speech that he was only defending objectivity and precise mathematical physics. Holten was defending a bad cause in vain.[6] Ørsted was silent. The cantankerous critic was more and more insistent reminding the public of the motto above the entrance to Plato's Academy: 'Let no one ignorant of geometry enter this academy!'. Yet, at the University of Copenhagen the physics professor delivered incorrect lectures not only on centrifugal force, but also on the calculation of the elliptical orbits of the planets. Section 211 of 'The Mechanical Part of Physics' was strikingly at odds with Newton's *Principia mathematica*; Ørsted was called scandalous, but even so the author remained silent.[7]

The anonymous reviewer, who signed himself θ [theta], was outraged. Tycho Brahe was his hero, and when his hero was the hero of all Danes, this must be due to some national hypocrisy, according to θ, for Tycho Brahe had been persecuted and condemned by his colleagues at the University, indeed his rival, seneschal Christoffer Valkendorf, managed to have a committee declare his studies not only useless, but outright harmful. Unfortunately, it was not the ingenious Tycho Brahe who spoke on behalf of Danish science, but his envious and incompetent adversaries. Now the critic saw a repetition of the old story as in Hans Christian Ørsted's case. He was considered the highest-ranking physicist in the country, but he did not even know the formula for centrifugal force let alone the elliptic orbit of the planets. Laplace and Brewster considered old Valkendorf a disgrace, but as late as the inauguration of the new University building in 1836 Valkendorf had been lauded in a cantata. But, had not he defrayed the expenses of a college? O, yes, but for the benefit of Tycho Brahe's paltry foes, not for his pupils. Had Ørsted repented his mistakes? No! Had the erroneous sections been rectified? No! Then θ pointed out some more mistakes.[8] Ørsted did not respond, but his quarrelsome adversary had made a few points and would not let them be silenced.

More than six months passed without succeeding in getting Ørsted to admit his blunders to the public. So, θ sent an open letter to the Vice-Chancellor asking six questions:

1. Is it true that the three false formulas are taught at the University?
2. Is it true that they are printed in textbooks used at the University?
3. Is it true that the three formulas mentioned are false and misleading?
4. Is it true that no respectable professor in Germany, France, and England would do anything but reject them?
5. Is it true that the equivalent true formulas are taught at the Military High School?
6. Is it true that the University has taken no steps to correct the errors?[9]

There was no answer. The more choleric the behaviour of θ (and his awareness of other people's errors seemed inexhaustible) the more phlegmatic the learned republic must have seemed to him. Therefore he started to threaten. So far he had been lenient only pursuing the truth, he wrote, but if the Vice-Chancellor showed no interest in defending the credibility of the University, he would not hesitate to internationalise the scandal, for instance by informing Arago, who would probably deprive Ørsted of his status as a foreign member of the Académie Royale des Sciences. '*Errare humanum est—perseverare diabolicum*'. Ørsted's and the Vice-Chancellor's silence had provoked this menace. Ørsted's textbook was now being translated into German.[10] Should Denmark's scientific honour be compromised there as well? Wicked tongues hinted that the anonymous reviewer was driven by a personal hatred against Ørsted, but they should bear in mind that it is never the person removing the cancer who deserves the blame. θ was merely guarding scientific integrity, nothing else.[11]

The campaign was sheer blackmail. Ørsted's public prestige risked being maligned at home and abroad. He was hurt, of course, but not knowing who was hiding behind the cowardly mask of anonymity he was reluctant to respond. However, the Vice-Chancellor had asked Hansteen for a report, which was published in *Fædrelandet*. As before, Hansteen emphasised that there were mathematical errors, but that they belonged in the department of trifles, and that the attack widely overshot the mark. This was a quite insufficient admission in the eyes of the critic, almost a kind of collegiate certificate of fame for the discoverer of electromagnetism. In this connection θ wanted to point out that Ørsted's theory of electromagnetism was wrong as well, but this had nothing to do with 'The Mechanical Part of Physics', of course. In addition, his *experimentum crucis* (ch. 44) had confirmed Ampère's mathematical theory rather than refuted it as was Ørsted's intention.[12] But no matter how often and how ruthlessly Ørsted's reputation as a physicist was slated, the anonymous reviewer did not succeed in getting his victim to appear in public and confess his guilt.

Finally, almost two years after the first attack, θ called upon his victim in Studiestræde and threw aside his mask. Behind it appeared the Icelander Thorleifur Gudmundson Repp, an old acquaintance of Ørsted, a failed master of arts and now a translator into English. He had been present in the auditorium in Nørregade in 1820 when the Students' Association was founded. He was bright, no doubt, and like Ørsted he had been awarded the gold medal twice for a philological and an aesthetical prize essay submitted anonymously, so cheating was out of the question. He had been a frequent guest at Bakkehuset, where he had helped Knud Lyne Rahbek study the Icelandic sagas, but Kamma had found his hot-headed and self-centred character rather trying. Ørsted recognised him easily at once, because he had expelled him from his own examination for the master's degree in 1826 due to scurrilities, a scandal that stuck to him for the rest of his life and which gave him a permanent sense of injured pride.[13] Ørsted had to admit that he was hurt by the attacks and that they had disturbed his sleeping, and urged Repp to terminate them, because he wanted peace to work. As already indicated by Hansteen and Holten, there were errors in the textbook and they would be corrected.[14] Ørsted's only apology was that he did not consider himself a mathematical physicist, but an experimental one as a matter of principle, because if physics was to be disseminated it had to be made understandable to people with no knowledge of differential and integral calculus. Now the seventy-year old Ørsted hoped to find peace.[15]

But all he had obtained soon turned out to be a fragile armistice. In August 1847 it broke down, when Repp sent Ørsted a letter complaining that the truce had been to the one-sided advantage of Ørsted. θ himself had not received what he had come for and what had kept his passion burning throughout his campaign. What Repp wanted was an academic degree, and

Ørsted had the power to confer it upon him. Now Repp had rescued Ørsted and Denmark from international scandal by not carrying out his threat, and so it was only fair that he was rewarded not only by the master's degree which had unjustly been withheld from him twenty years before, but also by the highest award the University could give, the doctorate, not because of the title itself which he could easily acquire abroad if he wanted to, but because of the right to lecture, the *jus docendi*, at the University. He generously offered Ørsted eight days within which to settle the matter. The cocksure Repp attached a draft of his letter to the Senate, in which he listed five reasons qualifying him for a doctorate:

1. He had been behind the establishment of 'Athenæum', the reading club, and thereby made books of sterling merit available to the people.
2. He had helped the Health Police by pointing out the dangerous state of drinking water.
3. He had founded a society to promote the cultivation of silk and written a history of silk production.
4. He had written a history of Danish law in English, and finally
5. Besides Latin he had a mastery of English, German, Danish and Icelandic.

And if this was not enough he had also helped several postgraduates write their doctoral dissertations. In brief: Repp was far more qualified for the doctorate than many to whom Ørsted had awarded it. Magnanimously, the psychopath promised Ørsted to keep the letter secret until the deadline of eight days had passed.[16]

So what did Ørsted do? He let *Fædrelandet* print Repp's letter uncensored crediting the reading public with sufficient intelligence to need no further enlightenment on the matter. A more biting revenge would be unimaginable. Ørsted also added the fact that Repp had been rejected as inspector of water supplies in Copenhagen. Repp was laid bare, and Ørsted stood out publicly as the incorruptible official, except that Repp's criticism of the mathematical level of the book (apart from the scurrilous tone) was justified. Ørsted admitted that he did not belong to the group of physicists who like Newton, Laplace, and Gauss found that physics should be treated mathematically. He could not advise anybody to base their study of physics on mathematics: 'this usually occasions a strange obscurity and confusion of the view on nature'. He had always emphasised that physics ought to be written in Danish and as plainly as possible to make it part of general education. In this respect he was pleased to say that he had shown new ways. If any mathematical inaccuracies had slipped in he regretted it.[17]

The next day Ørsted wrote a letter to the Danish Embassy in London asking for help in getting his reply to Repp's lampoons published should they find their way in to the British press. Repp was a translator into English, so that was where his threats were likely to show up. The diplomats set Ørsted's mind at ease by reassuring him that no newspaper of good repute would dream of publishing Repp's libels, and should they turn up in the gutter press Ørsted had better ignore it. To be on the safe side, Ørsted worked out a counter-attack that could be published at a short notice:

> This man has conceived a hatred bordering on madness against me, as I was the Vice-Chancellor of the University of Copenhagen twenty years ago, when Mr Repp was sent down

after a bad examination for a master's degree... By the way, I have never written any mathematical book or thesis; yet in popular works I have tried to convey the mathematical truths of physics in a way that would be easy to grasp by a large readership, endeavouring to describe the way of thinking in mathematical proofs rather than developing them in detail. Mr. Repp has snatched the deprecatory remarks he has heard from certain arch-mathematicians and used them with great lack of skill, but with all the artifice blind hatred can instil...[18]

Ørsted's apologia was never needed, but that year a *Tillæg til Naturlærens mechaniske Deel* Supplement to the Mechanical Part of Physics of thirty-four pages was published. The errors in section 211 and other places had been corrected.[19]

*

Still Ørsted was not left in peace to work for long. At the time of the culmination of the Repp case a vacancy as lecturer in mathematics at the Polytechnic Institute was to be filled. The board of directors was split. The professor in mathematics, C. Ramus (Schmidten's successor) recommended his assistant Adolph Steen, a polytechnic graduate, but a majority consisting of Director Ørsted and his loyal supporters, Zeise and Forchhammer to whom a new one should be added, namely C.G. Hummel, lecturer in engineering (succeeding Ursin and Dyssel), preferred C.V. Holten, also a polytechnic graduate in physics, and the amanuensis of the Director. It was unwise of them to turn down the candidate with the greatest expertise, particularly as the majority had one thing in common: none of them were mathematicians, apart from at the most elementary level. Steen, the rejected applicant, flew into a rage that a physicist of Ørsted's dynamical school was being appointed a lecturer in mathematics, and abused Hummel whose qualifications he had difficulties in spotting, and not only the mathematical ones.[20] The polytechnic students were complaining that mathematics instruction was deficient. How could an institution train engineers to construct water and gas works, establish railway and telegraph lines, and build bridges and steam engines, who had never been taught mathematical calculations? One of Steen's anonymous well-wishers exposed the Director and his supporters to public contempt in words not falling short of those used by Repp.[21] Like Ørsted, the majority considered mathematics to be only an auxiliary discipline and esteemed human qualifications more highly. Holten was reputed for his friendly temper, and Ørsted could trust his loyalty, while Steen was known to be self-assertive.

However, it was not only mathematics that was struggling in Ørsted's polytechnic department of the University. Technology itself was in dire straits. So, at any rate, readers of *Fædrelandet* were informed in a series of critical articles by the first graduate in applied science, J.F.C.E. Wilkens, now reader in technology.[22] Wilkens presented his criticism in a constructive spirit, suggesting that the Polytechnic Institute adjust its curriculum to the needs of industry instead of being 'vegetative'. His point was that the Polytechnic Institute was almost exclusively dedicated to scientific subjects and hence practically the faculty of science of the University, a self-contained state within the state, although it had been launched to serve industry. This criticism followed soon after the Scandinavian Science Conference in Copenhagen in 1847, when the staff of the Polytechnic Institute had chaired the various sections—except for Wilkens, because unlike the BAAS, the Scandinavian conference had no section for technology.

The Polytechnic Institute had become a self-contained ivory tower prioritising the individual student's personal and scientific education at the expense of industry's need for technical competence. This was what Wilkens meant to indicate by the term 'vegetative'. And the priorities were no mere coincidence. He found it to be encapsulated in the institution from the very beginning. Ursin's original proposal had aimed at walking on two legs, equality between theory and practice. For inscrutable reasons his proposal had landed on the table of the Board of the University, while the Directorate of Manufactories had no say in the matter. The Board had then requested a report from the primate of science, who had presented a proposal so cheap for the University and the Treasury that the Spartan Monarch Frederik VI could not refuse it. Ørsted had offered himself as director without a salary supported by already salaried, loyal assistants (Forchhammer, Zeise, and Schmidten); new staff had to put up with starvation wages, which forced them to take supplementary employment, Wilkens at the Military High School and Hummel as Inspector of Water Supplies in Copenhagen, for instance. The Polytechnic Institute was born into poverty, which was especially obvious when comparing its budget with that of Sorø Academy, paralysed in the midst of a reform while its funds kept growing. The Board of the Polytechnic Institute made a virtue out of its poverty by vegetating in 'the realm of the spirits'. The Institute could not afford a library. Teachers in technology and engineering had to empty their own pockets if they wanted to be updated with technical innovations. Workshops and the drawing office were lacking in the most basic equipment. The result was a one-sided bookish education. Members of staff were pressured to live within their means, realising that their posts did not exist for the sake of their private households, but the other way around.[23]

Since the Directorate of Manufactories had been excluded at the planning stage, the Polytechnic Institute had no contact with industry, nor had it made any during its first eighteen years. The conditions of admission were such that new students could not be recruited from guilds and industry. To get in one had to pass a test more comprehensive in mathematics than the *artium* and *philosophicum* at the University. How could an artisan possibly afford to prepare himself for such requirements? Only sons of well-to-do people could enter, which gave the polytechnic students a social profile and an academic identity that by itself made a gap between industry and the ivory tower. In addition, the polytechnic student, having climbed the ladder to obtain access, was little motivated to learn workshop skills from scratch. So how would familiarity with technical skills ever happen? The *Gewerbeschule* in Berlin that was said to have inspired Ursin and that Ørsted had visited twice was aimed at the needs of industry. Von Beuth was no scientist, and his institution was no annex to the University of Berlin. He was determined to recruit artisans having their wits about them and had shaped the admission requirements accordingly. Unlike Ørsted, Beuth's *Gewerbeschule* started on the workshop floor and approached the lecture hall and the library from there.[24]

The second point discussed by Wilkens was the duration of the education. Formally it was two years, as in the (other) faculties at the University, but actually it was longer, because exams took place after the fourth semester. Pretty soon this duration was prolonged, because the examination requirements turned out to be too ambitious, even when three hours a day for workshop training were cancelled in favour of bookish activities. The polytechnic education was organised in courses with fixed requirements. Every teacher prioritised his own subject at the expense of those of his colleagues, so work loads grew and extensions became standard.[25]

In comparison with similar institutions abroad science had almost supplanted technology at the Polytechnic Institute. This was reflected in the Board on which only the scientific staff (Ørsted Forchhammer, Zeise, and Ramus) had a seat, while Hummel (engineering and drawing), Poulsen (foreman of the workshop), and Wilkens (technology) had no say—hence Wilkens's articles. There was nothing with a bearing on industrial activities; the ambitious examination requirements were exclusively oriented towards mathematics and science. All experience showed, according to Wilkens, that large parts of the curriculum were soon forgotten when not applied. The disproportion between curriculum and applicability he found grotesque, and his opinion was substantiated by comparison with foreign countries, where specialisation was far more advanced, because they were oriented towards the concrete demands of the future. In Copenhagen there were two branches only: mechanics and chemistry, while in industrialising countries curricula were adapted to the rapidly developing technological division of labour (railways, telegraphs, steam ships, bridge building, road building, sewage systems, water supply, manufactories, etc.). The waste of resources was enormous in Denmark and it delayed technical progress.[26]

According to Wilkens his opinion was strongly supported by the pattern of employment of the sixty-six polytechnic postgraduates for the period 1832–47 compared to that of the polytechnic institution in the Duchy of Hanover, a city whose population was almost the same as the Danish.

In fifteen years the Polytechnic Institute had enriched Danish industry with two polytechnic graduates a year (half the graduates), while Hanover had contributed thirty a year (two thirds). Wilkens's point was not that Hanover was fifteen times as useful as Denmark, because the Polytechnic Institute had also trained workshop pupils, and some students had followed individual courses without intending to pass all examinations, such as Søren Hjorth, for instance, who became director of the railway company and gained an award for his electro-engine at the Great Exhibition in London in 1851. But despite these merits the statistics corroborated his main point that the Polytechnic Institute functioned as the faculty of science of the University. Its graduates qualified for official posts like those from other faculties, and not,

EMPLOYMENT OF POLYTECHNIC POSTGRADUATES IN	DENMARK	HANOVER
Private industry	23	405
Other practical industry	9	33
Non-industrial posts	30	222
Unknown	4	8
Total	66	668

as was intended or ought to have been intended, for carrying out technical assignments as needed by modern industry and infrastructure in Denmark.[27]

To remedy this miserable situation Wilkens proposed a comprehensive reform of the Institute. It was then divided into two: a number of workshops that the Board ignored and a scientific department being the equivalent of a faculty. It is not the intention to go into details as regards his proposal; the only aim is throw light on Ørsted's institution from a contemporary perspective. Suffice it to say that Wilkens proposed a two-stage polytechnic education aiming at fulfilling the demands of industry: a *Realskole* (ch. 49) for 14–19 year-olds and by extension a two-year polytechnic school where theory and practice weighed roughly the same. As both schools gave high priority to workshop training, Wilkens found it necessary to incorporate a series of industrial workshops in the scheme. This was precisely what Winstrup had offered when he fell foul of Ørsted in 1831. The radicalism of the proposal was that in future both the *Realskole* and the new polytechnic school would be administered by the Directorate of Manufactures, making Ørsted director of the existing scientific department only, which would continue as a High School under the auspices of the Board of the University.[28] From Ørsted's point of view Wilkens's scathing criticism could hardly have been published at a more inconvenient time. For it was precisely in October 1847 that he was putting the finishing touches to his application to the Board of the University for a considerable expansion of the premises and activities of the Polytechnic Institute.

*

Looking at the technological modernisation at that time the infrastructure was particularly important (railways, steamship routes, and telegraphs). In this process the Polytechnic Institute only had a modest, indirect role. Railways were built by the British entrepreneur Radford for a joint stock company, whose director, Søren Hjorth, true, had attended courses at the Institute, but it was not knowledge acquired there that qualified him for the leadership. Telegraph lines were also constructed with British technology, and the leadership of the company was in the hands of W.O.W. Lehmann, Orla's brother, who had graduated from the Military High School. Det Sjællandske Jernbaneselskab [The Zealand Railway Company] consulted Ørsted asking for his recommendation to the King on the construction of an electromagnetic telegraph line along the railway from Copenhagen to Roskilde. The company did not get what it asked for and for obvious reasons. The private railway company needed an electromagnetic telegraph to obtain a fast line of communication between stations in order to control the running of trains in both directions on the single-track railway. The government saw no interest in co-financing the railway, and when a few years later parliament discussed the issue, the civil war had broken out, and the State Telegraph was assigned to improve military communication and the collection of duties from ships (the Sound Dues). A few years later the Sound telegraph line between Ellsinore and Hamburg was established by means of a cable under the Great Belt and the Little Belt, but this technical solution was only possible after the successful development of a cable under the English Channel in 1850. Ørsted took no particular interest in helping the Zealand Railway Company. His one preoccupation was scientific, that is rapid communication from one meteorological station to the other in both north-south and east-west directions to study the weather and earth magnetism. To this end a thirty-kilometre section to Roskilde was insignificant. As far as the technical challenges were concerned he referred them to the entrepreneurs. His advice was to obtain offers from competing firms and to compare their prices, guarantees, and reputations. This was self-evident. If the railway company needed scientific advice he was prepared to help.

The electromagnetic telegraph is a good example with which to dismiss the myth Ørsted himself had been so successful in spreading, namely that science promotes technological development. But was not modern telegraphy electromagnetic, and was not Ørsted the discoverer of electromagnetism? Indeed, and Ørsted's name is mentioned in any jubilee publication to endow technology with scientific prestige. But the transformation from the discovery in 1820 to operational telegraphy did not capture Ørsted's interest.

The actual technological project to be carried out by the government concerned the improvement of the fresh water supply and a sewage system in Copenhagen. And when all streets and lanes in the city had to be dug up anyway, the city council decided to put gas pipelines under ground and establish a gasworks at the same time. Ørsted became a member of a committee set up in spring 1848.[29] The following year it arranged an international competition for a comprehensive solution to the water, gas, and sewage construction works of the city. When offers arrived Ørsted's life was fading. The new water supplies were not based on the artesian well borings that the Royal Danish Society had investigated during the previous twenty years but on traditional water reservoirs from lakes in the vicinity of the capital. Gas and sewage pipelines were constructed by foreign entrepreneurs like the railway and telegraph lines.

In retrospect, obviously, Ørsted's dynamical science was the embryo of the modern technological development of electrification and electronics; but in his day this perspective was a distant and unreal dream only and no motivating force in his scientific life. On the contrary, he looked towards the past and was preoccupied with proving the dynamical theory he had set out in *Ansicht*, his major work. His promises of technical progress had always been worded in general and non-committal terms calculated to release public funding for scientific research. They led to considerable scientific expansion at the University and Polytechnic Institute, epoch-making results and personal fame, but not to the technological progress he had indicated. The myth was bait and it got fish on the hook. His technological merit consisted of lectures for a generation of scientifically educated technologists whose primary achievement was the dissemination of physics and general education and only indirectly technical modernisation.

> Many a man now hit by the sting of death,
> was formerly a brother pious and kind.[1]

57 | 1848–9
Civil War and Free Constitution

O**N 20TH** January 1848 the President of the Royal Danish Society and the Society for the Dissemination of Science died and Ørsted lost an esteemed patron. We are talking about Christian VIII, of course. From that day on the absolute monarchy belonged to the past, and the future of Ørsted's scientific institutions was uncertain. Everybody realised that the successor, Frederik VII, was unsuitable to rule a country, and the liberals' demand for a constitution was therefore no longer utopian, but inevitable. The most controversial problem was the United Monarchy. The liberal Schleswig-Holsteiners would not put up with being separated from each other, but yearned to secede from the Danish kingdom, and on the other hand the Eider Danes would not relinquish Schleswig, as that would have left the Danish-speaking population in the German Federation. The hostile atmosphere Ørsted had experienced in Kiel a couple of years earlier threatened to degenerate into a regular civil war. Reason would no longer be able to prevent Ørsted's political house of cards from collapsing. A whipped up nationalism was about to undermine the United Monarchy. The national liberals would no longer settle for the abolition of an absolute monarchy, they were heading for a takeover of political power themselves.

Christian VIII's precarious health was broken when inspecting the crew of the corvette 'Valkyrien' in a biting January cold. When his physician-in-ordinary bled him to draw the inflammation out of his body the wound went septic and rapidly developed into blood poisoning that took the monarch's life. The next day Schouw and Clausen announced five political demands: a free constitution for the monarchy and Schleswig including (1) an elected representation with the right of granting supplies, (2) codetermination in deciding a biennial budget, (3) a shared legislative power with the King/executive power, (4) ministerial responsibility, and finally (5) public parliamentary debates.[2]

The most recent accession of a new king, Christian VIII, had been met with great expectations as to a change of constitution, but nothing had happened. This time the national liberals put their foot down by confronting Frederik VII with their five point plan. However, Schouw and Clausen were knocking on a door that was already open, or at least ajar. Some months before his death Christian VIII had clandestinely set up a committee to draft a new constitution attempting to keep the monarchy and the Duchies united by a common government which would meet in turn at Christiansborg and Gottorp Castle, and make the influence of the Assemblies of the Estates become more like a representative parliament. The duchies and the monarchy were to have the same number of representatives and by favouring the aristocratic element in the Duchies, the government attempted to split the Schleswig-Holstein movement. A.S. Ørsted was to work out the details with two other officials, and a few days later the constitutional rescript of 28 January was issued, including full freedom of the press with the addition of the sentence 'censorship can never be reintroduced'[3] that was soon to slide verbatim into the June constitution. The so-called January rescript was the monarchy's attempt at securing the continuing existence of the United Monarchy and at the same time offering constitutional concessions.

The January rescript satisfied neither the Schleswig-Holsteiners nor the Eider Danes. The former did not obtain the secession of Schleswig from the monarchy. Two factors contributed to whip up excitement further. The first was the February revolution in Paris. This time a workers' uprising brought about the fall of the Citizen King Louis Philippe. The revolution had a domino effect. The reactionary Count Metternich had to flee Vienna and in many German states demands for national unity were voiced. Friedrich Wilhelm IV decided to put up with the establishment of a common German Reichstag at St Paul's Church, Frankfurt. What started as a social revolution in France grew into a national revolution in Germany. In Denmark and the Duchies both national and constitutional claims flared up. The second factor was Orla Lehmann and his national liberal associates. He was abroad when the political scene changed. He rushed home and from the end of February his enraged rhetoric whipped up the excitement still more.[4]

Hans Christian Ørsted had withdrawn from politics ten years earlier, disappointed that the national liberals were polluting the public debate with their ambitions for power masked as constitutional claims rather than fighting for the rule of law. But he was particularly worried about their improvident and irresponsible nationalism, which had got out of control in rejoicing over Hjorth Lorenzen's provocation in the Schleswig Assembly of the Estates. Ørsted had found a new political sympathiser in the novelist A.M. Goldschmidt who had tried to dampen the ardour of the national liberals. Ørsted started subscribing to 'his spiritual and valuable journal *Nord og Syd*' ['North and South'].[5] Their common point of departure in Enlightenment ideals of religious tolerance and the rule of law based on reason made them sympathetic towards one another before they met face to face.

Goldschmidt, too, was suspicious of the motivation of the national liberals. Were they not fighting for their own political power rather than for the freedom of the people? He had been targeted by the censorship of the press himself, and as a Jew he was conscious of the importance of the protection of minority rights by the state. Censorship, Goldschmidt wrote trenchantly, is something the state inflicts upon the citizens, just as when a man finds out that his wife no longer loves him and contents himself by preventing her from real adultery. To Goldschmidt freedom is the crucial thing, and this could be safeguarded by the absolute monarchy just as well as by a constitution, as the January rescript had made clear. The national liberals were not interested in freedom for the entire population, but in achieving their own guardianship. The national liberals used the term 'constitution' as a magic formula believing that it offered the people a freedom they had never asked for. 'It would be rather difficult ... to present the people with a catechism, from which they can learn to love freedom, just as one cannot tell a man what to do or say to make him fall in love with a woman.'[6] Ørsted shared such opinions, they complied with the January rescript, and thus, of course, with the policy his brother pursued in the council of state.

We are now going to follow the course of the dramatic week from 18 to 24 March 1848 and the part the Ørsted brothers played in it, to the extent that the incomplete source materials will allow. On 18 March the Schleswig-Holstein movement arranged two conventions at Rendsborg on the edge of the Eider River: an open-air mass meeting and a united convention of the deputies of the Itzehoe and Schleswig Assemblies of the Estates. It was decided to send a delegation to the King in Copenhagen to demand the incorporation of the Duchy of Schleswig into the German Federation.

Two days later, on Monday 20th in the morning the SS Copenhagen arrived in the capital bringing the news of the Schleswig-Holstein demand. The Rendsborg resolution came as a shock to the government as well as to the Eider Danes. The committee of three (including A.S. Ørsted) working on the details of the January rescript convened and Frederik VII made it clear that he was not going to abandon the Duchy of Schleswig. For the time being the government decided to prepare for the mobilisation of the army in case the Schleswig-Holsteiners should choose to use military power to enforce their demand. The same day Orla Lehmann and the national liberal movement decided to summon a meeting of the city council of Copenhagen at six o'clock and to bring a public meeting forward by two days, to take place the same evening at eight in the Casino. The national liberals succeeded in obtaining the consent of both the city council and the public meeting to their Eider programme, as well as to the notorious resolution to the King 'not to drive the nation to the self-help of despair', a threat of revolution in case their demands for a new constitution and a new government dedicated to the Eider programme were not met.

On the same day Goldschmidt, writing in his journal *Nord og Syd*, realised the significance of the resolution: 'Nothing can save the Schleswig-Holsteiners as a party unless we Danes do wrong to the Schleswigers and by our threats induce them to take violent steps. Nothing can save the Ultra [the Eider] Danes as a party and provide them justice except when the Schleswig-Holsteiners bring about an uprising or foment trouble.' This meant that the struggle of both parties depended on a dramatic event, military or political, provoked by the adversary, which would in turn justify counter measures, military or political. Goldschmidt added a postscript to make it clear that this dramatic event had already occurred: 'an uprising is said to have been brought about. From this moment we consider it our duty to reach an agreement between all parties'.[7] Goldschmidt only had ambiguous rumours on which to base his analysis. He did not explicitly write a *violent* uprising, but depending on strategic interests 'uprising' could refer to the political demand, which was a reality, or a future outbreak of military action, which was not. In any circumstance, it was not far from thought to action.

On Tuesday 21st a Schleswig-Holstein deputation of five men departed from Kiel on the SS Skirner to Copenhagen to present the King with their demand. Of course, they were ignorant of the Casino meeting. They were at sea while the excited demonstrators marched in procession through the streets of Copenhagen to Christiansborg to find that the absolute monarchy had already surrendered. The January rescript had been withdrawn, the Government, including A.S. Ørsted, had been dismissed, and the army had been mobilised. This meant total victory for Orla Lehmann and his national liberal associates. Their Eider programme was endorsed as the official policy of the monarchy. Frederik VII appointed a new Government consisting of A.W. Moltke and the national liberal leaders, among them Orla Lehmann.

Goldschmidt summed up the fragile and cleverly manipulated background of the fatal events of the day in a pamphlet *Øieblikket* [The Moment]: 'The uprising has taken place, or what at the moment in relation to the two parties amounts to the same thing: everybody believes there has been an uprising'. The Schleswig-Holsteiners responded to exaggerated rumours to the effect that their adversaries had decided that 'Schleswig shall not be asked, but forced', and the Eider Danes responded similarly to rumours that the Schleswig-Holsteiners' 'thought had become action, and that they had risen up in arms against Denmark'. This was the rumour-mongering that caused the fall of the United Monarchy. 'In this way the Ultra Danish party won over the majority of public opinion and has been brought to the helm of state by the help of the people.'[8]

On Wednesday 22nd the Schleswig-Holstein deputation took lodgings at the Hotel d'Angleterre that was soon besieged by protesting Copenhageners crowding on Kongens Nytorv. The Schleswig-Holsteiners now met the representatives of quite another regime than that with which they had set out to present and negotiate their demands. Orla Lehmann represented the brand new March Government and immediately rejected the demand adopted by the Rendsborg meetings to incorporate Schleswig into the German Federation. But the Monarchy and the Duchies were not in a state of war. The German-minded chief negotiator Theodor Olshausen suggested a division of Schleswig according to language; referenda were to be held in each parish enabling the population to make up its own mind as to whether their parish should be Danish or German. Apparently, Lehmann misunderstood the proposal for he immediately started to put demarcations on a map, whereupon Olshausen had to point out to him that the idea was not to dictate a certain division, but that it be made according to a referendum (as actually happened in 1920). But the proposal of a negotiated solution did not meet the approval of Lehmann who brushed it aside as 'an amusing idea'.[9]

The same day Goldschmidt wrote a letter to H.C. Ørsted considering him the potential saviour of the United Monarchy and encouraging him to intervene to save the peace by placing himself at the head of 'all benevolent, but easygoing and impartial men to meet the events by determination and good will.' But why Ørsted of all people? Because, Goldschmidt said, 'Your soul is harmonious, you love beauty and have not forgotten the ethical in favour of the aesthetical; you love the arts and have preserved veneration for nature, and for the natural and healthy in life. You are an upright man and are regarded as one.'[10] Along with the letter followed Goldschmidt's pamphlet *Øieblikket* indicating that the national liberal party now occupying several posts in the March Government had been seen through. It was Goldschmidt's plan to gather sober-minded adherents of the United Monarchy in the hope of winning back the initiative from the Ultra Danes.

Ørsted responded that since the time when national rows had broken out he had entertained the same attitude as Goldschmidt, namely that:

> 'the different parts of the state should be kept together by the good deeds of a free constitution and by reciprocal justice of a free constitution, and that by seriously aiming at these benefits one would find a guarantee for the Danish and German nationalities, of which one should not violate the other... For the past few days I have thought of trying to do what has failed for others, but events have followed one upon another so rapidly that what seemed appropriate one day had to be halfway rejected the next and fully rejected the next day again. Still, yesterday morning [22 March when A.S. Ørsted was still in the Government], I seriously pondered a gathering of a group of like-minded men to pursue this end; but the news [of the

Fig. 157. Hans Olde's draft for an oil painting of the proclamation of the Provisional Government of Schleswig-Holstein in front of the Town Hall of Kiel 24 March 1848.

Olde's picture was made between 1912–17, when the German Kaiserreich was drawing to a close and shortly before the northern part of Schleswig was reunited with Denmark in 1920. The revolutionary aspect of the event of 24 March 1848 is understated in favour of the memorable festivity of the citizenship. In the middle Wilhelm Hartwig Beseler is reciting the proclamation of the provisional government in front of the town hall with its nationalistic slogan 'up ewig ungeteilt' [undivided forever]. To the right of Beseler stands the provost of the monastery Friedrich, Count of Reventlow-Preetz, leader of the landed aristocracy, and to the right in a light coat the Prince of Nør. The rejoicing crowd and the colour swinging students are kept at a distance. Only a little, simple boy has strayed right up to the well-dressed gentlemen. C. Degn 1994, 226–227. Schleswig-Holsteinische Landesbibliotek, Kiel. Photo by Dr Jens Ahlers.

formation of the March Government] that later came to my knowledge immediately crushed the idea.'[11]

Like his brother, H.C. Ørsted was committed to a free constitution according to the January rescript and to negotiations with the Schleswig-Holsteiners, but realised that he had missed the opportunity.

On 23 March the Schleswig-Holsteiners were received in audience by Frederik VII, who promised them an answer to their demand the next day. However, Goldschmidt did not accept Ørsted's resigned response for he immediately addressed Ørsted again by a letter that has gone missing, but to which he answered: 'After having learned more about your plan and considering that it deserves to be taken very seriously, I can think of nothing better than suggesting to you a discussion of the matter. You will find me at home tomorrow [24 March].'[12] Perhaps Goldschmidt had in mind a meeting between the moderates of both parties, where the rumour-mongering facilitated by the slow means of communication between Kiel and Copenhagen was revealed and real negotiations initiated. The risk of a civil war was too great to allow the two sober-minded intellectuals to simply stand by while political decisions were taken by nationalistic extremists on the basis of loose rumours that might turn into self-fulfilling prophecies. We do not have any sources to show what the two serious men discussed in Ørsted's study, or even if they met at all. At any rate no united action ensued.

On that Thursday, 23 March, the Copenhagen newspaper *Berlingske Tidende* reached Kiel and only now did the Schleswig-Holsteiners learn what had happened in the Danish capital two days earlier, namely that the absolute monarch had in fact abdicated. On Friday night at the town hall in Kiel the 'provisional government' of Schleswig-Holstein issued a proclamation serving to legitimise the uprising by the announcement that Frederik VII had now been rendered 'not free' to be their duke. The same day, 24 March, the Prince of Nør, the brother of the Duke of Augustenborg, and a group of students and soldiers from the garrison of Kiel took the train to Rendsborg by the Eider. The leader of the Schleswig-Holstein landed aristocracy, Fr. Reventlow-Preetz, the provost of the monastery, read to soldiers and officials the proclamation that the Danish Duke was henceforth 'not free', implying that the oath of allegiance could no longer be considered binding. Hence, they should have no scruples serving the provisional government of Schleswig-Holstein and joining the uprising. Most officers felt bound by the oath of allegiance to the King of Denmark, but they capitulated. Had the Schleswig-Holsteiners taken advantage of the human rights of the American constitution entitling citizens to throw down the gauntlet to a regent breaking the social contract? Or were they merely a group of judicial sophists determined to secede at any cost? People in Denmark and the Duchies were now so whipped up that the peace could no longer be kept. On 9 April the first battle at Bov, north of Flensborg was fought. The civil war was unstoppable. The German Reichstag in Frankfurt sent nine thousand auxiliary troops, the Prussian King an army of twelve thousand, while Oskar I, the Swedish-Norwegian King sent four thousand five hundred soldiers to fight for the Danish cause.

As usual the truth was the first victim of the war. In Germany no one heard of the villainy committed by Reventlow-Preetz when he released the soldiers of the garrison from their oath to the King of Denmark and hence legitimised the uprising. The Danish bishop Martensen drew attention to it in a pamphlet published in 1850 and sent by Ørsted to Schelling and to Humboldt, the advisor of Friedrich Wilhelm IV, asking him to present it to the King.[13] According to Ørsted, Martensen's pamphlet was a well-argued repudiation of the myth of the 'not free' status of the Danish King and the consequent mixing of religion and politics that was intended to legitimise the uprising of soldiers and the perjury of officials using quotations from the Bible.

Martensen stated that the Schleswig-Holstein demand of 18 March to incorporate the Duchy of Schleswig into the German Federation was the event that logically led to the perjury and the outbreak of the disastrous civil war.[14] For in the very formation of the provisional government the uprising and the will to fight were implicitly present. To transform the uprising against the Danish King into a civil war it was necessary to establish an adversary, which was not so easy, because both parties had sworn an oath of allegiance to the same King by the same colours. Hence, the Schleswig-Holsteiners circulated the myth of the 'not free' status of the King. Loyalty towards an oppressed King was a contradiction in terms. Regarding him as 'not free' not only released his subjects from their oath of allegiance, but also obliged them to liberate him from the rebels (the Danes). In this way they obtained nationalistic and diametrically opposed definitions of the terms 'loyalty' and 'rebel'.

Martensen, himself a Schleswiger by birth, sharply distanced himself from his German-minded clerical colleagues, because they had given religious legitimacy to the perjury and as a consequence eased the pangs of conscience of his fellow-countrymen, and even made it a moral duty to fight against the Ultra Danes and liberate their King. According to Martensen the King was not oppressed at all. The demand to have Schleswig incorporated into the German Federation was dealt with in exactly the same way by the January rescript and by Lehmann's resolutions. He found that 'the self-proclaimed rebellious government' had already started the process on 18 March. According to his nationalistic interpretation the definitive provocation rested totally on the shoulders of the Duchies. This view was a biased simplification underestimating the provocative measures taken by the formation of the Danish March Ministry. After all, the Schleswig-Holstein assembly of 18 March consisted of seven deputies of the Assemblies of Slesvig and Izehoe. A few of them were also members of the provisional government of 23–24 March.

Martensen's view was that Frederik VII had taken the decision to repudiate the Schleswig-Holstein deputation's claim as voluntarily as he had made the same decision a month and a half earlier. The Schleswig-Holsteiners, on the other hand, found that the Casino meeting had declared the Danish King 'not free' and thus had initiated an uprising against him the next day cemented by the formation of the March Government (for the sake of clarity called the Casino Ministry by the Schleswig-Holsteiners) who had then waged civil war, because until then the provisional government had been willing to negotiate. Hence the Schleswig-Holsteiners subsequently made efforts to show that 'the provisional government' was nothing but an amalgamation of the two assemblies of the estates of the Duchies in conformity with the January rescript, comparable to their previous administration as a unit by the German Chancellery. Moreover, they underlined that the demand of the provisional government was not an ultimatum, but a basis for negotiation they had come to Copenhagen to present. Their good will had been demonstrated by Olshausen in his talk with Lehmann about a possible division of Schleswig according to linguistic boundaries.

To Martensen it was not particularly interesting whether the uprising was violent or not. He saw the demand of 'the provisional government' to incorporate Schleswig into the German Federation as an uprising that would by necessity bring the Duchies into conflict with the United Monarchy and, if the King did not give in, to an armed one at that. The Schleswig-Holsteiners already knew that the King was opposed to giving up Schleswig. This was explicitly stated in the January rescript, and hence the political upheaval of 18 March was bound to be the first step leading to civil war.

Both views that have caused historians to quarrel ever since were built into Goldschmidt's analysis, namely that to the ultras of both parties it was only a question of providing legitimacy for their escalation of the conflict. Hence it is probable that what Goldschmidt wanted to talk to Ørsted about was a proposal for a meeting with moderate Schleswig-Holsteiners, for instance G.A. Michaelis, the medical doctor who had presided over the GDNÄ conference in Kiel in 1846, and with whom Ørsted had been in correspondence. Facing the imminent threat of civil war the two scientists might have reached a compromise as they had done as colleagues after the rowdy meeting in Kiel.[15] Michaelis and Ørsted had agreed to consider the row a misunderstanding. And the present escalation of the national conflict due to slow means of communication was a clear case in point.

Ørsted was very pleased with Martensen's effort. 'Your pamphlet seems to me a work of art profoundly rooted in clear feelings of truth and wise humanitarian love. Although you have wisely abstained from dealing with politics, you have still as a spokesman for religion and truth pointed to the appropriate kind of politics: that of honesty, reconciliation, and Christianity.'[16] Sending a copy to Humboldt he announced that he would soon receive the German translation of *The Soul in Nature*, in which he tried to draw certain philosophical and aesthetical conclusions from his research, similar to those of *Kosmos* that Humboldt had presented him with and in which Ørsted's scientific discovery was mentioned. The *Kosmos* volumes were jewels in his library.

*

Once again Ørsted's law of the pendulum had confirmed itself. Politics had swung from authoritarian monarchism to its opposite extreme: political and national chaos. How would one find political harmony in this acute situation? His maxim of reason as the highest good had been betrayed by his national liberal colleagues. The political reality was fundamentally changed and a fresh analysis was needed. If reason had failed, the political scene was now left in the hands of the people. A limited franchise as preferred by the national liberals would favour capitalists taking control of society and they were likely to feather their own nests. If the franchise was extended to the proletariat confusion, impotence, and envy were likely to ensue as long as the present level of education was maintained. Such was Goldschmidt's analysis in 'his spiritual and valuable journal *Nord og Syd*'. Goldschmidt's sober-minded strategy of enlightening the people was essentially similar to Ørsted's political view. They agreed that a free constitution was not a condition of freedom. It was the other way around; freedom is the condition under which a free constitution may work.

No doubt Ørsted had imagined that the common man was unaffected by the nationalist fanaticism of the Eider Danes, but had been manipulated into cheering. Here, too, an educational campaign was needed. Democratic rule demands more general education so that voters are not blinded by demagogues. Ørsted had experienced his colleagues in Kiel whipping up irrational nationalist leanings, and something similar now applied to his colleagues in Copenhagen. It was a disappointment to him that Orla, the son of the friend of his youth, Martin Lehmann, had become a demagogue, though this had been going on for some time. It was even worse that Schouw, his brother's and his own old loyal friend had become a supporter of the Eider policy of the Ultra Danes. Such examples made Ørsted doubt that reason was capable of unmasking political campaigns that demonised the opposition and whipped up feelings of fear. Religious revivalist preachers did the same thing and so did pseudoscientific tricksters, who took advantage of sick peoples' mental weaknesses.

Perhaps Ørsted had the March days in mind when he suggested that reason should dispel these irrational plots, for what else was there to trust? To the Royal Danish Society he proposed a prize for the best essay on the question: 'A thorough and impartial description of the means by which the spokesmen of parties consciously or subconsciously deceive the crowd'.[17] The March days had been a signal illustration of political manipulation. Ørsted had not given up his basic trust in education, integrity, and the prevalence of reason in the long run. He explained his proposal as follows:

> 'When the inappropriate means used by parties in religion, science, and politics make unspeakable contributions to confuse the crowd (not only the ignorant, but also some having acquired considerable enlightenment), it would be instructive to subject the means used by spokesmen for parties (sometimes out of delusion and sometimes consciously to disappoint the crowd) to scrutiny. Such scrutiny would examine the information on the basis of which the spiritual laws of nature could be deduced that determine the course of arguments, and the directions to follow to avoid self-deception, by which one might be induced to delude others, as well as to safeguard oneself against being seduced by others.'

The proposal was never carried out by the society. Four days later when Gitte was fifty-nine and her spouse, as he usually did, wrote a few congratulatory stanzas, they were set against a gloomy background, because 'Civil war will pull Denmark apart;//Swords are honed for fratricide'.[18]

On 5 June 1849 Frederik VII ratified the constitution the national assembly had approved, and in the autumn Ørsted's nephew and son-in-law, Peter Hasle, vicar at Asnæs in the county of Holbæk, was encouraged to be a candidate for the first general election for parliament. Having assessed the situation for himself, he asked his father-in-law for advice. He had various doubts, since he did not feel confident with politics and was nervous of jeopardising his good relations with the congregation, because if he crossed the programme of the Society of the Friends of the Peasantry, he would be exposed to public contempt in its newspaper, *Almuevennen*, widely read in the countryside and reflecting the views of the peasantry, a considerable number of whom had not yet been released from villeinage.[19]

Fig. 158. *The Brave Soldier*, a coloured lithograph illustrating Peter Faber's popular songs occasioned by the civil war. Peter Faber is also renowned for his Christmas songs; he was one of HCØ's polytechnic students later to become director of the State Telegraph Service. RL.

Fig. 159. The Ørsted family. Daguerreotype of 1849. In the back-row from the right:

1. Albert Nicolai Ø. (1829–1900), polytechnic candidate, later inspector at the Polytechnic Institute
2. Edvard August Scharling (1807–1866), Professor of chemistry at the University
3. Karen Scharling, née Ø. (1815–1892), married to EAS
4. Frederik Christian Emil Dahlstrøm (1815–1894), jurist, later Prefect of the Diocese of Ålborg
5. Anders Sandøe Ø. (1826–1905), later mayor at Vejle

In the front-row from the right:

1. Anders Sandøe Ø. (1778–1860), Minister of State, fired March 1848, later again Minister of State
2. Inger Birgitte Ø, née Ballum (1789–1875)
3. Hans Christian Ø. (1777–1851)
4. Sophie Wilhelmine Bertha Dahlstrøm, née Ø. (1821–1889)

The following members of family are missing: Anna Dorothea Marie Hasle, née Ø (1818–1850), Niels Søren Peter Hasle (1814–1890), pastor at Asnæs, Niels Christian Ø. (1816–1849), and Mathilde Elisabeth Ø. (1824–1906). The reason for the absence of the Hasle family must be the long journey to the capital. Niels Christian died 05.10.1849, and therefore I assume that the daguerreotype was taken after that date. The reason of Mathilde Ørsted's absence could be that the daguerreotype was intended as a present for her 25th birthday on 4 November 1849. DTM.

Ørsted answered promptly and constructively. One cannot study to become a clever politician, he began. If there is a more qualified candidate one should withdraw. All good men must learn political practice on the job in parliament. Most important is the education and uprightness of the candidate. The worst is an obsequious party tool. Hence, Hasle could comply with the request. 'If I were you I should have a talk with the local members of the Society of the Friends of the Peasantry telling them that you stand at the request of others.' The dominating issue of the constituency was farming. Ørsted felt there was a lot to be done to ameliorate the condition of the peasantry, such as the abolition of villeinage, the move to private ownership, alleviation of the lot of the poor, and the extension of good education to all citizens. To accomplish this the government had set up a commission whose proposals Ørsted would listen to and work for by peaceful means. He found much of the established order harmful to people and country, and a lot of conscientious work was needed to provide just reforms. If the leadership of the Society of the Friends of the Peasantry wanted to extort a personal promise in advance, he had better advise them to find another candidate. He would never consent to being in the pocket of special interests, but work first and foremost for the entire population. These were Ørsted's personal views, but whatever Peter Hasle made up his mind to do, his father-in-law wished him God's blessings.[20]

*

In 1848 the Ørsted-family did not travel for their summer holidays. After the departure of Christian VIII, Fredensborg was no longer available to them. Besides there was a civil war going on in Jutland so it was unsafe to go there. But the following summer there was a ceasefire, and all the family (Gitte and Hans Christian, Mathilde (24) Anders (23), and Nicolai (20) as well as uncle Anders) went to Rudkøbing to see their childhood home. Jens, their coach driver took them to Glorup, the manor house where Andersen, too, was a frequent guest. They walked in the park and took rides in the neighbourhood, for instance to the huge rock at Hesselager, which Christian VIII had allowed to be partially excavated, and to the manors of Egeskov and Brahe Trolleborg, whose beautiful surroundings deserved a more civilised owner according to Mathilde. From the town of Assens the family observed Schleswig, the territory the war was about, and wherever they came they found manors transformed into hospitals and filled with wounded soldiers. Many soldiers were already permanently resting in the cemetery.

At Hvidkilde manor house Søren, their foster-son, joined the family. They were ferried to the island of Tåsinge and on to Rudkøbing where honours were heaped upon the two brothers. They headed for the pharmaceutical yard they had left more than fifty years before. Most things were unchanged. A deputation invited them to dine at the newly built town hall. Rudkøbing was decorated with colours, and inside hung portraits of the Ørsted brothers festooned with flowers. They were welcomed by more than eighty fellow townspeople, toasts were proposed, cheers were raised, and speeches were given by the pastor for Hans Christian and by the local lawyer for Anders. A festive committee had planned excursions, balls, and other amusements, but there was no time for it all. Happy hosts accompanied the travellers to the jetty, once again the pastor thanked them for coming, and hurrahs were heard till the boat rounded the point of Tåsinge on its way to Svendborg.[21]

Showpieces from the poetic arsenal of the past…[1]

58 | 1849-50
The Soul in Nature

THE SOUL in Nature is an enigmatic title linking soul and nature, while the readership would probably expect a contrast, in which nature is seen as material as opposed to spiritual. Ørsted's own proposal for a title was 'Natural Science and the Education of the Spirit', which is more precise, but less striking. The book is Ørsted's philosophical testament, a collection of dialogues, speeches, essays, obituaries, and poems written over a period of more than forty years. There is neither introduction nor foreword. The reader will have to discover the connections between texts as he reads them. Ørsted never wrote a coherent dualistic philosophy. The book does not deal with science directly, but with the reflections of a scientist on the relevance of his work to aesthetics, politics, pedagogy, and religion. Scientific writings addressed to the community of scholars are absent. The Soul in Nature is written in a plain style without the use of technical terms, to reach a wide public. Ørsted himself is responsible for the selection of texts, unlike the rest of the nine volumes of his *Samlede og Efterladte Skrifter* [Collected and Posthumous Works], of which *The Soul in Nature* forms the first two; the seven last were published after his death in 1851 by Mathilde. Ørsted's intention with the book was no less than 'to prompt a change of the contemporary world view'.[2]

The paradoxical title refers to a modern dynamical science, a breakaway from the onesided mechanical and mathematical view of the past. The book opens the door to a comprehensive world view emanating from the new research disciplines which revolutionised science in his lifetime, such as electrochemistry, the theory of the polarisation of light, diamagnetism, earth magnetism, and thermoelectricity, but first and foremost, of course, his own electromagnetism. This dynamical science was driven by invisible polar forces not susceptible to being measured and weighed, and hence comparable to the noumenal forces of the human soul. This modern world view had been sparked off by Kant's critical philosophy, but Ørsted believed that the new dynamical insights had made important contributions to the fundamental issues Kant had wrestled with, especially as far as the philosophies of aesthetics and religion are concerned. Since the aim of dynamical science is to find the laws governing the intrinsic forces of nature, which are not immediately accessible by common sense, it is of the utmost importance that it be disseminated to the wider public. This enlightenment will show that human reason is capable of finding (not inventing) laws of nature, and that laws of nature are expressions of a superior reason. In short, *The Soul of Nature* is an optimistic book about the joy of scientific insight and its significance for general education.

This modern, dynamical science was not useful in the same sense as the mechanical, for it was difficult to see what kind of machinery it would yield, although the electromagnetic telegraph showed some potential. But Ørsted was more anxious to point out that dualism was the philosophy of science. It showed that human reason was capable of finding laws of nature, and that such laws of nature were true and expressions of a superior reason. When the knowledge of this modern science was disseminated, people would learn to distinguish between true and false and become emancipated from fear and superstition, and understand that art is a reflection of the beautiful in nature.

Ørsted emphasised the importance of having this philosophical testament published not only to a Danish, but also to a European readership. Up to then it was only his scientific research that had been published in German and French, while his dialogues and aesthetical essays were available only in Danish. In particular he wanted to have it published in Britain. In 1839 he had sent the German translation of 'The Airship' to John Herschel who, being a second generation immigrant, was among the few Britons capable of reading German. The book was accompanied by a poem for Herschel. 'Splendidly, thought and knowledge in you has joined in art, splendidly, from your innermost soul the wells of poetry gush out. From an admiring friend a small present is sent as a token of equality of mind, even if equality of force is wanting.'[3] He could hardly express any more clearly how he would like to be regarded by his admired colleague. In the same letter he asked Herschel to sit for the painter C.A. Jensen who brought the present to London.[4]

In 1841 Ørsted had arranged for Christian VIII to confer upon Herschel the knighthood of Dannebrog.[5] Ørsted readily admitted that orders and titles were a little ridiculous, but they might be defended by the view that they balanced the aristocracy's privileges of birth, because they were given for personal merit.[6] Herschel did not know if he was supposed to give the Danish King something in return or if he was expected to be received in an audience in Copenhagen to thank the King for it.[7] Neither, Ørsted replied, the only thing to happen was that his name would appear in the *Hof- og Statskalenderen* ['The Yearbook of the Court and State'].[8] But Herschel soon discovered that British court etiquette did not permit British people to wear foreign orders.[9] This rule, of course, also applied to Murchison, to whom Ørsted had handed the title of commander of Dannebrog after the BAAS meeting in Southampton, but Murchison intended to ignore it. Being a scientist he considered himself a citizen of the world, and he would be proud to publicly display his award from the Danish King who was so respected for his scientific insight. Murchison wore the commander's cross at the BAAS meeting in Oxford in 1847,[10] and Ørsted wrote to Herschel that if Murchison could wear it, so could he. If Herschel would now be so kind as to accept the order, Ørsted would see to it that he received it.[11] So, Ørsted used his status at court to nourish the vanity of a select group of foreign colleagues, possibly hoping for something in return.

In Southampton Ørsted had already entertained the idea of publishing his philosophical testament in English and to this end he had approached the London publisher John Murray recommended to him by Leonard Horner.[12] However Murray declined. So Ørsted asked Captain Edward Sabine to find another publisher, for instance for his essay 'All Existence a Dominion of Reason'. He fervently wished to be recognised as a natural philosopher (Whewell's newly coined word 'scientist' was not yet commonly used) in Britain, but he also wanted money, for his income shrank during the civil war. He told Sabine that he would have the translation made himself and promised to send a complete manuscript three months before the stipulated date of publication in Denmark. Sabine made contact with T.N. Longman, another London publisher, but he wanted to see the manuscript first. Now Ørsted had the dialogue 'On the Spiritual in the Material' translated, as it was intended to be the initial part. Longman, however, did not like it. Ørsted realized that this particular dialogue might give the impression that his book was too metaphysical, something the English were known to dislike, so he had 'The Fountain' translated to enhance the notion that his aesthetics of

Fig. 160. Sir John F.W. Herschel (1792–1871), mathematician and astronomer. Daguerreotype by Julia Margaret Cameron (1815–1879). Musée d'Orsay, Paris. RMN, Paris. Photo by René-Gabriel Ojéda.

nature was derived from laws of nature. But this dialogue did not open the door to Longman's firm, either.¹³

Now Ørsted asked Herschel for a favour. He sent him the translated parts of his manuscript together with Longman's negative response, asking him to help him find a publisher. If Herschel liked the dialogues he was welcome to publish them in a journal. Ørsted attached an introduction in English, written for the occasion, and asking for a royalty, yet only if Herschel thought it would be well received.¹⁴ Apparently, he did not think so, for in the margin of Ørsted's letter Herschel commented in pencil: 'nothing came out of this correspondence. I was unable to be of any use in this matter.'

Ørsted was on the verge of desperation. Was Herschel actually dissuading him from publishing the book? The war meant pecuniary losses; the students were drafted for military service, so lecture fees plummeted. For once Ørsted felt sorry for himself. He was now seventy-two, and too old to revise the manuscript, he wrote, and the translator was no scientist. What was the problem? His speeches and essays had always been favourably received in Germany, but this might be explained by the fact that what went down well with German *Naturphilosophen* was less palatable for British scientists, virtually all of whom abhorred metaphysics. Fair enough, but did not the British go too far? 'I'm not a supporter of German *Naturphilosophie*', Ørsted told Herschel:

> 'For half a century I have developed my own philosophy based on experimental physics. Niebuhr, who was decidedly critical of German metaphysics, already lauded my *Ansicht* forty years ago. You should realise that German metaphysicians would not approve of essays revealing their erroneous analogies and unsystematic methods. I have presented my own method of work in "First Introduction to Physics" in 1811... I do not deduce my conclusions from metaphysical axioms, but infer my philosophy from the laws of nature. Perhaps it was wrong of me to begin with the aesthetics of nature, but I am convinced that my views are original and have a certain profundity when compared to previous views. I tell you all this in order to get you to take a fresh look at my manuscript.'¹⁵

He never heard from Herschel again. Two months later he asked him to hand over the manuscript to Colding who happened to be in London.¹⁶ Ørsted had acknowledged his defeat. The correspondence shows the self-humiliating stubbornness with which Ørsted (quite unusually) fought for his reputation.

*

At the very end of 1849 the Danish version of *The Soul in Nature* became available in bookshops in Denmark. The book created a stir. The primate of the State Church, Bishop J.P. Mynster, saw it as an undisguised attack on the dogma he was employed to defend and he reacted vehemently. Molbech, too, was negative. Others such as Andersen, Hauch, and Goldschmidt praised it highly and read it as an epoch-making contribution to the cultural debate. In the course of spring 1850 the book appeared in German translation by K.L. Kannegiesser in Leipzig, introduced by P.L. Møller's biographical sketch and lithograph of the author. Andersen was travelling in Germany and told Mathilde that he was delighted to see his own *In Schweden* standing in German bookshop windows next to *Der Geist in der Natur*.[17] By contrast, *The Soul in Nature*, translated into English by the Horner sisters, went almost unnoticed in Britain.

Let us take a look at the Danish reviews first and begin with the negative ones. Molbech, whose jubilee publication had cooled the atmosphere between them since 1843, came forward with a proposal for a lecture on aesthetical theory to the Royal Danish Society of Sciences and Letters to take place on 7 March 1851, and Ørsted was looking forward to the discussion, but unfortunately he was prevented from taking part as he was by then on his deathbed.[18] Molbech's title was 'A Reflection on the Beautiful in Nature and its Relation to the Beautiful in the Arts and the Idea of Beauty'. In his *The Soul in Nature* Ørsted had argued in favour of a relationship between truth and beauty by showing that it is the laws of nature that determine the beauty of an object and hence that the sense-impression is less dependent on the imagination of the subject. By this argument Ørsted had underpinned his main point, that the sciences are capable of elucidating the riddle of aesthetics. While the humanities focused on the taste of the subject alone and rejected the objective source of the experience of beauty, Ørsted insisted that the latest discoveries within dynamical physics bridged the gap between the sensual and the metaphysical and were therefore indispensable to the understanding of aesthetics. Artists needed scientists, who should be given equal status in the learned republic and a greater role in the education of young people.

This was the aesthetics of nature that Ørsted considered his original contribution to cultural life and that Molbech saw it as his task to repudiate. As already mentioned, Ørsted could not be present. Molbech took a classical Platonist point of departure distinguishing between the real physical world, which he gladly left to the scientists and the ideal spiritual world, the domain of the humanities. The physical world would always remain an incomplete shadow world of the inaccessible transcendental world. Hence, Ørsted's bridge between the two worlds could only result in a faint reflection of the ideal in the sensual world. Against this background Molbech found that the whole idea of mimesis was a misunderstanding, a remnant from Antiquity (Aristotle) and the Enlightenment (Batteux). The beauty sought by artists does not exist in its pure noble form in nature, where determinism rules. Art is free and obeys no rules Molbech asserted, echoing Schelling. When nevertheless nature has a significant part in the arts, for instance for sculptors and landscape painters, they idealise nature. The creative process is not mimetic, but unfolds in complete freedom. Artistic beauty is rooted in the human genius taking advantage of its imagination to express the world of ideas. To Molbech the difference is obvious in music. True rhythm and notes are phenomena in time and space, but there is an insurmountable distance between the warbling of birds in the wild and the sound of an orchestra in the concert hall. A symphony is a free musical creation of ideas, whereas warbling birds merely obey laws of necessity. The truth of natural laws does not make them beautiful.[19]

In Molbech's eyes, science revolves around empirical realities and scientists look at nature in the same way as a farmer or a smith. They want to know how it works to exploit their knowledge technologically. Art on the other hand is above the phenomenal world and seeks its subject matter in the idea of nature. The artistic genius finds its ideal forms, colours and lights intuitively, and the taste of his audience is refined during a process of education. Like Rousseau, Molbech could not explain taste; science cannot do so either, he maintained, and will never be able to. Rousseau's *je ne sais quoi* does not express resignation, but is a precise characterisation of the essence of art. It has nothing to do with truth, as Ørsted seemed to think. Molbech did not say so himself, but if somebody had said that 'The Airship' (ch. 46) was really didactic verse and unbecoming poetry he would no doubt have agreed.

Nonetheless, when beauty does exist in nature this is due to pure coincidence, according to Molbech. The teleology of nature is not aesthetical, but functional. Humans may be handsome or ugly depending on taste and fashion, but from an organic point of view ugly humans function as well as handsome ones. An aesthetic of nature can only be aware of empirical realities, and if they live up to an ideal of beauty it is by a stroke of good luck. Consequently, aesthetics are not part of science. The beautiful and the functional emanate from two different kinds of reason, and Ørsted's talk of one comprehensive dominion of reason was sheer nonsense. There is a poetical form of beauty that is not true, for example unicorns, sphinxes, centaurs, satyrs, and angels. And in nature there are true realities that are decidedly unbecoming such as seals, sea lions, sperm whales, and octopuses. These are the examples suggested by Molbech. From the Galathea expedition he had become aware of the exotic species recently brought back from the depths of the sea, and against this background he concluded that whether these biological species are beautiful or ugly it is never their inherent purpose to live up to our aesthetical criteria.[20] Their appearance is functional in relation to their conditions of life. Seals and sea lions have their layer of fat to protect them against the sharp cold of icebergs, and certain species of fish have their splendid colours to be able to swim under cover of plants of the same colours.

The dying Ørsted was prevented from assessing these objections. But it is not difficult to see how he would have reacted, for his reflections on the relationship between the beautiful and the functional had resulted precisely in the view that the two notions were subsumed under the same reason. The functional *was* the beautiful. The swan is ugly in a hen run, but beautiful when floating in its right element. The same applies to bats, and seals and sea lions only seem ugly to us on rocks and ice floes, but as soon as they are swimming in the right element they become elegant. Molbech ignored all Ørsted's arguments in favour of a new aesthetics of nature derived from dynamical science, including the theory of the polarisation of light, probably because he did not accept let alone understand the noumenal character of the forces of nature. Ørsted's attempt to connect science and the humanities was conceived as a menace rather than as a gift. *The Soul in Nature* was an attempt to show that the arts and the sciences addressed the same subject, but from two different perspectives. The mechanical physics of the Enlightenment had had to capitulate to the challenges of aesthetics, and Romanticism had not succeeded in getting closer to a solution, but rather continued to obfuscate artistic beauty.

*

To the primate of the State Church, Bishop Mynster, *The Soul in Nature* was hostile to Lutheran Christianity.[21] Ørsted's dominion of reason was a denial of the Kingdom of God and of the dogma of the church that he was called to defend. Here two views on life collided: the dogma on the fall of man, salvation, and resurrection on the one hand and the teachings of Ørsted on the permanence of natural laws and human progress on the other. The structure of the irrecon-

Fig. 161. J.P. Mynster (1775–1854), Chaplain-in-Ordinary, Bishop of Zealand, member of the Board of the University and Grammar Schools, the Royal Danish Society, etc., whose criticism, 'Betragtninger over Aanden i Naturen' [Reflections on SiN] in Theologisk Tidsskrift [Theological Journal] terminated an otherwise lifelong friendship. RL.

cilable exchange of opinions is most clearly stated by Ørsted's rejoinder, which went as follows.[22]

Firstly, of course, the bishop could not accept (or perhaps understand) Ørsted's idea of the immutability of the laws of nature and the variability of nature. They were unequally armed for the debate because while Ørsted knew his Bible, Mynster was a mere amateur in science. Mynster could not understand how permanent laws of nature could produce changes of states of nature, that mountains could rise from the depths of the sea, that animal species could be extinguished and new ones emerge, and so on. But he did believe that Genesis and many dogmas were at odds with science, and hence he rejected the relevance of science to religious matters. God was not conceived by reason, according to Mynster, He could only be believed. There could be no smooth transition between faith and knowledge, but only a leap.

Secondly, Mynster was unable to accept that God should be subjected to the laws of nature. He rejected Ørsted's assertion that God and Reason are synonymous, believing that God is elevated above reason and has the power to suspend or change the laws of nature. God is capable of breaking the laws of nature, which he did as he sent his son to save the world, and when the world crucified him, made him rise from his grave to be reunited with his father in heaven, a faith he confessed at every Sunday's service. Ørsted argued that God is behind all laws of nature and that they are so wisely construed that they are equipped with unknown self-regulating mechanisms, like the centrifugal governor preventing the steam engine from running riot. The progress of civilisation is a confirmation of the dominion of reason and 'the secret plan of nature'. Hence divine reason is able to govern the world without resorting to miracles.

Thirdly, Mynster did not share Ørsted's ethical optimism that the arrow of history is hurrying towards more progressive civilisations. The bishop maintained the Christian philosophy of history according to which man had fallen and lost his paradisiacal innocence and therefore he would be able to fulfil his original destiny only upon having embraced his saviour. The Holy Scriptures testified to the belief that changes in the state of nature had happened in consequence of the fall of man, or perhaps even more than one fall, for after the expulsion of Adam and Eve from Paradise, the natural span of human life was several hundred years (Methuselah), but after the flood the laws of nature shortened Noah's life expectancy to the one we still know. Thanks to God's interference it will be changed again by the resurrection of the body to everlasting life.

Fourthly they resumed the classical discussion on the relation between faith and knowledge: did God reveal himself in the Holy Scriptures or in the book of nature? In *The Soul in Nature* Ørsted had fought superstition as being contrary to reason. As he had already argued thirty-five years before during the controversy on pantheism, the cultivation of science was essentially a religious worship. Ørsted had acquired a minimal conception of God beginning and ending with the conviction that the laws of nature were the only testimonies of the existence of Allreason or God. This was his complete concept of God. He was not a pantheist, but when people thought that he was, it was perhaps due to his assertion that we can know nothing about God's essence beyond what we read in the book of nature where he reveals himself. The concept of God represented by the church was of necessity man-made on the basis of interpretations of the Bible written throughout a long historical period. Ørsted found that the concept of God celebrated by the church was a projection of man's different cultures at different times. Hence the moral maxims of the church must of necessity derive from circular arguments using sources limited by place and time.

To Mynster the relation between faith and knowledge was different. Faith was paradoxical, and theology did not provide knowledge about God, but knowledge about what Christians believe. The mistake of deism, the religion of reason during the Enlightenment, was that much of what Ørsted believed that he knew, he only knew because he believed in authorities. But new research incessantly changes our knowledge, so that we know things today we did not know yesterday. Mynster referred to a paper on the relationship between science and belief he had given to the Royal Danish Society.[23]

Ørsted believed that his simple faith was a truth he had found, whereas Mynster's teachings were metaphysics invented by the church. He rejected the bishop's accusation of belief in authorities as merely polemical. Of course everybody cannot research everything first hand. Most things we need to learn from others, but this is not tantamount to blind subjection to authorities, but rather using one's reason in trusting others. Faith should not serve the purpose of saving us, but should enlighten us to find our calling and fulfil it. Even if Mynster did not say so explicitly, he considered Ørsted an atheist. Ørsted would have repudiated that label. He saw himself as a non-believer in the sense that he rejected the Lutheran concept of God and the rigid dogma of the state church. The rituals and sacraments of the state church were embedded in superstition and therefore a matter of indifference.[24] As he had heretically preached from the pulpit against Grundtvig, the cultivation of science is tantamount to religious worship. There was a widening gulf between Mynster's sermons on sin and salvation by grace of faith and Ørsted's gospel of natural science and reason about the True, the Good, and the Beautiful, only a stone's throw away in Studiestræde.

Andersen was obviously upset by Mynster's ruthless criticism of Ørsted. Since his arrival in Copenhagen in 1819 he had grown up with his patron's philosophy of religion, which had gradually permeated his own mental universe. While the Ørsted family were spending their summer holidays on Glorup and Langeland, Andersen had been travelling in Sweden for three months, and when he returned he settled in the manor of Glorup to write *In Sweden*, his account of the journey based on his diary notes. *In Sweden* is one of Andersen's most meticulously composed works, containing his reflections on Ørsted's *The Soul in Nature*, the second volume of which Ørsted was in a hurry to edit. Andersen could not dismiss Mynster's criticism with a single gesture. On the one hand the bishop was speaking on behalf of the state church, and on the other hand, Andersen wrote to Ørsted,

> 'I cannot understand our bishop;... I find it far more beneficial in one's faith also to know. Our Lord will approve of being seen through the reason he provided us with himself. I shall not go blindfolded to God; I shall keep my eyes open, look out and be aware, and if I do not arrive at another goal from that of the one who merely believes, at least my thought has been enriched.'[25]

Andersen tried to appear convinced, but inside he trembled with fear. He wrote himself out of his scruples at Glorup, and read aloud his manuscript in Studiestræde while conversing on *The Soul in Nature* with its author, rewriting his draft and adding new thoughts. The outcome was two chapters of *In Sweden* at which Ørsted nodded approvingly.

The first, 'En Historie' [A Story] is a romance that serves as a frame for a deeper purpose.[26] It begins with an idyllic description of nature: apple trees are flourishing, cats are purring, birds are twittering. Nature is generous and people are blissful. Church bells ring for worship. Suddenly the idyll changes. Inside God's house the vicar is preaching on the punishment of sin, chastising the infidels with whips and threatening them with the torments of the damned in metaphors reminiscent of Dante's *Inferno*. The contrast between nature and church is total, and so is the confusion of the minister's wife listening to the sermon. Back in the parsonage she asks her husband to explain why God, supposed to be infinite love, punishes so mercilessly. By now the plot of the romance is established. Andersen interpolates a dream, the minister's dream, but he only reveals at the end that it is actually a dream. In it the minister is sitting by his wife's deathbed, and a few days later he has buried her. Then she reappears as a ghost haunting him. She cannot find peace of mind unless her husband is willing to help her. He must go out in the world and pull out just one hair from an unforgivable sinner; this would

Fig. 162. Hans Christian Andersen. As opposed to Big Hans Christian (HCØ) Little Hans Christian (Andersen) was an expert at sitting quiet during the many seconds of exposure. Daguerreotype by unknown. OBM.

be no difficult task considering his infernal sermon. The minister goes to town where there is no lack of evil people committing the entire range of the seven deadly sins. He finds a haughty man in his palace, an avaricious man clinging to his gold, and an arsonist sitting in gaol. Facing these sinners the minister does not find it in his heart to tear out a single hair. So he returns home empty-handed asking God to handle the matter instead. Suddenly he wakes up. There right in front of him his wife is sitting, ready to be kissed. Outside the spring sun is shining and had children been present they would have turned somersaults. God has interceded and the story has a happy ending as romances must have. God has turned out to be Ørsted's loving divinity of nature and not the revenging god of the State Church. The prelate, who could have been modelled on Meisling or Mynster, was quite mistaken, and Ørsted's optimistic Providence carried the day.

The second story, 'Tro og Videnskab' [Faith and Science],[27] is apparently influenced by conversations in Studiestræde. It does not really fit the shape of a dialogue, but is more like a sermon construed dialectically, so one is led to think that Andersen is testing Mynster's sharpest arguments on Ørsted, who plays the same role as Alfred in his own dialogues. Andersen brings out one absurdity of Ørsted's optimistic dualism after the other. Does not science kill the faith in the immortality of the soul? Would it not be safer to stay in the naivety of the faith of one's childhood than to live in the dominion of reason? And what about prayers? They were Andersen's wings that carried him up to his comforting God, but does it make sense to pray to Reason? Is not Ørsted simply a haughty scientist believing that he can achieve the same wisdom as our Lord? If the erudite are correct that nature and people are governed by eternal laws of nature, do we not live then in a world totally predetermined. And if so, what is the use of folding one's hands to offer a prayer to God? Serious scruples like these made Andersen tremble. Was there any meaning in times of adversity if there was no longer the hope of immortality to cling to? Listening to the bishop all his worries could be epitomised as follows: 'Science lies as a stone on my grave, but it is my faith that makes it burst'.[28]

'It is not so!', Andersen hastens to say. Despite serious scruples, he acquiesces in Ørsted's maxim of the three aspects of reason (the True, the Good, and the Beautiful) in his album from 1833. Faith and science are not pointing to different directions, but to the same. God's spirit is revealed in his work of creation. We find his traces everywhere in nature, in acoustic figures for example. As the soul of man radiates from his eyes and the smile from his lips, God's threefold reason peeps out from life itself. The scientist scrutinises its impressions and the artist creates its ideas. The Biblical myth about the tree of knowledge is not worthy of credence. Why should we come to harm by becoming clever? The caricature of the scientist as a haughty person playing God is ill-founded in Ørsted's case. Little Hans Christian knew him too well. 'Science is like a chemical test showing that the gold is genuine', Andersen summed up, as trustingly as his patron.[29]

As well as Andersen, Goldschmidt wrote the most elaborate and promising review in *Nord og Syd*. He had been sitting at the fountains of the Tuileries in Paris enjoying *The Soul in Nature*. The book had provided him with an experience like the impression he had received the first time he saw the starry heavens through the binoculars of the Jesuit Observatory in Rome. To him the book anticipated a breach with Romanticism, an entirely new view of life and a new literary programme. In his eyes *The Soul in Nature* was a modern turning point that anticipated Georg Brandes' periodisation by two decades.[30] *Nord og Syd* like Brandes's *Hovedstrømninger i det 19. Aarhundredes Litteratur* [Main Currents of European Literature] set out to introduce a European context in which Danish literature and politics could be reflected. To Goldschmidt, Ørsted represented progress in European cultural life by basing a philosophy of life on modern science and a cosmopolitan attitude. Instead of Romanticism, mysticism, pietism, and nationalism Ørsted maintained the ideals of the Enlightenment: freedom, humanism, and tolerance. Freedom was justified by free research that does not acquiesce in well-established views, but challenges them, tries new methods, and presents results that nobody has seen before. Ørsted's humanism consisted in seeing the reasonable and divine in his fellow human beings, and in generously contributing to their education. And his tolerance, Goldschmidt wrote, was elevated above the common indifference in that he preserved his esteem for people with whom he disagreed while at the same time maintaining his disagreement with their standpoints. Adversaries should be persuaded by arguments rather than be coerced by power or humiliated by insults.

Goldschmidt saw Romanticism as an irrational reaction against the noble Enlightenment ideals of freedom, human rights, and tolerance. The Romanticists had had good reasons to combat the vulgarisation of these ideals, when the terror of the revolution had been turned against freedom and Christianity. But now Romanticism had become an irrational and nostalgic ideology, looking back to medieval mysticism, chivalrous tournaments, pietistic introversion, and asceticism as well as nationalistic self-conceit. When these trends emerged Goethe, Kant, and Humboldt had remained steadfast supporters of classicism and Enlightenment progress, and so had Ørsted in Denmark. This was the perspective in which Goldschmidt saw Ørsted's efforts in cultural life in favour of science and popular education, freedom of the press, the rule of law, aestheticism, and religion. In a long and contemporary account of European and Danish literature he sided with Ørsted's programme, in which people are elevated by science, while they feel its spiritual strength, observe its conquests, and get hope of victory, belief in perfectibility, and reassurance of purpose.[31]

Goldschmidt was totally at odds with the criticism that found scientific insight damaging to the enjoyment of the arts. To him knowledge of astronomy, acoustic figures, and the theory of light contributed to the pleasure of the beautiful in nature (the starry heaven) and the beautiful in the arts (symphonic music, imaginative literature, or landscape painting). He was delighted with the view of the high peaks of reason and of the expansion of the soul in the universe, the intuition of God as the consciousness of the laws of nature, for reason eliminates prejudices of coincidence and meaninglessness.

*

While Danish reviews of *The Soul in Nature* were directed towards the aesthetical and religious aspects of the book, Germans responded in particular to the philosophical. It was translated by K.L. Kannegiesser as *Der Geist in der Natur*[32] and reviewed in newspapers in May 1850 in Vienna, Leipzig, Dresden, Köln, Berlin, and Hamburg as well as in philosophical journals.[33] Some reviews were influenced by the civil war in Schleswig-Holstein. Reviewers juxtaposed Ørsted's views and German *Naturphilosophie*. They were compared to Humboldt's *Kosmos*, and in one case to Davy's *Consolations in Travel*, three examples of famous scientists drawing far-reaching philosophical conclusions from their studies, subtleties that the Germans yearned for, the English tolerated (if they were funny), but the French usually disliked. In a few months the first edition was sold out, and in October the second edition appeared. In general the reviews were positive, some very positive.

Some German reviewers were confused as to where to place *The Soul in Nature*.[34] What nobody in Denmark had pondered became the main issue in Germany. What was Ørsted's position vis-à-vis Kant, Fichte, and Schelling? Was he a *Naturphilosoph* or a Steffensian or ...? In *Allgemeine Monatsschrift für Litteratur* Dr O.L. Erdmann characterised Ørsted as partly a Kantian. And a certain Dr Frauenstädt continued to consider to which philosophical school he belonged. The dynamical lead was acceptable, but Ørsted offended against Kantian epistemology by using regulative ideas instead of following Kant's categorical scheme. When natural phenomena cannot be seen in time and space, the categories become useless, the reviewer thought. Hence Ørsted made a basic epistemological mistake when insisting on reason leading to science. Erdmann rejected Ørsted's statement that the reason of natural laws was the same as the reason of the scientist. However, Ørsted's discovery of electromagnetism had been acknowledged long before by the international research community, so what Erdmann really said was that Ørsted's erroneous philosophy had led to a correct physical result. The German reviewers considered the title to be an unsustainable philosophical postulate. A better title would have been 'The Nature in the Spirit' which would have indicated that the book, contrary to Kant's epistemology, dealt with a physical solution to a philosophical problem. But they did not mention that Ørsted had actually confirmed Kant's dynamical theory from *Metaphysische Anfangsgründe*, and that he was the only one who had verified it experimentally.

Ørsted's thought experiment said that the polar forces of electricity interchanged with those of magnetism. He did not claim to be saying anything about electromagnetism per se. On the contrary, like Kant he denied that any science could be produced about the noumenal, the only thing he talked about was the effect. Therefore, Ørsted's assertion in *The Soul in Nature* was exactly the opposite of what the two German critics accused him of asserting. They spoke on behalf of philosophy and saw Ørsted as a physicist who, while he had undeniably made a true discovery, had done so at the expense of philosophy.

From his Norwegian observation point, Christopher Hansteen did not look at it that way. When he received *The Soul in Nature*, he was reminded that in 1820, when Ørsted had made his discovery, he had reasoned in the same way. How could a priori knowledge be generated unless Ørsted's statement were true: 'If the laws of our reason were not in nature, we would not succeed in forcing them into it; if the laws of nature were not in our reason, we would not understand them'.[35] Presumably, all scientists thought like that, provided they were not too narrow-minded. Hansteen expressed his gratitude for the book with a quotation from his own writing:

> 'The internal invisible causes of the changes taking place in nature we call natural forces. [There must be] an inner activity by itself and it cannot be thought without an idea in advance determining its direction; for a completely undecided activity, an endeavour in all directions, cancels itself out. These ideas we call laws of nature..., and they could not diverge from the laws of our own thinking; if they did, all science would be in vain.'[36]

Mathilde immediately sent Kannegiesser's translation to London in order for Joanna Horner and her sister Leonora, both having grown up in Germany, to translate it for the British market. The two sisters even started to study Danish. The result was *The Soul in Nature*, dedicated to Mathilde.[37] It was published in January 1852 without being much noticed. The translation from Danish to German to English resulted in a rather poor edition. Charles Darwin read it amid his collections in Kent while arranging his notes from the Galapagos Islands. His comment was brief and merciless: 'Dreadful!'.[38] The two sisters hoped that to a certain degree they had fulfilled Mathilde's father's great wish to have his philosophy of nature made accessible to a British audience. They sent the book to Herschel, Murchison, and Lyell among others, but we do not know their reactions. The Horner sisters would have liked to translate more Ørstediana, but Mathilde found that many of her father's speeches and obituaries were less interesting for non-Danish readers.[39]

What he said about the good, the beautiful, and the true,
was of as much value to most people as a nutmeg would be to a cow![1]

59 | 1850–1
Big and Little Hans Christian's Modern Turning Point

WHILE THE Ørsted family toured Funen and Langeland, Andersen spent three months in Sweden, and on returning he settled himself at Glorup manor house to write his travel book *In Sweden*, in which he commented on the criticisms of Ørsted's *The Soul in Nature* by Bishop Mynster as mentioned in the previous chapter.

In his fairy tale *The Shadow* from 1847 he had used Ørsted as his model of 'the learned man', writing 'Everything went wrong for the learned man. Sorrow and trouble pursued him, and what he said about the good, the beautiful, and the true, was of as much value to most people as a nutmeg would be to a cow. At length he fell ill.' Andersen felt that Ørsted was concerned that his persistent attempts to arrange science, moral philosophy and aesthetics into a new synthesis with the title 'All Existence a Dominion of Reason' did not achieve the recognition he was hoping for. Drafts were read with a scepticism that did not correspond to the star treatment he was used to enjoying in the European research community. His misgivings became real when Herschel and Humboldt reacted half-heartedly, and Molbech, Mynster, and Grundtvig with hostility towards *The Soul in Nature*. Apart from Goldschmidt and Hauch, Andersen was the only admirer of Ørsted's philosophical testament. Big and Little Hans Christian became fellow-sufferers. The lukewarm response and harsh criticism (and from Grundtvig even ridicule, although he confessed he had not read it nor had any intention of doing so) was so disappointing that the scant recognition received was unable to make up for it. 'A nutmeg to a cow!'[2]

As years passed an ever-closer complementarity between the two gradually developed. What had begun in 1820 as a distinctive father-son relationship, in which big Hans Christian like Jonas Collin was the well-wisher and patron of little Hans Christian, grew into a profound artistic friendship that then developed even further and ended up as a relationship of equals. In the 1820s the unsettled and practically orphaned boy who felt so alone in the world had sought shelter in Studiestræde and found the door open to a home with a family life. There he was asked to dine, at first on Fridays later on Wednesdays, and there he received spiritual nourishment from around the table as well. In the 1830s little Hans Christian became the budding poet and storyteller in need of patronage and recommendations to gain travel grants for inspiration abroad. Ørsted used all his influence at court to favour Andersen, and perhaps even gambled with his esteem with Frederik VI by recommending the poet for a post at the Royal Library. At the same time Ørsted, having resumed poetry himself, hauled him over the coals when he received a piece of bad work like *Agnete and the Merman*, but conversely he also accorded the most unreserved praise when he received his pearls, such as *Tales, Told for Children*. In 1835 Ørsted had encouraged the poet with this unsurpassed declaration of confidence: 'If your novel *The Improvisatore* makes you famous, your fairy tales will make you immortal!'.[3] The combination of social insecurity and an insatiable urge to gain unconditional recognition made Andersen an extremely vulnerable artist. In the 1840s he felt safe and at home with the Ørsteds, who appreciated both factors. Big and Little Hans Christian were not names with which they addressed each other; they were only used when talked about by others, and they continued to say '*De*' to one another (as opposed to the intimate '*Du*'). Considering Ørsted's unforgettable praise it is no wonder that later, in an interview with his travelling companion Nicolai Bøgh, Andersen confessed, 'I think Ørsted is the man I have loved the most.'[4]

From a superficial point of view the differences between the two men are conspicuous, and it takes some imagination to understand what brought them together. Big Hans Christian could easily be Little Hans Christian's father, but as far as the sizes of their bodies are concerned, Big Hans Christian was the smaller who had to raise his head and look up to Andersen. Intellectually, Ørsted was a walking encyclopaedia, while Andersen's knowledge was scattered, at *philosophicum* he did not even know the phenomenon of electromagnetism. What Mrs Møller had been for the Ørsted brothers, Gitte became for Andersen. And what Manthey had been for the pharmacist's son from Rudkøbing, Big Hans Christian became to the cobbler's son from Odense. Whereas Ørsted was able to reciprocate Manthey's services, it is difficult to see what Andersen could do for Ørsted, particularly in the beginning. But by the 1840s their relationship had become equal. The weekly conversations around the dining table gave each of them a deeper appreciation of the other's way of thinking. Andersen became more and more infatuated with Ørsted's philosophy of life. It penetrated his artistic mind and was reflected in his stories, among others *The Ugly Duckling*, *The Bell*, *The Shadow*, and *The Drop of Water*. At the same time Ørsted was taken with Andersen's spontaneous ingenuity and secure grasp of form, and the direct, personal style in which he wrote his fairy tales exactly as he told them to Ørsted's children. Andersen's oral style could have no place in the professor's articles, although he was an ardent inventor of new words. The poet turned out to have developed a mature judgement about the genres of the scientist's works. His dialogues were not peopled with humans of flesh and blood, Andersen observed. Alfred and all the other figures were feeble personae, mere marionettes in Ørsted's hands, and being without lives of their own they did not work. The issues and the arguments were all correct and convincing, but what did it help if the interlocutors did not captivate the hearts of the audience? Literature must enthral people's emotions, otherwise it falls to the ground, and the dialogues and *The Airship* were boring. They would never become popular literature as intended. Just imagine! Andersen ventured to say this to Ørsted about his dearest works.[5] In the last five years of Ørsted's life they read everything they wrote aloud to one another before it went into the press. They also recited passages from the Bible, including Genesis, and they discussed religious questions, among them the immortality of the soul. But on top of their literary interests they were also very fond of one another as the human beings they were, each retaining his childish innocence. Their relationship was so confident that it easily survived even unpleasant 'truths', as long as they were sincere, which they were.

*

In August 1850 Andersen asked Ørsted to send him the second volume of *The Soul in Nature*. He was eager to learn how his patron had responded to Mynster's ruthless critique. In his reply Ørsted almost routinely comforted Andersen, advising him to ignore his critics. Not even Goethe and Schiller could escape ignoble attacks. This time Andersen's play *Ole Lukøie* was the target. It was staged at the Casino Theatre and more than two thousand tickets went like hot cakes. Ørsted asked him to write a non-fiction piece about the aesthetics of the fairy tale, and Andersen complied. *In Sweden* contained two essays supplementing Ørsted's aesthetics of nature and physico-theology from *The Soul in Nature*.

Like Plato and Kant Ørsted adhered to a dualistic world picture. On the one hand there was the earthly, timely, sensual, real nature consisting of empirical phenomena and man as part of nature. But on the other hand there was a higher world of ideas, too, governed by Almighty God who was behind the laws of nature constituting the noumenal world, the dominion of reason. The difference between Plato and Kant was that Plato's world of ideas was inaccessible, as it was only known via the shadows cast in the famous parable of the cave. In Kant's philosophy the noumenal in the sense of 'the thing-in-itself' is uncognisable, too, but even so there is 'something' that produces phenomenal effects in the physical world—electromagnetism for instance. The connection between the empirical world and the eternal ideas is established by reason. It is man's privilege and obligation to use his reason to emancipate himself from his self-imposed tutelage. It is Kant's merit to have analysed how reason interacts with the senses via the categories of the understanding ('Ariadne's Clew') to acquire true knowledge. When it comes to ethical action, however, reason must rest on a maxim, the categorical imperative, belonging to the realm of metaphysics. Kant endeavoured to bridge this dualism by means of an aesthetics consisting partly of physical phenomena (empirical nature) and partly of metaphysical elements united in the arts and leading man's imagination to eternal ideas.

Throughout his life Ørsted worked to consolidate this bridge. He believed he had made a breakthrough when, taking his point of departure in Kant, he succeeded in extending the natural sciences from their mechanical aspects to encompass also a dynamical area. The latter is the effect of fundamental forces, heat, light, electricity, and magnetism, uncognisable per se, and so noumenal. The discovery of electromagnetism affirmed this dynamical system, and a completely new dimension of science was opened up to scientific research. To Ørsted this discovery was a proof of the correspondence between phenomenal and noumenal reason, the bridge between the real and the ideal world. A new dominion of reason had emerged: the task of science was to deduce (noumenal) laws of nature from phenomena, and the task of the arts was to present the (phenomenal) experience of the ideas gained in the form of music, painting and poetry. 'We endeavour either to show the sensual world as a revelation of the world of reason, or to express the world of reason by means of the sensual.'[6]

This worldview had torn down the wall between the two realms or dominions of dualism. It also marked a division of labour between the two Hans Christians. Andersen responded to *The Soul in Nature* by wishing to indulge in the study of modern science.[7] At the same time he ventured to urge upon Ørsted the judgement that his poetry contained too much for reason and too little for the senses and imagination. Conversely, Big Hans Christian mildly reproached Andersen that his patriotic song 'I Danmark er jeg født, dér har jeg hjemme' [In Denmark I was born, she is my home] totally ignored the sciences, so he proposed a new verse highlighting the importance of scientific research and education. It has never been adopted.[8]

In *Poesiens Californien* [Poetry's California], one of the chapters in *In Sweden*, Andersen whole-heartedly embraced Ørsted's new worldview that implied a breach with the state church and the Romanticist aesthetics and was hence a modern turning point for science and the arts. Accordingly, the arts and the church should no longer romanticise the past and abide by superstition that made people stick to obsolete doctrines and silly mysticism, but should use allegories rooted in modern science, a gold mine of poetry. Andersen portrays a young poet entering a library to find material for his works. Two librarians offer their help: 'Superstition', an old crone with crutches, tarot cards, amulets and other witchcraft, and 'Science', a lovely youth with a lustrous sword and a fiery soul. 'I am the strongest in the region of romance!' the old crone says in recommendation of herself, 'Sing as a skald of what you see!'. The poet looks around observing gothic churches with stained glass windows, tumbledown baronial castles with drawbridges, a deep cellar with a basilisk, a monster with a lethal glance hatched from the egg of a cock, and a golden chalice on the altar table that had once served as a drinking cup of trolls. 'I give you a rich recompense!', the youth promises, 'Truth in the created, Truth in God!' He lifts his sword, cleaves in two the deep vaults in the churches and castles and stabs the basilisk to death. Instead the voice of science sounds all over the world. The age of miracles appears to have returned.

> Thin iron ties were laid over the earth, and along these the heavily-laden wagons flew on the wings of steam, with the swallow's flight; mountains were compelled to open themselves to the inquiring spirit of the age; the plains were obliged to raise themselves; and then thought was borne in words, through metal wires, with lightning's speed, to distant towns. 'Life! life!' it sounded through the whole of nature. 'It is our time! Poet, thou dost possess it! Sing of it in spirit and in truth!'.[9]

The choice was an easy one for Andersen, Hauch, and Goldschmidt, but Ingemann preferred the enigmatic Ossian and Walter Scott and 'Superstition', and Bishop Mynster refused to ask science for permission, as he phrased it, to love Homer's antiquated allegories.

Andersen recited from his manuscript in front of Ørsted, who had for a long time regarded Romanticism and the cathedral of Cologne with the same critical eyes, and in *The Soul in Nature* he had written about 'showpieces from the poetic arsenal of the past'.[10] Even before Little Hans Christian could walk, Ørsted in one of his first dialogues had slated mysticism. 'Poetry's California' was replete with their shared thoughts. But when Little Hans Christian arrived at the following passage he no doubt threw a nervous glance at the professor's expression: 'It is not our intention that the poet should versify scientific discoveries; the didactic poem, even in its best form, is and will always remain a mechanical puppet lacking fresh life'. This was a devastating judgement of Ørsted's most ambitious experiment in poetry. *The Airship* had been shot down (ch. 46). There is no indication that Ørsted took offence. He rejoiced in Andersen's company, also (or perhaps particularly) when he was relentlessly frank. Time and again he had suggested that independence was the first mark of reason. When Andersen had received the second volume of *The Soul in Nature*, including his response to Mynster, he answered from Glorup: 'I am pleased with your book, and also pleased with myself that it is so easy to read that it seems to me to be the result of my own thinking; reading it I seem to be able to say: "I should have said the same!" The truth in it has passed into and become part of myself...'[11] Andersen had received a letter from Humboldt with greetings to Ørsted and thanks for *Naturlehre des Schönen* ['Physics of Beauty'] translated by H. Zeise, the same man who translated Andersen's fairy tales into German. He made Ørsted happy by telling him that *The Soul in Nature* was displayed in the German bookshop in Rome, and Mathilde by telling her that in Germany it was placed next to *In Sweden*.[12]

The modern relationship between art and science that Ørsted and Andersen introduced in their books demanded an entirely new literary direction, a shift of paradigm, where the artist makes literature the expression of the true, the good and the beautiful and takes the part of the progress of science and technology. 'You have often been accused of lack of education', Ørsted one day said jokingly to Andersen, 'and perhaps you are the poet of ours who will accomplish the most for science!'.[13] Andersen was fascinated by Ørsted's comprehensive philosophy and liked to mediate it in his stories. The conquests of science were to impregnate all art forms, particularly landscape painting and imaginative literature. And the antiquated showpieces were to be relegated to the poetic arsenal of the past, for modern poetry and prose required new metaphors, and Ørsted himself was a provider of new words. Science was often described in textbooks and academic periodicals in mathematical language as experiments and analyses, but the arts taking their point of departure in the world of ideas demanded narrative forms of mediation in new genres.

*

The Galoshes of Fortune was an allegory in which people could go on long journeys in both time and space by wearing the galoshes.[14] Councillor Knap had been arguing about Ørsted's essay in the almanac *Old and Modern Times*; of course, the author favoured his own day, while the

Councillor preferred the times of King Hans as the noblest. When leaving the discussion for home, he put on by mistake the galoshes of fortune instead of his own and stepped directly into mud and mire in the street, because in the middle ages there were neither pavements nor sewers. A similar expression of Ørsted's belief in progress appeared in *The Thorny Road of Honour* where the fairy tale has become a realistic genre, or as Andersen put it, 'The story is very closely akin to reality; and it still has its harmonious explanation here on earth, while reality often points beyond the confines of life to the regions of eternity. The history of the world is like a magic lantern that displays to us, in light pictures upon the dark ground of the present, how the benefactors of mankind, the martyrs of genius, wandered along the thorny road of honour.'[15] We shall look at some more of Andersen's stories, so as to move towards an understanding of what it takes to transfer the noumenal in science and religion to a senses-based everyday account.

The Philosopher's Stone is staged in an idyllic fantasy setting in the Tree of the Sun, which has a huge crown, and a castle of crystal with a view over all the countries in the world. Inside the whole world is mirrored on the walls. Here lives the wisest man in the world (Big Hans Christian), the researcher perusing the book of truth, where everything is told about reason in its three dimensions, the true, the good and the beautiful—except for life after death. That page is empty, which worries the wise man. The Bible has given him a word of comfort on everlasting life, but he wants to read it in his own book, which is silent on the matter. This allegory corresponds to our knowledge of the two. Ørsted's dialogue *The Spiritual in the Material* breaks off where Sophie asks Alfred to explain immortality and receives the answer that this subject requires an entirely new talk that would have to wait for another time. The poet and the professor had been sitting in Studiestræde looking critically at parts of the Bible resulting in their rejection of the doctrine of original sin and salvation, which made Ørsted a heretic in Mynster's eyes, but not in Andersen's.

The wise man under the Tree of the Sun has five children, four sons and a daughter, and they all love each other very much. The number of five is due to each of them being richly endowed with one of the senses, the sons with sight, hearing, smell, and taste, and the daughter with a sense of touch, though in return she is blind. To orient herself she has the gift of a thread that helps her to feel her way. The sons with their strong senses are well equipped for empirical research. They can readily set out into the world to find the philosopher's stone symbolising the three aspects of reason. The father tells his children that as by studying nature one becomes aware of its hidden forces and reason and God, so by studying people one becomes aware of the existence of such a precious stone, a jewel of consciousness, though he can tell no more about it, only that it does really exist. This wording is obscure unless it is seen in the light of Ørsted's epistemology, 'Ariadne's Clew' helping the researcher to orient himself in the empirical reality, but not in religion, which by necessity must be reduced to a minimum. It is a waste of time to investigate the concept of God by metaphysical speculation. God is untraceable in time and space. And about that of which nothing can be said, one should remain silent.

The subtlety of Andersen's story is that he makes the four sons have the same dream, which also haunts the imagination of many people and perhaps did formerly for him as well, until Ørsted intervened. They dream that they set out into the world to search for the philosopher's stone, or in other words that their senses are capable of finding out what the true, the good and the beautiful are. And when they wake up they rush into the world in turn to make their dream real. It is not the father (alias Ørsted) who sends them out, for he already knows that this would be a vain effort. No, they do it on their own initiative, only to return home without success. Reason is not found in the real world, where one is disillusioned like the four brothers. One after the other returns crestfallen. They can neither see reason nor hear it, nor smell it, nor taste it. Instead they have met falsehood and malice and ugliness, sometimes personified as devils. To put the investigation in a totally absurd light one of the brothers ascends in a balloon (which the narrator makes fun of, asserting that it is not invented at all) and instead of reaching his objective the balloon gets caught on the spike of a weathercock on top of a church spire, and is spun round and round making him completely bewildered. 'They do not fare well!' the father says, looking lost at the empty page in the book of truth.

One night the blind daughter has a dream. She has gone out to look for her brothers. She does not find them but instead finds the philosopher's stone, which she holds tight in her hand to give it to her father. Then she wakes up, and realises that it is not the precious stone she holds in her hand but her spinning wheel, on which she has for a long time spun Ariadne's thread. She ties the thread firmly to the Tree of the Sun and sets out in the world with her fervent sensitivity. She finds the light of the rainbow, the warbling of birds, the smell of apple trees, but also hideous yells and cries, in short a mixture of true and false, of good and evil, of beauty and ugliness. Like her brothers she also meets the devil himself, who slyly makes a copy of her, so that nobody can find out who is the genuine girl and who is the copy. But she does not find the philosopher's stone. She then follows 'Ariadne's Clew' back to her father, yet not completely empty-handed, for in her palm she has caught specks of dust of the true, the good, and the beautiful that the wind has whipped up on her way. The dust escapes from her hand and floats down onto the empty page of the book of truth, where immortality should have been explained. The dust lands on the page forming one shining word, the word 'b-e-l-i-e-f'.

The story presents an experience rather than a solution, even if Andersen has given us the key to the allegories.[16] It is not a philosophical thesis, but a fairy tale, and these rarely gain by literal interpretations. Still, obviously Andersen has borrowed Ørsted's favourite metaphor, 'Ariadne's Clew', to show that epistemology has its limits vis-à-vis the noumenal, that there is a difference between science and faith, though they are not mutually exclusive, that the world does not in mysterious ways keep the philosopher's stone secret, that neither superstition nor empiricism is of any avail in religious matters, and that life contains so many fragments of the ideal that a person endowed with sensitivity can find impressions of the eternal in the temporal (to the extent to which we live up to the Ørsteds' maxim 'We should make our entire life an impression of the eternal reason that is revealed in us'), that immortality is a possibility to be believed, and finally that belief is not tied to the sacraments and doctrines of the church.

Usually literary critics see *The Bell* as Andersen's allegorical narrative of himself (the confirmand) and Ørsted (the prince), both infatuated with the distant ringing of a bell. Together they seek and find the bell on a walk in the country across fields, woods, and open waters. Other characters in the story are enticed by its sound, but give up on their way, because they lack the sense for nature. A confectioner puts up his tent halfway along to sell cakes and tea in the romantic environment, and some bread-and-butter students are also attracted by irrelevant matters such as writing tedious theses to acquire a diploma and a secure post. A bunch of confirmands also lose their way in the woods; they do not find the bell either, but some fool's bells, and give up. The poor confirmand and the noble prince remain. There is no indication that it is a question of a poet and a scientist finding the same goal although taking separate routes. It is the social, not the intellectual difference that is emphasised. The poor confirmand is probably an autobiographical invention. In 1844 Andersen paid a visit to Christian VIII on the island of Föhr, and one day he asked for a royal carriage and a coach and valet in red livery to be put at his disposal. As a young confirmand Andersen had been ill-treated by Reverend Tetens, who among other things had placed him at the back of the aisle of the church together with the poor confirmands of the chaplain. Now this prelate had obtained an office on Föhr, and Andersen would not miss the opportunity to see him. The King granted his request, and Andersen allowed the carriage and horses as well as coach and valet in red livery to park in the rural deanery while inflicting a cruel revenge on his clerical oppressor.[17]

In the story, written just after this incident, the poor confirmand with the short-sleeved jacket and clogs climbs the rocks and gazes across the sea at the sky in all the colours of the rainbow and countless bright stars after sunset. Then he catches sight of the prince. 'They ran towards each other and held each other's hands in the great church of nature and poetry, and above them rang the invisible, holy bell, happy spirits surrounded them, singing hallelujahs and rejoicing.'[18] There is a strong sense of the beautiful in nature combined with perseverance, which rewards the two with the same experience of nature. Andersen seems to say that experiences of nature have a holy character and are open to all whether they stem from the lowest or the highest social classes. The crucial prerequisite is that the experience of nature is the goal in itself and not some irrelevant side issue or even a means to other goals as it was to the confectioner and the students. The experience of nature like the cultivation of science is a form of religious worship in itself, as Ørsted said in his speech at the University in 1814. Big Hans Christian must have loved that point.

Andersen expounded Ørsted's programme in a range of stories. In *Thousands of Years From Now* modern technology has realised the American dream of seeing the entire old world in eight days by airship. In England passengers leave the flight to rush through the tunnel under the Channel to France, where the tourists re-embark to fly with Andersen, their experienced and knowledgeable tour guide, who tells them about all the famous authors, while they soar in the sea of air above the nations of Europe.[19] And in *The Muse of the New Century* he has sketched what the poetry of the future will contain. 'The showpieces of the poetic arsenal of the past' have been replaced by allegories of scientific discoveries and polytechnic inventions, from the forces of steam (*Master Bloodless*) via the airship of the Montgolfier brothers to Ørsted's electromagnetic telegraph and Daguerre's photography, without losing inspiration from the classical repertoire of music or the great literature of the world from Homer and Shakespeare and Molière. It is hard, of course, to prophesy in detail about the poetry of the future Muse, because she has only recently been born. However, she is not going to dope the readership with chloroform, but will offer a life-giving elixir in brief and clear prose and poetry, he promises.[20]

The so-called modern breakthrough is usually dated to around 1870, characterised by a literary paradigm shift from Romanticism to realism, from a religious to a secular worldview coloured by the progress of science (naturalism) and from nationalism to a cosmopolitan outlook. Perhaps it is worth considering a revision of this tradition and seeing the companionship between Little and Big Hans Christian (and Goldschmidt and Hauch) as the first step on the road to a modern turning point taking its point of departure in science and reason at the expense of the Romanticism of the Golden Age and with a scepticism vis-à-vis the doctrines of the state church.

Our Lord can bear to be seen through the reason he himself gave us.[1]

60 | 1849–51
Jubilee and Death

ON 5 October 1849 Ørsted's oldest son Christian died, hardly thirty-three years of age. The circumstances are unknown. He had become a graduate in forestry in 1836 and then district road engineer in Kronborg County. His father cared a great deal for him, for instance Christian had joined him at the Scandinavian Science Conference in Stockholm, and he recommended him to the King and to the prefect of Sorø County for a post as forest supervisor, but probably in vain. He must have died from some disease either in a hospital or at home, for in Gitte's ledger four rixdollars appear as payment to night nurses. Gitte had sewn a black gown out of twenty feet of merino and put black mourning crepe onto the hats the women of the family wore for the funeral.[2]

On 14 November 1849 Adam Oehlenschläger's seventieth birthday was celebrated on the Shooting Range on Vesterbro. Two hundred guests dined in the hall decorated with huge set pieces by Wilhelm Marstrand and Constantin Hansen. Clausen and Madvig were present as colleagues and representatives of the new democratic government, poets wrote songs, Andersen, Grundtvig, Hauch, Heiberg, Paludan-Müller, and Molbech were there, prelates Mynster, Martensen, and Tryde, too, as well as the Ørsted brothers, of course. Toasts were proposed and the atmosphere was uplifting; on a festive evening like this, where the homage was paid to the greatest skald of the North, old hostilities were put aside, Grundtvig and Clausen were briefly at peace, only Ingemann was absent, probably because Sorø was too far away.[3]

Oehlenschläger and the Ørsteds gathered together again on 21 December, when A.S. Ørsted's birthday was celebrated in Studiestræde as usual. A few days later Oehlenschläger fell ill, and he died on 20 January exactly two years after Christian VIII. Once again Bishop Mynster delivered the funeral sermon in a packed church draped in black and lit by chandeliers. The coffin was carried out to the accompaniment of J.P.E. Hartmann's funeral music, through the city gate and along the linden alley, passing the Liberty Memorial and the Shooting Range to Frederiksberg Cemetery, where he was laid to rest next to his parents and sister Sophie, whom he had buried thirty-two years earlier. A collection was arranged, and H.V. Bissen modelled a statue of the prince of poets that was cast in bronze and later erected at the entrance to The Royal Theatre.[4]

In March yet another sad event happened to the Ørsted family. Marie was the twin who had survived her sister Sophie in 1818. She had been married to Gitte's nephew, the theologian Peter Hasle, in 1843. Now she, too, was taken. Like her brother Christian she was only thirty-three. The family rented a carriage and drove the long and hard way to Asnæs vicarage, where Reverend Peter Hasle laid his wife to rest. Now Hans Christian and Gitte had lost both a son and daughter within six months, and altogether three out of their eight children.[5]

On 7 November 1850 it was the fiftieth anniversary of Ørsted becoming a university teacher, and he was celebrated for the entire day at Fasangaarden [The Pheasant Farm] in Frederiksberg Gardens, which was being refurbished for him to move in. Oehlenschläger had had it for his summer residence, but after his death Ørsted's many friends had managed to have it transferred to him and his family for life. It was arranged that Mathilde would look after the small garden. Four out of the many rooms had been furnished thanks to a national gift, and for the occasion Bissen had made a marble bust of Ørsted, which was later mass-produced in miniature in biscuit ware by the Royal Porcelain Factory. It was placed in the large garden room, where the reception took place. Only four women were present the whole day, Gitte and their three daughters, Karen Scharling, Sophie Dahlstrøm, and Miss Mathilde.[6] Ørsted was fêted against a sad background, and the mixed reception of *The Soul in Nature*, especially the devastating criticism by Mynster, overshadowed their former friendship and put a damper on the party.

A.S. Ørsted showed up with a set of silver cutlery to replace the iron set that had always been used in Studiestræde. A highly emotional letter alongside the gift reassured his brother that he was the best that Providence had ever given him. He thanked him for 'the beneficial influence on my entire spiritual development and mental direction' he had enjoyed for seventy years and for the comfort and support he had received at times of hardship.[7] Nobody had meant more to him than his beloved brother.

At midday the many well-wishers called. The weather was far better than expected on a Danish November day. Our protagonist had acquired a new toupee for the occasion. It was fixed by means of some steel wire that deflected a magnetic needle when he leaned forward to see what was happening. Heiberg had written a festive song saluting the man 'who followed the path of research to give us nature as our spiritual heritage'.[8] Forchhammer, far from being a great eulogist, listed all of Ørsted's scientific merits and handed him the contract for Fasangaarden, the family's new summer dwelling. Professor Madvig, the Minister of Education of the so-called

JUBILEE AND DEATH

November Government, did homage to Ørsted on behalf of the monarchy and pronounced him *Geheimekonferensraad*, the most distinguished title of the otherwise abolished rank system of the absolute monarchy, and a mark of honour that had never before been bestowed on a university teacher. Then a deputation from the University stepped forward presenting the professor with a doctoral ornament, whose Minerva head was framed in diamonds, and concluding the ceremony with a witty poem.

The Students' Association having been established in the same lecture hall and in the same week in which he had discovered electromagnetism thirty years earlier, pronounced him a member of honour.[9] The Free Masonic Order's Master of the Chair, Dr Carl Otto, handed Ørsted a diploma done in calligraphy making him an honorary member of the Zorobabel til Nordstjernen lodge [Z. of the North Star] as an appreciation of 'the lustre and glory you have spread in our two lodges as well as in the profane world by your name and your genuine Masonic character'.[10] Young polytechnic students and a deputation from the Industrial Association added yet another two eulogies to the series and more festive songs were sung. All deputations were served dinner, and at eight o'clock more than two hundred students turned up torches in hand. They formed a circle in the garden singing to their admired teacher: 'And the phenomenon will only be seen by sharp researchers' eyes as links in the chain of harmony. Therefore, true priest of nature! You accomplished so much. You have read the laws of nature and guessed the essential riddle…'.[11]

Gitte and his daughters were moved by the sight of the many guests of different social origin assembled to pay homage to the old patriarch. Ørsted stepped into the garden inviting

Fig. 163. HCØ's doctoral ornament framed by diamonds. A present from the University of Copenhagen in commemoration of his 50th jubilee as a university teacher. SNU.

645

everybody inside. There was little room for the two hundred students, but they all had a drink, and the old professor walked around sharing a toast with each individual. 'I shall do what I can', was the reply Andersen gleaned, when Gitte wanted to spare her spouse, asking the students to restrain themselves: 'No, no, my husband cannot possibly drink with all of you.'[12] The day of his jubilee was like a fairy tale.

*

On Saturday 1 March 1851 the entire Ørsted family was gathered at Nærum, 13 miles north of Copenhagen, for the birthday of Professor E.A. Scharling. This was a long and chilly drive by horse and carriage. The following morning Ørsted rose early to work in his study. There were so many thoughts he had to put on paper. His study was cold and so was he; his hands were trembling and could hardly keep his cup of tea still. He spilt tea on his papers. He had to go to bed again to get warm. He was adamant about rising again in the afternoon to proceed with his work. But it did not help, he had to stay in bed for the rest of the day and the next one as well. Then he let Professor C.E. Fenger, the family doctor be called for.

Fenger had often discussed medical problems with his patient. He had come to the conclusion that the medical profession was a hundred years behind. It ought to learn from the sciences, meaning in practice that the physician had to look for the laws of nature that govern the human body rather than become hypnotised by the symptoms of each individual. Only by understanding the statistically normal would physicians get to know and heal diseases, Fenger thought. 'The individual is everything lying beyond the rules and not comprised by them, and consequently the worst enemy of science; the individual exercises the same influence on medicine as a planet disobeying the laws of gravity would do on physics.'[13] This sounded like the outcome of Ørsted's deliberations on the aesthetics of physics, namely that it is the laws of nature of light, colours, and sound that determine the aesthetical sense impression and less the individual differences of taste. But even if Fenger agreed with Ørsted in putting the laws of nature above individual symptoms and tastes, he was right, unfortunately, in asserting that the science of medicine was a hundred years behind. Standing by the sickbed, he only had the usual dubious cures at his disposal: bloodlettings, emetics, laxatives, and sourdough under the soles of the feet.[14]

When Fenger had examined Ørsted he found problems with his lungs and ordered an emetic; the next day the patient was bled, and this did not help either, he prescribed Spanish flies that would bite wounds to drain pus and slime. As usual Ørsted was an optimist and calmed his family down by saying that he felt no pain and would recover soon. On the Friday he beckoned Scharling to come in as he quietly saluted through the open door. 'Please disturb me, I shall be well soon', he said to his son-in-law. 'I shall get up on Sunday!' he said to Andersen. But his powers dwindled, his voice weakened, and Scharling realised the direction he was going. On the Friday evening was the usual meeting of the Royal Danish Society, but the topic of the session was no routine matter. Molbech was talking about aesthetical theory, and Ørsted had so much wanted to listen and take part in the debate. On the Saturday this note appeared in *Berlingske Tidende* under the heading 'The Society for the Dissemination of Science': 'Geheimeconferentsraad H.C. Ørsted will be prevented from lecturing on Sunday, 9 March.'[15] On

Fig. 164. Glass containing Spanish flies, Dr Fenger's cure to extract HCØ's inflammation in March 1851. Medicinhistorisk Museion. Photo by Anders Olsen.

that day he was to have delivered the eighth in a series of Sunday lectures on the history of science that he had started just after his jubilee.[16] Instead on that morning the family gathered around his deathbed: Gitte, Karen, Sophie, Nicolai, Dahlstrøm, and Scharling. Although Andersen, too, was a kind of a son of the family, he sat down in the drawing room next to the study, ready to drop. It was horrific. He was in pain sitting passively and waiting for death to come. He walked to Collin's house. Rumours of Ørsted's sickness were known all over the city, and in the gateway in Studiestræde many friends of the Ørsted family were assembled. When he expired the servants were the first to know. Then Scharling went downstairs to tell the sad news.[17] Andersen walked back to Studiestræde. On the way he met Baron Holsten who told him that it was all over. Andersen dragged himself up the stairs. His eyes met Gitte calm, and pious. He went to the deathbed and wept, wept a lot.[18]

The next day Andersen was already putting his feelings of filial devotion, sorrow and admiration on paper: 'He was so true and very kind-hearted, he had a child's mind and was still the deep thinker..., like a fell seen from afar. The philosopher's stone was in that fell. He was so loving, upright, and pious, and yet so great. He will never be forgotten...'[19] Over the following days other commemorative poems were printed on the front page of *Fædrelandet* by Dahlstrøm,

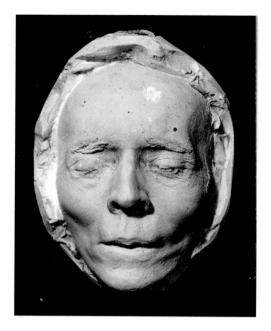

Fig. 165. HCØ's death mask in gypsum made by the sculptor J.A. Jerichau (1816–1883) (ch. 1). FBM.

Molbech, and many more. Romanticist views on arts and sciences were evident in Molbech's poem, where Ørsted's scientific life was delineated as pure mysticism on a par with Aladdin, who just happened by coincidence to rub the lamp, whereupon laws of nature were unveiled to the unprepared layabout. In Molbech's poem a hybrid between the Queen of the Night from *The Magic Flute* and the goddess Isis from the bowels of the Earth calls the merry son of nature, alias Hans Christian Ørsted, down to her through a fissure in the crust of the earth to present him with one of the secrets of nature that he is allowed to bring to light. Addressing the woman with the black veil this merry son of nature declares: 'I descended step by step on the ladder of the hidden forces [sic!], but nobody could say the word that would set my mind at peace. Now at the foot of the ladder I stand with the lamp of research in my hand and ask you, omniscient muse, please open your secret door for my spirit.' The Night agrees to reveal one secret of nature only, for if she reveals them all, it will be the end of the world....[20] Obviously, this is an allegory in the manner of 'the poetic arsenal of the past' and in total contrast to Ørsted's legendary diligence.

On 17 March a group of polytechnic students gathered in Studiestræde, when they laid a silver wreath on Ørsted's coffin and sang a song by Peter Faber, the polytechnic postgraduate and poet, who became the first director of the State Telegraph. They lit torches and carried their teacher to the ceremonial hall of the University draped in black and put the coffin on a catafalque surrounded by silver chandeliers. Here they sang the last of Faber's verses. During the night they kept vigil by the coffin. The next day a deputation from the grammar schools of the city placed another silver wreath on it, and in the midst of it a silver plate with engraved acoustic figures and a magnetic needle with the text: 'You taught me to obey'.

From the gallery the choir of the Students' Association sang a chorale by Carsten Hauch (text) and C.E.F. Weyse (music). Professor H.N. Clausen gave a speech portraying Ørsted as the University's eminently most interdisciplinary servant. 'True, Nature was the sanctuary, whose altar piece he served with priestly devotion early and late', but he was in scholarly touch with almost all colleagues, whether theologians, philosophers, linguists, historians, or literary critics. 'To communicate was the deepest urge of his character, to kindle the spark of the spirit.' His ability to captivate his audience had also struck Ritter and Hauch. Now Clausen commemorated the extinguished spark of the spirit. Ørsted's pedagogical talent did not consist in following rhetorical rules, but in being 'the flash of thought, a bolt of lightning safely hitting, sharply enlightening, electrically awakening... Like few men he has been the teacher of all Denmark.'[21]

Outside on Vor Frue Square a huge crowd of people was gathering, and the church bells tolled. The coffin was taken down the stairs and slowly (it took three quarters of an hour) and accompanied by funereal music it was carried behind the apse via Fiolstræde into Vor Frue Church. Rev. E.C. Tryde delivered the funeral sermon, not Bishop Mynster. Was he sick or hindered in other ways? No, he had not been asked, he said apologetically. Andersen was enraged. Should a friend be asked to speak for a friend?[22] True, Ørsted had not been a regular churchgoer; he had more frequently met the bishop at meetings or parties than in his church. Only once a year, the ritual minimum, did the Ørsteds take communion.[23] *The Soul in Nature* had provoked a breach between them after fifty years of friendship. Their civilised way of communicating had lately been only a facade. It was an intentional act of protest to let the coffin and the procession following it make a long detour round the bishop's palace thus turning its back contemptuously on the primate of the state church.

The church was draped in black as it had been at the funerals of Thorvaldsen and Oehlenschläger. Hartmann the organist, with the trumpeters, played the funeral march he had composed for Thorvaldsen a few years earlier. The choir sang a funeral hymn by Heiberg (text) and Niels W. Gade (music) and now Rev. Tryde delivered his funeral sermon. He portrayed Ørsted as 'an apostle feeling that it is not only important to see the light for oneself, but also to let it shine for others, not only to nourish great thoughts for oneself, but to express them in a language that is understandable by many.'[24] It was beautiful rhetoric, Andersen confessed to his diary, but it did not touch his emotions. He felt an urge to weep, but could not. Nor could Uncle Anders and the sons. Tryde dropped a hint that there were other religious opinions than Ørsted's. Andersen jumped from the church bench. Was his fatherly friend now to be accused of heresy in public? But the minister took the edge off his point by saying that Ørsted's religious views pointed towards the same goal as that of the State Church, and that he bridged the gap... thereby giving rise to a confused metaphor, for if the religious views pointed towards the same goal, there would be no gap to bridge.[25] Finally, the coffin was carried outside to the accompaniment of Hauch's chorale: 'Goodbye, You deep thinker, Great, rare friend of Nature! We are not likely to live to see the day when we meet your equal.'[26]

The newspapers wrote that the funeral procession following the coffin to Assistens Cemetery numbered two thousand people. At the grave the next of kin were allowed to be alone in their parting from him. Rev. Peter Hasle, his son-in-law, who had buried his Marie at Asnæs exactly a year before, spoke a word of comfort to the family. The patriarch they had lost was a role model for children, sons-in-law, daughters-in-law, and grandchildren. The dear departed was driven by his calling, he practised his ideals, he was zealous, incessantly finding new energies, the organiser entertaining grand visions, the patron generously giving help and advice, and the father encouraging his children to develop their talents, not to schoolmaster them, but to demonstrate from his own experience that personal growth is nourished from within. In the midst of all his fame he had preserved a childish innocence and warmth that opened his and Gitte's home to family, colleagues, and students, and if they had entered with troubles they walked away enlivened. On 21 December every year he invited the best forces of the nation to celebrate his brother's birthday. He and Gitte could not share everything, but their love was reciprocal. They read each other's thoughts before they were spoken, because they sought to establish a home for the true, the good, and the beautiful.[27]

On the day of his burial a commemoration was held in Stockholm. An obituary and some poems were recited against the background of a monument behind a transparent veil, inscribed with Ørsted's name framed by stars. At its foot sat mourning muses, and an orchestra played a funeral march by Beethoven.[28]

In the church of Frederiksborg Castle hangs the coat of arms made for the occasion of Ørsted's appointment to Knight of Dannebrog. Under the motto of 'Truth in Love' the spark of the spirit is framed by an electrical circuit and his author's quill writing 'the laws of nature are the laws of reason'. In his Masonic lodge a commemorative coat of arms was set up bearing this inscription:

Fig. 166. HCØ's tomb at Assistens Cemetery. The wreaths have been laid by various institutions founded by HCØ or attending the centenary of the discovery of electromagnetism in 1920. DTM.

Fig. 167. HCØ's Masonic coat of arms with his motto 'Truth in Love'. DFO.

Professor of Physics
Hans Christian Ørsted
Geheimeconferentsraad, Knight
Of Dannebrog, Knight of several Orders
And Dannebrogsmand,
Saw the light of the world at Rudkøbing
5777, 8, 14,
And the light of the Mason in the St. Johannes Lodge
Frederik to the Crowned Hope
5812, 1, 29.
At his 50th jubilee

> 5850, 11, 7,
> The St. Johannes Lodge,
> Zorobabel to the North Star,
> Honoured him by pronouncing him
> An Honorary Member
> And his mother lodge as Honorary Master.
> 5851, 3, 9,
> He went to the eternal East to
> View the solution of the riddles that
> Were still obscure to him.
> They were not many; for the Great Architect
> Had revealed to him the deepest secrets of
> The World Building; and the light
> He kindled for others
> Will shine lustrously till the last days.

On Good Friday, 18 April a mourning lodge was held in Zorobabel to the North Star, where Carl Otto, the phrenologist and Master of the Lodge, in front of the sarcophagus proclaimed to the assembled brothers: 'only rarely was he seen in our narrow lodge, but his works have been more numerous and more indefatigable in the great working temple of the world'. Another brother, the prison chaplain C.H. Visby said that

> as Ørsted had known to unite forces of nature that had so far been believed to be separate forces, magnetism and electricity, so in the evening of his life he had tried to show that the natural and the revealed religion were basically one in so far as 'the soul of nature' was also the spirit of God... When younger brothers see his coat of arms they shall be in awe of him and say: 'Ørsted, too, was a Mason', and the profane shall revere the lodge as noble, when they hear that Ørsted was working there.

Unlike the State Church, the Masonic Order approved the minimal concept of God that encapsulated Ørsted's scientific and religious beliefs.

> The entire teaching essentially consisted
> in taking Kierkegaard's *The Moment* in one hand
> and Ørsted's *The Soul in Nature* in the other,
> folding them together and binding them in one.[1]

61

Hans Christian Ørsted and the Golden Age in a Wider Perspective

Discussions of originality versus influence leading to the discovery of electromagnetism have dominated the historiography of Ørstedean science particularly during the past five decades. In this final chapter I shall focus on some of these discussions in the light of my own findings scattered throughout this biography. Ørsted's encounter with Kant's *Metaphysische Anfangsgründe der Naturwissenschaft* was a rude awakening and he based his doctoral dissertation in 1799 on these ideas, thus dissociating himself from the small Danish scientific community, who had derivatively imbibed French corpuscular theory. Soon afterwards he learned about the sensational discoveries made by Ritter and Volta, realising that this was good news for his dynamical project. He rushed to Oberweimar and discussed his ideas intensively with chemists and physicists working in the then nationally different communities of scholars in Germany and France (chs 11–16, 27–28). As I have tried to show throughout this biography Ørsted was adamant in resisting being categorized as a partisan of a specific school such as *Naturphilosophie* for instance at the expense of his independence (ch. 25). In his thinking he was open-minded and ready to learn from a variety of sources. He opposed doctrinaire extremism believing that the history of science shows a development in wide pendulum swings where the truth turns out to be found at the equilibrium, the golden mean (ch. 18). He urged French pre-positivists to learn from German philosophy and vice versa.[2]

Nevertheless, historians of science have argued persistently to categorize him as either (1) a contaminated Kantian/*Naturphilosoph* (in the negative sense) being cured by French pre-positivism (Meyer and Rosenfeld)[3], (2) a *Naturphilosoph*, i.e. a disciple of Schelling (in the positive sense) (Friedman, Wilson, and Caneva)[4], (3) a Kantian (also in the positive sense) (Gower, Shanahan, and myself)[5], or (4) both 2 and 3 (Stauffer, Williams, and Snelders).[6]

Meyer's introduction concluded that Ørsted had won international fame in spite of the fact that his discovery was nourished by metaphysical ideas construed by Kant, Schelling, and Ritter, lumped together as *Naturphilosophie*, some indefinable plague, from which he suffered as an adolescent, but from which he was graciously cured as an adult. In the same vein, Léon Rosenfeld, the Belgian Marxist physicist, rebukes Ørsted's unruly enthusiasm that fostered his juvenile mistakes. Fortunately, Rosenfeld concludes, he dropped his qualitative experiments and—perhaps subconsciously—learnt his lesson from sober-minded French science, represented by P.-S. Laplace in particular. But this is not so. Ørsted's career did not set out from wild speculation to end up in sober experiments. This juxtaposition makes sense to logical positivism only. As I have shown Ørsted was a meticulous experimenter already as a boy; he made experiments with Manthey, Ritter, Hermbstaedt, Vauquelin, and Charles; he acquired the best collection of scientific instruments and fitted up the best laboratory in Copenhagen. But he also realised that a fruitful experiment must be preceded by careful epistemological preparations.

The new trend in the history of science vindicated the virtues of *Naturphilosophie*, now in the strict sense of Schelling's early works, which were now appreciated as sound guidelines for scientific research. Following the footsteps of Stauffer, Williams, and Snelders, the articles by Friedman, Wilson, and Caneva moved Ørstedean historiography into the opposite extreme, because for them Kantian influence is ignored. Friedman went so far as to ascribe to Schelling 'the crucial step' that paved the way for the discovery of electromagnetism.[7]

This group of scholars takes for granted that influence is generally a decisive catalyst for any discovery and was so for Ørsted's. But given that the discovery was a source of amazement to everybody, one would not expect to find that its prerequisites were available from more or less readymade sources. Ørsted only began perusing Schelling's system in some depth in 1802; he was unimpressed, and even less impressed by his disciples. To be sure, Ørsted gave a paper imbued with *Naturphilosophie* in the Scandinavian Literary Society in 1804 with Steffens, Schelling's ardent disciple, in the audience. This paper was never reprinted, probably because he regretted its wild speculations, deduced from Steffens' geological works. By 1810 Schelling ceased working on *Naturphilosophie* and in the 1820s and '30s Steffens recognised that its speculations had proved futile for the progress of science, while at the same time paying full respect to Ørsted's discovery as above all a fruit of experimentation relatively independent of these ideas (ch. 52). At that time *Naturphilosophie* had long been generally used as a derogative term. But harsh criticism of *Naturphilosophie* already started, as we have seen, when it was being born in the very last years of the 18th century (ch. 10).

Wilson, Friedman, and Caneva argue that it was Schelling's philosophy that spawned and nurtured Ritter's electrochemical experiments. But it is the other way around! As convincingly established by Ostwald, Richter, Ørsted, Wetzels, and Dietzsch,[8] Ritter's imaginative and self torturing experiments with galvanism began with his supervision of Alexander Humboldt's attempts to solve the Galvani-Volta controversy in spring 1797. Ritter's work ended up showing that neither Galvani's theory of animal galvanism nor Volta's theory of contact between two metals caused the new phenomenon; instead he discovered that electricity is produced by a chemical process. Then Ritter communicated his experiments in 1797 to the Jena Naturforschende Gesellschaft in front of the early Romanticists (among them the well known Schlegel brothers, the Humboldt brothers, Schelling, Steffens, Tieck et al.). Ritter submitted his manuscript to a publisher in Leipzig who—uncomfortable with its daring scope—declined to print it. This caused a postponement of Ritter's *Beweis* that only saw the light of day the following year at his private expense.

In the meantime Schelling had already taken advantage of his sensational results in his *Ideas*. All members of the early Romanticist circle in Jena including Schiller and Novalis as well were fully aware that Ritter, the autodidact, had made the experimental contribution to Schelling's *Ideas*, and Schelling made no secret of it, although he was often accused of extrapolating wild speculations that found no proof in Ritter's work. As Ritter put it: 'They [Schelling, Steffens, and their followers] grasp a bit, but when they start their round dance it makes me dizzy'.[9] Ritter's discomfort with *Naturphilosophie* was due to Schelling's unqualified criticism of empiricism in general and in particular his insistence that speculation alone is conducive to scientific knowledge and therefore has a higher merit than experiment. The heart of the matter is that Schelling ranked *Naturphilosophie* above science and intended it to become the philosophy of objective idealism aiming at a reconciliation of man and nature. In Friedrich Schlegel's words: 'Schelling's *Naturphilosophie* was bound to stir much contradiction by aiming at devastating stark empiricism; fortunately, there is no reason to be afraid that his lack of scientific knowledge will prevail since at the same time Ritter has elaborated a physics of pure empiricism that by its rigorous method satisfies the strongest claims of scholarship.' This quotation formed the introduction to his journal *Europa* to which Ørsted contributed an article on recent developments of dynamical science.[10]

All of this became perfectly clear to Ørsted when on his first tour abroad in 1801 he headed directly to Oberweimar to encounter Ritter, the discoverer of electrochemistry, with whom he developed a close friendship that would last until Ritter's premature death nine years later. There can be no doubt that Ritter, the man and his works, meant far more to Ørsted than Schelling. There is widespread consensus that during Ritter's short life Ørsted was his closest friend and collaborator. During Ørsted's eighteen months' sojourn in Germany he spent the longest period face to face with Ritter, and most of the remaining time with the early Romanticists in Jena, in the Kränzchen salon, and in various natural philosophy societies and laboratories in Berlin (ch. 13). In Paris Ørsted fostered a plan to promote Ritter, at that time scarcely known outside the city walls of Jena, to world fame by winning for him the Napoleon prize awarded for important discoveries in galvanism (ch. 15).

Although Ritter died as early as 1810 Ørsted mentions him 38 times in his scientific works against eight times for Schelling. The equivalent numbers in his travel letters are 22 to 16. Ritter's letters to Ørsted fill 260 pages (Ørsted's letters to Ritter are lost). By contrast, he only met Schelling a couple of times, not for tête-à-tête talks, but mostly in the company of others. Ørsted sent him one letter, of which more in a moment. He received no reply. One would not expect such priorities from a scholar claimed to be under heavy influence of *Naturphilosophie*.

Andrew D. Wilson in his Introduction to Ørsted's scientific works (Princeton 1998) argues that Ørsted's dissertation on Kant's *Metaphysische Anfangsgründe der Naturwissenschaft* was first and foremost a critique of that book in two (or rather three) respects. Firstly, to be sure, Ørsted criticised Kant for not being entirely true to his own epistemology, because he sometimes substituted a priori judgement with empirical observation. This is correct, but did not make Ørsted less a Kantian, but rather, more Kantian than Kant himself. Kant's epistemology—by Ørsted often referred to as 'Ariadne's Clew'—guided Ørsted through his scientific and philosophical writings and justified his replacement of the atomic model of contemporary corpuscular theory in favour of his own dynamical project.[11]

Secondly, according to Wilson, Ørsted was dissatisfied with Kant's metaphysics of science because it mirrored the conventional Cartesian dichotomy between nature and mind, whereas Ørsted wanted a cosmic synthesis of the two that only Schelling could supply. Unfortunately, Wilson does not present any evidence of Ørsted's alleged desire, and I think it would be difficult for him to find it. Ørsted persistently talks about the relationship between nature and mind within the Kantian epistemology of things for us and things in themselves, and he is very explicit in arguing that natural philosophers can only find laws of nature that conform to human reason—and this is almost a truism. Embarking on a research project any scientist would take this precondition for granted. Ørsted never became a Schellingian *Identitätsphilosoph* (ch. 25).[12]

Thirdly, Ørsted should have complained that Kant saw no chance of chemistry ever moving from fragmentary empirical observations to becoming a true science. To be sure, Ørsted accepted with regret that this was an undisputable fact for the time being, but he devoted the rest of his life to change this state of affairs. In doing so there was little or no help from Schelling's *Naturphilosophie*. In fact the seven or eight times he mentioned Schelling in his works between 1799 and 1809 (when Schelling stopped his contributions to *Naturphilosophie*) this was more often than not to dissociate himself from the philosopher that Wilson considers his model. The overall reason for this dissociation is 'Schelling's speculations, because the empirical theorems that he adduces are often completely false'.[13] Although the ideas of Schelling and other *Naturphilosophen* are sometimes lucid, Ørsted repudiates Eschenmayer's and Schelling's theory 'as it is false from its first foundation'.[14] Conversely, in the preface to his dissertation Ørsted regrets that Schelling's excellent *Erster Entwurf einer Naturphilosophie* arrived too late for him to take it into account, but he also stresses the point that his theory of the cohesion of matter—although in accordance with Schelling's view—is not derived from his book. In turn Eschenmayer and Schelling slated Ørsted's *Ideen zu einer neuen Architektonik der Naturmetaphysik*. (ch 12).[15]

True enough, Ørsted sometimes praised Schelling's ideas: the clearest example stems from his lecture 'On the Harmony Between Electrical Figures and Organic Forms' (1805). However, this is the exception, not the rule. Steffens, Schelling's adherent, was in the audience and Ørsted probably wished to please him. True, Ørsted found inspiration in Schelling's idea of the effect of magnetical, electrical, and chemical forces in three different dimensions (length, breadth, and depth)[16], but as hypotheses only, and he had abandoned them before 1812, when he published his major work *Ansicht*, in favour of an alternative hypothesis of polar forces in equilibrium or conflict—derived from Ritter's electrochemistry.[17] His decisive idea that electricity might be propagated in undulatory manner owes nothing to Schelling. Otherwise Ørsted's positive remarks were put forward in order to moderate the devastating discomfort with *Naturphilosophie* expressed by Oehlenschläger in 1807 and by Grundtvig in 1814 (chs 24 and 29). Ørsted declined being labelled as an adherent of the new German school of *Naturphilosophie* just as he refused to dismiss it (ch. 25). His judgement was balanced and he always insisted that speculation (by which he meant Kant's regulative ideas) and experiment should go hand in hand—a method he called a thought experiment.

Does the evidence support Wilson's claim that Schelling's influence on Ørsted outweighs the close partnership between Ritter and Ørsted? I do not think so. He supports his claim by reference to Ørsted's letter of 18 January 1850, which he contends accompanied Hans Martensen's *Christian Dogmatics*.[18] Unfortunately, Ørsted sent Schelling quite another book, although by the same author, namely his *Sendschreiben an den Herrn Oberconsistorialrath Nielsen in Schleswig: Ein Wort über den Amtseid und die schleswig-holsteinische Geistlichkeit*.[19] Ørsted also sent this book to his friend Weiss and to court steward Alexander Humboldt to counter a widespread opinion in Germany that the Schleswig-Holsteinian insurgents were justified in breaching their loyalty towards the Danish King. Does it really matter whether Ørsted sent Schelling one book or another? I think so, because Wilson makes capital of Martensen's *Christian Dogmatics* which acclaimed *Naturphilosophie* by inferring that Ørsted, by sending him exactly that book, paid homage to Schelling as his master, whereas the heart of the matter is that Ørsted used Schelling as a messenger to dissuade the Prussian authorities from supporting the insurgents in Schleswig-Holstein.[20]

In his comprehensive article 'Kant—Naturphilosophie—Electromagnetism' M. Friedman aims at establishing Schelling as the natural link between Kant and Ørsted. The second edition of Schelling's *Ideas* (1803) is assigned the task of leading Ørsted beyond Kant towards his discovery of electromagnetism. Friedman's article is concerned with a much-needed attempt to elaborate Stauffer's theory that '*Naturphilosophie* in particular provided essential philosophical motivation for Ørsted's scientific discoveries'.[21] It seems to me, however, that the attribution of a key function to Schelling is at odds with the source material informative of the actual steps Ørsted took in pursuit of his dynamical project. Friedman argues that the 1803 edition of Schelling's *Ideas* is particularly important because in the supplement of this edition the author incorporated the results of Ritter's electrochemical experiments. But as already mentioned above Schelling's *Naturphilosophie* was from the very beginning spawned by Ritter's galvanic experiments presented to the Naturforschende Gesellschaft in Jena in 1797. It caused some discontent from members of the Jena circle to find Schelling's speculative ideas based on Ritter's ingenious experiments without giving him due credit.[22]

For chronological reasons alone Friedman's assertion that Ørsted learned about electrochemistry via Schelling's book from 1803 is untenable in so far as he had had full access to the originator two years earlier. Secondly, Ritter did not and could not make experiments with the Voltaic pile—as Friedman writes, before its invention in 1800. The point is that Ritter discovered the chemical effect of galvanism before the invention of the pile.[23] Thirdly, nowhere does Ørsted indicate his debt to Schelling's supplement of 1803. Remarkably, although Ørsted possessed all of Schelling's works, exactly the 1803 edition of the *Ideas* is missing in his library.[24] Indeed, Ørsted never mentioned it in his subsequent scientific writings, *Materialien* (1803), his unpublished 'Theory of Force' (1812), or *Ansicht/Recherches* (1812–13). Nor did Schelling ever become his friend. Nowhere is Schelling elevated to a position comparable to that of Ritter, whom Ørsted 'regarded as the creator of modern chemistry. His comprehensive ideas and his achievement undertaken with such great vigour and exertion, spread a great light in all directions.'[25]

In the remaining part of his article Friedman repeats his argument, now against historians of science such as Barry Gower, Timothy Shanahan and myself. Schelling alone was the man who taking his point of departure in Kant's metaphysics of natural science could develop it further by establishing the indispensable tripartite scheme (to which I shall return immediately) that ultimately paved the way for the discovery of electromagnetism. Again Friedman substantiates his view by reference to Ørsted's 1805 paper 'On the Harmony Between Electrical Figures and Organic Forms'. As I mentioned above this paper is a unique exception in wild speculation indulging Steffens' and Schelling's *Naturphilosophie*. I do not argue that Ørsted was not seduced by Schelling in this particular instance nor that Schelling had no impact at all. Ørsted was sometimes charged with partisanship of the new German school, but he resented being labelled as part of it as I have shown throughout this biography (chs 25, 52).

Friedman argues that without Schelling's improvement of Kant's metaphysics Ørsted could not have made his discovery. In his dissertation Ørsted concluded from Kant's metaphysics that Newton's atomic theory and as a consequence French corpuscular theory with its imponderable fluids was flawed, because it seemed to him that atomism was based on materialistic concepts amenable to mathematical calculation, but that these concepts defied empirical proof. The only sensible effect of matter was its resisting force manifested when trying to penetrate it. As an alternative Kant suggested a dynamical theory of matter. Ørsted wholeheartedly embraced Kant's analysis as well as its consequences, the most disheartening of which was that chemistry—at least for the time being—was incapable of becoming a natural science, because to be so would require concepts amenable to mathematical treatment; polar forces do not lend themselves to weight and measurement as did the concept of mass enabling the force of gravity to be calculated. Sadly, this insight, as Kant and Ørsted had to accept, pushed chemistry back to scratch.

Ørsted's response to this insight was to embark on his own dynamical project and he was soon encouraged to engage fully into it by Ritter's and subsequently Volta's discoveries. But he remained faithful to Kantian epistemology, which very few people knew let alone understood, and which Ørsted—for the benefit of the non-initiated majority—usually referred to as 'Ariadne's Clew'. He was immediately aware that Kant's epistemology operated with the dichotomy between noumenal and phenomenal principles. Fundamental forces per se are noumenal and thus undetectable, but nevertheless have to be assumed as a regulative idea that might lead to understanding the phenomenal effects. This according to Kant is the only purpose of noumenal and regulative ideas. Noumena alone can never become part of science, because they are metaphysical ideas defying empirical investigation. By contrast, men of science (and Kant was not a man of science) can conceptualize phenomena and explore them empirically. Science will always have to walk on two legs, an epistemological one and an experimental one. But the two must be kept strictly separate. What Ørsted criticized Kant for was exactly this: Kant had mingled empirical observations with his metaphysics (ch. 52).

Whereas for Friedman, Wilson, and Caneva Kant's relevance is tenuous and Schelling's concrete, for Gower, Shanahan, and myself it is the other way round. Moreover, Gower and Shanahan are perfectly justified in blaming Schelling for his negligence of experimental physics. There is no evidence that Schelling ever engaged in any experiment. When Ørsted remarks that 'it is well-known that Schelling, through speculation, has produced an attempt, which, as such, is of incalculable value, but the combined efforts of a great number of blessed geniuses are probably required for the accomplishment of this task',[26] Friedman takes it to mean an endorsement of Schelling's speculative method. Since Ørsted was an industrious experimenter himself, it almost suggests itself that Ørsted is being ironical. Schelling despised the empirical chaos of the manifold, but nevertheless snatched Ritter's epoch-making experimental results to fit them into his own system. By contrast Ritter and Ørsted both reasoned and experimented.

If Schelling's *Ideas* were so ground breaking—and Kant's metaphysics of natural science so inadequate—as Friedman (and Wilson and Caneva) claim, one must ask whether *Naturphilosophie* ever led to any (other) scientific discovery. The answer is no. Ørsted's discovery of electro-magnetism was not foreseen by any *Naturphilosoph* nor by anybody else (except the late Ritter) for that matter. Steffens recognised this when in 1829 he visited Ørsted in Copenhagen. *Naturphilosophie*, he now admitted, had proved to be barren speculation leading nowhere.[27]

To become convinced that Schelling exerted a direct influence on Ørsted's great discovery, a concrete scientific idea instrumental in bringing it about is needed. The most tangible suggestion presented by the three historians of science is Schelling's tripartite scheme already indicated above (note 16) which Ørsted mentioned once (in his 1805 paper) and never repeated as far as I am aware. This tripartite scheme has no place in Ørsted's *Theory of Force*, 1812, but in the epilogue he does refer to another scheme dealing with chemical forces for which Ritter is responsible: '[Ritter] can be regarded as the first to introduce into natural philosophy the truth that the forces, which previously we knew only under the names of electricity, galvanism, magnetism, are the same as the chemical forces'.[28] This quote is obviously different from Schelling's tripartite scheme mentioned in 1805; for instance directions of forces are totally absent. Finally, also in his *Theory of Force* Ørsted refers to a third scheme in which 'it is remarkable that there are only three main chemical classes, that is, three primary forms of the distribution of forces, just as there are only three dimensions in space'.[29] There is no hint that this is referring to Schelling, nor is there any indication of magnetism, so it is doubtful whether Ørsted has Schelling's scheme in mind at all. Indeed, his name is completely absent from this unpublished book and the same applies to his *Ansicht/ Recherches* (1812–13), for which it served as a draft. Nevertheless, Anja Skaar Jacobsen, the editor of *Theory of Force*, in a footnote ascribes the last quote to Schelling, supporting her claim on Caneva's text.[30] What is happening here is reading a historian's interpretation into the source material rather than writing history by an unbiased scrutiny of primary sources.

Since this is the only concrete proof of Schelling's influence offered by Friedman, Wilson, and Caneva it is hard to make sense of the suggestion. Caneva correctly admits that there were many tripartite schemes in the air at that time. One was Schelling's rather obscure scheme categorising the motion of three forces in three dimensions: length (magnetism), breadth (electricity), and depth (chemistry). A different electrochemical scheme was set up by Ritter earlier: three space-filling forces in combination (the galvanic pile), two space filling forces in conflict (the electrochemical series), and two space filling forces in combination (chemical compound). To validate their case Caneva suggests to identify 'specific concepts and explanatory strategies' represented in Schelling's opus, while disregarding 'Schelling's subtleties'.[31] Schelling's tripartite schema exemplifies the method of analogical thinking (the analogy between direction and motion of force), but it does not explain anything. The use of analogical method by *Naturphilosophie* was ridiculed by Kant's statement prompted by Herder's *Ideen*: 'They try to explain what they don't understand by something they don't understand at all.'[32] Schelling indulged in tripartite schemes waving his magic wand among analogical subtleties: first the inorganic 'Stufenfolge' (hierarchy): chemistry, electricity, magnetism; secondly the organic analogies: reproduction, [nervous] irritability and sensibility. In this way Schelling prepares his philosophy of identity between the inorganic and the organic by analogy. Reproduction is intensified chemistry, irritability is for the organisms what electricity is for certain metals, and thirdly sensibility is the ante state of magnetism. As in the inorganic realm the three forces are different manifestations of one fundamental force, the organic hierarchy is different metamorphoses of one fundamental principle of life.[33]

Friedman explains Schelling's concrete influence on Ørsted's discovery of electromagnetism as follows:[34] 'As Ørsted himself explains,[35] it appears that he arrived at his decisive experiment by beginning from Schelling's triadic schema and then adding the idea that electrical forces are much more "bound" (i.e. less manifest as static electricity or charge) in a galvanic circuit than they are in a static distribution of charge. From this point of view, current electricity is much closer, as it were, to magnetism, and so it is precisely here that we should seek the desired "attractive or repulsive effect".' But Ørsted's 'Observations on Magnetism' in *Ansicht* does not involve dimensions, let alone Schelling's or anybody else's tripartite scheme, at all, but refers mainly to the mathematical similarity between electricity and magnetism manifest in Coulomb's electrometer with which he experimented intensely in 1815, where both polar forces attract and repel along straight lines like gravity. The observations made in *Ansicht* independently led to his procurement of a more powerful galvanic apparatus 'in which electricity is found to be extremely bound', i.e. produces a strong effect.[36] By means of this apparatus Ørsted subsequently brought about his discovery.

*

The life of Hans Christian Ørsted is now over and the time has come to place his significance as a scientist and cultural personality in the Danish Golden Age in a wider perspective. The Golden Age is the time-honoured term for the period designating a cultural prosperity after a time of a decline during the war against Britain, the loss of Norway, and the subsequent economic crisis. Works of reference usually label the culture of the Golden Age as Romanticist as

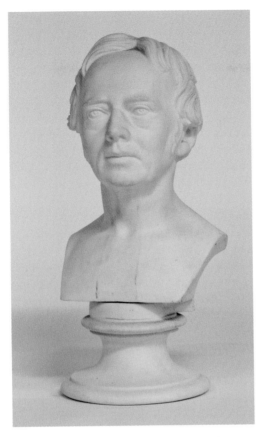

Fig. 168. Bust of HCØ by H.W. Bissen (1798–1868), manufactured in biscuit ware by the Royal Porcelain Factory, Copenhagen. DTM.

a reaction against the prepositivist ideology of the French Revolution. Traditionally, this Romanticist reaction is considered a parenthesis enduring from the late eighteenth century to about 1850 (or exactly the period of Ørsted's adult life). After him followed Modernism which resumed the Rationalism of the Enlightenment, playing a trump card at the so-called Modern Breakthrough around 1870. The credit for the Modern Breakthrough in literature and the arts is usually accorded to Georg Brandes, who made his first appearance on the scene of literary criticism in 1866 with a piece of seventy-six pages arguing against *Dualismen i vor nyeste Philosophie* [The Dualism of our Recent Philosophy]. In this booklet Ørsted is divided into two men, a scientist, whose international fame Brandes acknowledged, but hardly understood, and a leading person of culture, whose dualist philosophy he ridiculed. Brandes saw dualism as a kind of religious superstition, as modern man's vain attempt to live by homespun contrasts between knowledge and faith. Brandes caricatured the dualism of the Golden Age as follows: 'The entire teaching essentially consisted in taking Kierkegaard's *Øieblikket* [The Moment] in the one hand and Ørsted's *The Soul in Nature* in the other, folding them together and binding them in one.'[37] No reader of the two works can avoid seeing the platitude of this satire.

It has to be shown where Ørsted stood in relation to these epochal labels. The Danish authors who are most easily embraced by Romanticism, Baggesen, Grundtvig, and Ingemann were exactly those targeted by Ørsted when in his aesthetical criticism he distanced himself from 'the showpieces of the poetic arsenal of the past' (ch. 58). Ørsted did not look at medieval, gothic cathedrals, troubadours and mystics with nostalgic yearnings. Quite the contrary, he had visions of a modern architecture and a realistic art of poetry, music, and painting that would mimetically take advantage of the insight into natural laws that his dynamical project put at the disposal of artists. Apart from Ritter, the friend of his youth, whose experimental genius and neurotic temper at the same time was Ørsted's model and antithesis, the ideas of Romanticism did not appeal to Ørsted.

Against this background, it is hoped that it is clear by now that Romanticism is no appropriate epithet for Ørsted's ideas. The concept that more precisely characterizes the Golden Age is dualism, which has been suggested by Erik M. Christensen.[38] This epithet is now made increasingly justifiable by Ørsted's dynamical project. The dualism of the Golden Age is the basic philosophical structure with inherent tensions between the immanent and the transcendent levels of human life and the attempts to overarch them. In its simple form, dualism covers the philosophy of Plato, Neo-Platonism, Christian theology, and the German sources inspiring the Danish Golden Age: Kant, Fichte, Goethe, and Schiller. The pattern of dualist thinking allows a great deal of latitude, enabling various Romanticist figures such as Grundtvig, Mynster, and Molbech to feel at home there alongside the Ørsted brothers, creating a framework within which it makes sense to discuss and to disagree. To Ørsted (as with Plato and Kant) the model separates the phenomenal world from the noumenal. God can be left out of the framework, for religion has no place in science, neither in the metaphysics of nature nor of morality. The revealed Christianity of the state church is demythologized and its concept of God minimized, because we can know nothing about God, since He does not appear in time and space, although He is a possibility for believers as an irreducible concept of All-reason.

The Golden Age, most appropriately characterized as dualism, begins in 1798 with the Ørsteds taking charge of the editorship of *Philosophisk Repertorium* as the platform for the debate of Kant's critical philosophy. And the modern turning point did not begin with Brandes' lectures in 1870, but with the coordinated writings of Big and Little Hans Christian from 1849 to 1851 (*The Soul in Nature* and *In Sweden*) advocating a new world view impregnated with the idea of progress, scientific education, aesthetical awareness and enjoyment, reason, realism, and naturalism as well as a minimal concept of God.

In this way the Golden Age in general and the Ørsted brothers in particular are framed by French pre-positivism on one side and Brandesian positivism on the other. Georg Brandes was the person who tried to put down the Golden Age as a period in the history of ideas by rejecting the metaphysics of dualism. His simple method was to convert the two levels of dualism (the phenomenal and the noumenal) to knowledge and belief respectively. When Brandes so reduced Plato's philosophy, Kant's three critiques, and the metaphysics of dualism to religious belief pure and simple, he exposed himself as being not philosophically equipped to do away with dualism. By his account the Ørsteds' metaphysics of nature and aesthetics and the maxims of moral philosophy were put down not by solving their unresolved problem, no, the entire metaphysics was eradicated, when dualism was replaced by positivist monism.

As with the pre-positivists of the Enlightenment, all questions have scientific answers for Brandes. The metaphysics of nature, working with a concept of force such as Newton's gravity, or Ørsted's aesthetics of nature about the auditory experience of the sound waves of music, or the visual impressions of beams of light or colours, or the maxims of moral philosophy such as Kant's free will or Anders's categorical imperative, all have been reduced to metaphysical superstition by Brandes. These questions would either find their scientific answers, in which case they would be changed into knowledge, or they would remain metaphysical, that is religious nonsense deserving to be put down. In line with the French tradition from the positivists Auguste Comte and Hippolyte Taine, Brandes demands that all metaphysics is transformed into scientific insight or if that proves impossible, is thrown on the scrap heap. In his eyes dualism is a kind of intellectual spinelessness, a series of religious mantras lost on the empirical scientist's positive results, falling into wishful thinking, clerical lies, and empty speculation.

Brandes brought his arguments for denouncing dualism from French positivism. This was developed by Auguste Comte, who had graduated from the utilitarian École Polytechnique, a modern engineering school established in the aftermath of the French Revolution to serve the war efforts of Napoleon Bonaparte while useless faculties of theology and the humanities were closed; the Revolution was in no need of such good-for-nothings. To Laplace, a pre-positivist rejecting any kind of metaphysics, science and religion were rivals, and it was the task of science to wipe out the church as an institution.

*

Four aspects of Ørsted's view of physics distinguish him from so many other, but not all, physicists. First physics is a dualistic science consisting of two branches. The mechanical branch deals with material substances which are measurable and weighable as well as changeable because of external influences in time and space; this branch can be described in terms of material entities and mathematicised. The dynamical branch deals with internal forces in 'matter' causing chemical, electrical, magnetic, and other changes; these forces are not recognisable in time and space, but must exist nonetheless. Physicists are unable to examine these forces as such; only their effects are phenomenal. Forces are necessary auxiliary concepts that are not in themselves conducive to scientific knowledge. 'Matter' is put in inverted commas, because it only seems to be phenomenal. Like Kant, Ørsted talked about 'matter' as space-filling forces. As 'thing-in-itself' it is beyond recognition.

Secondly, and as an extension of this, Ørsted does not think that changes in the physical world, the research object of physics, acquire clarification and understanding by being mathematicised. On the contrary, mathematics is alienating, because it pretends to be clarifying and understanding nature, but in reality it only describes or calculates physical phenomena. Internal forces are noumenal and defy mathematical description. Ørsted demanded a visual and verbal physics as for instance acoustic figures, and the construction of new terms such as conflict, orbital lines, and electromagnetic field. Physical understanding means that the researcher sees physical change with his mind's eye and expresses it verbally.

Thirdly, scientific education is a significant part of general education, because the study of natural laws according to the method of thought experiments effectively sharpens one's sense of true and false. Science is a cognitive touchstone and natural laws have objective and universal existence. They are not human constructions or something the researcher invents, but realities he finds. Knowledge of them helps humans feel at home within nature.

Fourthly, Ørsted is convinced that the significance of physics, on top of its justification in itself as a cognitive activity, lies in aesthetics rather than in technology. Technical progress is of great importance for civilisation, but too closely related to materialism and capitalism and hence philistine. Far more important for man is the spiritual pleasure evoked by the arts and sciences. Ørsted did not acquiesce in Rousseau's gesture of resignation *'Je ne sais quoi'*, but found the beginning of an answer to artistic creativity and enjoyment in its mimetic relationship to the beautiful in nature. The dynamical branch of physics in particular shows that the beautiful in the arts consists in recently discovered laws of nature covering the waves of sound and light, the theory of the spectrum of colours, and others. Our fascination with music, painting, and poetry is due to the correspondence between the subjective experience and the objective reason of nature.

Consequently, Ørsted's dualism is not confined to the sciences, but is by extension the fundamental structure of his worldview, also comprising moral and aesthetical philosophy. He entered into a lifelong division of labour, together with his brother Anders, to concretise the dualistic project through reciprocal inspiration. Moral and legal philosophy became the domain of his brother, but when he was prevented from taking part in public debate, Hans Christian became the mouthpiece of their shared views on these and general cultural matters. They were both anxious to save the freedom and dignity of man by insisting on the distinction between homo phenomenon and homo noumenon. The first concept characterises man as a creation of nature, while the other expresses the creative talents of man within the sciences and the arts and his free will to choose his ethical maxim (Anders's categorical imperative) and his calling (to identify the divine idea in collaboration with other members of the learned republic).

Like Kant, the Ørsted brothers saw the greatest challenge of critical philosophy in man's profound wonder at the relationship between 'the starry heaven above us' and 'the moral law within us'. While laws of nature rule by relentless necessity, man is free to choose to take life into his own hands and live in moral freedom and responsibility. In other words: dualism. The question whether man is endowed with free will or not is not a scientific problem but a metaphysical one. Reason and free will are not organisms found in the human body by dissection. Nevertheless, we are free to use reason to make the choice of living *as if* we had a free will. People can experience themselves as being free to choose their calling, their intellectual or aesthetical use of their talent, their political and religious standpoints, in short: everything that gives man his identity and dignity.

NOTES

Chapter 1

1. C. Hauch 1852, *SES* 9, 159
2. HCØ 1844, 1
3. B. Dibner, 1961, 30–4
4. J. Aldini, 1804, 340. *Fortegnelse* no. 1165–1166
5. S. Stringari and R.R. Wilson, *Rendiconti Fisiche e Naturali* 2000, 115–36
6. W. Ostwald, 1895, 174
7. M. Faraday, *Annals of Philosophy* 1821, 195
8. J. Herschel 1831, 340
9. K. Meyer, *NS* I, xiii–clxvi. L. Rosenfeld 1979
10. T. Shanahan 1989, TL 126
11. G. Buchdahl 1986
12. R. Stauffer 1953 and 1957
13. P.L. Williams, *DSB* vol. x, 1974, 182–6
14. K. Caneva 1997 and 2007, A.D Wilson 1998 and 2007, M. Friedman 2007
15. M. Friedman 2007, 157 n.54
16. P. Guyer in S. Sedgwick, 2000, 19–53

Chapter 2

1. ØC 4
2. ØC 4
3. ØC 4
4. Aa. Schæffer 1978, 67–88
5. D-45, DTM
6. E. Schultz 1960
7. C. Meyer 1989, 25–7
8. ØC 102
9. ØC 4
10. E. Schultz 1969. R.R Vestergaard 1891
11. LHS journal matters 1783/379 and 1784/393
12. S. Jørgensen 1796, 122–23
13. J.E Jensen 1988, 36
14. HCØ 2002, 15–16
15. ASØ 1951, 23
16. L. Nørregaard 1784
17. C. Batteux 1773
18. HCØ 2002, 16

NOTES

Chapter 3

1. *NS* III, 193
2. AOe 1974, 68
3. H.N Clausen 1877, 30–5
4. H.N Clausen 1877, 30–5
5. H.N Clausen 1877, 30–5
6. H.N Clausen 1877, 8. ASØ 1951, vol. 1, 18
7. HCØ 2002, 17
8. E.C Werlauff 1968, 19–20
9. H.N Clausen 1877, 119–20
10. AOe 1974, 102–3
11. I. Kant 1798. B. Readings 1997, 57
12. *KUJ* 1793, 152–5
13. *Catalogus* 1786
14. *KUJ* 1793, 23–33
15. H. Rørdam 1879
16. DCC 1996, 41–7. KU vol. 2, 1993, 189–94
17. I.C Fabricius 1796, 25–31, 111–12,142–3
18. *KUJ* 1793, 58–71
19. H. Steffens 1844, 21–4
20. *KUJ* 1793, 90–3, 97–102
21. *KUJ* 1793, 90–3, 97–102
22. *KUJ* 1797, 106–10
23. *NS* III, 193. H.H Kjølsen 1965, 29–42
24. C. Christensen 1924–26, 66–92. C.C.A Gosch 1973. KU vol. 13, 1979
25. *Minerva* 1794, 297–326
26. A. Skaar Jacobsen, *Ambix* 2000
27. AOe 1974, 104f

Chapter 4

1. I. Kant 1913, vol. 5, 161–2
2. *KUJ* 1797, 31
3. AOe 1974, 68
4. S. Körner 1987, 168
5. H. Steffens 1844, 222
6. C. Hornemann 1795. J.H. Spleth was the external examiner at HCØ's doctoral dissertation in 1799, but had died before 1801. J.F.W. Schlegel was ASØ's tutor of law. In his prize essay ASØ often refers to Schlegel's journal *Astræa*, ASØ 1951, 27. Revd Jens Bindesbøll (the father of the future architect G. Bindesbøll, HCØ's travel companion in 1822–23) wrote the anonymous 'Forsøg til Forfatterens Charakteristik' [Essay to characterise the author, i.e. C. Hornemann] op.cit., i–xxxii
7. *SES* 7, 55
8. H. Steffens 1844, 222
9. I. Kant 1911, vol. 3, 12
10. I. Kant , 10

11. I. Kant 1911, vol. 3, 75: 'Ohne Sinnlichkeit würde uns kein Gegenstand gegeben, und ohne Verstand keiner gedacht werden. Gedanken ohne Inhalt sind leer, Anschauungen ohne Begriffe sind blind.... Nur daraus, daß sie sich vereinigen, kann Erkenntnis entspringen.'
12. P. Guyer 1999, 1
13. P. Guyer 1999, 1
14. ASØ 1798, 239
15. I. Kant 1913, vol. 5, 33–42
16. I. Kant 1911, vol. 4
17. I. Kant 1911, vol. 4
18. I. Kant 1911, vol. 4

Chapter 5

1. C. Batteux 1773, vol. 1, 147
2. P. la Huray 1981, 40
3. C. Batteux 1774
4. J.G. Schulzer 1792–94, vol. 3, 91
5. I.V.D Lühe 1979, 28–38
6. HCØ, *Minerva* 1797, 130
7. HCØ, *Minerva* 1797, 138–9
8. 'Professus grandia turget' [one, who pretends to know great things, proves turgid], Horace *Ars Poetica*, 27, was HCØ's motto. With his essay, anonymous, of course, followed an envelope marked with the motto. After the examination the envelope would be opened and the identity of the author come to light.
9. HCØ *Minerva* 1797, 159
10. Aa. Schæffer 1947, 577–90 og 1948, 486–8. K. Bærentsen and V. Gaunø Jensen 1977, 11–20
11. Aa. Schæffer 1947, 589 (my translation from the German)
12. K. Bærentsen and V. Gaunø Jensen 1977, 17–18
13. HCØ 1798, 238
14. HCØ 1798, 248. A.W Hauch 1794, reviewed by HCØ in HCØ 1798 and 1799. Hauch 1794 and Kratzenstein 1787 were quoted by F.A.C Gren 1797
15. F. Bull 1946, 37

Chapter 6

1. ASØ 1798, vol. 1, 310
2. NKS 3494, RL
3. NKS 3494, RL
4. A. Gamborg 1800
5. ASØ 1951, 37
6. ASØ 1951, 37
7. ASØ 1951, 37
8. ASØ's sojourn at Rudkøbing in the summer of 1798 was spent preparing the examen juridicum, *Philosophisk Repertorium*, vol. 1, 320 and *KUJ* 1799
9. ASØ 1951, vol. 1, 38
10. C.H. Pfaff 1798
11. I. Kant 1917, vol. 8, 95–116

12. AOe 1974, 67
13. ASØ 1951, vol. 1, 168–9
14. ASØ 1951, vol. 1 40–1

Chapter 7

1. I. Kant 1911, vol. 3, 75, cf. ch 4, note 11
2. HCØ, *Philosophisk Repertorium for den nyeste danske Litteratur*, vol. 2, 145–224
3. HCØ, *Philosophisk Repertorium for den nyeste danske Litteratur* , vol. 1, 123–6, 319–20 and 384, and vol. 2, 256
4. J.P Mynster 1854, 48
5. H. Steffens 1844, vol. 2, 228–9
6. J.P. Mynster, op.cit. 99
7. A.W. Hauch 1798
8. *NS* I, 33–78. *SSW*, 46–78
9. HCØ, 'Kemiske Breve' [Chemical Letters], in *Bibliothek for Physik, Medicin og Oeconomie* 1–2, vol. 14, 1798 and 3–4, vol. 16, 1799. The first of the four chemical letters was an introduction to Gren's dynamical chemistry, the second and third contained an exposition of Gren's theory of heat (mentioning Rumford's experiment against Lavoisier's 'caloric'), and the fourth dealt with combustion, calcination, and respiration according to Gren's antiphlogistics.
10. A.W. Hauch, op.cit., 770
11. *NS* III, 33 (not in *SSW*)
12. I. Newton 1687
13. I. Newton 1720, 375–6, 388
14. O. Knudsen 1987, 57
15. A.W. Hauch, op.cit., 629–30
16. A.W. Hauch, op.cit., 633–4
17. R. Fox 1974, 89–136 and 1996, 278–94. M. Tamny 1996, 597–609. S. Schaffer 1996, 610–26. S. Toulmin & J. Goodfield 1962. DCC 1995

Chapter 8

1. *NS* III, 45 (not in *SSW*)
2. *NS* III , 43 (not in *SSW*)
3. *NS* I, 79–105. *SSW*, 79–100. HCØ used the term 'metaphysics of nature' as a reference to Kant and not analogous to natural philosophy, which contemporarily was used in two meanings: partly as synonymous with science like 'natural philosophy' in English, and partly as the German notion *Naturphilosophie* fathered by F.W.J. Schelling. To avoid misunderstandings in the following I shall use the German term when I refer to Schelling's philosophy.
4. *NS* III, 49 (not in *SSW*)
5. *NS* I, 36 (*SSW*, 46–7)
6. 'filum Ariadneum', *NS* I, 9, *SSW*, 93. C. Hauch *SES* 9, 153
7. This position is repudiated by R.E Butts 1986, 195. Butts finds that the very metaphysical reflection on the concept 'fundamental force' is a valuable philosophical effort per se.
8. This position is sustained by Kant himself, cf. 1911, vol. 4, 523–5 and discussed by P. Kitcher 1986, 213
9. M. Friedman 1986, 56f. G.G Brittan 1986, 90f. H. Duncan 1986, 303. Finally, Gerd Buchdahl takes Kant's statement of the lack of mathematical constructibility of the dynamical theory with unruffled calm. The

negative result of Kant's analysis was what could be expected at his time, Buchdahl thinks. Modern atomic theory has acknowledged Kant's problem as a fruitful one for research. In nuclear physical theory, attraction and repulsion, for instance, are the combined fundamental forces on electrical charges of the atom and they explain the orbits around the nucleus of electrons and protons; cf. G. Buchdahl 1969, 553–6

10. I. Kant 1911, vol. 4, 564–5
11. I. Kant 1911, vol. 4, 471
12. *SSW* 93
13. H. Høffding 1909, 49
14. *SSW* 96
15. *SSW* 86
16. *NS* III 45, not in *SSW*
17. I. Kant 1911, vol. 4, 523–5
18. AOe 1974, vol. 1, 110

Chapter 9

1. LM>HCØ 16.11.1800, ØC 1–2
2. Census 1801, Copenhagen, Strand Kvarter, Matr. 45, 152
3. HCØ>LM 25.08.00 and 02.09.00, ØC 1–2
4. HCØ>LM 11.11.00, ØC 1–2
5. HCØ>LM 23.12.00, ØC 1–2
6. HCØ>LM 23.12.00, ØC 1–2
7. HCØ>LM 07.10.00, ØC 1–2
8. HCØ>LM ØC 1–2, and 28.10.00, LM>HCØ 20.10.00, ØC 1–2
9. LM>HCØ 16.11.00. ØC 1–2
10. LM>HCØ 16.11.00. ØC 1–2
11. LM>HCØ 16.11.00. ØC 1–2
12. Moreover, all indications of this relationship are censored away in MØ 1–2
13. N.G Bartholdy 2004, 9–37
14. SLS minutes of proceedings, NKS 768d, RL
15. SLS statutes, NKS 768d, RL
16. SLS minutes of proceedings, 34–5, NKS 768d, RL
17. HCØ>LM 01.07.00 and 12.07.00, ØC 1–2. Unfortunately M. Ørsted 1870 is inadequate as a sourcebook. Firstly, there are no clear criteria of selection, so the letters printed depend on the editor's whimsical discretion. Secondly, nearly all private affairs are censored away and thirdly, it is only noted haphazardly if and where omissions have been made.
18. LM>HCØ 25.07.00, ØC 1–2
19. Patronatsforestilling 01.11.00, quoted by G. Norrie 1939, vol. 3, 111–19
20. Patronatsforestilling 01.11.00, quoted by G. Norrie 1939, vol. 3, 111–19
21. P.C. Abildgaard>HCØ 22.08.00, MØ 1, 6
22. HCØ>LM 12.08.00, ØC 1–2
23. K. Meyer's assumption that the Duke and Moldenhawer worked against HCØ's career is thus without foundation. *NS* III, xiii and xv.
24. K. Meyer, *NS* III, xv and E. Andersen 1968, 22
25. *Universitets- og Skole-Annaler* 1806, 95–6 and 232
26. F.A.C Gren 1796

27. We do not know where HCØ lectured privately. Kjølsen's suggestion of Skidenstræde is guesswork and improbable, because LM and HCØ taught at the Academy of Surgery in Norgesgade and there is no evidence that the University's laboratory in Skidenstræde was used by HCØ
28. HCØ>LM 10.02.01, 03.03.01, 10.03.01, 24.03.01 and 31.03.01, ØC 1–2

Chapter 10

1. HCØ, *NS* I, 106, *SSW*, 101
2. O.H. Mynster 1794, 3. O. Bostrup 1996, 78
3. L. Galvani 1791, part iv
4. M. Pera 1988, 138ff.
5. A. Volta 1918–29, vol. 1, 574, JWR1800, vol. 1, part iii, 47
6. *DSB* vol. 14. Volta remarked on the voltaic pile: 'Its character of a perpetual motion machine may seem paradoxical and maybe it is inexplicable; nevertheless, it is a true reality that anybody can touch and feel with his own hands.'
7. A. von Humboldt 1797, vol. 2, 440. This second volume includes Ritter's 'Bemerkungen und Ergänzungen' [Comments and Supplements], cf. K. Poppe 1968, 83
8. *NS* I, 91
9. JWR 1798 containing among other things his speech of 29.10.1797 to the Scientific Society of Jena with the title 'Über den Galvanismus; die Entdeckung eines in der ganzen lebende und toten Natur sehr tätigen Prinzips.' [On Galvanism; the Discovery of a very Active Principle in the whole of Living and Dead Nature]
10. *NS* III, 31 (not in *SSW*)
11. JWR 1798, 173
12. HCØ>LM 28.10.00, ØC 1–2. HCØ read about JWR's galvanic experiments for the first time in *Intelligenzblatt der Allgemeinen Literaturzeitung*, Jena, 1800, No. 180, col. 1510
13. LM>HCØ 09.11.00, ØC 1–2
14. C.C. Rafn, *Nyt Bibliothek* 1801, vol. 1, 45
15. C.C. Rafn, *Nyt Bibliothek* 1801, vol. 1, 60. A. Skaar Jacobsen considers it a fact that JWR subscribed to Schelling's controversial and speculative *Naturphilosophie*, and that orthodox scholars in Scandinavia and Germany had high expectations that Hauch would guide him in the proper direction (*Ambix* 2000). The series of experiments mentioned above rather puts him in opposition to Schelling, whose views were unsubstantiated by experiments. The concept *Naturphilosophie* is used derogatively by Jacobsen. She does not present any evidence to support her claim, which is also at odds with Hauch's own judgement: 'I am aware that Ritter's view expressed with so much perspicacity will have the upper hand for a great many people.' A.W. Hauch, *Nyt Bibliothek* 1801, 152
16. C.C. Rafn, *Nyt Bibliothek* 1801, vol. 1, 60
17. C.C. Rafn, *Nyt Bibliothek* 1801, vol. 1, 28–68
18. C.C. Rafn, *Nyt Bibliothek* 1801, vol. 1, 31–4
19. EAS 1857, D-79, DTM
20. JWR 1798, 70. *Philosophical Transactions*, London 1799, 255–92
21. *Jahrbuch der Goethegesellschaft* 1921, (135–51) 143–4. Begging letter 13.10.00, 146–7, letter of thanks 10.04.01
22. HCØ>LM 03.03.01. ØS 1–2. *Nyt Bibliothek* 1801, 2. vol., 288
23. *Magazin für den neuesten Zuständen der Naturkunde* 1802, vol. 3, 412–17

Chapter 11

1. HCØ>SP 17.01.02, ØC 80, TL 39
2. LM>HCØ 19.12.01, ØC 1–2

3. LM>HCØ 15.01.02, ØC 1–2
4. *Fonden ad usus publicos*, protocol of letters No. 164 of 20.02.02
5. LM>HCØ 08.01.02, ØC 1–2
6. HCØ>LM 23.06.02, ØC 1–2
7. HCØ>SP 23.01.02, ØC 80, TL 41
8. HCØ>SP 14.10.02, ØC 80, TL 130
9. HCØ>SP 09.10.02, ØC 80, TL 129
10. HCØ>SP 20.10.02, ØC 80, TL 132
11. HCØ>SP 16.01.02, ØC 80, MØ I, 39
12. LM>HCØ 19.12.02 and 15.01.02. HCØ>LM 05.01.02, ØC 1–2
13. HCØ>SP 16.02.02, ØC 80, TL 56
14. Missing periods: 23.08.-31.08.01; 11.10.-29.11.01; 04.12.-27.12.01; 26.04.-09.05.02; 13.06.-24.06.02; 24.07.-13.08.02
15. e.g. HCØ>LM 03.10.01, 26.01.02, 16.02.02 and LM>HCØ 22.02.02, ØC 1–2
16. ØC 1–2
17. HCØ's notebook 1801–03, 1812, 1843, ØC 88
18. LM>HCØ 03.10.01, ØC 1–2
19. HCØ>SP 12.04.02, ØC 80
20. LM>HCØ 17.07.02, ØC 1–2
21. HCØ>SP 24.06.02, ØC 80. The quotation is not included in MØ I-II, TL 94
22. HCØ>LM 23.06.02, ØC 1–2, TL 92
23. HCØ's notebook 28, ØC 15
24. Bakkehusmuseets manuskriptsamling, vii, Letters to Kamma Rahbek, 12.07.02
25. Th. Bull 1954, 22. The letter is dated 05.08.05, the same year as B.A. Ørsted was engaged to G.J. Bull
26. HCØ's notebook 30–1, ØC 15
27. HCØ>SP 29.09.02, ØC 80, TL 123
28. HCØ>SP 14.05.02, ØC 80, TL75
29. HCØ>SP 11.05.02, ØC 80, TL 75
30. HCØ>SP 18.-19.05.02, ØC 80, TL 79. *For the Friends of Nature and Art*, 1997. R. Alex & P. Kühn 1995.
31. HCØ>SP 1.-2.06.02, ØC 80, TL 85
32. HCØ>SP 10.06.02, ØC 80, TL 88

Chapter 12

1. HCØ>SP 13.08.-04.09.02, ØC 80, TL 103
2. JWR>HCØ 26.07.09, C II 247
3. K.Richter 1988, 13–84. W. Benjamin 1979, 42. W. Wetzels 1973
4. C.v. Klinchowstroem 1921, 135–51
5. JWR>HCØ 21.02.02, C II 10
6. M.H Mendel 1802
7. JWR 1984, 67
8. HCØ>LM 14.08.01, ØC 1–2
9. HCØ>LM 04.12.01, ØC 1–2. TL 29
10. I. Mieck 1965, 345
11. Landesarchiv Berlin, Pr. Br. Rep. 42, Neue Folge vol. 2, Plankammer V, No. 3
12. *Nyt Bibliothek* vol. 3, 241–55, *NS* III, 77 (not in *SSW*)
13. HCØ>SP 03.12.01, ØC 80. TL 28–9

14. *Nyt Bibliothek* vol. 3, 241–55, *NS* III, 71–7 (not in *SSW*)
15. HCØ>LM 06.09.02, TL 113. My emphasis.
16. HCØ>SP 27.12.01, ØC 80
17. A.S Jacobsen 2000, 43–4
18. M.H Mendel 1802, 12
19. HCØ>LM 23.06.02 and LM>HCØ 17.07.02, TL 94
20. *NS* I, xvi–xviii
21. ØC 88, 6–9
22. ØC 88, 15–16
23. HCØ>LM 08.02.02, ØC 1–2, TL 52–4
24. LM>HCØ 17.07.02, ØC 1–2, MØ I 75
25. HCØ>LM 19.08.02, ØC 1–2, TL 99–102
26. HCØ 1803, *NS* I, 205–10
27. HCØ>LM 06.09.02, ØC 1–2, TL 107–12
28. J.J. Winterl>HCØ 18.09.02, C II 602
29. JWR>HCØ 28.10.02, C II 26
30. HCØ>SP 20.09.02, ØC 80, TL 120
31. M.H Mendel 1802, 7
32. K. Richter 1988, 66

Chapter 13

1. HCØ>SP 28.03.02, ØC 80, TL 66
2. HCØ>SP 23.09.01, ØC 80, TL 19
3. C.L Reinhold 1790–92
4. J. Baggesen 1831 covering the period 1790–1801
5. J.G. Fichte 1792
6. J.G. Fichte 1799
7. *KUJ* 1796, 20–1
8. *Die Horen* 1795
9. J.W Goethe 1986
10. F. Schiller 1972, 321–34. J.G. Fichte 1795 and 1800. Th. Ziolkowski 1990
11. P. Wilhelmy-Döllinger 2000
12. M.L. Davies 1995, 27
13. HCØ>SP 16.01.02, ØC 80, TL 37
14. HCØ>SP 23.-27.09.02, ØC 80, TL 122
15. F. Maurice 1997, 134
16. HCØ>SP 18.01.02, ØC 80, TL 40
17. F. Maurice 1997, 135
18. HCØ>SP 01.01.02, ØC 80, TL 32
19. HCØ>SP 20.03.02, ØC 80, TL , 63
20. HCØ>SP 10.01.02, ØC 80, TL, 35. HCØ>SP 13.04.02, ØC 80, TL , 68
21. HCØ's notebook, ØC 15
22. HCØ>LM 04.12.01, ØC 1–2, TL 30
23. F. Maurice 1997, 158–66
24. HCØ>SP 23.-27.09.02, ØC 80, TL 121–2

25. LM>HCØ 08.01.02, ØC 1–2, MØ I, 34–5
26. LM>HCØ 27.04.02, ØC 1–2
27. K. Fischer 1923, 190–3
28. HCØ>SP 29.01.02, ØC 80, MØ I, 43
29. J.G. Fichte 1801
30. HCØ>ASØ 07.02.02 and 16.02.02, ØC 1–2, TL 47–9, 56–8
31. J.H. Randall, Jr, 1965, vol. 2, 214–22
32. *Zeitschrift für spekulative Physik*, fasc. 4
33. HCØ>ASØ 16.02.02, ØC 1–2, TL 56–8
34. K. Fischer 1923, 36
35. K. Hufbauer 1982, 220–1
36. HCØ>SP 27.03.02, ØC 80, TL 64–5
37. T.H Broman 1996, 149
38. HCØ>SP 08.09.02, ØC 80
39. HCØ>SP 02.10.02, ØC 80, TL 114
40. HCØ>SP 16.01.02, ØC 80, TL 37–8
41. HCØ's notebook 22.09.01, ØC 15
42. F.Schlegel 1975, vol. 3, xxi
43. *Europa*, vol. 1, 50

Chapter 14

1. HCØ>SP 07.05.03, ØC 80. TL 174
2. HCØ>SP --.12.02, ibid., ibid., TL 138, 151
3. HCØ>SP 15.,23.03.03, ibid., ibid., TL 164, 167
4. Ground-plan of Passage des Petits-Pères, Archive of Musée Carnavalet, 29. J. Hillairet 1963–72, vol. 2, 264
5. J. Hillairet, 1963–72, vol. 2, 397
6. G. Poisson 1989, 40 (Conservatoire des Arts et Métiers), 112 (Panthéon), and 104 (Club des Jacobins)
7. *Histoire et Dictionnaire de la Révolution Française 1789–1799*, 1987
8. M. Crosland 1967, 13
9. B. Michael and H.-H. Schepp 1973, vol. 1, 130–5
10. M. Crosland 1967, 4–19
11. M. Crosland 1967, 155
12. HCØ>SP 13.04.03, ØC 80, TL 178
13. HCØ>SP --.12.02, ØC 80, TL 148
14. HCØ, fasc. 11, ØC 43
15. HCØ, fasc. 11, ØC 43
16. HCØ>SP --.12.02, ØC 80, TL 140
17. This must be an exaggeration, because at another place HCØ writes that Vauquelin had 40 listeners, which amount to (not 400, but) 160 louis d'or (about 960 Rixdollars). HCØ>SP --.12.02, ØC 80, extract in TL 147
18. More information on École Polytechnique in T. Bugge 1800
19. M. Crosland 1967, 192–208
20. HCØ's notebook 23.02.03, ØC 15. *Europa* 1803 vol. 1, NS I, 126, SSW 116,
21. M. Crosland 1967, 237
22. M. Crosland 1967, 246–7

23. *Europa* 1803, vol. 1, *NS* I, 126 and 130–1, *SSW* 116 and 118–19. HCØ>SP 01.06.03, ØS 80, MØ I 141–2, HCØ's travel diary 19.10.03, ØS 15
24. HCØ>SP 03.03 and 01.07.03, ØS 80, MØ I, 124 and 141–2
25. M. Crosland 1967, 56–79
26. W.D Wetzels 1973, 39–40

Chapter 15

1. HCØ>SP dec. 1802, ØC 80. TL 150
2. M. Crosland 1967, 20–3
3. HCØ>SP 13.05.03, 24.05.03, 04.06.03, 21.-31.08.03, 04.09.03, 14.09.03, and 16.09.03, ØC 80. TL 178–9, 182, 184, 203, and 204–5
4. M. Crosland 1967, 20–3
5. HCØ>SP dec. 1803, ØC 80. TL 150
6. JWR>HCØ 22.05.03. C II, 35–6
7. HCØ>SP 18.03.03, ØC 80. TL 166
8. HCØ>SP 13.05.03. ØC 80, TL 179
9. HCØ>LM 19.08.02 and LM>HCØ 04.09.02, ØC 1–2, TL 200–2
10. JWR>HCØ 20.05.03. CII, 34
11. JWR>Duke Ernst of Gotha 28.03.03, *Jahrbuch des Freien Deutschen Hochstifts* 1973, 190–223
12. HCØ's notebook 02., 07., and 10.06.03, ØC 15. HCØ>SP 10.06.03, ØC 80, TL 185
13. A.J Turner 1989
14. JWR>HCØ 20.06.03. C II, 44
15. JWR>Duke Ernst II of Gotha 28.03.03, *Jahrbuch des Freien Deutschen Hochstifts* 1973, 190–223
16. HCØ>LM 10.08.02, ØC 1–2, TL 197. The rate of exchange from French francs to Danish rixdollars deteriorated, and the calculation is subject to reservations, cf. LM>HCØ 16.07.03, ØC 1–2
17. JWR>HCØ 20.06.03, C II, 44–5
18. JWR>Frommann 29.07.03, K. Richter 1988, 131
19. JWR>Duke Ernst II of Gotha 14.06.02, *Jahrbuch des Freien Deutschen Hochstifts* 1973, 184–5
20. JWR>HCØ 15.06.03, C II, 41–4
21. HCØ>LM 04.09.03, ØC 1–2. TL 200–2
22. HCØ>LM 04.09.03, ØC 1–2. TL 200–2
23. 1.) *Journal de physique, de chimie, d'histoire naturelle et des arts*, vol. 57, 345–68, *NS* I 214–37. 2.) *Journal de physique, de chimie, d'histoire naturelle et des arts*, vol. 57 401–5, *NS* I, 237–2242. The experiment to prove the existence of the two electrical poles of the earth appears in the post-scriptum, 232–3. *Procès-verbaux . . .* , vol. 3, 1910–22, 11–15
24. JWR>HCØ 15.02.04, C II, 55
25. HCØ's notebook 02.11.03, 05.11.03, and 09.11.03. ØS 15. 1.) *Journal de physique, de chimie, d'histoire naturelle et des arts*, vol. 57, 345–68, NS I, 214–37. 2.) *Journal de physique, de chimie, d'histoire naturelle et des arts*, vol. 57 401–5, NS I, 237–42. 3.) *Journal de physique, de chimie, d'histoire naturelle et des arts*, vol. 57 406–9, NS I, 242–5. 4.) *Journal de physique, de chimie, d'histoire naturelle et des arts*, vol. 57 409–11, NS I, 245–8, (not in *SSW*)
26. HCØ>LM 06.10.03, ØC 1–2. TL 205–8
27. EAS in D-79, DTM. HCØ>LM 25.11.03, ØC 1–2, TL 212

Chapter 16

1. HCØ>LM 27.06.03, ØC 1–2, TL 188–9
2. P.C. Abildgaard>Th. Bugge 12.11.98 and 18.12.1798, NKS 287c, RL

3. *Nyt Bibliothek*, vol. 5, 273–4
4. HCØ>LM 04.12.01, ØC 1–2, TL 30
5. LM>HCØ 17.07.02, ØC 1–2, extract in MØ I, 74–6
6. HCØ>LM 06.09.02, ØC 1–2. HCØ *Materialien* 1803, NS I, 205–10, *SSW*, 162–5
7. LM>HCØ 16.07.03, ØC 1–2
8. *Nyt Bibliothek*, vol. 2, 158–74, NS III, 70 (not in *SSW*), cf. LM's comment: 'The author (HCØ) endeavours by various practical works that almost entirely engulf him to …', *Nyt Bibliothek*, vol. 3, 241–55, NS III, 76, note 1 (not in *SSW*)
9. HCØ>SP 21.-31.08.03, ØC 80, TL 198
10. HCØ>LM 06.09.02, ØC 1–2, TL 113
11. Aasheim, Bugge, Buntzen, and Hauch were owners of collections of scientific instruments, but they were private and not at the disposal to HCØ's lectures at the University
12. HCØ>LM 06.09.02, ØC 1–2, TL 113
13. LM>HCØ 18.12.02, ØC 1–2, extract in MØ I, 105–8
14. HCØ>LM 03.01.03, ØC 1–2, extract in MØ I, 108–10, TL 194–5
15. LM>HCØ 05.02.03, ØC 1–2.
16. HCØ>SP 25.08.03, ØC 80. A and C are excised by MØ, whereas B is included although slightly altered, MØ I, 148–50
17. HCØ>LM 27.06.03, ØC 1–2, TL 188–9
18. HCØ>LM 04.09.03, ØC 1–2, extract in MØ I, 155–7, TL 200–2
19. HCØ>LM 06.10.03, ØC 1–2, TL 208
20. HCØ's notebook 5–56, ØC 15
21. S. Miss 1996, 50–65
22. SP>B. Thorvaldsen 21.01.21, Thorvaldsens Museum
23. HCØ>LM 06.10.03, ØC 1–2, extract in MØ I, 160–3, TL 206

Chapter 17

1. AOe>HCØ 31.08.06, ØC 1–2, MØ I, 200
2. JWR>HCØ 15.02.04, ØC 1–2. This passage as well as other comments by JWR about HCØ's relationship with women are excised by M.C. Harding in C I-II, the rest of the letter C II, 53–7
3. JWR>HCØ 20.11.04, ØC 1–2. CII, 94
4. JWR>HCØ 15.02.04, ØC 1–2. CII, 94
5. JWR>HCØ 26.12.03, ØC 1–2. CII , C II, 46
6. R. Chenevix, *Annalen der Physik* 1804, 422–54
7. P.A. Heiberg>Rasmus Nyerup 31.08.04, S. Birket Smith 1884, 51–4
8. L. Engelstoft>Rasmus Nyerup 31.08.04, L. Engelstoft 1862, vol. 3, 318
9. *KLE* 26, 1804, 415
10. *KLE* 27, 1804, 430–1
11. HCØ>AOe 02.11.05, AOe 1945, vol. 1, 159
12. H. Steffens 1844, vol. 5, 125–31
13. AOe>HCØ 05.02.06, AOe 1945, vol. 1, 238
14. H. Steffens 1844, *Zeitschrift für spekulative Physik* I, 1800
15. H. Hultberg 1973, 46–51
16. H. Steffens 1844, vol. 5, 93
17. CH 1969, 27
18. H. Steffens 1844, vol. 5, 30–1

19. AOe 1945, vol. 1, 20, 27–53
20. H. Steffens 1844, vol. 5, 30–1
21. LHS journals 1783/379 and 1784/393, EA
22. HCØ>AOe 13.05.06, AOe 1945, vol. 2, 8–9
23. SØ>IBB 11.08.12, ØS 94
24. HCØ>E.H. Schimmelmann 23.05.05, NBD 19, RL
25. AOe>Crown Prince Frederik, after 30.07.05, AOe 1945, vol. 2, 97–100
26. HCØ>AOe 15.11.05, AOe 1945, vol. 2, 185
27. AOe>HCØ 02.12.05, AOe 1945, vol. 2, 185
28. HCØ>AOe 15.11.05, AOe 1945, vol. 2, 173
29. AOe>Director of the Royal Theatre 28.05.06, AOe 1945, vol. 2, 11
30. HCØ>AOe 23.01.07, AOe 1945, vol. 2, 154. HCØ>AOe 23.07.07, AOe 1945, vol. 2, 251–2. HCØ>AOe 26.05.09, AOe 1945, vol.2, 274
31. C. Hauch *SES* 9, 122–3
32. JWR>HCØ 08.02.05, ØC 1–2, C II, 97
33. LM>HCØ 17.07.02, ØC 1–2
34. LM>HCØ 04.09.02, ØC 1–2
35. *KUJ* 1795, 122, 190, and 196
36. *Nyt Bibliothek* 1803, 273. The original in ØC 59, fasc. xvi, dated Paris 26.03.03, so the journal's dating 1802 is erroneous.
37. LM>HCØ 07.05.03, ØC 1–2
38. LM>HCØ 05.02.03, ØC 1–2
39. *Annalen der Physik* 15, 1803 and 25, 1807
40. *Nyt Bibliothek* vol. 4, 1802, 203
41. *Fonden ad usus publicos* 1902, vol. 2, 76
42. *Fonden ad usus publicos* 1902, vol. 2, 128. In his book *Fra Skidenstræde til H.C. Ørsted Instituttet*, Copenhagen 1965, H.H. Kjølsen makes an attempt to identify the locations of HCØ's recidences, laboratories, and lecture halls from 1804 to 1851. His claim that HCØ would have used laboratorium chymicum in Skidenstræde for experimental lectures is improbable. Firstly, there is no written evidence for it, and secondly, HCØ could not possibly squeeze his many participants into this small laboratory of 25 square metres (in the winter term of 1805–06 there were 60 enrolled, of which 40 completed the course, HCØ>AOe 15.03.06, AOe 1945, vol. 1, 256). According to Kjølsen's own calculations the laboratory could only contain 10–15 people (p.32). Thirdly, we know that HCØ (as soon as he had been promoted professor in September 1806) moved from Collin's Court (JWR>HCØ 02.02.06, ØS 1–2) to the other side of the street to newer and larger premises in Thott's Palace (notes 36–7 above and HCØ>AOe 20.09.06).

Whereas Kjølsen claims the following sequence of HCØ's addresses: Skidenstræde: -1806; 52, Østergade: 1805-13; Thott's Palace: 1813-19; 35, Nørregade: 1819-; his documentable addresses are: 159, Norgesgade: 1804-06; 201-202, Norgesgade (Thott's Palace): 1806-1808; 68 (not 52), Østergade: 1808-12; 12, Gammel Strand: 1813; 38, Læderstræde: 1813-14; 401B, Dronningens Tvergade: 1815-1819; 35, Nørregade: 1819-1824; 106, Studiestræde: 1824-1851, cf. map of Copenhagen (ch 3)

Chapter 18

1. *SES* 5, 31–2. AOe 1964, 57
2. *catalogus præelectionum* 1809–10 and 1810–11

3. *NS* III 78–9 (not in *SSW*). According to L. Pearce Williams a misquotation from Schiller's *Columbus*: 'Mit dem Genius steht die Natur in ewigem Bunde. Was der Eine verspricht, leistet die Andere gewiss', DSB vol.x, 1974, 183
4. *NS* III 78
5. *KLE* 1804, 619–21
6. C. Hauch *SES* 9, 125–6
7. HCØ>AOe 15.03.06, AOe 1945, vol. 1, 256
8. HCØ>The Duke of Augustenburg c.1807, ØC 59, fasc. xviii, RL
9. HCØ>AOe 28.12.05, AOe 1945, vol. 1, 205. JWR>HCØ 11.02.05, ØC 1–2, C II, 103
10. JWR>HCØ 25.05.06, excised in C II, 175
11. JWR>HCØ 25.05.06, excised in C II, 175
12. JWR>HCØ 10.07.06, 19.01.07, 20.04.07, ibid., ibid., 180, 187, 188, 207
13. JWR>HCØ 25.05.06, ibid., ibid. 168
14. *NS* III 96–105, *SSW* 185–91
15. DCC 2000, 24–5
16. NKS 768d, RL
17. *SES* 8, 109
18. *NS* III 101–2, *SSW* 189–90
19. *NS* III 103, *SSW* 191
20. W.D Wetzels 1973, 76
21. *NS* III 114, *SSW* 198
22. *NS* III 115, *SSW* 199
23. *NS* III 156, *SSW* 284
24. T.S Kuhn 1970
25. AOe 1964, 57
26. *SES* 5, 31–2
27. *Fonden ad usus publicos* 1902, vol. 2, 226

Chapter 19

1. JWR>HCØ 01.05.04, ØC 1–2, CII, 66
2. *KLE* 1804, 399
3. *KLE* 1805, 319, and 1807, 268–9
4. ØC 43, fasc. ix, RL
5. JWR>HCØ 20.11.04, ØC 1–2, CII, 90
6. JWR>HCØ 16.-17.08.05, ØC 1–2, CII , 109
7. HCØ>AOe 13.06.06, AOe 1945, vol. 2, 30
8. ØC 43, fasc. ix, RL
9. JWR>HCØ 16.-17.08.05, ØC 1–2, C II, 111
10. *SSW* 286, 289
11. HCØ 1809, 147
12. H-40, H-42, DTM
13. *SSW* 291
14. A.S Jacobsen 2001, 184–218
15. KDVS announcement 1806
16. CF. Bucholz 1807, 336–44, *Nyt Bibliothek* 1806, 229–52

17. KDVS minutes 519/1808 and 585/1808
18. JWR>HCØ 10.07.06, ØC 1–2, C II, 176–7
19. W. Benjamin 1979, 42
20. JWR 1984, 31–9
21. E. Rehm 1974, 294
22. JWR>HCØ 06.09.05, ØC 1–2, C II, 130, Wesselhöft>HCØ 16.12.05, ibid[23] JWR>HCØ 02.12.05, ibid., ibid., 146–7
24. JWR>HCØ 10.07.06, ibid., ibid., 176
25. JWR>HCØ 26.07.09, ibid., ibid., 247 and 255
26. HCØ>AOe 22.10.05, AOe 1945, vol. 1, 141. JWR>HCØ 08.02.05, ØC 1–2, C II, 97
27. Kgl. Res. [Royal resolution] 05.09.06, PRO
28. Gjerlew>HCØ 20.01.08, ØC 1–2, partly in MØ I, 259
29. HCØ>Board of Directors, Univ. Cop., ØC 59, fasc. xvi, RL
30. HCØ Øc 59, fasc. ii, RL
31. HCØ Øc 59, fasc. ii, RL

Chapter 20

1. HCØ>AOe 08.09.07, ØC 1–2, MØ I, 220
2. A.C. Gjerlew>HCØ 19.02.08, ØC 1–2, op.cit. 267
3. AOe>Christiane Heger 23.11.06, AOe 1945, vol. 2, 135–7
4. J.G Fichte 1930, vol. 2, 462
5. SØ>J. Baggesen 11.08.07 in N. Bøgh 1881, 157
6. SØ>AOe 02.08.06, AOe 1945, vol. 2, 60–1
7. *Nær og Fjern* 4, 1875 No.168
8. Reventlow 1900–02, vol. 4, 257
9. AOe>HCØ 05.02.06, AOe 1945, vol. 1, 238–40
10. AOe 1850–51, vol. 2, 53
11. HCØ>AOe 23.07.07, AOe 1945, vol. 2, 65–6
12. P. Bagge 1970
13. HCØ>AOe 23.07.07, AOe 1945, vol. 2, 253
14. J.G Fichte 1845, vol. 6, 347–447
15. I. Kant 1917, vol. 8
16. HCØ>AOe 02.08.06, AOe 1945, vol. 2, 65–6
17. J.G Fichte 1845, vol. 5, (397–580), 476
18. P. Bagge 1970
19. *Fortegnelse over HCØ's efterladte Bogsamling* 1853, No.4126
20. J.G Fichte 1845, vol. 6, 385
21. *Aarbog for Universitetet og de lærde Skoler* 1807, vol. 2, 293–313
22. *Cancellitidende* 1807
23. T. Munch-Petersen 2007, 237–8
24. J.P Mynster 1854, 170–2
25. J.P Mynster 1854, 170–2
26. C. Hauch 1867, 210
27. HCØ>AOe 08./12.09.07, AOe 1945, vol. 2, 280–1
28. JWR>HCØ 08.10.07, C II, 207
29. S.D Thalbitzer 1966, 168

30. University Library Add 1289 a-b, file iv, FCS on SØ, fragment, RL
31. SØ>J. Baggesen 12.09.07, AOe 1945, vol. 2, 283–4
32. Notes by Miss B. Scharling, F-3, DTM
33. H. Callisen 1809, vol. 1, 638–42
34. HCØ>JWR 03.09.08, *NS* I 345
35. T. Bugge>Sjösten, NKS 287c, RL
36. Gjerlew>HCØ 25.11.07, ØC 1–2

Chapter 21

1. D-41, DTM
2. HCØ's notebook 05.-08.10.01, ØC 15, RL
3. E.F.F Chladni 1802
4. G.C Lichtenberg 1777
5. *NS* II 21 and 23, *SSW* 272–3
6. *Neues allgemeines Journal der Chemie*, vol. 3, 1807, 544–5, *NS* I, 343–4, *SSW* 261–2
7. *Magazin für den neuesten Zustände der Naturkunde*, vol. 9, 31–2, *NS* I, 261–2, *SSW* 180
8. *Neues allgemeines Journal der Chemie*, vol. 6, 1806, 292–302, *NS* I 267–73, *SSW* 181–4
9. *Magazin für den neuesten Zustände der Naturkunde*, vol. 9, 31–2, *NS* I, 261–2, *SSW* 180. *Bibliothèque Britannique* 30, 240 and 364–72; *NS* I 262–6, *SSW* 181–4. *Neues allgemeines Journal der Chemie*, vol. 6, 1806, 292–302, *NS* I 267–73, *SSW* 210–14. *KDVS Skrifter* 1810, *NS* II, 11–34. A.W. Hauch's PM 07.06.07, KDVS 543–1808
10. JWR>HCØ 20.05.03, ØC 1–2, C II, 40 and 90–1
11. JWR 1985, 268, cf. JWR>HCØ 31.03.09, ibid., ibid., 224
12. K. Meyer characteristically comments (*NS* I, xlii), 'it cannot be denied that the arguments employed in various places, especially to explain the generation of electricity, bear the stamp of being adapted so as to agree with a previously given result and of building not so much on mathematically or experimentally grounded facts as on hypotheses intuitively advanced. To French and English scientists, in particular, who were not infected with the phraseology of the German school of Nature Philosophy, the form must have been distasteful.'
13. KDVS 543–1808
14. KDVS 552-, 612-, and 621–1808, 14.11.08
15. SAK 1920, 350
16. HCØ>IBØ 09.06.23, ØS 1–2
17. HCØ>IBØ 14.08.23, ØS 1–2
18. HCØ>Ch. Wheatstone 12.06.23, C II, 590
19. This may refer to HCØ, 'Ueber die Art wie sich die Electricität forpflanzt (Ein Fragment)', *NS* I, 267–73, [On the Manner in Which Electricity is Transmitted (A Fragment)], *SSW* 210–14, which was translated into English from *Journal de Physique* and printed in *Nicholson's Journal*, 1806, or to HCØ's letter to Pictet, which was translated from *Bibliothèque Britannique* and printed in *The Philosophical Magazine* xxiv, 1806.
20. R.D. Tweney, *Physis*, 29, 1992, 149–64
21. L.P Williams 1965, 181

Chapter 22

1. *NS* II, 34, *SSW* 281
2. *NS* II, 34, *SSW* 281
3. HCØ's notebook 02.06.02, ØC 15, RL
4. N. Schiørring 1978, vol. 2, 66ff.

5. Account book, NKS 1011, RL
6. J.J Rousseau 1768, 227
7. J.J Rousseau 1768, 231
8. *SES* 5, 33–40
9. *Fortegnelse over Ørsteds efterladte Bogsamling* [Catalogue of Ørsted's Left Collection of Books], 1853. This catalogue does not contain his entire library, since parts of it, fiction in particular, was taken out by his widow. The catalogue numbers *c*.6,500 volumes. The following works and authors are relevant for his theory of music: The Encyclopaedia, Chladni, Euler, d'Alembert, Diderot, Rousseau, Gall, Lichtenberg, Kant, Fichte, Schelling, Ritter, Schlegel, Plato, Burke, Faraday, Herschel, Wheatstone, and S.Aa. Kierkegaard, the only Dane.
10. Of the five interlocutors included in 'On the Cause of the Pleasure Evoked by Tones' some were allotted parts in three new dialogues on the same topic that HCØ wrote thirty years later (in chronological order):
    1. 'On the Cause of Pleasure Evoked by Tones, 1808, *SES* 3, 67–99, *SiN*, 325–51 (under the title: 'The same Principles of Beauty exist in the Objects Submitted to the Eye and to the Ear') (Alfred, Valdemar, Herman, Felix, Julius)
    2. 'The Physical Effects of Tones, 1830, *SES* 3, 100–23, *SiN*, 352–71 (Alfred, Valdemar, Herman, Felix, Sophie)
    3. 'The Spiritual in the Material', 1839, *SES* 1, 3–36, *SiN* 1–27, (Alfred, Herman)
    4. 'On Symmetry and the Impressions of Beauty Evoked by it', 1839, *SES* 3, 124–54, not in *SiN*, (Alfred, Herman)
11. *SES* 3, 77, *SiN*, 333–4
12. E. Hanslick, who has a high regard of Ørsted's theory of music, erroneously ascribes Julius's attitude to HCØ, Hanslick, 1974, 110
13. *SES* 3, 91, *SiN* 345
14. *SES* 3, 69, 75, 79, 97–8, *SiN* , 326, 332, 335–6, 350–1
15. *SES* 3, 93, 80, *SiN* , 347, 336
16. L. Daston 2007, 235–46. DCC 2007, 115–34
17. *SES* 3, 121, *SiN*, 369
18. I. Kant 1913, vol. 5
19. DCC 2000, 18–31

Chapter 23

1. HCØ>Gjerlew 13.02.08, ØC1–2, MØ I, 261
2. HCØ>AOe 14.02.09, AOe 1945, vol. 3, 237. *Fonden ad usus publicos* 1902, 129.
3. HCØ>AOe 26.05.09, AOe 1945, vol. 3, 277
4. KDVS 612–1808, 31.10.08
5. e.g. HCØ>LM 06.09.03, jan. 03, and 01.04.15, ØC 1–2, RL
6. KDVS 698–1809
7. KDVS 707–1809
8. KDVS 857–1811
9. P. la Cour and J. Appel 1897, sections 211–212 and 273–274
10. H.N Clausen 1877, 22–3
11. FCS>SØ 30.08.11, F.C. Sibbern 1866, 49. Schumacher lodged with ASØ and SØ in their flat in Vestergade when he arrived in Copenhagen, SØ>HCØ 13. and 27.06.12, ØC 1–2, RL

12. CH>HCØ 02.04.15, C I, 109
13. KDVS 812–1811
14. KDVS 1909–1815
15. KDVS 747–1810, 811/812–1811. HCØ>SØ 07.07.12, ØC 1–2. *SSW* 310–92
16. KDVS 821 and 875–1811
17. KDVS 1854–1813
18. KDVS 1941–1815
19. KDVS 823–1811 (Berzelius, 1881–1815 (van Mons), 2037–1816 (Gay-Lussac), 2081–1817 (Werner), 2537–1821 (Davy), 2593–1821 (Erman, Brewster, Brera)
20. HCØ>JJB 25.02.11, C II, 8

Chapter 24

1. D-41, DTM
2. AOe 1945, vol. 1, 243 and vol. 2, 18, 31, and 195
3. Justitsarkiv. Rel. t. Kongen. Generalauditøren, parcel 17, ccxii, verdict No. 2, and Generalitets- og Kommissariats Koll. Kgl. ordrer og res. 1809, No. 249, PRO
4. AOe 1945, vol. 3, 306
5. NKS 1674, fasc. 17, RL
6. BT 28.06.11
7. F-32, DTM
8. E.C Werlauff 1968, 167–8
9. AOe>HCØ 05.02.06, AOe 1945, vol. 2, 232–3
10. AOe>HCØ 11.08.07, AOe 1945, vol. 2, 266
11. H. Steffens 1844, 244–5
12. Reventlow 1900–1902, vol. 5, 292
13. Reventlow 1900–1902, vol. 5, 248
14. Reventlow 1900–1902, vol. 5, 249
15. AOe 1945, vol. 2, 269
16. AOe 1945, vol. 2, 270–80
17. AOe>C.Heger 25.12.07, AOe 1945, vol. 3, 77–8. The breach between AOe and Steffens is dealt with by H. Hultberg 1973, 85, and by E. Thomsen 1950, 149, AOe 1945, vol. 3, 270–80
18. AOe>HCØ 11.08.07, AOe 1945, vol. 2, 273
19. AOe>HCØ 11.08.07, AOe 1945, vol. 2, 274
20. AOe>HCØ 11.08.07, AOe 1945, vol. 2, 276
21. *SES* 5, 97
22. *SES* 5, 54
23. HCØ>AOe 01.11.07, AOe 1945, vol. 3, 7
24. HCØ>AOe 01.11.07, AOe 1945, vol. 3, 9
25. HCØ>AOe 01.11.07, AOe 1945, vol. 3, 10
26. HCØ>AOe 01.11.07, AOe 1945, vol. 3, 12
27. HCØ>AOe 01.11.07, AOe 1945, vol. 3, 13
28. *SES* 5, 1–32, *SSW* 243–60
29. HCØ>AOe 06.1107, ØC 1–2, MØ I, 238
30. HCØ, ibid., 22, AOe, AOe 1945, vol. 2, 268
31. AOe 1945, vol. 4, 286
32. HCØ>AOe 06.11.07, AOe 1945, vol. 3, 23

33. HCØ>AOe 06.11.07, AOe 1945, vol. 3, 24–5
34. HCØ>AOe 06.11.07, AOe 1945, vol. 3, 23
35. J.L. Heiberg 1972, 26–7

Chapter 25

1. HCØ's scrapped draft for an early version of his 'Dialogue on Mysticism', in which the sentence is communicated by Ludvico to Francisco (names that HCØ later changed). ØC 73, fasc. i, RL
2. There are at least three drafts for this dialogue. Evidently, they have been corrected several times with different colours of ink. There is another draft for a dialogue between Ernst, Bartholomeus, and Alexander on 'What is Chemistry?' which ended up as a lecture. ØC 74, fasc. i–iii, and 43, fasc. v, RL
3. *SES* 5, 44–5. Not in *SiN*. Original manuscript in ØC 51, fasc. i, RL
4. HCØ>E.H. Schimmelmann 29.05.17, ØC 1–2, RL
5. *SES* 5, 45
6. *SES* 5, 50 and 61
7. *SES* 5, 56 and I.Kant, vol. 4, 564–5
8. *SES* 1, 18, *SES* 5, 92. DSB vol.x, 1974, 183
9. *SES* 1, 19
10. I. Kant 1911, vol. 4, ix
11. *SES* 5, 85–6
12. *SES* 5, 69
13. *SES* 5, 100–1
14. P. Guyer, 'Natural Ends and the End of Nature', in *Boston Studies in the Philosophy of Science*, vol. 241, 'Hans Christian Ørsted and the Romantic Legacy in Science', ed. by R.M. Brain, R.S. Cohen, and O. Knudsen, Dordrecht 2007, 75–96. *SES* 5, 104
15. which becomes obvious by comparing the first edition in *Skandinavisk Litteraturselskabs Skrifter* and the second one in *SES* 5
16. K. Fischer, vol. 7, 1923
17. JWR>HCØ 31.03.09, C II, 229
18. A.F. Gehlen>HCØ, C II, 367–70
19. E. Rehm 1974, 291–311. Personal information on the miniature from K. Richter, Weimar
20. JWR>HCØ 28.10.02, 20.05.03 and 16.–17.08.05, C II, 28, 32, 110
21. H. Berg and D. Hermann 1977, 83–113
22. JWR>HCØ 28.10.02, CII, 30
23. K. Richter 1988, 36–7
24. All letters from HCØ to JWR are lost in the latter's mess, or (less likely) Gehlen has failed to transfer JWR's archive to the Bavarian Academy, or (less likely still) the Academy has failed to keep it, E. Rehm 1974, 291–311
25. JWR>HCØ 16.–17.08.05, C II, 119
26. HCØ>LM 06.09.02 and 04.12.08, ØC 1–2, RL

Chapter 26

1. CH>HCØ 30.08.14, C I, 106
2. HCØ>AOe 20.09.06, ØC 1–2, RL, MØ I, 206
3. CH 1969, 26–32
4. CH>HCØ 1812, C I, 91–103

5. HCØ>CH 03.08.22, note 1, C I , 124. CH, *Magasin for Naturvidenskaberne*, 1824, vol. 4, 268–316 and 1825, vol. 5, 1–74
6. CH>HCØ 30.08.14, C I, 106
7. WCZ>HCØ 24.01.11, ØS 1–2
8. *Fonden ad usus publicos* 1902, 300–2
9. C. Hauch 1867, 218–19
10. C. Hauch 1867, 185, 200, 205, 207, and 222–6
11. C. Hauch 1867, 305
12. *Nyeste Skilderie af Kjøbenhavn* 1821, col. 89–90
13. *Fonden ad usus publicos* 1902, 340
14. C. Hauch in *SES* 9, 145
15. C. Hauch in *SES* 9, 150
16. ASØ 1015, vol. 1, 94–145
17. F.D.E Schleiermacher 1803
18. H. Høffding 1909, 49
19. ASØ 1815, vol. 1, 95
21. ASØ 1815, vol. 1, 95
21. ASØ 1815, vol. 1, 96
22. I. Kant 1907, xxiii–xxiv. HCØ *NS* I, 35, *SSW* 47, HCØ 1802
23. ASØ 1815, vol. 1
24. ASØ 1815, vol. 1, 105
25. *Catalogus* 1809–10 and 1810–11
26. ASØ 1815, vol. 1, 97
27. ASØ 1815, vol. 1, 98
28. HCØ *NS* III, 159, *SSW* 286
29. ASØ 1815, vol. 1, 99
30. HCØ, *NS* III, 163–4, *SSW*, 289
31. As to other views of ASØ's philosophy of ethics and law cf. D. Tamm 1980, 11–32, M. Blegvad 1980, 33–54. I follow S. Blandhol 2003, 153–210

Chapter 27

1. B.G. Niebuhr>HCØ 07.11.12, ØC 1–2, RL
2. N.G Bartholdy 2004, 9–37
3. HCØ, ØC 59, fasc. ii, RL
4. *Fonden ad usus publicos*, J. 1809–1817, No. 458 of 18.02.12 and 634 of 03.02.13, PRO
5. HCØ>SØ 18.08.12, MØ I, 301, TL 232
6. KU, Konsistorium [Senate] 1538–1881, copy-book 1810–13, 21 No. 1213, PRO
7. F. Fang 1932–33, 258
8. A. Hansen 1944, vol. 51, 267–77. SØ>HCØ 27.06.12, MØ I, 295
9. ASØ>HCØ 29.12.12, ØC 1–2, excised in MØ I, 308. SØ>I.B. Ballum 27.10.12, ØC 94
10. HCØ>SØ 29.05.12, ØC 94, excised in MØ I, 291, TL 228
11. HCØ's notebook 63, ØC 15
12. ASØ>HCØ 27.06.12, ØC 94
13. SØ>FCS 27.12.11, F.C. Sibbern 1866, 26
14. SØ>FCS 10.05.12, F.C. Sibbern 1866, 32

15. FCS>SØ 28.05.12, F.C. Sibbern 1866, 36
16. SØ>FCS 04.04.12, F.C. Sibbern 1866, 28
17. ASØ>FCS 01.09.08, NBD, RL
18. FCS>ASØ 01.10.11, F.C. Sibbern 1866, 51
19. FCS>SØ 17.01.12, F.C. Sibbern 1866, 59
20. FCS>SØ 16.07.12, F.C. Sibbern 1866, 73
21. FCS>SØ 17.01.12, F.C. Sibbern 1866, 58
22. *DBL* and *ADB*
23. HCØ>SØ 07.07.12, MØ I, 296–7, TL 230
24. HCØ>SØ 07.07.12, MØ I, 296, TL 230
25. HCØ>SØ 18.08.12, MØ I, 302, TL 232
26. *Fonden ad usus publicos* 1902, 175
27. B.G. Niebuhr>HCØ 07.11.12, ØC 1–2, RL
28. B.G. Niebuhr>HCØ 07.11.12, ØC 1–2, RL
29. HCØ>SØ 18.08.12, MØ I, 301–2, TL 232
30. V. Villadsen 2000, 120–1
31. HCØ>SØ 29.05.12, 09.06.12 (not preserved), 07.07.12, 18.08.12, SØ>HCØ 13.06.12 and 27.06.12. No letters to or from AOe have been preserved
32. HCØ>SØ 18.08.12, MØ I, 299–300, TL 234
33. SØ>I.B. Ballum 17.10.12, ØC 94
34. HCØ>ASØ 11.12.12, MØ I, 305, TL 236–7
35. HCØ>ASØ 11.12.12, MØ I, 304–5

Chapter 28

1. B.G. Niebuhr>HCØ 07.11.12, ØC 1–2, RL
2. HCØ>JJB 14.03.12, C I, 14
3. HCØ>FCS 17.03.12, Add. 1040, RL
4. H. Toftlund Nielsen 2000, 27–30. After his death in 2009 his copy went on auction and the RL considered buying it, but discovered that it already had a copy (20–47). So, two copies have actually survived. An annotation in the RL copy suggests that the remaining stock was pulped.
5. HCØ>FCS 24.06.12, Add 1040, RL
6. HCØ, *NS* II, 41, *SSW*, 313
7. HCØ 2003, 154–5
8. HCØ fasc. xviii ØC 59, RL
9. HCØ 2003, 253 (oxygen), 256 (hydrogen), 257 (carbon), 258 (nitrogen). J. Lund 1987, 101–14
10. KDVS 747–1810, 811-812-1811
11. *NS* II, 35–169, *SSW* 310–92
12. *NS* II, 171–7 (introduction only), not in *SSW*
13. M. Friedman 1992, 213–49, and 2007, 135–58
14. I. Kant 1920, 213.23–214.8
15. *NS* II, 41, *SSW*, 313
16. HCØ 1813, 238
17. *NS* II, 156–7, *SSW*, 384
18. *NS* II, 149–59, *SSW*, 379–85
19. *NS* II, 169, *SSW*, 392

20. HCØ>ASØ 25.02.13, ØC 1–2 (excised in MØ I, 311–14), TL 239
21. *Dansk Litteratur-Tidende* 1816, 574
22. HCØ>ASØ 25.02.13, ØC 1–2, MØ I, 312–13, TL 239
23. E.M. Chevreul, 'corps' in *Dictionnaire des Sciences Naturelle* 1818, vol. 10, 531–2
24. R. Holmes 1999, 117. T.H. Levere 1971, 134–5. A.S Jacobsen 2000, 195–7. The copy of *Ansicht* annotated by Coleridge is C.43a.17, British Museum
25. HCØ>Th. Thomson 05.07.16, MSS 1083A, Dibner Library. Reference taken from A.S Jacobsen 2000, 257
26. *Annals of Philosophy* vol. 13, 1819, 368–77, 456–63, and vol. 14, 47–50
27. Th. Thomson>HCØ 27.01.19, C II, 556–7

Chapter 29

1. Hebt er sich aufwärts,//und berührt mit dem Scheiten die Sterne,//Nirgends haften da de unsichern Sohlen,//und mit ihm spielen Wolken und Winde. A poem by J.W. Goethe quoted by NFSG 1815, 72
2. ASØ>HCØ 06.02.13, ØC 1–2, MØ I, 310
3. HCØ>ASØ 25.02.13, ØC 1–2, MØ I, 311–12, TL 238
4. NFSG 1812, 292–8
5. The first phase of the German Controversy on Pantheism began with F.H. Jacobi's *Über die Lehre des Spinoza in Briefen an den Herrn Moses Mendelssohn* [On the Teachings of Spinoza in Correspondence with Mr Moses Mendelssohn], which had involved Herder, Goethe, Hamann, and Kant. The second phase began in 1811 with Jacobi's *Von den göttlichen Dingen und ihrer Offenbarung* [On Divine Matters and their Revelation], a book that provoked Schelling to a sharp reply, *Denkmal der Schrift Jacobis von den göttlichen Dingen* [Remarks on Jacobi's Book on Divine Matters], published a year later. Jacobi was friendly with F. Reventlow, Emkendorf, J. Baggesen, B. Niebuhr, and M. Claudius, and thus close to some of the Danish King's German-speaking subjects
6. P. Franks 2000, 95–116. F. Beiser 1987
7. *ADB*, vol. 30, 583–4. F.H. Jacobi 1994, 338. W. Weischedel 1967. F.H. Jacobi 1811
8. HCØ 1814, 111–12
9. This example will suffice to show the polemical tone and erroneous quotations of the controversy: 'In the Retort of the Chronicle, p. 60, where firstly one is astonished by the reassurance that the Naturphilosophen consider freedom and necessity as irreconcilable opposites.' (HCØ 1814, 66). However, NFSG actually writes: 'The [Naturphilosophen] claim, that even if good and evil are so definitely against each other in life, they are still in a sense one. But as to necessity and freedom (accountability) on the other hand they claim the opposite implying that they cancel each other out. Consequently, they recognise an irreconcilable contrast, and so the entire philosophy of identity tears itself to pieces.' (NFSG 1813, 59–60)
10. FWJS 1804
11. HCØ 1814, 67, cf. FWJS 1812, 358–66
12. HCØ 1814, 67, cf. FWJS 1812, 358–66
13. S.J Gould 1999
14. NFSG 1815, 32–50 (FWJS), 38 (FWJS, 'Religion und Philosophie')
15. NFSG 1813, *passim*, e.g. 38, 58
16. HCØ 1814, 83–9
17. HCØ 1814, 129, 164–73. NFSG 1813, 160
18. HCØ 1814, 182, NFSG 1813, 137
19. HCØ 1814, 185, NFSG 1813 139–40
20. HCØ 1814, NFSG 1813

21. chs 6 and 26
22. *SES* 1, 185, *SiN* 139–40
23. ch. 20
24. *SES* 1, 188–9, *SiN*, 140–1
25. *SES* 1, 189, *SiN* , 141

Chapter 30

1. D-41, DTM
2. HCØ commented on this poem '"A Character for each of my Three Names": To understand this small poem one needs to know that as a little child I used to call myself Hans Christian Brave Fellow [Bravkarl]. To each of the two first names I have now added pet names, coincidentally in German.' D-41, DTM
3. SØ>S. Perbel 03.11.15: 'What has given me so much pleasure this summer is the little girl Gitte gave birth to in the middle of June. At first it made me quite uneasy to take care of the baby, but little by little I have calmed down, and now she is like my own child whom I can hardly bear not to see for one day.' D-95, DTM
4. AOe>C.Heger after 24.09.02, AOe 1945, vol. 4, 41, 43. The medical record exactly covering the third and fourth quarters of 1801 have gone missing. PRO. B. Wamberg suggests that a rheumatic fever has complicated the scarlet fever and caused a mistral stenosis (a stricture between the auricle and the ventricle of the heart), B. Wamberg 2002, 34
5. FCS about SØ. Fragment, Add. 1289, fasc. v, RL
6. JWR>HCØ 11.03.05, ØC 1–2
7. HCØ>AOe 05.08.06, ØC 1–2, MØ I, 200
8. FCS>SØ 15.11.15, FCS 1866, 80
9. SØ>FCS no date. Letters from SØ to FCS, Add. 1289b, RL
10. HCØ>I.B. Ballum 05.02.14, D-146, DTM
11. HCØ>I.B. Ballum 04.03., 12.03., 15.03., 26.03.14, D-147, D-44, D-148–149, DTM
12. HCØ's collection of physical and chemical instruments was located in rented premises at Thott's Palace from 1813–1819. When accompanied by experiments his lectures for army officers took place at Thott's Palace from winter 1813–14. So did his private lectures on 'Chemistry as a Science and its Application in Daily Life and for Artists, Artisans, and Manufacturers', etc. every Wednesday and Saturday evening from 6–8 o'clock (at 30 rixdollars) and on 'Experimental Chemistry in General' every Monday and Thursday evening from 6–8 o'clock (at 20 rixdollars). D-52, DTM. In December 1815 his collection of scientific instruments was transferred to the University, *Fonden ad usus publicos* 1902, 221
13. Aa. Schæffer 1964, vol. 71, 815
14. Extract from FCS's Notes on Sophie Ørsted, D-94, DTM
15. SØ>I.B. Ballum 01.02.14, ØC 94
16. T. Gyllembourg>A.C. Gjerlew 27.09.18, M. Borup 1947, 221
17. HCØ's application to the Board of the University, no date, ØC 59, fasc. xi
18. *SES* 9, 96

Chapter 31

1. SØ>S. Perbel 03.11.15, D-95, DTM
2. HCØ, ØC 59, fasc. ii, RL
3. HCA's notebook 63–72, ØC 15, RL
4. HCA's notebook 63–72, 90–100

5. HCØ's report, ØC 59, fasc. x, RL. Printed under the partly misleading title 'Proposal for a Reform of the Study of Physics', *NS* III, 191–200, not in *SSW*. Only half of the manuscript is preserved in the ØC
6. HCØ's recommendations, ØC 59, fasc. xviii, RL
7. HCØ, *NS* III, 199–200, not in *SSW*
8. A. Lomholt 1950, vol. 2, 510
9. KDVS 1852-1815
10. KDVS 1853/54-1815
11. KDVS 1854a, 1923-1815
12. KDVS 1892-1815
13. J. Collin 1815
14. E. Andersen made good the deficiency in 1965, cf. his *Mindeskrift* [Memorial Volume], 1968
15. Vedtægter [Statutes] for SLS, NKS 768d, RL
16. C. Molbech 1843, 354–5
17. H.N Clausen 1877, vol. 6
18. ASØ>HCØ 06.12.13. HCØ>ASØ 25.02.13, MØ I, 310–11
19. Ordensdekorationer [Decorations], etc., 05.08.15, ØC 81, fasc. iii, RL
20. Unregistered manuscript, fasc. F, DTM
21. SØ>S. Perbel 03.11.15, D-95, DTM
22. J. Engberg 1973, 102–20. L. Koch 1896, 74. M. Rubin 1895 (repro 1970), 298–300
23. Notes by Miss B. Scharling, F-3, DTM
24. HCØ>CH 19.05.15, C I, 110. HCØ's notes on his children, F-10, DTM. HCØ's oldest half-sister, born 1797 and also named after his mother, died 1808
25. SØ>S. Perbel 03.11.15, S-95, DTM
26. Song book used as ledger 1814–1818, NKS 1011, RL
27. HCØ's draft for application to the Foundation ad usus publicos, no date, ØC 59, fasc. xi, 1–2, RL. University of Copenhagen, Konsistorium [Senate], Copy book 1815–1817, 24, 1213–23, Nos. 346, 700, 721
28. HCØ's application 18.12.15, *Fonden ad usus publicos* 1902, vol. 2, 221, 227
29. BT 85, 14.10.14

Chapter 32

1. *SES* 6, 79
2. HCØ, BT 14.10.14. Cf. BT 03.11.15, 27.10.17 and *Nyeste Skilderie af Kjøbenhavn* 1819, col. 1415 and 1820, col. 1343
3. F. Barfod 1845, 11. H. Nielsen 1983, 41
4. HCØ, notes, among others 'Landoeconomisk Chemie' ['Agricultural Chemistsry'] 05.11.20, repeated 05.11.26, ØC 57, file xiv, RL
5. M.C Harding 1924, 236
6. ØC 34, 37, 57, files i–ii, 35, 62, 57, files iii–xiii, RL
7. HCØ>Admiralitetet 22.04.17, D-12, DTM
8. HCØ 2002, 31-2
9. HCØ 2002, 31
10. M.C Harding 1924, 226
11. M.C Harding 1924, 224
12. M.C Harding 1924, 225
13. HCØ>LM 16.07.16, ØC 1–2, RL

14. HCØ>J.S. Møsting 10.06.19, *Fonden ad usus publicos* 1902, 305
15. HCØ>J.S. Møsting 10.06.19, *Fonden ad usus publicos* 1902, 302
16. HCØ KDVS 2137-1817, *NS* II, 438
17. WCZ 1817
18. *SES* 8, 149–63. J.L Heiberg 1817
19. *NS* I, 345
20. *NS* I, 267–73
21. *Nyt Bibliothek* 1806, vol. 9:3, 268–75. *Neues allgemeines Journal der Chemie* 1806, vol. 6, 292–302, *Journal de physique, de chimie et d'histoire naturelle* 1806, vol. 62, 369–75. *Nicholson's Journal* 1806, vol. 15
22. *NS* II, 178
23. HCØ KDVS 1899-1815
24. HCØ KDVS 1899-1815, 13–14
25. KDVS *Oversigter* 1814–1815, 7–10. *NS* II, 432
26. HCØ>LM 30.09.15, ØC 1–2, RL
27. IBØ ledgers 1814–1818, NKS 1011, RL. IBØ ledger 10.01.19–0011.22, E-52, DTM
28. *NS* II, 206–11
29. HCØ KDVS *Oversigter* 1816–17, 7–9
30. HCØ KDVS *Oversigter* 1816–17, 304
31. O. Knudsen 2007, 387–98. *NS* II, 211–12. HCØ KDVS *Oversigter* 1917–18, 8–12. *NS* II, 438–41
32. *NS* II, 211
33. ASØ>FVI 30.06.17, Kongehusets arkiv [Royal Archive], PRO
34. J. Engberg 1973, 147–58

Chapter 33

1. HCØ 2002, 31
2. ASØ>M. Rogert 31.12.17, NBD xix, RL
3. HCØ, F-33, DTM
4. *Nyeste Skilderie af Kjøbenhavn* 17.02.18
5. *Veiviser for Kjøbenhavn*, 1818–19
6. P.M Boll 1990, vol. 2, 16
7. IBØ, NKS 1011, RL
8. JGF, KDVS *Oversigter* 1851, vol. 6, 156
9. DCC 1996, 180–92
10. The fee was two thousand rixdollars, which HCØ in a letter to Minister of Finance, S. Møsting, suggested reduced to half (probably to increase his credibility). Yet he asked J. Collin for a swift transfer of the amount, HCØ>J. Collin 26.08.19, NKS 1572, RL
11. HCØ>IBØ 17.09.18, ØC 1–2, RL
12. HCØ 1927, 3
13. HCØ and L. Esmarch 1819, *NS* III, 228–34, not in *SSW*
14. HCØ and L. Esmarch 1819, *NS* III , 234–9
15. HCØ>WCZ 03.11., 14.11.18 and 18.05., and 03.12.19, NKS 1686, RL
16. HCØ, F-10, DTM
17. IBØ>HCØ 17.02.23, ØC 1–2, RL
18. Fasc. F, DTM
19. IBØ paid 32 rixdollars for the deed for the grave, E-52, DTM

20. 2 Samuel, chapter 12, verses 15–24. C. Hauch, *SES* 9, 149
21. Notes by Miss B. Scharling, F-3, 7, DTM
22. Ledger from January 1819 to November 1822, 12.03., 15.03., 20.03., 29.05., 30.06., 12.07., 09.12.19, E-52, DTM
23. Notes by Miss B. Scharling, F-3, DTM
24. Kgl. Dir. for Univ.> KU Konsistorium 13.02.19, KU, Konsistorium Kopibog 1817–19, 25, 1213–24, No. 945, RRO
25. Vurderingsforretning 08.09.38, CCA
26. H.H Kjølsen 1965, 67
27. HCØ 1927, 15–18. HCØ and L Esmarch 1820, *NS* III, 274–5, not in *SSW*
28. HCØ 1927, 15–18. HCØ and L Esmarch 1820, *NS* III, 286
29. HCØ and L. Esmarch, *Statstidende* 38, 1821, *NS* III, 201–300
30. *Einladung zu einer Interessentschaft....*, 1819, RRO
31. H. Schæffer 1822
32. *Fonden ad usus publicos* 1902, vol. 2, 300
33. HCØ>IBØ 13.10.18, 18.08. and 28.08.19, ibid. HCØ>WCZ 03. and 14.11.18, NKS 1686, RL
34. HCØ>ASØ 11.05.23, ØC 1–2, RL
35. JGF>HCØ 31.05.23, ØC 1–2, RL C II, 343
36. HCØ's recommendation of JGF to FVI 11.12.20. *Fonden ad usus publicos* 1902, vol. 2, 284–6
37. HCØ and L. Esmarch 1820, *NS* II, 277, not in *SSW*
38. HCØ>IBØ 18.08.19, ØC 1–2, RL
39. *Annals of Philosophy* 16, 1820, 130f.

Chapter 34

1. HCØ 2002, 35
2. *Fonden ad usus publicos* 1902, vol. 2, 303–5 and 224–5
3. *Fonden ad usus publicos* 1902, vol. 2, 300–2
4. IBØ, E-62, DTM
5. M. Rubin 1895, 529–33
6. *SES* 6, 71–82. *Fonden ad usus publicos*, vol. 2, 306–7
7. *Fonden ad usus publicos*, vol. 2, 284–5
8. LHS journalsager 1820, EA
9. IBØ, E-52, DTM
10. J.A Dyssel 1820
11. *Hesperus* 1820, vol. 3, 312–21, KDVS *Oversigter* 1820–1821, 12–21, *NS* II 447–53, *SSW* 425–9. HCØ 1830, *NS* II 351–98, *SSW* 542–80, *NS* II 223–45, *SSW* 430–45
12. KDVS 2471–1820 and 2485–1820
13. DCC 1994, 28–65. *Nye Landoeconomiske Tidender* 1820, 370–85
14. KDVS 2478–1820
15. H.C.A Lund 1896, 51–2
16. H.C.A Lund 1896, 62
17. H.C.A Lund 1896, 57
18. H.C.A Lund 1896, 94
19. *Hesperus* 1820, vol. 2
20. H.C.A Lund 1896, 88
21. H.C.A Lund 1896, 84 and 89
22. H.C.A Lund 1896, 200

Chapter 35

1. HCØ 2002, 35
2. Ledger 10.01.19 to November 1822, E-52, DTM
3. AOe 1964, 57
4. JWR>HCØ 20.05.03, C II, 34
5. O. Knudsen 1987, 56–7
6. HCØ KDVS archive 1899/1815, KDVS *Oversigter* 1814–15, 9–10, SSW 395–6
7. HCØ>LM 30.09.15, ØC 1–2, RL
8. KDVS *Oversigter* 1816–17, 7–9
9. KDVS *Oversigter* 1816–17, 9
10. CH>HCØ 30.04.19, C I, 116–20
11. K. Meyer 1912, 27–9
12. ØC 23, folder viii and ix, RL—called 'Kladden' [the draft] and 'Supplement II' respectively by K. Meyer, *NS* I, lxxvi–lxxvii
13. *Dansk Litteraturtidende* 1820, No. 28, 447–8
14. Ledger 10.01.19 to Nov. 1822, E-52, DTM
15. A. Larsen 1920 has the original text as well as translations into French, Italian, German, English, and Danish
16. Ledger 10.01.19 to November 1822, E-52, DTM
17. ØC 23, RL
18. HCØ 2000, 35
19. HCØ in *Journal für Chemie und Physik, Bibliothèque universelle des sciences, Journal de physique, Annals of Philosophy*, SSW 431, cf. ch 28

Chapter 36

1. HG. von Schmidten>HCØ 21.03.22, ØC 1–2, MØ II, 25
2. JJB>HCØ 16.12.20, C I, 18
3. CSW>HCØ 02.08.20, C I, 266–70
4. CHP>HCØ 27.07.20, C II, 466–7
5. CSW>HCØ 02.08.20, C I, 270
6. CSW>HCØ 02.08.20, C I, 270
7. CSW>HCØ 02.08.20, C I, 270
8. HCØ>ASØ 09.01.23, ØC 1–2, MØ II, 41, TL 273
9. L.P. Williams 1980, 60
10. *Bibliothèque universelle*, vol. 14, 1820, 281–4, C II, 477
11. L.P Williams 1965, 143
12. L.P Williams 1989, 76–7
13. H.D. de Blainville>HCØ 03.11.20, C II, 271–4
14. H.D. de Blainville>HCØ 03.11.20, C II, 272. John Leslie and Sir David Brewster, both Edinburgh professors
15. GF>HCØ 23.10.20, MØ II, 12
16. C II, 305–8
17. *Dansk Litteraturtidende*, 45, 1820, 784
18. *Annals of Philosophy*, 1821, 195
19. C II 516

20. C II 385
21. H.C. von Schmidten>HCØ 21.03.22, ØC 1–2, MØ II, 22
22. HCØ>FVI 10.12.21, ØC 1–2, MØ II, 25
23. C II 409–10
24. M. Malmanger 2000, 88–9. J.C Dahl 1988, 122–3 and 128–9
25. CF>HCØ 31.10.20, ØS 1–2, MØ II, 16
26. H.P Giessing 1852, 134
27. H.G. von Schmidten>HCØ 21.03.22, ØC 1–2, MØ II, 25
28. SES 8, 86
29. Finansdep. Forestillinger og Resolutioner 1820, No. 277 of 12/17.10.20 incl. attachments, PRO
30. WCZ in *Tidsskrift for Naturvidenskaberne* 1822, 56
31. WCZ in *Tidsskrift for Naturvidenskaberne* 1822, 61–2
32. SES 8, 86

Chapter 37

1. HCØ>IBØ 16.06.23, ØC 1–2, RL, TL 335
2. KDVS 2528–1821
3. HCØ>IBØ 07.11.22, ØC 1–2, TL 246
4. HCØ>CH 03.08.22, C I, 124
5. CF>HCØ 24.12.22, ØC 1–2, MØ II, 33–5
6. HCØ>CF 05.05.23, ØC 1–2, MØ II, 68–72
7. A.C. Gjerlew>HCØ 26.-29.04.23, ØC 1–2
8. HCØ>IBØ 28.08.19, ØC 1–2
9. HCØ>IBØ 02.12.22, ØC 1–2, not in MØ II, TL 254
10. IBØ>HCØ 28.12.22, ØC 1–2, TL 266
11. HCØ>IBØ 02.12.22, ØC 1–2, TL 258
12. IBØ>HCØ 18.11.22, ØC 1–2, HCØ>IBØ 21.11.22, ØC 1–2, TL 252
13. HCØ>IBØ 08.01.23, ØC 1–2, TL 267–8
14. IBØ>HCØ 28.12.22, ØC 1–2, TL 266
15. IBØ>HCØ 09.05.23, ØC 1–2, TL 266,
16. IBØ>HCØ 27.01.23, ØC 1–2, TL 266
17. DCC 1996, 97–121
18. Carl of Hessen>HCØ 21.03.21, FMA
19. HCØ>IBØ 07.11. and 10.11.22, ØC 1–2, TL 247–8
20. HCØ>Carl of Hessen 08.03.23, FMA
21. Carl of Hessen>HCØ 26.03.23, FMA
22. HCØ>Carl of Hessen 08.03.23, FMA
23. HCØ>ASØ 11.05.23, ØC 1–2, MØ II, 73
24. Carl of Hessen>HCØ 26.03.23, FMA
25. HCØ>IBØ 15.12.22, ØC 1–2, TL 258
26. HCØ>IBØ 21.11. and 02.12.22, ØC 1–2, TL 251–5
27. HCØ>IBØ 02. and 10.12.22, ØC 1–2. TL 256. P. la Cour and J. Appel 1897, 469–70. HCØ>CF 05.05.23, ØC 1–2, MØ II, 68–72
28. J. Radkau 1989, 104
29. ch. 44

30. HCØ>IBØ 02.12.22, ØC 1–2, TL 253
31. J.W. Goethe, *Werke*, 1972, 2nd part, vol. 10, 523
32. J.W. Goethe>J.W. Döbereiner 20.10.20, J.W Goethe 1905, 313
33. J.W. Goethe, *Werke*, 1975, 3rd part, vol. 8, 272–3
34. HCØ>IBØ 14.01.23, ØC 1–2, TL 275–9
35. IBØ>HCØ 09.05. and 23.05.23, ØC 1–2, TL 275–9
36. HCØ>IBØ 15.12.22, ØC 1–2,, TL 261–3
37. J.W. Goethe>S. Boisserée 27.01.23, J.W Goethe 1907, 284
38. HCØ>ASØ 09.01.23, ØC 1–2
39. HCØ>IBØ 08.01.23, ØC 1–2, TL 268
40. HCØ>IBØ 25.01.23, ØC 1–2, TL 280

Chapter 38

1. HCØ>IBØ 23.02.23, ØC 1–2, TL 292
2. IBØ>HCØ 11.02.23, ØC 1–2, TL 292
3. HCØ KVDS *Oversigter* 1820–21, 21
4. *Procès-verbaux des séances de l'Académie*, vol. 7, 1910–22, 90
5. H.G. v. Schmidten>HCØ 21.03.22, ØC 1–2
6. O. Knudsen 1989, 6–35
7. JJB>HCØ 16.12.20, C I, 18–19. *NS* II, 230–2, *SSW*, 435–6
8. P. la Cour and J. Appel 1897, 463–5
9. HCØ>IBØ 12.02.23, ØC 1–2, MØ II, 50–1, TL 286–90
10. J.R Hofmann 1996, 144–53
11. *NS* II, 225, *SSW* 431–2
12. HCØ>IBØ 12.02.23, ØC 1–2, MØ II, 50–1, TL 286–90
13. *NS* II, 241–5, *SSW*, 443–5
14. HCØ>IBØ 23.02., 05.03.23, ØC 1–2, TL 294
15. HCØ>IBØ 07.02.23, ØC 1–2, TL 285
16. HCØ>IBØ 07.02.23, ØC 1–2, TL 285
17. HCØ>IBØ 07.02.23, ØC 1–2, TL 285
18. HCØ>CF 05.05.23, ØS 1–2, MØ II, 70–1, TL 311–4
19. HCØ>IBØ 05.03.23, ØS 1–2, MØ II, 70–1, TL 294
20. HCØ>IBØ 25.04.23, ØS 1–2, MØ II, 66, TL 309–10
21. HCØ>IBØ 12.02.23, ØS 1–2, MØ II, 48–9, TL 288–90
22. H.G. v. Schmidten>HCØ 21.03.22, ØS 1–2, MØ II, 28
23. HCØ>JGF 14.04.23, NBD, fasc. xvi, 22, RL. *Catalogus* 1823
24. HCØ>WCZ 14.04.23, NKS 1686, RL
25. R. Fox 1996, 278–94
26. C I-II
27. A Lomholt 1950, vol. 2, 386–428

Chapter 39

1. HCØ>IBØ 27.05.23, ØC 1–2, RL, TL 326
2. CF>HCØ 10.03.23, ØS 1–2,, MØ II, 67–8
3. KDVS 2537, 2541/1821, 08.03. and 23.03.21

4. T. Young>HCØ 31.05.21, D-2, DTM
5. RS Minutes of Council, vol. ix. 1811–1822, 243
6. A.L Macfie 1973
7. *Sketch of the Rise and Progress.... 1860*
8. HCØ>ASØ 11.05.23, ØC 1–2, MØ II, 72–6, TL 318–21
9. J. Newman 1990
10. J. Newman 1990
11. The last part was read 15.05.23. Journal Book of the RS, vol.xliii, 1819–1823, 615
12. HCØ>IBØ 15.06.23, ØC 1–2, TL 331–2
13. HCØ>IBØ 15.06.23, ØC 1–2, TL 331–2
14. HCØ>IBØ 15.05.23, ØC 1–2, TL 322
15. HCØ>IBØ 15.06.23, ØC 1–2, TL 333
16. HCØ>IBØ 16.06.23, ØC 1–2, TL 334
17. HCØ>IBØ 27.05.23, ØC 1–2, TL 328
18. A. Rupert Hall 1969, 1–2
19. IBØ>HCØ 09.06.23, ØC 1–2
20. HCØ>IBØ 28.06.23, ØC 1–2, TL 341
21. J. Cumming 1821. Ch. Babbage 1830
22. HCØ>IBØ 27.05.23, ØC 1–2, TL 328
23. L.P Williams 1965, 196
24. M. Berman 1971
25. HCØ>IBØ 15.05.23, ØC 1–2, TL 323–4
26. L.P Williams 1985, 82–104
27. M. Faraday, 1821, No.12, 186–7
28. DNB, xix
29. G. Cantor 1991, 44
30. A.C. Gjerlew>HCØ 26.-29.04.23 ØC 1–2
31. HCØ 1830, NS II, 359
32. HCØ>IBØ 15.05.23, ØC 1–2, TL 323–4
33. HCØ>IBØ 15.06.23, ØC 1–2, TL J.-B.-J. Delembre>HCØ 05.04.23, C II, 310–11
34. HCØ>IBØ 24.06., 08.07., 16.07., 25.07., 14.08.23, ØC 1–2, TL 335–6, 343, 349, 352–3
35. HCØ 1830, NS II 351–98
36. HCØ>IBØ 21.07.23, ØC 1–2, TL 352
37. HCØ>IBØ 28.06.23, ØC 1–2, TL 341
38. IBØ>HCØ 09.06.23, ØC 1–2, TL 341
39. A. Andersen Feldborg>AOe 23.07.23, AOe vol. 3, 1955, 236. AOe>Walter Scott 13.10.23, AOe vol. 3, 1955, 243–5. Walter Scott>AOe 16.01.24, AOe vol. 3, 1955, 245–8
40. HCØ>IBØ 08.07.23, ØC 1–2, TL 343
41. HCØ>IBØ 16.07.23, ØC 1–2, TL 343
42. HCØ>IBØ 25.07.23, ØC 1–2, TL 352–3
43. HCØ>IBØ no place and date; according to contents from Paris it must be the beginning of May 1823, and 27.05.23, ØC 1–2, TL 328
44. HCØ>IBØ 14.08.23., ØC 1–2, TL C. Hauch states (*SES* 9, 139–40) that the plan was conceived on his way *to* England, while E. Snorrason (1987, 127) suggests on board the steamer *from* England. Neither is true.

Chapter 40

1. C.C Gillespie 1980, 549 (my paraphrase)
2. IBØ>HCØ 09.06.23, ØC 1–2
3. F-3, 7, DTM
4. J.T Lundbye 1929, 128–9. B. Scharling's drawing, B-86, DTM
5. F. Jørgensen 1957–59, 4th series, vol. 5, 27–9
6. SES 5, 129–42
7. e.g academics C.H. Reinhardt, G. Forchhammer, J.F. Schouw, W.C. Zeise, J.L. Heiberg, and O. Bang, businessmen E. Erichsen and L.N. Hvidt, manufacturers I.C. Drewsen, J. Owen, and C.A. Brøndum as well as the brewers Ch. and J.C. Jacobsen (father and son).
8. HCØ's invitation 16.10.23. M.C Harding 1924, 23
9. EAS 1857, 53. M.C Harding 1924, 34
10. M.C Harding 1924, 91
11. This debate was by and large a repetition of the arguments brought forward in the complaint of WCZ's lectures in chemistry (ch. 34), cf. HCØ's reference to Chaptal's textbook: 'The courses teach the general principles of chemistry, i.e. the properties of gases, liquids, and metals, etc. and do not take their point of departure in the needs of a dyer or a tanner, a manufacturer of gunpowder or sulphuric acid.'
12. M.C Harding 1924, 38
13. M.C Harding 1924, 49
14. 'Among the artists following HCØ's lectures 1824, 25 were 13 painters: W. Bendz, D. Blunck, A. Küchler, M. Rørbye, J. Sonne, and Chr. Holm among them. (SNU archive, parcel 6, fasc. 060102), H. Tybjerg 1996, cf. ch. 56

Chapter 41

1. I. Kant 1917, vol. 8, 53–61
2. KDVS minutes 1819
3. J.D Herholdt 1826, 4
4. J.D Herholdt 1826, 13
5. K. Michelsen 1989
6. HCØ 'Fra hvor tidlig en Alder har jeg Erindringer?' [From what age do I have memories?], 13–18, ØC 4
7. J.D. Herholdt>HCØ 13.06.25, ØC 1–2, RL
8. K. Michelsen 1989, 2
9. ASØ *Juridisk Tidsskrift*, vol. 8, 1824, 131
10. ASØ *Juridisk Tidsskrift*, vol. 8, 1824, 144
11. ASØ *Juridisk Tidsskrift*, vol. 8, 1824, 132
12. ASØ *Juridisk Tidsskrift*, vol. 8, 1824, 127
13. F.G Howitz 1824, 20
14. HCØ>ASØ 07.02.02, MØ I, 45. HCØ>SP 27.04. and 09.05.02, MØ I, 58, 63
15. F.G Howitz 1824, 13
16. F.G Howitz 1824, 5, 11
17. *Skandinavisk Litteraturselskabs Skrifter* 1805, 426–65
18. F.G Howitz 1824, 5, 9, 11, 26, 112
19. C. Otto 1879, 205–7
20. H.G v. Schmidten 1829, 105–14
21. HCØ, monthly lecture No. 141 of 05.10.30, ØC 62, RL

Chapter 42

1. D.M Knight 2007, 429
2. H. Davy>HCØ 22.07.24, C II, 310
3. D.M Knight 2007, 429
4. H. Kragh and H.J. Styhr Petersen 1995, 35–7
5. HCØ, KDVS *Oversigter* 1822–23, 9–10, 1823–24, 13–14, *NS* II, 461–3, not in *SSW*
6. KDVS 2814–1824
7. D.M Knight 2007, 429
8. D.M Knight 2007
9. JJB>HCØ 03.07.24, C I, 26
10. F. Wöhler>JJB 08.10.24, JJB 1901, vol. 1, 10–15
11. M.C Harding 1924, 68–71
12. KDVS 2892–1825 of 18.02.25. 2898/99–1825 of 08.04.25
13. HCØ>CH 21.03.25, C I, 133. HCØ KDVS *Oversigter* 1824–25, 15–16
14. *Magasin for Naturvidenskaberne*, vol. 5, 1825, 176–7
15. F. Wöhler>HCØ 18.02.28, C II, 609
16. R. Keen 2005, 53–69. F. Wöhler>JJB 10.10.27, JJB 1901, vol. 1, 194–7
17. I. Fogh 1921, KDVS, vol. 3, 14, 1–17 and 15, 1–7. N.J Bjerrum 1926, 316–17. K. Meyer 1920, *NS* I, cxxiv. H. Kjølsen 1965, 102–21. T.A Bak 1987, 97. H. Kragh and H.J. Styhr Petersen 1995, 110–12
18. F. Wöhler 1827, 146–61
19. HCØ's notes for 118th monthly lecture of 08.01.28, ØC 35

Chapter 43

1. *SES* 2, 204, *SiN* 297
2. *SES* 9, 88
3. H.N Clausen 1877, 107–14, 129–36
4. ASØ *Juridisk Tidsskrift* vol. 12, 1825
5. F. Dahl 1929
6. FVI>F.J. Kaas 20.08.26. F. Dahl 1929, 156
7. FVI's rescript 21.09.26. F. Dahl 1929, 186–7
8. ASØ>F.J. Kaas 21.09.25. F. Dahl 1929, 186–7
9. F. Abrahamowitz 2000, 346
10. *SES* 2, 198–9, *SiN*, 292–3
11. *SES* 2, 206, *SiN* , 298
12. *SES* 6, 101
13. *SES* 6, 90
14. HCØ>C.E. Scharling, no date [1836], ØC 44, ii, RL
15. F-3, 11, DTM
16. *SES* 6, 92
17. *SES* 7, 115, 119
18. *SES* 6, 101
19. *SES* 9, 99–100
20. ASØ *Eunomia* 1815, vol. 1, 98
21. *SES* 6, 103–4
22. Matthew chapter 12, verse 33

Chapter 44

1. HCØ 1831, 854–7
2. F-10 and F-3, DTM
3. HCØ>IBØ 14.04.23, ØC 1–2, RL
4. IBØ>HCØ 12.03.23, ØC 1–2, RL
5. IBØ>HCØ 09. and 23.05.23, ØC 1–2, RL
6. HCØ>Klingberg 02.09. and 14.10.29, D unregistered, DTM
7. HCA SV 17, 77
8. HCA SV 95–6
9. J. Andersen 2003, 59–82
10. IBØ>HCØ 28.12.22, ØC 1–2, RL
11. HCA SV 17, 56
12. HCA, D-218, DTM
13. F-3, 9–12, DTM
14. A.V Nielsen 1949, 8–21 and 41–52, and 1950, 91–5
15. http://www.gdnae.de/werist/geschi.html
16. A. von Humboldt 1828, 8
17. A. von Humboldt 1828, 6–7
18. HCØ>IBØ 16.09.28, ØC 1–2, RL, TL 369–71
19. HCØ>IBØ 19.09.28, ØC 1–2, TL 371–5
20. HCØ>IBØ 19.09.28, ØC 1–2, TL 372. In the official congress report the lecture bears the title 'Remarks to the theory of HCØ's electromagnetic process, mainly in relation to the views of Mr Ampère, Paris on the same matter.' Bericht…1829.
21. HCØ 1829, 260–2
22. HCØ KDVS Oversigter 1929–30, 22–6, NS II, 479–81, SSW 539–41
23. Bericht…1831
24. HCØ ØC 4, RL. In the official congress report the title of the lecture is 'On the Mathematical Truth in Physics.'
25. HCØ 1831, 854–7
26. 'Calculer la situation respective de tous les êtres qui le composent'. H.G. von Schmidten>HCØ 21.03.22, MØ II, 26
27. KDVS 3519–1828. HCØ, KDVS Oversigter 1829, 12
28. HCØ>C.S. Weiss 'Briefwechsel über Atomistik und Dynamik' [Correspondence on Atomic and Dynamical Theory], C I, 335
29. HCØ>C.S. Weiss 'Briefwechsel über Atomistik und Dynamik' [Correspondence on Atomic and Dynamical Theory], C I, 289

Chapter 45

1. SES 2, 76, SiN 201
2. DCC 1986, 5–62
3. ch. 19
4. Universitetsdirektionen [Board of the University], journals, Ursin's proposal 1605/1827, HCØ's report 1155/1829, RRO
5. Universitetsdirektionen [Board of the University], journals, Ursin's proposal 1605/1827, HCØ's report, 1605/1827

6. J.T. Lumbye 1929, 70
7. M.F Wagner 1999, 306
8. A. Steen 1879, 9, 104
9. M.C Harding 1924, 41
10. DCC 1996, 41–52
11. Archive of the Polytechnic Institute, journals 1829–1842, 28, RRO
12. DCC 1996, 734–7
13. M.F Wagner 1999, 335–9
14. A. Paulinyi 1993, 24–9
15. DCC 1996, 168
16. G.F. Ursin 1830, No.15, 249–53
17. Cumming's patent, DCC 1996, 1
18. AOe 1860, 162–5
19. SES 2, 63–80, SiN, 192–204
20. SES 2, 78–9, SiN, 203–4
21. SES 2, 76, SiN, 201
22. O.J Rawert 1992, 120–8

Chapter 46

1. *SES* 9, 87
2. HCA, *SV* 16, 143
3. HCA, *SV* 2, 259
4. HCA, *SV* 17, 79
5. HCA, *SV*, 80
6. J. Andersen 2003, vol. 1, 193
7. HCA, *SV* 17, 78
8. HCA, *SV*, 82
9. H.N Clausen 1877, 139
10. P.V. Jacobsen 1899, 155
11. HCØ, ØS 4, RL
12. H.N Clausen 1877, 150
13. HCØ>JLH 23.04.29 and JLH>HCØ 24.04.29, MØ II, 92.99
14. SES 5, 189–99. HCØ>JLH 24,04.29, MØ II 92–4. *SES* 9, 76–7
15. *Maanedsskrift for Litteratur* 1, 1829, I, HCØ 424–33, II JLH, 433–56
16. G.C Byron 1986, 149
17. F. Paludan-Müller 1901, 105
18. HCØ>F. Paludan-Müller, no date, quoted from S. Kühle 1941, 98
19. *SES* 9, 96–8
20. *SES* 9, 98
21. HCA>HCØ 05.12.33, HCA 2000
22. HCA, *SV* 17, 79
23. HCA>HCØ 05.12.33, MØ II, 116–22
24. HCA, *SV* 10, 363–453
25. HCØ>HCA 08.03.34, HCA 2000
26. R. Stampe 1918, 2

27. M. Larsen 1956, 2nd part, 287–300
28. *SES* 9, 41
29. J. Steenberg 1969–79, vols.1–2, does not mention HCØ's sojourn, but a range of actors that A.W. Hauch as the previous head of the Royal Theatre invited to lodge in the buildings of the castle, then in a period of decline.
30. HCØ>K. Ørsted 23.07.36, D-122, DTM
31. HCØ>K. Ørsted 11.06.36, D-183. MØ II, 145
32. HCØ>MØ 23.04.36, D-189, DTM
33. CH>HCØ 18.12.36, MØ II, 153
34. *SES* 4, 23–4
35. J.F.W. Herschel>HCØ 29.07.39, C II, 392

Chapter 47

1. *SES* 5, 164
2. IBØ>EAS 07.11.35, D-231, DTM
3. JFS read parts of his prize essay to the KDVS, 2767–1823, replaced HCØ 18.05.27, 3378a-1827, was elected archivist, etc. 3664–1830, and re-elected 4326–1835, 5690–1845 6414–1850
4. H. Vammen 1966, 17–20
5. H. Vammen 1966, 21
6. O. Lehmann 1872, vol. 1, 68–9
7. HCØ>O.Lehmann 01.09.32, MØ 2, 108
8. HCØ>O.Lehmann 01.09.32, MØ 2
9. ωα *Maanedsskrift for Litteratur* vol. 5, 1831, 156–88 and vol. 7, 1832, 523–37
10. HCØ>O. Lehmann 03.09.32, MØ 2, 113
11. H. Jørgensen 1944
12. H. Jørgensen 1944
13. *Collegialtidende* 26.02.35
14. H. Jensen 1937–38
15. HCØ *Dansk Folkeblad* 1, 1835, *SES* 7, 27–38
16. HCØ *Dansk Folkeblad* 1, 1835, *SES* 7, 30
17. HCØ>F.C.E. Dahlstrøm 23.06.35, D-48, DTM
18. HCØ *Dansk Folkeblad* 1, 1835, *SES* 7, 36
19. HCØ *Dansk Folkeblad* 1, 1835, *SES* 7, 38
20. HCØ *Dansk Ugeskrift*, 1835, *SES* 7, 19–24
21. HCØ *Kjøbenhavns Flyvende Post* 1835, *SES* 4, 49–54
22. JLH>HCØ 03.11.35: 'The poem seems to me to be what such a poem ought to be: simple, clear, and heartfelt', ØC 1–2. CH>HCØ 18.12.36: 'Thank you for your beautiful poem that has brought pleasure to all friends of true liberty and poetry'
23. ASØ>HCØ 20.11.35, ØC 1–2
24. HCØ *Maanedsskrift for Litteratur* 13, 1835, *SES* 7, 1–16
25. HCØ *Maanedsskrift for Litteratur* 13, 1835, *SES* 7, 4
26. HCØ *Maanedsskrift for Litteratur* 13, 1835, *SES* 7, 12
27. E. Tryde *Dansk Folkeblad*, vol. 2, Nos. 2–3
28. DCC 1969, 22–32
29. Anonymous (Orla Lehmann) *Maanedsskrift for Litteratur*, vol. 7, 1832, 177–9. The editor 'had no misgivings publishing an interesting article on a subject of such common importance', although the author refused to reveal his identity. An elaborated account is given in DCC 1969

30. *Kjøbenhavnsposten* 169, 1836
31. HCØ *Dansk Folkeblad* 1836, *SES* 7, 99–102
32. HCØ ØC 4, C. Bræstrup>P.C. Stemann 11.03.35, DPB 1, No. 67, 95–6
33. HCØ>F.C.E. Dahlstrøm 23.06.35, D-48, DTM. HCA 1973
34. S. Nielsen 1954–57, 246–7
35. C. Molbech 1837
36. HCØ ØS 83
37. HCØ>E.D. Ehlers 12.01.39, unregistered, DTM

Chapter 48

1. EAS>HCØ 03.10.34, MØ II, 132
2. J.F.W. Johnston, *Edinburgh Journal of Science*, New Series, vol. 3, 1830
3. *Catalogus* 1824 et seq. The University and the Polytechnic Institute had a mutual course catalogue up till 1851. Before 1834 it was published in Latin only, thereafter also in Danish
4. HCØ manuscript 141, 05.10.30, ØC 62
5. H. Martensen>HCØ 24.11.35, MØ II, 142–4
6. EAS>HCØ 03.10.34, MØ II, 130–2
7. HCØ>EAS 20.12.34, D-226, DTM
8. HCØ>EAS 07.11.35, D-231, DTM
9. EAS>IBØ 25.04.35, D-228, DTM
10. HCØ>EAS 12.01.36, D-229, DTM
11. *Dansk Ugeskrift* 171, May 1835, 225–48
12. HCØ>EAS 20.06.35, D-230, DTM
13. M.C Harding 1924, 82
14. HCØ>WCZ 30.05.39, NKS 1686, RL
15. K. Meyer *NS* III, cxxxii–cxxxvii
16. P.C. Steman>KDVS 14.03.31, 3932-1831, KDVS
17. 3945-1831, KDVS
18. DCC 1996, 231–42
19. KDVS *Oversigter* 1832–33, 26–8
20. C. Molbech 1843, 482–3
21. A. Garboe 1961, 32–50
22. M.C Harding 1924, 53–8
23. HCA *SV* 9, 2004, 484–5
24. CF>HCØ 30.04.39, Kongehuset CVIII, 202, 1809–1847, Breve fra og om vidensk selskaber og institutioner [letters from and on scientific societies and institutions], No. 119, RRO
25. H. Ohlman 1990, 715. HCA *SV* 3. C. Wheatstone>HCØ 20.05.39, C II, 594
26. HCØ and IBØ>K. Ørsted 14.05.36, D-188, DTM
27. R. Berg 1938, 695–706. O.J. Rawert 1850, 136–43
28. *Industriforeningens Tidende* 1838, 66. C. Nyrop 1888, 32–3. Nyrop launched the myth of HCØ's industrial efforts (205–10), possibly in gratitude for HCØ's letters of recommendation to him when as a young instrument maker he applied for grants towards his grand tour

Chapter 49

1. *SES* 6, 128
2. *SES* 6, 130

3. NFSG, 1941–44, vol. 4, 232
4. HCØ>W. Whewell 26.08.1834, Add.Ms. a210.42, 04.09.1846, a210.43, of 20.06.1847, a210.44 og NFSG>W. Whelwell 28.06.1836, Add.Ms. a205.53, Wren Library, Trinity College, Cambridge
5. NFSG, 1983, 78–88
6. HCØ, *SES* 4, 131–42
7. numerous references to Plato, e.g. HCØ, *SES* 2, 127, 129, 130, *SES* 5, 135, *SES* 5, 164, *SES* 2, 65–80
8. HCØ, *SES* 6, 109–56
9. NFSG, 1941–44, vol. 4, 216
10. O. Korsgaard and L. Løvlie, 2003, 9–36
11. G. Nissen, 1968, 49–67
12. HCØ, *SES* 2, 130
13. HCØ, *SES* 7, 177
14. P.J Jensen, 1963, vol. 2, 187
15. E. Spang-Hansen, P.J Jensen, 1963, vol. 2, 219–21 and 231–3
16. K.H Gad, 1833
17. HCØ, *Maanedskrift for Litteratur*, 1836, vol. 16
18. NFSG, 1941–44, vol. 4, 367
19. K. Hørby, 1967, 59–84
20. NFSG, 1941–44, vol. 4, 372–5
21. NFSG, 1941–44, vol. 4, 232
22. NFSG, 1983, 36
23. NFSG, 1983, 22
24. *Dansk Ugeskrift*, 171, 1835, 225–48
25. HCØ, *Dansk Ugeskrift* 2/19, 1842, 281–94
26. NFSG, 1983, 48
27. HCØ, *SES* 7, 97–102
28. HCØ>Carsten Hauch 30.06.1842, MØ 2, 173
29. K. Hørby, 1967, 77
30. DCC, 1999, 39
31. J.T Lundbye, 1929, 89

Chapter 50

1. *Förhandlinger* 1840, 33
2. *Förhandlinger* 1840, 33
3. HCØ>CVIII 12.07.40, ØC 1–2
4. JJB>HCØ 24.07.40, C I, 44–5
5. L. Zeuthen 1866, 78
6. H. Steffens>HCØ 04.11.40, MØ II, 166–7
7. HCØ>Mansa 01.07.42, D-26, DTM
8. HCØ>JJB 17.09.40, C I, 46–7. JJB>HCØ 04.01.41, C I, 48–9. HCØ>JJB 22.05.41, C I, 50–1
9. *Förhandlinger*, s.a.
10. *SES* 3, 155–69
11. HCØ>IBØ 15.07.42, ØC 1–2
12. HCØ>CVIII 26.07.42, MØ II, 168–72
13. CVIII>HCØ 07.08.42, ØC 1–2. The snide comment is excised by MØ II, 172

14. JJB>HCØ 08.08.42, C I, 60–1
15. M. Jensen 1963, 70
16. M. Rubin 1895, 302–7
17. HCØ>CH 04.05.44, C I, 205–7. HCØ' travel diary 1846, 11, ØS 83
18. F. Barfod 1844, 15
19. HCØ and IBØ>their son Anders 30.07.44, D-27, DTM
20. *Forhandlinger* 1847
21. *SES* 1, 55–75, *SiN*, 41–55
22. HCØ>CH 05.01.45, C I, 214. CH>HCØ 12.02.45, C I, 215–21
23. *BT* 15.07.47
24. *BT* 17.07.47
25. *BT* 22.07.47
26. *BT* 19.07.47
27. *BT* 22.07.47

Chapter 51

1. HCØ *Dansk Ugeskrift* II, vol. 4, 1844, *SES* 7, 77
2. *SES* 8, 23–6
3. KDVS *Oversigter* 1839, 4–5
4. H. Jensen 1934, vol. 2, 218
5. D. Bruun>M.P. Bruun 10.12.39, *DPB* I, No. 244. J. Paludan-Müller>C. Paludan-Müller 12.12.39, *DPB* I, No. 245. I.C. Drewsen>J.F. Schouw 02.01.40, *DPB* I, No. 255
6. *SES* 9, 90
7. Kongehuset 202, CVIII 1809–1847. Letters from and on scientific societies and institutions, No. 119, Reports from HCØ 1839–47, RRO. CF>HCØ 11.03.38 and 02.04.39, ØC 1–2
8. 'HCØs senere biografi efter 1828' [Supplement to HCØ's biography after 1828], ØC 4
9. CF>HCØ 22.11.37, ØC 1–2
10. *Forhandlingerne ved Studentersamfundets Stiftelse* 1840, iii-viii
11. *Forhandlingerne ved Studentersamfundets Stiftelse* 1840, 10
12. P.E. Lind>HCØ 02.12.43, ØC 1–2, MØ II, 208–10
13. 'HCØs senere biografi efter 1828' [Supplement to HCØ's biography after 1828], ØC 4
14. 'HCØs senere biografi efter 1828' [Supplement to HCØ's biography after 1828], ØC 4
15. 'HCØs senere biografi efter 1828' [Supplement to HCØ's biography after 1828], ØC 4
16. HCØ>CVIII 16.05.40, MØ II, 162–6
17. DCC 1969, 66–101
18. JFS *Dansk Ugeskrift*, II, vol. 7, 1847, 97–104
19. HCØ's travel diary 1846, ØC 83
20. HCØ *Dansk Folkeblad*, vol. 1, 1936. *SES* 7, 41–58
21. HCØ *Dansk Ugeskrift* II, vol. 4, 1844. *SES* 7, 68

Chapter 52

1. *Fædrelandet* 13, 16.01.50
2. HCØ>CM 01.04.40, NKS 2336, RL
3. *SES* 8, 99–114
4. *CM* 1843, 431

5. HCØ KDVS 5416–1843. HCØ>CVIII 01.07. and 20.11.43. Kongehuset 202, CVIII 1809–1847. Letters from and on scientific societies and institutions, No. 119, RRO
6. CM 1843, 431, xvi
7. HCØ>P.O. Brøndsted 03.05.42, NKS 1545, RL
8. HCØ, KDVS *Oversigter* 1835, 18
9. HCØ>CVIII 27.12.39, 20.11.43. Kongehuset 202, CVIII 1809–1847. Letters from and on scientific societies and institutions, No. 119, RRO
10. A. Lomholt 1960, vol. 3, 184–8. HCØ>W. Weber 31.12.39, C II, 567
11. P.W. Lund>CF 14.02.37, KDVS 4575–1837. T. Wolff 1994, 156–63
12. *SES* 8, 107
13. *SES* 2, 151–92, *SiN* 257–88
14. *SES* 2, 185
15. *SES* 2, 166
16. *SES* 2, 167
17. *NS* III, 151–90, 158, *SSW* 282–309, 285
18. *SES* 2, xi og 190
19. H. Steffens 1841, 25–42
20. H. Steffens 1841, 33
21. *SES* 8, 105
22. *SES* 8, 113, *SES* 2, 153
23. KDVS 4712–1839. *NS* II, 528–9, *SSW* 607–9
24. W. Marstrand 1929, 17–27. P.F Dahl 1972, xiii–xxxv. K. Meyer, *NS* III, lxxiv-lxxviii
25. KDVS 5438–1843 and 5504–1843
26. K. Caneva, HSPS, 1997, 28:1, 61, 68, 86
27. Caneva completely ignores Ørsted's dissertation and his reading of Kant's 'Metaphysical Principles of Natural Science' (ch. 8), according to which mechanical motions are phenomena, while fundamental forces are noumenal. This is the lesson Ørsted deduced from Kant and which became the foundation of the entire dynamical research programme. As Ørsted stresses in his *Ansicht* 'It goes without saying that we do not presume to have explained anything with this definition. For the moment, we leave completely open what these forces are, and whether they are independent or merely modifications of other forces or of substances. We merely adopt the usage of referring to each active property as a force.' *NS* II, 73, *SSW* 332
28. K. Caneva, 1997a, 97
29. HCØ 1843, *NS* III, lxxvi
30. HCØ, KDVS *Oversigter* 1848, 92
31. V. Marstrand 1929, 21
32. V. Marstrand 1929, 57

Chapter 53

1. J. Herschel 1831, 340
2. HCØ>CH 02.01.42 and 23.05.43, C I
3. A. Meyer-Abich 1967, 171
4. A. v. Humboldt 1845, vol. 1
5. HCØ>IBØ --.07.43, ØC 1–2, partly in MØ II, 177–8, TL 387
6. HCØ>IBØ --.07.43, ØC 1–2, partly in MØ II, 177–8, TL 387
7. A. v. Humboldt>HCØ 23.07.43, C II, 411–12

8. HCØ>IBØ 28.07.43, ØC 1–2, TL 409–410
9. HCØ>IBØ 01.08.43, ØC 1–2, partly in MØ II, 188, TL 402
10. DCC 1996, 507–24
11. HCØ's travel notes 1843, 12–13, ØC 91, RL
12. HCØ>IBØ 06.08.43, ØC 1–2, MØ II, 192, TL 406. HCØ>IBØ 08.08.43, ØC 1–2, MØ II, 197, TL 411
13. HCØ>IBØ 29.07.43, ØC 1–2, MØ II, 183–7, TL 397–400
14. HCØ>IBØ 06.08.43, ØC 1–2, MØ II, 191–2, TL 405–406
15. HCØ>IBØ 08.08.43, ØC 1–2, MØ II, 198, TL 412
16. HCØ>IBØ 10.08.43, ØC 1–2, MØ II, 199, TL 413

Chapter 54

1. SES 3, 147
2. SAK 1968 (A72), 49
3. D. Jørgensen 2003, 262–9
4. SES 1, 24
5. SES 3, 127
6. SES 3, 127–8
7. SES 1, 39–54
8. SES 3, 146–7. SES 3, 212–13
9. F. Bremer>HCØ 04.07.49 and HCØ>F.Bremer, s.d., MØ 2, 264–7
10. F. Bremer>HCØ 04.07.49 and HCØ>F.Bremer, s.d., MØ 2, 214–15
11. F. Bremer>HCØ 04.07.49 and HCØ>F.Bremer, s.d., MØ 2, 151
12. SES 3, 146–7. The peacock is included both in HCØ's dialogue and HCA's fairy tale.
13. HCA SV 2, 286, 288, 291
14. HCA SV 2, 197
15. HCA SV 2,, 432, http://hca.gilead.org.il/
16. HCA SV 3, 382, http://www.andersen.sdu.dk
17. HCA SV 3, 26–8
18. HCØ, 'Sytten Forelæsninger om det Skjønne' [Seventeen Lectures on the Beautiful], 8th lecture, ØC 48
19. HCØ 1845
20. E. Hannover: C.W.Eckersberg., 1898, C.V. Petersen, 1933–34, 114–22; P.J. Nordhagen 1968, K. Monrad 1989
21. I. Ydstie 1999, 25
22. M.C Harding 1924, 49
23. HCØ>Direktionen for SNU [The Society for the Dissemination of Science 20.05.49, SNU Archive, parcel 9, file 090201. H. Tybjerg, 1996

Chapter 55

1. Daily News 18.09.46, ØC 85
2. HCØ>IBØ 30.07.46, ØC 1–2
3. HCØ's travel notes for 1846, 5–7, ØC 83
4. HCØ>EAS 29.07.46, MØ II, 226–8. HCØ>IBØ 30.07.46, ØC 1–2
5. HCØ>IBØ 19.08.46, ØC 1–2
6. HCA, SV 3, 27
7. R.D Altick 1978, 375–89
8. HCØ's travel diary 1846, 45–6, ØC 83
9. HCØ's travel diary 1846, 46–7

10. HCØ's travel diary 1846, 50–1. R.D Altick 1978, 141–62
11. E. Nystrøm 1913, 226–7
12. MØ's diary from the tour of 1846, 45–6, F-39, DTM
13. MØ's diary from the tour of 1846 52 and MØ II, 238–9
14. MØ's diary from the tour of 1846 54 and MØ II, 239
15. MØ's diary from the tour of 1846 59–60
16. MØ's diary from the tour of 1846 56 and MØ II, 240–1
17. MØ's diary from the tour of 1846 56–7 and MØ II, 241–2
18. MØ's diary from the tour of 1846, 70
19. CF>HCØ 02.04.39, 202 Kongehuset CVIII, 1809–1847, Breve fra og om vidensk selskaber og institutioner. [Letters from and on learned societies and institutions], No. 119, PRO. HCØ *Erindringsblad* [Memoir] 1840. J. Herschel>HCØ 04.12.41, C II, 397–8
20. MØ's diary from the tour of 1846, 63, F-39, DTM
21. MØ's diary from the tour of 1846 68, MØ II, 244
22. *Daily News* 11.09.46
23. MØ's dairy from the tour of 1846, 85, F-39, DTM. MØ II, 250
24. D. Knight 1986, 134
25. R.M. MacLeod 1983, 55–90. D. Knight 1986, 128–48
26. *Daily News* 18.09.46, ØC 85
27. HCØ's travel diary for 1846, 71, ØC 83. MØ II, 245
28. HCØ's travel diary for 1846 78
29. Report of the sixteenth Meeting of the BAAS 1846, 1847, Notices, 2–3. NS II, 416–18. Amtlicher Bericht 1847, 192–3
30. *Daily News* 18.09.46, ØC 85
31. *Daily News* 18.09.46, ØC 85 and HCØ's travel diary for 1846, 80–2, ØC 83

Chapter 56

1. *NS* III, clxii
2. HCØ 1844
3. *Kjøbenhavnsposten* 248, 23.10.44
4. *Kjøbenhavnsposten* 261, 18.11.44
5. HCØ>CH 09.10.44 and 12.02.45, C II, 211 and 216–19
6. *Kjøbenhavnsposten* 271, 18.11.44
7. *Kjøbenhavnsposten* 282, 30.11.44
8. *Kjøbenhavnsposten* 135, 14.06.45
9. *Kjøbenhavnsposten* 142, 23.06.45
10. HCØ 1851
11. *Kjøbenhavnsposten* 165, 19.07.45
12. *Kjøbenhavnsposten* 191–2, 19.-20.08.45
13. HCØ>K.L. Rahbek 19.11.25, NKS 1455c, file xvi, RL. DBL
14. HCØ 1847
15. *Fædrelandet* 180, 03.08.47
16. T.G. Repp>HCØ 31.07.47, ØS 1–2
17. *Fædrelandet*, 180, 03.08.47
18. R. Bielke>HCØ 14.08.47, ØS 1–2

19. HCØ 1847, 9
20. A. Steen 184
21. *Fædrelandet* 196, 21.08.47. O. Pedersen 1987, 142–66
22. J.F.C.E. Wilkens *Fædrelandet* 254, 255, 257, 258, 259, 1847. M.F Wagner 1999, 371–92.
23. J.F.C.E. Wilkens *Fædrelandet* 206, 01.09.47. Cd. (anonymous) M.F Wagner 1999, 209, 04.09.47. M.F Wagner 1999, 211, 07.09.47
24. J.F.C.E. Wilkens, M.F Wagner 1999, 254, 26.10.47
25. J.F.C.E. Wilkens, M.F Wagner 1999, 255, 27.10.47
26. J.F.C.E. Wilkens, M.F Wagner 1999, 257, 29.10.47
27. J.F.C.E. Wilkens, M.F Wagner 1999 258, 30.10.47
28. J.F.C.E. Wilkens, M.F Wagner 1999 259, 01.11.47
29. V. Christensen 1912, 553–6

Chapter 57

1. *SES* 4, 38
2. H.N. Clausen and J.F. Schouw 1848
3. L. Koch 1896, 190
4. H. Vammen 1988, vol. 88.2, 253–79
5. HCØ>A.M. Goldschmidt 29.01.48, NKS 4252, RL
6. DCC 1969, 102–7
7. A.M Goldschmidt 1850, vol. 2, 111–12
8. A.M Goldschmidt 1850, vol. 2, 112
9. C. Degn 1995, 226–31
10. A.M. Goldschmidt>HCØ 22.03.48, A.M. Goldschmidt 1963, 135–6
11. HCØ>A.M. Goldschmidt 23.03.48, A.M. Goldschmidt 1963, 136
12. HCØ>A.M. Goldschmidt 23.03.48, A.M. Goldschmidt 1963, 137
13. HCØ>A. v.Humboldt, no date, C II, 412–14. HCØ>J.W.F. Schelling 18.01.50, D-55, DTM
14. H.L Martensen 1850
15. HCØ>G.A. Michaelis 06.10.46, G.A. Michaelis>HCØ 10.10.46, C II, 437–40
16. HCØ>H.L. Martensen 09.01.50, MØ II, 161–2, misdated to 09.01.40
17. *SES* 9, 74–5
18. *SES* 4, 17
19. P. Hasle 18.09.49, MØ II, 269–71
20. HCØ>P.Hasle 18.09.49, MØ II, 271–5
21. MØ's travel diary 1849, F-40, DTM

Chapter 58

1. *SES* 1, 100, *SiN*, 73, where the quotation has been translated into 'in the magnificent ancient dress of past days'
2. *SES* 2, 3, *SiN*, 143. It should be noted that *The Soul in Nature* is not a translation of the same texts that appeared in *Aanden i Naturen*. *The Soul in Nature* also contains texts from *SES* vols.3, 5, and 7
3. *SES* 4, 23–4, not in *SiN*
4. HCØ>J. Herschel 26.04.39, HS 13.163, RS
5. HCØ>J. Herschel 11.06.41, HS 13.165, RS
6. HCØ>J. Herschel 04.07.41, HS 13.166, RS

7. J. Herschel>HCØ 10.10.41, C II, 396–7
8. HCØ>J. Herschel 16.11.41, HS 13.167, RS
9. HCØ>R.J. Murchison 28.11.46, C II, 456–7
10. R.J. Murchison>HCØ 13.12.46 and 05.02.47, C II, 457–9
11. HCØ>J. Herschel 20.06.47, HS 13.170, RS
12. HCØ's travel diary 1846, 13.09.46, MØ II, 247
13. HCØ>E. Sabine 06.10.48, C II, 508–10
14. HCØ>J. Herschel 25.08.49, HS 13.171, RS
15. HCØ>J. Herschel 12.10.49, HS 13.172, RS, C II 402–4
16. HCØ>J. Herschel 28.12.49, HS 13.173, RS
17. HCA>MØ 01.09.51, HCA 2000, 597
18. HCØ>CM 25.02.51, NKS 2336, RL
19. G. Forchhammer, KDVS *Oversigter* 1851, 43–62
20. T. Wolff 1994, 156–63
21. J.P Mynster 1850, 16
22. *SES* 2, 39–49, *SiN* 172–81
23. J.P. Mynster 1821
24. J.P Mynster 1850, 57
25. HCA>HCØ 03.08.50, HCA 2000, 583
26. HCA *SV*15, 64–7
27. HCA *SV*15, 99–102
28. HCA *SV*15, 99
29. HCA *SV*15, 102
30. J. Andersen suggests something similar in his *Andersen. A Biography*, vols. I–II, 2003, 190, but argues that Andersen in his novel *To be or not to be* dissociates himself from their mutual programme.
31. A.M Goldschmidt 1850, vol. 5, 639–69 and 691–726. G. Brandes 1919, vol. 2, 121 only vouchsaves HCØ one sentence, viz. in a section on HCA: 'How grateful he [HCA] was for every kind of instructions is shown by his relationship with HCØ, who as an aestheticist, however, was heavily prejudiced.'
32. HCØ, *Der Geist in der Natur*, Leipzig 1850
33. *Beilage zum Morgenblatte der Wiener Zeitung*, 60, 18.05.50
    Leipziger Zeitung, 232, 20.08.50
    Die Grenzboten. Zeitschrift für Politik und Literatur, 35, 23.08.50
    Deutsche Reform—politische Zeitung für das constitutionelle Deutschland, 1152, 11.10.50
    Dresdner Journal und Anzeiger, 289, 16.10.50
    Hamburger literarische und kritische Blätter, 116, 13.11.50
    Kölnische Zeitung, 294, 08.12.50
34. O.L. Erdmann *Allgemeine Monatschrift für Litteratur* 1, 1850 393–400 I, Frauenstädt O.L. Erdmann *Allgemeine Monatschrift für Litteratur*, 2, 20–9
35. HCØ *SES* 1, 24, *SiN*, 18
36. CH>HCØ 14.05.50, CII 233–4
37. HCØ, *The Soul in Nature*, 1852
38. Darwin's note book 11.05.52. C. Darwin 1988, vol. 4, 488. I am indebted to Prof. David Knight for this reference
39. MØ>J. Horner 14.05. and 11.08.52, NKS 2500, RL

Chapter 59

1. HCA, *SV* 2, 2003, 417
2. NFSG, *Danskeren* 1850
3. HCA>H. Wulff 16.03.35, HCA 1959, vol. 1, 211
4. N. Bøgh 1887
5. HCA, SV 17, 11
6. *SES* 9, 41
7. HCA>Carl Alexander of Weimar, December 1850, HCA 2000, 591
8. HCA, *SV* 17, 2007, 117–18
9. HCA, *SV* 15, 117–22, http://www.online-literature.com/hans_christian_andersen/pictures-of-sweden/26/
10. *SES* 1, 100, *SiN*, 73
11. HCA>HCØ 03.08.50, HCA 2000, 583
12. HCA>MØ 01.09.51, HCA 2000, 597
13. A. Grum-Schwensen 2001, 139
14. HCA, *SV* 1, 2003, 460–85. HCØ, *Almanakken* 1835, SES 2, 133–49, *SiN*, 244–56
15. HCA, *SV* 2, 2003, 138–49, http://hca.gilead.org.il/thorny_r.html
16. HCA>S. Læssøe 01.01.59, HCA 2000, 692
17. G. Brandes 1919, vol. 2, 125–6. The story is not told by HCA in 'The Fairy Tale of my Life' (*SV* 17–18) as stated by GB.
18. HCA, *SV* 1, 2003, 369–72, http://hca.gilead.org.il/bell.html
19. HCA, *SV* 2, 2003, 72–72
20. HCA, *SV* 2, 2003, 382–7

Chapter 60

1. HCA>HCØ 03.08.50, HCA 2000, 583
2. IBØ's ledger 08.10.49, E-62, DTM
3. K. Arentzen 1879, 495–518
4. HCØ>N.C.L. Abrahams 02.02.50, D-31, DTM
5. IBØ's ledger 09.04.50, E-62, DTM
6. MØ>J. Horner 31.12.50, D-10, DTM
7. ASØ>HCØ 07.11.50, MØ II, 296–7
8. J.L. Heiberg, *Til Hans Christian Ørsted den 7de November 1850*
9. Studenterforeningens Seniorat>HCØ 07.11.50, MØ II, 297
10. Letter of 5850.11.6 (Masonic time, i.e. 06.11.50], D-35, DTM
11. *Til HCØ 07.11.50 fra Studenter*
12. HCA>H. Wulff 10.11.50, HCA 2000, 589
13. C.E. Fenger, *Ugeskrift for Læger*, 1839
14. M.A Skydsgaard 2006
15. *BT* 08.03.51
16. HCØ, Søndagsforelæsninger [Sunday Lectures] 1850–1851, ØS 49, xiv, RL
17. EAS's Notes on HCØ, F-11, DTM
18. HCA 1974, vol. 4, 21
19. HCA, *Fædrelandet* 11.03.5120.03.51
20. *Hans Chr. Ørsteds Jordefærd* 1851, 49

21. *Hans Chr. Ørsteds Jordefærd* 1851, 14–18
22. HCA 1974, vol. 4, 21
23. EAS's Notes on HCØ, F-11, 11, DTM
24. *Hans Chr. Ørsteds Jordefærd* 1851, 19–29
25. HCA 1974, vol. 4, 21
26. *Hans Chr. Ørsteds Jordefærd* 1851, 9
27. *Hans Chr. Ørsteds Jordefærd* 1851, 30–5
28. *Fædrelandet* 31.03.50

Chapter 61

1. G. Brandes 1866, 67
2. TL 292
3. K. Meyer, 'The Scientific Life and Works of H.C. Ørsted', NS I xiii–clxvi, L. Rosenfeld 1979
4. M. Friedman 2007, A.D Wilson 1998, 2007, K.L Caneva 1997, 2007
5. B. Gower 1973, T. Shanahan 1989, D.C Christensen 1995
6. R. Stauffer 1953, 1957, P.L Williams 1974, H.A.M. Snelders 1990, A. Skaar Jacobsen 2000
7. M. Friedman 2007 155–7
8. W. Ostwald, 'Johann Wilhelm Ritter, Rede, gehalten auf der ersten Jahresversammlung der Deutschen Electrochemischen Gesellschaft am 5. Oktober 1894 in Berlin', in J.W. Ritter, *Fragmente aus dem Nachlasse eines jungen Physikers*, ed. by Steffen and Birgit Dietzsch, Hanau 1984, pp.321–43. *Der Physiker des Romantikerkreises Johann Wilhelm Ritter in seinen Briefen an den Verleger Carl Friedrich Ernst Frommann*, ed. by Klaus Richter, Weimar 1988. W.D. Wetzels, *Johann Wilhelm Ritter: Physik im Wirkungsfeld der deutschen Romantik*, Berlin/N.Y. 1973
9. JWR>F.K. von Savigny 13.07.1803 in C. Klinkowstroem, 'Drei Briefe von Johann Wilhelm Ritter.' *Der grundgescheute Antiquarius* 1, Munich 1921, 125
10. *Europa, eine Zeitschrift*, vol. 1, 1803, 50
11. K. Nielsen and H. Andersen, *Boston Studies in the Philosophy of Science* No. 241, 2007, 97–114
12. SES 1, 18
13. SSW 77
14. SSW 71
15. HCØ 1802, 7 and 12
16. SSW 190
17. SSW 342–9
18. A.D Wilson 2007, 1–11
19. H. Martensen 1850
20. HCØ>FWJS 18.01.50, D-55, DTM: 'This matter is not merely a Danish concern. Germans, Danes, Swedes, Norwegians, and Dutch form together the greatest German-Gothic star placed in the middle of Roman and Slav peoples, and they are bound together through numerous leagues between states and families of language. To stir hatred between them is pernicious for everybody.'
21. M. Friedman 2007, 149–50
22. K. Richter 1988, 36, W.D Wetzels 1973, 24
23. M. Friedman 2007, 145, n.29
24. *Fortegnelse....*1853
25. SSW 313
26. SSW, 199

27. *SES* 8, 105–7
28. HCØ, 1812, 303
29. HCØ 290–1
30. HCØ , 291, n.36 referring to K.L Caneva, 1997b.
31. K.L. Caneva 1997 10
32. W.D. Wetzels, 76
33. W.D. Wetzels , 5
34. M. Friedman 2007 154, n.48
35. *SSW* 378–9
36. *SSW* 379
37. G. Brandes 1866, 67
38. E.M Christensen 1966, 11–45. V. Andersen 1916, 109–40

ARCHIVAL MATERIAL & BIBLIOGRAPHY

Archival Material

Bakkehusmuseets manuskriptsamling, Frederiksberg Kommunebibliotek
vii, breve til [letters to] Kamma Rahbek

British Museum
C.43a.17, Coleridge's annotated copy of HCØs *Ansicht*

Danmarks Tekniske Museum, Helsingør
(register and card index)

A, 1–44	møbler, inventar [furniture etc.]	
B, 1–128	billeder, buster, m.v. [pictures, busts, etc.]	
C, 1–30	medaljer [orders and medals]	
D, 1–247	breve, digte [letters, poems]	(D-79, E.A. Scharling's notes on HCØ)
E, 1–70	personlige minder [personal memories]	(E-52f., IBØ's ledgers)
F, 1–45	slægtsoptegnelser [family relations]	
	(F-3, Birgitte Scharling's notes of her mother's account)	
	(F-39–40, MØ's travel diaries 1846, 1849)	
G, 1–	fysiske instrumenter [physical instruments]	
H, 1–179	HCØ's værker [works]	
J, 1–75	HCØ's biografier [biographies]	
K, 1–155	HCØ's bibliotek [library]	
L, 1–99	diverse [sundry]	

Dibner Library, Smithsonian Institution, Washington D.C.
MSS 1083A, HCØ>Th.Thomson, 05.07.16

Erhvervsarkivet, Aarhus
Landhuusholdningsselskabets journalsager [The Royal Danish Agricultural Society's journals]

Det Kgl. Bibliotek [RL]
Collection of manuscripts
Re letters from and to HCØ see electronic register of letters
Add.1040, Breve fra HCØ til FCS [letters from HCØ to FCS]
Add.1289b, Billetter og breve fra SØ til FCS [notes and letters from SØ to FCS]
Add 1289 a-b, læg iv FCS om SØ, brudstykke [FCS on SØ, fragment]
NKS 287c, iv, Thomas Bugge's archive
NKS 1011 Songbook used as ledger 1814–1818
NKS 768d Minutes of the SLS, statutes and list of members 1800
NKS 2500, MØ>J. Horner

NKS 3494, Gamborg, A., 'Den philosophiske Moral—dicteret af Prof. Gamborg', [The Philosophical Morality—dictated by Prof. Gamborg] pp.30

Grundtvigarkivet [The Grundtvig Archive], fasc. 355, iv, F.M. Knuth>N.F.S. Grundtvig

Ørstedsamlingen [The Ørsted Collection]

(register at the reading room of the Collection of Manuscripts, the following references do not appear in the register)

1–2, breve til og fra HCØ [letters to and from HCØ]

1–2, vii, koncept til svar på rundskrivelse fra JFS til Clausen, Gad, Ørsted, Delblanc af 11.07.1837 [draft of reply to circular letter from JFS to.... of 11.07.1837]

4, 'Fra hvor tidlig en Alder har jeg Erindringer?' [From how early an age do I have memories?]

4, 'Til HCØs senere Biographie efter 1826' [Addition to HCØ's later biography after 1826]

4, manus til tale på Studentersamfundets møde, udat. [draft of .HCØ's speech to the Students' Union, s.a.]

4, udkast til konsistorial erklæring i–ii, udat. [draft of declaration to the Senate i–ii, s.a.]

15, Rejsedagbog. Korte Bemærkninger [Travel notes. Short entries]

23, viii–ix, Elektro-magnetisme—af K. Meyer kaldt henholdsvis 'Kladden' og 'Supplement II' [electro-magnetism—'the draft' and 'supplement II' respectively by K. Meyer]

35, månedsforelæsning af 08.01.1828 [monthly lecture of 08.01.1828]

43, v, kladde til dialog 'Hvad er Chemie?', som endte som forelæsning og essay [draft of the dialogue 'What is Chemistry?' ending up as a lecture and an essay]

43, ix, 'Noget om Lærebøger', udat. manus [Remarks on textbooks, manuscript s.a.]

43, xi, 'Forslag til Oplysningens Udbredelse hos os. De Pariser Athenéer' [Proposal on the diffusion of enlightenment in Denmark. The Parisian Athenaea]

44, ii, HCØ>C.E. Scharling

48, 17 Forelæsninger om Det Skjønne [Lectures on the Beautiful]

49, xiv, søndagsforelæsninger 1850–51. 'Naturlærens Historie' [Sunday lectures 1850–51. 'The History of Physics']

51, i, orig. manus til 'Samtale over Mysticismen' [original manuscript to 'Dialogue on Mysticism']

57, xiv, div. unummererede og udaterede noter, bl.a. 'Landoeconomisk Chemie 05.11.1820, gjentaget 05.11.1826' [sundry unnumbered and undated notes, among others 'Agricultural Chemistry 05.11.1820, repeated 05.11.1826']

59, ii, 'Forslag til en practisk Læreanstalt for den eksperimentale Naturvidenskab', udat., (efter 1808) [proposal for a practical institute for experimental science, s.a., after 1808]

59, x, Forslag, udat. (1813) [Proposal, s.a., 1813]

59, xi, HCØs udat. ansøgn. til Univ. Dir. [HCØ's undated application to the Board of the University]

59, xvi, PM til Univ. Dir., udat., (c.1808) [PM to the Board of the University, s.a., c.1808]

59, xvi, 'Brev fra Doctor Ørsted i Paris til Professor Manthey' 1803 (ikke 1802 som angivet i *Nyt Bibliotek for Physik, Medicin og Oeconomie*, 1803, 273) ['Letter from Dr Ørsted in Paris to Professor Manthey', 1803 (not 1802 as stated in...)]

59, xviii, Kladde til Betænkning om det physiske Studium ved Kjøbenhavns Universitet, udat., (c.1807) [Draft of report on the study of physics at the University of Copenhagen, s.a., c.1807]

62, manus til Maanedsforelæsning 141, 05.10.1830 [manuscript for monthly lecture 141 of 05.10.1830]

74, i–iii og v, forsk. kladder til 'Samtale over Mysticismen' [sundry drafts of 'Dialogue on Mysticism']

80, (+ 82) 'Rejsebreve i Dagbogsform' (MØs betegnelse, dvs. HCØ>SP) 1801–04 ['Travel letters in diary form' -MØ's description, i.e. HCØ>SP, 1801–04]

81, iii, ordensdekorationer, mv. [decorations, etc.]

83, rejsedagbog 1846, 'Einige Tischreden in der Kieler Versamlung der GDNÄ' [travel diary 1846, 'Some after-dinner speeches in the Kiel Assembly of the GDNÄ']

83, papirer vedk. Trykkefrihedsselskabet udat. [papers concerning the Society for the Right Use of the Freedom of the Press, s.a.]

85, *Daily News*, London 1846

Beilage zum Morgenblatte der Wiener Zeitung 60, 18.05.50

Leipziger Zeitung 232, 20.08.50

Die Grenzboten. Zeitschrift für Politik und Literatur 35, 23.08.50

Deutsche Reform—politische Zeitung für das constitutionelle Deutschland 1152, 11.10.50

Dresdner Journal und Anzeiger 289, 16.10.50

Hamburger literarische und kritische Blätter 116, 13.11.50

Kölnische Zeitung 294, 08.12.50

88, Rejsebemærkninger 1801–03, 1812 [travel notes 1801–03, 1812]

91, Rejsebemærkninger 1843 [travel notes 1843]

94, Breve til og fra HCØ [letters to and from HCØ]

102, Rudkjøbing Kirkeprotokol [Rudkøbing church register]

Det Kgl. Danske Videnskabernes Selskab, Copenhagen [KDVS]

Mødeprokol 1800- [minutes 1800–]

Hovedarkivet [General archive]

(mødeprotokollens nr.-år, f.x. 543–1808 ref. t. journalsag i hovedarkivet, hvor den dog ikke altid findes) [number and year of the minutes, e.g. 543–1808 referring to journal items in the general archive, where it cannot always be found]

1899–1815, 'Over Loven om de electriske Kræfters Udbredelse' ['On the Law of the Propagation of Electrical Forces']

kartotek over HCØs forelæsninger i KDVS, som er optaget i 'Oversigter' [register of HCØ's lectures in KDVS, included in the annual transactions]

Specialarkivet, Artesisk brøndboring 1831–34, 1835– [Special archive, Artesian Well Boring 1831–34]

Københavns Stadsarkiv, Copenhagen

Vurderingsforretning 08.09.1838 [valuation]

Københavns Universitets specialebibliotek [prize essays]

Christensen, Dan Ch., 'En Analyse af hovedpunkter i forfatningsdiskussionen i Danmark 1830–1849', Københavns Universitets prisspørgsmål i historie (typewritten) 1969

Michelsen, William, 'H.C. Ørsteds Stilling i dansk Aandsliv' Københavns Universitets prisspørgsmål i nordisk filologi (handwritten) 1940

Vammen, Hans, 'J.F. Schouw som politiker', Københavns Universitets prisspørgsmål i historie (typewritten) 1967

Landesarchiv Berlin

Grundplan af Hermbstaedts laboratorium, Pr. Br. Rep. 42, Neue Folge vol.ii, Plankammer V, nr.3 [ground plan of Hermbstaedt's laboratory]

Landsarkivet for Sjælland

Folketællingen 1801, København, Strand Kvarter, Matr. 45, 152 [census 1801]

Københavns Skiftekommission 1800–1801, Forseglingsprotokol 3/17 [administration of estate]

Musée Carnavalet, Paris
Grundplan af rue des Petits-Pères, Paris [ground plan of rue des Petits-Pères, Paris]

Niels Bohr arkivet, KU, Copenhagen
(registrant over SNUs arkiv) [register of the archive of the Society for the Dissemination of Science in Denmark]
Selskabet for Naturlærens Udbredelses arkiv, pak.6, læg 060102 og pk.9, læg 130703, HCØ>Direktionen for SNU 18.11.1848 [HCØ> the Board of the SNU]

Oxford University Museum archive
The History of the Building of the Museum, box 2.4. Correspondence concerning donations for statues

Rigsarkivet, Copenhagen [Public Record Office, Copenhagen] [RA, PRO]
Finanscoll. j.nr.1532, 1815A, HCØ og Jeppe Smith>Finanskollegiet
Finansdep. Forestillinger og Resolutioner 1820, nr. 277 af 12./17.10.1820 m. bil. [recommendations and resolutions, including attachments]
Fonden ad usus publicos, brevprotokol, nr.164 [letter book]
Fonden ad usus publicos, J. 1809–1817, nr. 458 og 634
Generalitets- og Kommissariats Koll. Kgl. ordrer og res. 1809, nr. 249 [royal orders and resolutions]
Justitsarkiv. Relationer til Kongen. Generalauditøren, pk.17, Rel. ccxii, dom nr.2 [communications to the King]
Kgl. res. 05.09.06
Kgl. res. 14.09.15
Kongehuset, FVI, ASØs brev af 30.06.17
Kongehuset CVIII 1809–1847, 202, Breve fra og om videnskabelige selskaber og institutioner, nr. 119, HCØ>CVIII 08.01.42 og 09.09.47 [Letters from and on scientific societies and institutions]
Københavns Universitet, Konsistorium 1538–1882, Kopibog 1810–13, 21 nr. 1213 [The Senate's letter book]
Københavns Universitet, Konsistorium Kopibog 1817–19, 25, 1213–24, nr.945, Univ.dir.>KU Konsistorium 13.02.1819 [The Senate's letter book]
Den polytekniske Læreanstalts arkiv. Journalregistratur 1829–1842 og journalsager 28, 1830 [journal items, the Polytechnic Institution]
Universitetsdirektionen, Journalsager, Ursins forslag 1605/1827, HCØs forestilling 1155/1829 [the Board of the University, journal items, Ursin's proposal 1605/1827, HCØ's recommendation 1155/1829]

Royal Society, London
Minutes of Council, vol. ix, 1811–1822, 243
Journal Book of the Royal Society, vol. xliii, 1819–1823, 615
HCØ>John Herschel 25.08.1849, HS 13.171
HCØ>John Herschel 12.10.1849, HS 13.172
HCØ>John Herschel 28.12.1849, HS 13.173

Den store danske Frimurerordens arkiv, Copenhagen [FMA]
HCØ>Carl af Hessen, 08.03.1823
Sørgelogen 5851.4.18 (18.04.1851) [mourning lodge]

Thorvaldsens Museums arkiv, Copenhagen
SP>Bertel Thorvaldsen

| ARCHIVAL MATERIAL & BIBLIOGRAPHY

Wren Library, Trinity College, Cambridge
Add.Ms. a 210.42, Add.Ms. a 210.43, Add. Ms.a 210.44, HCØ>W. Whewell
Add.Ms. a 205.53, N.F.S. Grundtvig>W. Whewell

Books and Articles

Abrahamowitz, Finn, *Grundtvig. Danmark til Lykke*, Copenhagen 2000
Agardh, C.A., Tale ved åbningen af det første skandinaviske naturforskermøde, *Förhandlingar vid det af skandinaviska naturforskare och läkare hållna möte i Götheborg år 1839*, Gothenburg 1840
Aldini, J., *Essay Théorique et Experimental sur la Galvanisme*, Paris 1804
Alex, Reinhard & Peter Kühn, *Schlösser und Gärten um Wörlitz*, Leipzig 1995
Allgemeine Deutsche Biographie
Altick, R.D., *The Shows of London*, Cambridge, Mass. 1978
Amtlicher Bericht über die 24. Versammlung Deutscher Naturforscher und Aerzte in Kiel im September 1846, ed. by G.A. Michaelis and H.F. Scherk, Kiel 1847
Andersen, Einar, *Thomas Bugge. Et Mindeskrift i anledning af 150 årsdagen for hans død 15. januar 1815*, Copenhagen 1968
——, *Dagbøger*, vol.iv, 1851–60, ed. by Tue Gad, Copenhagen 1974
——, *samlede værker. Eventyr og Historier I, 1830–1850*, ed. by Klaus P. Mortensen, Copenhagen 2003
——, *samlede værker. Eventyr og Historier II, 1852–62*, ed. by Klaus P. Mortensen, Copenhagen 2003
——, *samlede værker, Eventyr og Historier III, 1863–75*, ed. by Klaus P. Mortensen, Copenhagen 2003
——, *samlede værker, Blandinger 1822–1875*, 'Anmeldelse', ed. by Klaus P. Mortensen, Copenhagen 2004
——, *samlede værker, Skuespil I, 1822–1834*, ed. by Klaus P. Mortensen, Copenhagen 2005
——, *samlede værker, Selvbiografier I*, 'Levnedsbog 1832', ed. by Klaus P. Mortensen, Copenhagen 2007
——, *samlede værker, Selvbiografier II–III*, 'Mit Livs Eventyr', ed. by Klaus P. Mortensen, Copenhagen 2007
——, samlede værker, *Rejseskildringer II, 1851–72*, 'I Sverrig', ed. by Klaus P. Mortensen, Copenhagen 2007
——, *Astrid Stampes Billedbog*, Copenhagen 2003
Andersen, H.C. og Henriette Wulff. En brevveksling, vol.i, Odense 1959
Andersen, Jens, *Andersen. en biografi*, vol.i–ii, Copenhagen 2003
Andersen, Vilhelm, *Tider og Typer. Goethe II*, Copenhagen 1916
Anonym, 'Om de praktiske Videnskaber ved Universiteterne', *Minerva* 1794
Anonym, *Sketch of the Rise and Progress of the Royal Society Club*, London 1860
Arentzen, Kristian, *Adam Oehlenschläger. Literaturhistorisk Livsbillede*, Copenhagen 1879
Babbage, Charles, *Reflections on the Decline of Science in England*, London, 1830
Bagge, Povl, 'Akademikerne i dansk politik i det 19. århundrede. Nogle synspunkter.', *Historisk Tidsskrift*, 12.rk., vol.iv, hf.3, 1970
Baggesen, K. & A. (udg.), *Aus Jens Baggesen Briefwechsel mit K.L. Reinhold und F.H. Jacobi*, vol.i–ii, Leipzig 1831
Bak, Thor A., 'Kemikeren', i *Hans Christian Ørsted*, ed. by F.J. Billeskov Jansen, Egill Snorrason og Chr. Laurtiz-Jensen, Isefjordsværket 1987
Bang, Ole, *I kast med dampen og elektriciteten. Søren Hjorth—en dansk pioner 1801–1870*, Copenhagen 1982
Barfod, Fr., 'Hans Christian Ørsted', *Skandinavisk Folkekalender*, 1845

Bartholdy, Nils G., 'Friedrich Münter—videnskabsmand og frimurer', *Acta masonica scandinavica*, Copenhagen 1998
——, 'Kronprins Frederiks (VI) og hans svoger hertugen af Augustenborgs forhold til frimuereriet', *Acta masonica scandinavica*, vol.ix, Copenhagen 2004
Batteux, Charles, *Indledning til de Skiønne Konster og Videnskaber ved Hr. Batteux, oversat og formered ved Jens Hvas*, vol.i–iv, Copenhagen 1773
Batteux, Charles, *Principes de la littérature*, 5.ed., Paris 1774, reprint by Slatkine Reprints, Genève 1967
Beiser, Frederick, *The Fate of Reason, German Philosophy from Kant to Fichte*, Harvard 1987
Bekjendtgørelse fra det Kongelige Videnskabernes Selskab, 1806
Benjamin, Walter, *Deutsche Menschen. Eine Folge von Briefen*, Leipzig/Weimar 1979
Berg, Hermann & Dietrich Germann, 'Ritter und Schelling—Empirie oder Spekulation', *Die Philosophie des jungen Schelling*, ed. by Erhard Lange, Weimar 1977
Berg, R., 'Da Industriforeningen blev til', *Historiske Meddelelser om København*, 1938
Bericht über die Versammlung deutscher Naturforscher und Ärtzte in Hamburg ved die Geschäftsführer J.H. Bartels og J.C.G. Fricke, Hamburg 1831
Berlin, Isaiah, *The Roots of Romanticism*, ed. by Henry Hardy according to stencil manuscript with the more modest title 'Some Sources of Romanticism' (BBC's arkiv), London 1999
Berman, Morris, *Social Change and Scientific Organisation: the The Royal Institution, 1799–1810*, The John Hopkin's University (Ph.D.thesis) 1971
Bjerrum, N.J., 'Die Entdeckung des Aluminiums', *Zeitschrift für angewandte Chemie* 39, 1926
Boll, Per M., 'W.C. Zeise—liv og levned', i *Historisk-kemiske skrifter*, 2, Copenhagen 1990
Borup, Morten, *Johan Ludvig Heiberg, I. Barndom og Ungdom 1791–1825*, Copenhagen 1947
Bostrup, Ole, *Dansk Kemi 1770–1807. Den kemiske revolution*, Copenhagen 1996
Brandes, Georg, *Dualismen i vor nyeste Philosophie*, Copenhagen 1866
——, *Samlede Skrifter* I, 2nd ed., Copenhagen 1919
Breve fra H.C. Andersen, ed. by C.St.A. Bille & Nicolaj Bøgh, Copenhagen 2000
Breve fra og til N.F.S. Grundtvig, ed. by Georg Christensen og Stener Grundtvig, vol.ii, Copenhagen 1826
Breve fra P.V. Jacobsen, ed. by Julius Clausen, Copenhagen 1899
Breve og Aktstykker vedr. Johan Ludvig Heiberg, ved Morten Borup, vol.i, 1806–1825, Copenhagen 1946
Breve til Hans Christian Andersen, ed. by C.St.A. Bille og Nikolaj Bøgh, Copenhagen 1877
Breve til og fra Sibbern, ed. by C.L.N. Mynster, Copenhagen 1866
Breve til og fra Adam Oehlenschläger, ed. by H.A. Paludan, Daniel Preisz & Morten Borup with the participation of Louis Bobé og Carl S. Petersen, vol.i–iv, Copenhagen 1945
Breve til og fra H.C. Ørsted, ed. by Mathilde Ørsted, vol.i–ii, Copenhagen 1870
Briefwechsel zwischen J. Berzelius og F. Wöhler, ed. by O. Wallach, vol.i, Leipzig 1901
Brittan, Gordon B., jr. 'Kant's two grand Hypotheses', 'Kant's philosophy of physical science: Metaphysische Anfangsgründe der Naturwissenschaft 1786–1986, *The University of Western Ontario series in philosophy of science*, vol.33, Dordrecht 1986
Broman, Thomas H., *The Transformation of German Academic Medicine 1750–1820*, Cambridge 1996
Brown, G.I., *Scientist, Soldier, Statesman, Spy—Count Rumford*, Sutton Publishing 2001
Buchdahl, Gerd, "'Special Metaphysics' and Metaphysics of Science", "Kant's philosophy of physical science: Metaphysische Anfangsgründe der Naturwissenschaft 1786–1986, *The University of Western Ontario Series in Philosophy of Science*, vol.33, Dordrecht 1986
Buchdahl, Gerd, *Metaphysics and the Philosophy of Science*, Oxford 1969

Bucholz, C.B., 'Prüfung des Winterlschen Systems; enthaltend eine Untersuchung des neusten Verfahrens Winterls die Andronie zu gewinnen', *Journal für die Chemie und Physik*, 1807, 336–44

Bugge, Thomas, *Thomas Bugge's Journey to Paris in 1798 and 1799*, Copenhagen 1800

Bull, Francis, *Tradisjoner og minner*, Oslo 1946

Bull, Th. (udg.), *Et Dansk-Norsk Hjem belyst ved Brev fra den Bull-Ørstedske Familiekrets*, Oslo 1954

Butts, Robert E., 'The Methodological Structure of Kant's Metaphysics of Science', 'Kant's philosophy of physical science: Metaphysische Anfangsgründe der Naturwissenschaft 1786–1986, *The University of Western Ontario Series in Philosophy of Science*, vol.33, Dordrecht 1986

Byron, Lord, 'Don Juan', Canto II 1819, 194, *Lord Byron The Complete Works*, vol.v, ed. by Jerome J. McGann, Oxford 1986

Bærentsen, Kurt & Jensen, V.Gaunø, 'Om den unge H.C.Ørsted', *Theriaca. Samlinger til Farmaciens og Medicinens Historie*, 1977

Bøgh, Nicolaj, *Fra Oehlenschlægers Kreds. Biografier*, Copenhagen 1881

——, *Illustreret Kalender for Danmark*, Copenhagen 1887

Callisen, Heinrich, *Physisk Medizinske Betragtninger over Kiøbenhavn*, vol.i, Copenhagen 1809

Caneva, K.L., (1997a) 'Colding, Ørsted, and the meaning of force', *Historical Studies in the Physical and Biological Sciences*, 28:1, 1997, 1–138

——, (1997b) 'Physics and *Naturphilosophie*: A Reconnaissance', *History of Science*, 35, 1997, 35–106

——, 'Ørsted's Presentation of Others', 'Hans Christian Ørsted and the Romantic Legacy in Science. Ideas, Disciplines, Practices', ed. by Robert M. Brain, Robert S. Cohen and Ole Knudsen, *Boston Studies in the Philosophy of Science*, 241, Dordrecht 2007, 273–338

Cantor, Geoffry, *Michael Faraday. Sandemanian and Scientist*, London 1991

Catalogus lectionum a professoribus in universitate regia hafniensi a kalendis inde octobris anni mdcclxxxvi habendarum

Catalogus praelectionum in universitate regia havniensis per semestre aestivum a kalendis maji mdcccxxiv habendarum

Chaptal, J.A., *Chemien anvendt paa Kunster og Næringsdrift af Grev J.A. Chaptal, forhen Minister for det Indre i Frankrig. En Oversættelse, gjennemseet og forsynet med Anmærkninger ved H.C. Ørsted. Professor og Ridder af Dannebrogen. Første Deel, indeholdende de to første Dele af Originalen*, Copenhagen 1820

Chenevix, Richard, 'Remarques sur un ouvrage intitulé Materialien zu einer Chemie des neunzehnten Jahrhunderts, ou Matériaux..., publiés par Dr. J.B. Oersted, Regensburg/Ratisbonne 1803, *Annalen der Physik*, ed. by L.W. Gilbert, Halle 1804

Chladni, E.F.F., *Die Akustik*, Leipzig 1802

Christensen, Carl, *Den danske Botaniks Historie*, vol.i, Copenhagen 1924–26

Christensen, Dan Ch., 'En Analyse af hovedpunkter i forfatningsdiskussionen i Danmark 1830–1849', prize essay in history issued by the University of Copenhagen (stencil), Copenhagen 1969

——, 'Naturvidenskabelig og teknologisk udvikling i socioøkonomisk sammenhæng. En analyse af H.C. Ørsteds virke', *Humanistisk Årbog*, vol.ii, 5–62, RUC 1986

——, 'Romantikkens natursyn. H.C. Ørsteds fald og oprejsning'. *Guldalderhistorier. 20 nærbilleder af perioden 1800–1850*, ed. by Bente Scavenius, Copenhagen 1994

——, *Det Moderne Projekt. Teknik og Kultur i Danmark-Norge 1750–(1814)–1850*, Copenhagen 1996 (doctoral thesis)

——, 'The Ørsted-Ritter Partnership and the Birth of Romantic Natural Philosophy', *Annals of Science* 52, 1995, 153–85

——, 'Plovkonkurrencen på Strandmøllen 1820. Den teknologihistoriske konflikt bag svingplovens indførelse i Danmark', *Søllerødbogen* 1994, 28–65
——, 'Aqva Metropolis', *Memento Metropolis*, Copenhagen 1996, 231–42
——, 'Grundloven og Det moderne Projekt', i *Grundlovens Danmark. Samfund og kultur før og efter 1849*, ed. by Søren Bitsch Christensen, Aarhus 1999
——, 'Fysik som Kunstart', *Krydsfelt. Kunst og Videnskab i Guldalderen*, ed. by Mogens Bencard, Copenhagen 2000
——, 'Ørsted's Concept of Force & Theory of Music', 'Hans Christian Ørsted and the Romantic Legacy in Science. Ideas, Disciplines, Practices', ed. by Robert M. Brain, Robert S. Cohen and Ole Knudsen, *Boston Studies in The Philosophy of Science, 241*, Dordrecht 2007, 115–34
Christensen, Erik M., 'Guldalderen som idéhistorisk periode: H.C. Ørsteds optimistiske dualisme, *Guldalderstudier. Festskrift til Gustav Albeck den 6. juni 1966*, Aarhus 1966
——, 'Den blicherske tolkning?', *Kritik. Tidsskrift for litteratur, forskning, undervisning*, Copenhagen 1967/4, 20–44
——, 'En fortolkning af "Højt fra Træets grønne Top". Verifikationsproblemet ved litteraturvidenskabelig meningsanalyse belyst i praksis*, Copenhagen 1969, 84–120
——, 'Det Uudsigelige', *Edda. Nordisk tidsskrift for litteraturforskning*, Oslo 1990, 291–7
——, 'H.C. Andersen og den optimistiske dualisme', *Andersen og verden. Indlæg fra den første internationale H.C. Andersen-konference 25.-31. august 1991*, Odense 1993, 177–91
Christensen, Villads, *København i Kristian den Ottendes og Frederik den Syvendes Tid 1840–1857*, Copenhagen 1912
Clausen, H.N. & J.F. Schouw, *Ved Thronskiftet 1848*, Copenhagen 1848
——, *Optegnelser om mit Levneds og min Tids Historie*, Copenhagen 1877
Collin, Jonas, [mindetale over Bugge], *Nye Landoeconomiske Tidender*, 1815
Cour, Poul la & Jacob Appel, *Historisk Fysik*, vol.ii, *Den nyere Naturforskning*, Copenhagen 1897
Crosland, Maurice, *The Society of Arcueil, A View of French Science at the Time of Napoleon I*, London 1967
——, *Gay-Lussac. Scientist and Bourgeois*, Cambridge 1978
Cumming, James, 'The Effects of the Galvanic Fluid on the Magnetic Needle', *Transactions of the Cambridge Philosophical Society* 02.04.21
Dahl, Frantz, *Frederik VI og Anders Sandøe Ørsted i 1826*, Copenhagen 1929
Dahl, Johan Christian, 1788–1857, *Jubileumsutstilling 1988, Nasjonalgalleriet*, Oslo 1988
Dahl, Per F., *Ludvig Colding and the Conservation of Energy Principle. Experimental and Philosophical Contributions*, N.Y. 1972
Dansk Biografisk Leksikon
Dansk Litteraturhistorie, vol.ii, 'Fra Oehlenschläger til Kierkegaard' by Gustav Albeck et al., Copenhagen 1965
Danske politiske breve fra 1830'rne og 1840'rne, ed. by Povl Bagge, Povl Engelstoft and Johs. Lomholt-Thomsen, vol.i–iv, Copenhagen 1945–58
Daston, Lorraine, 'Ørsted and the Rational Unconscious', 'Hans Christian Ørsted and the Romantic Legacy in Science. Ideas, Disciplines, Practices', ed. by Robert M. Brain, Robert S. Cohen and Ole Knudsen, *Boston Studies in the Philosophy of Science, 241*, Dordrecht 2007
Davies, Martin L., *Identity or History? Marcus Herz and the End of the Enlightenment*, Wayne State University Press, s.a.
Degn, Christian, *Schleswig-Holstein—eine Landesgeschichte*, Neumünster 1995

Dibner, Bern, *Oersted and the Discovery of Electromagnetism*, Burndy Library, Norwalk, Connecticut 1961.

Dictionnaire des sciences naturelles, vol.x, 1818

Dictionary of National Biography

Duncan, Howard, 'Kant's Methodology', 'Kant's philosophy of physical science: Metaphysische Anfangsgründe der Naturwissenschaft 1786–1986, *The University of Western Ontario Series in Philosophy of Science*, vol.33, Dordrecht 1986

Dyssel, J.A., *Nogle Bidrag til et Skilderie af Justizen og de Undergivnes Behandling ved Hs. Majestæt Kongens Livkorps, og et lidet Tillæg af Engelbreth*, Copenhagen 1820

Efterladte Papirer fra den Reventlowske Familiekreds, vol.v, ed. by Louis Bobé, Copenhagen 1902

Elers Collegium 1691–1941, ed. by Elersianersamfundet, Copenhagen 1942

Engberg, Jens, *Dansk Guldalder eller oprøret i Tugt,- Rasp- og Forbedringshuset*, Copenhagen 1973

Engelstoft, Lauritz, *Udvalg af Laurids Engelstofts Skrifter*, ed. by C.F. Allen, vol.iii, Copenhagen 1862

Erdmann, O.L., 'Philosophie. Empirische und speculative Naturbetrachtung', *Allgemeine Monatsschrift für Litteratur*, vol.i, 1850

Erindringsblad til de ved Selskabet for Naturlærens Udbredelse foranstaltede Foreviisninger af physiske og chemiske Forsøg, i Aaret 1840

Fabricius, I.C., *Über Academien insonderheit in Dännemark*, Copenhagen 1796

Fang, Fanny, 'Knudsens Gaard—Kroghs Gaard', *Fra Kjøbenhavns Amt*, 1932–33

Faraday, Michael, 'New electro-magnetic apparatus', *Quarterly Journal of Science*, 12, 1821, 186–7

Fenger, C.E., 'Den numeriske Methode', *Ugeskrift for Læger*, 1839

Fichte, J.G., 'Versuch einer Kritik aller Offenbarung' (1792), *Sämmtliche Werke*, vol.v, ed. by J.H. Fichte (1845, reprint) Berlin 1965

——, 'Über die Bestimmung der Menschen' (1795), *Sämmtliche Werke*, vol.i, ed. by J.H. Fichte (1845, reprint) Berlin 1965

——, 'Appellation an das Publikum über die durch ein Kurfürstl. Sächs. Konfiskationsreskript ihm beigemessenen atheistischen Äusserungen. Eine Schrift, die man zu lesen bittet, ehe man sie confiscirt' (1799), *Sämmtliche Werke*, vol.i, ed. by J.H. Fichte (1845, reprint) Berlin 1965

——, 'Über die Bestimmung der Gelehrten' (1800), *Sämmtliche Werke*, vol.i, ed. by J.H. Fichte (1845, reprint) Berlin 1965

——, 'Sonnenklarer Bericht an das grössere Publikum, über das eigentliche Wesen der neuesten Philosophie. Ein Versuch, die Leser zum Verstehen zu zwingen' (1801), *Sämmtliche Werke*, vol.ii, ed. by J.H. Fichte (1845, reprint) Berlin 1965

——, 'Über das Wesen des Gelehrten und seine Erscheinungen im Gebiete der Freiheit' (1805), *Sämmtliche Werke*, vol.vi, ed. by J.H. Fichte, Berlin 1845

——, 'Die Anweisung zum seligen Leben, oder auch die Religionslehre' (1806), *Sämmtliche Werke*, vol.v, ed. by J.H. Fichte, Berlin 1845

——, 'Die Grundzüge des gegenwärtigen Zeitalters' (1806), i *Sämmtliche Werke*, vol.vii, ed. by J.H. Fichte, Berlin 1846

Fichte, J.G., Briefwechsel, *Kritische Gesamtausgabe*, vol.ii, ed. by Hans Schulz, 1930

Fischer, Kuno, *Geschichte der neuern Philosophie*, vol.vi, 'Fichtes Leben, Werke und Lehre', Heidelberg 1923

——, *Geschichte der neuern Philosophie*, vol.vii, 'Schellings Leben, Werke und Lehre', Heidelberg 1923

Fogh, I., 'Über die Entdeckung des Aluminiums durch Oersted im Jahre 1825', *Matematisk-fysiske Meddelelser 3*, KDVS, Copenhagen 1921

Fonden ad usus Publicos. Aktmæssige Bidrag til Belysning af den Virksomhed, ed. by RA, vol.ii, 1801–1826, Copenhagen 1902

For the Friends of Nature and Art. The Garden Kingdom of Prince Franz von Anhalt-Dessau in Age of Enlightenment, ed. by Kulturstiftung Dessau, Wörlitz 1997

Forchhammer, J.G., *Rejse til Færøerne 1821*, ed. by Ad. Clément, Copenhagen 1927

Forhandlinger ved de Skandinaviske Naturforskeres fjerde Møde i Christiania den 11–18 Juli 1844, Christiania 1847

Forhandlinger ved de skandinaviske Naturforskeres andet Møde, der holdtes i Kjøbenhavn fra den 3die til den 9de Juli 1840, Copenhagen 1841

Forhandlingerne ved Studentersamfundets Stiftelse, ed. by M. Hammerich og A.Fr. Krieger, Copenhagen 1840

Fortegnelse over H.C. Ørsteds efterladte Bogsamling nr.4126, Copenhagen 1853

Fox, Robert, 'Laplacian Physics', *Companion to the History of Modern Science*, ed. by R.C. Olby, G.N. Cantor, J.R.R. Christie, and M.J.S. Hodge, London/N.Y. 1996, 278–94

—— 'The rise and fall of Laplacian physics', *Historical studies in the Physical Sciences*, 4, 1974

Fra, J.L. 'Heibergs Ungdom', *Memoirer og Breve xxxvii*, ed. by Julius Clausen and P.Fr. Rist, Copenhagen 1972

Franks, Paul 'All or nothing: systematicity and nihilism in Jacobi, Reinhold, and Maimon', *The Cambridge Companion to German Idealism*, ed. by Karl Ameriks, Cambridge 2000

Friedman, Michael, 'Metaphysical Foundations of Newtonian Science', 'Kant's philosophy of physical science: Metaphysische Anfangsgründe der Naturwissenschaft 1786–1986, *The University of Western Ontario Series in Philosophy of Science*, vol.33, Dordrecht 1986

——, *Kant and the Exact Sciences*, Cambridge, Mass. 1992

——, 'Kant—Naturphilosophie—Electromagnetism', 'Hans Christian Ørsted and the Romantic Legacy in Science. Ideas, Disciplines, Practices, ed. by Robert M. Brain, Robert S. Cohen and Ole Knudsen, *Boston Studies in the Philosophy of Science* 241, Dordrecht 2007, 135–58

Frauenstädt, I, 'Philosophie. Ueber das Verhältniß der Physik zur Metaphysik', *Allgemeine Monatschrift für Litteratur*, 2, 1850

Förhandlingar vid det af skandinaviska naturforskare och läkare hållna möte i Götheborg år 1839, Gothenburg 1840

Förhandlingar vid de Skandinaviska Naturforskarnes tredje Möte i Stockholm den 13–19 Juli 1842, Stockholm s.a.

Gad, K.H., *Hvor skal jeg sætte min Søn i Skole, naar han ikke skal studere? Hvor kan den kommende Købmand, Fabrikant eller dannede Haandværker erholde den første Undervisning?*, Copenhagen 1833

Galvani, Luigi, *De viribus electricitatis in motu musculari commentarius*, (disp.), Bologna 1791, vol. iv, *Commentary on the Effects of Electricity on Muscular Motion* transl. by Margaret Glover Foley, Notes and a Critical Introduction by I. Bernard Cohen, Norwalk, Burndy Library 1953

Gamborg, A., 'Forslag til at forbedre Fuglenes Sang i vore Skove', *Scandinavisk Museum*, 1, Copenhagen 1800

Garboe, Axel, *Geologiens Historie i Danmark. Forskere og Resultater*, vol.i–ii, Copenhagen 1961

Giessing, H.P., *Kong Christian den Ottendes Regjeringshistorie*, Copenhagen 1852

Gillespie, Charles C., *Science and polity in France at the end of the old regime*, Princeton, N.J. 1980

Giversen, Søren (udg. og overs.), *Den ukendte Gud. Hermes-skrifterne*, Copenhagen 1983

Goethe, J.W., 'Dichtung und Wahrheit', *Sämtliche Werke, Briefe, Tagebücher und Gespräche*, 1. Abteilung, 14, Berlin 1986

——, *Goethes Werke*, ed. by Sophie von Sachsen, 1. Abteilung, vol.i, 86–7, Tokyo 1975 ('Trost in Tränen')
——, *Goethes Werke*, ed. by Sophie von Sachsen, 2. Abteilung, Autobiographische Schriften, vol.x,
——, *Goethes Werke*, ed. by Sophie von Sachsen, 3. Abteilung, vol.viii, 272–3, Tokyo 1975 (HCØ and Bindesbøll)
——, *Goethes Werke*, ed. by Sophie von Sachsen, Hamburg 1972
——, *Goethes Briefe*, 4. Abteilung, vol.33, Weimar 1905
——, *Goethes Briefe*, 4. Abteilung, vol.36, Weimar 1907
Goldschmidt, A.M., *Nord og Syd*, vol.v, Copenhagen 1850
Gosch, C.C.A., *Udsigt over Danmarks zoologiske Literatur*, 2nd Series, vol.i, Copenhagen 1973
Gould, Stephen J., *Rocks of Ages. Science and Religion in the Fullness of Life*, N.Y. 1999
Gower, Barry, 'Speculation in Physics: The History and Practice of *Naturphilosophie*', *Studies in the History and Philosophy of Science 3*, 1973, 301–56
Gren, F.A.C., *Grundriß der Chemie*, Halle 1796
Gren, F.A.C., *Grundriß der Naturlehre*, 3.ed., Halle 1797
Grum-Schwensen, Ane, 'H.C.Ørsted og digtekunsten', i *Krydsfelt. Kunst og Videnskab i Guldalderen*, ed. by Mogens Bencard, Copenhagen 2000
——, 'H.C. Andersen og familien Ørsted', *Fynske Minder*, 2001
Grundtvig og Ingemann. Brevvexling 1821–1859, ed. by Svend Grundtvig, Copenhagen 1882
Grundtvig, N.F.S., *Kort Begreb af Verdens Krønike i Sammenhæng*, Copenhagen 1812
——, *Imod den lille Anklager, det er Prof. H.C. Ørsted, med Beviis for at Schellings Philosophie er uchristelig, ugudelig og løgnagtig*, Copenhagen 1815
——, *Værker i Udvalg 1–10*, ed. by Georg Christensen and Hal Koch, Copenhagen 1941
——, *Krønikens Gienmæle*, Copenhagen 1813
——, *Mands Minde 1788–1838. Foredrag over det sidste halve Aarhundredes Historie holdte 1838*, ed. by Svend Grundtvig, Copenhagen 1877
——, *To Dialoger om Højskolen*, ed. by DCC, Kvanløse 1983
——, *Statsmæssig Oplysning. Et udkast om samfund og skole*, ed. by K.E. Bugge and Vilhelm Nielsen, Copenhagen 1983
Guyer, Paul, 'Introduction: The starry heavens and the moral law', *The Cambridge Companion to Kant*, ed. by Paul Guyer, Cambridge 1999
——, 'The Unity of Nature and Freedom: Kant's Conception of the System of Philosophy', *The Reception of Kant's Critical Philosophy. Fichte, Schelling, and Hegel*, ed. by Sally Sedgwick, Cambridge 2000
——, 'Natural Ends and the End of Nature', 'Hans Christian Ørsted and the Romantic Legacy in Science. Ideas, Disciplines, Practices', ed. by Robert M. Brain, Robert S. Cohen and Ole Knudsen, *Boston Studies in The Philosophy of Science*, 241, Dordrecht 2007, 75–96
Hall, A. Rupert, *The Cambridge Philosophical Society. A History 1819–1969*, Cambridge 1969
Hansen, Arnold, 'Da Apoteker Søren Christian Ørsted forsøgte at oprette et Laboratorium', *Archiv for pharmaci og chemi*, vol.li, 1944, 267–77
Hanslick, Eduard, *The Beautiful in Music. A Contribution to the Revisal of Musical Aesthetics*, (transl. from 7th ed., Leipzig 1885) N.Y. 1974
Hansteen, Christopher, *Til Fots til Bergen Anno 1821. Av Christopher Hansteens Erindringer*, Oslo 1969
Harding, M.C., *Correspondance de H.C. Ørsted avec divers savants*, vol.i–ii, Copenhagen 1920
——, *Selskabet for Naturlærens Udbredelse. H.C. Ørsteds Virksomhed i Selskabet og dettes Historie gennem hundrede Aar*, Copenhagen 1924
Hauch, A.W., *Begyndelsesgrunde til Naturlæren*, vol.i–ii, Copenhagen 1794, 2nd ed. 1798

———, *Det Physiske Cabinet eller Beskrivelse over de til Eksperimental-Physiken henhørende vigtigste Instrumenter tillligemed Brugen deraf*, vol.ii, Copenhagen 1836

Hauch, Carsten, 'Hans Christian Ørsteds Levnet', 1852, SES 9, 109–83

———, *Minder fra min Barndom og min Ungdom*, Copenhagen 1867

Haward, Birkin, *Oxford University Museum. Its architecture and art*, Oxford 1991

Heiberg, J.L., *De poeseos dramaticæ genere hispanico, et præsertim de Petro Calderone de la Barca principe dramaticorum*, Copenhagen 1817

Hein, Piet, 'Teknoti og Kultisme', *Menneskesag. En Piet Hein antologi udg. i anledning af hans 70-års dag*, Bondes Bogtryk, 1975

Henriksen, Aage, *Den Rejsende. Otte kapitler om Baggesen og hans tid*, Copenhagen 1961

Herholdt, J.D., *Udtog af Professor Herholdts Dagbøger over Rachel Hertz' Sygdomme i Aarene 1807–26*, Copenhagen 1826

Herschel, John, *Preliminary discourse on the Study of Natural Philosophy*, London 1831

Hillairet, Jacques, *Dictionnaire historique des rues de Paris*, vol.ii, Paris 1963–72

Histoire et dictionnaire de la Révolution française 1789–1799, ed. by J. Tulard, J.-F. Fayard & A. Fierro, Paris 1987

Hofmann, James R. *André-Marie Ampère*, Cambridge 1996

Holmes, Richard, *Coleridge. Early Visions*, London 1999

Hornemann, C., *Christian Hornemanns efterladte philosophiske Skrifter*, Copenhagen 1795

Howitz, F.G., *Om Afsindighed og Tilregnelse, et Bidrag til Phychologien og Retslæren*, Copenhagen 1824

Hufbauer, Karl, *The Formation of the German Chemical Community (1720–1795)*, Berkeley/Los Angeles/London 1982

Hultberg, Helge, 'Den unge Henrich Steffens 1773–1811', *Københavns Universitets festskrift*, Copenhagen 1973

Humboldt, Alexander von, *Versuch über die gereizte Muskel- und Nervefaser*, vol.ii, Berlin 1797, incl. J.W. Ritters 'Bemerkungen und Ergänzungen', jvfr. Kurt Poppe, 'Über das wissenschaftliche Leben Johann Wilhelm Ritters', J.W. Ritter, *Fragmente aus dem Nachlass eines jungen Physikers* (selection) Stuttgart 1968

———, *Rede gehalten bei der Eröffnung der Versammlung deutscher Naturforscher und Ärtzte in Berlin am 18ten September 1828*, Berlin 1828

———, *Kosmos. Entwurf einer physischen Weltbeschreibung*, vol.i–iv, Berlin 1845–58

Høffding, Harald, *Danske Filosofer*, Copenhagen 1909

Hørby, Kai, 'Grundtvigs højskoletanke og Sorø Akademis reform 1842–1849', *Årbog for dansk skolehistorie*, 1967, 59–84

Jacobsen, Anja Skaar, 'A.W. Hauch's Role in the Introduction of Antiphlogistic Chemistry into Denmark', *Ambix* 47/2, 2000

———, *Between Naturphilosophie and Tradition. Hans Christian Ørsted's Dynamical Chemistry*, (Ph.D.thesis) Aarhus 2000

———, 'Spirit and Unity: Ørsted's Fascination by Winterl's Chemistry', *Centaurus*, 43, 2001

Jensen, Hans, *De danske Stænderforsamlingers Historie 1830–1848*, vol.ii, Copenhagen 1934

———, 'Vi alene vide-. Et kildekritisk Bidrag til Vurdering af dansk Enevælde', *Historisk Tidsskrift*, 10th Series, vol.iv, 1937–38

Jensen, J.E., *Historiske huse og gårde i Rudkøbing*, Nationalmuseet, Copenhagen 1988

Jensen, Magnus, *Norges historie. Unionstiden 1814–1905*, Oslo/Bergen 1963

Jensen, Povl Johs., 'Madvig som filolog', *Johan Nicolai Madvig—et Mindeskrift*, vol.ii, Copenhagen 1963

Johnston, James F.W., 'Scientific Men and Institutions in Copenhagen', *Edinburgh Journal of Science, New Series*, vol.iii, Edinburgh 1830

Jørgensen, Dorthe, *Skønhedens metamorfose. De æstetiske idéers historie*, Copenhagen 2003

Jørgensen, Frank, 'Københavnske foreninger 1820 til 1848', *Historiske Meddelelser om København*, 4th Series, vol.v, 1957–59

Jørgensen, H., *Trykkefrihedsspørgsmaalet i Danmark 1799–1848*, Copenhagen 1944

Jørgensen, S. *Efterretninger om Rudkiøbing Kiøbstæds nuværende Tilstand med Hensyn til min Embedsforvaltning sammesteds*, Odense, 1796

Jørgensen, Tormod, *Anton Frederik Tscherning*, Copenhagen 1938

Kant, I., 'Beantwortung der Frage: Was ist Aufklärung?', *Kants gesammelte Schriften*, ed. by Die Königlich Preussische Akademie der Wissenschaften, vol.v, Berlin 1928

——, 'Grundlegung zur Metaphysik der Sitten', *Kants gesammelte Schriften*, ed. by Die Königlich Preussische Akademie der Wissenschaften, vol.iv, Berlin 1911

——, 'Idee zu einer allgemeinen Geschichte in weltbürgerlicher Absicht' (1784), *Kants gesammelte Schriften*, ed. by Die Königlich Preussischen Akademie der Wissenschaften, vol.viii, Berlin 1917

——, 'Kritik der reinen Vernunft', 2.udg. (1787), *Kants gesammelte Schriften*, ed. by Die Königlich Preussische Akademie der Wissenschaften, vol.iii, Berlin 1911

——, 'Kritik der praktischen Vernunft', *Kants gesammelte Schriften*, ed. by Die Königlich Preussische Akademie der Wissenschaften, vol.v, Berlin 1913

——, 'Kritik der Urtheilskraft', *Kants gesammelte Schriften*, ed. by Die Königlich Preussischen Akademie der Wissenschaften, vol.v, Berlin 1913

——, 'Die Metaphysik der Sitten', 1.del, 'Metaphysische Anfangsgründe der Rechtslehre' og 2.del 'Metaphysische Anfangsgründe der Tugendlehre', *Kants gesammelte Schriften*, ed. by Die Königlich Preussische Akademie der Wissenschaften, vol.vi, Berlin 1907

——, 'Metaphysische Anfangsgründe der Naturwissenschaft', *Kants gesammelte Schriften*, ed. by Die Königlich Preussischen Akademie der Wissenschaften, vol. iv, Berlin 1911

——, 'Der Streit der Fakultäten' (1798), *Kants gesammelte Schriften*, ed. by Die Königlich Preussische Akademie der Wissenschaften, vol.vii, Berlin 1917

——, 'Über die Macht des Gemüths durch den bloszen Vorsatz seiner krankhaften Gefühle Meister zu werden', *Kants gesammelte Schriften*, ed. by Die Königlich Preussische Akademie der Wissenschaften, vol.viii, Berlin 1917

Krarup, Per, 'Forholdet til Skolen', *Johan Nicolai Madvig—et Mindeskrift*, vol.i, 218–19

Keen, Robin, *The Life and Work of Friedrich Wöhler (1800–1882)*, ed. by Johannes Büttner, Nordhausen 2005

Kierkegaard, Søren, *Samlede Værker*, 2nd ed., Copenhagen 1920

Af Søren Kierkegaards Efterladte Papirer, ed. by H.P. Barfod og H. Gottsched, Copenhagen 1869

Kitcher, Philip, 'Projecting the Order of Nature', 'Kant's philosophy of physical science: Metaphysische Anfangsgründe der Naturwissenschaft 1786–1986, *The University of Western Ontario Series in Philosophy of Science*, vol.33, Dordrecht 1986

Kjølsen, Hans H., *Fra Skidenstræde til H.C. Ørsted Instituttet*, Copenhagen 1965

Klinckowstroem, Graf Carl von, 'Goethe und Ritter, Mitteilungen aus dem Goethe- und Schiller-Archiv', *Jahrbuch der Goethegesellschaft* 8, 1921, 135–51

Knight, David, *The Age of Science. The Scientific World-view in the Nineteenth Century*, London 1986

—— 'The Spiritual in the Material', 'Hans Christian Ørsted and the Romantic Legacy in Science. Ideas, Disciplines, Practices', ed. by Robert M. Brain, Robert S. Cohen og Ole Knudsen, *Boston Studies in the Philosophy of Science* 241, Dordrecht 2007, 417–32

Knudsen, Ole, 'Fysikeren', *Hans Christian Ørsted*, ed. by F.J. Billeskov Jansen, Egill Snorrason og Chr. Laurtiz-Jensen, Isefjordsværket 1987

———, 'Elektromagnetisme 1820-1900', *Studier i Elektromagnetismens historie* (doctoral thesis) Aarhus 1989

———, 'Ørsted's work on the compressibility of liquids and gases, and his dynamic theory of matter', 'Hans Christian Ørsted and the Romantic Legacy in Science. Ideas, Disciplines, Practices', ed. by Robert M. Brain, Robert S. Cohen og Ole Knudsen, Boston Studies in the Philosophy of Science 241, Dordrecht 2007, 387-98

Koch, Hal, 'Introduction to N.F.S. Grundtvig', *Værker i Udvalg*, ed. by Georg Christensen and Hal Koch, vol.ii, xxi-xxiii, and vol.iii, vii-viii, Copenhagen 1941-42

———, *Den danske kirkes historie, 1800-1848*, ed. by Hal Koch and Bjørn Kornerup, vol.vi, Copenhagen 1954

Koch, L., *Anders Sandøe Ørsted*, Copenhagen 1896

Korsgaard, Ove & Lars Løvlie, 'Indledning' til *Dannelsens Forvandlinger*, ed. by Rune Slagstad, Ove Korsgaard and Lars Løvlie, Oslo 2003, 9-36

Kragh, Helge & Hans Jørgen Styhr Petersen, *En nyttig videnskab. Episoder fra den tekniske kemis historie i Danmark*, Copenhagen 1995

Kratzenstein, C.G., *Vorlesungen über die Experimentalphysik*, 6th ed., Copenhagen 1787

Kuhn, Thomas S., *The Structure of Scientific Revolutions*, 2nd ed., Chicago 1970

Kühle, Sejer, *Frederik Paludan-Müller*, part 1, Copenhagen 1941

Københavns Universitet 1479-1979, Almindelig historie 1788-1936, vol.ii, ed. by Leif Grane and Kai Hørby, Copenhagen 1993

Körner, S., *Kant*, London 1987

Larsen, Martin, 'H.C. Andersen og Grímur Thomsen', i *Anderseniana* 1956, 2nd half-vol., Copenhagen 1956

Lehmanns, Orla, *Efterladte Skrifter*, vol.i, ed. by Hother Hage, Copenhagen 1872

Levande vatn—fossen som natur og symbol, ed. by M. Malmanger et al., Baroniet Rosendal 1999

Levin, Poul, 'H.C. Andersen og Naturen', *Egne og Stæder. Æstetiske Studier*, Copenhagen 1899, 131-52

Lichtenberg, G.C., *Über eine neue Methode, die Natur und die Bewegung der elektrischen Materie zu erforschen*, Göttingen 1777

Lomholt, Asger, *Det kongelige danske Videnskabernes Selskab 1742-1942*, vol.ii, Copenhagen 1950

Lund, H.C.A., *Studenterforeningens Historie 1820-70*, Copenhagen 1896

Lund, Jørn, 'Sprogmanden', *Hans Christian Ørsted*, ed. by F.J. Billeskov Jansen, Egill Snorrason, Chr. Lauritz-Jensen, Isefjordsværket 1987

Lundbye, J.T., *Den polytekniske Læreanstalt 1829-1929*, Copenhagen 1929

Lundgreen-Nielsen, Flemming, 'Grundtvigs københavnske aftenskole. Omkring den folkelige forening "Danske Samfund" 1839-43', *Guldalderhistorier*, ed. by Bente Scavenius, Copenhagen 1994, 100-9

Lühe, Irmela von der, *Natur und Nachahmung. Untersuchungen zur Batteux-Rezeption in Deutschland*, Bonn 1979

Macfie, A.L., *The Crown & Anchor Tavern. The Birthplace of Birkbeck College*, 1973

MacLeod, Roy, M., 'Whigs and savants: reflections on the reform movement in the Royal Society, 1830-48', *Metropolis and Province. Science in British culture 1780-1850*, ed. by Ian Inkster and Jack Morell, London 1983

Malmanger, Magne, 'På sporet av den norske natur', *Kistefos-Museet*, ed. by Nina Sørlie, Oslo 2000

Marstrand, Vilhelm, 'Ingeniøren og Fysikeren Ludvig August Colding: Mindeskrift i Anledning af Den polytekniske Læreanstalts Hundredaarsfest', *Ingeniørvidenskabelige Skrifter A, 20*, Copenhagen 1929

Martensen, H.L., *Sendschreiben an den Herrn Oberconsistorialrath Nielsen in Schleswig: Ein Wort über den Amtseid und die schleswig-holsteinische Geistlichkeit*, Copenhagen 1850

Mattheson, Johann, *Der vollkommene Capellmeister: Studienausgabe im Neusatz des Textes und der Noten*, ed. by Friederike Ramm, Bärenreiter 1999

Maurice, Florian, *Freimaurerei um 1800. Ignaz Aurelius Feßler und die Reform der Großloge Royal York in Berlin*, Tübingen 1997

Mendel, M.H. (udg.), *D.Johann Christian Oersteds Ideen zu einer neuen Architektonik der Naturmetaphysik nebst Bemerkungen über einzelne Theile derselben*, Berlin 1802

Meyer, C., '"Genopdukket" kisteplade fra 1700-tallet, oprindeligt hjemmehørende i Sct.Peders kirke i Slagelse', *Årbog for Historisk Samfund for Sorø Amt*, 1989

Meyer, Kirstine, 'The Scientific Life and Works of H.C. Ørsted', NS I, Copenhagen 1920, xiii–clxvi

——, 'H.C. Ørsteds Arbejdsliv i det danske Samfund', NS III, Copenhagen 1920, xi–clxvi

——, *Dansk Maal og Vægt fra Ole Rømers Tid til Meterloven*, Copenhagen 1912

Meyer-Abich, Adolf, *Alexander von Humboldt*, Hamburg 1967

Michelsen, Knud, *Synålejomfruen og lægevidenskabens menneskeopfattelse*, Copenhagen 1989

Michelsen, William, 'H.C. Ørsteds Stilling i dansk Aandsliv' (prize essay in Nordic philology, stencil) 1940

Mieck, Ilja, 'Sigismund Friedrich Hermbstaedt (1760 bis 1833) Chemiker und Technologe in Berlin', *Technikgeschichte*, 32, 1965

Miss, Stig, 'En billedhugger under uddannelse', *1996 København som kulturby*, Thorvaldsens Museum, Copenhagen 1996

Molbech, Christian og Nikolai Grundtvig Severin Grundtvig. En Brevvexling, samlet af Chr. K.F. Molbech og ed. by L. Schrøder, Copenhagen 1888

Molbech, Christian, *Det Kongelige Danske Videnskabernes Velskabs Historie i dets første Aarhundrede 1742–1842. Udarbeidet efter Kilderne*, Copenhagen 1843

Munch-Petersen, Thomas, *Defying Napoleon, how Britain Bombarded Copenhagen and Seized the Danish Fleet in 2007*, Stroud, Gloucestershire, 2007

Music and Aesthetics in the Eighteenth and Early-Nineteenth Centuries, ed. by Peter le Huray and James Day, Cambridge 1981

Mynster, J.P., 'Udvikling af Begrebet Tro', *KDVSs philosophiske og historiske Afhandlinger*, vol.i, Copenhagen s.a.

—— 'Bemærkninger ved Skriftet "Aanden i Naturen", første og anden Deel', *Nyt Theologisk Tidsskrift*, Copenhagen 1850

——, *Meddelelser om mit Levnet*, Copenhagen 1854

Mynster, Ole H., 'Estne Electricitat signum destinctivum vitæ et mortis animalis?', *Physicalsk, oeconomisk og medicochirurgisk Bibliothek for Danmark og Norge*, vol.iii, 1794

Newman, John, *Somerset House. Splendour and Order*, London 1990

Newton, Isaac, *Opticks or A Treatise of the Reflections, Refractions, Inflections & Colours of Light*, (4th ed. London, 1720), N.Y. 1979

——, *Philosophia Naturalis Principia Matematica*, London 1687

Nielsen, Axel V., *Nordisk Astronomisk Tidsskrift*, 1949, 8–21, 41–52 and 1950, 91–5

Nielsen, H. Toftlund, 'H.C. Ørsteds "Chemie"', *Dansk Kemi*, 8:3, 2000, 27–30

Nielsen, Helge, 'Det classenske Biblioteks karakter og benyttelse', *Ugeskrift for Jordbrug* 41, 1983

Nielsen, K. and Andersen, H., 'The Influence of Kant's Philosophy on the Young H.C. Ørsted', 'Hans Christian Ørsted and the Romantic Legacy in Science. Ideas, Disciplines, Practices', ed. by Robert

M. Brain, Robert S. Cohen and Ole Knudsen, *Boston Studies in The Philosophy of Science*, 241, Dordrecht 2007, 97–114

Nielsen, Sigurd, 'Selskabet for Trykkefrihedens rette Brug', *Historiske Meddelelser om København*, 4th Series, vol.iv, Copenhagen 1954–57

Nissen, Gunhild, 'Fra dannelsesdiskussionen i 1830ernes og 1840ernes skoledebat', *Årbog for dansk skolehistorie*, 1968, 49–67

Nogle Blade af J.P. Mynsters Liv og Tid, ed. by C.L.N. Mynster, Copenhagen 1875

Norrie, Gordon, *Af Medicinsk Facultets Historie*, vol.iii, Copenhagen 1939

Nyrop, C., *Industriforeningen i Kjøbenhavn 1838–1888. En historisk Oversigt*, Copenhagen 1888

Nyström, Ejler, *Offentlige Forlystelser i Frederik den Sjettes Tid*, Copenhagen 1913

Nørregaard, L., *Natur-Rettens første Grunde*, Copenhagen 1784

O'Dwyer, Frederick, *The Architecture of Deane and Woodward*, Cork University Press, s.a.

Ostwald, Wilhelm, *Electrochemistry: History and Theory*, Leipzig 1895

Ostwald, Wilhelm, *Vorträge und Reden*, Leipzig, 1904

Otto, C., *Af mit Liv, min Tid og min Kreds. En autobiografisk Skildring*, Copenhagen 1879

Pais, Abraham, *Niels Bohr og hans tid i fysik, filosofi og samfundet*, Copenhagen 1996

Paludan-Müller, Fr., *Dandserinden—Et Digt i tre Sange*, Copenhagen 1901

Paulinyi, Akos, 'Machine Tools in the Transfer Policy', *European Historiography of Technology. Proceedings from the TISC-Conference in Roskilde*, ed. by DCC, Odense 1993, 24–9

Pera, Marcello, 'Radical Theory Change and Empirical Equivalence, The Galvani-Volta Controversy', *Revolutions in Science, Their Meaning and Relevance*, ed. by William R. Shea, Science History Publications/U.S.A. 1988

Pedersen, Olaf, 'Det længere perspektiv', *Hans Christian Ørsted*, ed. by F.J. Billeskov Jansen, Egill Snorrason, Chr. Lauritz-Jensen, Isefjordsværket 1987, 142–66

Petersen, Richard, *Henrik Steffens. Et Livsbillede*, Copenhagen 1881

Pfaff, C.H., *Brown's System der Heilkunde. Nach der lezteren Englischen Ausg. übersetzt, und mit einer kritischen Abhandl. über die Brownischen Grundsätze begleitet von Chrph. Heinr. Pfaff. Nebst einer tabellarischen Uebersicht des Brownischen Systems von Sam. Lynch*, 1796

Pihl, Mogens, 'Kirstine Meyer', *Fysisk Tidsskrift* 1943, 175–91

Pihl, Mogens, 'H.C. Ørsted. Foredrag holdt på Danmarks Tekniske Museum d. 23. marts 1970', *Fysisk Tidsskrift*, 1970

Poisson, Georges, *Paris au temps de la Revolution 1789–1799*, Paris 1989

Politik und Schule von der Französischen Revolution bis zur Gegenwart, ed. by B. Michael and H.-H. Schepp, vol.i, Frankfurt am Main 1973

Kurt Poppe, 'Johann Wilhelm Ritter und Ernst II., Herzog von Sachsen-Gotha etc. Zwei unbekannte Briefe aus den Jahren 1802–03', *Jahrbuch des Freien Deutschen Hochstifts*, Frankfurt a.M. 1973

Procès-verbaux des séances de l'Académie des Sciences tenues depuis la fondation de l'Institut jusqu'au mois d'août 1835, vol.i–x, Hendaye 1910–22

Radkau, Joachim, *Technik in Deutschland vom 18. Jahrhundert bis zur Gegenwart*, Frankfurt 1989

Randall, John Herman Jr., *The Career of Philosophy*, vol.ii, 'From the German Enlightenment to the Age of Darwin', New York/London 1965

Rawert, O.J., *Kongeriget Danmarks Industrielle Forhold fra de ældste Tider indtil Begyndelsen af 1848*, reprint, 1992

Readings, Bill, *The University in Ruins*, Cambridge, Mass. 1997

Rehm, Else, 'Über den Tod und die letzten Verfügungen des Physikers Johann Wilhelm Ritter', *Jahrbuch des Freien Deutschen Hochstifts*, Frankfurt a. M. 1974, 291–311

Reinhold, Carl Leonh., *Briefe über die Kantische Philosophie*, vol.i–ii, Leipzig 1790–92

Report of the sixteenth Meeting of the British Association for the Advancement of Science, 1846

Reumert, Elith, *Sophie Ørsted*, Copenhagen 1920

Richter, Klaus (ed.), *Der Physiker des Romantikerkreises Johann Wilhelm Ritter in seinen Briefen an den Verleger C.F.E. Frommann*, Weimar 1988

Ritter, J.W., *Beweis, dass ein beständiger Galvanismus den Lebensprozess in Thierreich begleite*, Weimar 1798

Ritter, J.W., *Fragmente aus dem Nachlass eines jungen Physikers. Ein Taschenbuch für Freunde der Natur. Herausgegeben von J.W.Ritter*, ed. by S. & B. Dietzsch, Hanau 1984

Rosenfeld, Léon, 'Oersted's scientific Concepts', 'Selected Papers of Léon Rosenfeld', ed. by Robert S. Cohen og John J. Stachel, *Boston Studies in the Philosophy of Science 21*, Dordrecht 1979

Rousseau, J.J., *Dictionnaire de musique*, Paris 1768

Rubin, Marcus, *Frederik VIs Tid fra Kielerfreden til Kongens Død*, Copenhagen 1895 (reprint 1970)

Rørdam, Holger, *Fra Universitetets Fortid*, Copenhagen 1879

Schaffer, Simon, 'Newtonianism', *Companion to the History of Modern Science*, ed. by R.C. Olby, G.N. Cantor, J.R.R. Christie and M.J.S. Hodge, London/N.Y. 1996, 610–26

Scharling, E.A., *Bidrag til at oplyse de Forhold, under hvilke Chemien har været dyrket i Danmark*, Copenhagen 1857

Schelling, F.W.J., *Ideas for a Philosophy of Nature*. Translated by E. Harris and P. Heath, Cambridge 1988

——, *Philosophische Untersuchungen über das Wesen der menschlichen Freiheit und die damit zusammenhängenden Gegenstände*, Stuttgart 1804

——, *Denkmal der Schrift des Herrn Jacobi von den göttlichen Dingen, und der ihn in derselben gemachten Beschuldigung eines absichtlich täuschenden Lügeredenden Atheismus*, Tübingen 1812

Schiller, Fr., 'The Nature and Value of Universal History: An Inaugural Lecture' [1789], *History & Theory*, vol.xi, 3, 1972, 321–34

——, 'Über die ästhetische Erziehung des Menschen, in einer Reihe von Briefen', *Die Horen* 1795

Schiørring, Niels, *Musikkens Historie i Danmark*, vol.ii, 1750 til 1870, Copenhagen 1978

Schlegel, Fr., 'Charakteristiken und Kritiken II (1802–1829)', *Kritische Friedrich-Schlegel-Ausgabe*, ed. by Ernst Behler et al., vol.iii, Zürich 1975

Schleiermacher, F., *Grundlinien einer Kritik der bisherigen Sittenlehre*, Berlin 1803

Schmidten, H.G. von, 'Om Phrenologien i Anledning af Dr. Ottos phrenologiske Tidsskrift, December 1828', *Maanedskrift for Litteratur*, vol.i, 1829, 105–14

Schrøder, Vibeke, *Tankens Våben. Johan Ludvig Heiberg*, Copenhagen 2001

Schultz, E., *H.C.Ørsted's Slægt*, consultation by Dir. H.C. Ørsted, 1960

Schæffer, Herman, *Et andet Slags Bekjendtgjörelse indeholdende Beviis for at Hr. Professor H.C. Örsted og Justitsraad Esmarch ikke have forstaaet sig paa at bedömme Skipper Coulthards Forslag, betreffende den af ham formeentlige fundne, 50 p'Ct haltig, Jernmalm og Steenkul; … Copenhagen 1822*

Schæffer, Aage, 'Søren Christian Ørsted 1750–1822. Apoteker i Rudkøbing, Sorø, Ringsted, København', *Theriaca. Samlinger til Farmaciens og Medicinens Historie*, 1978

——, 'Om H.C.Ørsteds Uddannelse og Farmaceutiske Eksamen', *Archiv for Pharmaci og Chemi*, 1947

——, 'H.C.Ørsteds farmaceutiske Lærebrev', *Archiv for Pharmaci og Chemi*, 1948

——, 'Nogle bemærkninger om medicinforsyningen på københavnske sygehuse i ældre tid, navnlig vedrørende Almindelig Hospital', *Archiv for pharmaci og chemi*, 1964

Shanahan, Timothy, 'Kant, Naturphilosophie, and Ørsted's Discovery of Electromagnetism: A Reassessment'. *Studies in History and Philosophy of Science* 20, 1989, 287–305
Sibbern, F.C., *Skrivelse til Digteren Ingemann i Anledning af Huldregaverne. Fra en Ven i Kjøbenhavn*, Copenhagen 1831
Skydsgaard, Morten A., *Ole Bang og den sidste hippokratiske Medicin*, (Ph.D. thesis), Copenhagen 2004
——, *Ole Bang og en brydningstid i dansk medicin*, Copenhagen 2006
Smith, S. Birket (udg.), *Til Belysning af litterære Personer og Forhold i Slutningen af det 18de og Begyndelsen af det 19de Aarhundrede. En Samling Breve*, Copenhagen 1884
Snelders, H.A.M., 'Oersted's Discovery of Electromagnetism', *Romanticism and the Sciences*, ed. by A. Cunningham and N. Jardine, Cambridge 1900
Snow, C.P., *Two Cultures*, London 1959
Spang-Hansen, E., 'Madvig og Københavns Universitet', *Johan Nicolai Madvig—et Mindeskrift*, vol.ii, Copenhagen 1963, 219–21 and 231–3.
Stampe, Rigmor, *H. C. Andersen og hans nærmeste omgang*, 1918
Stauffer, R.C., 'Persistent Errors regarding Oersted's Discovery of Electromagnetism', *Isis* 44:4, 1953, 307–10
——, 'Speculation and Experiment in the Background of Oersted's Discovery of Electromagnetism', *Isis* 48:1, 1957, 33–50
Steen, Adolph, *Mathematiken som Læregjenstand ved Høiskoler, især med Hensyn til den polytechniske Læreanstalt og et Misgreb ved samme*, Copenhagen 1847
——, *Den polytekniske Læreanstalts første halvhundrede Aar 1829–1879*, Copenhagen 1879
Steenberg, Jan, *Fredensborg Slot. Monumenter og Minder*, vol.i–ii, Copenhagen 1969–79
Steffens, Henrich, 'Recension der neueren naturphilosophischen Schriften des Herausgebers', *Zeitschrift für spekulative Physik* 1, 1800
——, 'Over Naturphilosophiens Forhold til den empiriske Naturvidenskab', *Forhandlinger ved de skandinaviske Naturforskeres andet Møde, der holdtes i Kjøbenhavn fra den 3die til den 9de Juli 1840*, Copenhagen 1841
——, *Was ich erlebte. Aus der Erinnerung niedergeschrieben*, vol.i–v, 2nd ed., Breslau 1844
Stringari, S. and Wilson, R.R., 'Romagnosi and the Discovery of Electromagnetism', *Renconditi Lincei, Scienze, Fisiche e Naturali*, Serie 9, vol. 11, 2000, 115–36
Strøm, Elin, *Naturhistorieselskabet i København 1789–1804*, Oslo 1841
Sulzer, J.G., *Allgemeine Theorie der schönen Künste, in einzelnen, nach alphabetischer Ordnung der Kunstwörter aufeinanderfolgenden Artikeln abgehandelt*, vol.i–iv, 2nd ed., Leipzig 1792–94
Thalbitzer, Sophie Dorothea, '*Grandmamas Bekiendelser*', *Memoirer og Breve IV*, ed. by J. Clausen and P.Fr. Rist, Copenhagen 1966
Tamny, Martin, 'Atomism and the Mechanical Philosophy', *Companion to the History of Modern Science*, ed. by R.C. Olby, G.N. Cantor, J.R.R. Christie and M.J.S. Hodge, London/N.Y. 1996
Thomsen, Ejnar and Brix, Hans, *Københavns Universitets Festskrift*, Copenhagen 1941
Thomsen, Ejnar, 'Omkring Oehlenschlägers tyske Quijotiade', *Københavns Universitets Festskrift*, Copenhagen 1950
Toulmin, S. & Goodfield, J., *The Architecture of Matter*, London 1962
Tryde, E., 'Hvor er Kjernen af Folket?', *Dansk Folkeblad*, vol.ii, nr. 2–3
Turner, A.J., *From Pleasure and Profit to Science and Security. Etienne Lenoir and the transformation of precision instrument-making in France 1760–1830*, The Whipple Museum of the History of Science, Cambridge 1989

Tybjerg, Hans, *Omkring Købkes Frederiksborg Slot ved aftenbelysning*, Copenhagen 1996

Tychsen, N., 'Mit Levnedsløb eller en kort Fortælling om hvorledes det er gaaet mig indtil jeg kom her paa Biergstaden Kongberg', *Theoretisk og Praktisk Anviisning til Apothekerkunsten*, vol.i, ed. by J.F. Bergsøe, Copenhagen 1804

Tyndall, J., *Sound*, N.Y. 1871

Tweney, Ryan D., 'Stopping Time: Faraday and the Scientific Creation of Perceptual Order', *Physis*, 29,1992, 149–64

Veiviser for Kjøbenhavn, 1818–1819

Vestergaard, R.R., *Bidrag til Elers' Collegiums Historie*, Copenhagen 1891

Villadsen, Villads, 'Rendezvous in Paris', i *Krydsfelt. Kunst og Videnskab i Guldalderen*, ed. by Mogens Bencard, Copenhagen 2000

Volta, Alessandro, 'On the Electricity excited by the mere Contact of Conducting Substances of Different Kinds', in Alessandro Volta, *Le opere*, Milano 1918–1929, vol. i, transl. of J.W. Ritter and edition of *Beyträge zur näheren Kenntnis des Galvanismus und der Resultate seiner Untersuchung*, vol.i, part 3, Jena 1800

Wagner, Michael F., *Det polytekniske Gennembrud. Romantikkens teknologiske konstruktion 1780–1850* (doctoral thesis), Aarhus 1999

Wamberg, Bodil, *Sophies hjerte*, Copenhagen 2001

Vammen, Hans, 'J.F. Schouw som politiker', (prize essay in history, stencil), Copenhagen 1967

——, '"Schouw er velsignet… ." En professor og hans guldaldernetværk', i *Krydsfelt. Ånd og natur i Guldalderen*, ed. by Mogens Bencard, Copenhagen 2000

——, 'Casino 1848', *Historisk Tidsskrift*, vol.88, 2nd. Series, Copenhagen 1988, 253–79

Weischedel, Wolfgang, *Streit um de göttlichen Dinge. Die Auseinandersetzung zwischen Jacobi und Schelling*, Darmstadt 1967

Wergeland, Nicolai, *Mnemosyne. Et Forsøg paa at besvare den af Kongl. Selskab for Norges Vel fremsatte Opgave om et Universitet i Norge*, Christiania (Oslo) 1811

Werlauff, E.C., 'Erindringer af mit Liv', *Memoirer og Breve*, ed. by Jul. Clausen and P.Fr. Rist, Copenhagen 1968

Wetzels, Walter D., *Johann Wilhelm Ritter: Physik im Wirkungsfeld der deutschen Romantik*, Berlin/N.Y. 1973

Wilhelmy-Döllinger, Petra, *Die Berliner Salons*, Berlin/N.Y. 2000

Wilkens, J.F.C.E., 'Den polytekniske Læreanstalts Historie, Læreplan og Virksomhed', *Fædrelandet* Nos. 254, 255, 257, 258, 259, 1847

——, 'Om Embedsmændenes Forpligtelser, et par Ord til Forsvar', *Fædrelandet* 211, 07.09.1847

——, 'Den polytekniske Læreanstalts Historie, Læreplan og Virksomhed', *Fædrelandet* 254, 26.10.1847

Williams, L. Pearce, *Michael Faraday*, N.Y. 1965

——, *The Origins of Field Theory*, N.Y. 1966

——, 'Oersted, H.C.', *Dictionary of Scientific Biography*, ed. by C.C. Gillespie, vol. x, N.Y. 1974, 182–6

——, 'Faraday and Ampère: A Critical Dialogue', i *Faraday Rediscovered. Essays on the Life and Work of Michael Faraday 1791–1867*, ed. by David Gooding and Frank A.J.L. James, 1985

——, 'André-Marie Ampère', *Scientific American*, vol.260, 1, 1989

Wilson, Andrew D., 'Introduction', *Selected Scientific Writings of Hans Christian Ørsted*, transl. and ed. by K. Jelved, A.D. Jackson, and O. Knudsen, Princeton 1998, xv–xl

——, 'The Way from Nature to God', 'Hans Christian Ørsted and the Romantic Legacy in Science. Ideas, Disciplines, Practices', ed. by Robert M. Brain, Robert S. Cohen and Ole Knudsen, *Boston Studies in The Philosophy of Science*, 241, Dordrecht 2007, 1–12

Wolff, Torben, 'Korvetten Galatheas rejse omkring jorden. En Ekspedition med videnskabelige og politisk-økonomiske formål', *Guldalderhistorier*, ed. by Bente Scavenius, Copenhagen 1994, 156–63
Wöhler, F., 'Ueber das Aluminium', *Annalen der Physik und Chemie* 11, 1827, 146–61
Ydstie, Ingebjørg, 'Brusende vannfald og mektige kaskader. Fossen fra typologi til topografi', i *Levende vatn—fossen som natur og symbol*, ed. by Baroniet Rosendal 1999
Zeise, W.C., *De vi corporum alcalinorum materias regno organico peculiares transmutandi, annexis quibusdam de transmutatione chemica in genere*, Copenhagen 1817
——, 'Om det hos os oprettede offentlige chemiske Øvelses-Laboratorium', i *Tidsskrift for Naturvidenskaberne*, ed. by Hornemann, Reinhardt and Ørsted, 1822
Zeuthen, Ludvig, *Mine første 25 Aar (1805–1830)*, Copenhagen 1866
Ziolkowski, Th., *German romanticism and its institutions*, Princeton 1990
Oehlenschlägers Levnet fortalt af ham selv, vol.i, Copenhagen 1974
Oehlenschläger, Adam, *Langelands-Reise i Sommeren 1804*, ed. by Povl Ingerslev-Jensen, Copenhagen 1972
——, Adam Oehlenschläger, 'En Gammel Mand med en Perspektivkasse', *Sanct Hansaften-Spil*, ed. by M. Baumann Larsen, Copenhagen 1964
——, 'Cantate opført i Anledning af den polytechniske Lære-Anstalts Indvielse, ved Kongens høie Nærværelse den 5te November 1829', *Oehlenschlägers Poetiske Skrifter*, ed. by F.L. Liebenberg, vol.xxi, Copenhagen 1860, 162–5
——, *Erindringer*, vol.ii, Copenhagen 1850–51
Ørsted, A.S., *Forsøg til en rigtig Fortolkning og Bedømmelse over Forordningen om Trykkefrihedens Grændser*, part 1, Copenhagen 1801
——, *Af mit Livs og min Tids Historie*, Copenhagen 1951
——, *Over Sammenhængen mellem Dydelærens og Retslærens Princip. Et academisk Prisskrift*, parts i–ii, Copenhagen 1798
——, 'Et Par Ord i Anledning af den foranstaaende Afhandling', *Juridisk Tidsskrift*, vol.viii, Copenhagen 1824
——, 'Behøver den danske Kirkeforfatning en omsiggribende Forandring?', *Juridisk Tidsskrift*, vol.xii, 1825
Ørsted, H.C., 'Forsøg til en Besvarelse af det, for Aaret 1796, ved det kiøbenhavnske Universitet, udsatte Priisspørgsmaal i Æsthetiken: Hvorledes kan det prosaiske Sprog fordærves ved at komme det poetiske for nær? og hvor ere Grændserne mellem det poetiske og det prosaiske Udtryk?', *Minerva* 1797
——, 'Svar, paa det ved Kjøbenhavns Universitet, i Aaret 1797 udsatte medicinske Priisspørgsmaal: om Modervandets Oprindelse og Nytte', *Bibliothek for Physik, Medicin og Oeconomie*, 3, 1798
——, 'Grundtrækkene af Naturmetaphysiken', *Philosophisk Repertorium*, Copenhagen, 1799, NS I 33–78
——, 'Naturphilosophiens første Grunde. I Anledning af Hauchs Naturelære', *Philosophisk Repertorium*, vol.ii, Copenhagen 1799
——, 'Kemiske Breve', 1–2, i *Bibliothek for Physik, Medicin og Oeconomie*, vol.14,'98, and 3–4, vol.xvi, Copenhagen 1799
——, 'Recension over Begyndelsesgrunde til Naturlæren', 2nd part, Copenhagen 1799, *Kjøbenhavnske Lærde Efterretninger* 51, Copenhagen 1799, NS III 33–49
——, *Dissertatio de forma metaphysices elementaris naturæ externæ*, Copenhagen 1799, NS I 79–105
——, 'Forsøg og Bemærkninger over den galvaniske Elektricitet af Doktor H.K. Ørsted', *Nyt Bibliothek for Physik, Medicin og Oeconomie*, vol.i, Copenhagen 1801, NS I 106–9
——, *Materialien zu einer Chemie des neunzehnten Jahrhunderts*, ed. by Dr. Johann Christian Oersted, Erstes Stück, Regensburg 1803, NS I 205–10
——, 'Udtog af et Brev fra Doct.Ørsted, til Professor Manthey', *Nyt Bibliothek for Physik, Medicin og Oeconomie*, 3, Copenhagen 1803, NS III 77

——, 'Uebersicht der neuesten Fortschritte der Physik', *Europa*, vol.i, 1803, NS I 112–31

——, 'Expériences sur un appareil a charger d'électricité par la colonne électrique de Volta, par M.Ritter, a Jéna, présentées a l'Institut National par J.C.Orsted, docteur a l'université de Copenhague', *Journal de physique, de chimie, d'histoire naturelle et des arts*, vol.lvii, Paris 1804, NS I 214–37

——, 'Expériences avec la pile électrique faites par M.Ritter, a Jéna; communiquées par M.Orsted', *Journal de physique, de chimie, d'histoire naturelle et des arts*, vol.lvii, Paris 1804, NS I 237–42

——, 'Expériences sur la magnétisme par M.Ritter, a Jéna, communiquées par Orsted, docteur a l'université de Copenhague, *Journal de physique, de chimie, d'histoire naturelle et des arts*, vol.lvii, Paris 1804, NS I 242–5

——, 'Expériences sur la lumière par M.Ritter, a Jéna, communiquées par Orsted, docteur a l'université de Copenhague, *Journal de physique, de chimie, d'histoire naturelle et des arts*, vol. lvii, Paris 1804, NS I 245–8

——, 'Indbydelse til Physiske og Chemiske Forelæsninger', *Kjøbenhavns Lærde Efterretninger*, 39, 1804, 619–21, NS III 78–9

——, 'Schreiben des Hrn. Dr. Oersted zu Kopenhagen an Hrn I.W. Ritter zu Jena, Chladni's Klangfiguren in elektrischer Hinsicht betreffend', *Voigts Magazin*, vol.ix, 31–2, NS I 261–2

——, 'Ueber die Art, wie sich die Electricität fortpflanzt. (Ein Fragment)', *Gehlens Journal* 6, 1806, 292–302, NS I 267–73

——, 'Lettre de Mr. Orsted, Prof. de phil. a Copenhague, au Prof. Pictet, sur les vibrations sonores', *Bibliothèque Britannique*, 30, 240, 364–72, NS I 262–6

——, 'Forsøg foranlediget af nogle Steder i Winterls Skrifter', *Nyt Bibliothek for Physik, Medicin og Oeconomie*, 1806, 229–52

——, 'Ueber die Art, wie sich die Electricität fortpflanzt', 1806, NS I 267–73

——, 'Ueber die Klangfiguren', *Gehlens Journal* 3, 1807, NS I 343–4

——, Videnskaben om Naturens almindelige Love, vol.i, Copenhagen 1809

——, 'Forsøg over Klangfigurerne', *KDVS Skrifter*, 1810, NS II 11–34

——, 'Første Indledning til den almindelige Naturlære', Copenhagen 1811, NS III 151–90

——, 'H.C. Ørsted's Theory of Force. An unpublished textbook in dynamical chemistry' (1812), ed. and transl. by Anja Skaar Jacobsen, Andrew D. Jackson, Karen Jelved, and Helge Kragh, *Historisk-filosofiske Meddelelser 86, Det Kongelige Danske Videnskabernes Selskab*, Copenhagen 2003

——, *Ansicht der Chemischen Naturgesetze durch die neueren Entdeckungen gewonnen*, Berlin 1812, NS II 35–169

——, *Recherches sur l'identité des forces chimiques et électriques*, Paris 1813

——, 'Forslag til reform af det fysiske studium', NS III 191–200

——, *Tentamen nomenclaturæ chemicæ omnibus linguis scandinavico-germanicis communis, prolusionis loco scripsit M. Johannes Christianus Ørsted, Anniversaria in memoriam reipublicæ sacræ et litterariæ cum universæ tum danicæ nostræ restauratæ celebranda indicit regiæ universitatis hauniensis rector cum senatu academico*, (dvs. Universitetets årsskrift), Copenhagen 1814, NS II 178–205

——, *Imod den store Anklager*, Copenhagen 1814

——, 'Ueber das Gesetz der Elektrischen Anziehung',1814, NS II 178

——, 'Bemerkungen hinsichtlich auf Contactelectricitaet', 1817, NS II 206–11

——, 'Ueber die Zusammendrueckung des Wassers', 1817, NS II 211–12

—— 'Bornholm. Dagbøger ført paa Reiserne i 1818 og 1819', ed. by Ad. Clément, *Muséum de minéralogie et de géologie de l'université de Copenhague, Miscellanées*, 9, Copenhagen 1927

——, 'Beretning om en Undersøgelse over Bornholms Mineralrige, udført 1818 efter kongelig Befaling giennem Rentekammeret af Professor H.C. Ørsted og Justitsraad L. Esmarch', Copenhagen 1819, NS III 201–47

——, *Experimenta circa effectum conflictus electrici in acum magneticam*, Copenhagen 1820

——, *La découverte de l'électromagnétisme faite en 1820 par J.C.Ørsted*, ed. by Absalon Larsen, Copenhagen 1920

——, 'Briefwechsel [m. S. Weiß] über Atomistik und Dynamik' (1828–29), CI

——, 'Thermo-electricity', *The Edinburgh Encyclopædia*, ed. by David Brewster, Edinburgh 1830, NSII 351–98

——, 'Betrachtungen ueber den Electromagnetismus', NS II 223–45

——, 'Ueber den Magnetismus des electrischen Stroms', *Isis—Encyclopädische Zeitung von Oken*, 1829, col. 260–2

——, 'Bemærkninger til den af HCØ opstillede teori om den elektromagnetiske proces, hovedsagelig i henseende til de over samme genstand fremførte synspunkter af Hr. Ampère, Paris.' *Bericht über die Versammlung deutscher Naturforscher und Ärtzte in Berlin* ved die Geschäftsführer A.v. Humboldt und H. Lichtenstein, Berlin 1829

——, 'Ueber die Verschiedenheit des physicalischen Vortrages von dem mathematischen, auch wenn beyde dieselben Wahrheiten darstellen', *Isis—Encyclopädische Zeitung von Oken*, 1831, col.854–7

——, *Erindringsord til Forelæsninger over Lyset*, Copenhagen 1835

——, *Luftskibet*, Copenhagen 1836

——, *Tale holden i de Skandinaviske Naturgranskeres første Møde i Kjøbenhavn*, Copenhagen 1840

——, *Naturlærens mechaniske Deel*, Copenhagen 1844

——, *Naturlehre des Schönen*, oversat af H. Zeise, Hamburg 1845

——, *Tillæg til Naturlærens mechaniske Deel*, Copenhagen 1847

——, *Aanden i Naturen*, Copenhagen 1849

Ørsteds, Hans Christian, Jordefærd, Copenhagen 1851

——, *Samlede og efterladte Skrifter*, vol.i–ix, ed. by Mathilde Ørsted, Copenhagen 1851–52 (SES)

Ørsted, Mathilde, *Breve til og fra H.C. Ørsted*, Copenhagen 1870 (MØ I–II)

H.C. Ørsteds rejsebreve, ed. by K. Jelved and Andrew D. Jackson, Det Kongelige Danske Videnskabernes Selskab, Copenhagen 2011

The Travel Letters of H.C. Ørsted, ed. by K. Jelved and Andrew D. Jackson, The Royal Danish Academy of Sciences and Letters, Copenhagen 2011 (TL)

——, *H.C. Ørsteds Naturvidenskabelige Skrifter I–III*, ed. by Kirstine Meyer, Copenhagen 1920, (NS I–III)

Selected Scientific Works of Hans Christian Ørsted, translated and edited by Karen Jelved, Andrew D. Jackson, and Ole Knudsen, with an introduction by Andrew D. Wilson, Princeton University Press, 1997 (SSW)

Ørsteds, H.C., *Selvbiografi*, ed. by Anja Skaar Jacobsen and Svend Larsen, Aarhus 2002

H.C. Ørsted's Theory of Force. An unpublished textbook in dynamical chemistry, edited and translated by Anja Skaar Jacobsen, Andrew D. Jackson, Karen Jelved, and Helge Kragh, Historisk-filosofiske Meddelelser 86, The Royal Danish Society of Sciences and Letters, Copenhagen 2003

TIDSSKRIFTER (contemporary with HCØ)

Annals of Philosophy, London 1819–20

Berlingske politiske og Avertissements Tidende, 1814, 1847, 1851

Bibliothèque universelle, vol.xiv, 1820

Cancellitidende, 1807
Collegialtidende, 1821, 1835
Dansk Folkeblad 1835–
Dansk Litteratur-Tidende, 1811, 1816, 1820
Dansk Ugeskrift 1832–
Danskeren, ed. by N.F.S. Grundtvig, 1851
Eunomia, eller Samling af Afhandlinger, henhørende til Moralphilosophien, Statsphilosophien, og den Dansk-Norske Lovkyndighed, part 1, 1815
Fædrelandet, 1847, 1850–51
Hesperus. For Fædrelandet og Litteraturen, vol.iii, Copenhagen 1820
Historical Studies in the Physical and Biological Sciences
History of Science 1997
Industriforeningens Tidende, 1838
Intelligenzblatt der Allgemeinen Literaturzeitung, Jena 1800
Isis
Journal de physique, de chemie et d'histoire naturelle, 1806, 1820
Journal für Chemie und Physik, 1820
KDVS Oversigter 1814–51
Kjøbenhavns Flyvende Post, 1835
Kiøbenhavns Universitetsjournal, ed. by Jac. Baden, Copenhagen 1793
Kiøbenhavns Lærde Efterretninger, Copenhagen 1798–99, 1804
Kjøbenhavnsposten 1844, 1845
Magasin for Naturvidenskaberne, vol.iv–v, 1824–25
Magazin for Kunstnere og Haandværkere, 5, 1830
Magazin für den neuesten Zustände der Naturkunde, 3, 1802
Maanedskrift for Litteratur 1829–
Neues allgemeines Journal der Chemie, ed. by A.F. Gehlen, 1806
Nicholson's Journal 1806
Nord og Syd, vol.ii, v
Nye Landoeconomiske Tidender, 1820
Nyeste Skilderie af Kjøbenhavn, 1818–1820
Nyt Bibliothek for Physik, Medicin og Oeconomie, 1801–02, 1806
Nær og Fjern 4, 1875
The Philosophical Magazine, xxiv, 1806
Philosophical Transactions of the Royal Society, London 1800
Philosophisk Repertorium for den nyeste Literatur, 1798–99
Rendiconti Lincei, Scienze, Fisiche e Naturali, 2000
Det skandinaviske Litteraturselskabs Skrifter, 181
Statstidende, 1821
Studies in the History and Philosophy of Science, 1973
Tidsskrift for Naturvidenskaberne, 1822–
Universitets-og Skole-Annaler, ed. by L. Engelstoft, Copenhagen 1806
Zeitschrift für spekulative Physik, 4.hf.
Aarbog for Universitetet og de lærde Skoler, 1807, 1809

INDEX OF NAMES

Aasheim, A.N., Norwegian-Danish professor, 62, 78, 133, 175
Abich, R.A., German physicist, 306
Abildgaard, P.C., Danish veterinary professor, 78, 156
Agardh, K.A., Swedish bishop, 520–1, 532
Aischylos, Greek playwright, 189
Albert, British prince, 14, 597
Aldini, Italian writer, 5
Alembert, J. le Rond d', French mathematician and philosopher, 138
Althaus, German phycisist, 349
Ampère, A.-M., French physicist, 349, 351–3, 377–81, 383, 386–7, 400, 447–9, 451, 557, 565, 602
Andersen, H.C., Danish story teller, 11, 13, 237, 408, 442–3, 465, 472–4, 493, 498, 504, 521–3, 571, 574–5, 577, 588, 624, 629–32, 634–43, 647–9
Arago, D.F.J., French physicist, 349, 351–4, 358, 361, 378–80, 383–4, 386–7, 565
Archimedes, Greek technologist, 453
Aristotle, Greek philosopher, 14, 47, 476

Baader, F., German writer, 174
Babbage, Ch., British mathematician and writer, 393, 395, 400, 590, 594–6
Baden, J., Danish professor, 31, 124
Baggesen, J., Danish writer, 76–7, 122, 124, 135, 204–5, 233–4, 249, 283–4, 662
Ballum, Anne Marie, IBØ's sister, 407, 440–1, 537
Ballum, Bolette Sophie, IBØ's sister, 440
Ballum, Inger Birgitte, see Ørsted, I.B.
Banks, J., British scientist, president of RS, 391, 413, 561
Barfod, Fr., Danish writer, 526, 531
Bardenfleth, Danish scientist, 77
Barlow, P., British physicist, 395, 397
Batsch, German professor, 123
Batteux, Ch., French aestetician, 23, 47–8, 217, 569
Baumgarten, A.G., German aestetician, 48
Beck, J.S., German physicist, 349

Becker, G., Danish chemist and professor, 174, 193
Beethoven, L. van, German composer, 215, 650
Bendavid, L., German scientist, 117, 121, 130
Bendz, W., Danish painter, 578
Benjamin, W., German writer, 191
Bernstorff, A.P., Danish foreign secretary, 331, 488
Bernstorff, J.P.E., his uncle, 488
Berthollet, C.-L., French chemist, 64–5, 138–9, 141, 143–4, 147, 150, 158, 178, 226, 265, 268
Berzelius, J.J., Swedish chemist, 226, 248, 265, 268, 270–1, 314, 317, 339, 349–50, 378, 384, 417, 426, 442, 517, 519–26, 532–3
Beuth, P.C.W. von, German minister, 370, 459, 565–6, 605
Bindesbøll, M.G.B., Danish millwright and architect, 360, 369, 371, 374–6, 388–9, 403, 579–80
Biot, J.-B., French scientist, 139, 147–9, 152–3, 265, 349, 351, 382, 386
Birkbeck, G., British physician, 212, 395
Bissen, H.V., Danish sculptor, 644
Blainville, H.D. de, French physicist, 349, 353–4, 386
Blücher, German general, 541, 565
Blumenbach, J.F., German physiologist, 50, 98
Blunck, D. C., Danish painter, 578, 580
Bohn, ASØ's housekeeper, 442–3
Bohr, N., Danish scientist, 13
Boisgiraud, J.P.T., French scientist, 349, 386
Boisserée, S., German art historian, 372–4, 408, 585
Bonpland, A., French explorer, 561
Bornemann, M.H., Danish jurist, 420
Borring, A.D., see Ørsted, A.D.
Bowring, J., British whig politician, 410
Brahe, Tycho, Danish astronomer, 3, 162, 336, 356, 601
Brandes, G., Danish literary critic, 661–3
Brandis, J.D., German physician, 442, 585
Brecknell, S., British art historian, 12
Bremer, Fr., Swedish writer, 573–75

733

INDEX OF NAMES

Brera, V.L., Italian medical professor, 226
Brewster, David, Scottish physicist, 226, 354, 400
Brongniart, A., French geologist, 426
Brougham, H., British whig politician, 408, 410
Brown, J., Scottish physician, 56–7, 110
Bruun, C. M., Danish geographer, 60
Brünnich, M.T., Danish mining engineer, 35
Buch, C.L. von, German geologist, 349
Bucholz, C.F., German chemist, 190
Bugge, Th., Danish professor, secretary of KDVS, 7, 26, 31, 40, 63, 78–9, 98, 158, 177–8, 181, 190, 206, 290–2, 343, 352, 444–5, 545–6
Bull, G.J., Norwegian jurist, HCØ's brother-in-law, 103, 169–70, 247, 527
Buntzen, Th., Danish physician, 168–9, 175–6
Byron, G.G., Lord, British poet, 287, 471–2
Bøgh, N., Danish writer, 635

Cagliostro, A., Italian count and charlatan, 65
Callisen, H., Danish physician and professor, 98, 206, 262, 416
Caneva, K., American historian of science, 6, 239, 556–7, 654, 659–60
Canova, A., French sculptor, 388
Carl, Prince of Hessen, Danish-Norwegian Freemasonic Grand Master, 254, 364–8, 376
Carl XIV Johan, Count Bernadotte, King of Sweden-Norway, 520, 523, 525–6, 560
Carlisle, A., British physicist, 85–6
Caroline Amalie, Queen of Denmark, 358, 425, 442, 535, 538
Caroline Mathilda, Queen of Denmark-Norway, 202
Carstensen, G., Danish lieutenant, founder of Tivoli and Casino, 529
Carus, C.G., German painter and Naturphilosoph, 446, 568, 578
Cavendish, H., British chemist, 61
Champollion, J.F., French egyptologist, 367
Chaptal, J.A.C., French chemist and Home secretary, 139, 145, 300, 328–9
Charles, J.A.C., French chemist and baloonist, 85, 137, 140–1, 157, 179, 226, 388, 478, 654
Chenevix, R., Irish chemist, 167–8, 190, 398
Chevreul, M.E., French chemist, 271–3, 380, 383, 386

Children, J.G., British electrician, secretary of RS, 308, 395
Chladni, E.F.F., German jurist, physicist, and musician, 188, 207–10, 215, 217, 222, 244, 569
Chrétien, G.L., French physionotracist, 160
Christensen, D.Ch., Danish historian, 654
Christensen, E.M., Danish professor, 662
Christian III, King of Denmark-Norway (1536–59), 247
Christian IV, King of Denmark-Norway (1588–1648), 20
Christian VI, King of Denmark-Norway (1730–46), 247
Christian VII, King of Denmark-Norway (1769–1808), 195, 202
Christian VIII, Prince Christian Frederik, King of Denmark (1839–48), 254, 349, 358–9, 361, 370, 376, 385, 387, 390, 393, 408, 412–3, 425, 460, 475, 502, 508, 515, 521–3, 525, 529, 534–9, 542, 545, 549, 566, 563, 567, 579, 581, 609–10, 620, 622, 646
Classen, J.F., Danish manufacturer, 104, 328
Classen, P.H., director of the Classen Trust, 104, 328
Claudet, A., French-British daguerreotypist, 592
Clausen, H.N., Danish theologian, 31, 224, 431–4, 438, 469, 486, 493, 505, 522, 529, 531–2, 609–10, 643, 649
Colding, L.A., Danish polytechnician, 552–9, 624
Coleridge, S.T., British writer, 272
Collin, J., Danish civil servant, 55, 176, 201, 291, 361, 413–14, 442–3, 456, 462–3, 474, 545, 635
Columbus, C., Italian explorer, 5, 178, 357, 480, 565
Comte, A., French sociologist, 663
Condorcet, M.J.A.N., French mathematician and philosopher, 139
Confiliachi, Italian physicist, 349
Conté, N.J., called Nicollet by Prince Carl/ HCØ, French egyptologist, 367–8
Cook, J., British explorer, 413, 561
Copley, G., British scientist, 355
Correggio, A.A., Italian painter, 567
Coulomb, C.A., French physicist, 5, 64, 139, 147, 153, 300–4, 331, 340–1, 351–2, 380, 554
Coulthard, D., British-Danish entrepreneur, 315–16, 322–3
Cousin, V., French philosopher, 384, 386

Cumming, J., British chemist, 395
Cuvier, J.-L.-N.-F., French zoologist, 139–40, 179, 249, 361, 385

Dahl, I.C., Norwegian painter, 106, 309, 358, 567–8, 578
Dahlstrøm, F.C.E., HCØ's son-in-law, 491, 619, 647
Dalton, J., British chemist, 272
Darbes, J., Danish painter, 130–1
Darwin, Ch., British naturalist, 8, 14, 394–5, 633
David, J.-L., French painter, 262
David, C.N., Danish economist, 485
Davy, H., British chemist, president of RS, 226, 265, 268, 270–1, 297, 339, 349, 351, 355, 361, 390–3, 395–9, 402, 409–10, 424–8, 535, 632
Delambre, J.-B. J., French astronomer, 358
Delamétrie, French writer, 130,155
Dietzsch, S. & B., German historians of science, 654
Dulong, P.L., French physicist, 380, 384, 386
Dyssel, I.A., Danish student of medicine, HCØ's amanuensis, 327, 329–31, 333, 342–3, 359, 414, 426, 498
Döbereiner, J.W.D., German chemist, 349, 371
Döllinger, I.D., German anatomist, 133

Eckersberg, C.W., Danish painter and professor, 26, 180, 216, 250, 262, 344, 362–3, 522, 578–81
Engelstoft, L., Danish professor, 166–7, 389
d'Èpinay, L.-F.-P., French woman writer, 159
Erdmann, O.L., German chemist, 567, 632
Erdmannsdorff, Fr. W. von, German landscape architect, 103–4
Erman, P., German physicist, 226, 260, 349–51, 409, 565
Eschenmayer, A.K.A.E., German physician and philosopher, 656
Esmarch, L., Danish jurist and surveyor, 304–5, 315–16, 322, 336, 341, 343
Euclid, Greek mathematician, 14
Euler, L.E., German mathematician, 383
Euripides, Greek playwright, 476, 564
Ewald, Johs., Danish writer and poet, 71

Faber, P., Danish polytechnician, 648
Fabricius, I.C., Danish professor, 36

Faraday, M., British physicist, 5, 212–13, 355, 395–9, 451, 565, 591–6
Fenger, C.E., Danish physician, 646–7
Ferdinand, King of Two Sicilies, 358
Ferguson, A., Scottish philosopher, 40
Fessler, I.A., German freemason, 131
Fichte, J.G., German philosopher, 9, 40, 55–8, 107, 122–6, 131–2, 170, 176, 178, 196–202, 228, 232, 252, 258–9, 274, 283, 334, 436, 469, 483, 491, 509, 541, 632, 662
Forchhammer, J. G., Danish chemist, geologist, and professor, 13, 314–16, 322–4, 327, 329–30, 342, 349, 356, 364, 389, 414, 426, 442, 455, 458, 498–9, 501, 503, 505, 519, 532, 585–99, 605–6
Forster, G., German explorer, 561
Fourcroy, A.-F., French chemist, 139–40, 155
Fourier, J.B.J., French physicist, 370, 386, 388
Franklin, B., American physicist, 147
Fraunhofer, J., German instrument maker, 374
Frederik VI (1808–39), King of Denmark (and Norway–14), Crown Prince Frederik (1784–1808), 151, 172, 174, 189, 195, 202, 225–6, 229–31, 254–55, 261, 289–90, 306, 308–10, 315, 327, 334, 357, 359, 374, 400, 411, 432–4, 440, 443, 459, 481–2, 484–6, 500, 520, 534–5, 538, 605
Frederik Christian, Duke of Augustenborg, Chancellor, 34, 41, 48, 78–9, 97, 117, 124, 157,9, 167, 169, 175–6, 191, 231, 254, 289, 292–4, 299–300, 410
Frederik VII, King of Denmark (1848–63), 429, 609–10, 612–14, 617
Fresnel, A.J., French physicist, 349, 380–1, 386, 582
Friedländer, D., German editor, 131
Friedländer, M., German physician, 130, 144, 150
Friedman, M., American historian of science, 6, 654, 657–60
Friedrich Franz, Duke of Anhalt-Dessau, 104–5
Friedrich der Grosse, King of Prussia (1740–86), 371
Friedrich Wilhelm III, King of Prussia (1797–1840), 128, 259, 447
Friedrich Wilhelm IV, King of Prussia (1840–61), 549, 560–2, 567, 610, 614–15
Frommann, C.F.E., German publisher, 109, 123, 152

INDEX OF NAMES

Gachet, madame de, French emigrant, 110, 165
Gade, N.W., Danish composer, 649
Galen, Greek physician, 50
Galilei, G., Italian mathematician and physicist, 14, 67, 278, 435
Gall, F.J., German phrenologist, 133, 311, 420
Galvani, L., Italian anatomist, 80–1, 109, 655
Gamborg, A., Danish professor, 31, 52
Gamst, H.C., Danish technologist, 505
Gau, F.C., German architect, 389
Gauss, C.F., German mathematician, 445, 451, 518, 547–9, 561
Gay-Lussac, L.-J., French chemist, 139, 226, 351, 358, 382–3, 386, 400
Gehlen, A.F., German editor, 191, 242
Gilbert, L.W., German editor, 167, 191, 349–50
Gjerlew, A,C., Danish theologian, 100, 135, 152, 170, 206, 227, 281, 361, 372, 399
Gluck, W., Austrian composer, 421
Goethe, J.W., German writer, 58, 81, 91, 109–10, 122–4, 146, 170, 173, 188, 196, 228, 230, 234, 256–9, 261, 265–6, 271, 283, 296, 361, 371–4, 387, 401, 447, 570–1, 573, 585, 636, 662
Goldschmidt, A.M., Danish writer, 610–14, 616, 624, 631–2, 634, 642
Goldsmith, O., British poet and novelist, 599
Gower, B., British historian of science, 6, 654, 658
Gren, F.A.C., German chemist, 62, 79, 81, 248
Grimm, J., German linguist, 250, 567
Grimm, W., German linguist, 250, 567
Grundtvig, N.F.S., Danish theologian, hymnist, and historian, 10, 274–81, 313, 431–3, 506–16, 518, 520–2, 526, 530, 536, 539, 543, 591, 634, 643, 657, 662
Guizot, F.P.G., French prime minister, 587
Gunnerus, E., Norwegian bishop and professor, 37
Guyton de Morveau, L.-B., French chemist, 139, 141–2, 153
Gyllembourg, Th., née Buntzen, Danish writer, 126, 227–8, 235, 286, 442–3, 470–1
Göttling, J.F.A., German chemist, 122

Hachette, J.N.P., French physicist, 349, 386
Hallé, J-N., French medical scientist, 147, 153
Haller, A. von, German physiologist, 50

Hannover, E., Danish art historian, 577
Hansen, C.F., Danish painter and architect, 36
Hansen, C., Danish painter, 578–80, 643
Hansen, H., HCØ's grandfather, merchant, 16
Hansteen, Chr., Norwegian mathematician and physicist, 169, 225, 247–8, 270, 306, 343, 330, 349, 361, 382, 428, 442, 445, 457, 518, 520–3, 526–8, 532–3, 547–9, 561, 600–2, 633
Harding, M.C., Danish historian of science, 414
Harsdorff, C.F., Danish architect and professor, 72, 74
Hartmann, J.P.E., Danish composer, 216, 644
Hasle, N., Danish shipping agent, 441
Hasle, P., Danish vicar, HCØ's son-in-law, 617–19, 649
Hauch, A.W., Danish Court Steward, 36–7, 61–5, 70, 79, 87–90, 93, 97, 167, 202, 213, 215, 248–9, 254, 334, 345, 350, 364, 413, 442, 475, 501, 595
Hauch, Carsten, Danish zoologist, novelist, and poet, 11, 179, 203, 248–9, 330, 479, 523, 540, 570, 575, 624, 634, 642–4, 649
Haüy, R., French crystallographer, 139, 147, 226, 361
Hegel, G.W.F., German philosopher, 118
Heger, C., Christiane's and Kamma's brother, 170
Heger, H., Danish brewer and amateur astronomer, Christiane's and Kamma's father, 34, 203
Heidenrich, Danish goldsmith and thief, 168
Heiberg, J. L., Danish writer, 11, 215–16, 227, 235–6, 300, 330, 442, 466, 469–70, 479, 483, 643, 649
Heiberg, J. L., Danish actress, wife of JLH, 470
Heiberg, P.A., Danish writer, father of JLH, 22, 228
Hein, Piet, Danish architect and poet, 11
Herder, J.G., German theologian and philosopher, 110, 183, 508, 510, 541, 660
Herholdt, J.D., Danish professor, 416–18
Hermansen, Karen, HCØ's mother, see Ørsted, Karen
Hermbstaedt, S.F., German chemist, 112, 114–17, 139, 157, 260, 288, 359, 654
Herschel, F.W., British astronomer, 61, 91, 357
Herschel, J.F.W., British astronomer and physicist, 5, 7, 212, 304, 357, 393, 395, 425, 428, 480, 504, 561, 565, 579, 585, 591, 596–7, 622–4
Hertz, R., Danish patient, 416–20

Herz, H. J., German salon hostess, 126–8, 130, 169, 258, 260, 467, 566
Herz, M., German physician, her husband, 126–7
Hetsch, G.F., Danish professor, 455
Hippocrates, Greek physician, 14
Hjorth, S., Danish railway director, 438
Hjort-Lorenzen, P., Danish politician, 539
Hofmann, J.R., British historian of science, 381
Holberg, L., Danish-Norwegian playwright and philosopher, 27, 444, 549, 564
Holm, Chr., Danish painter, 578
Holmblad, Danish dyer, 225
Holten, C.V., Danish polytechnician, 414, 442, 498, 600
Homer, Greek historian, 520, 641
Horace, Roman poet, 520
Hornemann, Chr., Danish student, 41, 53, 60
Hornemann, J.W., Danish professor, 290
Horner, L., Scottish geographer, 591, 623
Horner, J., his daughter, 597, 633
Horner, L., his daughter, 597, 633
Hostrup, J., Danish writer, 530
Howard, L., British meteorologist, 582
Howitz, F.G., Danish physician and phrenologist, 310, 415–23
Humboldt, A. von, German geographer, 81, 123, 139, 250, 358, 445–7, 547, 561–4, 567, 582, 616, 632, 654, 657
Humboldt, W. von, his brother, 123, 250, 567
Hume, D., Scottish philosopher, 40–1, 59
Hummel, C.G., Danish polytechnician, 414, 498, 505, 554, 605–6
Huygens, Chr., Dutch physicist, 383
Hvidt, L.N., Danish entrepreneur, 400
Händel, G.F., German-British composer, 447
Höpp, J.P., Danish civil servant, 482, 485

Iffland, German playwright, 129
Ingemann, B.S., Danish hymnist and novelist, 662

Jacobi, F.H., German writer, 275
Jacobsen, A. Skaar, Danish historian of science, 659
Jacobsen, H.J., Danish pharmacist, 190
Jacobsen, I.C., Danish brewer, 12

Jacobson, L.L., Danish physician, 345
Jensen, C.A., Danish painter, 579, 622
Jerichau, J.A., Danish sculptor, 12–13, 591
Jessen, P.W.J., German psychiatrist, 541
Johnston, J.F.W., Scottish professor, 497
Jones, mayor of Southampton, 593
Joule, J.P., British physicist, 552, 558
Jørgensen, Danish accountant, 98
Jørgensen, S., Danish judge, 23

Kall, N.C., Danish professor, 8, 407
Kannegiesser, K.F.L., German translator, 624, 632
Kant, I., German philosopher, 4, 6–9, 30–1, 33, 40–6, 51–58, 59–65, 66–71, 117–22, 124–6, 132, 143, 178, 183–4, 189, 199–200, 215, 220, 228, 238–40, 244, 251–3, 265–8, 274, 279, 308, 338–9, 351, 377, 380–1, 398, 420, 423, 436, 450–1, 474, 509, 541, 550–2, 558, 570–1, 577, 622, 632, 636, 653–56, 658, 660, 662–4
Karsten, K.J.B., German metallurgist, 117, 133, 369, 567
Kastner, W.G., German chemist, 190
Kepler, J., German astronomer, 67
Kierkegaard, S.Aa., Danish philosopher and writer, 511, 570, 661
Kierulff, A., Danish chief of police, 422
Klaproth, M.H., German chemist, 260, 116, 131, 369
Kleist, E.G. von, German physicist, 149
Klingberg, H.M.V., surgeon at the Royal Danish Navy, 407, 440–1
Knight, G., British scientist, 271, 308
Knudsen, O., Danish historian of science, 269, 340
Kotzebue, A., German playwright, 125, 129, 334
Kratzenstein, C.G., Danish professor, 31, 37, 158, 174–5, 193
Krumm, J., Norwegian chemist, 255, 262
Krøyer, H., Danish student, 333
Kuhlau, Fr., Danish composer, 215–16
Küchler, A., Danish painter, 578, 580
Købke, Chr., Danish painter, 577, 580–1, 583
Køster, L. Schack, Danish pharmacist, 427
Kaas, Fr. J., Danish minister of state, 294, 308–10, 419, 433

INDEX OF NAMES

Lamarck, J.-B., French zoologist, 139
Laplace, P.S. de, French mathematician and physicist, 8, 61, 63, 65, 138–9, 143, 145, 147–8, 153, 184, 223, 267, 329, 349, 351–2, 382–4, 386–7, 409, 450, 654, 663
Lasteyrie, C.-Ph., French salon host, 137
Lavoisier, A.-L., French chemist, 61–62, 64, 87–8, 119–20, 138, 141, 190, 224, 272, 347, 383
Lehmann, M., Danish student, civil servant, 135, 152, 227
Lehmann, O., Danish politician, 135, 484–5, 488–90, 493, 505, 610–17
Lehmann, W.O.W., Danish engineer, 607
Leibniz, G.W., German mathematician and philosopher, 14
Lenoir, E., French instrument maker, 152
Leslie, J., Scottish mathematician and natural philosopher, 354
Lessing, G.E., German writer, 40, 126, 129
Lichtenberg, G.C., German physicist and writer, 188, 210–1, 244
Lind, K.C., Danish dyer, 440
Lind, P.E., theologian, HCØ's nephew, 537
Linnæus, C., Swedish naturalist, 14
Locke, J., British philosopher, 40–42, 59
Louis Philippe, King of France (1830–48), 587, 610
Lumbye, H.C., Danish musician and composer, 529
Lund, P.W., Danish naturalist, 519, 591
Lundbye Th., Danish painter, 577–80
Luther, M., German theologian and reformer, 334, 432
Lyell, Ch., British geologist, 535, 592

Madvig, J.N., Danish professor, 511, 545
Manthey, J.G.L., Danish professor and pharmacist, 7, 50–1, 71–6, 79, 97–8, 102, 112, 114, 117–19, 121, 131, 154, 156–62, 169, 174–5, 181, 190, 227, 244, 254, 300, 344, 349, 364, 389, 410, 654
Marat, J.-P., French revolutionary, 138, 478
Marezoll, J.G., German theologian, 32, 288
Marstrand, W., Danish painter, 556–8, 580, 643
Martensen, H.L., Danish theologian, 438, 442, 498, 522, 615–16, 643, 657
Marum, M. van, Dutch physicist, 152

Mayer, J.R., German physicist, 552, 558
Maxwell, J.C., Scottish mathematician and physicist, 5
Meisling, S., Danish headmaster, 443, 465
Mendel, M.H., German physician, 112, 117–18, 121
Mendelssohn, M., German writer, 126, 130
Mendelssohn-Bartholdy, J.L.F., German composer, 447
Mesmer, A., Austrian quack, 65, 81
Metternich, C. von, Austrian minister of state, 447, 482, 610
Meyer, K., née Bjerrum, Danish historian of science, 6, 11, 654
Michaelis, G.A., German professor, 616
Mitscherlich, E., German chemist, 567
Molbech, Chr., Danish historian, 503, 544–6, 625–6, 634, 643, 648, 662
Moldenhawer, D.G., Danish professor, 31
Molière, J.-B., French playwright, 644
Moll, G., German physicist, 349
Moltke, A.G., Danish court steward, 35
Moltke, A.W., Danish minister of state, 612
Moltke, O., Danish minister of state, 482, 485
Monge, G., French technologist, 139, 145
Monro, A, British sculptor, 12
Mons, J.B. van, Dutch physicist, 226, 297
Montaigne, M. de, French essayist, 40
Montgolfier, J. M., French baloonist, 476–8
Montgolfier, J. E., his brother, 476–8
Mozart, W.A., Austrian composer, 81, 106–7, 220, 527
Munch-Ræder, Norwegian jurist, 538
Murchison, R., British geologist, 527, 592, 596–7, 622–3
Müller, F.H., Danish pharmacist, 438
Murray, J., British publisher, 623
Mynster, J.P., Danish bishop, 60, 203, 389, 438, 443, 479, 491, 521, 545, 626–9, 634, 643–4, 649, 662
Mynster, O. H., Danish professor, 80, 290
Münter, Fr., Danish bishop, 31, 78, 293, 328, 408, 431
Møller, Engelke Cathrine, née Ørsted, HCØ's aunt, 17–18, 26–7, 33, 70, 103, 168–9, 172, 205–6, 228, 284, 295, 363, 527, 564
Møller, J., Danish professor, 333–4, 407
Møller, J.P., Danish painter, 579

Møller, P.L., Danish writer, 624
Mönchgesang, D., JWR's wife, 165–6, 180, 192, 242, 247
Møsting, J.S. von, Danish minister of finance, 297–8, 315, 327, 359, 413, 522

Napoléon Bonaparte, French general and emperor, 75, 81–2, 85, 109, 138–9, 145, 147–8, 191–2, 197, 202, 208, 230, 259, 328, 358, 366, 371, 377, 384, 394, 525, 565, 589
Neef, C.E., church fatherGerman physicist, 349
Nelson, H., British admiral, 79
Newton, I., British mathematician and phycisist, 14, 42, 44, 48, 63–65, 67, 91, 118, 183, 302, 358, 391, 409, 450, 476, 558, 570–1, 579, 588, 601
Nicholson, W., British physicist, 85–86
Nicollet, see Conté
Niebuhr, B.G., German historian and civil servant, 258–61
Niebuhr, C., Danish explorer, his father, 259, 366
Nielsen, H. Toftlund, Danish chemist, 264
Novalis (Hardenberg, Fr. von), German writer, 109, 112, 125, 188, 235, 243, 272
Nyerup, R., Danish professor, 166, 333
Nørregaard, A., Danish jurist, 23

Oeder, G.C., German-Danish professor, 36
Oehlenschläger, A., Danish poet and playwright, 11, 19–20, 32, 40, 56–58, 70–1, 76–7, 102, 167–9, 171–4, 179, 185, 188, 190, 193, 196, 198, 203–4, 221, 229–37, 247, 249, 280, 336, 374, 401–2, 408, 441–2, 444, 460, 463, 467, 469–70, 474, 522–3, 535, 538, 549, 566, 643–4, 649, 657, 664
Oehlenschläger, Christiane, née Heger, AOe's wife, 103, 172, 196, 202, 235
Oehlenschläger, G., AOe's and SØ's father, organist, 56, 102–3, 198, 408, 442
Oken, L., German biologist, 446, 449
Oldenburg, Chr., German-Danish wig maker, 15, 21–2, 543
Oldenburg, his Danish wife, 21–2
Olshausen, Th., Schleswig-Holsteinian politician, 612–6
Olufsen, C., Danish professor, 291

Origen, Greek father of the church, 277
Oskar I, King of Sweden-Norway, 525–6, 528, 614
Ostwald, F.W., German chemist and historian of science, 5, 654
Otto, C., Danish phrenologist, 310–1, 420–2, 425, 645, 652

Pais, A., American historian of science, 13
Paludan-Müller, Fr., Danish writer and poet, 442, 471, 643
Panzer, J.H.L., German physicist, 207
Paracelsus, (Ph.A.Th.B. von Hohenheim), German iatrochemist, 50
Pauelsen, E., Norwegian painter, 576
Pepys, W.H., British physicist and instrument maker, 395, 397
Perkins, J., American inventor, 306, 395, 400, 588
Perthes, W., German bookseller, 191
Petersen, the HCØ-family's private teacher, 317, 407
Petit, A.-T., French physicist, 349, 383–4, 386
Petrarch, F., Italian poet, 477
Peymann, H.E., Danish general, 202
Pfaff, C.H., German chemist and physicist, 57, 314, 329, 349–50
Pictet, M.-A.T., Swiss physicist, 349, 351
Pingel, Danish carpenter, 320, 407
Plato, Greek philosopher, 134, 185, 189, 200, 214, 217–18, 279, 281, 284, 490, 511, 600, 636, 662
Poggendorff, J.C., German physicist, 349
Poisson, S.-D., French chemist, 386
Posch, German medallist, 369
Prangen, F.E. von, Danish colonel, 458, 462
Prechtl, J.J., Austrian physicist, 349, 378
Priestley, J., British chemist, 61, 271
Probsthein, C.D., Danish painter, 102–3, 161
Probsthein, S., HCØ's first fiancée, 75–6, 79, 100–3, 112, 156, 158–62, 165, 215, 281–2, 286, 389
Puy, E. du, French-Danish composer, 215

Rahbek, Kamma, née Heger, 169–70, 172, 232, 442–3, 602
Rahbek, K.L., Danish professor, her husband, 174, 202, 442, 602
Rambuteau, C.-P., French count, mayor, 587

INDEX OF NAMES

Ramus, C., Danish professor, 545, 554, 557, 606
Ranke, L. von, German historian, 260
Raphael (Raphaello Sanzio), Italian painter, 567
Ravnholdt, Danish executioner, 422
Rawert, O.J., Danish painter and historian of technology, 463
Reinhardt, J.C.H., Danish zoologist, 345, 519
Reinhold, K.L., German philosopher, 58, 122, 259
Repp, T.G., Icelandic-Danish translator, 600–4
Retzius, A.A., Swedish anatomist, 532
Reventlou-Preetz, Fr., Schleswig-Holsteinian provost, 614
Reventlow, C.D.F., Danish minister of state, 36, 98
Richter, J.B., German chemist, 117, 121, 654
Ridolfi, Italian physicist, 349, 351
Riisbrigh, B., Danish professor, 24, 32, 41–2, 44, 53
Ritter, J.W., German physicist, 11, 71, 81–8, 108–12, 119–21, 123, 130, 144, 146–55, 157, 165–6, 174, 178–82, 184–5, 188–9, 191–3, 204, 211, 220, 232, 242–3, 268, 270, 281, 304–5, 327, 331, 338–41, 355, 357, 374, 381, 410, 547, 552, 588, 649, 653–5, 657, 662
Rive, C.G. de la, Swiss physicist, 349, 351
Robinson, H. Crabb, British student, 272
Robespierre, A.B.J., French dictator, 144
Rogert, Mathilde, married to ASØ, 312, 317, 443
Romagnosi, G.D., Italian amateur physicist, 5
Rose, V., German pharmacist, 121, 130
Rosenfeld, L., Belgian physicist, 654
Rosing, M., Norwegian-Danish actor, 170
Rousseau, J.-J., French philosopher and composer, 152, 159, 215–17, 447, 570, 625, 664
Rubens, Danish civil servant, 296, 320
Rumford, British count, previously B. Thompson, American physicist, 265
Rømer, O., Danish mathematician and astronomer, 152, 343, 387
Rørbye, Martinus, Danish painter, 578, 580
Röschlaub, A., German professor, 128
Rosenørn Lehn, O., Danish count, 264

Sabine, E., British explorer, 592, 623
Savart, F., French physicist, 349, 351, 386
Saxtorph, M., Danish professor, 50, 78
Say, J.B, French political economist, 361
Scharling, C.E., Danish theologian, 438, 499
Scharling, E.A., Danish chemist, HCØ's son-in-law, 413, 438, 442, 481, 498–501, 503, 523, 532, 530, 619, 646
Scheele, C.V., Swedish chemist, 91
Scherer, A.N., German chemist, 133
Schelling, J.F.W., German philosopher, 6–7, 9, 58, 60, 118, 123, 132, 134, 167–8, 178, 181, 183, 190, 199, 218–19, 232–5, 238–43, 259, 274–7, 280, 374, 549–52, 556, 563, 615, 632, 654, 660
Schifter, A., Danish admiral, 501–2, 522
Schiller, Fr., German writer, 40, 58, 100, 110, 123–4, 126, 129, 178, 197, 220, 228, 240, 259, 271–2, 447, 480, 570–1, 636, 655, 662
Schimmelmann, Ch., EHS's wife, 173–4, 192
Schimmelmann, E.H., Danish minister of finance, 97–8, 124, 158, 169, 172–4, 176, 197, 228, 230, 254, 297, 315, 328, 331, 413
Schinkel, K.F., German architect, 370, 565
Schlegel, A.W., German writer, 53, 58, 106–7, 125, 131, 133–4, 218, 250, 259, 272, 655
Schlegel, C., A.W. Schlegel's wife, 134
Schlegel, Fr., German writer, 53, 106–7, 125, 128, 133–7, 150, 152, 218, 243, 250, 259, 272, 655
Schlegel, J.E., Danish professor, 47
Schlegel, J.W.F., Danish jurist, 53, 491
Schleiermacher, F.D.E., German theologian, 173, 250, 258
Schmidt-Phiseldek, C.G.F. von, Danish professor, 291
Schmidten, H.G. von, Danish mathematician, 328, 330, 354, 358, 377, 382, 389, 423, 426, 442, 450, 458, 466, 468, 498, 523, 605
Schouw, J.F., Danish botanist and politician, 483, 486, 491, 493, 500, 505, 519, 521–2, 529, 532, 545, 582, 609–10
Schumacher, H.C., Danish astronomer, 224, 356, 442, 444–5, 535, 546–7
Schweigger, J.S.C., German editor, 248, 349, 351, 374, 377, 428
Schäffer, H., Danish judge, 323
Scott, W., Scottish novelist, 401, 599
Sedgwick, A., British geologist, 395

INDEX OF NAMES

Seebeck, Th. J., German physicist, 349–51, 361, 369–70, 374, 377, 387, 400, 565
Seip, J. A., Norwegian historian, 482
Serres, M. de, French professor and translator, 262
Shakespeare, W., British playwright, 58, 599, 641
Shanahan, T., American historian of science, 6, 654, 658
Sibbern, F.C., Danish philosopher, 224, 256, 258, 262, 265, 283, 286, 296–7, 312, 318, 335, 491, 532, 566
Simon, P.L., German chemist, 121, 341
Smith, A., Scottish philosopher, 40
Snelders, H.A.M., Dutch historian of science, 654
Snow, C.P., British physicist and novelist, 11
Socrates, Greek philosopher, 217
Sonne, J., Danish painter, 578–80
Spurzheim, German phrenologist, 420
Stauffer, R., American historian of science, 6, 654
Steen, A., Danish mathematician, 604
Steenstrup, J.J.S., Danish professor, 532
Steenstrup, P., Danish mining engineer, 102
Steffens, H., Norwegian-Danish-German naturalist, 'the Emperor', 35, 37, 60, 158, 167–9, 172, 174, 176, 178, 181–4, 191, 193, 197–8, 230–3, 240, 243, 300, 316, 444, 462, 468, 521–3, 549–52, 556, 564, 655–6, 658
Steffens, J.H., officer, HS's brother, 176
Stemann, P.C., Danish minister of state, 482, 485–6, 490, 501
Struensee, Fr., Danish usurpator, 36
Suhr, O.B., merchant, 13
Sulzer, J.G., Swiss aesteticist, 47

Taine, H., French literary critic, 663
Taube, A., Swedish countess, 236
Taube, G., her brother, 235
Tetens, N., Danish philosopher, 40
Thalbitzer, S., wife of Danish merchant, 204
Thénard, P., French chemist, 139–141, 383, 386
Thomson, Th., Scottish chemist, 272–3, 324, 349, 354
Thorlacius, B., Icelandic-Danish professor, 290
Thorlacius, G., Icelandic-Danish headmaster, 290
Thorvaldsen, B., Danish sculptor, 76, 161–2, 473, 563, 580, 649

Tieck, J.L., German writer, 125, 137, 232, 272, 564, 655
Tode, J.C., Danish professor, 74, 78
Treviranus, L.C., German physiologist, 182
Trommsdorff, J.B., German chemist, 100
Tryde, E.C., Danish provost, 522, 643, 649
Tscherning, A.F., Danish officer, 360, 500, 505, 516

Ursin, G.F., Danish mathematician, 414, 453–62, 468, 565, 605

Valkendorf, Chr., Danish founder of college, 601
Vammen, H., Danish historian, 484
Vauquelin, L.N., French chemist, 139–42, 157–8, 226, 654–5
Veit, D., née Mendelssohn, Fr. Schlegel's wife, 107, 127–8, 135
Viborg, E., Danish professor, 290
Victoria, Queen of the British Empire, 12, 596–7
Visby, C.H., Danish prison chaplain, 652
Voigt, J.K.W., German chemist, 122
Volta, A., Italian physicist, 71, 80–5, 87, 109, 147, 149, 152, 165, 178, 208, 564, 653–4

Wagner, M., Danish historian of technology, 458
Wallich, A.W., scene painter, 579
Warberg, O., Danish astronomer and industrial spy, 175, 316, 322
Watt, J., British inventor, 400
Weber, C. M. von, German composer, 371, 547–9
Weiss, C.S., German crystallographer, 117, 191, 258, 260, 349–51, 369, 371, 442, 446
Welhaven, J.S., Norwegian writer, 489
Wellesley, Arthur, Duke og Wellington, British general, 203, 565, 589
Wergeland, H., Norwegian writer, 526
Werner, A.G., German mine director, 191, 386, 504
Wessel, Casper, Danish mathematician and geodesist, 546
Westrumb, J.F., German chemist, 119
Weyse, C.E.F., Danish composer, 649
Wetzels, W.D., American-German historian of science, 654

741

Wheatstone, Ch., British physicist, 212, 393, 395, 425
Whewell, W., British mathematician, 395, 425, 592
Wieland, Ch.M., German playwright, 40
Wilke, J.K., Swedish physicist, 149
Wilkens, J.F.C.E., Danish polytechnician, 464, 498, 505, 605–7
Williams, L.P., American historian of science, 6, 654
Wilson, A.D., American historian of science, 6, 654–7, 659
Winstrup, O.J., Danish mechanic and foreman, 454–5, 457–8
Winterfeldt, J.B., Danish admiral, 291
Winther, Chr., Danish writer, 332–3
Winterl, J.J., Hungarian-Austrian chemist, 116–17, 119–20, 141, 144, 157, 167, 178, 185, 190–1, 238, 244, 268, 339, 345, 552
Wleugel, P.J., Danish naval officer, 336, 343
Wolff, Chr., German philosopher, 23, 59
Wollaston, W.H., British physicist, 212, 349, 355, 395, 398–9, 402
Worm, Danish executioner, 422
Wöhler, Fr., German chemist, 426–30

Yelin, J.K., German physicist, 349
Young, Th., British physicist and Egyptologist, 349, 356, 367, 390, 395, 582, 59

Zeise, W.C., Danish chemist, 248–9, 300, 306, 314, 317, 324, 327, 329–30, 335, 342, 345, 359, 361, 389, 394, 408, 414, 417, 426, 442, 455, 459–60, 498, 500–1, 525, 605–6
Zimmermann, E.A.W., German physicist, 306

Ørsted, Anders Sandøe, HCØ's brother, jurist, and minister of state, 'The Asse-child', 3, 8, 11, 21–2, 28, 31, 35, 41, 46, 52–8, 70–1, 76, 101–2, 118, 131, 157–9, 161, 169–70, 197, 203, 228, 250–3, 255, 262–3, 265, 274, 279, 282–4, 292–4, 296, 308–10, 312–13, 316–17, 327, 361, 384, 415–23, 431–3, 440–4, 482, 484–7, 490–3, 521, 609, 619–20, 643–4
Ørsted, Anders Sandøe, HCØ's son, 560, 585, 619–20

Ørsted, Anne Dorothea, née Borring, HCØ's stepmother, 170, 172, 255
Ørsted, A.N., 560, 619–20
Ørsted, Anna Dorothea **Marie**, HCØ's daughter, 319, 362, 407, 492, 619, 644
Ørsted, Barbara Albertine, called **Tine**, HCØ's sister, 103, 169–70, 172, 206, 527
Ørsted, Barbara Albertine, née Witth, HCØ's grandmother, 17
Ørsted, Benedicte, HCØ's aunt, 17, 20
Ørsted, Christian Sørensen, HCØ's grandfather, 17
Ørsted, Hans Christian, Danish physicist, 'the Professor with the Nose' not indexed
Ørsted, Herman, HCØ's brother, apprentice, 255, 263
Ørsted, Inger Birgitte, født Ballum, called **Gitte**, HCØ's wife, 172, 206, 227, 255, 285–7, 294–6, 304, 312–3, 316–7, 322, 349, 361–4, 372, 377, 384, 393–5, 400–4, 408, 415, 440–1, 444, 474, 481, 491–2, 560, 568, 578, 617, 619–20, 643–47, 650
Ørsted, Jacob Albert, HCØ's brother, pharmacist, 76, 169–71, 228, 256, 263, 318, 363
Ørsted, Jacob Albert, HCØ's uncle, vicar, 17–18, 20
Ørsted, Karen, née Hermansen, HCØ's mother, 16–20
Ørsted, Karen, HCØ's daughter, 294–5, 317, 319, 362, 403, 407–8, 415, 440, 443–4, 446, 472, 492, 498–500, 560, 619, 644, 647
Ørsted, Lauritz Gerhard, HCØ's uncle, clothier, 17
Ørsted, Mathilde, née Rogert, ASØ's second wife, 361, 415
Ørsted, Mathilde Elisabeth, HCØ's daughter, 10, 101, 159, 161, 415, 472, 504, 585–99, 619–20, 633, 644
Ørsted, Niels Christian, HCØ's son, called **Christian**, 319, 362, 407, 440, 502, 523, 585, 619, 643, 647
Ørsted, Niels Randulph, HCØ's brother, 'the Fixer', 76, 102, 169, 172, 228–30, 263
Ørsted, Sophie, née Oehlenschläger, ASØ's wife, 'the Powderkeg', 128, 165, 277–9, 281, 284, 331–6,

344, 346, 390–393, 436–7, 442–4, 447–52, 501, 505–10, 513, 524, 526, 556–9, 566–7
Ørsted, Sophie Wilhelmine Berta, HCØ's daughter, called **Sophie**, 362, 407, 472, 492, 504, 619, 644, 647

Ørsted, Søren Christian, called little Søren, HCØ's fosterson, 284, 318, 362–4, 407–8, 440, 523, 620
Ørsted, Søren Christian, HCØ's father, pharmacist, 16–21, 102, 170–1, 256, 336